Sustainable Nanotechnology

Sustainable Nanotechnology

Strategies, Products, and Applications

Edited by

Yashwant V. Pathak
University of South Florida Health
Tampa, USA

Govindan Parayil
University of South Florida
Tampa, USA

Jayvadan K. Patel
Sankalchand Patel University
Gujarat, India

WILEY

This edition first published 2022

Registered Office
John Wiley & Sons, Inc., 111 River Street, Hoboken, NJ 07030, USA

Editorial Office
111 River Street, Hoboken, NJ 07030, USA

For details of our global editorial offices, customer services, and more information about Wiley products visit us at www.wiley.com.

Wiley also publishes its books in a variety of electronic formats and by print-on-demand. Some content that appears in standard print versions of this book may not be available in other formats.

Library of Congress Cataloging-in-Publication Data

Names: Pathak, Yashwant Vishnupant, editor. | Parayil, Govindan, 1955– editor. | Patel, Jayvadan K., editor.
Title: Sustainable nanotechnology : strategies, products, and applications / edited by Yashwant V. Pathak, University of South Florida Health, Tampa, USA, Govindan Parayil, University of South Florida, Tampa, USA, Jayvadan K. Patel, Sankalchand Patel University, Gujarat, India.
Description: First edition. | Hoboken, NJ : John Wiley & Sons, Inc., 2022. | Includes bibliographical references and index.
Identifiers: LCCN 2021050548 (print) | LCCN 2021050549 (ebook) | ISBN 9781119649977 (hardback) | ISBN 9781119650300 (adobe pdf) | ISBN 9781119650317 (epub)
Subjects: LCSH: Nanotechnology. | Green chemistry. | Sustainable development.
Classification: LCC T174.7 .S87 2022 (print) | LCC T174.7 (ebook) | DDC 620/.5–dc23/eng/20211208
LC record available at https://lccn.loc.gov/2021050548
LC ebook record available at https://lccn.loc.gov/2021050549

Cover Design: Wiley
Cover Image: © KTSDESIGN/SCIENCE PHOTO LIBRARY/Getty Images

Set in 9.5/12.5pt STIXTwoText by Straive, Pondicherry, India

10 9 8 7 6 5 4 3 2 1

MIX
Paper from responsible sources
FSC www.fsc.org FSC® C013604

Dedication – Professor Viness Pillay (1970–2020)

This book is dedicated to a leading pharmaceutical sciences researcher and academic, Professor Viness Pillay, who passed away peacefully on the morning of 24 July 2020 after a lengthy illness. He was just four days away from his 50th birthday.

A Personal Professor of Pharmaceutics, Prof. Pillay held the Department of Science and Innovation (DSI) – National Research Foundation (NRF) of South Africa Research Chair in "Pharmaceutical Biomaterials and Polymer-Engineered Drug Delivery Technologies" and also served as the founding director of the Wits Advanced Drug Delivery Platform (WADDP) Research Unit for more than two decades.

Prof. Pillay spent his career making a substantial contribution to pharmaceutical sciences research. His work transcended in producing several inventions in the design of advanced drug delivery systems, biomaterials, nanomedicines, and de novo tissue engineering and bio-inks for 3D printing.

He had an outstanding insight and ability to identify important therapeutic challenges and present research results that questioned conventional wisdom. This led him to developing several inventions including the world's fastest dissolving matrix for the onset of rapid drug action in the human body, a neural device for therapeutic intervention in spinal cord injury and novel wound-healing technologies. He has also pioneered numerous molecular modeling paradigms as a first-in-the-field of pharmaceutics including his very own PEiGOR Theory – Pillay's Electro-influenced Geometrical Organization–Reorganization Theory published in the International Journal of Pharmaceutics.

His research was always at the forefront and led him to publish more than 300 research articles in ISI-accredited international journals, 45 book chapters, editor of 2 books, and 21 granted patents. He was also the recipient of numerous prestigious national and international accolades.

Capacity development and transformation played a central role in the work of Prof. Pillay, so much so that he has developed the largest cohort of talent in the pharmaceutical sciences in South Africa. This was a fitting testament to his servant heart that always spoke to doing what was right and what was good for people – a living motto of, "the world can always use one more kind and compassionate person." He had a brilliant poise of producing not only great science but scientists too.

Throughout his career Prof. Pillay was greatly loved as an influential researcher, teacher, prodigy, friend, mentor, life coach, supervisor, and confidant of more than 120 postgraduate students and postdoctoral fellows from nine different countries as well as colleagues and friends from around the world. In 2019, he received the prestigious NRF Award as the Champion of Research Capacity Development and Transformation at Public Science and Higher Education Institutions in South Africa. He loved to love people, and this was strongly felt by his students who have gone on to successful careers making a difference as leaders in some of the world's best-run pharmaceutical companies, academic institutions, and research organizations.

His work brought together outstanding individuals from many different academic disciplines and is an inspiration to his students and colleagues. In addition to using his powerful intellect, creativity, and communication skills to make research contributions, he has used these gifts to also bring about new thinking and models of cohort mentorship and team-based research. He created science, synthesized ideas, provided and challenged viewpoints, developed careers, redefined pharmaceutical innovation, garnered accolades, and above all, made researchers – all with an unshakeable attitude and indomitable humility.

A visionary, a giant of science, and a truly wonderful person, gone too soon. His colossal knowledge, inspirational leadership, and legacy has made a positive impact on all those who knew him and will always be carried forward to ease such a great loss and pain.

Contents

List of Contributors

Samson A. Adeyemi
Wits Advanced Drug Delivery Platform Research Unit
Department of Pharmacy and Pharmacology
School of Therapeutic Sciences
Faculty of Health Sciences
University of the Witwatersrand
Johannesburg
South Africa

P. Sreeramana Aithal
College of Management & Commerce
Srinivas University
Mangalore
India

Shubhrajyotsna Aithal
Faculty
College of Engineering & Technology
Srinivas University
Mangalore
India

Norma Alcantar
University of South Florida (USF) College
of Engineering
USF
Tampa
FL
USA

Umi Athiyah
Faculty of Pharmacy
Universitas Airlangga
Surabaya
Indonesia

Baishakhi Bairagi
Department of Zoology
University of Kalyani
Kalyani Nadia
West Bengal
India

Pratyusha Banerjee
Department of Zoology
University of Kalyani
Kalyani Nadia
West Bengal
India

Sunita Chaudhary
Arihant School of Pharmacy and Bio-Research Institute
Gandhinagar
Gujarat
India

Vishal Chavda
Institute of Pharmacy
Nirma University
Ahmedabad
Gujarat
India

Yahya E. Choonara
Wits Advanced Drug Delivery Platform Research Unit
Department of Pharmacy and Pharmacology
School of Therapeutic Sciences
Faculty of Health Sciences
University of the Witwatersrand
Johannesburg
South Africa

Ilaria Corsi
Department of Physical
Earth and Environmental Sciences
University of Siena and INSTM Local Unit
Siena
Italy

Richa Dayaramani
Khyati College of Pharmacy
Palodiya
Ahmedabad
India

Kangkan Deka
Department of Pharmaceutical Sciences
Dibrugarh University
Dibrugarh
Assam
India

Manopriya Devisetty
University of South Florida (USF) College of Engineering
USF
Tampa
FL
USA

Tasneem Fatma
Department of Biosciences
Jamia Millia Islamia
Central University
New Delhi
India

Andrea Fiorati
Department of Chemistry, Materials,
and Chemical Engineering "G. Natta"
Politecnico di Milano and INSTM Local Unit
Piazza Leonardo da Vinci
Milano
Italy

Giacomo Grassi
Department of Physical
Earth and Environmental Sciences
University of Siena and INSTM Local Unit
Siena
Italy

Dewi Melani Hariyadi
Faculty of Pharmacy
Universitas Airlangga
Surabaya
Indonesia

Bibhuti Bhusan Kakoti
Department of Pharmaceutical Sciences
Dibrugarh University
Dibrugarh
Assam
India

Manjir Sarma Kataki
Department of Pharmaceutical Sciences
Dibrugarh University
Dibrugarh
Assam
India

Nida Jamil Khan
Department of Biosciences
Jamia Millia Islamia
Central University
New Delhi
India

Ridita Khan
University of South Florida (USF) College of Engineering
USF
Tampa
FL
USA

Lalit Kumar
Department of Pharmaceutics
Manipal College of Pharmaceutical Sciences
Manipal Academy of Higher Education
Manipal
Karnataka
India

Pradeep Kumar
Wits Advanced Drug Delivery Platform
Research Unit
Department of Pharmacy and Pharmacology
School of Therapeutic Sciences
Faculty of Health Sciences
University of the Witwatersrand
Johannesburg
South Africa

Samarth Kumar
Formulation-R & D
Sun Pharmaceutical Industries Ltd.
Vadodara
Gujarat
India

Saliha B. Kurt
Faculty of Science and Arts
Department of Chemistry
Terzioglu Campus
Canakkale
Turkey

Giulia Liberatori
Department of Physical
Earth and Environmental Sciences
University of Siena and INSTM Local Unit
Siena
Italy

Shirleen Miriam Marques
Department of Pharmaceutics
Manipal College of Pharmaceutical Sciences
Manipal Academy of Higher Education
Manipal
Karnataka
India

Layla Muraisi
Taneja College of Pharmacy
University of South Florida
Tampa
FL
USA

Debjani Nath
Department of Zoology
University of Kalyani
Kalyani Nadia
West Bengal
India

Prachi Pandey
Babaria Institute of Pharmacy
BITS Edu campus
Vadodara
Gujarat
India

Govindan Parayil
Patel College of Global Sustainability
University of South Florida
Tampa
Florida
USA

Komal Parmar
ROFEL, Shri G.M. Bilakhia College of Pharmacy
Vapi
Gujarat
India

Anand Patel
Department of Chemistry
Uka Tarsadia University
Bardoli
Gujarat
India

Bhavin Patel
Department of Chemistry
Uka Tarsadia University
Bardoli
Gujarat
India

Jayvadan K. Patel
Faculty of Pharmacy
Nootan Pharmacy College
Sankalchand Patel University
Visnagar
Gujarat
India

Manish Patel
L M College of Pharmacy
Gujarat Technological University
Ahmedabad
Gujarat
India

Nishith Patel
Arihant School of Pharmacy and Bio-Research Institute
Gandhinagar
Gujarat
India

Vimal Patel
Institute of Pharmacy
Nirma University
Ahmedabad
Gujarat
India

Yashwant Pathak
Taneja College of Pharmacy
University of South Florida
Tampa
FL
USA

Adjunct Professor
Faculty of Pharmacy
Airlangga University
Surabaya
Indonesia

Viness Pillay
Wits Advanced Drug Delivery Platform Research Unit
Department of Pharmacy and Pharmacology
School of Therapeutic Sciences
Faculty of Health Sciences
University of the Witwatersrand
Johannesburg
South Africa

Carlo Punta
Department of Chemistry, Materials,
and Chemical Engineering "G. Natta"
Politecnico di Milano and INSTM Local Unit
Piazza Leonardo da Vinci
Milano
Italy

Twishi Puri
Department of Biology
Worcester State University
Worcester
MA
USA

Kavyashree Puttananjegowda
University of South Florida (USF) College of Engineering
USF
Tampa
FL
USA

Ananya Rajkumari
Department of Pharmaceutical Sciences
Dibrugarh University
Dibrugarh
Assam
India

Anugrah Ray
Department of Zoology
University of Kalyani
Kalyani Nadia
West Bengal
India

Laura Riva
Department of Chemistry, Materials,
and Chemical Engineering "G. Natta"
Politecnico di Milano and INSTM Local Unit
Piazza Leonardo da Vinci
Milano
Italy

Puspendu Roy
Department of Zoology
University of Kalyani
Kalyani Nadia
West Bengal
India

Mehtap Sahiner
Canakkale Applied Science
Department of Fashion Design
CanakkaleOnsekiz Mart University
Terzioglu Campus
Canakkale
Turkey

Nurettin Sahiner
Faculty of Science and Arts
Department of Chemistry
Canakkale Onsekiz Mart University
Canakkale
Turkey

Department of Chemical and Biomolecular
Engineering
University of South Florida
Tampa
FL
USA

Department of Ophthalmology
Morsani College of Medicine
University of South Florida
Tampa
FL
USA

William Serrano-Garcia
University of South Florida (USF) College of Engineering
USF
Tampa
FL
USA

Bianca Seufert
University of South Florida (USF) College of Engineering
USF
Tampa
FL
USA

Jigar Shah
Institute of Pharmacy
Nirma University
Ahmedabad
Gujarat
India

Pranav Shah
Maliba Pharmacy College
Uka Tarsadia University
Bardoli
Gujarat
India

Tabassum Siddiqui
Department of Biosciences
Jamia Millia Islamia
Central University
New Delhi
India

Sylvia Thomas
University of South Florida (USF) College of Engineering
USF
Tampa
FL
USA

Preface

Nanotechnology, if applied strategically and carefully, can provide sustainable solutions to many global challenges and lead us toward sustainable development. Sustainable innovations derived from nanotechnology applications can help us achieve several of the 17 sustainable development goals (UN-SDGs), a set of forward-looking goals for the world to adopt in order to achieve economic, social, and environmental sustainability. The SDGs were formally adopted by the United Nations in 2015, which garnered universal support from all UN member states. The agenda of actions to make the SDGs achievable by 2030 exhorts all nations adopt appropriate policy changes to their respective national development goals and plans (Source: United Nations Development Programme: https://www.undp.org/content/undp/en/home/sustainable-development-goals/background/). The major challenges the world is facing now, such as extreme weather, droughts, wildfires, devastating floods, sea-level rise, loss of biological diversity, recurrence of global pandemics, and increasing poverty and income/wealth inequality within and between nations, if not addressed immediately, will lead to an uncertain future for the present and future generations.

The world is facing challenges on all three dimensions of sustainable development – economic, social, and environmental. More than one billion people are still living in extreme poverty, income, and wealth inequality. The covid-led shut down of the global economy is felt most acutely in developing nations, and hundreds of millions of people who had escaped poverty over the past two decades are now being pushed back into poverty. At the same time, unsustainable consumption and production patterns over many decades have resulted in huge economic and environmental costs as the externalities associated with these unsustainable practices have been left unresolved. These externalities in the form of carbon dioxide pollution from fossil fuels and billions of tons of plastics and other consumption wastes have ended up in our oceans, landfills, and natural landscapes, to name a few major problems. Life on our planet is under threat from these and other sources of environmental degradation.

Achieving sustainable development will require concerted global actions to deliver on the legitimate aspirations for further economic and social progress, requiring economic growth and employment while strengthening environmental protection. Sustainable development must be inclusive and should take special care of the needs of the poorest and most vulnerable. Strategies need to be ambitious, action oriented, collaborative, and capable of adapting to different levels of development.

That means we need to systemically change consumption and production patterns, which, *inter alia*, will entail significant price corrections; encourage preservation of natural capital; reduce inequality; and strengthen economic governance.

(Taken from World Economic and Social Survey 2013: Sustainable Development Challenges https://sustainabledevelopment.un.org/content/documents/2843WESS2013.pdf)

Continuation of the current development strategies will not be sufficient to achieve sustainable development. Moreover, relying on "business as usual" scenarios presents clear risks, because evidence is mounting against the status quo such that:

a) The impact of climate change threatens to escalate in the absence of adequate safeguards, and there is a need to promote integrated and sustainable management of natural resources and ecosystems and take mitigation and adaptation action in keeping with the principle of common but differentiated responsibilities;
b) Hunger and malnourishment, while decreasing in many developing countries, remain persistent in other countries, and food and nutrition security continues to be an elusive goal for too many;
c) Income inequality within and among many countries has been rising and has reached an extremely high level, invoking the specter of heightened tension and social conflict;
d) Rapid urbanization, especially in developing countries, calls for major changes in the way in which urban

development is designed and managed as well as substantial increases in public and private investments in urban infrastructure and services;

e) Energy needs are likely to remain unmet for hundreds of millions of households, unless significant progress in ensuring access to modern energy services is achieved;
f) Recurrence of financial crises needs appropriate measure of prevention, and the financial system has to be redirected toward promoting access to long-term financing for investments required to achieve sustainable development.

One of the solutions for addressing these challenges is the "Need for inclusive strategies and technology innovation." In recent years, nanotechnology applications in major industries and occupations have grown significantly.

Nanotechnology has shown significant potential for resolving global challenges for sustainable development in many areas, such as water and sanitation, environmental pollution, soil degradation, health care challenges, food and agriculture, animal welfare, and deterioration of biological diversity, among others.

In this volume, we focus nanotechnology applications in three areas:

1) Nanotechnology and eco-friendly biobased products;
2) Nanotechnology and eco-friendly manufactured products;
3) Models and ways to understand the life cycle and risk assessment of manufactured nanomaterials and nanoengineering with special reference to global sustainability.

Given the enormous scientific and technological advances in this area, it will be impossible to cover all aspects of nanotechnology applications in one volume. Nevertheless, nanotechnology, which is expanding fast, will definitely have impact on environment and sustainability. We focus, specifically, on how the researches and innovations in nanotechnology will affect the environment. The book consists of several chapters authored by experts who are working in this area of research and development. We strived to cover as many areas as possible where nanotechnology made lasting impact, with particular focus on these impacts positively affecting global sustainability.

We would like to extend our sincere thanks and express our gratitude to all the authors who have contributed the chapters to this book; without their support, we would not have completed this book within the timeframe that we had set out at the outset. We would also like to extend our sincere appreciation to John Wiley & Sons, their editorial, production, and marketing teams who spared no efforts to make this book project a success.

We would like to thank our families who supported our efforts in bringing out this book. Our thanks also go to our respective home institutions that allowed us the time and freedom to complete this task.

We believe this book will be an excellent reference volume for people who are working in the area of nanotechnology and sustainability.

We believe we did our best to make sure the book is free of errors and faults; however, if you the reader find any mistakes or factual errors, we request you to bring them to our notice, so that we can correct and update the content in the next edition.

With best wishes to our readers

Yashwant V. Pathak, Govindan Parayil, and Jayvadan K. Patel

Foreword

There have been enormous developments in the area of nanotechnology with its application to deliver antimicrobial properties in hand washes, bandages, and socks, and zinc or titanium nanoparticles are the active UV-protective elements in modern sunscreens. With several innovations a significant interest has developed among governments, industries, and academicians.

Sustainability is now one of the greatest challenges in nanotechnology research. In addition to natural resources, there is a need for social and economic resources. The principle of sustainability has three major pillars: economy, society, and the environment, being translated to profit, people, and planet.

Nanotechnology provides a promise of positively contributing to global sustainability challenges, by providing interesting solutions such as water purification, energy efficiency, cheaper versions for solar energy, and solar panel development. In addition, nanotechnology plays an important role in health care through nanomedicine, nanorobots, nano health care challenges, nano-pharmacogenomics solutions to chronic diseases, nanoscaffolds, noninvasive surgeries, and so on.

Some challenges posed by nanotechnology are developing sustainable nanotechnology, to deal with potential ecological and health effects of nanotechnology on human health and to design nanotechnology-based products with minimal environment adverse impact.

This book by Dr(s) Yashwant Pathak, Govindan Parayil, and Jayvadan Patel addresses many of these issues and provides nanotechnology-based research priorities for global sustainability to improving sustainability and methods for development of various biobased products using sustainable nano biocomposites. Some chapters have also glimpses of novel approaches to design eco-friendly material based on natural nanomaterials.

I congratulate the editors to bring out this compilation addressing sustainable nanotechnology their strategies, products, and applications.

I am sure that the readers will find this book very useful for improving their understanding about sustainable nanotechnology and to plan further research in the field of nanotechnology.

With best wishes to the editors and all the chapter authors.

Prof. (Dr.) Ramesh K. Goyal
Vice Chancellor
Delhi Pharmaceutical Sciences and Research
University, India

1

Nanotechnology-Based Research Priorities for Global Sustainability

Twishi Puri[1], Yashwant Pathak[2,3], and Govindan Parayil[4]

[1] *Department of Biology, Worcester State University, Worcester, MA, USA*
[2] *Taneja College of Pharmacy, University of South Florida, Tampa, FL, USA*
[3] *Faculty of Pharmacy, Airlangga University, Surabaya, Indonesia*
[4] *Patel College of Global Sustainability, University of South Florida, Tampa, FL, USA*

1.1 Introduction

Sustainable development is not a newly constructed idea. Since the 1970s, conversations about this particular field have been fueled by the fear that the world cannot sustain an immense growth in human population and economic activities without damaging the environment and our social fabric. Concerns that at some point in the not too distant future, we will reach a "limit to growth" prompted policy-makers to urgently seek solutions for global sustainability, a concept that is so vast that it is subject to various interpretations. The vastness of the topic, however, still generates three aspects that global sustainability can affect. There is a global consensus that is agreed upon by the general public and the business community at large that environmental responsibility, economic efficiency, and social equity are necessary conditions for attaining global sustainability. Environmental responsibility, as the name suggests, is concerned with the conservation of resources – i.e. food, water, energy, etc. and safe methods of disposal of waste products of all forms. Likewise, economic efficiency focuses on resource production and meeting the demands of the market place. It is concerned with open trade and no inhibition in terms of the allocation of resources for production. Social equity, on the other hand, is concerned with the distribution of resources based on the productivity of an individual or an organization. In other words, social equity allows people to make decisions and ensures that the rewards that they get are based upon their efforts [1]. So, why address sustainability? Today, concerns for maintaining sustainability have increased, especially among the younger generation. Concerns about the availability of natural resources and the safe and efficient disposal of the by-products of our production and consumption practices have added urgency to transitioning to a sustainable future. Consequently, finding solutions to these concerns have focused on using innovative science and technology applications. Therefore, applying nanotechnological research to maintain global sustainability has become a priority [2].

Nanotechnology refers to understanding and control of the material at the nanoscale. For reference, a nanometer is one billionth of a meter. The field was introduced almost half a century ago and, over time, it has established itself as an active research area [3]. It incorporates nanoscale science, engineering, and technology – three very useful fields with various applications. At the nanoscale, materials show unusual biological, physical, and chemical properties. In fact, according to quantum theory, nanomaterials, with size being within the range of 1 and 250 nm, lie between the quantum effects of atoms, molecules, and the bulk properties of materials. This nanoscale is known as the "no-man's-land" where their properties are controlled by the phenomenon that has its own critical dimensions. The structure of nanoparticles can be manipulated to produce materials with desired properties. Using nanomaterials with these unusual properties gives us an opportunity to enhance existing technology with profound features that have technical, economic, and societal implications [4]. Advocates of nanotechnology claim that the combinations of nanotechnology with various fields such as information technology, biotechnology, and cognitive sciences produce far-reaching advances. In terms of global sustainability, nanotechnology's influence in various areas can change the future of our efforts for sustainability [5].

There are many fields in which incorporating nanotechnology can lead us to global sustainability. The three focus

Sustainable Nanotechnology: Strategies, Products, and Applications, First Edition.
Edited by Yashwant V. Pathak, Govindan Parayil, and Jayvadan K. Patel.

areas are the environment, the economy, and society. There are various fields in which nanotechnological research has already affected the growth. This chapter focuses on the fields of medicine, food, environment, health, and industry. These fields fall perfectly under the three focus areas of global sustainability. The nanotechnology-based research done under these fields has not only enhanced them but also made them safe and sustainable.

1.2 Medicine

Medicine and nanotechnology, for the most part, go hand in hand. Whether it is the field of surgery or drug delivery, nanotechnological research has been very much involved in revolutionizing medicine. The growing interest in medical applications of nanotechnology has resulted in the emergence of the field popularly known as nanomedicine. Nanomedicine refers to applying nanoscale biotechnology to medicine. It allows us to use nanotechnology to improve the human biological system as well as create powerful tools for treating human diseases. In terms of sustainability, nanomedicine's aim is to improve the overall quality of life by working at a molecular level to target diseases and formulate treatments [6].

1.2.1 Nano Oncology

Cancer is the result of uncontrolled cell division and has the tendency to spread to other regions of the body. Healthy cells can be converted to tumor cells with the right combination, or in this case, the wrong combination of genes and environmental factors. According to the statistics published by the Global Cancer Observatory (GLOBOCAN), in 2018, there were 18.1 million new cancer cases and 9.6 million cancer deaths, the leading cause being lung, bowel, prostate, and female breast cancer [7].

Over the past several decades, nanotechnology has made magnificent contributions to oncology, not just in terms of diagnosis but also regarding drug delivery for treatment. Specifically, in cancer therapy, the use of nanomaterials has allowed the development of targeted drug delivery, enhanced the properties of therapeutic molecules, and developed a sustainable or stimulus-triggered drug delivery [8]. There are a lot of factors involved in the effective management of cancer treatment, one of which is early detection. In order to detect uncontrolled growth, pathologists use cancer biomarkers. According to the US Food and Drug Administration (FDA), biomarkers are "any measurable diagnostic indicator that is used to assess the risk or presence of disease" [9]. Every cell type in the body has unique molecular features and characteristics.

Cancer cells, or other cells in response to the presence of abnormal growth in the body, release biomolecules that are different from the biomolecules released from healthy cells. These biomolecules are defined as biomarkers and can be used to define the molecular definition of cancer [10]. Examples of biomarkers include genes; gene products; specific cells; enzymes; hormones present in blood, urine, tissues, and other bodily fluids; proteins or protein fragments; and DNA- or RNA-based fragments [11, 12].

There are several existing methods of detection available including:

a) the Papanicolaou test to detect cervical cancer and mammography for breast cancer detection for women,
b) prostate-specific antigen (PSA) test for a blood sample of men to detect prostate cancer,
c) occult blood test for colon cancer detection, and
d) endoscopy, X-rays, ultrasound imaging, CT scans, and MRI are used for various detection purposes.

However, there are many limitations to the current methods. Furthermost, these methods are not always successful at detecting cancer at early stages. In addition, they are neither affordable nor available to many people who require them. The priority should be to discover new methods of detections that are accessible when needed. For detection, nanomaterials' physical, optical, and electrical properties are quite useful. Over the years, the development of nanomaterials such as quantum dots, gold nanoparticles (GNPs), carbon nanotubes (CNTs), magnetic nanoparticles, gold nanowires, and many others works to lessen the limits of the standard methods of detection and increase the precision of detection [12].

1.2.1.1 Gold Nanoparticles

In comparison with other nanomaterials, the nanostructure of metallic nanoparticles is most flexible due to the synthetic control of their shape, size, structure, composition, assembly, and encapsulation, along with the tenability of their optical properties. Within these metals, GNPs are extremely useful in biomedical applications because their preparation time is shorter and the process is simpler than the others. Gold nanospheres can be prepared by reducing auric acid with different concentrations of sodium citrate for size variation. In addition, the citrate capping on the gold particles can be replaced with biomolecules such as DNA, peptides, and antibodies; they form covalent and noncovalent bonds with GNPs [13].

There are many applications of GNPs in cancer imaging and cancer therapy. The combination of GNPs and dynamic light scattering, a technique usually used for nanoparticle size analysis, enabled the analysis and detection of chemical and biological target analytes such as proteins, DNA, viruses, carbohydrates, and chemical and environmental toxins [14, 15].

With regard to treatment, photothermal therapy (PTT) using GNPs can initiate a hyperthermic physiological response in the tumor [16, 17]. The GNPs are able to convert light into heat, which can "melt" the targeted tumor. Gold, in particular, is useful in PTT because of its specificity. They can be administered to a local tumor area, which decreases the chances of its distribution to healthy cells, a problem with the conventional cancer treatment methods [18]. They can also be used to prevent migration of tumors to other areas of the body. In 2017, a study showed that GNPs targeting the nuclear membrane of cancer cells can increase nuclear stiffness and prevent cell migration and invasion. The nanoparticles trapped at the nuclear membranes can lead to overexpression of lamin A/C protein that leads to cell stiffness [19].

1.2.1.2 Quantum Dots

Histological assessment of solid tumors includes imaging and biopsy, and in most cases, surgery is performed to remove the primary tumor and evaluate the surrounding lymph nodes. Visible, fluorescent, and radiolabeled small molecules have been used as contrast agents to improve detection during real-time intraoperative imaging, but unfortunately, the current dyes lack the tissue specificity, stability, and signal penetration needed for optimal performance. Graphene quantum dots are used in cancer-targeted drug delivery. It was recorded that the mean survival time of tumor-bearing mice can be extended by 2.5 times when treated with Qdots [20]. Semiconductor quantum dots having superior optical properties are well-established fluorescent imaging probes.

Compared to conventional small molecule dyes, their size, high stability, non-photobleaching, and water solubility made them a unique fluorophore. At the same time, there have been major concerns regarding their potential nanotoxicity because high-quality Qdots often contain heavy metal elements [21, 22]. Newly emerged theranostic drug delivery system using quantum dots helped in a better understanding of the drug delivery mechanism inside the cells. Nanoscale quantum dots, with unique optical properties, have been used for the development of theranostics. Surface-modified quantum dots and their applications became widespread in bioimaging, immune histochemistry, tracking intracellular drug, and intracellular molecules target [23]. Chemotherapy or PTT is always inefficient due to their inherent limitations, but their combination for the treatment of cancers has attracted great interest during the past few years. A promising theranostic agent, black phosphorus quantum dots (BPQDs), due to its excellent photothermal property, extinction coefficient, and good biocompatibility and biodegradability, hold great potential for cancer treatment. However, the rapid degradation of BP with oxygen and moisture causes the innate instability that is the Achilles' heel of BP, hindering its further applications in cancer theranostics. The BPQDs-based drug delivery system exhibited pH- and photo-responsive release properties, which could reduce the potential damage to normal cells. The in vitro cell viability study showed a synergistic effect in suppressing cancer cell proliferation [24, 25]. Studies show that nanoplatform of BPQDs camouflaged with a platelet membrane (PLTm) carrying hederagenin (HED) significantly enhances tumor targeting and promotes mitochondria-mediated cell apoptosis and autophagy in tumor cells [26].

1.2.1.3 Carbon Nanotubes

CNTs are one of the unique one-dimensional nanomaterials discovered by SumioIijima in 1991. CNTs can be functionalized via different methods to perform their specific functions and received more and more attention in biomedical fields. It is because of their unique structures and properties, including high aspect ratios, large surface areas, rich surface chemical functionalities, and size stability on the nanoscale range [27, 28]. Being attractive carriers and mediators for cancer therapy, they have also been applied as mediators for PTT and photodynamic therapy to directly destroy cancer cells without severely damaging normal tissue.

CNTs are becoming one of the strongest tools that are available for various other biomedical fields as well as for cancer therapy [29]. CNTs are used as nanocarrier transporters to transport anticancer drugs, genes, and proteins for chemotherapy that makes them effective in delivering biomolecules and drugs [30, 31]. They have the ability to enter cells, and this behavior is independent of cell type and functional group at their surface.

Research shows a variety of chemically functionalized CNTs have the ability of biocompatibility with the biological environment. The behavior of the material can be regulated making them a useful tool for all kinds of diagnosis and therapeutic as well as drug delivery applications [32, 33]. Besides CNTs, MnO⍰nanotubes, platinum nanoparticles, paclitaxel-loaded riboflavin, and thiamine-conjugated multiwalled CNTs showed promising potential in the treatment of cancer [34–37].

1.2.2 Drug Delivery

The current drug delivery system for cancer treatments may have shown positive results, but they definitely have negative consequences as well. These drugs are toxic, have poor tissue selectivity, and continuously using them may increase resistance against them [38]. The toxicity of a drug is always a concern because most drugs are not tissue specific, i.e. they have the same effect on normal and abnormal cells. The use of nanotechnology makes it possible to have a specifically targeted drug delivery system. From size manipulation to creating various delivery vehicles, nanotechnology can help specifically to target and assemble in tumor cells [39].

1.2.2.1 Metal-Based Drug Delivery

There are several metal-based nanodrugs that have shown great success in treatments; however, they are known to produce large quantities of toxic and other harmful substances [40]. A way to lower the negative consequences of nanometal drugs is to modify their properties. For example, the study of biodegradable iron stents and cobalt-chromium stents in porcine coronary arteries of juvenile pigs showed to have the potential to reduce chronic inflammation and premature recoil. Another study shows that modifying biocompatible and monodispersed iron oxide superparamagnetic nanoparticles with the combination of folic acid (FA) and polyethylene glycol (PEG) increased the affinity of nanoparticle uptake by targeted cells [41].

For antitumor therapy, two-dimensional molybdenum disulfide (MoS_2) nanosheets have proved to be a good photothermal agent. An extensive study has shown that soybean phospholipid encapsulated MoS_2 nanosheets have shown good photothermal conversion performance and photothermal stability. In addition, soybean phospholipids can be found in nature, so the cost of obtaining it from natural resources is always lower than synthetically developing it. The reason MoS_2 nanosheets do not necessarily carry a drug to cancerous cells; however, as a photothermal agent, they can absorb near-infrared reflectance (NIR) light and convert it into heat, which then can be transported to tumor cells. This process will bring the temperature to the critical temperature of 42 °C and result in efficient cell death. This was tested on mice with breast tumor growth by intravenously and intratumorally injecting soybean phospholipid molybdenum disulfide (SP-MoS_2) nanosheets. Both methods showed suppression of growth of the tumor [42].

Metallic nanodrugs can also be used as antiseptics or for antimicrobial purposes. For example, hydrophilic metallic silver nanoparticle (AgNP) nanocomposites composed of a polymer matrix of *N*-vinylpyrrolidone (poly [VT-*co*-VP]) have various uses in medicine especially as an ingredient in burn medicine. The study shows these metallic nanocomposites exhibiting antimicrobial activities toward Gram-negative and Gram-positive bacteria. Additionally, silver has shown to have more antimicrobial effect on Gram-negative bacteria due to better linkage between the silver nanocomposites and the hydrophilic channels present in the outer membrane of Gram-negative bacteria. In addition to antimicrobial properties, silver nanocomposites have shown to not precipitate or shrink when stored in an aqueous environment for four months. This is due to the stability of functional groups in the nanocomposites. It is proposed that these silver nanocomposites can be used for the treatment of various infectious diseases and can be quite useful after surgeries in which the major problems can be caused by exposure to bacteria [43].

1.2.2.2 Biotechnology-Based Drug Delivery

There are various developments in drug delivery systems based on combinations of biomacromolecules and nanoparticles. Since drug delivery is popular in cancer treatment, most of the developments have occurred in oncology. For the treatment of malignant melanoma, folate-decorated cationic liposomes have been developed as nonviral vectors of hypoxia-inducible factor 1-α siRNA (HIF-1α siRNA). Hypoxia-inducible factor 1-α is a transcription factor that responds to hypoxic stress and could be a potential target in malignant melanoma therapy. When HIF-1α is upregulated, transcription is activated that results in angiogenesis. Small interfering RNA (siRNA) are pieces of double-stranded RNA that can interfere with the translation of proteins and inhibit angiogenesis when used against HIF-1α. The double-stranded RNA alone did not achieve the antiangiogenesis activity, thus HIF-1α siRNA vector is an excellent vehicle that can load siRNA and protect it from degradation [44]. Another method of delivering anticancer drugs, such as quercetin, is a lecithin-based mixed polymeric micelle. Although quercetin (Que) is a well-known and successful anticancer drug, its low solubility and low oral bioavailability (BA) hinders its use in clinical settings. A micelle as a delivery system is quite useful in this case because its hydrophobic core and hydrophilic shell provide a safe passage for low soluble drugs. To increase the solubility of drugs, the more hydrophobic material is added to the micelle, which increases the space in the hydrophobic core and provides more space for drugs to be solubilized. Lecithin is a hydrophobic mixture of organic phospholipids that help in the absorption of drugs. In this case, lecithin helps increase the BA of Que. These micelles not only are able to increase drug solubility and BA but also, due to their nanosize, are able to enter and gather in tumor sites [45].

Even though most of the drug delivery methods are focused on cancer treatments, some specialize in other problematic areas of the human body. Skin is the largest organ of the body and the stratum corneum is the main barrier that drugs need to penetrate to get into the deeper layer of the skin. In the case of antifungals, drugs should be able to get through this layer but are not always able to do so. The development of nanosized colloidal carriers can be used as vehicles for drug delivery. Studies done on naftifine-loaded microemulsion colloidal carriers showed that the carriers were an effective way of delivering naftifine, an antifungal drug, to deeper layers of the skin. Additionally, the method of delivery was shown to have low levels of cytotoxicity [46].

1.2.3 Biosensors

Biosensors are tools used to detect and analyze biological elements. Conventional biosensors have their advantages, but they also exhibit several limitations. Nanotechnology, however, eliminates the limitations of conventional methods. In fact, as the material dimensions are minimized, the applicability of biosensors is improved [47]. There are several nanoparticle-based biosensors that can help detect pathogenic viruses. For example, the development of quantum dots-based imaging and capturing systems for selective capturing and detection of the HIV in whole blood. It is a dual-stain imaging system for the detection of HIV1 gp120 envelope glycoprotein. It is also capable of obtaining countable imaging. This system can work with 10 μl of a blood sample, is portable, and is highly cost-effective compared to other methods [48]. For the detection of multiple viruses, the fluorescence characteristic of AgNPs is quite useful for optical biosensors. Using silver nanoclusters, biosensors have been prepared for the detection of specific DNA sequences of HIV, hepatitis B virus (HBV) and human T-lymphotropic virus type I (HTLV-I) gene. Before binding to the DNA sequence, these nanoclusters exhibited high fluorescence activity. After attaching to the sequence, however, the fluorescence intensity decreased, allowing the target sequences to be detected [49]. Additionally, magnetic nanoparticles have also been utilized for virus detection. Amino functional carbon-coated magnetic nanoparticles, for example, have been used to distinguish hybridization of HBV nucleic acids [50, 51].

MicroRNAs (miRNAs) are small noncoding RNAs that regulate gene expression by inhibiting translation and play a role in RNA degradation [52–54]. Currently, miRNA detection has several limitations such as sensitivity and selectivity [55]. Additionally, the current methods of miRNA detection, including northern blotting, microarray analysis, and RT-PCR, are high in cost, complicated to handle, and do not produce stable results [56]. Tumor-derived miRNA-141 deregulation in human plasma is an important biomarker for blood-based detection of various cancers, such as prostate [57], colons [58], and ovarian [59]. An electrochemical immunosensor composed of modified gold electrodes, reduced graphene oxide, and CNTs has been developed for the detection of the miR-141 gene. With this method, the detection limit was down to 10fM [60]. Similar to miRNA-141, miRNA-155 is another biomarker for diagnosis of diffuse large B-cell lymphoma [61]. Oligo-hybridization-based electrochemical biosensors can be used for its detection. These biosensors utilized GNPs on sheets of graphene oxide situated on glassy carbon electrodes. This particular biosensor exhibited higher selectivity as it was able to distinguish between complementary target miRNAs, three-base mismatch, and noncomplementary miRNAs.

It also was capable of directly detecting miR-155 in plasma without any need for sample preparation, extraction, and amplification [62]. Another example of a useful biomarker is miRNA-21, the most frequently upregulated miRNA in breast cancer and can also be used for early diagnosis and drug development for cardiovascular disease [63]. An electrochemical biosensor based on a metal ion functionalized titanium phosphate nanospheres is a sensitive and selective tool for detecting miR-21. The addition of cadmium ion to titanium phosphate nanosphere exhibited improvement of the electrochemical signals by five times [64, 65].

1.3 Food and Agriculture

Advocates of global sustainability recognize and emphasize the importance of sustainable development of agriculture and food [66]. Currently, agriculture is one of the largest causes of global environmental change. The process of food production alone contributes to 30% of the global greenhouse gas emissions [67], occupies 40% of the land [68], uses 80% of the freshwater [69], and is one of the largest factors contributing to species extinction [70]. There are several aspects of the food sector that can be enhanced to prevent and minimize the negative consequences of the current system. Using nanotechnology in food and agriculture can not only increase the safety of the product and protect the environment but can also be used to improve the mechanisms of food distribution.

1.3.1 Fertilizers

Chemical-based conventional fertilizers may have worked previously but have been deemed unsustainable since the beginning of the green revolution [71]. Current fertilizers cause a loss of nutrients. Nitrogen is an essential mineral required for the growth of crops. It is lost through processes of nitrate leaching, denitrification, and ammonia volatilization. This loss of nitrogen not only affects plant growth but also contributes to pollution, global warming, and causes a huge economic loss [72]. This and other problematic results of conventional fertilizers can be prevented by using nanofertilizers. Nanofertilizers can enhance nutrient use efficiency by causing a higher uptake of nutrients. This is accomplished by the smaller surface area of nanomaterials, which is known to increase the nutrient surface interaction. Nanomaterials can also be used to enhance the results of conventional fertilizers. Slow-release chemical-based fertilizers, for example, can be coated with nanoparticles and significantly reduce the results of nitrate leaching and denitrification [73]. Compared to conventional fertilizers, nanofertilizers have many advantages. Where conventional fertilizers work rapidly, nanofertilizers feed the crops gradually in a controlled manner. They are highly effective in nutrient absorption by plants and result in a lower loss of essential nutrients. This is due to their nanosized pores and their ability to utilize various ion channels within the plants. In addition, polymer-coated fertilizers are able to avoid contact with the soil and water due to the coating encapsulating the nanoparticles. This ensures that the nutrients and minerals are available for the plant to uptake when it is ready to do so. This also minimizes the unnecessary loss of nutrients [74].

1.3.2 Application in Food Science

Due to public apprehension and the regulatory agencies not reaching upon agreement, the application of nanotechnology in food preparation does not have any worldwide applicable rules and, hence, is still in the developmental phase [75–77]. There are various aspects needed to be explored in the relationship between nanotechnology and food preparation. Application, for one, is certainly a matter of discussion [78]. Nanotechnology can have various possible applications in food science. For example, nanomaterials can be used in food products to increase its freshness and improve its taste. This can be done through methods of nanoencapsulation. SiO_2 nanomaterials, for example, can act as carriers for flavors in food [79]. In addition, the functionality of these applications should also be the focus development of this field. What are the reasons for using nanomaterials? Protection against biological deterioration, chemicals, and enhancement of physical properties can be few of the reasons. Furthermore, the safety of using nanomaterials over conventional materials and methods should be assessed, for example, what are the results of these applications in in vivo and in vitro experiments, or do nano-enhanced food contribute to food allergies? [78]

Healthy and sustainable food is certainly a goal for the future, especially to achieve global sustainability. Sustainability of food is a bit tricky because food has a shelf life and for most items, it is not very long. A reason for that is biological pathogens. Silver nanoparticles and nanocomposites have been identified as antimicrobials by the US FDA. The Ag^+ ions in AgNPs are responsible for binding to cause morphological changes and generation of reactive oxygen species by binding to membrane proteins of bacteria. This causes damage to cells and death due to oxidative stress [80, 81]. Even though some nanomaterials are used to cause oxidative stress, there is some that act as antioxidant carriers. A developed example of this is SiO_2 gallic acid nanoparticles that contain a high capacity of 2,2-diphenyl-1-picrylhydrazyl radicals, a compound used to measure antioxidant activity [82]. Nano-delivery materials can also be used to increase the BA of bioactive compounds in food products. Depending on the type of bioactive compound, there are many nano-carries available. For instance, coenzyme Q10 is a lipophilic compound and is not very soluble in water, which is the cause of its low BA. A lipid-free nano-CoQ10 system is modified with various surfactants, which improve the solubility and BA of coenzyme Q10 in oral administration [83].

1.3.3 Food Packaging

Similar to its contribution to food science, nanotechnology's contribution to food packaging focuses on increasing the shelf life of food and improving its safety. Currently, polymer-based materials, synthetic and organic, are widely used in biomedical sciences and agriculture. However, polymers alone cannot achieve the required performance of an innovative packaging; hence, nanomaterials such as CNTs, nano clay, and biocomposites have been used to manipulate the polymers and improve their performance [83]. In food packaging, the combination of polymers and nanomaterials has developed intelligent and active packaging systems. The basis of each packaging system is the mimicking of biological processes, which preserves the integrity of the package and foods in food chain systems [84].

1.3.3.1 Intelligent Packaging

An intelligent packaging system focuses on monitoring the conditions and quality of food products during the distribution and storage stage of the supply chain and intends on delivering this information to the consumer of the product. Intelligent packaging systems can be distributed in four categories: data carriers, quality indicators, sensors, and others such as organic light-emitting diodes (OLEDs) and holograms [84]. Nano-based communication devices such as Radio Frequency Identification (RFID) tags and a barcode with wireless sensors could be used to provide product authenticity, anti-theft, anti-counterfeiting [85], and product traceability [86]. For instance, a wireless RFID sensor tag consisting of two planar inductor-capacitor resonators to monitor relative humidity has shown a sensitivity range of 20–70%.

This system could be integrated into the RFID sensors of conventional packaging machines [87]. ZnO nanoparticle and polyvinylpyrrolidone (PVP)-based luminescent films also showed the capability of sensing the status of food substrates by the intensity of their luminescence [88]. In addition, the use of quantum dots and graphene to develop a chip-based sensor to detect oil samples exhibits the capability to differentiate between various oil samples in a laboratory setting. Compared to the conventional methods of high-performance liquid chromatography (HPLC), this system was able to discriminate between eight different unknown oil samples with an accuracy of 92.5% [89].

1.3.3.2 Active Packaging

In active packaging systems, food products, packaging materials, and the environment are interacting together to extend the shelf life, quality, and safety of the products [85]. Active packaging systems can be categorized as scavengers, blockers, releasers, and regulators [84]. The focus here is to protect the food products from harmful microbes [90], excess moisture [91], and excess oxygen [92]. Active antimicrobial packaging systems are combinations of antimicrobial agents and nanomaterials. For example, antimicrobial nanofibrous films of polyvinyl alcohol-b-cyclodextrin with cinnamon essential oil performed well in suppressing the growth of *Staphylococcus aureus* and *Escherichia coli* [93]. Substitution of cinnamon essential oil with lemongrass and oregano essential oils exhibited suppression of *Salmonella enteritidis* in ground beef placed in a sterile plastic bag for six days [94]. For reduction of the rate of oxygen transmission, ascorbic and iron powders or copper chloride can be used as catalysts for oxygen scavenging thermoplastic starch films. This method reduces the oxygen transmission rate from 20.9% to 1% in 15 days at 80% relative humidity (RH) [93]. Additionally, for the modification of atmosphere packaging, oxygen scavenging polyethylene terephthalate (PET) films, PET-aluminum oxide coatings, polylactic acid films, and oriented polypropylene (o-PP) films were deposited with palladium layers through vacuum deposition on the silicon oxide layer. This method was also able to significantly reduce the oxygen transmission rate [91]. Similarly, low-density polyethylene (LDPE) films combine with activated carbon and sodium erythorbate exhibited the oxygen concentration absorbance rate of 80% [92].

1.4 Human Health and the Environment

Presently, environmental sustainability is recognized as one of the biggest issues fueled by and affecting humankind. Directly related to the continuous increase in population, the constant deteriorating environmental health is holding us back from achieving global sustainability. Moreover, almost every aspect of sustainability discussed in this chapter either is dependent upon the environment or contributes to its condition [95]. Although there are many aspects of the environment that can be improved to prevent any more damage, the three discussed here are air, water, and energy. The influence of nanotechnology can enhance these three facets of the environment in a safe manner and minimize the effects of any future anthropogenic activities.

1.4.1 Water Purification

Clean water is an essential part of global sustainability. The constantly increasing the global human population increases the demand for clean water; however, population increase is also a factor in the growth of industries, a huge factor in the decrease in water quality. Making clean and affordable water accessible to people is still a challenge today [96]. There are various conventional methods used today for water treatment and purification. Nanomaterial-based water purification methods, however, not only improve the quality of water but also extend purification treatments to remote areas without electricity [97]. Many nanomaterials used in nano oncology and for drug delivery are also utilized in water treatments. For example, CNTs, in this case acting as nano adsorbents, are better alternatives of activated carbon because they are able to absorb organic chemicals more efficiently than activated carbon [98].

Nanomembranes are another method of removing microparticles from water. These membranes are composed of nanofibers and, when combined with metal oxide nanoparticles, can intensify membrane surface hydrophilicity, water permeability, and fouling resistance [96]. Nanocatalysts, such as zero-valent metal, semiconductor materials, and bimetallic nanoparticles, are used in purification treatment to amplify reactivity and degradation of contaminants such as pesticides and herbicides [99]. Studies show that silver nanocatalysts, N-doped TiO_2 and ZrO_2 nanoparticles are successful in the degradation of contaminants in water [100]. Similar to nanocatalysts, nanostructured catalytic membranes also have higher rates of decomposition and selectivity. These membranes require less contact time, can be scaled for commercial purposes, are composed of homogeneous catalytic sites, and allow multiple reactions to take place simultaneously [101].

1.4.2 Air Purification

Air is a unique essential resource needed by living beings. Unlike water, living without air is quite impossible for aerobic species. With water, part of the problem is that it is inaccessible to some remote areas of the world. With air the situation is different because there is an abundance of breathable air, the only problem is the quality of it. As the global population increases and continues to spread, the result is the settlement of factories and other industrial buildings and an increase in automobile use, which contributes to the poor quality of breathable air.

Some examples of nanotechnology used in air purification methods are CNTs, GNPs, and nanocatalysts. CNTs have a small pore structure and large surface area of functional groups, which can be manipulated through optimum chemical or thermal treatment. These characteristics allow CNTs to be highly efficient in trapping perilous substances from the air [102]. Unlike CNTs, GNPs have shown to exhibit converting characteristics. For example, when combined with titanium dioxide, GNPs are able to convert sulfur dioxide present in polluted air into sulfur [103]. Nanocatalysts also exhibit converting characteristics. The surface area of these catalysts is large enough for chemical reactions to take place. These reactions are able to convert the harmful gases produced by automobiles and factories into safe gases.

1.4.3 Energy

Water and air are essential to the survival of most living beings, but for humans, energy has become as important. As much as we promote the use of alternative sources for energy, most of it is still produced from fossil fuels. This has a huge impact on the environment and is actually a barrier to achieving sustainability. Fossil fuels may have been an excellent source for many of our needs, but it certainly has proved to not be a reliable source. First, fossil fuels are considered to be nonrenewable resources, i.e. at one point we would have to deal with their depletion and find reliable alternative resources for our needs. Additionally, this resource is not readily available to everyone because of its uneven distribution – a huge concern for those who are not able to access it [104, 105]. There are many alternative sources of energy available, most popular being solar energy, but they have not been used at a large scale due to issues with converting nonrenewable sources to renewable energy and minimizing energy loss [106]. The use of nanotechnology in the energy sector is able to provide clean energy in a cost-effective way by developing both conventional and renewable energy sources.

1.4.3.1 Energy Conversion

Even though there are many forms of energy available to us, we cannot directly use them. In everyday life, the most useful forms of energy for us are electrical and heat energy, but they can only be generated through conversion from other forms of energy. In reality, all forms of energy, including nuclear energy, come from the sun, hence solar energy is the common source. This energy can be converted to chemical, heat, wind/hydro, and mechanical energy, all of which can be ultimately converted to electrical energy, the most common form of energy used [107]. Every day, the sun releases a huge amount of energy, making solar energy abundant and cost-free. This energy is released in the form of heat and radiation. Using photovoltaic cells, sunlight can be converted into solar energy that can then be converted into other forms of energies [108]. Conventional solar cells have two drawbacks; they are expensive to produce and their efficiency is rarely above 20%. This is mainly due to the energy of the photons in the cell being larger than the bandgap energy, the right amount of energy needed for the solar cell to work. To correct this, solar cells are enhanced with quantum dots. Due to their variation in size, quantum dots are able to produce various bandgaps that allow the photons that previously had larger energy to pass through. In addition, solar cells made of nanomaterials, such as nanocrystalline silicon, are capable of increasing the efficiency of solar cells by 40–50% [109].

1.4.3.2 Energy Production

Some of the already discussed nanomaterials can be used to make already established energy production mechanisms more effective. To make solar cells more cost-efficient and effective, organic materials such as CNTs have been used. Organic materials have quite a few advantages over inorganic materials in the case of solar cells. Organic semiconductors exhibit a high absorbance coefficient, which allows the photons to be absorbed within a thin layer of the solar cells. This marginally decreases the cost of solar cell production. Additionally, organic materials have shown to be more efficient with increasing temperature, whereas inorganic materials have exhibited a loss of efficiency when the temperature increases. Even though the use of CNTs in polymer-based solar cells have very limited efficiency, for commercial use, it is more desirable due to its low cost and various applications [110].

Nanotechnology is utilized in unconventional methods of energy production in many ways. In the case of windmills, nanotechnology can be used to enhance the operation and efficiency of it. The use of CNTs in making rotor blades results in higher fatigue resistance, shear strength, and fracture toughness [111]. Nanocomposites can also be used to coat the blade to prevent it from damage from the weather and other environmental factors. For example, superhydrophobic nanocomposites containing titanium power have been shown to reduce water adhesion strength, which means in extremely cold temperatures the water will not stick to the rotor blades and possibly damage them [112]. Nanocoatings and nanocomposites can also be used for corrosion protection in hydropower systems and for drilling equipment of geothermal, oil, and gas systems [113].

1.4.3.3 Energy Storage

Sustainability of energy includes more than just safe production and conversion; being able to store it for later use is quite important. Nanotechnology's influence in energy storage can make it a safe and cost-effective process in addition to sustainability. The simplest form of energy storage and one that most people are familiar with are batteries [114]. Most of the electronic devices that are used today are portable, which has increased the demand for an energy storage unit that is high density yet lightweight. This can be done by using nanocrystalline separator plates in batteries, which not only allow more storage of energy than conventional methods but also make the battery lightweight due to their foam-like structure [115].

A safer alternative to fossil fuel-generated energy is using hydrogen as an energy carrier. Hydrogen has shown the potential to hold a tremendous amount of energy and can be converted into other energy forms without releasing any harmful emissions. Various nanomaterials, especially carbon based, are good candidates for hydrogen storage due to their high absorbency, high specific area, pores, and low-mass density [116]. Combination of single-walled CNTs and BH_3 may work as a reversible hydrogen storage system and allow storage and release of hydrogen. This makes it optimal for hydrogen-based fuel cells that could be used in vehicles.

1.5 Industry

Most applications of nanotechnology highlighted in this chapter are somewhat directly related to the health and sustainability of the human body. Industries can contribute to the deteriorating condition of the environment by emitting harmful gasses. It can also have a negative influence on the health of living beings through the emission of harmful gases and particles. Nanotechnology, however, is not limited to just those applications. The use of nanomaterials in various industries can produce safe materials and minimize their negative consequences. It can also increase the cost-efficiency of the materials and make the industry economically sufficient.

1.5.1 Automotive

The data on the ownership of automobiles continues to climb as the influence of industrialization continues to spread. Along with the increase in the automotive industry comes an increase in fuel consumption, greenhouse gases, and resource usage. This in return increases the demand and cost of fuel and resources. Even though car manufacturers do put in the effort to minimize the negative consequences of automobiles and increase its efficiency, measures to achieve sustainability are rarely implemented [117]. With the introduction of nanotechnology, new opportunities to make the industry safe and sustainable have arisen. The combination of car engineering and nanotechnology has influenced change in each part of the car. For instance, the improvisation of nanomaterials such as carbon black and silica in car tires results in lower rolling resistance, abrasion resistance, friction, and extended tire life and safety; decrease in weight; and overall a superior performance. In addition, brominated isobutylene-co-para-methyl styrene elastomer-based nanoclay has proved to increase the air retention of tires by 50% in comparison with halobutyl rubbers. CNTs have also been used to enhance the tensile strength and tear strength of the tires [118]. In terms of thermal performance, nanofluids have the potential to improve the cooling rates of the engine by increasing efficiency, decreasing the weight, and making the thermal management systems more simple. They can also be added to fuel additives, coolants, engine oils and greases, and brake fluids [119–121].

1.5.2 Construction

The construction industry is a significant contributor to the world's economy. According to the Global Construction Perspectives and Oxford Economics, China, United States, and India will experience an 85% growth in construction by 2030 [122]. It is important to consider enhancing the materials and functional properties of the construction [123]. Concrete, a predominantly used material in construction, is composed of several ingredients that have their own disadvantages. Nanomaterials can be used to alter and improve the properties of concrete. For example, adding nanosilica to concrete can improve its resistance to segregation, increases its strength of hardening, prevents calcium leaching, and decreases the ability to absorb water. In addition, using fiber sheets containing nanosilica particles and hardeners can increase the strength and durability of existing structures [124]. Some of the already mentioned nanomaterials, such as CNTs, can also be used to improve the mechanical properties of concrete [125]. During concrete production, the soil is endangered due to the carbon emission caused by the process. To prevent this, nano-aggregate, such as C–S–H gel, can be added to concrete. This gel is able to breathe carbon dioxide into carbon and oxygen and decrease the amount of carbon emission [126]. Steel is another vital component of the construction industry due to its properties such as strength, corrosion resistance, welding ability, and low cost. The American Iron Steel Institute and the US Navy have developed steel with higher strength than usual by adding nanomaterials such as carbon nanoparticles and CNTs. This also makes the steel more cost-effective [127].

1.6 Further Training

Although nanotechnology has various applications in the path to global sustainability, it has its risks and limitations that would need to be sorted out before any further development. The enhancement of agricultural methods with the help of nanomaterials has been discussed earlier, but its contribution to the food sector also generates some major risk factors. For one, the toxicology assessments for nanomaterials may not be sufficient enough. The current data from traditional assessments rely on mortality and sublethal endpoints. These tests are also time consuming, costly, and do not relay the data regarding delayed toxicity. The data from one of these assessments may generate a result of low toxicity, but how is that affecting the human body and the environment, in the long run, is not predictable from current methods of testing. Some have suggested using genomic and proteomic techniques for a faster and cost-effective assessment of long-term toxicity. However, these techniques do require the state-of-the-art instrumentation [128]. When it comes to nanomaterials, the simple concentration and exposure time are not the only factors that determine its toxicity [129]. The unique properties of nanoparticles, such as size, morphology, and chemistry, could affect their toxicity. In addition, the functional groups and other contaminants present on the surface of these materials can also induce significantly greater toxicity effects than pure nanomaterials alone. These factors contribute to the necessity of redefining the risk assessments for any further nanotechnological developments [130].

1.7 Conclusion

The development of sustainability-focused nanotechnology plays a major role in enhancing and improving the current mechanisms and technology and leads to the path of global sustainability. The use of nanoparticles in sectors such as medicine, food, environment, and industry certainly overcomes limitations, improves the field, makes it more cost-effective, and minimizes the negative results that contribute to poor human and environmental health. For human health certain nanoparticles, such as CNTs, quantum dots, and nanosilica, have been used in medicine and agriculture to develop safer and cost-effective technologies. For the environment, nanoparticles have been used to improve the air and water quality, minimize the negative consequences of the automotive and construction industry, and enhance the process of energy conversion, a field that will need to be developed further due to the constant decline in natural resources. Using nanoparticles certainly comes with the risk of toxicity, which will require further testing to create a set of centralized regulations for their use around the world.

The research done so far on nanoparticles is excellent in creating a fascinating path to a safer, cleaner, and more economically sufficient planet; however, as mentioned before, further assessments of the long-term effects of the involvement of nanotechnology certainly needs to be done to make sure that the mistakes that were made before under the greed of operating new technology are not repeated.

References

1 Rodrigue, J.-P. (2020). Chapter 4: transportation, energy and environment. In: *The Geography of Transport Systems*, 456. New York: Routledge.

2 Mata, T.M., Caetano, N.S., and Martins, A.A. (2015). Sustainability evaluation of nanotechnology processing and production. *Chemical Engineering Transactions* 45: 1969–1974.

3 Sargent, J.F. (2014). The national nanotechnology initiative: overview, reauthorization, and appropriations issues. In: *National Nanotechnology Initiative Overview & Strategic Plan* (ed. J.F. Sargent Jr.), 1. Congressional Research Service.

4 Roco, M.C., Mirkin, C.A., and Hersam, M.C. (2011). Nanotechnology research directions for societal needs in 2020: summary of international study. *Journal of Nanoparticle Research* 13: 897–919.

5 Wang, G. (2018). *Nanotechnology: The New Features*. arXiv.

6 Shrestha, M. (2012). Nanotechnology to revolutionize medicine. *Journal of Drug Delivery and Therapeutics* 2: 156–161. https://doi.org/10.22270/jddt.v2i5.302.

7 Globocan Observatory W (2019). Cancer Today - World. *International Agency for Research on Cancer* 876: 2018–2019.

8 Shi, S., Zhou, M., Li, X. et al. (2016). Synergistic active targeting of dually integrin αvβ3/CD44-targeted nanoparticles to B16F10 tumors located at different sites of mouse bodies. *Journal of Controlled Release* 235: 1–13.

9 Ilyin, S.E., Belkowski, S.M., and Plata-Salamán, C.R. (2004). Biomarker discovery and validation: technologies and integrative approaches. *Trends in Biotechnology* 22: 411–416.

10 Goossens, N., Nakagawa, S., Sun, X., and Hoshida, Y. (2015). Cancer biomarker discovery and validation. *Translational Cancer Research* 4: 256–269.

11 Rhea, J.M. and Molinaro, R.J. (2011). Cancer biomarkers: surviving the journey from bench to bedside. *Medical Laboratory Observer* 43: 10–12, 16, 18.

12 Choi, Y.E., Kwak, J.W., and Park, J.W. (2010). Nanotechnology for early cancer detection. *Sensors* 10: 428–455.

13 Huang, X., Jain, P.K., El-Sayed, I.H., and El-Sayed, M.A. (2007). Gold nanoparticles: interesting optical properties and recent applications in cancer diagnostics and therapy. *Nanomedicine* 2: 681–693.

14 Liu, X., Dai, Q., Austin, L. et al. (2008). A one-step homogeneous immunoassay for cancer biomarker detection using gold nanoparticle probes coupled with dynamic light scattering. *Journal of the American Chemical Society* 130: 2780–2782.

15 Zheng, T., Pierre-Pierre, N., Yan, X. et al. (2015). Gold nanoparticle-enabled blood test for early stage cancer detection and risk assessment. *ACS Applied Materials & Interfaces* 7: 6819–6827.

16 Dreaden, E.C., Austin, L.A., MacKey, M.A., and El-Sayed, M.A. (2012). Size matters: gold nanoparticles in targeted cancer drug delivery. *Therapeutic Delivery* 3: 457–478.

17 Singh, P., Pandit, S., Mokkapati, V.R.S.S. et al. (2018). Gold nanoparticles in diagnostics and therapeutics for human cancer. *International Journal of Molecular Sciences* 19: 1–16.

18 Vines, J.B., Yoon, J.H., Ryu, N.E. et al. (2019). Gold nanoparticles for photothermal cancer therapy. *Frontiers in Chemistry* 7: 1–16.

19 Ali, M.R.K., Wu, Y., Ghosh, D. et al. (2017). Nuclear membrane-targeted gold nanoparticles inhibit cancer cell migration and invasion. *ACS Nano* 11: 3716–3726.

20 Iannazzo, D., Pistone, A., Salamò, M. et al. (2017). Graphene quantum dots for cancer targeted drug delivery. *International Journal of Pharmaceutics* 518: 185–192.

21 Zhao, M.X., Zhu, B.J., Yao, W.J., and Chen, D.F. (2016). Therapeutic effect of quantum dots for cancer treatment. *RSC Advances* 6: 113791–113795.

22 McHugh, K.J., Jing, L., Behrens, A.M. et al. (2018). Biocompatible semiconductor quantum dots as cancer imaging agents. *Advanced Materials* 30: 1706356.

23 Ranjbar-Navazi, Z., Eskandani, M., Johari-Ahar, M. et al. (2018). Doxorubicin-conjugated D-glucosamine- and folate- bi-functionalisedInP/ZnS quantum dots for cancer cells imaging and therapy. *Journal of Drug Targeting* 26: 267–277.

24 Luo, M., Cheng, W., Zeng, X. et al. (2019). Folic acid-functionalized black phosphorus quantum dots for targeted chemo-photothermal combination cancer therapy. *Pharmaceutics* 11: 242.

25 Chen, B.Q., Kankala, R.K., Zhang, Y. et al. (2020). Gambogic acid augments black phosphorus quantum dots (BPQDs)-based synergistic chemo-photothermal therapy through downregulating heat shock protein expression. *Chemical Engineering Journal* 390: 124312.

26 Shang, Y., Wang, Q., Wu, B. et al. (2019). Platelet-membrane-camouflaged black phosphorus quantum dots enhance anticancer effect mediated by apoptosis and autophagy. *ACS Applied Materials & Interfaces* 11: 28254–28266.

27 Sahoo, N.G., Bao, H., Pan, Y. et al. (2011). Functionalized carbon nanomaterials as nanocarriers for loading and delivery of a poorly water-soluble anticancer drug: a comparative study. *Chemical Communications* 47: 5235–5237.

28 Iijima, S. (1991). Helical microtubules of graphitic carbon. *Nature* 354: 56–58.

29 Son, K.H., Hong, J.H., and Lee, J.W. (2016). Carbon nanotubes as cancer therapeutic carriers and mediators. *International Journal of Nanomedicine* 11: 5163–5185.

30 Elhissi, A.M.A., Ahmed, W., Hassan, I.U. et al. (2012). Carbon nanotubes in cancer therapy and drug delivery. *Journal of Drug Delivery* 2012: 1–10.

31 Xie, J., Teng, L., Yang, Z. et al. (2013). A polyethylenimine-linoleic acid conjugate for antisense oligonucleotide delivery. *BioMed Research International* 2013: 1–7.

32 De Faria, P.C.B., Dos Santos, L.I., Coelho, J.P. et al. (2014). Oxidized multiwalled carbon nanotubes as antigen delivery system to promote superior CD8+ T cell response and protection against cancer. *Nano Letters* 14: 5458–5470.

33 Hassan, H.A.F.M., Diebold, S.S., Smyth, L.A. et al. (2019). Application of carbon nanotubes in cancer vaccines: achievements, challenges and chances. *Journal of Controlled Release* 297: 79–90.

34 Liu, C., Wang, D., Zhan, Y. et al. (2018). Switchable photoacoustic imaging of glutathione using MnO 2 nanotubes for cancer diagnosis. *ACS Applied Materials & Interfaces* 10: 44231–44239.

35 Berber, M.R., Elkhenany, H., Hafez, I.H. et al. (2020). Efficient tailoring of platinum nanoparticles supported on multiwalled carbon nanotubes for cancer therapy. *Nanomedicine* 15: 793–808.

36 Singh, S., Mehra, N.K., and Jain, N.K. (2016). Development and characterization of the paclitaxel

loaded riboflavin and thiamine conjugated carbon nanotubes for cancer treatment. *Pharmaceutical Research* 33: 1769–1781.

37 Layton, E., McNamar, R., Hasanjee, A.M. et al. (2017). The effects of single-walled carbon nanotubes on cancer cell migration using a pancreatic tumor model. In: *Biophotonics Immune Responses XII* (ed. W.R. Chen), 7. The International Society for Optics and Photonics (SPIE).

38 Qiao, H., Zhu, Z., Fang, D. et al. (2018). Redox-triggered mitoxantrone prodrug micelles for overcoming multidrug-resistant breast cancer. *Journal of Drug Targeting* 26: 75–85.

39 Zhang, X. (2018). Mini-review: nanotechnology forms of drug delivery. *Journal of Drug Delivery and Therapeutics* 8: 275–277.

40 Ravichandran, S. (2010). Green chemistry – a potential tool for chemical synthesis. *International Journal of ChemTech Research* 2: 2188–2191.

41 Jahangirian, H., Lemraski, E.G., Webster, T.J. et al. (2017). A review of drug delivery systems based on nanotechnology and green chemistry: green nanomedicine. *International Journal of Nanomedicine* 12: 2957–2978.

42 Li, X., Gong, Y., Zhou, X. et al. (2016). Facile synthesis of soybean phospholipid-encapsulated MoS2 nanosheets for efficient in vitro and in vivo photothermal regression of breast tumor. *International Journal of Nanomedicine* 11: 1819–1833.

43 Pozdnyakov, A.S., Emel'yanov, A.I., Kuznetsova, N.P. et al. (2016). Nontoxic hydrophilic polymeric nanocomposites containing silver nanoparticles with strong antimicrobial activity. *International Journal of Nanomedicine* 11: 1295–1304.

44 Chen, Z., Zhang, T., Wu, B., and Zhang, X. (2016). Insights into the therapeutic potential of hypoxia-inducible factor-1α small interfering RNA in malignant melanoma delivered via folate-decorated cationic liposomes. *International Journal of Nanomedicine* 11: 991–1002.

45 Chen, L.C., Chen, Y.C., Su, C.Y. et al. (2016). Development and characterization of self-assembling lecithin-based mixed polymeric micelles containing quercetin in cancer treatment and an in vivo pharmacokinetic study. *International Journal of Nanomedicine* 11: 1557–1566.

46 Erdal, M.S., Ozhan, G., Mat, M.C. et al. (2016). Colloidal nanocarriers for the enhanced cutaneous delivery of naftifine: characterization studies and in vitro and in vivo evaluations. *International Journal of Nanomedicine* 11: 1027–1037.

47 Bellan, L.M., Wu, D., and Langer, R.S. (2011). Current trends in nanobiosensor technology. *Wiley Interdisciplinary Reviews, Nanomedicine and Nanobiotechnology* 3: 229–246.

48 Kim, Y.G., Moon, S., Kuritzkes, D.R., and Demirci, U. (2009). Quantum dot-based HIV capture and imaging in a microfluidic channel. *Biosensors & Bioelectronics* 25: 253–258.

49 Cao, Q., Teng, Y., Yang, X. et al. (2015). A label-free fluorescent molecular beacon based on DNA-Ag nanoclusters for the construction of versatile Biosensors. *Biosensors & Bioelectronics* 74: 318–321.

50 Altay, C., Senay, R.H., Eksin, E. et al. (2017). Development of amino functionalized carbon coated magnetic nanoparticles and their application to electrochemical detection of hybridization of nucleic acids. *Talanta* 164: 175–182.

51 Mokhtarzadeh, A., Eivazzadeh-Keihan, R., Pashazadeh, P. et al. (2017). Nanomaterial-based biosensors for detection of pathogenic virus. *Trends in Analytical Chemistry* 97: 445–457.

52 Keshavarzi, M., Darijani, M., Momeni, F. et al. (2017). Molecular imaging and oral cancer diagnosis and therapy. *Journal of Cellular Biochemistry* 118: 3055–3060.

53 Keshavarzi, M., Sorayayi, S., JafarRezaei, M. et al. (2017). MicroRNAs-based imaging techniques in cancer diagnosis and therapy. *Journal of Cellular Biochemistry* 118: 4121–4128.

54 Jafari, S.H., Saadatpour, Z., Salmaninejad, A. et al. (2018). Breast cancer diagnosis: imaging techniques and biochemical markers. *Journal of Cellular Physiology* 233: 5200–5213.

55 Guo, J., Yuan, C., Yan, Q. et al. (2018). An electrochemical biosensor for microRNA-196a detection based on cyclic enzymatic signal amplification and template-free DNA extension reaction with the adsorption of methylene blue. *Biosensors & Bioelectronics* 105: 103–108.

56 Liu, C., Chen, C., Li, S. et al. (2018). Target-triggered catalytic hairpin assembly-induced core-satellite nanostructures for high-sensitive "off-to-on" SERS detection of intracellular MicroRNA. *Analytical Chemistry* 90: 10591–10599.

57 de Souza, M.F., de Syllos Cólus, I.M., Fonseca, A.S. et al. (2017). Abstract 1476: Cell-free miR-141 as a molecular marker for prostate cancer metastasis. https://cancerres.aacrjournals.org/content/77/13_Supplement/1476 (accessed 1 August 2020).

58 Cheng, H., Zhang, L., Cogdell, D.E. et al. (2011). Circulating plasma MiR-141 is a novel biomarker for metastatic colon cancer and predicts poor prognosis. *PLoS One* 6: e17745.

59 Iorio, M.V., Visone, R., Di Leva, G. et al. (2007). MicroRNA signatures in human ovarian cancer. *Cancer Research* 67: 8699–8707.

60 Tran, H.V., Piro, B., Reisberg, S. et al. (2014). An electrochemical ELISA-like immunosensor for miRNAs detection based on screen-printed gold electrodes modified with reduced graphene oxide and carbon nanotubes. *Biosensors & Bioelectronics* 62: 25–30.

61 Fang, C., Zhu, D.X., Dong, H.J. et al. (2012). Serum microRNAs are promising novel biomarkers for diffuse large B cell lymphoma. *Annals of Hematology* 91: 553–559.

62 Azimzadeh, M., Rahaie, M., Nasirizadeh, N. et al. (2016). An electrochemical nanobiosensor for plasma miRNA-155, based on graphene oxide and gold nanorod, for early detection of breast cancer. *Biosensors & Bioelectronics* 77: 99–106.

63 Cheng, Y., Ji, R., Yue, J. et al. (2007). MicroRNAs are aberrantly expressed in hypertrophic heart: do they play a pole in cardiac hypertrophy? *The American Journal of Pathology* 170: 1831–1840.

64 Cheng, F.F., He, T.T., Miao, H.T. et al. (2015). Electron transfer mediated electrochemical biosensor for microRNAs detection based on metal ion functionalized titanium phosphate nanospheres at attomole level. *ACS Applied Materials & Interfaces* 7: 2979–2985.

65 Shabaninejad, Z., Yousefi, F., Movahedpour, A. et al. (2019). Electrochemical-based biosensors for microRNA detection: nanotechnology comes into view. *Analytical Biochemistry* 581: 113349.

66 Pretty, J., Sutherland, W.J., Ashby, J., and Auburn, J. (2010). Taylor & Francis Online: The top 100 questions of importance to the future of global agriculture. *International Journal of Agricultural Sustainability* 8 (4): 219–236.

67 Vermeulen, S.J., Campbell, B.M., and Ingram, J.S.I. (2012). Climate change and food systems. *Annual Review of Environment and Resources* 37: 195–222.

68 Foley, J.A., DeFries, R., Asner, G.P. et al. (2005). Global consequences of land use. *Science* 309: 570–574.

69 Viala, E. (2008). Water for food, water for life a comprehensive assessment of water management in agriculture. *Irrigation and Drainage Systems* 22: 7–9.

70 Tilman, D., Clark, M., Williams, D.R. et al. (2017). Future threats to biodiversity and pathways to their prevention. *Nature* 546: 73–81.

71 Albanese, A., Tang, P.S., and Chan, W.C.W. (2012). The effect of nanoparticle size, shape, and surface chemistry on biological systems. *Annual Review of Biomedical Engineering* 14: 1–16.

72 Haverkamp, R.G. and Marshall, A.T. (2009). The mechanism of metal nanoparticle formation in plants: limits on accumulation. *Journal of Nanoparticle Research* 11: 1453–1463.

73 Ma, X., Geiser-Lee, J., Deng, Y., and Kolmakov, A. (2010). Interactions between engineered nanoparticles (ENPs) and plants: phytotoxicity, uptake and accumulation. *Science of the Total Environment* 408: 3053–3061.

74 Subramanian, K.S., Manikandan, A., Thirunavukkarasu, M., and Rahale, C.S. (2015). Nano-fertilizers for balanced crop nutrition. In: *Nanotechnologies in Food Agriculture* (ed. M. Rai, C. Ribeiro, L. Mattoso and N. Duran), 69–80. Cham Springer.

75 Bieberstein, A., Roosen, J., Marette, S. et al. (2013). Consumer choices for nano-food and nano-packaging in France and Germany. *European Review of Agricultural Economics* 40: 73–94.

76 Ravichandran, R. (2010). Nanotechnology applications in food and food processing: innovative green approaches, opportunities and uncertainties for global market. *International Journal of Green Nanotechnology: Physics and Chemistry* 1: 72–96.

77 Coles, D. and Frewer, L.J. (2013). Nanotechnology applied to European food production – a review of ethical and regulatory issues. *Trends in Food Science and Technology* 34: 32–43.

78 Durán, N. and Marcato, P.D. (2013). Nanobiotechnology perspectives. Role of nanotechnology in the food industry: a review. *International Journal of Food Science and Technology* 48: 1127–1134.

79 Dekkers, S., Krystek, P., Peters, R.J.B. et al. (2011). Presence and risks of nanosilica in food products. *Nanotoxicology* 5: 393–405.

80 Morones, J.R., Elechiguerra, J.L., Camacho, A. et al. (2005). The bactericidal effect of silver nanoparticles. *Nanotechnology* 16: 2346–2353.

81 Kim, J.S., Kuk, E., Yu, K.N. et al. (2007). Antimicrobial effects of silver nanoparticles. *Nanomedicine: Nanotechnology, Biology and Medicine* 3: 95–101.

82 Deligiannakis, Y., Sotiriou, G.A., and Pratsinis, S.E. (2012). Antioxidant and antiradical SiO2 nanoparticles covalently functionalized with gallic acid. *ACS Applied Materials & Interfaces* 4: 6609–6617.

83 Zhou, H., Liu, G., Zhang, J. et al. (2014). Novel lipid-free nanoformulation for improving oral bioavailability of coenzyme Q10. *BioMed Research International* 2014: 793879.

84 Mlalila, N., Kadam, D.M., Swai, H., and Hilonga, A. (2016). Transformation of food packaging from passive to innovative via nanotechnology: concepts and critiques. *Journal of Food Science and Technology* 53: 3395–3407.

85 Dobrucka, R. and Cierpiszewski, R. (2014). Active and intelligent packaging food-research and development-a review. *Polish Journal of Food and Nutrition Sciences* 64: 7–15.

86 Dasgupta, N., Ranjan, S., Mundekkad, D. et al. (2015). Nanotechnology in agro-food: from field to plate. *Food Research International* 69: 381–400.

87 Feng, Y., Xie, L., Chen, Q., and Zheng, L.R. (2015). Low-cost printed chipless RFID humidity sensor tag for intelligent packaging. *IEEE Sensors Journal* 15: 3201–3208.

88 Sarapulova, O., Sherstiuk, V., Shvalagin, V., and Kukhta, A. (2015). Photonics and nanophotonics and information and communication technologies in modern food packaging. *Nanoscale Research Letters* 10: 229.

89 Liu, H., Fang, G., Deng, Q., and Wang, S. (2015). A triple-dimensional sensing chip for discrimination of eight antioxidants based on quantum dots and graphene. *Biosensors & Bioelectronics* 74: 313–317.

90 Wen, P., Zhu, D.H., Wu, H. et al. (2016). Encapsulation of cinnamon essential oil in electrospunnanofibrous film for active food packaging. *Food Control* 59: 366–376.

91 Yildirim, S., Röcker, B., Rüegg, N., and Lohwasser, W. (2015). Development of palladium-based oxygen scavenger: optimization of substrate and palladium layer thickness. *Packaging Technology and Science* 28: 710–718.

92 Joven, R., Garcia, A., Arias, A., and Medina, J. (2015). Development of an active thermoplastic film with oxygen scavengers made of activated carbon and sodium erythorbate. *Packaging Technology and Science* 28: 113–121.

93 Mahieu, A., Terrié, C., and Youssef, B. (2015). Thermoplastic starch films and thermoplastic starch/polycaprolactone blends with oxygen-scavenging properties: Influence of water content. *Industrial Crops and Products* 72: 192–199.

94 De Oliveira, T.L.C., de Araújo Soares, R., and Piccoli, R.H. (2013). A Weibull model to describe antimicrobial kinetics of oregano and lemongrass essential oils against *Salmonella enteritidis* in ground beef during refrigerated storage. *Meat Science* 93: 645–651.

95 Arora, N.K. (2018). Environmental sustainability—necessary for survival. *Environmental Sustainability* 1: 1–2.

96 Sushma, D. and Richa, S. (2015). Use of nanoparticles in water treatment: a review. *International Research Journal of Environmental Sciences* 4: 103–106.

97 Savage, N. and Diallo, M.S. (2005). Nanomaterials and water purification: opportunities and challenges. *Journal of Nanoparticle Research* 7: 331–342.

98 Prachi, Gautam, P., Madathil, D., and Brijesh Nair, A.N. (2013). Nanotechnology in waste water treatment: a review. *International Journal of ChemTech Research* 5: 2303–2308.

99 Zhao, X., Lv, L., Pan, B. et al. (2011). Polymer-supported nanocomposites for environmental application: a review. *Chemical Engineering Journal* 170: 381–394.

100 Chaturvedi, S., Dave, P.N., and Shah, N.K. (2012). Applications of nano-catalyst in new era. *Journal of Saudi Chemical Society* 16: 307–325.

101 Choi, H., Zakersalehi, A., Al-Abed, S.R. et al. (2014). Nanostructured titanium oxide film- and membrane-based photocatalysis for water treatment. In: *Nanotechnology Applications for Clean Water: Solutions for Improving Water Quality*, 2e (ed. A. Street, R. Sustich, J. Duncan and N. Savage), 123–132. William Andrew.

102 Kaushik, B.K. and Majumder, M.K. (2015). Carbon nanotube: properties and applications. *SpringerBriefs in Applied Sciences and Technology* 1: 17–38.

103 Morelli, J. (2011). Environmental sustainability: a definition for environmental professionals. *Journal of Environmental Sustainability* 1: 2.

104 Li, S., Feng, K., and Li, M. (2017). Identifying the main contributors of air pollution in Beijing. *Journal of Cleaner Production* 163: S359–S365.

105 Martins, F., Felgueiras, C., and Smitková, M. (2018). Fossil fuel energy consumption in European countries. *Energy Procedia* 153: 107–111.

106 Dos Santos, M.C., Kesler, O., and Reddy, A.L.M. (2012). Nanomaterials for energy conversion and storage. *Journal of Nanomaterials* 2012: 3127–3171.

107 Liu, F., Wang, W., Wang, L., and Yang, G. (2013). Working principles of solar and other energy conversion cells. *Nano Energy* 2: 3–10.

108 Sharma, S., Jain, K.K., and Sharma, A. (2015). Solar cells: in research and applications—a review. *Materials Sciences and Applications* 6: 1145–1155.

109 Serrano, E., Rus, G., and García-Martínez, J. (2009). Nanotechnology for sustainable energy. *Renewable and Sustainable Energy Reviews* 13: 2373–2384.

110 Alturaif, H.A., ALOthman, Z.A., Shapter, J.G., and Wabaidur, S.M. (2014). Use of carbon nanotubes (CNTs) with polymers in solar cells. *Molecules* 19: 17329–17344.

111 Geng, Y., Liu, M.Y., Li, J. et al. (2008). Effects of surfactant treatment on mechanical and electrical properties of CNT/epoxy nanocomposites. *Composites. Part A, Applied Science and Manufacturing* 39: 1876–1883.

112 Arianpour, F., Farzaneh, M., and Kulinich, S.A. (2013). Hydrophobic and ice-retarding properties of doped silicone rubber coatings. *Applied Surface Science* 265: 546–552.

113 Maksimović, M. and Forcan, M. (2020). The role of nanotechnology in revolutionizing energy sector. *International Journal of Electrical, Electronics and Computer* 3: 211204609.

114 Sunseri, C., Cocchiara, C., Ganci, F. et al. (2016). Nanostructured electrochemical devices for sensing, energy conversion and storage. *Chemical Engineering Transactions* 47: 43–48.

115 Arivalagan, K., Ravichandran, S., Rangasamy, K., and Karthikeyan, E. (2011). Nanomaterials and its potential applications. *International Journal of ChemTech Research* 3: 534–538.

116 Sudan, P., Züttel, A., Mauron, P. et al. (2003). Physisorption of hydrogen in single-walled carbon nanotubes. *Carbon N Y* 41: 2377–2383.

117 Stoycheva, S., Marchese, D., Paul, C. et al. (2018). Multi-criteria decision analysis framework for sustainable manufacturing in automotive industry. *Journal of Cleaner Production* 187: 257–272.

118 Felix, D.G. and Kumar, D.G.S. (2014). Nano particles in automobile tires. *IOSR Journal of Mechanical and Civil Engineering* 11: 7–11.

119 Bhogare, R. and Kothawale, B. (2013). A review on applications and challenges of nano-fluids as coolant in automobile radiator. *IJSRP Org.* 3: 1–11.

120 Ekengwu, I.E., Utu, O.G., and Okafor, C.E. (2019). Nanotechnology in automotive industry: the potential of graphene. *Iconic Research and Engineering Journals* 3: 31–37.

121 Malani, A.S., Chaudhari, A.D., and Sambhe, R.U. (2015). A review on applications of nanotechnology in automotive industry. *International Journal of Mechanical and Mechatronics Engineering* 10: 36–40.

122 Perspectives, G.C. (2015). Economics O: Global Construction 2030| A Global Forecast for Construction Industry to 2030. Construction Market.

123 Shariati, S., Abedi, M., Saedi, A. et al. (2017). Critical factors of the application of nanotechnology in construction industry by using ANP technique under fuzzy intuitionistic environment. *Journal of Civil Engineering and Management* 23: 914–925.

124 Said, A.M., Zeidan, M.S., Bassuoni, M.T., and Tian, Y. (2012). Properties of concrete incorporating nano-silica. *Construction and Building Materials* 36: 838–844.

125 Bigley, C. and Greenwood, P. (2003). Using silica to control bleed and segregation in self-compacting concrete. *Concrete* 37: 43–45.

126 D'Alessandro, A., Materazzi, A.L., and Ubertini, F. (ed.) (2020). *Nanotechnology in Cement-Based Construction.* Jenny Stanford Publishing.

127 Singh, A., Sangita, and Singh, A. (2015). Overview of nanotechnology in road engineering. *Journal of Nano- and Electronic Physics* 7: 02004-1–02004-6.

128 Krewski, D., Acosta, D., Andersen, M. et al. (2010). Toxicity testing in the 21st century: a vision and a strategy. *Journal of Toxicology and Environmental Health. Part B, Critical Reviews* 13: 51–138.

129 Schrand, A.M., Dai, L., Schlager, J.J., and Hussain, S.M. (2012). Toxicity testing of nanomaterials. *Advances in Experimental Medicine and Biology* 745: 58–75.

130 Hristovski, K.D. and Markovski, J. (2018). Environmental, health, and safety risks associated with nanotechnology. *Metallurgical and Materials Engineering* 24: A158–A161.

2

The Road to Sustainable Nanotechnology: Challenges, Progress, and Opportunities

Sunita Chaudhary[1], Nishith Patel[1], and Jayvadan Patel[2]

[1] *Arihant School of Pharmacy and Bio-Research Institute, Gandhinagar, Gujarat, India*
[2] *Nootan Pharmacy College, Sankalchand Patel University, Visnagar, Gujarat, India*

2.1 Introduction

Advances in nanotechnology are generating novel nanomaterials (NMs) with extraordinary characteristics that can enrich and enhance the applicability of NMs in various sectors. As new uses are being explored on a daily basis in various diversified areas such as medicine, agriculture, automotive, and energy, it poses numerous challenges to environmental sustainability [1]. Application of NMs are improving energy conservation and increasing efficiency, productivity and profits of industrial and health sectors with lower environmental impact [2]. Using NMs in current technology can improve economic and environmental aspects with various applications. Simultaneously, it is accepted that the use of NMs increases the various challenges in human safety and regulations in various countries [3].

In 2011, a group of researchers, students, and government staff from a multidisciplinary area came together to check the effect of science and engineering on developing a new path with sustainable future. This panel of scientists had done research on nanotechnology and correlated it to various aspects of sustainability such as society, environment, and economy [4]. The Sustainable Nanotechnology Organization (SNO) has come into existence with the aim to support the advances in sustainable nanotechnology. It also promotes the progression of scientific work in the field of nanotechnology for safety, environment, and health. However, every developing technology is required to be balanced between the benefits for human and its undesired effects on environment and life [5].

As advances in nanotechnology continue, the research is progressing on many issues regarding sustainability for next decades. Recent advances in nanotechnology report worldwide challenges in various applications such as purification of water, management of greenhouse effect, and manufacturing using green chemistry [6]. Because of the significant physicochemical properties of NMs, it gives us more effective solution for sustainable technologies. In comparison with bulk materials, NMs have effective larger surface area. The synthesis of NMs with incorporation of functional group can increase their affinity toward a given compound with improved efficiency. NMs can also give us an opportunity to enhance them with improved optical, magnetic, and catalytic properties [7]. For instance, in the case of semiconductor quantum dots, the variation in fluorescence emission capacity depends strongly on the difference in particle size. Nanoscale materials have different physicochemical and biological properties, which meet many growing needs of society. By exploring the novel development of NMs with new properties, it can be possible to enhance the performance of materials significantly, in terms of achievement of more product performance using less material [8, 9]. Development of NMs using required approach having ecofriendly characteristics is need of todays industries like textiles, paints and biomedical [10]. The available methods used for preparation of nanostructured materials are ultrasonication, reverse-phase micelle, microwave assisted techniques, and deposition of vapour by chemical and physical means [11]. Nano metal oxides such as Al_2O_3, TiO_2, and SiO_2 have been efficiently obtained from the nature and is one of the best examples of such techniques [12]. Today, the available solar cells having effectiveness of 10–25% are made up of silicon [13]. The cost of a solar cell can be reduced by improving its efficiency, which makes it more economically competitive [14]. By using nanostructured surface for reflectivity in solar cell devices, the antireflection ability may be enhanced

Sustainable Nanotechnology: Strategies, Products, and Applications, First Edition.
Edited by Yashwant V. Pathak, Govindan Parayil, and Jayvadan K. Patel.

effectively using NMs having diameter less than the wavelength of incident light. To increase the usage of solar light, antireflection NMs are patterned with active components, and thus the cost can be reduced [15]. Also, light-capturing capacity can be additionally increased by light trapping, so it requires less amount of material to absorb solar light, and the cost can be reduced [16].

NMs can be used to solve various challenges in the environment, e.g. it can be used to restore the polluted soil with cleaning of water and air, and it also can reduce the impact of chemical manufacturing using nanoscale catalyst [17, 18]. Therefore, various opportunities are offered by nanotechnology in manufacturing novel NMs that improves living conditions using advanced techniques in various fields.

Despite significant development and increased utilization in many marketed products, NMs and their technology are still facing challenges in energy, environment, health, and safety (EHS). One of the principal challenges that comes up with nanotechnology is its influence on environment and its toxicity to humans. The cause for toxicity in humans is due to the availability of NMs with different properties that might lead to adverse drug reaction. NMs undertake biodegradation in their working atmosphere. This might lead to intracellular changes and gene modifications. If the gene alteration is undesirable, then it might prove hazardous for human beings. NMs are not always environmentally safe. They sometimes enter the ecology's food chain and cause changes. So, it is necessary first to understand the life cycle of NMs with their activity in the environment. It is necessary to focus on the size, structure, and reactivity of NMs in environmental systems [19].

2.2 Road to Sustainability in Nanotechnology

To have sustainability in nanotechnology, there is a need to (Figure 2.1)

1) address the accountability of NMs and their related toxicity.
2) strengthen large-scale manufacturing of NMs and minimize possible risks involved in it.
3) develop new capabilities for sustainable environment and health.
4) have a robust regulatory guideline and support.

2.2.1 Accountability of Nanomaterials and Related Toxicity

The developments in NMs are evolving as an inevitable part of daily life. With increase in the use of nanostructured material, it will eventually augment its exposure to the environment and to all those living and nonliving things that are made up of natural NMs called atom. Although the overall exposure by NM will be very less currently, in future with surge in the acceptance, the likelihood of exposure will increase. The effects of nanocomponents on environment are unknown as it largely depends on exposure and toxicity. Even after the selection of relatively less toxic material, the chance of harmful effects is unpredictable. Ironically, the unique properties of nanoparticles (NPs) that convey significant characteristics bring toxicity as well [20]. For instance, a naturally occurring oxide of titanium (TiO_2), which is frequently used in the preparation of

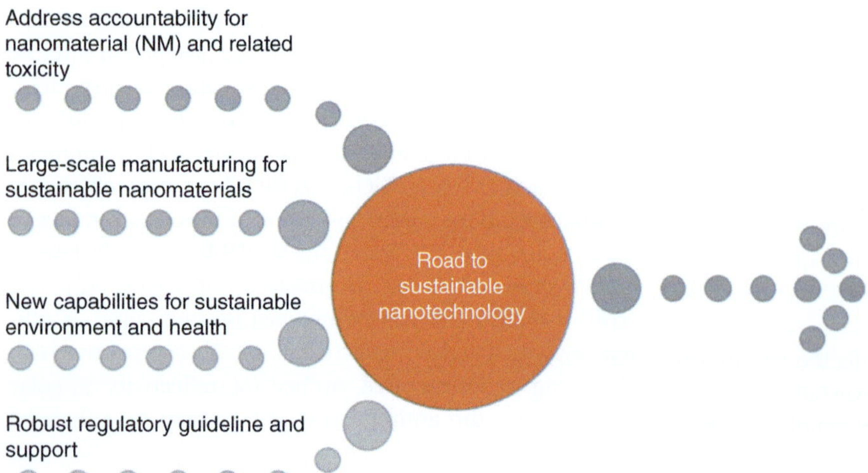

Figure 2.1 Road to sustainability in nanotechnology.

NMs, in the presence of sunlight, these TiO_2 NPs can be toxic to some freshwater organisms even at parts-per-billion levels due to uptake and in situ photogeneration of reactive oxygen species [21, 22]. The severity of toxicity of NMs depends on the physicochemical properties, viz., size, surface area, structure, shape, stability, and surface energy [20]. These make an imperative need to assess the toxic effect based on composition and physicochemical properties of NMs.

2.2.1.1 Size

The characteristic that brings the unique performance of NMs is its nanoscale structured engineering. At the same time, this physicochemical property creates opportunity for increased interaction with biological tissues at molecular level having similar size and structure. This same principle has caused pharmaceutical companies to formulate highly efficient and intelligent drug delivery systems for several diseases, which was not possible in their conventional size formulation. Not only in health care but also in all other sectors, these NMs have improved their characteristics such as strength, weight, appearance, efficiency, and durability. A research for pharmacological action based on NP size shows that NPs < 50 nm diffuse quickly to tissues of living things and bring potential toxicity to those tissues, while NPs>50 nm taken up by a cleaning system of mammalians, reticuloendothelial system (RES), and the RES organs, viz., liver, spleen, and lymph nodes, will become the target of oxidative stress [23]. Few other researchers show that particles <10 nm get deposited in the tracheobronchial of the lung, while all the other <100 nm particles are deposited all over the lungs and cause respiratory adverse effects [24–26]. Several other toxic effects, such as mitochondrial perturbation by silica [27], damage to nervous system [28], endothelial dysfunction [29], generation of neoantigens [30], and immune toxicity [31], are reported with limited clinical evidence. The uptake and interaction in biological tissues observed previously and substances generally regarded as safe now show adverse responses. The NPs generated during manufacturing may get inhaled, and ultrafine particles (<100 nm) induce pulmonary inflammation, oxidative stress, and distal organ involvement or get absorbed through the lungs and can create toxicity in vital systemic circulation. As the size reduces, it increases surface area and finally enhances capability to react with oxygen. Due to increased reaction with oxygen, it enhances inflammation, fibrosis, cytotoxicity, oxidative injury, and carbon deposition in lungs [32]. It is the sole responsibility of the researcher to involve toxicology scientists and closely monitor the toxicity of NPs during each development stage until robust regulatory guidelines based on size and surface area become available. One also needs to take care until the airborne NP hazard has been appropriately assessed; this risk should be managed by taking steps to avoid large quantities of these NPs becoming airborne.

2.2.1.2 Surface Area

The surface area increases with the reduction in the size of the same quantity of any material. NPs, although made from nontoxic materials, become hazardous, as the material developed reactivity at molecular level. The toxic effect of few such particles does not seem hazardous, but if more surface area becomes available, it will further add on risk. Thus, surface area also requires attention and monitoring for the toxicity study along with the size. As size reduces, it increases surface area and finally enhances capability to react with oxygen. Due to increased reaction with oxygen, it increases inflammation, fibrosis, cytotoxicity, oxidative injury, and carbon deposition in lungs of mineral particles, quartz, titanium dioxide, asbestos, and carbon black despite the same materials being inert at macro or higher size range[32]. Toxicity in biological systems generated by NPs is predominantly through the formation of oxidative responses and consequent formation of free radicals. Free radicals are hazardous as they oxidize the lipid and damage the DNA with inflammatory responses. The data also shows that the smaller the size, the more able it is toward the formation of reactive oxygen species [20]. Every NP thus closely needs to be monitored for its new developed chemical reactivity, and it should be done from its development phase.

2.2.1.3 Surface

The surface of NMs is a unique property, and the surface phenomenon seems to be a new direction in developing nontoxic NPs. The groups present on the surface or coating can change the physical, chemical, magnetic, electrical, and optical properties of NPs, which invariably can alter in vivo solubility, partition coefficient, distribution, pharmacokinetics, and accumulation [32–34]. The presence of oxygen on the surface can lead to the generation of reactive oxygen species that may lead to toxicity [24, 35]. In pharmaceuticals, the effect of NPs based on conformation has always been of interest for the formulation developer. The endocytic system in the body can actively remove any nanosize foreign material from the body, and same observed with drug loaded NP and cannot perform their desire role. The reason could be negatively charged cell membrane. Against such cell membrane, positive, negative, or neutral charge NPs cannot behave in the same manner. Similarly, NPs with lipophilic surface will have more affinity and eventually be quickly absorbed through the lipid bilayer membrane of the cell, while NPs coated with hydrophilic polymers such as polyethylene glycol (PEG), polyvinylpyrrolidone (PVP), and dextran avoid interaction due to the hydrophobic cell surface [16]. The surface energy of NPs alters the capacity to diffuse through the blood–brain barrier. The data shows that the anionic energy on the surface of NPs with the size range of 50–500 nm is transported through all layers of the skin, in contrast to the neutral or cationic NPs that do not have such similar characteristics [20].

2.2.1.4 Shape

The study has shown that the membrane diffusion of NMs is deviated based on its surface structure design. Yet, the type of surface of the NMs and their relationship with membrane diffusion are far from being broadly understood [36, 37]. Classically, interactions between NMs and the tissues of living things are driven by the chemical functionalities on the surface in addition to their shape and size. The importance of the surface can also be enlightened from a well-known phenomenon of naturally occurring peptides that show the inability to perform their fate of getting diffused into the cell membrane if they remain in random coil configuration rather than specific helical structure [36]. Different shapes, viz., fiber, spheres, tubes, ring, and planes, have been assessed to achieve their potential characteristics, including their adverse effects. Nickel NPs used in electronic application while testing toxicity based on size demonstrated that the change in configuration has more toxicity rather than the size [37]. The data shows that endocytosis of globular particle is quicker when compared to cylinder-shaped particles in nanoscale [38]; further toxicity of globular NPs is based on whether their configuration is homogenous or heterogeneous [39], and if they are other than spherical, then they will quickly show movement in systemic circulation, with possible biological consequences [40].

2.2.1.5 Composition and Crystalline Structure

For the preparation of NMs, various metals, polymeric materials, and bioceramics have been used. For medical purpose, phospholipids, PEG, and natural polymers have been used for formulation. It is the phospholipid that makes NMs compatible with the human tissue as the cell walls are made up of the same phospholipid. The composition of ethosomes has made it possible for them to enter through the skin from the space smaller than their own diameter by deforming the structure. One important report on the effect of composition of NPs on few species having vital role in trophic levels showed that the nanosilver and nanocopper with their soluble forms caused toxicity in all tested organisms, whereas TiO_2 of the same dimensions did not cause any toxicity issues [41]. Crystal form of NPs also influences the toxicity, and it has been reported that crystalline TiO_2NPs show toxicity in the absence of light including oxidative DNA injury, whereas NPs of metastable form of the same material with the same size and chemical composition do not show such toxicity [25]. In one more report, the cytotoxicity was previously claimed due to the size and then due to ultrahigh reactivity of NMs itself [42]. Several such materials of composition, viz., metals, aluminum oxide, gold, copper oxide, silver, zinc oxide, iron oxide, and titanium oxide; nonmetals, such as carbon and silica; and polymeric materials, have shown toxicity not only in animals and humans but also in nature [43].

2.2.2 Large-Scale Manufacturing of Sustainable Nanomaterials

The cutting-edge technological applications and characteristic advantage of NMs are due to their physical properties, composition, and colloidal stability (if in liquid form), and at the same time these factors are vulnerable from the view point of environment and health; so, henceforth, sustainable processes for the large-scale productions are desirable. The situation is that the scaling up of NPs/NMs on large scale, despite tuning nanoscale features, has become a technological barrier for the development. Apart from scale-up, concerns are raised for the toxic manifestations of NMs through their varied mechanisms for both environmental and health issues, although no clinically pertinent toxicity with their mechanism has yet been established that can prove them hazardous over their expediency. Sometimes, the methods of detection of toxicity and models used for the same are conflicting and inconsistent. So, based on few experimental models, judging more valuable NPs as more toxic to biological systems or vice versa is inappropriate [43].

Large-scale manufacturing aims for superiority, desired nanoproduct stipulations, desired physicochemical constraints, and sterility requirements. For such mass production of NMs, the selection of methods depends on the following factors:

1) Type of approach used;
2) type of NMs; and
3) regulatory requirements for production.

2.2.2.1 Type of Approach

Broadly, the tuning of particle size for nanoscale is being carried out by two approaches: top-down (TD) and bottom-up (BU). In TD approach, there is diminution of larger materials into smallest possible size, while in the BU approach there is assembly, aggregation, or formation of NPs, atom by atom, through precipitation or growth of nuclei. In TD method, pulverization can be achieved by means of impact (such as hammer mill) and attrition alone or impact and attrition (ball mill and fluid energy mill) together. This approach requires high energy requirement and fewer steps. The main issue with such impact- and/or attrition-based mill is unavailability of narrow size range. Still narrow size range can be achieved, but that will bring huge amount of waste.

BU approach is well known for its customization in design such that it reduces waste production, but this method uses organic solvents. Even after completion of the manufacturing process, these organic solvents remain in the system and need additional step for their removal, and thus, toxicity of residual solvents always remains a threat in NMs manufactured by such methods.

The use of organic solvents can be replaced with few nonorganic solvents. Such green methods use supercritical fluids (supercritical CO_2, ethyl alcohol, or water) that bring extremely pure NMs. The greener methods require exclusive pressurized apparatus with further successive steps [44]. Green methods for the production of NMs can address sustainability issues scalable for large manufacturing, cost-effective, versatile, and tunable nonagglomerated nanoclusters and overcome the key barrier for the progress of large-scale manufacturing of NMs.

2.2.2.2 Types of Nanomaterials

The essentiality of a sustainable approach to nanotechnology is becoming more and more urgent in the past few decades while many questions concerning all the steps of nanomanufacturing are unanswered [45]. Apart from issues concerning scale-up and large-scale manufacturing, materials-related issues need to be addressed that are selected based on the type of NMs being produced. The use of nanotechnology in the field of pharmacy involving inert material, pharmaceutical active ingredients, and solvents should not have any adverse events during manufacturing, sterilization, or after consumption by patients, while consumers do not require to take those precautions in regard to nanotechnology. Among all the issues, the first issue concerning the starting materials or substances, their nature, energy cost, and sources is very important. The second issue concerns the methods of synthesis, e.g. if NPs generate greenhouse gases and other pollutants in air, or if there is high energy consumption. Next concern is the requirement for extra capitals in the usage of NMs and nano-enabled products that may damagingly affect the environment.

2.2.2.3 Regulatory Requirements for Production

For any production unit, the requirement related to in-house qualification is very important, especially in pharmaceuticals, where every new product needs to be qualified from certain regulatory bodied as well as a good manufacturing practice (GMP) certification for scale-up and manufacturing process. Apart from these aspects, there is always search for cost-effective excipients and materials, flexible and uninterrupted production, and process setup. Overall, the industrial production should produce a product of high quality without contamination and free from elemental impurities residual solvent.

Further being expensive, manufacturing may involve accidental exposure that affects the environment and human health. The most common human effect that arises, in this case, are breathing issues and skin rashes. Yet, these issues can be minimized by appropriate awareness for possible hazards, and source of defensive apparel and microfiltered air throughout the production unit are necessary to avoid these consequences. Already discussed in the previous section about the physicochemical properties that make NPs attractive for numerous applications are indeed the properties that are the source of concern as they affect biokinetics and drug activity. The toxic effect flashes are inconsistent as some NPs are comparatively gentle, but others are possibly treacherous. Similarly, the environmental hazards that usually depend on its biodegradability show no surety of safe products and may disperse through the earth or aquatic ecosystems, such as unwanted penetration through cell membranes, the blood–brain barrier, fetuses or infants as the result of breast milk ingestion, placental wall, and causing pregnancy problems, inducing irritation into testicles cells, and being intrusive in sperm production by damaging DNA in germ cells [4]. As the formulation's backbone is similar with respect to size, the material of composition surface charges and more surface area becoming available to react with several such permeations far from target have been reported. Such phenomenon is being risen frequently and need to be taken care of for sustainable large-scale production. The regulatory bodies need to have separate guidelines for all nanotechnology-oriented production houses. These manufacturers produce NMs on a large scale. A large number of workers and operators must be involved at different stages of the production who are then prone to skin or tissue penetration; thus for handling, extreme precaution is required.

As one of the characteristics of NMs is their unique size range, they need a number of ultrapurification techniques in mass production. In addition, other ingredients including enzymes, precursors, and catalysts need to be removed if present at the end of the product. In pharmaceuticals, removal of pyrogens and sterilization is also carried out with filtration. Most importantly, the biological activities of such effluents and NMs are unknown and unpredictable. All these effluents also need some techniques to avoid nonbiodegradable NMs being excreted with them, so polluting water, earth, or air can be avoided.

Apart from all the abovementioned manufacturing issues, several conventional issues for industrial sustainability also remain. For example, for 1 g of finally purified and sterilized product, almost 15–20 l of solvent may be used. For large-scale production, several liters of potable and sterile water are required. The source of water in industries is mostly groundwater. For both potable and sterile water preparation, ultrapurification, reverse osmosis, distillation, and condensation require a cumulative high energy consumption. Several methods of production of NMs require organic solvents as an important part of manufacturing. Removal of them with extra precautions is required to avoid known toxicity related with residual solvents. Filtration of air should also be performed to avoid the interference of particles from outside the reactor with the chemical synthesis. The energy intensity required for a clean room with a class ranging from 1 to 10 is $1.017\,\mathrm{kWh\,cm^{-2}\,year^{-1}}$ [46].

Taking this into consideration, many organizations have widened the toxicology studies. These studies are not only expensive but also time consuming. One more concern regarding these toxicology studies is about the testing protocol, as no specific guidelines for the toxicology study of NMs are available. These studies are being carried out on cell lines or cell culture models; further assay techniques are then used to detect toxicity on these models. The high reactive nature of NMs due to their unique property also hinders with the assay materials or with detection systems. Such hindrance makes all the results and data questionable as they are inconsistent and conflicting. The regulatory bodies need to intervene with such studies and should coordinate with them globally. Such initiative can record the data of all studies globally so that all arising questions can be answered; most importantly, this will minimize the cost and time for determination of global toxicity record, which is the actual need of the hour.

2.3 Development of New Capabilities for Sustainable Environment and Health

The real task of sustainability of NMs commences with taking an insight into the complete lifespan of them. Throughout the life cycle of NMs, to control the toxicity one needs to take a glance at the life cycle of NMs and possible toxicity. It is very important that one must look into the fact that NMs that are made for betterment of performance are becoming toxic at which stage? This information will also be helpful for the development of regulatory guidelines to control toxicity. Every nanotechnology-based product can be categorized for possible critical toxicity attributes at different stages of life cycle as shown in the image (Figure 2.2).

2.3.1 Life cycle and Expulsion

2.3.1.1 Raw Material

Actions associated with the possession of different resources, materials, and carrying raw materials to handling services.

2.3.1.2 Processing and Product Development

Handling of required resources by various steps including mixing, separation, refinement, and if required, allowing them to react. Prepare them for the production stage; store or transport the processed materials to production capabilities.

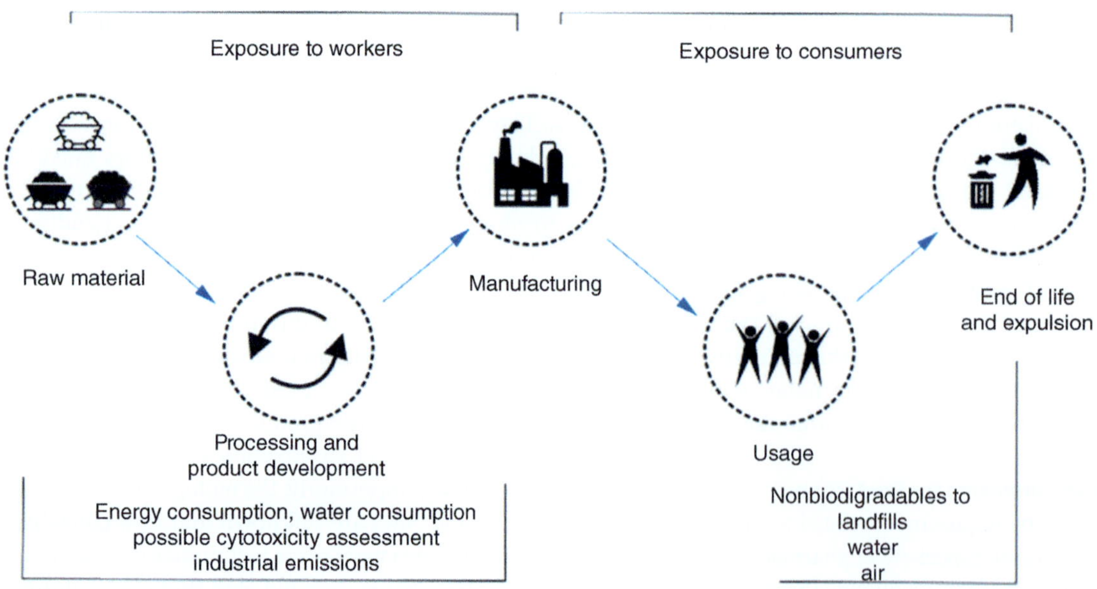

Figure 2.2 Life cycle assessment of nanotechnological products.

2.3.1.3 Manufacturing

Production, storage and transport, or transport and storage until it reaches to consumer.

2.3.1.4 Usage

By consumer, usage, storage, and maintenance up to specified period.

2.3.1.5 End of Life and Expulsion

Expulsion at the end of life span, which may face disposal, transportation, recycling, or incineration.

Assessment of the sustainability of NPs at each stage of life cycle is intended to lay the foundation for creating a decision-backing structure via constant updates. Further efficient control at each stage of life cycle could allow the growth of eco-friendlier goods.

2.3.2 Water Purification and Reuse

Nowadays, society and the global economy is facing the most critical problem, which is availability of clean water. Continuous supply of potable water for humans and in food processing, in energy generation, and in many more industries is a major challenge in many parts of the world. The combination between nanotechnology and water science technology revolutionizes the advancement in water treatment, purification, and recycling technologies [47]. Today, the available techniques for water purification are reverse osmosis or vapor compression. The available technologies for treatment of water have many limitations to provide sufficient quality to meet the requirements of humans and the environment. Hence, it is mandatory to reuse water after its exit from industry; also, we can recycle it, and it can be repurposed. Swaminathan et al. have developed nanofluidic systems that consist of membranes containing nanocapillary scale of 1 to 100 nm composed of polycarbonate sheet with measurement of 6–10 μm. The authors emphasized the effect of pore geometry, charge density, and surface charge to regulate the movement of ions or charged particles in nanopores. The authors focused on the addition of nanocapillary membrane into nanofluidic system that can manage the transport of molecules from one chemical environment to another [48]. Park et al. describe nanofibers as membrane for nanofiltration, which are made up of polyvinylidene fluoride (PVDF) and polyethyleneimine (PEI) for fabrication using electrospinning techniques with improved performance of nanofiltration for purification of water [49]. Usually, nanoadsorbents are very effective in eliminating pollutants from wastewater. Ali studied the benefit of using nanoadsorbents for treatment of wastewater due to its special properties such as very small size with larger surface area having lots of active sites for binding, which gives ease of separation [50]. Today, polymeric nanoadsorbents are more promising due to their excellent adsorption capacity and thermal stability in all pH range [51].

2.3.3 Food Science Technology

Recent advances in nanotechnology have made diverse applications in food technology possible, which include food safety, food additives, nanodelivery systems, biosecurity, and nanotoxicity. Health benefit and quality of food is a major concern for consumers, without reducing nutrition benefits. Due to the availability of many required elements with less-toxic effect of NMs, the demand of NMs is continuously growing in food science [52]. In nanoencapsulation of various components such as vitamins, lycopene, and lutein, many naturally occurring colors can be used as a tool for the delivery of various additives. By nanoencapsulation, spoilage and microbial contamination can be prevented in the food product. It also improves shelf life while improving safety of the food product [53]. Bratovčić et al. have emphasized on the improvement of sustainability of agrifood using nanotechnology by controlling microorganisms for decreasing the wastage of food and increasing the safety of food [54].

Fernandez et al. have worked with carotene protection using nanofibers of zein by encapsulation to increase its stability to oxidation with the help of nontoxic solvents such as dimcthyl formamide and chloroform, which makes it sustainable [55]. In nanoencapsulation, there is a direct contact of NMs with food during consumption, e.g. silicon dioxide (SiO_2) is mostly used as a carrier for fragrance of food products [56]. Many lipid-based formulations that are nanoencapsulated were developed to increase the effectiveness of antioxidants with its bioavailability and solubility [57]. One study showed that nanosized edible coating is a versatile solution for extension of shelf life of food and minimization of spoilage [58].

2.3.4 Sustainability of Aquaculture

Nanotechnology can revolutionize fisheries and aquaculture industry by rapid detection of disease with enhancement of various drugs absorption capacity of fish. Many approaches focus on the prevention of disease in aquatic species. For instance, the use of oil emulsion in vaccine may reduce side effects to a minimum. There is an availability of many inorganic nanocarriers based on carbon or calcium for fish vaccine. Carbon-based inorganic materials have good stability but have many limitations. Because of its variable chemical properties, it can be used as a carrier for vaccine delivery [59]. Liu et al. demonstrated the use of nanopolyplexes for DNA vaccine carrier in fish, which is made up of oleoyl-carboxymethyl-chitosan in hyaluronic acid for its more physiological stability. NMs used for fish vaccine should be safe because fish are eaten by humans who eventually consume the NMs that remain intact in fish.. Generally, NMs used for vaccine for fish need to be safe for both fish and humans [60].

2.3.5 As Nanobiopesticide

Nowadays, there is an increase in environmental hazard due to insecticides, and limitation in the synthesis of novel, safe, and effective molecules has increased the demand of research in this field. Nanosilica with amorphous property can be an effective carrier for better functional pesticide. Insects protect themselves from dehydration using cutical lipids and prevent death of the pest. By the mechanism of physical absorption, the nanosilica absorbed into the cuticular lipid causes killing of insects and pest. As per World Health Organization (WHO), the use of amorphous form of silica is safe for oral consumption for humans [61, 62]. Spinosad is a newly developed environmentally friendly biopesticide with limitation of having short duration of action and less potency in farm. Zhang et al. worked with novel nanospinosad that uses porous NPs for loading of spinosad. This nanospinosad exhibits rough surface texture, which makes it unique for high adhesion behaviors of pest, which also gives high mortality rate of pest with improved photostability in fields [63].

2.3.6 Conservation of Work of Art

Nowadays, work of art is mainly made up of organic materials such as acrylic polymer, which are difficult to remove during cleaning. However, this organic polymer needs to be removed from work of art composition. Nanotechnology comes up with versatile solutions that provide unique solutions by decreasing degradation of work of art materials and also can achieve long-term sustainability of artistic heritage. This approach can be beneficial for unstable materials that are used by modern artist. Some frequently occurring issues in conventional method of reestablishment can be overcome by nanoproducts such as nanofluids that consist of microemulsions [64]. Kolman et al. emphasize on an approach that combines two materials for canvas preparation such as silica NPs treated with polyelectrolyte and nanofibrils made up of cellulose. It was seen that using this combination had strengthening effect for fibrous material such as paper and also improvement in stiffness compared to individual components [65].

2.3.7 Plant Protection Using Nanofibers

Pheromones are a chemical substance generally released by female insects in the environment to attract male insects. Now, these pheromones are used in farm to protect crops from insects and as an alternative to chemical hazards of insecticide to soil. Generally, these pheromones are spread in the farm to prevent mating of insects, thus ultimately reducing their population by confusing male insects [66]. Hellmann et al. have studied the use of nanofibers as carriers using polyamides and cellulose acetate for pheromones. The ideal requirement of nanofibers for this purpose is high porosity and large surface. It was observed that solubility of pheromones is higher in cellulose acetate and gives more loading compared to polyamide nanofibers. The release of pheromones can be observed up to 30–50 days, depending on the material used, which minimizes washout due to rain with long-lasting effect [67]. Ponmurugan et al. had carried out a study of red root rot disease on tea plant. They compared nano-CuO with bulked CuO for antifungal effect on tea plant, which is a disease caused by *Poria hypolateritia* fungus. They found that nano-CuO gives more reduction in fungal growth with high yield in terms of leaves [68].

2.3.8 Management of Greenhouse Effect

Since past 10 years, increase in pollution from vehicles and industries has increased emission of greenhouse gases such as CO_2, which leads to global warming. Although many alternative sources are being developed that are non-CO_2 emitting, the world will continue to use substantial amounts of fossil fuels in the future. The upcoming alternative for excess CO_2 released in the atmosphere is the use of carbon capture and storage using nanotechnology, which gives efficient and cost-effective separation of CO_2 [69]. Initially, conventional membranes with polymeric structures were used for separation of CO_2 due to its low cost and ease of handling [70]. Idem et al. recently studied thermally rearranged polymers that are aromatic polymers containing heterocyclic rings. These polymers have excellent chemical and thermal stability due to the presence of microporous structures that are formed by thermal rearrangement in solid state, which makes it unique for transportation of gas for separation [71].

2.3.9 Materials Supply and Utilization

Materials supply and its proper utilization is needed for industries in electronics due to their rapid growth. These electronic industries produce many e-wastes, which are a serious hazard to the environment. Electronic materials generally contain resin and silica. Even resin contains 80%

of silica. A huge amount of silica present in e-waste is wasted. So, this silica from e-waste is used to prepare porous molecular sieves that contain NMs. Thus, recycling resin from e-waste can be a sustainable approach for the environment [72]. Liou et al. synthesized mesoporous silica with high purity and larger surface from resin ash having a pore size of 3 nm. This approach can minimize the disposal problem of electronic items, which is again a benefit for sustainable environment [72].

2.3.10 Encapsulation of Fertilizers

Fertilizers are generally needed in the soil for nutritional supplement for better growth of plants. Nowadays, mostly farmers use NPK fertilizer, which consists of nitrogen, phosphorous, and potassium. Due to the high solubility of NPK fertilizer, it is mostly lost in the environment due to rain.

This approach of addition of fertilizer becomes costly to farmers and harmful to humans and the environment. Due to the advancement in nanotechnology nanofertilizers have been developed, which have the potential to increase the nutritional value of plants. It also has the tendency of slow delivery because the nutrients coated with NMs can control delivery [73]. One study showed that the presence of biosensor in nanofertilizers can control delivery depending on the need of the soil and the growth rate of plant. Singh et al. compared the bulk NPs of zinc oxide on the growth of cabbage and cauliflower. It was observed that compared to bulk ZnO particles, nanozinc particle is more effective in terms of germination growth and protein content [74].

2.3.11 Regulatory Development

The properties of NMs, i.e. extremely high chemical reactivity due to nano size range, large surface area, and narrow size range, make them hazardous to the environment and human health. The therapeutic NMs, which are prepared from all known inert materials, could become penetrable or react with the non-targeted site and molecule, respectively. On one hand, the toxic manifestation of several NMs is still unknown, and on the other hand, the large-scale production technology is growing at its peak. The exposure of NMs by such mass production industry can create unprecedented event. To anticipate and minimize unfavorable consequences or unintended consequences and to have future public acceptance of nanotechnology-based innovations and development, robust regulatory guideline and support are required today.

Appreciable initiatives are taken by different countries/states till date to regulate nanotechnology, NMs, and the production thereof, including Europe, United States, Asia, and all over the world.

The European Commission has actively reacted on the issue, and under its guidance during 2011–2013, a detailed tactical research outline was published by the associates of the European Union (EU) NanoSafety Cluster with a strategic goal for future exploration on the secure utilization and safeguarded applications of all formulated NMs. The timeline for this documentation was kept for 10 future years starting from 2015. The document focused on delivering a key action ahead in the development of safe and sustainable NMs. Shortly after, the regulatory body of the European Chemicals Agency (ECHA) developed massive advice on the regulatory risk assessment under the European Community Regulation on chemicals and their safe use (REACH) for NMs and published it in April 2012. The European Union funded project BILAT USA 4.0 is one of the milestone events in the development of regulatory guidelines globally. In this event, one of the major concerns was the cooperation for nanosafety between the United States and EU. The result of the event was the identification of several prospective tools and five components for dedicated and accurate evaluations of human and environmental hazards caused by nanotechnology that includes improved testing assays, pertinent endpoints, technical awareness, risk assessment classification for NMs under development or in usage, and harmonized methods to toxicity evaluation.

The US government has its own initiative for nanotechnology that is active since 2000 called the National Nanotechnology Initiative (NNI). NNI actively looks into all matters concerning the sustainability of NMs. The issues related to EHS implications of NMs are specifically taken care by two different groups, Nanotechnology Environmental and Health Implications (NEHI) and Nanoscale Science, Engineering, and Technology (NSET). The first one is the working group of the Nanoscale Science, and the second one is the subcommittee of the National Science and Technology Council. Under NEHI, the US government has been funding since 2005 for EHS-related research.

A persistent initiative taken by a conference in 2011 and the result of the conference was establishment of an organization – SNO – as discussed in previous section. The objective of SNO was set to support sustainability in nanotechnologies and support and promote further development of nano-based formulations, taking into considera-tion health and environment. The SNO also aimed at the safe use of nanoproducts along with providing support for policy and decision-making. The establishment of such organization made its impact all over the world.

An organization in Singapore working since 2004 was registered in 2007 as Asia Nano Forum (ANF) society to promote development, awareness, and safety of nanotechnology. One of the objectives of ANF is to encourage and harmonize standardization and well-being of nanotechnology assessments. Including Singapore, 13 different organizations from Japan, Korea, Malaysia, Australia, Philippines, Taiwan, Thailand, and Vietnam are working under one roof for the sustainable development of nano-based products. Similarly, in Canada, Switzerland, India, and Australia, several government, individual, and autonomous bodies are working hard to address the issues of development, sustainability, and promotion for end users. Although individual organizations are working in their best possible way, it is the need of the hour to unite all the efforts and to address issues arising in development, sustainability, and promotion for end users. Such united efforts can not only uniformly report on single desk but also minimize the time required for recording unknown toxicity of all NMs and related products.

2.3.11.1 The Possible role of Standards in United Regulation

1) Development of procedures for life cycle assessment (LCA) of nanotech-based ingredients, devices, and products.
2) Investigate and release risk assessment tools in the field of nanotechnologies and update it from time to time.
3) Development and delivery of protocols for robust and unique toxicology study to measure toxicity of all different NMs on a single scale.
4) Record all toxicological data on a single platform and update it from time to time.
5) Identify and release robust industrial health protocols related to nanotechnologies and include all possible hazards related to the same occupational health hazards list.
6) Support regulation in the area of nanotechnologies.

2.4 Conclusion

Nanotechnology has appeared as a flexible technical solution to global sustainability challenges. The economic growth in the field of nanotechnologies will lead to an increased variety and increased volumes of engineered nanocarriers that are produced. Certainly, there are challenges for sustainability of nanotechnology itself. Threat of unknown and irreversible toxicity to human and environment is the major issue among them. The use of a life cycle assessment method, combined with risk assessment, to understand the potential problems and to implement green nanomanufacturing methods that are less troublesome to the environment and human health, is currently in progress. Considering the efforts made by different organizations all over the world, there is an opportunity to consider new universally agreed free-of-bias regulatory standards, with special emphasis on the uniform assessment techniques. As a long-time requirement, a harmonized initiative is also required for the toxicology testing of each NM and products based on it and for reporting them on a single platform. In the near future, with such harmonized efforts, nanotechnology will predictably include sustainability standards to reduce its impacts on the environment and human health using sustainable manufacturing practices and employing possible alternatives.

References

1 Eason, T., Meyer, D.E., Curran, M.A., and Upadhyayula, V.K. (2011). Guidance to facilitate decisions for sustainable nanotechnology.

2 Wiek, A., Foley, R.W., and Guston, D.H. (2014). Nanotechnology for sustainability: what does nanotechnology offer to address complex sustainability problems? In: *Nanotechnology for Sustainable Development* (ed. M.S. Diallo, N.A. Fromer and M.S. Jhon). Cham: Springer International Publishing.

3 Hutchison, J.E. (2016). The road to sustainable nanotechnology: challenges, progress and opportunities. *ACS Sustainable Chemistry & Engineering* 4 (11): 5907–5914.

4 Falsini, S., Bardi, U., Abou-Hassan, A., and Ristori, S. (2018). Sustainable strategies for large-scale nanotechnology manufacturing in the biomedical field. *Green Chemistry* 20 (17): 3897–3907.

5 Roco, M.C. (2005). Environmentally responsible development of nanotechnology. *Environmental Science & Technology* 39 (5): 106A–112A.

6 Brinker, C.J. and Ginger, D. (2011). Nanotechnology for sustainability: energy conversion, storage, and conservation. In: *Nanotechnology Research Directions for Societal Needs in 2020*, 261–303. Springer.

7 Sohail, M.I., Waris, A.A., Ayub, M.A. et al. (2019). Environmental application of nanomaterials: a promise

to sustainable future. *Comprehensive Analytical Chemistry* 87: 1–54.

8 Li, C. and Yue, Y. (2014). Fluorescence spectroscopy of graphene quantum dots: temperature effect at different excitation wavelengths. *Nanotechnology* 25 (43): 435703.

9 Iavicoli, I., Leso, V., Ricciardi, W. et al. (2014). Opportunities and challenges of nanotechnology in the green economy. *Environmental Health* 13: 78.

10 Cui, Q., Hernandez, R., Mason, S.E. et al. (2016). Sustainable nanotechnology: opportunities and challenges for theoretical/computational studies. *The Journal of Physical Chemistry B* 120 (30): 7297–7306.

11 Allianz-Aktiengesellschaft. (2005). Opportunities and risks of nanotechnologies: report in Co-operation with the OECD International Futures Programme. Allianz Center for Technology.

12 Yuliarto, B., Septiani, N.L.W., Kaneti, Y.V. et al. (2019). Green synthesis of metal oxide nanostructures using naturally occurring compounds for energy, environmental, and bio-related applications. *New Journal of Chemistry* 43 (40): 15846–15856.

13 Almosni, S., Delamarre, A., Jehl, Z. et al. (2018). Material challenges for solar cells in the twenty-first century: directions in emerging technologies. *Science and Technology of Advanced Materials* 19 (1): 336–369.

14 Nozik, A.J., Beard, M.C., Luther, J.M. et al. (2010). Semiconductor quantum dots and quantum dot arrays and applications of multiple exciton generation to third-generation photovoltaic solar cells. *Chemical Reviews* 110 (11): 6873–6890.

15 Oh, J., Yuan, H.-C., and Branz, H.M. (2012). An 18.2%-efficient black-silicon solar cell achieved through control of carrier recombination in nanostructures. *Nature Nanotechnology* 7 (11): 743–748.

16 Garnett, E. and Yang, P. (2010). Light trapping in silicon nanowire solar cells. *Nano Letters* 10 (3): 1082–1087.

17 Ghernaout, D., Alghamdi, A., Touahmia, M. et al. (2018). Nanotechnology phenomena in the light of the solar energy. *Journal of Energy, Environmental & Chemical Engineering* 3: 1–8.

18 Sharma, P. and Bhargava, M. (2013). Applications and characteristics of nanomaterials in industrial environment. *Research and Development (IJCSEIERD)* 3 (4): 63–72.

19 Ray, P.C., Yu, H., and Fu, P.P. (2009). Toxicity and environmental risks of nanomaterials: challenges and future needs. *Journal of Environmental Science and Health. Part C, Environmental Carcinogenesis & Ecotoxicology Reviews* 27 (1): 1–35.

20 Gatoo, M.A., Naseem, S., Arfat, M.Y. et al. (2014). Physicochemical properties of nanomaterials: implication in associated toxic manifestations. *BioMed Research International* 2014: 1–8.

21 Bar-Ilan, O., Louis, K.M., Yang, S.P. et al. (2012). Titanium dioxide nanoparticles produce phototoxicity in the developing zebrafish. *Nanotoxicology* 6 (6): 670–679.

22 Bar-Ilan, O., Chuang, C.C., Schwahn, D.J. et al. (2013). TiO_2 nanoparticle exposure and illumination during zebrafish development: mortality at parts per billion concentrations. *Environmental Science & Technology* 47 (9): 4726–4733.

23 De Jong, W.H., Hagens, W.I., Krystek, P. et al. (2008). Particle size-dependent organ distribution of gold nanoparticles after intravenous administration. *Biomaterials* 29 (12): 1912–1919.

24 Donaldson, K. and Stone, V. (2003). Current hypotheses on the mechanisms of toxicity of ultrafine particles. *Annali dell'Istituto Superiore di Sanità* 39 (3): 405–410.

25 Gurr, J.-R., Wang, A.S.S., Chen, C.-H., and Jan, K.-Y. (2005). Ultrafine titanium dioxide particles in the absence of photoactivation can induce oxidative damage to human bronchial epithelial cells. *Toxicology* 213 (1): 66–73.

26 Asgharian, B. and Price, O.T. (2007). Deposition of ultrafine (NANO) particles in the human lung. *Inhalation Toxicology* 19 (13): 1045–1054.

27 Guo, C., Wang, J., Jing, L. et al. (2018). Mitochondrial dysfunction, perturbations of mitochondrial dynamics and biogenesis involved in endothelial injury induced by silica nanoparticles. *Environmental Pollution* 236: 926–936.

28 Krzysztof, S., Magdalena, C., Magdalena, M.-K. et al. (2019). Toxicity of metallic nanoparticles in the central nervous system. *Nanotechnology Reviews* 8 (1): 175–200.

29 Cao, Y. (2018). The toxicity of nanoparticles to human endothelial cells. *Adv. Exp. Med. Biol.* 1048: 59–69.

30 Hannon, G., Lysaght, J., Liptrott, N.J., and Prina-Mello, A. (2019). Immunotoxicity considerations for next generation cancer nanomedicines. *Advanced Science* 6 (19): 1900133.

31 Chen, L., Liu, J., Zhang, Y. et al. (2018). The toxicity of silica nanoparticles to the immune system. *Nanomedicine* 13 (15): 1939–1962.

32 Nel, A., Xia, T., Mädler, L., and Li, N. (2006). Toxic potential of materials at the nanolevel. *Science* 311 (5761): 622–627.

33 Yin, H., Too, H.P., and Chow, G.M. (2005). The effects of particle size and surface coating on the cytotoxicity of nickel ferrite. *Biomaterials* 26 (29): 5818–5826.

34 Gupta, A.K. and Gupta, M. (2005). Cytotoxicity suppression and cellular uptake enhancement of surface modified magnetic nanoparticles. *Biomaterials* 26 (13): 1565–1573.

35 Sayes, C.M., Fortner, J.D., Guo, W. et al. (2004). The differential cytotoxicity of water-soluble fullerenes. *Nano Letters* 4 (10): 1881–1887.

36 Verma, A. and Stellacci, F. (2010). Effect of surface properties on nanoparticle–cell interactions. *Small* 6 (1): 12–21.

37 Ispas, C., Andreescu, D., Patel, A. et al. (2009). Toxicity and developmental defects of different sizes and shape nickel nanoparticles in zebrafish. *Environmental Science & Technology* 43 (16): 6349–6356.

38 Champion, J.A. and Mitragotri, S. (2006). Role of target geometry in phagocytosis. *Proceedings of the National Academy of Sciences of the United States of America* 103 (13): 4930–4934.

39 Lee, M.-K., Lim, S.-J., and Kim, C.-K. (2007). Preparation, characterization and in vitro cytotoxicity of paclitaxel-loaded sterically stabilized solid lipid nanoparticles. *Biomaterials* 28 (12): 2137–2146.

40 Kim, S.T., Chompoosor, A., Yeh, Y.-C. et al. (2012). Dendronized gold nanoparticles for siRNA delivery. *Small* 8 (21): 3253–3256.

41 Griffitt, R.J., Luo, J., Gao, J. et al. (2008). Effects of particle composition and species on toxicity of metallic nanomaterials in aquatic organisms. *Environmental Toxicology and Chemistry* 27 (9): 1972–1978.

42 Meng, H., Chen, Z., Xing, G. et al. (2007). Ultrahigh reactivity provokes nanotoxicity: explanation of oral toxicity of nano-copper particles. *Toxicology Letters* 175 (1): 102–110.

43 Bahadar, H., Maqbool, F., Niaz, K., and Abdollahi, M. (2016). Toxicity of nanoparticles and an overview of current experimental models. *Iranian Biomedical Journal* 20 (1): 1–11.

44 Türk, M. and Erkey, C. (2018). Synthesis of supported nanoparticles in supercritical fluids by supercritical fluid reactive deposition: current state, further perspectives and needs. *The Journal of Supercritical Fluids* 134: 176–183.

45 Karn, B. and Sadik, O. (2013). Sustainability and nanotechnology. *ACS Sustainable Chemistry & Engineering* 1 (7): 685.

46 Sartor, D., Piette, M.A., Tschudi, W., and Fok, S. (2000). *Strategies for Energy benchmarking in Cleanrooms and Laboratory-type Facilities*. Berkeley, CA: Lawrence Berkeley National Lab.

47 Cloete, T.E. (2010). Nanofiltration for water and wastewater treatment. In: *Nanotechnology in Water Treatment Applications* (ed. I. Koyuncu and M. Cakmakci). Caister Academic Press.

48 Swaminathan, V.V., Gibson, L.R., Pinti, M. et al. (ed.) (2014). Ionic transport in nanocapillary membrane systems. In: *Nanotechnology for Sustainable Development*, 2014e. Cham: Springer International Publishing.

49 Park, S.-J., Cheedrala, R.K., Diallo, M.S. et al. (2012). Nanofiltration membranes based on polyvinylidene fluoride nanofibrous scaffolds and crosslinked polyethyleneimine networks. *Journal of Nanoparticle Research* **14**: 884.

50 Ali, I. (2012). New generation adsorbents for water treatment. *Chemical Reviews* 112 (10): 5073–5091.

51 Kunduru, K.R., Nazarkovsky, M., Farah, S. et al. (2017). 2 - Nanotechnology for water purification: applications of nanotechnology methods in wastewater treatment. In: *Water Purification* (ed. A.M. Grumezescu), 33–74. Academic Press.

52 Roselli, M., Finamore, A., Garaguso, I. et al. (2003). Zinc oxide protects cultured enterocytes from the damage induced by *Escherichia coli*. *The Journal of Nutrition* 133 (12): 4077–4082.

53 Kashyap, P.L., Xiang, X., and Heiden, P. (2015). Chitosan nanoparticle based delivery systems for sustainable agriculture. *International Journal of Biological Macromolecules* 77: 36–51.

54 Bratovčić, A., Odobašić, A., Ćatić, S., and Šestan, I. (2015). Application of polymer nanocomposite materials in food packaging. *Croatian Journal of Food Science and Technology* 7 (2): 86–94.

55 Fernandez, A., Torres-Giner, S., and Lagaron, J.M. (2009). Novel route to stabilization of bioactive antioxidants by encapsulation in electrospun fibers of zein prolamine. *Food Hydrocolloids* 23 (5): 1427–1432.

56 Dekkers, S., Krystek, P., Peters, R.J.B. et al. (2011). Presence and risks of nanosilica in food products. *Nanotoxicology* 5 (3): 393–405.

57 Mozafari, M.R., Flanagan, J., Matia-Merino, L. et al. (2006). Recent trends in the lipid-based nanoencapsulation of antioxidants and their role in foods. *Journal of the Science of Food and Agriculture* 86 (13): 2038–2045.

58 Flores-López, M.L., Cerqueira, M.A., de Rodríguez, D.J., and Vicente, A.A. (2016). Perspectives on utilization of edible coatings and nano-laminate coatings for extension of postharvest storage of fruits and vegetables. *Food Engineering Reviews* 8 (3): 292–305.

59 Sahdev, P., Ochyl, L.J., and Moon, J.J. (2014). Biomaterials for nanoparticle vaccine delivery systems. *Pharmaceutical Research* 31 (10): 2563–2582.

60 Liu, Y., Wang, F.-Q., Shah, Z. et al. (2016). Nano-polyplex based on oleoyl-carboxymethyl-chitosan (OCMCS) and hyaluronic acid for oral gene vaccine delivery. *Colloids and Surfaces. B, Biointerfaces* 145: 492–501.

61 Kah, M. and Hofmann, T. (2014). Nanopesticide research: current trends and future priorities. *Environment International* 63: 224–235.

62 Barik, T.K., Sahu, B., and Swain, V. (2008). Nanosilica—from medicine to pest control. *Parasitology Research* 103 (2): 253.

63 Zhang, J., Brown, G., Fu, J. et al. (2020). Nanobiopesticides: silica nanoparticles with spiky surfaces enable dual adhesion and enhanced performance. *EcoMat* 2 (2): e12028.

64 Baglioni, P. and Chelazzi, D. (2013). *Nanoscience for the Conservation of Works of Art*. Royal Society of Chemistry.

65 Kolman, K., Nechyporchuk, O., Persson, M. et al. (2018). Combined nanocellulose/nanosilica approach for multiscale consolidation of painting canvases. *ACS Applied Nano Materials* 1 (5): 2036–2040.

66 Diallo, M. and Brinker, C.J. (2011). Nanotechnology for sustainability: environment, water, food, minerals, and climate. In: *Nanotechnology Research Directions for Societal Needs in 2020*: Science Policy Reports, 221–259. Dordrecht: Springer.

67 Hellmann, C., Greiner, A., and Wendorff, J.H. (2011). Design of pheromone releasing nanofibers for plant protection. *Polymers for Advanced Technologies* 22 (4): 407–413.

68 Ponmurugan, P., Manjukarunambika, K., Elango, V., and Gnanamangai, B.M. (2016). Antifungal activity of biosynthesised copper nanoparticles evaluated against red root-rot disease in tea plants. *Journal of Experimental Nanoscience* 11 (13): 1019–1031.

69 Kim, S. and Lee, Y.M. (2014). Thermally rearranged (TR) polymer membranes with nanoengineered cavities tuned for CO_2 separation. In: *Nanotechnology for Sustainable Development*, 2014e (ed. M.S. Diallo, N.A. Fromer and M.S. Jhon). Cham, Cham: Springer.

70 Baker, R.W. (2002). Future directions of membrane gas separation technology. *Industrial and Engineering Chemistry Research* 41 (6): 1393–1411.

71 Idem, R., Wilson, M., Tontiwachwuthikul, P. et al. (2006). Pilot plant studies of the CO_2 capture performance of aqueous MEA and mixed MEA/MDEA solvents at the University of Regina CO_2 capture technology development plant and the Boundary Dam CO_2 capture demonstration plant. *Industrial & Engineering Chemistry Research* 45 (8): 2414–2420.

72 Liou, T.-H. (ed.) (2014). Recovery of silica from electronic waste for the synthesis of cubic MCM-48 and its application in preparing ordered mesoporous carbon molecular sieves using a green approach. In: *Nanotechnology for Sustainable Development*, 2014e. Cham: Springer International Publishing.

73 Qureshi, A., Singh, D., and Dwivedi, S. (2018). Nano-fertilizers: a novel way for enhancing nutrient use efficiency and crop productivity. *International Journal of Current Microbiology and Applied Sciences* 7 (2): 3325–3335.

74 Singh, N., Amist, N., Yadav, K. et al. (2013). Zinc oxide nanoparticles as fertilizer for the germination, growth and metabolism of vegetable crops. *Journal of Nanoengineering and Nanomanufacturing* 3 (4): 353–364.

3

Opportunities and Challenges for Green and Eco-Friendly Nanotechnology in Twenty-First Century

P. Sreeramana Aithal[1] and Shubhrajyotsna Aithal[2]

[1] *College of Management & Commerce, Srinivas University, Mangalore, India*
[2] *Faculty, College of Engineering & Technology, Srinivas University, Mangalore, India*

3.1 Introduction

Identifying problems as challenges and involving in solving them is the nature of innovators in society. In this process, the technology is used as a tool. Most of the current problems of human beings related to basic needs, advanced wants, and dreamy desires can be solved using twenty-first century technologies, which include nanotechnology (NT) and information communication and computation technology (ICCT) [1]. Though the word NT is first used by a Japanese Professor Norio Taniguchi of Tokyo University of Science in 1974 conference to describe the characteristics of a thin film of the order of a nanometer, the idea is originally proposed by Richard Feynman in 1959 in his speech on there's plenty of space at the bottom at the American Physical Society meeting. Thereafter in 1981, Eric Drexler independently used the word NT in his scholarly publication. The discovery of scanning tunneling microscope in 1980 and the invention of fullerenes in 1985, supported for the initial development of NT. In 1986, Eric Drexler published a book on Engine of Creation and Nanosystems, which proposed the idea of nanoscale devices using NT. Further development of semiconductor nanocrystals, metal oxide nanoparticles, nano-quantum dots, and the invention of the atomic force microscope further fuelled the growth of NT as a general-purpose technology [2]. After that NT became research topics of all areas of basic sciences and applied sciences due to its potential ability and advantages of solving scientific, engineering, and industrial problems [3]. At the beginning of the twenty-first century, the developments in NT are accelerated due to its wide acceptance as frontier technology development area, and many postgraduate and research programs were started through country governments intensified research funding.

Even though NT involves manipulating matter on an atomic, molecular, and supramolecular scale, the particular technological goal is of precisely manipulating atoms and molecules for the fabrication of macroscale products, also now referred to as molecular NT. NT being a general-purpose technology encompassing four generations of products with increasing structural and dynamic complexity as (i) passive nanostructure (ii) active nanostructures, (iii) nanosystems, and (iv) molecular nanosystems. It is predicted that by 2020, the increasing integration of nanoscale science and engineering knowledge and of nanosystems promises mass applications of NT in industry, medicine, and computing and in better comprehension and conservation of nature. NT's rapid development worldwide is supporting this anticipation of its expected progress of transforming the society and the future of living beings. However, the slow pace of actual progress against predicted roadmap has been re-examined and based on more planned efforts to be made by country governments and NT research organizations. The modified timeline of NT industry progress is predicted by Aithal et al. [4, 5] based on the last 15 years development trend and it is listed in Table 3.1. The delay is mainly due to the fear and caution of many scientists, organizations, and country governments on fast acceptance for the NT commercialization process. However, through currently developed and proven green and eco-friendly NT processes further accelerates the nanotechnology growth and to stick the timeline as perpredicted in Table 3.1.

Sustainable Nanotechnology: Strategies, Products, and Applications, First Edition.
Edited by Yashwant V. Pathak, Govindan Parayil, and Jayvadan K. Patel.

Table 3.1 Anticipated timeline of nanotechnology innovations [5].

Nanotech generations	Development	Examples	Prediction by Roco and coworker [6]	Prediction by Aithal and Aithal [5]
First generation	Passive nanostructures E.g., coatings, nanoparticles, nanostructured metals, polymers, ceramics	Nanomaterials, including nanotubes and nanolayers	2000–2005	2000–2015
Second generation	Active nanostructures E.g., 3D transistors, amplifiers, targeted drugs, actuators, adaptive structures	Change their state during use, responding in predictable ways to the environment	2006–2010	2016–2020
Third generation	Nanosystems E.g., guided assembling; 3D networking and new hierarchical architectures, robotics	Assemblies of nanotools work together to achieve a final goal	2011–2015	2021–2035
Fourth generation	Molecular nanosystems E.g., molecular devices "by design," atomic design, gene therapy	Involves the intelligent design of molecular and atomic devices, leading to unprecedented understanding and control over the basic building blocks of all natural and man-made things	2016–2020	2036–2050
Fifth generation	Singularity	Growth rate in NT applications becomes infinite	2020–2025	Beyond 2050

As per the Table 3.1, NT is in the transition from its second generation to the third generation. The active nanostructures like 3D transistors, amplifiers, targeted drugs, actuators, adaptive structures using metal-nanomaterials, metal-nanocomposites, semiconductor nanocomposites, and nano-quantum dots are achieved and the next generation of nanosystems to fabricate Guided assembling systems, 3D networking systems, and new hierarchical architecture systems and robotics using nanosystems are approaching from coming years. The developed nanosystems for the next industrial revolution called industry 5.0 with the objective of total automation and mass customization can be achieved only if the fear of adopting NT systems in industries. By means of developing and promoting green and eco-friendly nanosystems [7, 8], the technology can spread at an accelerated speed, and many more components, devices, and systems will be commercialized from different industries. NT can become a part of ideal technology that can solve all problems of society optimally, provided the potential risks are taken care of.

In this chapter, we have analyzed the benefits and risks of NT and suggested how to use green nanotechnology (GNT) models to solve the potential risks. Such an attempt on greening nanotechnology applications will add value to the products and processes that are related to basic needs, advanced wants, and dreamy desires of the people. The discussions are based on the systematic analysis of the opportunities and challenges using standard frameworks on primary industry sector, secondary industry sector, tertiary industrial sector, and quaternary industry sector to interpret the importance of the advent of GNT in society.

3.2 Related Works

GNT deals with environmentally friendly processes of preparation, large-scale manufacturing, and industrial use of nanomaterials by minimizing environmental degradation and potential risks of health hazards. A systematic review of important related works published recently on various underlying areas of GNT is listed in Table 3.2 by identifying issues of environmental degradation. Based on the review and current understanding of various issues of environmental degradation, the opportunities and challenges of using NT as green and eco-friendly technology are discussed and analyzed.

3.3 Objectives and Research Methodology

To understand the possibility of using emerging NT as green and eco-friendly technology by means of the following objectives:

1) To identify green processes for nanomaterial preparation, handling, and conversion into nanosystems.
2) To discuss the opportunities and challenges of GNT in the primary industry sector including agriculture, forestry, mining, fisheries, etc.

Table 3.2 Review on research scholarly publications in green nanotechnology-related areas.

S. No.	Area	Issue and outcome	References
1	Environmental challenges of production of nanomaterials	Challenges in toxicity and environmental risks of nanomaterials	Ray et al. [9]
		Risks of nanomaterials in aquatic and terrestrial environments	Batley et al. [10]
		Environmental risk assessment of nano-TiO_2, nano-Ag, nano-ZnO, CNT, and fullerenes	Coll et al. [11]
		Potential health risks of nanomaterials used in biomedical applications	Lanone and Boczkowski [12]
		Risks of nanomaterials used in regenerative medicine, delivery systems, theragnostic, and therapy	Medina-Reyes et al. [13]
		Frameworks and tools for risk assessment of manufactured nanomaterials	Hristozov et al. [14]
2	Green production of nanomaterials	Green production of carbon nanomaterials	Rezaei and Kamali [15]
		Green nanotechnology of Au particles from plant extracts	Geraldes et al. [16]
		Green manufacturing of ultrapure engineered nanomaterials	Ortiz de Zárate et al. [17]
		Green synthesis of iron nanoparticles	Saif et al. [18]
		Green approach for the production of zinc oxide nanoparticles	Al-Dhabi and Arasu [19]
		Eco-friendly approaches for green synthesis of silver nanoparticles	Bhosale et al. [20]
3	Nanomaterials for agriculture and food industries	Role of nanotechnology in agriculture with special reference to the management of insect pests	Rai and Ingle [21]
		Novel environment-friendly crop improvement strategies	Yashveer et al. [22]
		Nanomaterials in food and agriculture: safety concerns and regulatory issues	Jain et al. [23]
		Nanosensors applications in agriculture and food industry	Omanović-Mikličanina and Maksimović [24]
		Nanomaterials in plant protection	Mazzaglia et al. [25]
		Nanomaterials for food packaging	Huang et al. [26]
		Nanotechnology in precision agriculture	Duhan et al. [27]
4	Nanomaterials for potable water and clean environment industry	Ideal water purifier system using nanotechnology	Aithal and Aithal [28]
		Nanotechnology for water treatment – a green approach	Patanjali et al. [29]
		Synthesis and applications of biogenic nanomaterials in drinking and wastewater treatment	Gautam et al. [30]
		Remediation of water and wastewater by using engineered nanomaterials	Bishoge et al. [31]
		Nanotechnology applicability in industrial wastewater treatment	Kamali et al. [32]
		New generation nanomaterials for water desalination	Teow and Mohammad [33]
		Engineered nanomaterials for water treatment and remediation	Adeleye et al. [34]
		Sustainable development of environment using green nanomaterials	Sivaraj et al. [35]

(*Continued*)

Table 3.2 (Continued)

S. No.	Area	Issue and outcome	References
5	Nanomaterials for renewable energy	Nanotechnology innovations and business opportunities in renewable energy sector	Aithal and Aithal [36]
		Concept and characteristics of ideal energy system	Aithal and Aithal [37]
		Realization opportunity of ideal energy system using nanotechnology	Aithal and Aithal [38]
		Nanomaterial used in clean energy technologies	Moore et al. [39]
		Green nanotechnology of trends in future energy	Guo [40]
		Nanotechnology for achieving green-economy through sustainable energy	Pandey [41]
		Nanomaterials for energy conversion and storage	Choi et al. [42]
		Polydopamine-inspired nanomaterials for energy conversion and storage	Qu et al. [43]
6	Nanomaterials for infrastructure and construction	Nanomaterials in cementitious composites	Adesina [44]
		Reinforcement efficiency of low-cost graphite nanomaterials in high-performance concrete	Peyvandi et al. [45]
		Nanomaterials to enhance microstructure and mechanical properties of concrete	Kwalramani and Syed [46]
		Nanotechnology as a preventive engineering solution to highway infrastructure failures	Ugwu et al. [47]
		Nanotechnology innovations for the sustainable buildings of the future	Sev and Ezel [48]
		Energy savings and sustainable construction	Oke et al. [49]
7	Nanomaterials for cosmetics	Use of nanomaterials in cosmetics	Yapar and Inal [50]
		Current role of nanomaterials in cosmetics	Srinivas [51]
		Present situation and future of nanomaterials in cosmetics	Masunaga [52]
		Gold nanomaterials in consumer cosmetics nanoproducts	Cao et al. [53]
		Role of nanostructured materials in cosmetics	Bilal and Iqbal [54]
		Silver nanoparticles in cosmetics	Gajbhiye and Sakharwade [55]
		Material engineering and nanotechnology for improving sports performance and equipment	Shalaby and Saad [56]
8	Nanomaterials for transportation and automobiles	Potential applications of nanotechnology in transportation	Mathew et al. [57]
		Economic and environmental implications of using nanocomposites in automobiles	Lloyd and Lave [58]
		Nanotechnological innovations and business environment for the Indian automobile sector	Aithal and Aithal [59]
		Nanotechnology applications in future automobiles	Wallner et al. [60]
		Nanotechnology in automobile industry for efficiency enhancement	Gurjar and Tyagi [61]
		Nanotechnology in transportation vehicles: an overview	Shafique and Luo [62]
		Applications of carbon nanotubes in automobiles	Krishnan et al. [63]
9	Nanomaterials for consumer applications industry	Nanomaterials in consumer products: a challenging analytical problem	Contado [64]
		Nanomaterials in consumer products	Hansen et al. [65]
		Nanotechnology in the real world: redeveloping the nanomaterial consumer products inventory	Vance et al. [66]

Table 3.2 (Continued)

S. No.	Area	Issue and outcome	References
10	Nanomaterials for electronics and computer industry	MEMS and nanotechnology research for the electronics industry	Pak [67]
		Inorganic nanomaterials for printed electronics	Wu [68]
		Application of carbon nanotubes (CNT) on the computer science and electrical engineering	Moghaddam et al. [69]
		Polymer nanocomposites and their applications in electronics industry	Tyagi and Tyagi [70]
		The impact of carbon nanotubes and graphene on electronics industry	Vargas-Bernal et al. [71]
		Potential impact of nanomaterials in information and communication technologies	Anuhya and Eunice [72]
		Graphene-reinforced polymeric nanocomposites in computer and electronics industries	Kardanmoghaddam et al. [73]
		Fully integrated graphene and carbon nanotube interconnects for gigahertz high-speed CMOS electronics	Chen et al. [74]
		Carbon nanomaterials for non-volatile memories	Ahn et al. [75]
		3D assembly of graphene nanomaterials for advanced electronics	Le Ferrand et al. [76]
11	Nanomaterials for health and pharmaceutical industry	Impact of nanomaterials on health and environment	Thomas et al. [77]
		Nanotherapeutics – product development along the "nanomaterial" discussion	Wacker [78]
		Occupational safety and health criteria for responsible development of nanotechnology	Schulte et al. [79]
		Effective drug delivery system of biopolymers based on nanomaterials and hydrogels	Gopi et al. [80]
		Biopharmaceutics and therapeutic potential of engineered nanomaterials	Liang et al. [81]
12	Nanomaterials for space industry	Engineered nanomaterials in aerospace	Arepalli and Moloney [82]
		Carbon nanomaterials on a space station board	Rizakhanov et al. [83]
		Nanotechnology safety in the aerospace industry	Haynes and Asmatulu [84]
		Space electric propulsion systems based on smart nanomaterials	Levchenko et al. [85]
		Bio-inspired hierarchical nanomaterials for space applications	Carpinteri and Pugno [86]
		Potential space applications of nanomaterials	Novikov and Voronina [87]

3) To discuss the opportunities and challenges of GNT in secondary industry sector including (i) potable water sector, (ii) environment cleaning sector, (iii) food and food processing sector, (iv) renewable energy sector, (v) construction industry sector, (vi) consumable goods industry sector, (vii) automobiles industry, (viii) medical equipment's and drug synthesis, etc.
4) To discuss the opportunities and challenges of GNT in service industry sector including (i) transportation and space industry, (ii) telecommunication industry, (iii) entertainment industry (iv) education and research, (v) electronic and photonics industry, (vi) healthcare industry, etc.
5) To analyze the technological solution to realize the 17 sustainable development goals (SDG) of the United Nations.

The current status of NT research in different industry sectors is studied through a systematic literature review and the nanomaterial prepared using green and eco-friendly methods and applications in some of the prominent industry sectors along with challenges and opportunities of GNT are also discussed. This technology analysis methodology is used to analyze the possibility of realization of the SDG of the United Nations.

3.4 Global Sustainable Development Goals

NT and ICCT are combinedly called twenty-first century technologies and are potentially realize the SDG of United Nations Member States announced in 2015 with a slogan of action to end poverty, to protect the planet, and to ensure peace and prosperity by the year 2030. Table 3.3 lists the SDG and suitable 21st technologies to achieve them to balance social, economic, and environmental sustainability. To realize each SDG based on the objective of an issue as maximum or minimum, suitable technologies are selected and proposed. It is argued that NT and ICCT, individually or combinedly, are capable to achieve global sustainable goals to a larger extent within the given target of 15 years [88].

3.5 Concept and Characteristics of Ideal Technology

Technology is a tool to solve many problems in society. The concept of ideal technology is a hypothetical technology that can solve all problems of human beings and provide luxury and comfort in life without affecting the society and environment. Ideal technology should have characteristics in order to elevate the quality of life to a unique level with perfect equality so that every human being in this universe should lead a happy and comfortable life and realize the so-called concept of heaven on earth. Based on various factors which decide the ideal technology system characteristics, a model consisting of input conditions, output conditions, environmental conditions, and system requirements is developed [89]. The input properties are (i) Manipulate the fundamental nature of matter to provide solutions to the basic and advanced problems of mankind. (ii) In-expensive and self-reliable in terms of resources to make it attractive to be used by people/countries of varied economical situations. (iii) Ubiquitous so that the technology provides solutions and services at anytime, anywhere, any amount of time to the users. (iv) Affordable to everybody so that it uses common materials available in nature and manipulates effectively to the need of a human being at an affordable cost. The output properties are (i) Solve basic needs like food, drinking water, renewable energy, clothing, shelter, health, and a clean environment. (ii) Provide comfort life to the users by providing solutions to their desires. (iii) Equality: ideal technology provides equal opportunity and similar solutions to every user irrespective of their gender, religion, background, education, economic status, and country of origin. (iv) Automation: ideal technology automates

Table 3.3 SD goals and suitable technologies to achieve them.

Goal No.	Focus issue	Goal by 2030	Objective	Technology
1	Reduce poverty	To zero	Minimize	NT
2	Reduce hunger	To zero	Minimize	NT
3	Health and well-being	To everyone	Maximize	NT
4	Quality higher education	To everyone	Maximize	ICCT
5	Gender equality	To everyone	Maximize	ICCT
6	Clean water and sanitation	To everyone	Maximize	NT
7	Affordable renewable energy	To everyone	Maximize	NT
8	Decent employment	To everyone	Maximize	NT and ICCT
9	Sustainable industrialization	Everywhere	Maximize	NT and ICCT
10	Reduced inequalities among countries	To zero	Minimize	—
11	Safe and sustainable cities and communities	Everywhere	Maximize	ICCT
12	Ensure sustainable production and consumption	Everywhere	Maximize	NT and ICCT
13	Combat on climate change	To zero	Minimize	NT
14	Conserve ocean and marine resources	Everywhere	Maximize	NT
15	Protect life on land	Everywhere	Maximize	NT
16	Ensure peace and justice	To everyone	Maximize	ICCT
17	Global partnership for sustainability	To highest	Maximize	ICCT

all processes in every type of industry to avoid human interference in work/control in order to provide an expected output based on programming. (v) Immortality is the ultimate goal of ideal technology so that it can create an avenue for a deathless situation or enhancement of the human life span. The System Requirement Properties are (i) General purpose technology to support all fields and problems of human and living beings on the earth. (ii) Self-directed, self-controlled, and self-regulated so that technology can control itself in order to achieve its goal. (iii) Easy, simple, quick, and user friendly to solve all types of problems and to provide a quick ideal solution. (iv) Scalable so that it is used for solving small and simple problems to large and complex problems of life. (v) Omnipotent to identify and solve problems and provide comfortability to human beings and feeling him like God. (vi) Exploring new opportunities to improve and explore comfortability and further leisure in the life of people. (vii) Infinite potential for further development of life in the universe. The environment/external properties are (i) Maintain a clean environment through its processes and avoid the footprint of processes, while achieving specific functions. (ii) Infinite business opportunities by creating new products/services with ideal characteristics. (iii) Adaptive to any situation to achieve the stated goal. (iv) No side effects such that it should be safe for users and the environment. Any technology which has the above properties/characteristics is considered as ideal technology and conventional technologies have serious drawbacks/limitations in terms of the above properties [89]. One of the properties of ideal technology is sustainability and zero green gas emissions to the environment, i.e. ideal technology is green technology (GT). Every technology can be made sustainable by adding green components so that they can avoid environmental degradation and converted into green technologies to provide a clean environment for future generations.

As discussed earlier, NT is expected to solve both the basic needs and comfort want of human beings. The basic needs of human beings are food, drinking water, energy, cloth, shelter, health, and environment, and the comfort wants are realizing automation in every field, space travel, and expanded lifespan, and so on. NT is the manipulation of matter on an atomic, molecular, and supramolecular scale. Planned and controlled development of NT leads to environmental sustainability and hence can be used as GT. GNT is evolving as a general-purpose technology due to its applications in all areas of society. Hence in the advanced form, it will have a significant impact on almost all industries and all areas of society by offering better built, longer-lasting, cleaner, safer, and smarter products for the home, for communications, for medicine, for transportation, for agriculture, and for every industry, in general. Thus, by controlled utilization of NT for environmental sustainability, it can be developed as GNT technology for sustainable development.

3.6 Contribution of Universal Technologies on Achieving Sustainability Developmental Goals

It is argued that the dream of realizing Global SDG of the United Nations is possible through the proper use of technologies. In this process, two mega technologies, NT and ICCT, have potential abilities and if used systematically by every participating country can reach the goal.

NT, being a mega technology with many branches including Nanomaterials Development Technology, Nanomechanics Technology, Nanoelectronics Technology, Nanophotonics Technology, Nanobiotechnology, and Nanomedicine, is considered as a general-purpose technology of the twenty-first century. NT has expected to change the rules of development games in many areas including agriculture and food industry, drinking water systems, efficient automobiles, renewable energy systems, high-speed optical computers, low-cost durable shelters, embedded intelligence, space vehicles, health and medical solutions, etc.

Similarly, ICCT being a mega technology supports many innovative general-purpose technologies which are going to change the business models of almost every industry. ICCT underlying technologies are supporting total automation of primary, secondary, tertiary, and quaternary industries by creating artificial intelligence to replace human beings totally, ubiquitous 3D printing through internet of things (IoT) and cloud computing, optimum business model creation through data science and business analytics, online education, retailing, entertainment, social connections through virtual and augmented reality, and high-speed processing of information supports the total transformation of society by changing the lifestyle and comfortability levels of individuals. ICCT allows human beings to become ubiquitous and total automated products and services for individuals and hence acts as a pillar of social, technological, and economical transformation.

These two technologies are capable to transform human life by offering more and more comfortability. As discussed earlier, every human being (i) needs nutritious food, clean drinking water, clean air, affordable shelter, energy from renewable sources, and good health as basic needs, (ii) many products and services as advanced wants for comfortability and satisfaction, (iii) dreams to realize many individual and collective desires to enjoy and get happiness to acquire the status of super-human with three desired abilities as (i) ubiquitous, (ii) omnipotent, and (iii) immortal. Both NT and ICCT can work together to realize the basic needs, advanced wants, dreamy desires at an affordable cost in the near future [90]. Since, based on our prediction of the ability of these two technologies to convert humans into super-humans in the process of transforming society by serving and solving everyone's problems, they are collectively called Universal technology [91]. This integration of ICCT and NT into Universal technology allows us to solve all the above three kinds of problems in society.

3.7 Risks Associated with Nanotechnology

Though NT has innumerable benefits including improved manufacturing methods, improved environment, and water purification systems, efficient renewable energy systems, physical systems property improvement, and performance enhancement, optimization of health problems through nanomedicine, better food production methods, and enhanced nutrition in food, large-scale infrastructure auto-fabrication through self-replicating machines, etc., if not handled properly with creating proper awareness and precautions may have potential disadvantages in terms of risks to the health of living beings, environment, social life, and economy of the countries. Some of the issues related to health, environment, social, economic, and a newly predicted effect called green goo are discussed below:

3.7.1 Health-Related Risks

Nanomaterials are expected to show toxicity effects that are not associated with larger particles. For example, even inert elements like gold become highly active at the nanometer dimensions. Size is a key factor in determining the potential toxicity of a particle. However, it is not the only important factor. Other properties of nanomaterials that influence toxicity include chemical composition, shape, surface structure, surface charge, aggregation and solubility, and the presence or absence of functional groups of other chemicals. The inhaling of nanoparticles due to their size may mainly cause the toxic effect of damage to the lungs and sometimes they may reach to the bloodstream and are predicted to cause heart problems. The ingestion of nanoparticles into the human body may also become toxic and lead to various diseases including colon cancer, Crohn's disease, arrhythmia, asthma, lung cancer, autoimmune diseases, neurological disease, etc. Only little is known about the ill-effects of nanoparticles on the living body and further study, as well as precautions, are essential.

3.7.2 Environmental Related Risks

The unused nanoparticles or waste nanomaterials during synthesis may agglomerate into larger particles or longer chains with modified physical and chemical properties, which may expose to the environment and may enter the human body and spread toxicity. Unused silver nanoparticles if mishandled may contaminate sewage sludge and affect the microorganisms of soil of agricultural fields. The silver nanoparticles show a toxic effect on fishes and other marine animals in the ocean. Silver nanoparticles at high concentrations may be toxic at high concentrations. Thus, it is assumed that nanotechnological products, processes, solutions, and different applications may affect significantly to the environment and climate. Thus, nanoparticles are likely to be more toxic due to their particle size, surface charge, and characteristics compared to bulk materials and hence may pose a risk to the environment.

3.7.3 Social Risks

NT supports new and easy solutions to many problems in agricultural, food processing, renewable energy, and healthcare sectors and removes many existing jobs through improved and automated technology used in manufacturing and service sectors which contribute to loss of manufacturing and agricultural jobs. Such mass loss of jobs in primary and secondary sectors creates social inequalities. NT will enable micro supercomputers on a very small scale, detection of minute amounts of substances, rapid analysis of genomes, and implantation of microchips into humans may lead to a darker side of violation of privacy. Though NT supports surveillance using nano-sensors extremely small cameras, people be afraid of the security and privacy of individuals by tracking their location and their instantaneous behavior. Such a negative perception of NT in society may result in questionable marketing decisions and hindrance in the speed of technology acceptance.

3.7.4 Economic Risks

NT supports huge agricultural production, artificial food at low cost, renewable energy for everyone, low-cost shelter, long life automobiles, low-cost healthcare services, and these innovations in the society leads to economic market crashes due to potential lower demand to oil and gas resources as well as due to crashed market for precious metals like silver, gold, or diamonds, etc. due to artificial reproduction of such things using molecular manipulation techniques. All this leads to crash of economic market and hence many industry performances.

3.7.5 Predictive Green Goo

Another potential danger predicted recently due to NT advances is that with time progress with NT advancement, a stage may reach where nanobots will become commonplace in society, and with artificial intelligence technology, these nanobots may develop their own intelligence and replicate in an uncontrolled manner such a way that one day the earth may overrun by these nanobots. This hypothetical situation is called a gray goo effect. Alternatively, one day, there is a risk on the entire planet that may be overrun by nanoengineered organisms called green goo.

The above NT risks are in turn hindering the progress and investment of financial resources on NT research by many countries' governments.

3.8 How Nanotechnology Can Be Made Green and Eco-Friendly

Ideally, nanomaterial development should be incorporate a safety-by-design approach, as there is a marketing edge for nano-enabled products with a reduced potential impact on health and the environment. Such GNT solutions play a major role in realizing SDG and eliminate the threat of the technification of development processes. GT is an environmental healing technology that reduces environmental damages created by the products and technologies for peoples' conveniences. It is believed that GT promises to augment farm profitability while reducing environmental degradation and conserving natural resources GTs are sustainable technologies which will not create footprint when used for various processes/applications [92]. GTs support the use of natural organic resources and avoid the production of green gasses. They also consume less resources and do not support to increase the entropy of the universe. GTs do not support any kind of environmental degradation. They support the automation of every process and hence avoid human intervention. Since they do not support environmental degradation and contribute to creating the footprint, they are sustainable, improve the lifestyle of the people, and contribute to human comfortability. The major technologies used in the present day like aircraft technology, automobile technology, biotechnology, computer technology, telecommunication technology, education technology, internet technology, renewable energy technology, atomic and nuclear technology, NT, space technology, etc. can be made green using the principle of GT [93–102]. NT predicted as to be pioneering technology of the twenty-first century, if modified as a GT, will be accepted by every user and play an important role in solving problems of society at both basic and advanced levels. The objectives of GNT in some of the basic and advanced fields of society are listed in Table 3.4.

There are many green synthesis protocols which use green chemistry principles for the preparation of nanoparticles and hence nanomaterials compared to conventional methods [103, 104]. This include

1) Nonhazardous naturally occurring materials to use as starting material using bottom-up approach.
2) Recyclability and reuse of magnetic nanoparticles in nano-catalysis applications.
3) Metal nanoparticles can be prepared using natural antioxidant agents like polyphenols from Tea or wine or agricultural residues.
4) Nanoparticles prepared using mild reaction conditions in the facile synthesis display reduced toxicity and are suitable for environmental remediation applications.
5) Metal nanoparticles with antibacterial activities can be synthesized using biogenetic reduction by plants using reducing agents involved include various water-soluble metabolite compounds.
6) Silver and gold nanoparticles which have applications in many industries are prepared using plant-based green chemistry preparation principles.
7) Many crystalline inorganic compounds are prepared using bottom-up low-temperature methods such as hydro/solvothermal synthesis, template-assisted approaches, nucleation, and growth in solution/suspension, microemulsion, miniemulsion, etc.
8) Microbial synthesis of nanoparticles using bacteria, fungi, and viruses; phototrophic eukaryotes, including plants, diatoms, and algae; heterotrophic human cell lines and some other biological agents fall under green synthesis of nanoparticles as eco-friendly, cost-effective, and simple approaches.
9) Microwave-assisted organic synthesis methods are used to prepare metal nanoparticles.

Hence, nanostructures and nanocomposites of metals and metal oxides like, Au, Ag, Al, Eu, Co, Pd, Pt, Fe, C_{60}, CdS, ZnO, Bi_2O_3, TiO_2, $NiFe_2O_4$, etc. are prepared using bottom-up methods like sol–gel method or chemical reduction methods using natural sources like plants, fungi, etc. These green chemistry preparation processes of nanomaterials and nanostructures boost the GNT movement and give confidence to industries to promote NT-based products and services.

Table 3.4 Objectives of green nanotechnology in various areas of society.

S. No.	Area	Objectives of green nanotechnology
1	Agriculture	To avoid environmental degradation in nanotechnology supported agricultural processes including pest control
2	Food processing	To eliminate poisonous contents in food and to avoid green gas emission and environmental degradation in all food packaging processes which are supported by green nanotechnology
3	Potable water	To develop large scale filters for water purification and seawater desalination through green nanotechnological processes without environmental degradation
4	Sustainable energy	To develop green nanotechnological processes for harvesting potential natural energy sources to generate required energy for human civilization without degrading environment
5	Consumer products	To produce a variety of new generation consumer products using green nanotechnology without side effects and without degrading environment in any manner during production, packaging, and in actual use by consumers
6	Automobiles	To produce energy efficient, zero emissions, durable automobiles using renewable energy processes based on green nanotechnology
7	Construction	To build environmentally friendly, energy efficient, smart buildings with the help of green nanotechnological processes
8	Industrial automation	To develop industrial processes which are environmentally friendly, no green gas emission, recyclable waste products using green nanotechnology
9	Computer and information communication	To develop and utilize environmentally friendly, recyclable electronic, and computer components which use renewable energy and efficient performance using green nanotechnology
10	Education	Use of green nanotechnology in all hardware required in education services
11	Health	Use of green nanotechnology with green processes in all health and medical services
12	Aircraft and space travel	Use of green energy and green nanomaterials and environmentally friendly nanotechnological processes in air and space travel

3.9 Green Nanotechnology in Primary Industry Sector

The primary industry sector in the economy includes all industries that are involved in the production and extraction of raw materials from nature such as farming, mining, oil and natural gas, forestry, fishing, etc. The primary industry sector constitutes a larger portion of the economy in the entire world with more contributions from developing countries. GNT principles and processes have immense advantages and benefits in the primary industry sector compared to its constraints. Table 3.5 lists opportunities and challenges for GNT-based innovations in the primary industry sector.

3.10 Green Nanotechnology (GNT) in Secondary Industry Sector

The secondary industry sector in an economy includes secondary processing of raw materials as inputs into various usable products using different types of machines. The output of the secondary industries is manufactured or assembled finished products which are tangible in nature. GNT principles and processes have direct implications of business performance in the secondary industry sector. Table 3.6 lists opportunities and challenges for GNT-based innovations in some of the secondary industry sectors.

3.11 Green Nanotechnology in Tertiary Industry Sector

The tertiary industry sector in an economy includes business which offers various services to consumers. These services are usually intangible in nature and produce high gross domestic products (GDP) and employment. GNT affects the service industry sector both directly and indirectly. GNT principles and processes have implications in future performances in this industry sector. Table 3.7 lists opportunities and challenges for GNT-based innovations in the tertiary industry sector.

Table 3.5 Green nanotechnology-based innovations in primary industry sector.

S. No.	Natural resources industries	Opportunities (O) and challenges (C) of green nanotechnology
1	Agriculture	O: Use of nanofertilizers, nanopesticides, nanobiosensors, and nano-enabled remediation are used in precision forming and biotic and abiotic remediation, for controlled release of nutrients to targeted soils, soil biota, soil organic matter, and plant morphological and physiological responses, aimed to obtain their fullest biological efficacy without overdosage Nano-sensors and nano-remediation methods may detect and remove environmental contaminants C: There is limited knowledge concerning nanomaterial biosafety, adverse effects, fate, and acquired biological reactivity once dispersed into the environment, which requires further scientific efforts to assess possible nano-agricultural risks Lack of adequate risk management strategies for workers, occupational safety practices, and policies, as well as to develop a responsible regulatory consensus
2	Forestry	O: Green nanotechnology has the ability to reduce carbon footprints of petroleum-based products by means of renewable forest-based nanocellulose. Nanocellulose is considered important material for research and development of plastics, coatings, sensors, electronics, automobile body and aerospace materials, medical implants, and body armor so that future day plastics, cellular telephones, medical implants, body armor, and flexible displays will be produced as forest products C: To achieve improvement in the performance-to-weight ratio of paper and packaging products through green nanotechnology to create features such as optical, electronic, barrier, sensing thermal, and surface texture
3	Mining	O: Use of nanosized vessels to recover valuable minerals that end up in the waste Use of green nanomaterials like graphene coatings on drill bits that perform borehole drilling to increase effectiveness and longevity Use of green nanomaterial for lubrication for all the mechanical parts heavy-duty machinery to work optimally Use of green nanotechnology allows us to isolate gold from raw materials in a selective manner instead of using cyanide C: To prove that the nanoparticles prepared through the green route used in mining do not have side effects for mining workers and the environment
4	Fisheries	O: Green nanotechnology can revolutionize fisheries and aquaculture industry with new tools like rapid disease detection, to speed up the absorption of drugs like hormones, vaccines, and nutrients to fish, and by using antibacterial nanocoatings, shelf life of fish, and shellfish may be improved. Green supply chain using nanotechnology can decrease supply chain time between origin to destination Nanotechnological applications in fisheries also include antibacterial surfaces in the aquaculture system, nano delivery of veterinary products in fish food using porous nanostructures, and nanosensors for detecting pathogens in the water including removing microbes, organic chemicals, and metals C: Proper methods of measurement of environmental effects and the surveillance of nanomaterials in products, especially food such as fish fillets, are needed
5	Oil and natural gas	O: Green nanotechnology provides nanomaterials to be used as drilling fluids and enhanced oil recovery in addition to other applications including cementing and well stimulation to enhance well productivity C: Economic feasibility of nanoparticle to be used and their commercial availability. To know the hazardous nature of nanoparticles on health, environment, and safety, and predicted severe health issues

Table 3.6 Nanotechnology-based manufacturing innovations in secondary industry sector.

S. No.	Manufacturing industries	Opportunities (O) and challenges (C) of green nanotechnology
1	Potable water sector	O: Green nanotechnology can be used to convert impure water and seawater into potable water. The nanofilters made by nanomembranes can remove all kinds of water contaminants including turbidity, oil, bacteria, viruses, and organic contaminants from impure water or salt from seawater
		C: Implementation of nanotechnology-enabled alternatives systematically using optimum nanomembranes for the conversion of impure water into potable water throughout the world in a fixed timeframe
2	Environment cleaning sector	O: Green nanotechnology products, processes, and applications are capable to clean degraded environments including air cleaning, water cleaning, and sound cleaning and controls climate change by reducing greenhouse gases and hazardous wastes
		C: Implementation of systematically designed low-cost renewable energy supported nanotechnology-based environment purifying system in every country within a fixed timeline
3	Food and food processing sector	O: Use of green nanotechnology in food protection and delivery to targeted sites, improving food flavor, to encapsulate nutrients such as vitamins, adding antibacterial green nanoparticles into food for enhancement of shelf life, sensing the contamination, improving food storage, tracking, training, brand protection, etc.
		C: Identifying the potential harm of nanomaterials to human beings due to added green nanomaterials to food and food packaging applications
4	Renewable energy sector	O: Use of green nanotechnology for renewable energy generation, transmission, storage, efficient lighting, and energy management systems at low cost
		C: Identifying optimum nanomaterial for a particular application, reduction of cost toward zero, improving efficiency toward 100%, optimization of storage properties of nanotechnology-based storage device, etc.
5	Construction industry sector	O: Green nanotechnology allows to improve the properties of construction materials including cement with the addition of nanoparticles will lead to stronger, more durable, self-healing, air purifying, fire resistant, easy to clean, optimum heat and noise insulation, and quick compacting concrete
		C: Challenges include unknown environmental, health and safety risks, uncertainty concerning the market, and consumer acceptance
6	Consumer goods industry sector	O: Green nanotechnology has made an impact on fast consumer goods like textile and fabrics, cosmetics and skin cares, sporting goods, cleaning products, furniture, home appliances, etc. in terms of durability, production cost, enhanced features, security, etc.
		C: Challenges include technology transfer, government approvals, consumer acceptance and awareness, negative propaganda and lobby of existing conventional manufacturers, etc.
7	Automobile industry	O: Green nanotechnology supported lightweight but stronger automobile components, increased performance with long mileage, durable tires, self-repairing, long-life batteries, renewable energy through nanopaints, which lead to cleaner, quieter, and more pleasant automobiles
		C: Commercialization of green nanomaterials, nanocomponents, and nanosystems related to automobiles. Country government support to create awareness among automobile manufacturers and customers
8	Medical equipment and drug synthesis	O: Green nanotechnology supports to revolutionize drug manufacturing, targeted drug delivery, medical diagnostics, regenerative medicines
		C: Worldwide acceptance of new drugs, treatment procedures, and regulatory practices take time for global usage
		Monitoring side effects and attitudes of medical practitioners also hinder the medical treatments in the health science regime
9	Electrical, electronics, and computer industry sector	O: Green nanotechnology based high speed and miniature-sized communication devices and computation devices, high-density memory chips, nano-sensors, etc. for ubiquitous communication, computation, embedded wearable electronics, and entertainment
		C: Complexity involved in fabricating nanoelectronics devices and the resistance of many companies to shift from silicon-based electronics to molecular nanomaterials-based devices

Table 3.6 (Continued)

S. No.	Manufacturing industries	Opportunities (O) and challenges (C) of green nanotechnology
10	Aerospace and defense sector	O: Green nanotechnology supports miniaturized drones or a swarm of artificial bees to provide additional awareness and visibility. The miniaturized bots equipped with artificial intelligence support give information on the battlefield situations. Hence, GNT with nanosatellites, nano-battlesuit, nanosensors, nano-drones, nanosystems planted in human bodies, and nano-nuclear chemical and biological weapons will give the upper hand in defense and aerospace sector against conventional technologies
		C: Technology transfer, skilled human resource, huge initial investment, awareness at decision-making level, procrastination of decisions

Table 3.7 Nanotechnology-based service innovations in the tertiary industry sector.

S. No.	Service industries	Opportunities (O) and challenges (C) of green nanotechnology
1	Advertising industry	O: Green nanotechnology provides special effect paints and displays which change their color at different light intensity levels and hence at a different time of the day
		C: Commercialization of such technology, cost against existing systems/models, and durability are yet to be tested
2	Education industry	O: Green nanotechnology as a career option, improving and innovations in educational technology through higher quality and low-cost internet as well as display devices leading to ubiquitous online education
		C: Challenges include, complexity involves in technology and initial investment cost
3	E-commerce industry	O: Green nanotechnology supports the identification of counterfeit goods. Certified QR codes and tracking devices supported by nanotechnology can be utilized for product packaging
		C: Creating awareness on the use of such technology with identity benefits to many products consumes time
4	Entertainment industry	O: Green nanotechnology supports to improve the efficiencies of digital entertainment instruments and their durability. It also improves the speed and reachability of internet signals for high speed online video games
		C: Cost and creating awareness are two major challenges for nanotechnology-based entertainment services
5	Fashion industry	Green nanotechnology embedded fabrics can be designed to resist liquids, fight off wrinkles, quick drying, and breathe. Also, for the killing of microbes in cloths, coating that repels water and stain-producing liquids, antistatic nanoparticles to discharge accumulated static charge, new designs, and patterns on fabrics and fashion equipment, etc.
		C: Awareness creation and reachability of GNT features in the industry
6	Financial services industry	O: Huge investment in mega-technology will facilitate the banking sector and drive economic growth. The financial industry will have a key role in the transfer of technology from research centers to various industries. GNT provides technological support for authentic and secured financial transactions
		C: Slow technology transfer, delay in investment decisions, financial constraints for start-ups, effective utilization of government budgets, etc.
7	Healthcare industry	O: Disease control by means of disease diagnostics, prophylactics, and treatment of diseases. Nanoprobes and nanosensors have the potential for prevention and control of diseases. GNT-based organ regeneration and lifespan expansion are also possible
		C: Use of GNT in health care may raise concern on regulation, transparency, patient privacy, consent, etc.
8	Hospitality industry	O: Food preservation with original taste; self-cleaning of floors, walls, fabrics, and furniture; bacteria-repellant bathroom surfaces; bed sheets that resist wear and soil and adjust for comfort or a pillowcase that glows when a guest reads in bed
		C: Cost of technology during the initial investment time is certainly high

(Continued)

Table 3.7 (Continued)

S. No.	Service industries	Opportunities (O) and challenges (C) of green nanotechnology
9	Insurance industry	O: Insurance industry sees GNT as a big opportunity for its future survival and growth. Nanotechnology companies might adopt insurance coverage to reduce their risks and liability for new futuristic business in every industry sector
		C: The fundamental difficulty in making risk assessment unless quantifiable statistical data are readily available, and hence probability and severity are difficult to calculate
10	Print and media industry	O: GNT has applications from printing inks to digital printing processes, videography, wearable audio recording devices, and high speed online electronic and optical communication device
		C: Slow technology transfer and higher cost for early entrants
11	Online services industry	O: GNT supports online ubiquitous services through 5G and future 6G technology where it can offer audio, video, smell, taste, and touch feelings of products and services for online selections. Using smart and artificial intelligence-enabled computers, mass customization of online services is possible
		C: Slow penetration to the market due to early breathing problems
12	Tourism industry	O: GNT-based attractive display screens at tourism centers, airports, and various other locations to provide quick information
		C: Cost factor until mass usage
13	Security services	O: Tagging and tracking, monitoring, advancing sensors technology, improved RFID technology in body armor, combating fraud with nanoparticle-based inks
		C: Integrating nanotechnology with ICCT for specific services
14	Coating services industry	O: Coating service, based purely on nano coating on surfaces/devices for specific purposes
		C: Attractive for large scale coatings only based on cost–benefit analysis
15	Event management industry	O: GNT-based electronic decorations, waste management, food and beverage quality management, music service, clothing service, etc.
		C: Coordination with technology provider and event management team
16	Smart city services industry	O: Green nanomaterial enabled the network to provide a backbone for smart city communications using 5G technology. Inclusion of nanotechnology in smart city solutions along with Information technology solutions, power plant, water treatment, road infrastructure, air pollution, etc.
		C: Delay in the realization of smart cities by governments in developed and developing countries

3.12 Green Nanotechnology in Quaternary Industry Sector

The quaternary industry sector in the economy includes the activities based on the intellectual or knowledge-based part of the economy. This sector is found in only most advanced countries in which through research and development, latest ICCT, and typically includes services such as information generation and sharing, information technology (communication and computing), consultation, education, research and development, and other knowledge-based services supported by technology. The important ICCT underlying technologies which work with GNT to provide intelligent services are artificial intelligence, 3D printing, cloud computing, IoT, quantum computing, information storage technology, mobile business technologies, and online education technologies. GNT being general-purpose technology of the twenty-first century supports ICCT underlying technologies as another set of general-purpose technologies to develop super-intelligent machines and super-human beings when these technologies get saturated [90].

3.13 Challenges in Managing Nanotechnology Innovations

NT offers a plethora of new materials for different industries and industry sectors. Though public perception is generally supportive of NT, some risk analysis of the potential long-term effects of green routed manufactured nanomaterials in human food is required. In different industry sectors, the exposure risk would relate to different aspects and it is a challenge to the scientist and engineers as well as local governments while managing the NT innovations.

This also needs to arrange some public awareness to explain the uses of NT by all industry sectors to explain the benefits and risks to the consumers. The exact number of available nanomaterials naturally available as well as man-made in the environment including the oceans are not understood and the fate and behavior of manufactured nanomaterials in important systems, such as the oceans, are poorly understood. The design of nanomaterials should be according to the principles of GNT that would complement and support current regulations of the government and to address the predicted risk while fostering the sustainable development of NT as GT.

3.14 How Green Nanotechnology Is Different and Secured to Achieve All 17 Sustainable Development Goals of UN

While discussing about opportunities and challenges of GNT and nanomaterials as general-purpose technology to solve many problems in the primary industry sector, secondary industry sector, tertiary industry sector, and quaternary industry sector, it is found that NT is a boon to mankind and gives incredible power to human beings to improve the comfortability and quality of life. While comparing the United Nations SDG with GNT opportunities, 12 goals out of 17 goals can be realized by using GNT. Five goals can be realized using ICCT underlying technologies and one goal of reduced inequalities among Countries is not directly related to technological innovations. The possibility of using GNT processes/NT systems in solving 17 goals of global sustainable development is shown in Table 3.8.

The NT solutions in different industry sectors are planned by ensuring nanoscale materials are designed and developed with human health and the global environment in mind, and hence further fears on sustainability due to degradable environmental concerns are unnecessary for future years. To accelerate the development of GNT, countries should consider the following steps:

1) Educating the people and entrepreneurs to create a supportive environment in society to accept new technology products.
2) Assessment of GNT implications to gauge the trade-off between benefits and constraints for nanoproducts and their existing counterparts.
3) Develop a performance standard globally for GNT products for the producers and users.
4) Branding green nano products and services globally by creating awareness in every country to use and to support to realize the SDG of the United Nations.
5) Country governments should encourage the production and use of GNT products and services by providing tax relaxation and other financial incentives for the initial few years. This will accelerate the penetration of GNT products and services development and usage in the country.
6) Government support for open innovation without patent rights leads to wide production and marketing of GNT products and services by many companies in different industry sectors. A systematic rewarding policy should be developed to honor such an open innovation.
7) Accelerated Focus on general-purpose technology development through national technology policy to provide more resources and encouragement for researchers and investors.
8) Government and Nongovernment organizations (NGO) efforts on creating awareness programs on the advantages of green nano products and risk prevention strategies through the design of safer and green processes that make them.

The proactive policies of government and industry sectors to encourage investors and development of skilled human resources. The responsible approach of government, industries, organizations, and individuals in developing green and eco-friendly NT as a technological tool will enable more sustainable products and processes for the next industrial revolution to realize the United Nations SDG.

3.15 Conclusion

Converting NT into green and eco-friendly NT by means of using bottom-up preparation techniques of green chemistry allows us to decrease the risks associated with it to various industrial applications and resembles the many characteristics of ideal technology [89]. Being a general-purpose technology with characteristics like pervasiveness, improvement, and innovative opportunities, NT has its roots and branches in almost all parts of science and technology applications in society. The fear of adverse effects of nanomaterials on user health and environment is also possible to take care by choosing green synthesis methods at room temperature processes. The United Nations identified 17 SDG in the year 2015, for global prospectus of humanity as systematic development objectives to be realized by 2030 with a timeframe of 15 years. These SDG goals can be realized using two general-purpose technologies of the twenty-first century that include NT and ICCT, in which NT has a major role to support 12 goals for realization. The opportunities and challenges of GNT in major part of primary, secondary, tertiary, and quaternary industry sectors encourages and accelerates the growth and acceptance of technology by enhancing investments and support NT usage by every country.

Table 3.8 SD goals, suitable technologies and focus on different industry sector to achieve them.

Goal No.	Focus issue	Technology	Focus of technological solutions in different industry sectors
1	Reduce poverty	NT	Green nanotechnology in primary sector with focus on agriculture industry
2	Reduce hunger	NT	Green nanotechnology with a focus on food preservation, transportation, and even preparation of artificial food
3	Health and well-being	NT	Green nanotechnology in healthcare and environment cleaning
4	Quality higher education	ICCT	Education technology using ICCT and nanotechnology in the tertiary sector
5	Gender equality	ICCT	Awareness and equal opportunity creation using ICCT-based education
6	Clean water and sanitation	NT	Green nanotechnology in primary and secondary industry sectors
7	Affordable renewable energy	NT	Green nanotechnology for highly efficient renewable energy generation and storage
8	Decent employment	NT and ICCT	Green nanotechnology together with ICCT created skilled jobs and employment
9	Sustainable industrialization	NT and ICCT	Green nanotechnology together with ICCT provides industry 4.0 and industry 5.0 based mass customization and total automation in all industry sectors
10	Reduced inequalities among countries	—	Through technology transfer and cooperation between the countries through free trade
11	Safe and sustainable cities and communities	ICCT and NT	Green nanotechnology together ICCT supports to create smart cities integrated industrial facilities
12	Ensure sustainable production and consumption	NT and ICCT	Green nanotechnology together with artificial intelligence and IoT supports sustainable production and consumption of essential commodities
13	Combat on climate change	NT	Green nanotechnology supports to control environmental degradation and provides the optimum solution to clean the environment and helps to combat climate change
14	Conserve ocean and marine resources	NT	Green nanotechnology is capable enough to clean the ocean to conserve marine resources
15	Protect life on land	NT	Green nanotechnology in the healthcare sector, strong shelter, and pollution-free clean environment through nanotechnology can protect life on the planet
16	Ensure peace and justice	ICCT	Communicate and collaborate using ICCT for peace and prosperity
17	Global partnership for sustainability	ICCT	Communicate and collaborate using ICCT for resource sharing

References

1 Aithal, P.S. and Aithal, S. (2019). Management of Universal Technologies & their Industry Implications. *Proceedings of International Conference on Emerging Trends in Management, IT and Education 1* (2): 318–328. ISBN No.: 978-87-941751-2-4. DOI: https://doi.org/10.5281/zenodo.3559719.

2 Aithal, P.S. and Aithal, S. (2018). Study of various general-purpose technologies and their contribution towards developing sustainable society. *International Journal of Management, Technology, and Social Sciences (IJMTS)* 3 (2): 16–33.

3 Aithal, P.S. (2016). Nanotechnology innovations & business opportunities: a review. *International Journal of Management, IT and Engineering (IJMIE)* 6 (1): 182–204.

4 Aithal, P.S. and Aithal, S. (2015). Managing anticipated breakthrough technologies of 21st century - a review. *International Journal of Research & Development in Technology and Management Sciences* 21 (6): 112–133.

5 Aithal, P.S. and Aithal, S. (2016). Nanotechnology innovations and commercialization – opportunities, challenges & reasons for delay. *International Journal of Engineering and Manufacturing (IJEM)* 6 (6): 15–25.

6 Renn, O. and Roco, M. (2006). White paper no.2 Nanotechnology risk governance. International Risk Governance Council, Geneva, Switzerland. p. 103.

7 Jaiswal, S. and Mojahid, A. (2020). Innovation in society through green\eco-friendly technology. *Journal for Modern Trends in Science and Technology* 6 (01): 37–43.

8 Aithal, P.S. and Aithal, S. (2016). Opportunities & challenges for green technology in 21st century. *International Journal of Current Research and Modern Education (IJCRME)* 1 (1): 818–828.

9 Ray, P.C., Yu, H., and Fu, P.P. (2009). Toxicity and environmental risks of nanomaterials: challenges and future needs. *Journal of Environmental Science and Health Part C* 27 (1): 1–35.

10 Batley, G.E., Kirby, J.K., and McLaughlin, M.J. (2013). Fate and risks of nanomaterials in aquatic and terrestrial environments. *Accounts of Chemical Research* 46 (3): 854–862.

11 Coll, C., Notter, D., Gottschalk, F. et al. (2016). Probabilistic environmental risk assessment of five nanomaterials (nano-TiO_2, nano-Ag, nano-ZnO, CNT, and fullerenes). *Nanotoxicology* 10 (4): 436–444.

12 Lanone, S. and Boczkowski, J. (2006). Biomedical applications and potential health risks of nanomaterials: molecular mechanisms. *Current Molecular Medicine* 6 (6): 651–663.

13 Medina-Reyes, E.I., Garcia-Viacobo, D., Carrero-Martinez, F.A., and Chirino, Y.I. (2017). Applications and risks of nanomaterials used in regenerative medicine, delivery systems, theranostics, and therapy. *Critical Reviews in Therapeutic Drug Carrier Systems* 34 (1): 35–61.

14 Hristozov, D., Gottardo, S., Semenzin, E. et al. (2016). Frameworks and tools for risk assessment of manufactured nanomaterials. *Environment International* 95: 36–53.

15 Rezaei, A. and Kamali, A.R. (2018). Green production of carbon nanomaterials in molten salts, mechanisms and applications. *Diamond and Related Materials* 83: 146–161.

16 Geraldes, A.N., da Silva, A.A., Leal, J. et al. (2016). Green nanotechnology from plant extracts: synthesis and characterization of gold nanoparticles. *Advances in Nanoparticles* 5 (3): 176–185.

17 Ortiz de Zárate, D., García-Meca, C., Pinilla-Cienfuegos, E. et al. (2020). Green and sustainable manufacture of ultrapure engineered nanomaterials. *Nanomaterials* 10 (3): 466–480.

18 Saif, S., Tahir, A., and Chen, Y. (2016). Green synthesis of iron nanoparticles and their environmental applications and implications. *Nanomaterials* 6 (11): 209–235.

19 Al-Dhabi, N.A. and Arasu, M.V. (2018). Environmentally-friendly green approach for the production of zinc oxide nanoparticles and their anti-fungal, ovicidal, and larvicidal properties. *Nanomaterials* 8 (7): 500–513.

20 Bhosale, R.R., Kulkarni, A.S., Gilda, S.S. et al. (2014). Innovative eco-friendly approaches for green synthesis of silver nanoparticles. *International Journal of Pharmaceutical Sciences and Nanotechnology* 7 (1): 2328–2337.

21 Rai, M. and Ingle, A. (2012). Role of nanotechnology in agriculture with special reference to management of insect pests. *Applied Microbiology and Biotechnology* 94 (2): 287–293.

22 Yashveer, S., Singh, V., Kaswan, V. et al. (2014). Green biotechnology, nanotechnology and bio-fortification: perspectives on novel environment-friendly crop improvement strategies. *Biotechnology and Genetic Engineering Reviews* 30 (2): 113–126.

23 Jain, A., Ranjan, S., Dasgupta, N., and Ramalingam, C. (2018). Nanomaterials in food and agriculture: an overview on their safety concerns and regulatory issues. *Critical Reviews in Food Science and Nutrition* 58 (2): 297–317.

24 Omanović-Mikličanina, E. and Maksimović, M. (2016). Nanosensors applications in agriculture and food industry. *Bull ChemTechnol Bosnia Herzegovina* 47: 59–70.

25 Mazzaglia, A., Fortunati, E., Kenny, J.M. et al. (2017). Nanomaterials in plant protection. *Nanotechnology in Agriculture and Food Science* 7: 115–133.

26 Huang, Y., Mei, L., Chen, X., and Wang, Q. (2018). Recent developments in food packaging based on nanomaterials. *Nanomaterials* 8 (10): 830.

27 Duhan, J.S., Kumar, R., Kumar, N. et al. (2017). Nanotechnology: the new perspective in precision agriculture. *Biotechnology Reports* 15: 11–23.

28 Aithal, S. and Aithal, P.S. (2018). Concept of ideal water purifier system to produce potable water and its realization opportunities using nanotechnology. *International Journal of Applied Engineering and Management Letters (IJAEML)* 2 (2): 8–26.

29 Patanjali, P., Singh, R., Kumar, A., and Chaudhary, P. (2019). Nanotechnology for water treatment: a green approach. In: *Green Synthesis, Characterization and Applications of Nanoparticles* (eds. A. Shukla and S. Iravani), 485–512. Elsevier.

30 Gautam, P.K., Singh, A., Misra, K. et al. (2019). Synthesis and applications of biogenic nanomaterials in drinking and wastewater treatment. *Journal of Environmental Management* 231: 734–748.

31 Bishoge, O.K., Zhang, L., Suntu, S.L. et al. (2018). Remediation of water and wastewater by using engineered nanomaterials: a review. *Journal of Environmental Science and Health, Part A* 53 (6): 537–554.

32 Kamali, M., Persson, K.M., Costa, M.E., and Capela, I. (2019). Sustainability criteria for assessing

nanotechnology applicability in industrial wastewater treatment: current status and future outlook. *Environment International* 125: 261–276.
33 Teow, Y.H. and Mohammad, A.W. (2019). New generation nanomaterials for water desalination: a review. *Desalination* 451: 2–17.
34 Adeleye, A.S., Conway, J.R., Garner, K. et al. (2016). Engineered nanomaterials for water treatment and remediation: costs, benefits, and applicability. *Chemical Engineering Journal* 286: 640–662.
35 Sivaraj, R., Salam, H.A., Rajiv, P., and Rajendran, V. (2015). Green nanotechnology: the solution to sustainable development of environment. In: *Environmental Sustainability*, 311–324. New Delhi: Springer.
36 Aithal, P.S. and Aithal, S. (2016). Nanotechnology innovations & business opportunities in renewable energy sector. *International Journal of Engineering Research and Modern Education (IJERME)* 1 (1): 674–692.
37 Aithal, P.S. and Aithal, S. (2018). The concept & characteristics of ideal energy system and its realization constraints. *International Journal of Applied Engineering and Management Letters (IJAEML)* 2 (2): 127–137.
38 Aithal, S. and Aithal, P.S. (2018). The realization opportunity of ideal energy system using nanotechnology based research and innovations. *International Journal of Advanced Trends in Engineering and Technology* 3 (2): 1–15.
39 Moore, E.A., Babbitt, C.W., Gaustad, G., and Moore, S.T. (2018). Portfolio optimization of nanomaterial use in clean energy technologies. *Environmental Science and Technology* 52 (7): 4440–4448.
40 Guo, K.W. (2012). Green nanotechnology of trends in future energy: a review. *International Journal of Energy Research* 36 (1): 1–17.
41 Pandey, G. (2018). Nanotechnology for achieving green-economy through sustainable energy. *Rasayan Journal of Chemistry* 11: 942–950.
42 Choi, J.W., Wang, D., and Wang, D. (2016). Nanomaterials for energy conversion and storage. *ChemNanoMat* 2 (7): 560–561.
43 Qu, K., Wang, Y., Vasileff, A. et al. (2018). Polydopamine-inspired nanomaterials for energy conversion and storage. *Journal of Materials Chemistry A* 6 (44): 21827–21846.
44 Adesina, A. (2020). Nanomaterials in cementitious composites: review of durability performance. *Journal of Building Pathology and Rehabilitation* 5 (1): 1–9.
45 Peyvandi, A., Soroushian, P., Farhadi, N., and Balachandra, A.M. (2018). Evaluation of the reinforcement efficiency of low-cost graphite nanomaterials in high-performance concrete. *KSCE Journal of Civil Engineering* 22 (10): 3875–3882.
46 Kwalramani, M.A. and Syed, Z.I. (2018). Application of nanomaterials to enhance microstructure and mechanical properties of concrete. *International Journal of Integrated Engineering* 10 (2): 98–104.
47 Ugwu, O.O., Arop, J.B., Nwoji, C.U., and Osadebe, N.N. (2013). Nanotechnology as a preventive engineering solution to highway infrastructure failures. *Journal of Construction Engineering and Management* 139 (8): 987–993.
48 Sev, A. and Ezel, M. (2014). Nanotechnology innovations for the sustainable buildings of the future. *World Academy of Science, Engineering and Technology International Journal of Civil, Environmental, Structural, Construction and Architectural Engineering* 8 (8): 886–896.
49 Oke, A.E., Aigbavboa, C.O., and Semenya, K. (2017). Energy savings and sustainable construction: examining the advantages of nanotechnology. *Energy Procedia* 142: 3839–3843.
50 Yapar, E.A. and İNAL, Ö. (2012). Nanomaterials and cosmetics. *İstanbul ÜniversitesiEczacılıkFakültesiDergisi* 42 (1): 43–70.
51 Srinivas, K. (2016). The current role of nanomaterials in cosmetics. *Journal of Chemical and Pharmaceutical Research* 8 (5): 906–914.
52 Masunaga, T. (2014). Nanomaterials in cosmetics--present situation and future. *Yakugakuzasshi: Journal of the Pharmaceutical Society of Japan* 134 (1): 39.
53 Cao, M., Li, J., Tang, J. et al. (2016). Gold nanomaterials in consumer cosmetics nanoproducts: analyses, characterization, and dermal safety assessment. *Small* 12 (39): 5488–5496.
54 Bilal, M. and Iqbal, H. (2020). New insights on unique features and role of nanostructured materials in cosmetics. *Cosmetics* 7 (2): 24.
55 Gajbhiye, S. and Sakharwade, S. (2016). Silver nanoparticles in cosmetics. *Journal of Cosmetics, Dermatological Sciences and Applications* 6 (1): 48–53.
56 Shalaby, M.N. and Saad, M.M. (2020). Advanced material engineering and nanotechnology for improving sports performance and equipment. *International Journal of Psychosocial Rehabilitation* 24 (10): 2314–2322.
57 Mathew, J., Joy, J., and George, S.C. (2019). Potential applications of nanotechnology in transportation: a review. *Journal of King Saud University-Science* 31 (4): 586–594.
58 Lloyd, S.M. and Lave, L.B. (2003). Life cycle economic and environmental implications of using nanocomposites in automobiles. *Environmental Science & Technology* 37 (15): 3458–3466.

59 Aithal, P.S. and Aithal, S. (2016). Nanotechnological innovations & business environment for Indian automobile sector: a futuristic approach. *International Journal of Scientific Research and Modern Education (IJSRME)* 1 (1): 296–307.

60 Wallner, E., Sarma, D.H.R., Myers, B., et al. (2010). Nanotechnology applications in future automobiles (No. 2010-01-1149). SAE Technical Paper. pp. 1–12. DOI: https://doi.org/10.4271/2010-01-1149.

61 Gurjar, B.S. and Tyagi, P. (2015). Applications of nanotechnology in automobile industry for efficiency enhancement and energy saving-a review. *International Journal of Interdisciplinary Research* 2 (3): 1–7.

62 Shafique, M. and Luo, X. (2019). Nanotechnology in transportation vehicles: an overview of its applications, environmental, health and safety concerns. *Materials* 12 (15): 2493.

63 Krishnan, A., Shandilya, S., Balasubramanya, H.S., and Gupta, P. (2020). A review on applications of carbon nanotubes in automobiles. *International Journal of Mechanical Engineering and Technology* 11 (1): 204–210.

64 Contado, C. (2015). Nanomaterials in consumer products: a challenging analytical problem. *Frontiers in chemistry* 3 (48): 1–20.

65 Hansen, S.F., Baun, A., Michelson, E.S. et al. (2009). Nanomaterials in consumer products. In: *Nanomaterials: Risks and Benefits* (eds. I. Linkov and J. Steevens), 359–367. Dordrecht: Springer.

66 Vance, M.E., Kuiken, T., Vejerano, E.P. et al. (2015). Nanotechnology in the real world: redeveloping the nanomaterial consumer products inventory. *Beilstein Journal of Nanotechnology* 6 (1): 1769–1780.

67 Pak, Y.E. (2001). MEMS and nanotechnology research for the electronics industry. In: *Smart Structures and Materials 2001: Smart Electronics and MEMS*, vol. 4334 (ed. V.K. Varadan), 23–29. International Society for Optics and Photonics.

68 Wu, W. (2017). Inorganic nanomaterials for printed electronics: a review. *Nanoscale* 9 (22): 7342–7372.

69 Moghaddam, H.K., Maraki, M.R., and Rajaei, A. (2020). Application of carbon nanotubes (CNT) on the computer science and electrical engineering: a review. *International Journal of Reconfigurable and Embedded Systems* 9 (1): 61–82.

70 Tyagi, M. and Tyagi, D. (2014). Polymer nanocomposites and their applications in electronics industry. *International Journal of Electronic and Electrical Engineering* 7 (6): 603–608.

71 Vargas-Bernal, R., Herrera-Pérez, G., and Tecpoyotl-Torres, M. (2019). The impact of carbon nanotubes and graphene on electronics industry. In: *Advanced Methodologies and Technologies in Digital Marketing and Entrepreneurship* (ed. M. Khosrow-Pour), 382–394. IGI Global.

72 Anuhya, K.H. and Eunice, T.G. (2016). Potential impact of nanomaterials in information and communication technologies. *Indian Journal of Research in Pharmacy and Biotechnology* 4 (6): 267–270.

73 Kardanmoghaddam, H., Maraki, M., and Rajaei, A. (2020). Graphene-reinforced polymeric nanocomposites in computer and electronics industries. *FactaUniversitatis, Series: Electronics and Energetics* 33 (3): 351–378.

74 Chen, X., Akinwande, D., Lee, K.J. et al. (2010). Fully integrated graphene and carbon nanotube interconnects for gigahertz high-speed CMOS electronics. *IEEE Transactions on Electron Devices* 57 (11): 3137–3143.

75 Ahn, E.C., Wong, H.S.P., and Pop, E. (2018). Carbon nanomaterials for non-volatile memories. *Nature Reviews Materials* 3 (3): 1–15.

76 Le Ferrand, H., Chabi, S., and Agarwala, S. (2020). 3D assembly of graphene nanomaterials for advanced electronics. *Advanced Intelligent Systems* 2 (5): 1–16.

77 Thomas, S.P., Al-Mutairi, E.M., and De, S.K. (2013). Impact of nanomaterials on health and environment. *Arabian Journal for Science and Engineering* 38 (3): 457–477.

78 Wacker, M.G. (2014). Nanotherapeutics—product development along the "nanomaterial" discussion. *Journal of Pharmaceutical Sciences* 103 (3): 777–784.

79 Schulte, P.A., Geraci, C.L., Murashov, V. et al. (2014). Occupational safety and health criteria for responsible development of nanotechnology. *Journal of Nanoparticle Research* 16 (2153): 1–17.

80 Gopi, S., Amalraj, A., and Thomas, S. (2016). Effective drug delivery system of biopolymers based on nanomaterials and hydrogels—a review. *Drug Desigining* 5 (129): 2169–0138.

81 Liang, X.J., Chen, C., Zhao, Y. et al. (2008). Biopharmaceutics and therapeutic potential of engineered nanomaterials. *Current Drug Metabolism* 9 (8): 697–709.

82 Arepalli, S. and Moloney, P. (2015). Engineered nanomaterials in aerospace. *MRS Bulletin* 40 (10): 804–811.

83 Rizakhanov, R.N., Polyanskiy, M.N., Malinovskaya, O.S., and Tsvetkova, E.V. (2012). CVD facility for the formation of carbon nanomaterials on a space station board. *Fullerenes, Nanotubes and Carbon Nanostructures* 20 (4–7): 482–486.

84 Haynes, H. and Asmatulu, R. (2013). Nanotechnology safety in the aerospace industry. In: *Nanotechnology Safety* (ed. R. Asmatulu), 85–97. Elsevier.

85 Levchenko, I., Xu, S., Teel, G. et al. (2018). Recent progress and perspectives of space electric propulsion systems based on smart nanomaterials. *Nature Communications* 9 (1): 1–19.

86 Carpinteri, A. and Pugno, N. (2008). Bio-inspired hierarchical nanomaterials for space applications. *Journal of the British Interplanetary Society* 61 (8): 290–294.

87 Novikov, L.S. and Voronina, E.N. (2017). Potential space applications of nanomaterials. In: *Protection of Materials and Structures from the Space Environment* (ed. J. Kleiman), 139–147. Cham: Springer.

88 Nilsson, M., Griggs, D., and Visbeck, M. (2016). Policy: map the interactions between sustainable development goals. *Nature* 534 (7607): 320–322.

89 Aithal, P.S. and Aithal, S. (2015). Ideal technology concept & its realization opportunity using nanotechnology. *International Journal of Application or Innovation in Engineering & Management (IJAIEM)* 4 (2): 153–164.

90 Aithal, P.S. and Aithal, S. (2018). Nanotechnology based innovations and human life comfortability –are we marching towards immortality? *International Journal of Applied Engineering and Management Letters (IJAEML)* 2 (2): 71–86.

91 Aithal, P.S. and Madhushree, L.M. (2019). Emerging trends in ICCT as universal technology for strategic development of industry sectors. In: *IT and Computing for All the Domains and Professionals: The Emergence of Computer and Information Sciences* (eds. P.K. Paul, A. Bhuimali, K.S. Tiwary and P.S. Aithal), 1–26. New Delhi: New Delhi Publishers, ISBN: 978-93-88879-66-8.

92 Aithal, P.S. and Aithal, S. (2016). Nanotechnology innovations and commercialization – opportunities, challenges & reasons for delay. *Proceedings of National Conference on Changing Perspectives of Management, IT, and Social Sciences in Contemporary Environment* 14: 1–12, ISBN 978-93-5265-6523.

93 Sridhar Acharya, P. and Aithal, P.S. (2015). Innovations in effective management of energy using green technology. *International Journal of Conceptions on Management and Social Sciences* 3 (2): 18–22.

94 Aithal, P.S. and Jeevan, P. (2016). Strategic rethinking of management education: green MBA model. *International Journal of Management, IT and Engineering (IJMIE)* 6 (1): 55–73.

95 Aithal, P.S. and Rao, P. (2016). How service industries can transform themselves into green business industries. *International Journal of Management Sciences and Business Research (IJMSBR)* 5 (4): 150–158.

96 Han, W. and Liu, L.C. (2009). Discussion on green education in universities. *Journal of Daqing Normal University* 1 (1): 39–45.

97 Wu, G. (2011). A new concept of green education: the cultivation model for successful and practical talents. *International Forum of Teaching & Studies* 7 (1): 45–48.

98 Aithal, P.S. (2015). Concept of ideal business & its realization using e-business model. *International Journal of Science and Research (IJSR)* 4 (3): 1267–1274.

99 Aithal, P.S. (2015). Mobile business as an optimum model for ideal business. *International Journal of Management, IT and Engineering (IJMIE)* 5 (7): 146–159.

100 Aithal, P.S. and Aithal, S. (2015). An innovative education model to realize ideal education system. *International Journal of Scientific Research and Management (IJSRM)* 3 (3): 2464–2469.

101 Rao, P. and Aithal, P.S. (2016). Green education concepts & strategies in higher education model. *International Journal of Scientific Research and Modern Education (IJSRME)* 1 (1): 793–802.

102 Boye, J.I. and Arcand, Y. (2013). Current trends in green technologies in food production and processing. *Food Engineering Reviews* 5 (1): 1–17.

103 Varma, R.S. (2012). Greener approach to nanomaterials and their sustainable applications. *Current Opinion in Chemical Engineering* 1 (2): 123–128.

104 Silva, L.P., Reis, I.G., and Bonatto, C.C. (2015). Green synthesis of metal nanoparticles by plants: current trends and challenges. In: *Green Processes for Nanotechnology* (eds. V. Basiuk and E. Basiuk), 259–275. Cham: Springer.

4

Improving the Sustainability of Biobased Products Using Nanotechnology

Shirleen Miriam Marques and Lalit Kumar

Department of Pharmaceutics, Manipal College of Pharmaceutical Sciences, Manipal Academy of Higher Education, Manipal, Karnataka, India

4.1 Introduction

In recent times, the growing environmental concerns and depletion of petroleum reserves have increased our dependence on biobased materials in various applications, which has led to growing concern for the conservation and sustainability of these bioresources [1]. A sustainable bioeconomy can be realized by reducing the pressure on biological resources and by availing biowaste resources [2]. The quest for development of these biobased products is on the rise due to their low cost, accessibility, sustainability, biodegradability, flexible processability, low environmental burden, and reduction in the emission of greenhouse gases [3–5]. The European bioeconomy generates a turnover estimated at around €2 trillion that employs more than 17 million people [6]. Enhanced productivity, effective recycling, capital availability, technology innovations, and market opportunities are various aspects that are required for a bioeconomy to thrive [5]. The usefulness of biobased products in producing sustainable materials has expanded to various fields such as environmental, biomedical, automotive, construction, food packaging, drug delivery systems, and waste management [4].

Nanotechnology has emerged as the emblem of innovation in the twenty-first century. It can be defined as the size reduction process in which large-sized particles are deployed and controlled in the nanoscale range [7]. Over the past few decades, a myriad of applications of nanoscience and nanotechnology have been explored by a large panel of scientific researchers and scholars. This inquisitiveness could be attributed either to the reinvestigation of scientific fields by taking into consideration the nanoscale as an important platform for enriching our knowledgebase or from the remarkable invention of new instruments that have democratized access to the nanoscale [8]. The ambition of nanotechnology has resulted in the continuous exploration of an increasing number of new nanomaterials which have found use in a variety of industries ranging from health care to engineering materials [9]. Due to the ever-growing concern of environmental problems such as air and water pollution, the utilization of nanostructured materials has been recently suggested for the area of environmental remediation [10, 11]. The goal of this method is to not use nanomaterials in the direct form but instead as building blocks to develop nanoporous microdimensional systems and overcome the ecological and health threats that are linked with the use of nanoscale materials [10]. The use of single-function environmental nanomaterials in current and novel environmental treatment technologies is eventually approaching its threshold limits. Therefore, smart nanomaterials with environmentally approachable functionalities have been recognized to upgrade the current environmental technologies. Intelligent nanomaterials are known to be applicable in different scenarios to achieve the best results by normal design of their constructions and functionalities [11].

4.2 Biopolymers

Due to the rapid increase in environmental pollution caused by synthetic products obtained from petrochemicals in addition to the waste generated from the plastics industry, there has been an urgent need to shift to more sustainable materials [3, 12]. Biopolymers are macromolecules that are obtained from a wide variety of plants and

Sustainable Nanotechnology: Strategies, Products, and Applications, First Edition.
Edited by Yashwant V. Pathak, Govindan Parayil, and Jayvadan K. Patel.

microorganisms [3]. Over the past few years, the development and commercialization of biopolymers obtained from renewable sources has led to the replacement of many petroleum-based materials. Biopolymers possess a wide array of environmental benefits [13]:

i) Decrease in the depletion of nonrenewable fossil fuel resources.
ii) Biopolymers are degradable, thus causing less environmental harm than their petroleum-based counterparts.
iii) They are devoid of carbon and sustainable, since they are obtained from natural resources.
iv) Biopolymers decrease the emission of carbon and carbon dioxide concentration in the atmosphere.

The utilization of biopolymers in conjunction with nanotechnology has resulted in groundbreaking work in numerous applications, e.g. healthcare, water purification, food packaging, and the energy sector [3]. A few of the major biopolymers along with their sources are depicted in Figure 4.1. Some of the biopolymers used for various applications have been briefly discussed in Table 4.1.

4.2.1 Starch

Owing to its abundance and renewability starch has emerged as a popular biopolymer in the past decade, for biomedical and food-packaging applications. Structurally, starch comprises two polysaccharides, i.e. amylose and amylopectin [23]. Starch can be used as a replacement for the petroleum-derived polymers since it negates the carbon footprint of traditional resins in addition to being degradable and can also be used in conjunction with a compostable substance without impeding the degradation process. Plastics derived from starch are also gaining importance due to the development of newer resin grades, their capability to blend with other biopolymers, and a growing number of suppliers. Moreover, the utilization of starch-based plastics has been rapidly increasing in the pharmaceutical and medical industries due to its low toxicity, biodegradability, and mechanical strength [24]. Starch for bioplastics is primarily obtained from corn, although attempts are being made to evaluate the possible utilization in bioplastics for starches derived from other sources such as potato, wheat, rice, barley, oat, and soy [25].

4.2.2 Cellulose

Cellulose comprises one of the most abundantly available biopolymers present in the world and is naturally present in all plants. It has found use in a multitude of applications owing to its biocompatibility, low density, excellent mechanical strength, and cost-effectiveness [26, 27]. Cellulose is a linear homopolymer composed of

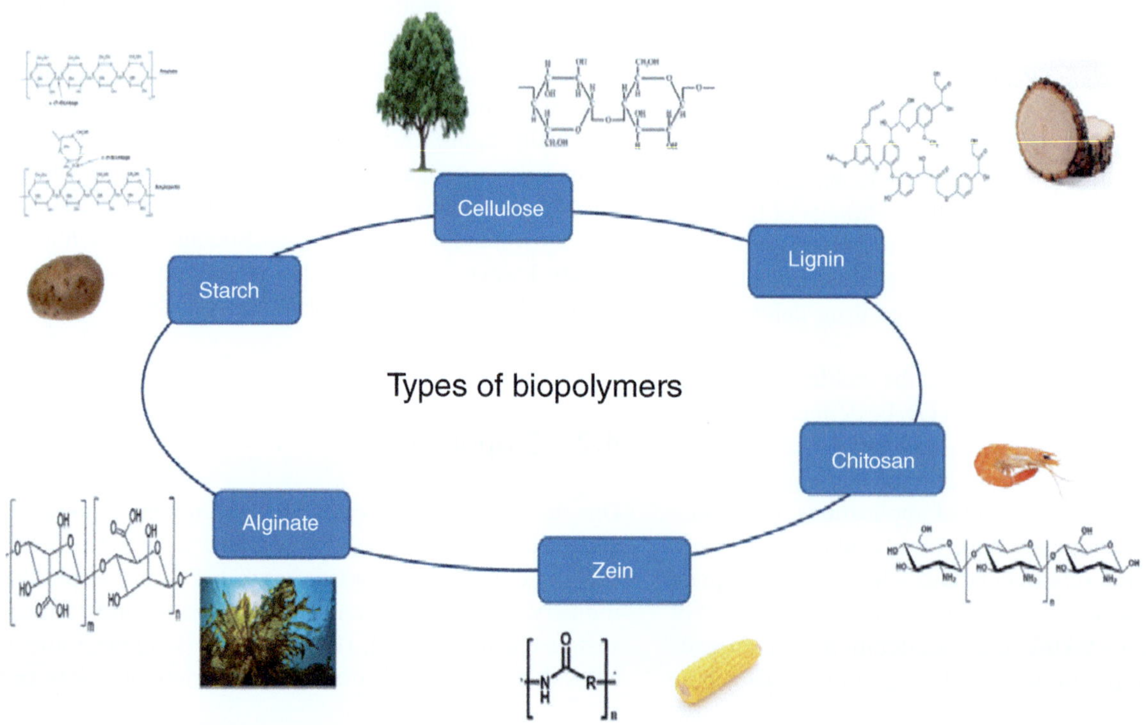

Figure 4.1 Types of biopolymers and their source of origin.

Table 4.1 List of some biopolymer formulations.

Biopolymer	Formulation	Application	References
Polylactic acid	Nanofilm	Antimicrobial packaging	[14]
Pectin	Solid lipid nanoparticles	Encapsulation of bioactives	[15]
Polylactic acid	Nanocomposite	Antibacterial food packaging	[16]
Chitosan	Nanoparticles	Ocular drug delivery	[17]
Polylactic acid	Nanofibrous scaffolds	Anticancer drug delivery	[18]
Chitosan	Nanoparticles	Active packaging	[19]
Zein	Composite nanoparticles	Encapsulation of bioactives	[20]
Nanocrystalline cellulose	Solid dispersion	Tablet excipient	[21]
Chitosan	Nanoparticles	Drug delivery	[22]
Starch	Film	Food packaging	[23]

D-anhydro-glucopyranose units, which are bonded by β-1,4-glycosidic linkage. Biocomposite materials comprising cellulose fibers, nanocellulose, and derivatives of cellulose as fillers are suitable biosustainable alternatives to produce high-quality polymer composites and functional polymeric materials [28]. Due to the exceptional feature and numerous applications of nanofibers, the use of cellulose nanofibers has been increased in the field of material science, including biomedical, filtration process, chemical synthesis, textiles, and electronics [29].

4.2.3 Polylactic Acid

Polylactic acid (PLA) has gained relevance as a sustainable biopolymer in various applications owing to its biodegradability, biocompatibility, mechanical strength, and ease of processing. Formulating blends of PLA with other polymers provides various alternatives to improve associated properties or to produce novel PLA polymers/blends for target applications. The biomedical applications, e.g. drug delivery, tissue engineering, implants, and sutures, are some of the areas where the use of PLA blends has been widely explored [30]. In 2019, the production volume of PLA was approximately 1.9×10^5 tons [31]. Polylactic acid is mainly synthesized using ionic polymerization technique [3].

4.2.4 Chitin and Chitosan

Chitin and chitosan are the most ubiquitous natural polysaccharides present on the earth. Chitin is a deacetylated chitosan which exists in three α-, β-, and γ-chitins polymorphic forms, which is attributed to the presence of strong hydrogen bonds between the amide and carbonyl groups of adjacent chains [32]. An enormous amount of crustacean shell waste obtained from crab and shrimp is a by-product of the seafood industry that is generated every year and can be used to produce chitin, which in turn can be deacetylated to produce chitosan. Thus, the production of chitin and chitosan is relatively cost-effective as compared to other biopolymers as they are extracted from a biowaste product using energy-efficient methods [33]. Chitin and chitosan are biopolymers with excellent bioactive properties, for example biodegradability, nontoxicity, biocompatibility, hemostatic activity, and antimicrobial activities [34].

4.2.5 Polyhydroxyalkanoate (PHA) and Polyhydroxybutyrate (PHB)

Polyhydroxyalkanoate (PHA) is a group of biodegradable linear polyester biopolymers and is produced by bacteria in nutrient deficiency and excess carbon supply by fermentation of sugars and lipids that build up in the cell as a carbon and energy storage source. Polyhydroxybutyrate (PHB), which is one of the most important members of the biodegradable thermoplastic PHA family, is a homopolymer of 3-hydroxybutyrate. It possesses high melting point, high degree of crystallinity, and poor permeability to water, oxygen, and carbon dioxide [13]. Various microorganisms, for example bacteria and fungi, and some plants synthesize and catabolize PHA biopolyesters [12]. PHB is thermoplastic and biodegradable, which has made it a potential candidate in the industrial sector for food-packaging applications with greater water vapor barrier and oxygen barrier properties than polyethene terephthalate and polypropylene [13].

4.3 Nanotechnology in Environmental Remediation

Nanotechnology harbors a myriad of strategies that can be leveraged to address environmental pollution. Nanoremediation exploits the potential of nanoparticles and nanomaterials for environmental remediation and has been suggested as a method for cleaning up existing pollution, ameliorating manufacturing methods to prevent generation of new pollution, and preserving ecosystems that suffer from the increase in human population, pollution, and urbanization [35]. The contamination of landfills, oil fields, and manufacturing and industrial sites by hazardous substances needs to be amended as it presents a grave threat to the ecosystem as well as the health and well-being of mankind [36]. Nanoremediation, in comparison to the current *in situ* remediation methods such as thermal treatment, pump-and-treat, and chemical oxidation containing bioremediation, presents advantages of being less costly, more effective, and also contributes to environmental, social, and economic sustainability. The properties of nanoparticles, that is high reactivity and surface area, are particularly beneficial in eliminating a wide range of environmental contaminants that include organohalogenated compounds, hydrocarbons, and heavy metals [7, 8]. It is of great importance that the materials used to combat pollution are not pollutants themselves, thereby acting as potential candidates for the remediation of pollution [37].

Increases in the environmental concerns due to soil, water, and air pollution have enhanced the pursuit to invent newer techniques to address the issue of environmental remediation. An enormous amount of research has been directed toward solving the problem of air pollution. A major cause of air pollution are a class of compounds known as volatile organic compounds (VOCs) that are released by anthropogenic sources such as chemical factories, paper industry, agricultural operations, and pharmaceutical industries into the atmosphere. Even when their concentrations are within permissible limits they possess the capacity to be extremely toxic, malodorous, and irritant. Cellulose has caught the attention of many researchers for applications in environmental remediation due to its availability, biodegradability, biocompatibility, and also since it can be easily manipulated to alter its physical and/or chemical features by modifying the carbinol functionalities. Guerra and coworkers [38] determined the performance of cellulose nanocrystals decorated with amine functionalities in capturing aldehyde VOCs. The modified cellulose nanocrystals were found to have superior properties which could be attributed to surface-to-volume ratio, which meant that there were highly reactive amine sites present on the surface per unit volume. The results obtained were crucial to fabricate the cellulose-derived gas-capturing materials with optimum physicochemical properties. With the success of these initial studies, the development of related amine-modified cellulose nanocrystal materials functionalized with ethylene diamine and Tris was carried out. Results revealed that small amine coats, ethylene diamine, and Tris produced materials that were efficient at the capture of aldehyde VOCs as compared to the cellulose nano-crystals-poly(ethyleneimine) (CNC-PEI) material. Present efforts are geared toward scale-up and validation of materials in the industrial settings [38].

4.4 Methods to Assess Sustainability of Nanotechnology

4.4.1 Life Cycle Assessment (LCA)

The burgeoning of nanotechnology in various sectors such as medicine, automotives, energy, and agriculture is bringing about revolutionary transformation, but these developments have also made the question of sustainability a key issue. Although the benefits of nanomaterials and nanoproducts are seemingly obvious, there is a need to understand the potential risks that these products may pose to human and environmental health. Hence, as nanotechnology progresses from the laboratory to the global marketplace, complementary measures have to be taken to resolve environmental, health, and safety issues that would prevent us from foreseeing social and economic benefits.

LCA has recently emerged as a powerful comprehensive tool to evaluate the environmental impacts of nanotechnology [39]. The existence of ISO standards for LCA has strengthened its scientific reputation and widespread use. It is an approach used to realize the upstream and downstream repercussions of nanotechnology on the environment and has motivated researchers to pursue promising applications while prohibiting the adverse effects [40]. LCA includes all facets in the life cycle of a product such as procurement of raw materials, production process, usage of products, and postuse management including recycling and disposal [39, 40]. Although the LCA of nanotechnology is still in its early stages, the development of nanomaterials can be done by keeping sustainability in mind through the use of cleaner or greener methods during the entire life cycle of a nanoproduct [39].

4.4.2 Case Study

Bartolozzi et al. [10] applied the LCA for optimization and scale-up of cellulose nanosponge synthesis to be utilized for the purpose of soil and water remediation. Through the application of this eco-design tool as well as taking into consideration the potential repercussions on the environment throughout the scale-up period the practicality of production of a sustainable and environment friendly material was assessed. Also, the global environmental impacts due to optimization of the energy consumption were assessed through the aid of a simulated scale-up model.

Initially, a laboratory-scale analysis was performed to recognize the main stages in the production process that have the greatest environmental impacts. Based on the analysis carried out changes were proposed in the washing solvent and temperature at the labscale itself. A second investigation was conducted during the optimization process to determine the reduction in environmental impact achieved. Although the energy-consuming processes also significantly add onto the global environmental impact, it could not be modified at the labscale. Thus, a simplified scaled-up production process was modeled, wherein the optimization of the energy-consuming stages was conducted. A final scale-up analysis was conducted to confirm the environmental benefits obtained.

The production process of CNS comprises mainly of three steps as depicted in Figure 4.2:

i) Oxidation of cellulose.
ii) Cellulose nanofibers synthesis by sonication.
iii) Cellulose nanosponge production by cross-linking the cellulose nanofibers in the presence of b-PEI.

The application of LCA at such an early stage in the research process proved to be very beneficial to determine the stages having the greatest environmental impact, thereby enabling an optimal eco-design of the production process. The results of the environmental impact calculated for this study were found to be two orders of magnitude higher than that compared to materials employed in industrial applications for soil and water remediation. Therefore, current analysis indicated that a simplified scale-up model as that suggested in the current study could result in providing a reduction of about two orders of magnitude in the environmental impact during the laboratory synthesis stage [10].

During the lab-scale synthesis, the energy-consuming processes and the final washing procedures were found to be the major steps contributing to the environmental impact. Optimization of the washing procedure was conducted at the labscale itself by replacing methanol with water as a solvent and by decreasing the solvent temperature. A simplified scale-up model of the cellulose nanosponge synthesis was conducted for optimization of the energy-consumption process that contributed majorly to the environmental impacts. A reduction of about two orders of magnitude was achieved with regard to the lab-scale synthesis, as determined by LCA. Therefore, through the above example the effectiveness of LCA in the early stages of research in reducing the environmental impact and improving the sustainability of green nanotechnology was emphasized [10].

4.4.3 Risk Assessment

A major apprehension regarding nanomaterials is associated with the question of whether the properties exclusive to nanoparticles, for example increase in the surface-to-volume ratio and greater surface reactivity, result in any significant impact on the ecosystem that is unlike and that is caused by conventional pollutants [40, 41]. Due to the rapid progress of nanotechnology in the recent years,

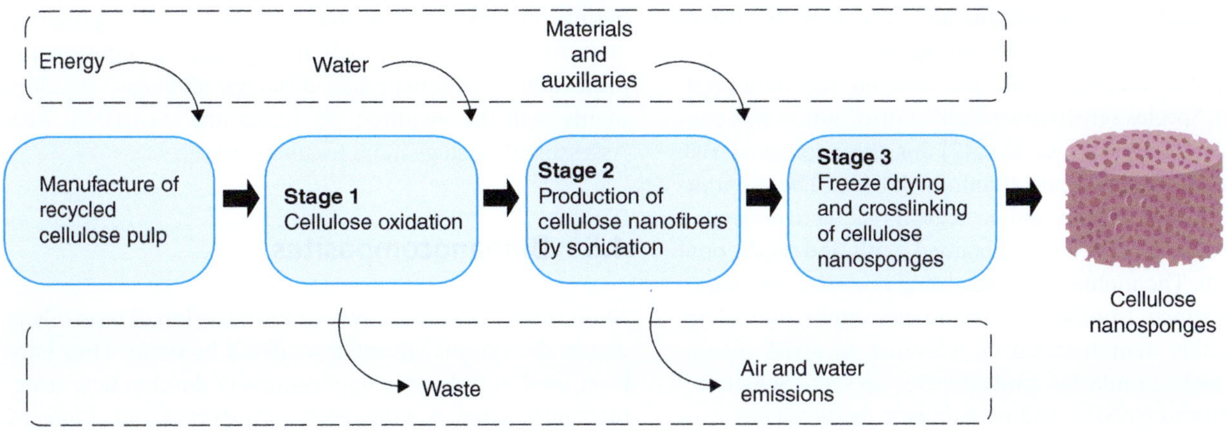

Figure 4.2 Production steps of cellulose nanosponges.

concern regarding its safety and repercussions on the environment has been increasing due to the lack of a suitable risk assessment method [42]. As the emergence of nanotechnology has been increasing over a wide spectrum of applications, there is a growing need to identify the balance between the potential benefits and the risks. The advantage of a risk assessment-based technique is its ability to be incorporated into the decision-making process of an organization or regulatory body and to bridge the gap linking risk assessment and risk management. Methodologies employed in risk assessment should be relevant to the present times as well as be adequately adaptable to permit incorporation of future technical and scientific knowledge in the development of nanoparticles [43, 44].

Grouping and risk-assessment strategies have been applied for nanomaterials [44]. It is based on grouping the nanomaterials depending on their physicochemical properties to design a testing procedure applicable to the entire group. Generic grouping of the materials based on properties such as dissolution rate, aspect ratio, and chemical reactivity is done to bring attention to issues that require additional observation or to read-across. Read-across is also considered when specific endpoint data is required from a specific nanomaterial [44]. Structured methods such as risk-ranking approaches have been used by Grieger et al. [45] who employed a risk-ranking technique including qualitative and quantitative data pertaining to human health risks of workers and soldiers caused due to engineered nanomaterials.

Another project that was undertaken to assess the environmental, health, and safety risks was the SUN project based on "Sustainable Nanotechnologies." It aims at monitoring the risks during the entire life cycle of manufacturing of the nanomaterials and eventually consolidate the data obtained into rules and specifications for sustainable manufacturing of nanoproducts. The SUN project also characterizes the nanomaterials released from products in the case studies during various life cycle stages for use in ecotoxicological and behavior studies, thereby focusing not only on the nanomaterials but also on the nanoproduct [46]. Species sensitivity-weighted distribution was conducted by Semenzin et al. [47] for the ecological risk assessment of nanoscale titanium dioxide. The environmental quality criteria and ecological risk were estimated and compared to results obtained with the traditional approach. The inclusion of weighting criteria such as species relevance, trophic level abundance, and nanotoxicity data quality demonstrated its relevance as a risk assessment tool. Similarly, probabilistic Species Sensitivity Distributions (pSSDs) and probabilistic predicted environmental concentrations (PECs) were used to quantify the environmental risks of five engineered nanomaterials [nano-TiO_2, nano-Ag, nano-ZnO, carbon nanotubes (CNT), and fullerenes] in water, soil, and sediments. The risk assessment results obtained during this study allowed a systematic assessment of the environmental risks caused by nanomaterials by concern of material/compartment combinations where the main possibility for effects with projected environmental concentrations is expected [48].

Moreover, it has been realized that certain aspects of current regulatory risk assessment (RA) methods are not appropriate to imitate the risk caused by nanomaterials. Inadequacy includes description of relevant material characteristics, a dearth of appropriate exposure models, a dearth of knowledge on which species are generally affected by nanomaterials, and how to include such elements in the risk characterizations process [49].

The existing strategies of risk assessment do not take into consideration the possible physicochemical changes that nanomaterials undergo during their life cycle, thereby requiring newer and more efficient methods of risk assessment. Moreover, nanoparticles comprise various primary particles, aggregates, and agglomerates of different proportions, thus making the assessment process very resource intensive. A risk assessment strategy comprising two phases has been introduced within the EU FP7 project called Managing Risks of Nanoparticles (MARINA) [50]. The Phase 1 constitutes a problem-framing stage wherein a base set of information is collected, related exposure situations are determined, and the scope for Phase 2 is identified. The severity of a relevant exposure scenario is identified by information on exposure, kinetics, and/or risk, which are involved as distinct pillars that contain precise tools. Phase 2 comprises a stage of risk characterization, identification of data requirement, and unified collection and evaluation of data on three areas, till satisfactory information is attained to achieve on probable risks in a relevant exposure scenario [49]. A fourth pillar, i.e. risk characterization, is defined which contains risk assessment tools. This strategy defines flexible and effective methods for data gathering and risk assessment, which is important to confirm the protection of nanomaterials, although additional developments will be required to make the MARINA Risk Assessment Strategy more feasible [50].

4.5 Bionanocomposites

Nanocomposites have captured the attention of researchers due to the various advantages offered by them. They have been used in industrial applications to develop new structures and materials with versatile properties and enhanced flexibility. Moreover, nanocomposites vary from the bulk polymer as they possess a greater volume of interfacial matrix attributed to the larger surface area. Despite the wide spectrum of advantages offered by nanocomposites they were found to have some constraints with respect to their properties, and thus, researchers came up with the concept of replacing polymers with biopolymers to produce "bionanocomposites" [51]. They can be defined as materials that constitute particles with at least one dimension between 1 and 100 nm and a component of biological origin such as biopolymers. Fillers based on inorganic nanomaterials have been explored as an interface with biological species for use in nanocomposites along with biopolymers. Various materials such as clay minerals, layered double hydroxides, carbonates, metal oxides, and carbonaceous nanomaterials have been investigated to fabricate nanocomposite materials in conjunction with biopolymers that have different effect on its mechanical, transport, or swelling properties [51, 52]. Biopolymers such as polylactic acid, polycaprolactone, polysaccharides, and proteins have been explored for use in bionanocomposites along with layered silicates of the smectite group that is used in regenerative medicines, drug delivery, tissue engineering, electronics, and food-packaging applications, though current studies display that the usage of microfibrous clay minerals, for example sepiolite and palygorskite, results in a motivating strengthening of polymer as well as biopolymer matrices in the development of clay–polymer nanocomposites [53]. Figure 4.3 depicts the type of bionanocomposites. The composition and application of various bionanocomposites have been briefly discussed in Table 4.2.

4.6 Methods of Production of Bionanocomposites

4.6.1 Solvent Casting

The solvent casting method relies on a solution system wherein the solvent is able to resolve the polymer in addition to dispersing the nanoparticles. The solvent also facilitates the polymer chains' mobility that results in intermingling of the nanoparticle polymer and solvent. On evaporation of the latter, the polymer which enters between the layers of the nanomaterial layers remains intact, thereby resulting in the formation of the nanocomposite. The solvent casting method is suitable for water-soluble polymers. However, it also requires large amount of organic solvents, which increases the cost and toxicity [64].

4.6.2 *In Situ* Polymerization

In situ intercalative polymerization results in swelling of the layered silicates present in liquid monomers or monomer solutions. Polymerization is initiated between the intercalated sheets when the monomers begin to migrate to the layered silicates. Further, the reaction is initiated by the supply of heat and radiation in conjunction with diffusion of initiators, organic initiator, or catalyst through a cation exchange before the layered silicates begin to swell. This method results in the formation of long polymeric chains within the clay galleries [65].

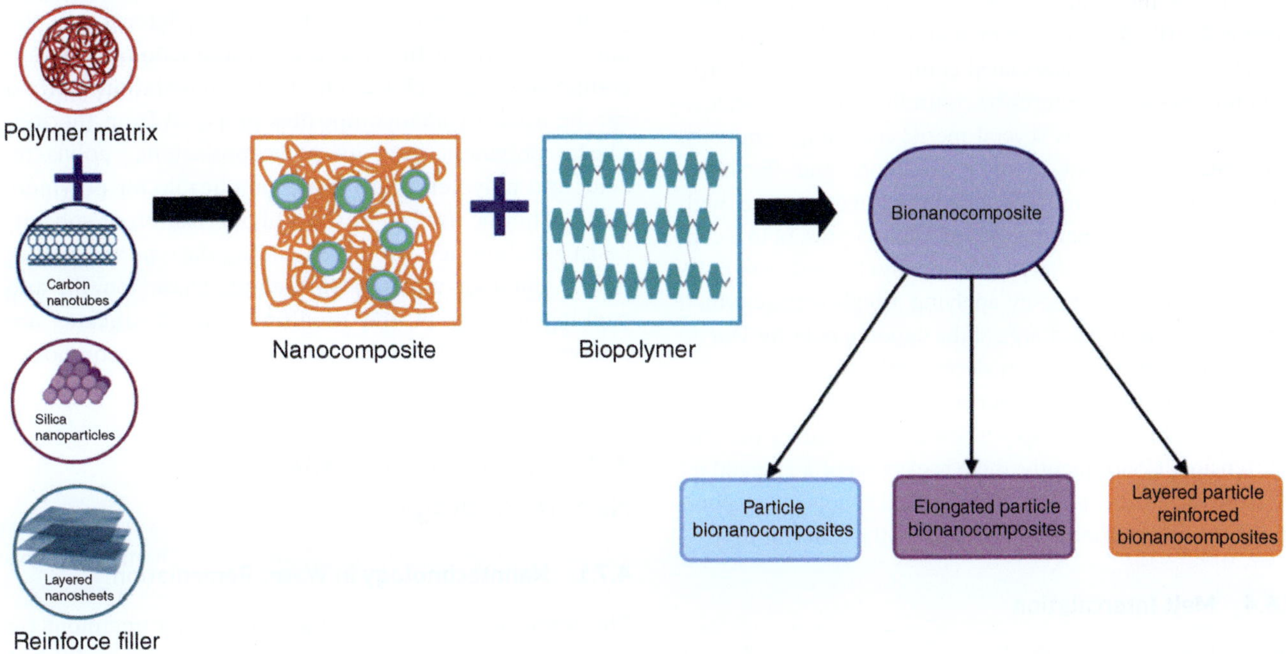

Figure 4.3 Types of bionanocomposites. *Source:* Modified from Saini et al. [53].

Table 4.2 Composition and applications of some bionanocomposites.

Continuous phase	Nanodimensional phases	Applications	Enhanced properties	References
Potato starch	Starch nanocrystals	Food packaging	Thermal stability and mechanical strength	[54]
Maize starch	Chitin nanowhiskers	Antibacterial packaging	Antimicrobial and mechanical properties	[55]
Whey protein isolate	Zein nanoparticles	Food packaging	Barrier and mechanical properties	[56]
Whey protein isolate	Montmorillonite	Food packaging	Mechanical strength	[57]
Cellulose	Nanofibers	Water purification	Flexibility and electromagnetic suppression	[58]
Alginate	Mesoporous silica	Drug delivery	Mechanical properties and pharmacological potential	[59]
Chitosan	Cellulose nanocrystals	Drug release	Mechanical performance and cytocompatibility	[60]
Chitosan	Iron (oxyhydr)oxide	Water purification	Mechanical properties	[61]
Chitosan	Zinc oxide nanoparticles	Dye removal	Thermal stability	[62]
Poly (3-hydroxybutyrate-*co*-3-hydroxyhexanoate)	Cellulose nanocrystals	Packaging	Thermal stability and mechanical properties	[63]

4.6.3 Electrospinning

Electrospinning is a very convenient technique that results in the production of various types of continuous ultrafine nanofibers by forcing a molten polymeric solution through a small orifice to produce a Taylor cone [64, 65]. Electrospun fiber mats are generally obtained by supplying a high voltage for the generation of an electric field between the orifice and the target like a rotating mandrel/disk or a mesh plate. Different polymer solutions such as polylactic acid, poly (ethyleneoxide) (PEO), and poly(methyl methacrylate) (PMMA) as well as a range of associated composites can be electrospun into fibrous structures by controlling the polymer/cellulose ratio [65]. It has several merits including simplicity, economical, fast, vast material selection, and flexibility. Electrospinning is composed mainly of three parts, i.e. high voltage supply, polymer/composite solution present in a capillary tube, and a collector. An electrically charged jet of polymer solution is obtained by applying a high voltage which generates an electrical charge of the opposite polarity. Due to this, there is formation of an electric field, thereby electrifying the surface of the polymer solution droplet that causes repulsion, which encourages a power that disables the surface tension. Consequently, there is ejection of a charged jet from the tip of the capillary tube followed by solvent evaporation that causes nanofiber formation on the collector [65].

4.6.4 Melt Intercalation

The melt blending method is one of the most proficient methods for the preparation of nanocomposites due to the absence of solvents, compatibility with industrial processes, environment friendliness, and cost effectiveness. This method involves annealing the polymers and nanomaterials together in a molten state above the softening point of the polymer at high shear using an extruder, mixer, or ultrasonicator followed by exhaustive mixing that effects in a thermal motion of the polymer chains and assembling of the molten mass into layers of the nanomaterials to produce both delaminated and intercalated polymer nanocomposites. The compatibility of polymer matrices and layered silicates determines the amount to which the intercalated and exfoliated nanocomposites are attained. The melt intercalation method can be used for nanocomposites prepared from thermoplastic polymers such as polycaprolactone, polylactic acid, and polyolefin and is also applicable for polymers that cannot be used in *in situ* polymerization or solvent-casting techniques. Factors including the type and nature of the polymer and the surfactant, chain length, and packing density of the modified layered silicates are known to affect the characteristics of the nanocomposite structure [64].

4.7 Applications of Biobased Nanotechnology

4.7.1 Nanotechnology in Water Remediation

The applications of nanotechnology in water treatment are presenting newer perspectives to overcome traditional treatment technology limitations. The generation of sludge and hazardous materials through the traditional methods of water treatment requires further treatment, thereby incurring additional energy and cost. The techniques based on nanotechnology are gaining superiority over the conventional methods in environmental remediation and are also gaining importance at the industrial scale due to lowered production costs [66, 67]. A majority of the biopolymers that are produced by bacteria under restrictive settings store carbon and energy resources. The mechanical and barrier properties of these biopolymers improve on incorporation of various nanofillers that are biocompatible and biodegradable and make it useful over a wide spectrum of applications [3].

The modification of conventional water purification membranes with nanocellulose warrants special attention as a biobased product for selective adsorption of contaminants from industrial and drinking water. The nature and availability of the functional groups that act as adsorption sites determine the adsorption capacity, whereas the structure of pores in the membrane was found to determine the accessibility to the adsorption sites [68]. Nanocellulose is an effective bioadsorbent that has tremendous strength and flexibility together with a high surface area and adaptable surface chemistry especially when used as composites [26, 69]. Cellulose nanofibers and nanocrystals derived from nanocellulose, with customized surface properties, have been produced by an extensive range of facile hydrolysis and catalytic process methods. Due to the presence of cationic or anionic moieties it has the ability to scavenge heavy metal and organic pollutants such as dyes, oils, and pesticides, from aqueous solutions. Hybrid membranes of nanocellulose along with other nanoparticles are expected to make a breakthrough in water purification methods due to their improved pore structure, mechanical stability, and adsorption capacity [69]. Nanofiltration and tight ultrafiltration are among the foremost methods for decontamination of industrial and domestic water to supply clean water which is scarce in many parts of the world [70]. Different types of nanocellulose such as bacterial cellulose, wood-derived cellulose, TEMPO-oxidized cellulose nanofibrils, and cellulose nanocrystals have been recently described as a substitute for organic solvent nanofiltration membranes due to their exceptional mechanical properties, low thermal expansion coefficients, high optical transparency, and good gas-barrier properties [70].

Cellulose nanocrystals that are obtained naturally are environment friendly and biodegradable and hence suitable for the production of nanocomposites. Zero-valent iron nanoparticles (nano-ZVI) reinforced with cellulose nanocrystals were prepared by Bossa et al. for the reduction of contaminants and *in situ* remediation of groundwater. The role of the cellulose nanocrystals is to induce stabilization and to increase the surface reactive sites on the nano-ZVI by creating a porous mesh around its surface. Nano-ZVI is known to adsorb a wide variety of groundwater contaminants such as halogenated hydrocarbons, azo dyes, hexavalent chromium, and heavy metal ions [71].

An entirely biobased membrane was prepared with cellulose nanofibers by a simple nanopaper production method and was further dip coated with cellulose nanocrystals. Results of the scanning electron microscopy indicated that the cellulose nanocrystals were well dispersed over the layer of the cellulose nanofibers. The membrane thickness, tensile strength, and tensile modulus were higher due to the cellulose nanocrystals dip coated onto the membrane as compared to the previous studies where cellulose microfibers were used [68]. The membranes prepared displayed an excellent capacity for the removal of silver, copper, and ferric/ferrous ions from industrial wastewater. Moreover, optimizing the size of the pores in the support layer and reducing the thickness of the different layers may make these membranes suitable for commercial applications [68]. Karim et al. demonstrated the usefulness of a nanocomposite biobased cellulose membrane to capture metal ions present in industry effluents. A high-flux nanocellulose composite with a layered cellulose membrane was prepared using vacuum filtration followed by hot pressing. Mechanical stability was provided by the support layer, whereas the functional layer caused immobilization of metal ions from contaminated water by adsorption. The porous matrix of microfibers in conjunction with a thin functional layer of nanocellulose produced a membrane with high flux and maximum functionality. Biodegradation studies conducted in effluent water and soil indicated the potential of these membranes for industrial applications where durability and possible biodegradation at the end of life are crucial, economical, and efficient [72].

The utilization of nanocellulose for impregnation of electrospun mats was one of the first examples of membranes based on functionalized nanocellulose for water purification. Impregnating electrospun mats with different ratios of nanocrystals presents a prospect for production of membranes with isotropic or anisotropic structures [69, 73]. The presence of copper in industrial wastewater streams is increasing the concern regarding its removal using an efficient method with a low ecological footprint as well as improving the quality of drinking water [74, 75]. Traditional methods of water purification such as continuous membrane processes and batch-wise adsorbents application have been used to overcome this problem. However, they pose several limitations such as poor permeance and disposal of saturated adsorbents [75]. Sheikhi and coworkers attempted to use electrosterically stabilized nanocrystalline cellulose (ENCC) as a nanoremediation adsorption technique for the elimination of copper from water. Currently, attention has been increased toward emerging sustainable and environmentally friendly nanoadsorbents to eradicate heavy metal ions from aqueous media. ENCC is a sustainable and eco-friendly method to eradicate heavy metal compounds and is prepared from wood fibers through the oxidation of periodate/chlorite and also possesses a high charge content and colloidal stability. Through this work, a copper elimination capacity of $\sim 185\,\mathrm{mg\,g^{-1}}$ was attained due to the presence of highly charged dicarboxylic cellulose polyanions extending from the ENCC. Thus, these characteristics in addition to biorenewability and fast, simplistic, and cost-effective removal of copper ions make ENCC an ideal sustainable method for the purification of wastewater [74]. Another method for removal of copper was accomplished through the use of a nanopaper ion exchanger incorporated with phosphorylated cellulose nanofibrils. The efficiency of these phosphorylated nanopaper ion-exchange membranes was dependent on their ability to reject copper and its permeability. Moreover, the ability of the nanopaper membranes to be recycled and reused was also demonstrated, thereby making it a sustainable solution for environmental remediation [75].

The use of chitin nanocrystals embedded onto an electrospun cellulose acetate fiber mat was also suggested by Goetz et al. to produce membranes with a lower abiotic fouling tendency for water treatment. Significant rise in the mechanical strength and modulus of electrospun cellulose acetate fiber membrane could be due to the coating of chitin nanocrystals on the individual fibers that were joined together at the junctions by webs formed by the chitin nanocrystals. Also, there was a decline in the development of biofilm and abiotic fouling due to deposition of the chitin nanocrystals on the membrane and the surface chemistry of chitin nanocrystals as well as the interactions between the surfaces of the cellulose acetate membrane and *Escherichia coli* cells. Thus, impregnation of chitin nanocrystals onto the cellulose acetate network produced membranes with high flux with prospective use to purify water from food industries contaminated with biologic and organic contaminants [73].

Nitrates which are a major portion of fertilizers used to improve crop yield have also been posing a major health threat due to their presence in potable water which deteriorates the quality of drinking water. The rapid intensification of nitrates in the environment warrants special attention due to their adverse side effects and their ability to cause eutrophication and acidification [76]. An attempt was made to produce a nanopaper ion exchanger derived from nanofibrillated cellulose to scavenge nitrates present in water. The nanocellulose ion exchanger was designed to operate in flow-through conditions and was produced from fiber sludge, which is a by-product generated during paper production. Cationic quaternary trimethylammonium groups were used to modify the cellulose nanofibrils, and their efficiency was assessed based on their permeance and nitrate adsorption. The dynamic filtration experiments demonstrated that the nitrates were effectively adsorbed onto the cellulose nanopaper and could be thus eliminated from drinking water. The nanofibrillated ion exchanger possessed adsorption properties comparable to its commercial counterparts in addition to having the benefit of reduced contact time [76].

4.7.2 Nanotechnology in Food Packaging

Food packaging plays a key role in the preservation of processed or fresh food. It ensures that there is no deterioration in the quality of food from the time of harvest or manufacture till its availability to shoppers [33, 77]. The use of traditional packing methods using nonbiodegradable substances has been rapidly escalating the strain on the environment due to the difficulties in disposal, slow degradation, depletion of aquatic life due to ocean pollution, and the unavailability of collection and recycling facilities in numerous countries [78]. Therefore, the drawbacks of nonrenewable materials have led to replacing noncompostable packaging resources with more renewable, sustainable, and eco-friendly substances [23].

Nanocomposites are a variability of nanoparticle-reinforced polymers, which have been established to improve the packaging materials, that usually contain up to 5% w/w nanoparticles with clay nanoparticle composites and possess enhanced barrier properties (80–90% decrease) and have been utilized in the manufacturing of beer bottles, carbonated drinks, and edible oils. Recently, the US Food and Drug Administration (USFDA) has also permitted nanocomposites contact with food [33]. Ternary polymeric films containing polylactic acid were prepared by Yang et al., which were developed by dispersing cellulose nanocrystals and lignin nanoparticles in different concentrations. Increase in the crystallinity values for the ternary system by incorporating the two lignocellulosic nanofillers resulted in nanocomposites with a higher strength and modulus values and also had enhanced nucleation and crystal growth as compared to the neat polylactic acid and polylactic acid binary systems [16].

Food packaging based on antimicrobials is also emerging as a promising style of active packaging, which relies on incorporating natural antimicrobial materials into food packaging, thereby increasing the shelf life of the product as well as ensuring food safety. Hosseini et al. prepared bioactive films comprising fish gelatin and nanoparticles based on chitosan, blend infused with oregano essential oil. In addition, nanocomposites possessing antimicrobial properties are extremely helpful to attenuate the expansion of postprocessing contaminant microorganisms, thereby regulating any undesirable change in the quality of food. The bioactive nanoparticle films were found to possess antimicrobial function against common food pathogens such as *Staphylococcus aureus*, *Listeria monocytogenes*, *Salmonella enteritidis*, and *E. coli* [79]. Wen et al. also reported a novel antimicrobial packaging material that had prospective use in active food packaging. It was prepared by dispersing cinnamon essential oil/β-cyclodextrin inclusion complex into polylactic acid nanofibers through the electrospinning technique, and the resultant polylactic acid/cinnamon/β-cyclodextrin nanofilm was found to have greater antimicrobial activity in comparison to the polylactic acid/cinnamon essential oil nanofilm, thereby proving that it could extend the shelf life of pork [14].

Whey protein isolate (WPI) is among the plethora of proteins that are obtained from milk as a by-product in the manufacturing of cheese or casein. In comparison to fossil fuel-based synthetic films, WPI has excellent film-forming and gas-barrier properties. Solution casting was used to prepare whey protein isolate-based bio-nanocomposite films incorporated with zein nanoparticles. The addition of zein nanoparticles considerably enhanced the vapor barrier and mechanical strength of the WPI without compromising the elongation of the films, thus proving to be potential food-packaging materials [56].

Starch, a biopolymer, is one of the foremost promising candidates for food-packaging applications, particularly because of its engaging combination of low cost, extensive accessibility, high purity, nontoxicity, biodegradability, and environmental compatibility. Fabra et al. developed nanobiocomposites of thermoplastic corn starch with bacterial cellulose nanowhiskers by direct melt-mixing, and they were characterized in terms of morphological, optical, barrier, and tensile properties. The moisture and barrier properties of the nanobiocomposites were found to be improved by coating the films with electrospun poly(3-hydroxybutyrate) fibers, thereby opening new avenues for the industrial packaging sector [80].

4.7.3 Nanotechnology in Biomedical Applications

Significant headway in the use of natural polysaccharides for controlled drug delivery has been achieved. Polysaccharides can be easily assimilated to nontoxic products by the body, are eco-friendly, cost-effective, and possess desirable characteristics such as biocompatibility and renewability with low protein and cell adhesion [81, 82]. In recent years, polysaccharide-based hydrogels have also acquired substantial value over synthetic polymers as controlled drug delivery systems in pharmaceutical science [81, 83]. Anirudhan and coworkers reported a novel drug delivery system of 5-fluorouracil, an anticancer drug, encapsulated in a cellulose-grafted polymethacrylic acid succinyl cyclodextrin, which proved to be a promising carrier that exhibited sustained and controlled release of 5-fluorouracil to the target site [81]. Ghayempour et al. worked on tragacanth gum, which is a natural and safe material to formulate zinc oxide nanoparticles. Cytotoxicity studies indicated lower cytotoxicity due to the use of tragacanth as a noncytotoxic biomaterial on the synthesized urchin-like zinc oxide nanorod arrays [82]. Kono and Teshirogi formulated a hydrogel based on carboxymethyl chitosan and carboxymethyl β-chitosan using carbodiimide as a cross-linker. The chitosan hydrogels loaded with aspirin exhibited a sustained release of the drug, thus possessing the potential of functioning as a sustainable eco-friendly material with controlled drug-release characteristics [84].

Polysaccharide-based hydrogels have also been deliberated upon as a prime method for drug delivery to the colon. Hosaini and coworkers reported nanosize hydrogels for indomethacin prepared with different cross-linkers such as glutaraldehyde, polyvinyl alcohol, and diglycidyl ether using the swelling–diffusion method. The hydrogels were based on tragacanth biopolymer that contained reactive sites for modification and cross-linking [83].

Conventional cross-linked elastomeric particles are synthesized from synthetic rubber latex which are obtained from petroleum products and are not biodegradable. Hence, Wang and coworkers reported the synthesis of a biodegradable cross-linked elastomer nanoparticle that was found to be suitable to modify the characteristics of biodegradable brittle elastomer particles such as polylactic acid and polyhydroxyalkanoate and was also found to be useful for various biomedical applications [85].

Chitosan is one of the most sought after biopolymers that is extensively used in medical and pharmaceutical industries due to its abundant availability, low cost, low toxicity, capability to enhance membrane permeability, biodegradability, and biocompatibility [17, 86]. Zhang et al. carried out the preparation and characterization of naringenin-loaded chitosan particles for ophthalmic delivery. The negative charge of the cornea and conjunctiva results in their interaction with the positively charged chitosan nanoparticles, thereby prolonging the residence time and increasing the concentration of the drugs. Thus, these chitosan-based nanocarriers were found to be advantageous over the conventional methods of drug delivery due to the ease of administration, enhanced ocular bioavailability, prolonged drug release, and reduction in the dosing frequency [17]. Nanocomposite hydrogels have perked the interest of many researchers due to their superior properties over pure polymer hydrogels. Yadollahi et al. developed chitosan zinc oxide nanocomposite hydrogel beads by *in situ* generation of zinc oxide nanoparticles. Based on the data obtained, the prepared chitosan–zinc oxide nanocomposite hydrogels were found to be prospective candidates for the controlled delivery of drugs [86]. Rudzinski et al. utilized chitosan and PEGylated chitosan to formulate nanoparticles to treat colon cancer in humans; 3 nmol of siRNA was encapsulated in the nanoparticles by the ionic gelation method, and the data obtained indicated that the chitosan and PEGylated chitosan nanoparticles were successfully targeted to the colon cells, thereby inhibiting a protein responsible for tumor progression [87].

Biodegradable polymers such as chitosan, polylactic acid, polycaprolactone, alginate, and gelatin have also gained interest in pulmonary drug delivery due to their favorable attributes such as better encapsulation, controlled release of the drug for a prolonged period of time, and resistance to systemic degradation. Chitosan has found widespread use as a biopolymer for delivery to the lungs of various drugs, genes, and vaccines due to its reduced toxicity, encapsulation, and targeting. Due to the negative charge of the sialic acid residues of the alveolar macrophages, the positively charged chitosan nanoparticles are effectively targeted to the lungs. Rawal et al. developed rifampicin-loaded chitosan nanoparticles for local targeted delivery to the lungs [88]. Nanoparticles based on a hydrophobic derivative of chitosan, i.e. octanoyl chitosan, were also formulated by Petkar et al. The resultant nanoparticles were used to encapsulate rifampicin for pulmonary delivery in tuberculosis [89].

Researchers in veterinary science have also been exploring the use of biopolymers as a sustainable option for drug delivery. In particular, hydrogels based on starch are gaining attention in biomedical drug delivery since they are water swellable, biocompatible, and biodegradable. Gok et al. aimed to synchronize the estrus cycle in ewes by formulating a modified starch-based vaginal tablet containing progesterone. From the data obtained, the vaginal tablet based on biocompatible wheat starch, containing 75 mg progesterone, was proposed to be a potential substitute for the commercially available sustained-release products that are utilized for estrus synchronization [90]. Gao et al. also formulated carboxymethyl chitosan nanoparticles for the oral delivery of extracellular products of *Vibrio anguillarum* for the oral vaccination of turbots. Therefore, these biobased nanoparticles were found to be effective carriers of the oral vaccine to enhance both adaptive and innate immunity in fish [91].

The use of electrospun nanofibrous scaffolds has been widely explored for various applications, for example bone, wound dressing, and drug delivery system. Polylactic acid, polylactic-*co*-glycolic acid, and polycaprolactone are materials that are preferred for fabricating nanofibrous scaffolds due to their biodegradability and processability [92]. Anaraki et al. formulated a multiwalled carbon nanotube electrospun nanofibrous scaffold of polylactic acid/polyethylene glycol by the electrospinning process. The nanofibrous scaffolds were used to encapsulate doxorubicin hydrochloride, an anticancer drug that exhibited sustained release for more than 30 days [18]. Chen et al. worked on nanofiber scaffold for instances wherein antithrombogenicity and accelerated endothelialization are preferred. The emulsion electrospinning technique was used to encapsulate heparin and vascular endothelial growth factor into the core of poly(L-lactic acid-*co*-ε-caprolactone) core–shell nanofibers and was found to have immense potential in the fabrication of vascular grafts [93]. Electrospun poly(vinyl pyrrolidone)–nanopoly(lactic-*co*-glycolic acid) core–shell composite nanofibers of flurbiprofen axetil were also reported by Zhu et al. The formulated composite nanofibers were suggested to be an optimistic idea for the delivery of drugs, owing to their remarkably decreased burst-release profile, excellent cytocompatibility, and antiadhesion activity [94].

4.7.4 Nanotechnology in Agriculture

The global population is expected to hit a whooping eight billion in 2024, which will cause an escalation in the food demand, thereby putting an immense pressure for food supply from the agriculture sector. The current worth of the agri-food sector is estimated at US$5 trillion, which will continue to surge higher, together with a 70% rise in caloric demand and a 100% increase in crop demand. To meet this ever-growing demand for food, researchers are now exploring the field of nanotechnology to address the issues of food security, sustainability, and nutritional safety along with increasing the agricultural productivity [95]. Popular food industries such as H.J. Heinz, Nestle, Hershey, Unilever, and Kraft are now exploring nanotechnology-based food and food packaging wherein the nanofood sector is expected to cross the US$20 billion mark by 2020 [95].

Over the past few decades the use of conventional pesticides and fertilizers in agriculture has been contributing to environmental damage. Therefore, to combat the use of these harmful chemicals, a more sustainable approach based on biopolymers such as polysaccharides is being developed. These biopolymer derived formulations could be applied for controlled release preparations and to reduce the toxic effects of agrochemicals, thereby resulting in their safe management [96]. Nowadays, there has been a paradigm shift from the use of conventional to more nano-based fertilizers. Abdel-aziz and coworkers formulated nanocarriers based on chitosan with encapsulated nitrogen, phosphorus, and potassium for foliar uptake by wheat plants. Results obtained from the study indicated that the wheat plants treated with the nanochitosan fertilizer on sandy soil had a greater harvest index, crop index, and mobilization index compared to the plants treated with normal fertilizers. Therefore, nanofertilizers were found to enhance the productivity and plant growth, thereby creating new opportunities in agricultural production [97].

Nanoparticles produced from biopolymer polysaccharides such as chitosan and cyclodextrin were also developed by Campos and coworkers for the delivery of carvacrol and linalool. The nanocarriers were prepared by the ionic gelation method. The resultant nanosystem offered a dual function by encapsulating hydrophilic molecules within the chitosan matrix, whereas the hydrophobic molecules could be transported in the cyclodextrin cavities, thus providing perspective for a wide variety of applications apart from agriculture. There was a significant increase in the acaricidal and oviposition activities of carvacrol and linalool encapsulated in the nanoparticles, whereas the unencapsulated molecules possessed effective repellent activity. The nanocarriers developed were thus found to be better alternatives for the delivery of botanical compounds in agriculture and control of agricultural pests as well as contribute to environmental remediation [98].

Recently, nanotechnology has spiked up interest as a sustainable method for enhancing the efficiency of herbicides as well and combating their adverse effects on the ecosystem. Various attempts have been made to develop herbicide carrier systems that specifically act on weeds, without harming nontarget organisms. Nanosystems based on biopolymers such as polysaccharides are now coming to the forefront as an effective way of delivering herbicides. Chitosan, which is a biodegradable polymer obtained from deacetylation of chitin, has been highlighted as a carrier for agrochemicals and plant micronutrients [99]. Maruyama and coworkers proved that encapsulating the herbicides enhanced their action and lowered their toxicity. Nanoparticles of imazapic and imazapyr were formulated using biopolymers such as alginate/chitosan and chitosan/tripolyphosphate, and their physicochemical stability was determined. Real-time polymerase chain reactions were used to determine the effect of the loading herbicides in nanoparticles on microorganisms present in the soil. The results obtained from the study proved that encapsulation of herbicides within the nanoparticles decreased their toxicity, thereby proving it to be a potential candidate for agro-food applications in the future [99].

Mohammadi et al. also used chitosan to prepare nanoparticles to enhance the antimicrobial efficacy as well as the stability of *Cinnamomum zeylanicum* essential oil against *Phytophthora drechsleri* which is the causative agent of rotting in cucumber (*Cucumis sativus*). The fruits coated with the *C. zeylanicum* essential oil chitosan nanoparticles were found to be more firm, maintained their color and water content, and also revealed lower microbial counts during the storage period in comparison to the essential oil. Thus, the biobased nanoparticle coatings were found to be a sustainable method to extend the postharvest shelf life of cucumbers [100]. An easy dripping technique was used by Perez and Francois to prepare macrospheres of chitosan and chitosan-starch blends to obtain controlled release of fertilizers. The formulated macrospheres which were formed by ionotropic gelation and neutralization were later incorporated into a hydrogel using a cross-linking agent such as sodium tripolyphosphate solution. Prepared hydrogels were found to be an appropriate option for the controlled release of fertilizers in the agrochemical industry [101].

Botanical repellents exhibit superiority over the conventional methods to reduce our dependence on synthetic pesticides and in soil remediation, but poor stability along with rapid degradation hampers their successful use in the field. Therefore, nanosystems based on biopolymers, which are environment-friendly carriers, can be used for the protection of compounds susceptible to degradation, impart desired release properties, and decrease the toxicity caused to humans and the ecosystem. Oliveira and coworkers prepared zein nanoparticles for the delivery of the essential oils of citronella (geraniol and R-citronellal). Nanoparticles based on biopolymers such as zein could be utilized as an effective environment-friendly carrier system for the protection of essential oils against premature degradation, exhibit desirable release properties, and attenuate the toxicity caused to the environment and to human health. Hence, this nanotechnology-based product offered a better choice for the application of botanical repellents in agriculture and also resulted in reduced toxicity, prevented premature degradation, and provided efficient pest control [102].

4.7.5 Nanotechnology in Dye Removal

Increase in the environmental pollution due to industrial effluents is gaining global concern due to its repercussions on human and environmental health. Treating discharges from the textile dyeing industry is one of the most challenging tasks [103]. In the textile industry, annually about 7×10^5 tons of dyes and pigments are produced, and 10–15% of dyes which have not been used in the dyeing process are released into the water systems [103, 104]. Approximately 100 000 dyes are currently available commercially [105]. Synthetic dyes are usually stable and nonbiodegradable.

Moreover, some dyes have also been found to be carcinogenic and teratogenic, which pose a serious threat to the water ecosystem as well as humans [103]. There have been various approaches for the removal of these harmful dyes through the use of biobased nanomaterials. A highly efficient and recyclable bioadsorbent was prepared by Song et al. by utilizing a keratin sponge matrix incorporated with cellulose nanocrystals for the deletion of dyes present in aqueous solutions. The study indicated that the adsorption power of the keratin bioadsorbent was highly enhanced due to the incorporated cellulose nanocrystals nanofiller. Additionally, the bioadsorbent had excellent packed-bed column operation performance, thereby suggesting its use for industrial dye wastewater treatment [103].

Yue et al. also developed an adsorbent for removal of dyes by reinforcing amino-terminated hyperbranched polymer and beta-cyclodextrin (β-CD) onto cotton fibers. The study revealed that the reinforced adsorbent was successfully able to remove the Congo red and methylene blue dyes and that the adsorption isotherms and kinetic studies were in accordance with the Freundlich model and pseudo-second-order model, respectively [104].

Biobased nanocomposites proved to be another feasible approach for the removal of dyes. Vilela et al. designed a zwitterionic nanocomposite membrane with cross-linked poly(2-methacryloyloxyethyl phosphorylcholine) and bacterial nanocellulose. The study aimed at assessing the removal of two water-soluble dyes, i.e. methylene blue and methyl orange, from water. Results revealed that the dyes were successfully eliminated under the experimental conditions reaching a maximum dye absorption of 4.4–4.5 $mg\,g^{-1}$. Thus, these nanocomposites proved to be a promising approach for the fabrication of antibacterial biobased adsorbent membranes for effluent treatment containing anionic and cationic dyes [106].

A novel method was also developed by Tavakolian through the use of hairy nanocellulose for dye removal from wastewater. The use of hairy nanocellulose has been gaining momentum because it is biorenewable, environmentally friendly, can exclusively be directed for particular applications, and is also produced by a cost-effective procedure. ENCC, which is a negatively charged form of hairy nanocellulose, was used for the removal of methylene blue, which is a cationic dye from wastewater. The electrosterically stabilized nanocrystalline exhibited an uptake of 1400 $mg\,g^{-1}$ of methylene blue, which was significantly greater to any other reported value. Thus, the use of hairy nanocellulose proved to be a proficient method for adsorbing harmful dyes from industrial wastewater [107].

Tang et al. prepared an aerogel sorbent of cellulose nanofibrils for the removal of contaminants from wastewater. Nanomaterials extracted from plant cellulose, such as cellulose nanofibrils, can be easily transformed to an aerogel having a controllable density. The cross-linked networks of the cellulose aerogel had high porosity and were found to effectively adsorb methyl orange and copper (II) ions. Thus, the cellulose nanofibrils-based aerogels were believed to be attractive avenues as commercial adsorbents due to their low cost, sustainability, greater adsorption capacity, and ease of fabrication [108].

Figure 4.4 depicts some of the applications of biobased nanoproducts. Some of the commercial biobased nanoproducts are listed in Table 4.3.

4.8 Expert Opinion

With the rapid dwindling of fossil-fuel/petrochemical reserves and increased anxiety regarding the adverse impacts on the environment, researchers have been exploring ways to switch over to natural resources for sustainable development. Recently, nanotechnology has been heralded as a prime technology, holding many prospects for future markets. Combining the aforementioned notions, the idea of nanomaterials doped with biopolymers has aroused the interest of many scientists, to convert to a more eco-friendly lifestyle. The ample advantages of biobased products including their abundance, cost-effectiveness, biodegradability, and biocompatibility have resulted in its use in various applications such as soil and water remediation, agriculture, energy-storage devices, food-packaging applications, and drug delivery. There have also been phenomenal advances in the application of nanocomposites in the biomedical field such as skin tissue engineering, guided bone regeneration, bone-healing process, wound healing, and wound acceleration. But the road to development is never a straight one, and all potential solutions come with a price.

With respect to biopolymers, to make the idea of "sustainability" more perceivable, they have to be economically superior to their synthetic counterparts. The rigorous processing, extensive use of solvents, and higher energy demands create a major hurdle for its commercial production. Thus, efforts need to be directed to achieve greater functionality and scalable production. Also, despite all the in-depth research, very less information is available about the fate of nanomaterials that are released into the environment and their impact on human health. Due to the greater surface-to-volume ratio, the biological activity of these particles is likely to vary from larger particles, and although there is a possibility that they may cause toxic effects when inhaled or in the blood circulation, there is no conclusive data to indicate the same. The many uncertainties regarding nanomaterials necessitate

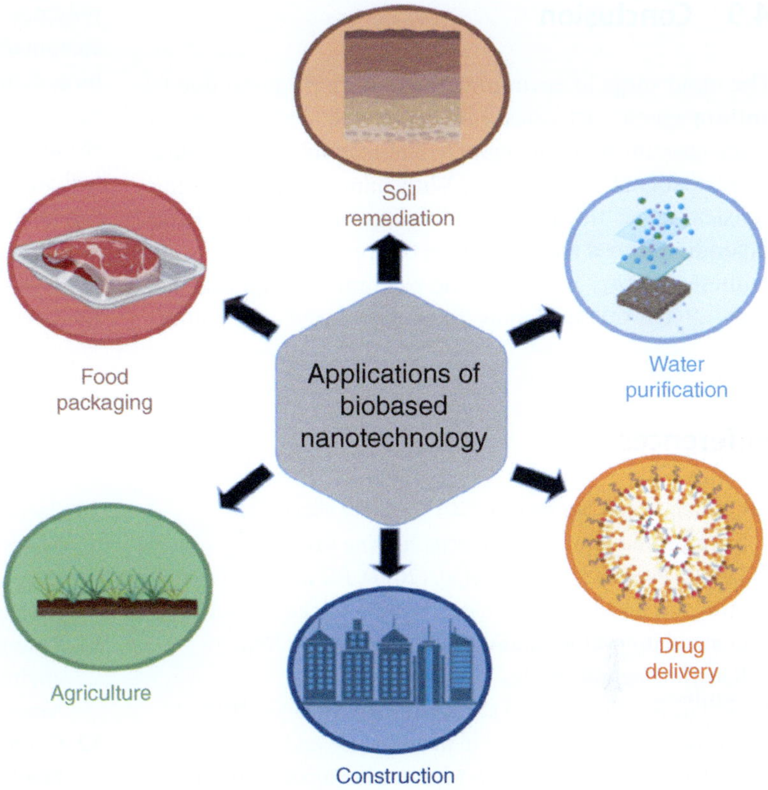

Figure 4.4 Applications of biobased nanotechnology.

Table 4.3 Commercial products of some biobased nanoproducts.

Commercial product	Content	Application	Producer/supplier
Master Nano Chitosan Organic Fertilizer	Liquid chitosan, organic acid, salicylic acid, and phenol compounds	Agriculture	Pannaraj Intertrade, Thailand
TAG NANO (NPK, PhoS, Zinc, Ca, etc.) fertilizers	Proteino-lacto-gluconate chelated with micronutrients, vitamins, probiotics, seaweed extracts, and humic acid	Agriculture	Tropical Agrosystem India (P) Ltd, India
Nano Green	Extracts of corn, grain, soybeans, potatoes, coconut, and palm	Agriculture	Nano Green Sciences, Inc., India
Curran®	Nanocellulose	Additive to paints and coatings, cosmetics, food, etc.	CelluComp Ltd, United Kingdom, Scotland
NF moisture Nanocell Mask, Mist and Serum	Nanocellulose	Cosmetics	Asia Nano Group

the need for strict guidelines for the responsible and safe manufacture of engineered nanomaterials. The existing tools of risk assessment for nanotechnology sustainability are also highly uncertain, require diverse evaluation criteria, present challenges in prediction, and are a more "relative" form of assessment, which is more applicable for products that are to be introduced or already available on the market.

Nevertheless, the current trajectory of development toward creating a more sustainable planet for future generations is a goal that should be pursued to build a more intelligent, greener, and prosperous economy. With the blossoming advancement of nanotechnology and higher demands by humans for medicine, food, and industrial production, the applications of biobased products have an immense opportunity of development in the near future.

4.9 Conclusion

The rapid surge in environmental pollution caused due to anthropogenic activities has motivated researchers to investigate the domain of green nanotechnology to create a more sustainable ecosystem. With nanotechnology being considered as the future technology, its use has been expanding over a wide spectrum of applications due to its multidisciplinary nature. In the past, much emphasis has been given to developing products based on natural polymers that are ubiquitous, biodegradable, and promote sustainability. Very recently, nanoproducts based on biopolymers have been utilized for a variety of applications such as water and soil remediation, agriculture, energy storage, and construction owing to their enhanced mechanical strength and barrier properties. Remarkable progress has been achieved in the production of biobased nanomaterials as well as nanocomposites, although there is a need to develop more facile and efficient processes for commercial production.

References

1 Catalan, J. and Norppa, H. (2017). Safety aspects of bio-based nanomaterials. *Bioengineering* 4 (4): 94.

2 Bell, J., Paula, L., Dodd, T. et al. (2017). EU ambition to build the world's leading bioeconomy—uncertain times demand innovative and sustainable solutions. *New Biotechnology* 40: 25–30.

3 Mohan, S., Oluwafemi, O.S., Kalarikkal, N. et al. (2016). Biopolymers – application in nanoscience and nanotechnology. In: *Recent Advances in Biopolymers* (ed. F.K. Perveen). PA: InTechOpen.

4 Attaran, S.A., Hassan, A., and Wahit, M.U. (2015). Materials for food packaging applications based on bio-based polymer nanocomposites: a review. *Journal of Thermoplastic Composite Materials* 30 (2): 143–173.

5 Brodin, M., Vallejos, M., Opedal, M.T. et al. (2017). Lignocellulosics as sustainable resources for production of bioplastics – a review. *Journal of Cleaner Production* 162: 646–664.

6 Ronzon, T., Santini, F., M'Barek, R. (2015). The Bioeconomy in the European Union in numbers. Facts and figures on biomass, turnover and employment. European Commission, Joint Research Centre, Institute for Prospective Technological Studies, Spain, 4p.

7 Luo, Y., Wang, Q., and Zhang, Y. (2020). Biopolymer-based nanotechnology approaches to deliver bioactive compounds for food applications: a perspective on the past, present, and future. *Journal of Agricultural and Food Chemistry* 68: 12993–13000.

8 Dudefoi, W., Villares, A., Peyron, S. et al. (2018). Nanoscience and nanotechnologies for biobased materials, packaging and food applications: new opportunities and concerns. *Innovative Food Science and Emerging Technologies* 46: 107–121.

9 Endes, C., Camarero-Espinosa, S., Mueller, S. et al. (2016). A critical review of the current knowledge regarding the biological impact of nanocellulose. *Journal of Nanbiotechnology* 14: 78.

10 Bartolozzi, I., Daddi, T., Punta, C. et al. (2019). Life cycle assessment of emerging environmental technologies in the early stage of development. *Journal of Industrial Ecology* 24 (1): 1–15.

11 Chang, J., Zhang, L., and Wang, P. (2018). Intelligent environmental nanomaterials. *Environmental Science: Nano* 5: 811–836.

12 Cardozo, J.R.G., Martinez, A.L.M., Perez, M.Y., and Londono, G.A.C. (2016). Production and characterization of polyhydroxyalkanoates and native microorganisms synthesized from fatty waste. *International Journal of Polymer Science* 2016: 6541718.

13 Rajan, K.P., Thomas, S.P., Gopanna, A., and Chavali, M. (2018). Polyhydroxybutyrate (PHB): a standout biopolymer for environmental sustainability. In: *Handbook of Ecomaterials* (eds. L. Martinez, O. Kharissova and B. Kharisov), 1–23. PA: Springer.

14 Wen, P., Zhu, D., Feng, K. et al. (2016). Fabrication of electrospun polylactic acid nanofilm incorporating cinnamon essential oil/β-cyclodextrin inclusion complex for antimicrobial packaging. *Food Chemistry* 196: 996–1004.

15 Xue, J., Wang, T., Hu, Q. et al. (2018). Insight into natural biopolymer-emulsified solid lipid nanoparticles for encapsulation of curcumin: Effect of loading methods. *Food Hydrocolloids* 79: 110–116.

16 Yang, W., Fortunati, E., Dominici, F. et al. (2016). Synergic effect of cellulose and lignin nanostructures in PLA based systems for food antibacterial packaging. *European Polymer Journal* 79: 1–12.

17 Zhang, P., Liu, X., Hu, W. et al. (2016). Preparation and evaluation of naringenin-loaded sulfobutylether-β-cyclodextrin/chitosan nanoparticles for ocular drug delivery. *Carbohydrate Polymers* 149: 224–230.

18 Anaraki, N.A., Rad, L.R., Irani, M., and Haririan, I. (2015). Fabrication of PLA/PEG/MWCNT electrospun nanofibrous scaffolds for anticancer drug delivery. *Journal of Applied Polymer Science* 132 (3): 41286.

19 Caro, N., Medina, E., Diaz-Dosque, M. et al. (2016). Novel active packaging based on films of chitosan and chitosan/quinoa protein printed with chitosan-tripolyphosphate-thymol nanoparticles via thermal ink-jet printing. *Food Hydrocolloids* 52: 520–532.

20 Dai, L., Zhou, H., Wei, Y. et al. (2019). Curcumin encapsulation in zein-rhamnolipid composite nanoparticles using a pH-driven method. *Food Hydrocolloids* 93: 342–350.

21 Emara, L.H., El-Ashmawy, A.A., Taha, N.F. et al. (2016). Nano-crystalline cellulose as a novel tablet excipient for improving solubility and dissolution of meloxicam. *Journal of Applied Pharmaceutical Science* 6 (02): 32–43.

22 Debnath, S.K., Saisivam, S., Debnath, M., and Omri, A. (2018). Development and evaluation of chitosan nanoparticles based dry powder inhalation formulations of prothionamide. *PLoS One* 13 (1): e0190976.

23 Shahabi-Ghahfarrokhi, I., Goudarzi, V., and Babaei-Ghazvini, A. (2019). Production of starch based biopolymer by green photochemical reaction at different UV region as a food packaging material: physicochemical characterization. *International Journal of Biological Macromolecules* 122: 201–209.

24 The utility of starch-based plastics. https://www.greendotbioplastics.com/starch-based-plastics (accessed 22 September 2020).

25 Kusuwandi, B. (2016). Nanotechnology in food packaging. In: *Nanoscience in Food and Agriculture 1* (eds. S. Ranjan, N. Dasgupta and E. Lichtfouse), 151–183. PA: Springer.

26 Mishra, R.K., Sabu, A., and Tiwari, S.K. (2018). Materials chemistry and the futurist eco-friendly applications of nanocellulose: status and prospect. *Journal of Saudi Chemical Society* 22: 949–978.

27 Dassanayake, R.S., Acharya, S., and Abidi, N. (2018). Biopolymer-based materials from polysaccharides: properties, processing, characterization and sorption applications. In: *Advanced Sorption Process Applications* (ed. S. Edebali). PA: InTechOpen.

28 Shaghaleh, H., Xu, X., and Wang, S. (2018). Current progress in production of biopolymeric materials based on cellulose, cellulose nanofibers, and cellulose derivatives. *RSC Advances* 8: 825–842.

29 Jahanbaani, A.R., Behzad, T., Borhani, S., and Darvanjooghi, M.H.K. (2016). Electrospinning of cellulose nanofibers mat for laminated epoxy composite production. *Fibers and Polymers* 17 (9): 1438–1448.

30 Singhvi, M.S., Zinjarde, S.S., and Gokhale, D.V. (2019). Polylactic acid: synthesis and biomedical applications. *Journal of Applied Microbiology* 127: 1612–1626.

31 Jem, K.J. and Tan, B. (2020). The development and challenges of poly (lactic acid) and poly (glycolic acid). *Advanced Industrial and Engineering Polymer Research* 3: 60–70.

32 Philibert, T., Lee, B.H., and Fabien, N. (2017). Current status and new perspectives on chitin and chitosan as functional biopolymers. *Applied Biochemistry and Biotechnology* 181: 1314–1337.

33 Priyadarshi, R. and Rhim, J.W. (2020). Chitosan-based biodegradable functional films for food packaging applications. *Innovative Food Science and Emerging Technologies* 62: 102346.

34 Singh, R., Shitiz, K., and Singh, A. (2017). Chitin and chitosan: biopolymers for wound management. *International Wound Journal.*

35 Corsi, I., Fiorati, A., Grassi, G. et al. (2018). Environmentally sustainable and ecosafe polysaccharide-based materials for water *Nano*-treatment: an eco-design study. *Materials* 11: 1228.

36 Corsi, I., Winther-Nielsen, M., Sethi, R. et al. (2018). *Ecofriendly* nanotechnologies and nanomaterials for environmental applications: key issue and consensus recommendations for sustainable and ecosafe nanoremediation. *Ecotoxicology and Environmental Safety* 154: 237–244.

37 Guerra, F.D., Attia, M.E., Whitehead, D.C., and Alexis, F. (2018). Nanotechnology for environmental remediation: materials and applications. *Materials* 23: 1760.

38 Guerra, F.D., Campbell, M.L., Attia, M.F. et al. (2018). Capture of aldehyde VOCs using a series of amine-functionalized cellulose nanocrystals. *ChemistrySelect* 3: 1–8.

39 Tsuzuki, T. (2014). Life cycle thinking and green nanotechnology. *Austin Journal of Nanomedicine & Nanotechnology* 2: 1.

40 Gallagher, M., Allen, C., Buchman, J.T. et al. (2017). Research highlights: applications of life-cycle assessment as a tool for characterizing environmental impacts of engineered nanomaterials. *Environmental Science: Nano* 4: 276–281.

41 Peijnenburg, W., Praetorius, A., Scott-Fordsmand, J., and Cornelis, G. (2016). Fate assessment of engineered nanoparticles in solids dominated media - current insights and the way forward. *Environmental Pollution* 218: 1365–1369.

42 Scott-Fordsmand, J.J., Peijnenburg, W., Amorim, M.J.B. et al. (2016). The way forward for risk assessment of nanomaterials in solid media. *Environmental Pollution* 218: 1363–1364.

43 Collier, Z.A., Kennedy, A.J., Poda, A.R. et al. (2015). Tiered guidance for risk-informed environmental health and safety testing of nanotechnologies. *Journal of Nanoparticle Research* 17: 155.

44 Oomen, A.G., Bleeker, E.A.J., Bos, P.M.J. et al. (2015). Grouping and read-across approaches for risk assessment of nanomaterials. *International Journal of Environmental Research and Public Health* 12: 13415–13434.

45 Grieger, K.D., Redmon, J.H., Money, E.S. et al. (2015). A relative ranking approach for nano-enabled applications to improve risk-based decision making: a case study of Army materiel. *Environment Systems & Decisions* 35: 42–53.

46 Nowack, B., Boldrin, A., Caballero, A. et al. (2016). Meeting the needs for released nanomaterials required for further testing - the SUN approach. *Environmental Science & Technology* 50: 2747–2753.

47 Semenzin, E., Lanzellotto, E., Hristozov, D. et al. (2015). Species sensitivity weighted distribution for ecological risk assessment of engineered nanomaterials: the n-TiO_2 case study. *Environmental Toxicology and Chemistry* 9999: 1–16.

48 Coll, C., Notter, D., Gottschalk, F. et al. (2016). Probabilistic environmental risk assessment of five nanomaterials (nano-TiO_2, nano-Ag, CNT and fullerenes). *Nanotoxicology* 10 (4): 436–444.

49 Scott-Fordsmand, J.J., Peijnenburg, W.J.G.M., Semenzin, E. et al. (2017). Environmental risk assessment strategy for nanomaterials. *International Journal of Environmental Research and Public Health* 14 (10): 1251.

50 Bos, P.M.J., Gottardo, S., Scott-Fordsmand, J.J. et al. (2015). The MARINA risk assessment strategy: a flexible strategy for efficient information collection and risk assessment of nanomaterials. *International Journal of Environmental Research and Public Health* 12: 15007–15021.

51 Urvoas, A., Valerio-Lepiniec, M., Minard, P., and Zollfrank, C. (2017). What are Bionaocomposites? In: *Bionanocomposites* (eds. C. Aime and T. Coradin), 1–7. PA: Wiley.

52 Pande, V.V. and Sanklecha, V.M. (2017). Bionanocomposite: a review. *Austin Journal of Nanomedicine & Nanotechnology* 5 (1): 1045.

53 Saini, R.K., Bajpai, A.K., and Jain, E. (2018). Fundamentals of bionanocomposites. In: *Biodegradeable and Biocompatible Polymer Composites* (ed. N.G. Shimpi), 351–377. PA: Elsevier.

54 Sessini, V., Arrieta, M.P., Kenny, J.M., and Peponi, L. (2016). Processing of edible films based on nanoreinforced gelatinized starch. *Polymer Degradation and Stability* 132: 157–168.

55 Zhang, Y.Q.S., Jing Yu, J., Yang, J. et al. (2016). Effects of chitin nano-whiskers on the antibacterial and physicochemical properties of maize starch films. *Carbohydrate Polymers* 147: 372–378.

56 Oymaci, P. and Altinkaya, S.A. (2016). Improvement of barrier and mechanical properties of whey protein isolate based food packaging films by incorporation of zein nanoparticles as a novel bionanocomposite. *Food Hydrocolloids* 54: 1–9.

57 Wakai, M. and Almenar, E. (2015). Effect of the presence of montmorillonite on the solubility of whey protein isolate films in food model systems with different compositions and pH. *Food Hydrocolloids* 43: 612–621.

58 Gopakumar, D.A. (2017). Nanocellulose based functional constructs for clean water and microwave suppression. *Materials.*

59 de Lima, H.H.C., Kupfer, V.L., Moises, M.P. et al. (2018). Bionanocomposites based on mesoporous silica and alginate for enhanced drug delivery. *Carbohydrate Polymers* 196: 126–134.

60 Rao, K.M., Kumar, A., and Han, S.S. (2017). Polysaccharide based bionanocomposite hydrogels reinforced with cellulose nanocrystals: drug release and biocompatibility analyses. *International Journal of Biological Macromolecules* 101: 165–171.

61 He, J., Bardelli, F., Gehin, A. et al. (2016). Novel chitosan goethite bionanocomposite beads for arsenic remediation. *Water Research* 101: 1–9.

62 Kumar, S., Krishnakumar, B., Sobral, A.J.F.N., and Koh, J. (2019). Bio-based (Chitosan/PVA/ZnO) nanocomposites film: thermally stable and photoluminescence material for removal of organic dye. *Carbohydrate Polymers* 205: 559–564.

63 Zhou, J., Ma, X., Li, J., and Zhu, L. (2019). Preparation and characterization of a bionanocomposite from poly (3-hydroxybutyrate-co-3-hydroxyhexanoate) and cellulose nanocrystals. *Cellulose* 26: 979–990.

64 Ucankus, G., Ercan, M., Uzunoglu, D., and Culha, M. (2018). Methods for preparation of nanocomposites in environmental remediation. In: *New Polymer Nanocomposites for Environmental Remediation* (eds. C.M. Hussain and A.K. Mishra), 1–28. PA: Elsevier.

65 Mousa, M.H., Dong, Y., and Davies, I.J. (2016). Recent advances in bionanocomposites: preparation, properties, and applications. *International Journal of Polymeric Materials* 65 (5): 225–254.

66 Lofrano, G., Libralato, G., and Brown, J. (eds.) (2017). *Nanotechnologies for Environmental Remediation*, 1–25. Switzerland, PA: Springer.

67 Liu, L., Gao, Z.Y., Su, X.P. et al. (2015). Adsorption removal of dyes from single and binary solutions using a cellulose-based bioadsorbent. *ACS Sustainable Chemistry & Engineering* 3 (3): 432–442.

68 Karim, Z., Claudpierre, S., Grahn, M. et al. (2016). Nanocellulose based functional membranes for water cleaning: tailoring of mechanical properties, porosity and metal ion capture. *Journal of Membrane Science* 514: 418–428.

69 Voisin, H., Bergstrom, L., Liu, P., and Mathew, A.P. (2017). Nanocellulose-based materials for water purification. *Nanomaterials* 7: 57.

70 Mautner, A., Lee, K.Y., Tammelin, T. et al. (2015). Cellulose nanopapers as tight aqueous ultra-filtration membranes. *Reactive and Functional Polymers* 86: 209–214.

71 Bossa, N., Carpenter, A.W., Kumar, N. et al. (2017). Cellulose nanocrystal-zero valent iron nanocomposites for groundwater remediation. *Environmental Science: Nano* 4 (6): 1294–1303.

72 Karim, Z., Mathew, A.P., Kokol, V. et al. (2016). High-flux affinity membranes based on cellulose nanocomposites for removal of heavy metal ions from industrial effluents. *RSC Advances* 6: 20644–20653.

73 Goetz, L.A., Jalvo, B., Rosal, R., and Mathew, A.P. (2016). Superhydrophilic anti-fouling electrospun cellulose acetate membranes coated with chitin nanocrystals for water filtration. *Journal of Membrane Science* 510: 238–248.

74 Sheikhi, A., Safari, S., Yang, H., and Van de Ven, T.G.M. (2015). Copper removal using electrosterically stabilized nanocrystalline cellulose. *ACS Applied Materials & Interfaces* 7 (21): 11301–11308.

75 Mautner, A., Maples, H.A., Kobkeatthawin, T. et al. (2016). Phosphorylated nanocellulose papers for copper adsorption from aqueous solutions. *International journal of Environmental Science and Technology* 13: 1861–1872.

76 Mautner, A., Maples, H.A., Sehaqui, H. et al. (2016). Nitrate removal from water using a nanopaper ion-exchanger. *Environmental Science: Water Research & Technology* 2: 117–124.

77 Kusuwandi, B. (2017). Environmental friendly food nano-packing. *Environmental Chemistry Letters* 15: 205–221.

78 Fortunati, E., Luzi, F., Yang, W. et al. (2018). Bio-based nanocomposites in food packaging. In: *Nanomaterials for Food Packaging* (eds. C. MAPR, J.M. Lagaron and V. AAM), 71–110. PA: Elsevier.

79 Hosseini, S.F., Rezaei, M., Zandi, M., and Farahmandghavi, F. (2016). Development of bioactive fish gelatin/chitosan nanoparticles composite films with antimicrobial properties. *Food Chemistry* 194: 1266–1274.

80 Fabra, M.J., Lopez-Rubio, A., Ambrosio-Martin, J., and Lagaron, J.M. (2016). Improving the barrier properties of thermoplastic corn starch-based films containing bacterial cellulose nanowhiskers by means of PHA electrospun coatings of interest in food packaging. *Food Hydrocolloids* 61: 261–268.

81 Anirudhan, T.S., Nima, J., and Divya, P.L. (2015). Synthesis, characterization and in vitro cytotoxicity analysis of a novel cellulose based drug carrier for the controlled delivery of 5-fluorouracil, an anticancer drug. *Applied Surface Science* 355: 64–73.

82 Ghayempour, S., Montazer, M., and Rad, M.M. (2015). Tragacanth gum biopolymer as reducing and stabilizing agent in biosonosynthesis of Urchin-like ZnO nanorod arrays: a low cytotoxic photocatalyst with antibacterial and antifungal properties. *Carbohydrate Polymers* 136: 232–241.

83 Hosaini, M., Hemmati, K., and Ghaemy, M. (2016). Synthesis of nanohydrogels based on tragacanth gum biopolymer and investigation of swelling and drug delivery. *International Journal of Biological Macromolecules* 82: 806–815.

84 Kono, H. and Teshirogi, T. (2015). Cyclodextrin-grafted chitosan hydrogels for controlled drug delivery. *International Journal of Biological Macromolecules* 72: 299–308.

85 Wang, Q., Zai, Y., Yang, D. et al. (2016). Bio-based elastomer nanoparticles with controllable biodegradability. *RSC Advances* 6: 102142–102148.

86 Yadollahi, M., Farhoudian, S., Barkhordari, S. et al. (2016). Facile synthesis of chitosan/ZnO bio-nanocomposite hydrogel beads as drug delivery systems. *International Journal of Biological Macromolecules* 82: 273–278.

87 Rudzinski, W.E., Palacios, A., Ahmed, A. et al. (2016). Targeted delivery of small interfering RNA to colon cancer cells using chitosan and PEGylated chitosan nanoparticles. *Carbohydrate Polymers* 147: 323–332.

88 Rawal, T., Parmar, R., Tyagi, R., and Butani, S. (2017). Rifampicin loaded chitosan nanoparticle dry powder presents: an improved therapeutic approach for alveolar tuberculosis. *Colloids and Surfaces, B: Biointerfaces* 154: 321–330.

89 Petkar, K.C., Chavhan, S., Kunda, N. et al. (2018). Development of novel octanoyl chitosan nanoparticles for improved rifampicin pulmonary delivery: optimization by factorial design. *AAPS PharmSciTech* 19: 1758–1772.

90 Gok, M.K., Ozgumus, S., Demir, K. et al. (2016). Development of starch based mucoadhesive vaginal drug delivery systems for application in veterinary medicine. *Carbohydrate Polymers* 136: 63–70.

91 Gao, P., Xia, G., Bao, Z. et al. (2016). Chitosan based nanoparticles as protein carriers for efficient oral antigen delivery. *International Journal of Biological Macromolecules* 91: 716–723.

92 Zhao, W., Li, J., Jin, K. et al. (2016). Fabrication of functional PLGA-based electrospun scaffolds and their applications in biomedical engineering. *Materials Science and Engineering: C* 59: 1181–1194.

93 Chen, X., Wang, J., An, Q. et al. (2015). Electrospun poly(L-lactic acid-co-ε-caprolactone) fibers loaded with heparin and vascular endothelial growth factor to improve blood compatibility and endothelial progenitor cell proliferation. *Colloids and Surfaces, B: Biointerfaces* 128: 106–114.

94 Zhu, T., Chen, S., Li, W. et al. (2015). Flurbiprofen axetil loaded coaxial electrospun poly(vinyl pyrrolidone)–nanopoly(lactic-co-glycolic acid) core–shell composite nanofibers: Preparation, characterization, and anti-adhesion activity. *Journal of Applied Polymer Science* 132 (22): 1–10.

95 Sampathkumar, K., Tan, K.X., and Loo, S.C.J. (2020). Developing nano-delivery systems for agriculture and food applications with nature-derived polymers. *iScience* 23 (5): 101055.

96 Campos, E.V.R., Oliveira, J.L.D., Fraceto, L.F., and Singh, B. (2015). Polysaccharides as safer release systems for agrochemicals. *Agronomy for Sustainable Development* 35: 47–66.

97 Abdel-Aziz, H.M.M., Hasaneen, M.N.A., and Omer, A.M. (2016). Nano chitosan-NPK fertilizer enhances the growth and productivity of wheat plants grown in sandy soil. *Spanish Journal of Agricultural Research* 14 (1): 1–9.

98 Campos, E.V.R., Proenca, P.L.F., Oliveira, J.L. et al. (2018). Chitosan nanoparticles functionalized with β-cyclodextrin: a promising carrier for botanical pesticides. *Scientific Reports* 8: 2067.

99 Maruyama, C.R., Guilger, M., Pascoli, M. et al. (2016). Nanoparticles based on chitosan as carriers for the combined herbicides Imazapic and Imazapyr. *Scientific Reports* 6: 19768.

100 Mohammadi, A., Hashemi, M., and Hosseini, S.M. (2015). Chitosan nanoparticles loaded with *Cinnamomum zeylanicum* essential oil enhance the shelf life of cucumber during cold storage. *Postharvest Biology and Technology* 110: 203–213.

101 Perez, J.J. and Francois, N.J. (2016). Chitosan-starch beads prepared by ionotropic gelation as potential matrices for controlled release of fertilizers. *Carbohydrate Polymers* 148: 134–142.

102 Oliveira, J.L., Campos, E.V.R., Pereira, A.E.S. et al. (2018). Zein nanoparticles as eco-friendly carrier systems for botanical repellents aiming sustainable agriculture. *Journal of Agricultural and Food Chemistry* 66: 1330–1340.

103 Song, K., Xu, H., Xu, L. et al. (2017). Preparation of cellulose nanocrystal-reinforced keratin bioadsorbent for highly effective and recyclable removal of dyes from aqueous solution. *Bioresource Technology* 232: 254–262.

104 Yue, X., Huang, J., Jiang, F. et al. (2019). Synthesis and characterization of cellulose-based adsorbent for removal of anionic and cationic dyes. *Journal of Engineered Fibers and Fabrics* 14: 1–10.

105 Crini, G., Torri, G., Lichtfouse, E. et al. (2019). Dye removal by biosorption using cross-linked chitosan-based hydrogels. *Environmental Chemistry Letters* 17 (4): 1645–1666.

106 Vilela, C., Moreirinha, C., Almeida, A. et al. (2019). Zwitterionic nanocellulose-based membranes for organic dye removal. *Materials* 12: 1404.

107 Tavakolian, M., Wiebe, H., Sadeghi, M.A., and Van de Ven, T.G.M. (2020). Dye removal using hairy nanocellulose: experimental and theoretical investigations. *ACS Applied Materials & Interfaces* 12: 5040–5049.

108 Tang, J., Song, Y., Zhao, F. et al. (2019). Compressible cellulose nanofibril (CNF) based aerogels produced via a bioinspired strategy for heavy metal ion and dye removal. *Carbohydrate Polymers* 208: 404–412.

5

Improving Sustainable Environment of Biopolymers Using Nanotechnology

Manish Patel[1] *and Jayvadan Patel*[2]

[1] *L M College of Pharmacy, Gujarat Technological University, Ahmedabad, Gujarat, India*
[2] *Nootan Pharmacy College, Sankalchand Patel University, Visnagar, Gujarat, India*

5.1 Introduction

Biopolymers are polymeric biomolecules that contain monomeric units that are covalently bonded to shape greater particles. The prefix "bio" signifies they are biodegradable materials made by living creatures. A wide kind of materials generally obtained from natural sources like microorganisms, plants, or trees are regularly portrayed utilizing the expression "biopolymer". Materials delivered by synthetic chemistry from biological/natural sources like vegetable oils, sugars, fats, resins, proteins, amino acids, at that point on likewise can be portrayed as biopolymer [1].

When contrasted with synthetic polymers which have a less difficult and more irregular structure, biopolymers are complex atomic gatherings that receive exact and characterized 3D shapes and structures. This is one significant property that makes biopolymers dynamic atoms in vivo. Their characterized shape and structure are surely keys to their working function. For instance, hemoglobin would not have the option to convey oxygen in the blood in the event that it was not collapsed in a quaternary structure. The primary property that recognizes biopolymers from nonrenewable energy source determined polymers is their sustainability, particularly when joined with biodegradability. Biodegradable biopolymers from renewable resources have been incorporated to give options in contrast to fossil-fuel-based polymers. They are mainly obtained from starch, sugar, natural fibers, or other natural biodegradable segments in different formulation quality. The biopolymers are debased by an introduction to microorganisms in the soil, compost, or marine residue. Besides, exposing biodegradable biopolymers to garbage removal by using their attribute of being degradable by the microscopic organisms especially bacteria in the ground fundamentally diminishes the release of CO_2 contrasted with conventional incineration. In this way, the utilization of biodegradable biopolymers from the perspective of global warming prevention. Lately, with the basic circumstance of the worldwide condition worsening because of global warming, the development of systems with maintainable utilization of materials has been accelerated from the perspective of adequately utilizing limited carbon resources and saving limited energy resources. Moreover, the expense of oil feedstocks has risen drastically and there is a rising purchaser enthusiasm for utilizing "green" (or renewable resources) as the reason for consumer items [2].

One of the rapidly developing materials areas in the last few years has been the manufacturing of polymers from sustainable resources also can be said as renewable resources. Their advancement is filled by the potential these polymers hold to supplant fossil-fuel-based polymers. The primary explanations behind this drive can be summed up as follows:

1) insufficient fossil-fuel resources;
2) pricing uncertainty of fossil fuels;
3) the commitment of the petroleum derivative as a feedstock to environmental change;
4) its infrequent role as a political weapon; and
5) its relationship with the garbage discarding issue made by the fossil-fuel derivative determined polymers.

Sustainable Nanotechnology: Strategies, Products, and Applications, First Edition.
Edited by Yashwant V. Pathak, Govindan Parayil, and Jayvadan K. Patel.

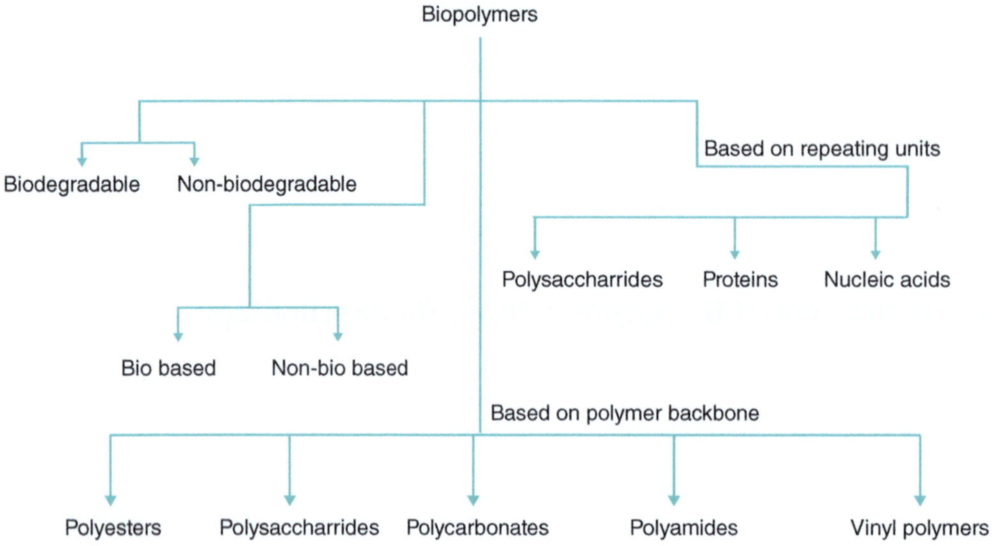

Figure 5.1 Biopolymer classes [3].

5.2 Different Class of Biopolymers

Biopolymers are classified into various ways depending on various scales. Depending on their degradability, biopolymers can be partitioned into two major classes, in particular biodegradable and nonbiodegradable, and then again, into bio-based and non-bio-based biopolymers (Figure 5.1). Based on their polymer general structure, biopolymers can be categorized generally into the following classes: polyesters, polysaccharides, polycarbonates, polyamides, and vinyl polymers [2]. These classes are again subdivided into a few subgroups dependent on their origin. Biopolymers can be characterized, based upon the character of the repetition unit they are made of, into three gatherings: (i) polysaccharides are made of sugars (for example cellulose present in plants cell), (ii) proteins are made of amino acids (for example myoglobin found in muscle tissues), and (iii) nucleic acids are made of nucleotides (for instance, DNA). Depending on their use, biopolymers can be named bioplastics, biosurfactant, bio detergent, bioadhesive, biofloculant, and more [3].

Biopolymers are categorized according to different sizes, in various ways. Biopolymers can be divided into two large classes based on their degradability, namely biodegradable and nonbiodegradable, and alternatively, into bio-based and non-bio-based biopolymers (Figure 5.1). Biopolymers can be loosely divided into the following classes based on their polymer backbone: polyesters, polysaccharides, polycarbonates, polyamides, and vinyl polymers. Once again, these groups are divided into many subgroups based on the source. Biopolymers can be classified into three categories according to the structure of the repeating unit they are made of (i) polysaccharides are composed of sugars (e.g. cellulose found in plants), (ii) amino acid proteins (e.g. myoglobin found in muscle tissues), and (iii) nucleic acids consist of nucleotides (DNA, the genetic material of the organism in question). Biopolymers can be categorized, depending on the application, as bioplastics, biosurfactant, bio detergent, bioadhesive, biofloculant, and more.

5.3 Sources and Preparation of Biopolymers

Polymeric biomaterials are polymers that are chemically inferred or modified, intended for various applications. Designing the production of novel biopolymers in plants is giving their manufacturing process a truly biorenewable approach. Like all-polymer firms, these polymers are additionally supplied in bulk and subsequently formed for specific end-use. Microorganisms additionally play a significant job in manufacturing a vast variety of biopolymers, for example, polysaccharides, polyesters, and polyamides which go from viscous solutions for plastics (Table 5.1). Their physical properties are subject to the structure and atomic weight of the polymer. The properties of different biopolymers derived by the guide of microorganisms can be custom fitted by the genetic control of microorganisms, subsequently making it reasonable for high-esteem clinical application, for example, tissue building and drug delivery.

Biopolymers that are created with the assistance of microorganisms require certain supplements and also, controlled ecological conditions. They are delivered either

Table 5.1 The preparation and application of alginate manufacturing method and its composite nanoparticles [4].

Polymer	Method of fabrication	Particle size (mm)	Remarks
Alginate	Control of the gel formation of alginate by calcium ions	250–850	Evaluation for the drug-loading capacity with doxorubicin as a model drug
	Gelation in the presence of calcium ions and further cross-linking with poly-L-lysine	—	Nanosponges are new antisense oligonucleotide carrier system for specific delivery to lungs, liver, and spleen
	Gelation in presence of calcium ions and further Eudragit E-100	200–950	In vitro release study revealed sustained release of gliclazide from gliclazide loaded alginate nanoparticles
	Modified coacervation or ionotropic gelation method	205–572	Optimization of mucoadhesive nanoparticulate carrier systems for prolonged ocular delivery of the drug
Sodium alginate and BSA	Emulsion solidification method	166 ± 34	Determination of the kinetic parameters of 5-FU sodium alginate BSA nanoparticles metabolism
Calcium alginate	Water-in-oil microemulsion followed by calcium cross-linking of glucuronic acid units	80	Examination of the nanoparticles for their potency as carriers for gene delivery
Alginate and chitosan	Ionotropic pregelation of alginate with chloride followed by complexation between alginate and chitosan	764–2209	Monitor the complexation calcium of contrary charged polyelectrolytes as insulin nanoparticulate carriers
	Induction of a pre-gel and with calcium counters chitosan ions, followed by polyelectrolyte complex coating with chitosan	850 ± 88	Development of an oral insulin delivery system having mild formulation conditions, high insulin entrapment efficiency for gastrointestinal release
	Induction of a pre-gel and with calcium counters, chitosan ions, followed by polyelectrolyte complex coating with chitosan	750	In vivo evaluation of the pharmacological activity of the insulin-loaded nanoparticles

straightforwardly by means of fermentation or on the other hand by synthetic polymerization (i.e. by chemical method) of monomers, which are then manufactured through maturation. Most biopolymers are biocompatible with no unfavorable impacts on biological systems. The process of manufacture of biopolymers from bacterial sources is accepted to be either because of their defense mechanism or as storage material [3]. They can be debased naturally, microorganisms, and enzymes so it can be reabsorbed in nature. Biopolymers or natural plastics are a type of plastics got from renewable biomass sources, for example, vegetable oil, corn starch, pea starch, etc. (Figure 5.2). By moving the concentrate more into the biopolymers, the protection of fossil assets and a decrease in CO_2 outflows can be accomplished consequently by promoting sustainable development. Among the microorganisms, algae fill in as a great feedstock for plastic creation as a result of their high yield and the capacity to develop in a range of environmental conditions. The utilization of algae opens up the chance of using carbon and killing ozone harming substance emissions from production lines or power plants.

Algae-based plastics have been an ongoing pattern in the time of bioplastics contrasted with traditional techniques for using feedstocks of corn and potatoes as plastics. While algae-based plastics are in their earliest stages, when they are into commercialization they are probably going to discover applications in a wide scope of businesses. As of now, microbial plastics are considered as a significant source of polymeric material that has an extraordinary potential for commercialization. They can change the flow properties of liquids, stabilize suspensions, flocculate particles, encapsulated materials, and produce emulsions [5].

5.4 Biopolymer Nanoparticles

Biopolymer nanoparticles were originally designed for the use of albumin and nonbiodegradable polymers, such as polyacrylamide and poly (methyl acrylate). The danger of persistent toxicity due to intracellular and additional tissue overburdening of nondegradable polymers was subsequently regarded as a significant restriction for the systemic

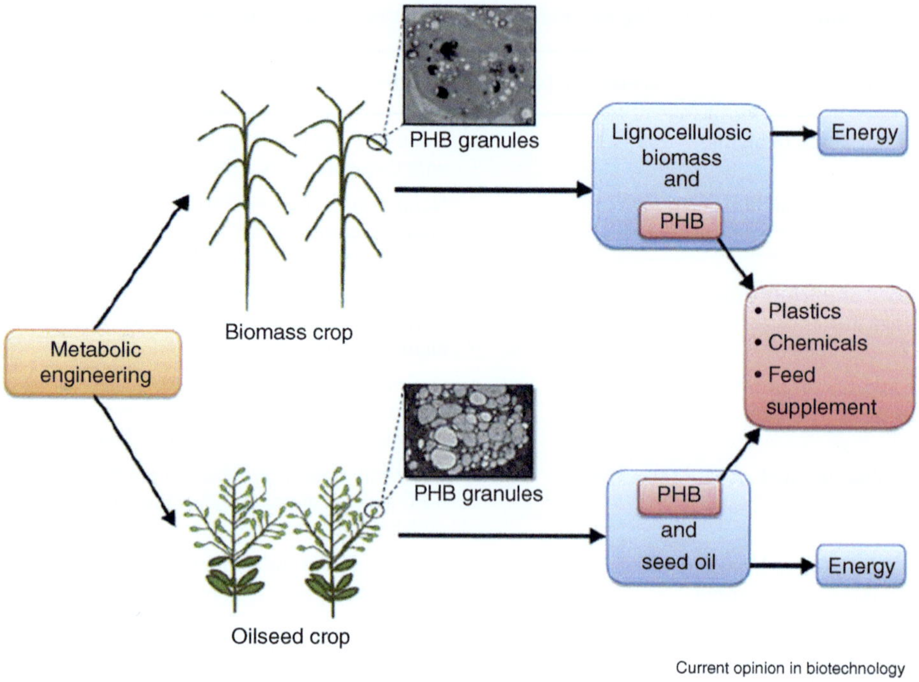

Figure 5.2 Preparation of polyhydroxybutyrate (PHB) in both biomass and oilseed crops. *Source:* K. D. Snell, et al., (2015), Fig 01, Pg 03 / with permission of ELSEVIER.

route of polyacrylamide and poly (methyl acrylate) nanoparticles administration in humans. As a result, synthetic biodegradable polymers like polyalkylcyanoacrylate, poly (lactic-*co*-glycolic acid), and polyanhydride were developed for the form of nanoparticles that had a lot of consideration [6–8].

The therapeutic capability of these colloidal biodegradable systems has been explored for various applications. These systems may also be concerned about toxicological concerns because of the extraordinarily significant reports described in the literature. The administration of hydrophilic atoms based on bio nanoparticles, such as peptides, proteins, and nucleic acids (oligonucleotides and qualities), has been shown to have extraordinary therapeutic potential, is further limited. This drawback is primarily due to the fact that the polymers that form these nanoparticles are primarily hydrophobic, although they are mature in proteins, peptides, and nucleic acids. This causes difficulties in effectively encapsulating the drug and protecting it against enzymatic degradation [9]. This examined the composition of nanoparticles using more hydrophilic, natural materials. It was felt years ago that we needed to build biodegradable nanoparticles (liposome, virus-like particle (VLP), protein, etc.) as successful drug delivery devices. In addition to the general benefits of nanoparticles, biopolymer nanoparticles, in particular, offer many advantages, including the simplicity of their preparation from well-understood biodegradable polymers and their high stability in, as well as during the storage of biological fluids. Nanoparticles made from biodegradable polymers such as proteins and polysaccharides can serve as efficient drug delivery vehicles for sustainable, controlled, and targeted release, with the goal of improving the therapeutic effects and also reducing the side effects of formulated drugs [10, 11].

5.4.1 Protein Nanoparticles

The first naturally occurring substance used to produce nanoparticles was two proteins, albumin and gelatin [12]. Among the colloidal systems, protein-based systems are very promising because they are biodegradable, less immunogenic, and nontoxic; they have higher in vivo and storage stability, are relatively simple to prepare and track size distribution, and can be scaled up in their development [10, 11, 13]. Furthermore, the protein-based nanoparticles give various possibilities for surface modification and covalent drug attachment due to the given primary structure of the proteins.

5.4.1.1 Albumin

Linked to its various functions and uses, albumin, a protein found in blood plasma, has always been a fascinating molecule [14]. Albumin is a protein that is biodegradable, biocompatible, and less immunogenic. In the circulatory system, the primary function of albumin is to aid transport, metabolism, and the distribution of exogenous and endogenous ligands. It is also capable of acting as an effective extracellular antioxidant and providing protection against free radicals and other harmful chemical agents. These specific albumin properties have provided a leading position in drug therapy from ancient times [13]. The study suggests the use of modified serum albumin as a selective agent for tumor detection and/or therapy or as a toxic compound delivery method to remove *Mycobacterium tuberculosis* via the delivery of receptor-mediated drugs. Thus, the nanotechnology period has used the well-established properties of albumin, both human serum albumin (HSA) and bovine serum albumin (BSA), for various purposes such as nanoparticular drug (antibodies, interferon-gamma, antiviral compounds) targeting carriers, anticancer therapeutic enhancer, adapted vehicles for drug delivery through the brain to the central nervous system and even across blood–brain barriers [6, 14].

5.4.1.2 Collagen

Collagen in a vertebrate is considered as a structural building material and the most abundant mammalian protein comprising 20–30% of total body proteins. Collagen has a particular sequence of structure, scale, and amino acids those results in triple helix fiber formation. Because of its excellent biocompatibility, biodegradation, and availability, collagen is considered a useful biomaterial. In addition, its capacity to change paved the way for its numerical applications in the manufacture of nanoparticles. These modifications involve the addition of other proteins, such as elastin, fibronectin, and glycosaminoglycan, resulting in the enhancement of its physicochemical and biological properties as well as the regulation of biodegradability and subsequent release of ligand through the use of such interlinking agents as glutaraldehyde, formaldehyde, ultraviolet, and gamma radiation. The collagen-dependent biodegradable nanoparticles are thermally stable, readily sterilizable, can be taken up by the reticuloendothelial system, and allow increased absorption of drug molecules into the cells [15, 16].

5.4.1.3 Gelatin

Gelatin is a natural water-soluble macromolecule which is the product of collagen heat dissolution and partial hydrolysis. There are two types of gelatin: type-A gelatin is obtained by isoelectric-point (pI) acid treatment of collagen between 7.0 and 9.0, while type-B gelatin is produced by alkaline hydrolysis of collagen with pI between 4.8 and 5.0. Gelatin has many advantages over other synthetic polymers including nonirritability, biocompatibility, and biodegradability, making it one of the promising materials as a carrier molecule. It is a natural, nontoxic, noncarcinogenic, low-immunogenic, and antigenic macromolecule. Gelatin has many surface functional groups that aid with chemical cross-linking and derivatization. These advantages led to its application for the synthesis of nanoparticles for drug delivery over the last 30 years [17].

5.4.1.4 Silk Protein

Sericin and fibroin nanoparticles. Silk fibers consist mainly of fibroin and sericin in which the structural protein, fibroin, and sericin are enveloped by the gum-like sticky protein. Table 5.2 describes preparation and application of silk protein nanoparticles [4].

5.4.1.4.1 Fibroin

It is made up of more than 70% of cocoon, the "core" proteins and is also known as hydrophobic glycoprotein. This insoluble protein is made almost entirely from the amino acids glycine, alanine, and serine that lead to the formation of an antiparallel-pleated layer in the fibers. Fibroin is semi-crystalline and consists of two phases: one is the shape of a solid crystalline base, and the other is noncrystalline. Also, silk fibroin is histocompatible, less immunogenic, and less toxic. Silk fibroin can be transformed into different types including gels, fibers, membranes, scaffolds, hydrogels, and nanoparticles [18]. Soil fibroin matrices with high specific surface area, high porosity, good biocompatibility, and biodegradability have comprehensive biomaterial and drug delivery applications. For many years, the soil has been used as a suture material, and it has been identified as a biopolymer that evokes minimal foreign body response. In addition, silk-based biomaterials are highly biocompatible with different cell types and encourage cell growth and proliferation (Figure 5.3).

5.4.1.4.2 Sericin

Sericins – hydrophilic glycoproteins that act as a "glue" – make up 20–30% of the coconut. These hot water-soluble proteins contain various polypeptides ranging in weight from 24 to 400 kDa and have an exceptionally high serine content (40%) along with substantial glycine content (16%). Sericin consists of 35% sheet and 63% random coil and has no helical material – thus, its partly unfolded state. Apart from the general benefits of protein nanoparticles, sericin nanoparticles may also provide some other benefits of the sericin's inherent property. These include antioxidants and antitumor action; enhancing the bioavailability of such elements as Zn, Mg, Fe, and Ca; and suppressing coagulation when sulfated; sericin is nontoxic to fibroblast cells. The content of methionine and cysteine in silk sericin is an important factor for cell growth and collagen synthesis. Water-soluble silk sericin has no immunogenicity and is also like silk fibroin a biocompatible macromolecular

Table 5.2 Preparation and application of silk protein nanoparticles [4].

Protein	Method of fabrication	Particle size (mm)	Remarks
Silk fibroin	Precipitation using water-miscible protonic and polar aprotonic organic solvents	35–125	Globular insoluble particles well dispersed and stable in aqueous solution
	Precipitation using water-miscible protonic and polar aprotonic organic solvents	50–120	Matrix for immobilization of L-asparaginase
	Microemulsion	167	Color dye-doped silk fibroin nanoparticles
	Conjugated covalently with insulin using cross-linking reagent glutaraldehyde capillary microdot technique	40–120	Insulin–silk fibroin nanoparticles bioconjugates
		<100	Sustained and long-term therapeutic delivery of curcumin to breast cancer cells
	Desolvation	150–170	Cellular uptake and control release studies
Silk sericin	Conjugation of sericin with activated PEG	200–400	Overcomes its problem of instability in water and insolubility in organic solvents
	Sericin–PEG self-assembled through hydrophobic interactions delivery	204	Self-assembled nanostructures for immobilization and drug
	Sericin–polymethacrylate core-shell nanoparticles by graft copolymerizing technique	100–150	Potential biomedical application as delivery systems
	Self-assembled silk sericin/poloxamer nanoparticles. Self-assembled silk sericin nanostructures	100–110	Nanocarriers of hydrophobic and hydrophilic drugs for targeted delivery
			Fractal self-assembly of silk protein sericin

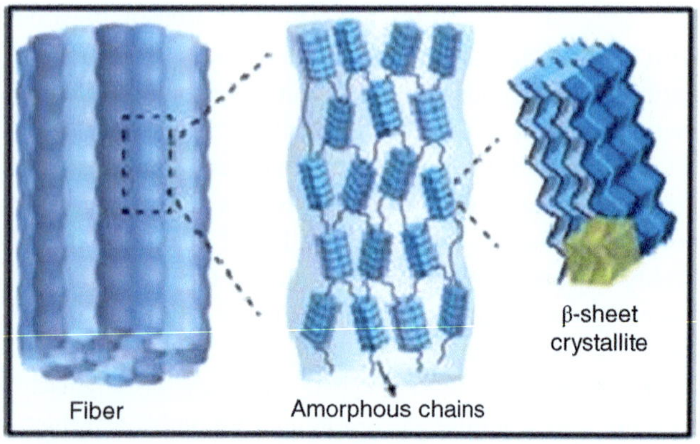

Figure 5.3 Fibroin is assembled from nanofibril units which crystal network consists of β-sheet crystallites dispersed within an amorphous matrix. *Source:* T. P. Nguyen et al., (2019), Fig 01, Pg 03 / MDPI / CC BY 4.0.

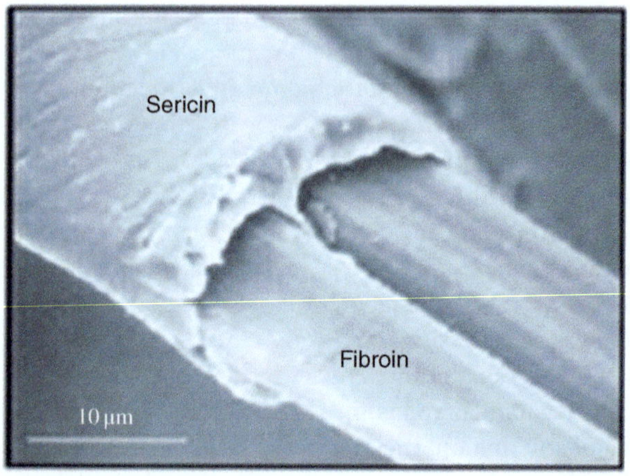

Figure 5.4 Main proteins of silkworm silk fibers are fibroin and sericin [18].

protein. An examination of the macrophage silk protein response indicates that sericin does not usually manifest inflammatory activity when present in a soluble form. Nayak and Kundu recently concluded that sericin facilitates the healing of wounds without inducing any inflammation [18, 19] (Figure 5.4).

5.4.1.5 Keratin

Keratins are a group of structural proteins rich in cysteine which exhibit a high mechanical strength due to a large number of disulfide bonds. Keratin was recently used as a nanosuspension that results in ultra-thin, transparent keratin coatings to investigate the action of the proliferation of in vitro cells as a possible coating material for standard cultivation. The nanosuspension keratin coatings can provide an affordable alternative to materials such as collagen or fibronectin. Keratin nanosuspension, if further explored, can also find applications in tissue engineering [18, 20].

5.4.2 Polysaccharide Nanoparticles

Nanoparticles derived from polysaccharides and nanostructured surfaces contribute to improving the biocompatibility of cell toxic material along with modern immobilization methods that are currently being developed for modern pharmaceutical formulations derived from bionanoparticles. Nanoparticles from naturally occurring polysaccharides were developed for peptide, protein, and nucleic acid administration [21].

5.4.2.1 Alginate

Alginate is a water-soluble, naturally occurring, a linear unbranched polysaccharide derived from brown seaweed. Alginate consists of two uronic acid forms, -L-guluronic acid and D-mannuronic acid. The monomeric units are split into three ways: alternating blocks of guluronic and mannuronic residues, guluronic acid blocks, and mannuronic acids. Alginate has been identified as a mucoadhesive, biocompatible, nonimmunogenic material under normal physiological conditions and undergoing dissolution and biodegradation. Alginate solubility in water depends on the cations related to it. Sodium alginate is water soluble because of calcium induces gel formation [21]. Alginate can also form complexes with polycations such as polyenimine (PEI), chitosan, or simple peptides such as polylysine and polyarginine, in addition to the interaction with calcium. Carboxylic uronic acid groups carry negative alginate charges. Chitosan endows nanoparticles with a positive surface charge, prolongs the contact time of the active ingredients with the epithelium, and improves absorption through the paracellular transport pathway via the near junctions. It is easy to obtain micro- and nanoparticles alginate by inducing calcium ion gelation [22]. This property can be used to produce a pre gel composed of very small aggregates of gel particles, followed by the addition of an aqueous polycation solution to form a complex polyelectrolyte coating. A cationic natural polymer, poly-L-lysine (PLL) was used to combine with alginates to prepare nanoparticles. However, if administered, PLL is toxic and immunogenic. Selected recently as an alternative cationic polymer, chitosan (CS). Table 5.1 displays the preparation and applications of alginate and its composite nanoparticles [4].

5.4.2.2 Chitosan

Chitosan is the naturally occurring second-abundant polysaccharide. Chitosan is made of D-glucosamine (deacetylated unit) and *N*-acetyl-D-glucosamine(acetylated unit), which are naturally distributed. It is produced through the deacetylation of chitin extracted from crab shells, shrimps, and krill. Commercially available chitosan is 66–95% deacetylated and it has an average molecular weight of between 3.8 and 2000 kDa. Chitosan is linear, hydrophilic, positively charged, and has the properties of mucoadhesive. Because of its excellent biocompatibility and biodegradability, it is an excellent biopolymer for the preparation of microparticles and nanoparticles [23]. It is degraded in vivo by lysozyme. Furthermore, the amino groups impart a high charge density on the molecule and are readily available with acids for chemical reactions and salt formation. Chitosan is soluble in different acids and can also interact with polyons in order to form complexes and gels. These characteristics are exploited in the manufacture of nanoparticles based either on the spontaneous formation of complexes between chitosan and polyons like DNA or on the gelation of a chitosan solution dispersed in an emulsion of water in oil [24] (Figure 5.5).

5.5 Fabrication Methodologies

There are three popular methods of preparing nanoparticles based on proteins and polysaccharides, namely emulsification, desolvation, and coacervation. They have been complemented very recently by the electrospray drying technique [12].

5.5.1 Emulsification

The theory of formation of nano-emulsions is based on the spontaneous emulsification that occurs when mixing an organic and aqueous phase. Organic phase is a homogeneous solution of tar, lipophilic surfactant, and water-miscible solvent, whereas the aqueous phase is hydrophilic surfactant and water. This method may be defined as the dissolution of hydrophobic substances in an organic solvent that is further emulsified at very high shear with an aqueous solution. This means very small droplets (50–100 nm) are produced. After emulsification, the organic solvent is removed by evaporation, resulting in strong dispersions of solid nanoparticles. The main drawback of this method is a need to apply and then remove organic solvent. On top of that, organic solvent residues can cause toxic problems [26] (Figure 5.6).

Figure 5.5 Chemical structure of chitosan biopolymer [25].

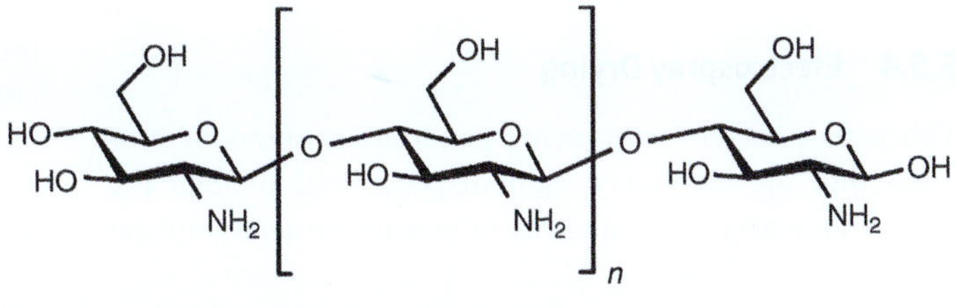

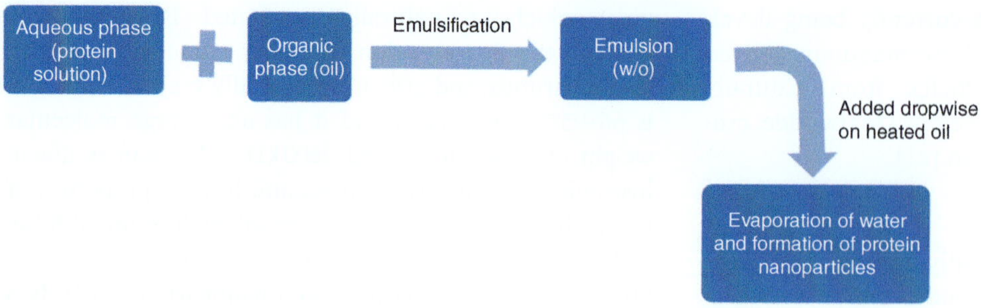

Figure 5.6 Process describing evaporation.

5.5.2 Desolvation

A different method for manufacturing nanoparticles which required a slow addition to the protein solution of a desolvation factor, such as natural salts or alcohol. The desolvation factor alters the tertiary protein structure [27]. When the critical level of desolvation is achieved, protein clump will be formed which will result in the formation of nanoparticles while cross-linking with a chemical substance (e.g. glutaraldehyde). Then a variation in the desolving method was done as a two-step process for the synthesis of gelatin nanoparticles. In the first stage of desolvation, decanting eliminates the low molecular gelatin fractions present in the supernatant, and in the second step, large molecular fractions present in the sediment are redissolved and then dissolved again at pH 2.5. Then the resulting particles can be separated easily by centrifugation and redispersion [4].

5.5.3 Coacervation

The method of coacervation is similar to the method of desolvation; it uses mixing of the aqueous protein solution with organic solvent such as acetone or ethanol to create tiny coacervates. The addition of the cross-linking agent, such as glutaraldehyde, etc., limits these coacervates. The difference of methods of coacervation and desolvation is the different parameters that influence the manufacturing process to give the nanoparticles the desired property. These parameters include initial protein concentration, temperature, pH, cross-linker concentration, agitation velocity, protein/organic solvent molar ratio, and organic solvent addition rate.

5.5.4 Electrospray Drying

The process of electrospraying generates sections of proteins that are relatively monodisperse and biologically active. This approach involves preparing protein solution by dissolving the dry powder in a solution that is electrosprayable. Dispersion of the solution accompanied by evaporation of the solvent leaves dry residues deposited on appropriate deposition substrates. This method had prepared insulin nanoparticles sized between 88 and 110 nm. Even higher nanoparticles output rate increases their scale. The biological activity of the nanoparticles based on the electrosprayed proteins is not affected by the process conditions (Table 5.3).

5.6 Nanotechnology and Biopolymer Composites with a Sustainable Approach

Large-scale use of biopolymers requires enhancing biopolymers' technological performance with regard to petroleum-based polymers, without seriously sacrificing environmental performance. The recent developments in silicate nanocomposites coated with biopolymers are very promising in this respect [29].

Varied fibers and platelets have long been used as reinforcements to enhance the mechanical properties of polymeric materials. The use as reinforcements of layered silicate clays is relatively recent, starting in the early 1990s. The widely used polymer composite clays are montmorillonite (MMT), hectorite, and saponite, which belong to the general phyllosilicate family of 2 : 1. Their crystal structure consists of either aluminum or magnesium hydroxide layers composed of two silica tetrahedrals fused to an edge-shared octahedral sheet. Individual layers are approximately 1 nm thick, and lateral platelet dimensions may vary from 30 nm to several microns, depending on the particular silicate, and are therefore referred to as nanoclays. When mixed with polymers, these nanometer-thick layers are able to exfoliate into individual platelets. Exfoliated silicate nanoclays have a high aspect ratio (defined as the diameter to platelet thickness ratio) of 10–1000, and therefore a small weight percentage of nanoclays properly spread

Table 5.3 Some biopolymers, their sources, and preparation [7, 28].

Polymer	Method/source	Example of bacteria used for synthesis
Hybrid plastics	Adding denatured algae biomass to petroleum-based plastics like polyurethane and polyethylene as fillers	Filamentous green algae, Cladophorales
Cellulose-based plastics	Biopolymer of glucose	30% of the biomass produced after extraction of algal oil is known to contain cellulose
Poly-lactic acid (PLA)	Polymerization of lactic acid	Bacterial fermentation of algal biomass
Bio-polyethylene	Ethylene produced from ethanol, by a chemical reaction called cracking. Ethanol derived from natural gas or petroleum	Bacterial fermentation of algal biomass
Poly esters	Biomass	Bacteria like *Alcaligenes eutrophus*, *E. coli*, etc.

throughout the polymer provides a much larger surface area for polymer filler interactions than traditional micron-sized reinforcements do. Nanoclays, in their pristine form, are naturally hydrophilic in nature, resulting in very poor dispersion in organic polymers. The modification of the surface chemistry of nanoclays by ion-exchange reactions with cationic surfactants, such as quaternary alkyl ammonium or alkyl phosphonium, has been shown not only to improve dispersion by making organophilic nanoclays, But also to improve polymer adhesion by providing functional groups that can react with the polymers and, in some cases, by initiating monomer polymerization. Organically modified montmorillonite (OMMT) clays with different specifications are manufactured commercially under the trade name CloisiteR by Southern Clay Products Inc. [30]

Three main techniques have been used in the preparation of multi-layer silicate polymers.

1) Polymer intercalation from solution, where the organically modified layered silicates (OMLS) is first swollen and mixed with the polymer solution in a solvent such as chloroform or toluene. The polymer chains intercalate and displace the solvent inside the silicate interlayer, forming a solvent-removing nanocomposite.
2) In situ intercalative polymerization accompanied by polymerization, where the organically transformed clay is swollen within the liquid monomer or a monomer solution.
3) Melt-intercalation in which a mixture of polymer and OMLS is rinsed statically or under shear above the polymer softening stage.

Melt intercalation is evolving as a conventional technique since it is compatible with current industrial processes such as extrusion and molding by injection. Furthermore, melt intercalation is deemed environmentally preferable to the other approaches due to the absence of organic solvents and related pollution [31].

The strong interfacial interactions between the polymer and organophilic layered silicates, the availability of large surface areas for such interactions due to high aspect ratios, and the good dispersion achieved with organomodifiers result in significant improvements in the technical performance of nanoclay composites. Mechanical properties such as tensile strength and nanoclay composite modulus, for example, dramatically increase with relatively low clay load by weight of less than 5%. The strengthening of polymers by nanoclays also increases the temperature of heat distortion, which is the temperature at which a polymer sample deforms under a specified load. The gas barrier properties of nanocomposites are stronger than those of pure polymers, firstly because nanoclays create a labyrinth of tortuous pathways that hinder the movement of gas molecules through the polymer resin, and secondly because local molecular level transformation in the polymer matrix often reduces permeability in the presence of silicate galleries and organic modifications. The thermal stability of polymeric materials is usually studied through thermogravimetric analysis, where weight loss following degradation at high temperature is monitored due to the formation of volatile products. Nanoclay dispersion increases polymer thermal stability, as the silicate layers serve as barriers to the flow of these volatile gaseous by-products and associated heat.

Although the impact of nanoclays on the above performance measures is positive, the direction of the impact on biopolymers' biodegradability is not unambiguous, as the mode of attack by compost microorganisms differs depending on the characteristics of nanocomposite and organomodifier and the compost conditions. Quaternary alkyl ammonium salts, the most widely used organic nanoclay modifiers, are considered biodegradable under acceptable waste treatment conditions and thus are not likely to adversely affect the biodegradability. Nevertheless, some studies report that biodegradation is delayed due to higher barrier properties of nanoclay composites, while some studies do not notice any variations in 60-day biodegradation between pristine biopolymers and their nanoclay composites [32].

The significant improvements in the properties of biopolymer nanoclay composites compared to pristine biopolymers have been documented in a number of experimental studies. These modifications include increased weight, increased storage modules, lower heat distortion, lower gas permeability, and flammability, and, frequently, greater biodegradability. Nevertheless, the observed degree of these improvements is a complex phenomenon depending on the polymer, nanoclay, clay filling, organomodifier, compatibilizers, process parameters, degree and nature of clay dispersion, morphology, and test conditions. Moreover, most studies report changes over a variety of these parameters, and typically these relations are not linear. For example, mechanical properties such as tensile strength increase initially with the increase in clay loading and start to fall after a critical loading stage. Similarly, the percentage of storage modules improved depends on the range of temperatures measured. As a result, the findings of different studies can hardly be summarized and concluded that biopolymer – nanoclay composites have enhanced technical performance against pristine biopolymers under appropriate conditions, making them more competitive with petroleum-based polymers and traditional composites [29].

5.7 Characterization of Nanoparticles

Using such advanced microscopic techniques as scanning electron microscopy (SEM), transmission electron microscopy (TEM) and atomic force microscopy (AFM), nanoparticles are typically characterized by their scale, morphology, and surface charge. The average diameter of the particles, their size distribution and charge influence the physical stability and distribution of nanoparticles in in-vivo studies [33]. Techniques for electron microscopy are very useful in determining the overall form of polymeric nanoparticles, which can determine their toxicity. The surface charges of the nanoparticles influence both the physical stability and redispersibility of polymer dispersion and their effects in vivo [3].

5.7.1 Particle Size

Nanoparticles have a significant role in drug release and drug targeting. Particle size has been shown to have an effect on drug release. Smaller particles have greater surface area. As a result, much of the drug loaded onto them would be exposed to the surface of particles leading to rapid release of the substance. By comparison, drugs disperse slowly inside larger particles [34]. As a disadvantage, smaller particles tend to accumulate during nanoparticle dispersion storage and transport. Hence, a balance exists between a small size and optimum nanoparticles stability.

Another important application of nanoparticles is the immobilization of enzymes in the field of biosensors. Reducing the size of carrier materials usually improves the performance of immobilized enzymes, because the smaller particles provide a broad surface area for enzyme surface attachment and reduce the resistance to substratum diffusion. In addition, nanoparticles' physical characteristics, such as increased diffusion and particle motility, can affect the intrinsic catalytic activity of the attached enzymes. Particular size can also influence the degradation of polymers. For example, the degradation rate of poly(lactic-*co*-glycolic acid) has been found to rise with the in vitro particle size. As discussed below, there are many methods for assessing the size of the nanoproducts [35].

5.7.1.1 Dynamic Light Scattering

Photon-correlation spectroscopy (PCS) or dynamic light scattering (DLS) is the fastest and most common method of determining particle size today. DLS is commonly used to assess the size of Brownian nanoparticles in nano- and submicron-ranged colloidal suspensions. In Brownian motion, reflecting monochromatic light (laser) onto a solution of spherical particles causes a Doppler change as the light reaches the moving object, shifting the wavelength of the incoming light. This alteration is related to particle size. The size distribution can be extracted and a description of the motion of the particle in the medium can be given, the diffusion coefficient of the particle is calculated and the autocorrelation function is used. This method has several advantages: the experiment is quick, is nearly automatic, and requires no extensive experience. In addition, this approach has modest cost of developing. The benefit of using dynamic scattering is the possibility of examining samples containing large ranges of species with widely different molecular masses (e.g. a native protein and different aggregate sizes) and detecting very small quantities of the higher mass species ($<0.01\%$ in certain cases). PCS describes the average particle size and polydispersity index (PI) which is a set of particle size measurements within measured samples. Observing larger particles as opposed to smaller particles shows that the larger particles travel more slowly than the smaller ones for a given temperature [34, 36].

5.7.1.2 Nanoparticle Tracking Analysis

A minor change to the nanoparticle tracking analysis (NTA) photon-correlation spectroscopy is a technique developed by Nano-Sight Ltd to determine the size distribution profile of small particles in a liquid suspension. The technique is used in conjunction with an ultramicroscopy which allows visualization of small particles moving under Brownian motion in liquid suspension. Using the Stokes–Einstein equation, computer software is then used to record motion of particles and then calculate their hydrodynamic radius afterward. Furthermore, because the samples need minimal preparation, the time needed to process one sample is considerably reduced [36, 37].

5.7.2 Particle Morphology

Nanoparticles' size and morphology exert a profound influence on the physical and chemical properties which determine their interaction with the environment and biological systems. There are also methods for examining nanoparticular morphology. Other parameters such as morphology or surface roughness of nanoparticles are also calculated by microscopic techniques such as SEM, TEM, and AFM, along with particle size and distribution study [35].

5.7.2.1 Scanning Electron Microscope

Nanoparticles solution should first be processed into a dry powder for SEM characterization, which is then placed on a sample holder and then coated with a conductive metal, such as gold, using a sputter coater. The sample is then analyzed with a focused, fine electron beam. The sample is then scanned with a guided, fine electron beam. The surface properties of the sample are derived from the secondary electrons released from the surface of the sample. The nanoparticles must be able to withstand vacuum, which can damage the polymer by the electron beam. The mean size obtained by SEM is comparable to the results obtained by dispersing dynamic light [38].

5.7.2.2 Transmission Electron Microscope

TEM works on a different concept than SEM but also brings in the same data type. TEM sample preparation is complex and time consuming due to its need to be ultra-thin for transmitting electrons. The dispersion of nanoparticles is deposited on assistance grids or films. Nanoparticles are fixed using either a negative staining substance such as phosphotungstic acid or derivatives, uranyl acetate, etc., or plastic embedding content to escape the instrument vacuum and encourage handling. Alternate approach is to expose the sample to liquid nitrogen temperatures after incorporation into vitreous ice. Sample surface characteristics are obtained by passing a beam of electrons through an ultra-thin tube, communicating with the tube as it moves through [38].

5.7.2.3 Atomic Force Microscopy

AFM is just another instrument used at the atomic level to classify a variety of surfaces, including nanoparticles, and it is one of the key types of microscopes for scanning probes. The primary benefit of AFM is its ability to image nonconductive samples without any special treatment, enabling the imaging of delicate biological and polymeric nanostructures and microstructures. AFM requires a limited preparation of samples and can be carried out under ambient conditions. The images of the sample surface are given by scanning with a sharp probe across its surface, and then tracking and documenting tip-sample interactions [36, 38].

5.7.3 Particle Stability

The colloidal stability of nanoparticles is studied by zeta potential. This ability is an indirect indicator of the charge on the surface [35]. It corresponds to potential difference between the shear surface and the outer Helmholtz plane. Laser Doppler anemometry is the tool used to calculate the potential of the zeta. It is based on the assessment of particle velocity by the shift induced in the interference fringe, which is created by the intersection of two laser beams. The mobility of electrophoresis is then converted into zeta potential. Many colloidal particles have potentially negative zeta values ranging from around −100 to −5 mV. Because of strong electrostatic repulsion, surface charges avoid the agglomeration of polymer nanoparticles dispersions, thus improving nanoparticle stability. The zeta potential may also provide details about the type of material encapsulated or coated on the surface inside the nanocapsule.

5.7.4 Particle Structure

Analysis of structural changes in the free protein sample and protein nanoparticles is vitally important in order to understand the nature of the protein modifications in terms of validation, folding, chemical bonding, etc.

5.7.4.1 X-ray Diffraction

One of the techniques for this purpose is X-ray diffraction (XRD) which is the primary method to examine the structure of crystalline materials, from atomic arrangement to crystallite size and imperfections. XRD also analyzes the production of the nanoparticles' phase composition, crystallite size and shape, lattice distortions and faults, composition variations, orientation, and in situ structure. The XRD pattern is usually obtained by illumination the sample with a source of X-rays (Copper K line) with a wavelength of 1.54 Å and by scanning the diffraction within a certain angle range 2.

5.7.4.2 Fourier Transform Infrared Spectroscopy

Another method to complement XRD is Fourier transform infrared spectroscopy (FTIR). The benefit of FTIR over crystallographic techniques is its capability to provide information about the structural features of proteins in solution with greater spatial and temporal resolution. The sample used for characterization normally consists of lyophilized nanoparticles in minutes. The fundamental theory governing this is that the bonds and bond groups vibrate at characteristic frequencies. A molecule exposed to infrared rays, at wavelengths typical of that molecule, absorbs infraround energy. The FTIR analysis is performed by lighting the sample with a modulated infrared spectroscopy (IR) beam. The sample transmission and reflection of the infrared rays at different wavelengths is converted into an IR absorption map, which is then analyzed and matched in the FTIR library with known signatures of identified material [39].

5.7.4.3 Cellular Uptake and Cytotoxicity

Besides these nanoparticles characterizations, it is important to evaluate their cellular uptake and cytotoxicity, both in vitro and in vivo. The cellular absorption of nanoparticles is calculated by labeling nanoparticles with fluorescent tags such as fluorescence isothiocyanate (FITC) followed by incubating these fluorescence-tagged nanoparticles with cells and visualizing them under confocal laser scanning microscope. Analysis of cytotoxicity is normally performed by incubating nanoparticles with cells and performing 3-(4,5-dimethylthiazol-2-yl)-2, 5-diphenyl tetrazolium bromide (MTT) assays, a common method for assessing the biomaterial toxicity based on mitochondrial activity. The theory underlying this is the reduction of yellow to purple formazan in living cells. Followed by the addition of a solubilizing solution to dissolve the insoluble violet formazan liquid into a color solution (usually either dimethyl sulfoxide, acidified ethanol solution, or a detergent sodium dodecyl sulfate solution in diluted hydrochloric acid). The absorbance of this colored solution is then quantified using a spectrophotometer at a given wavelength (depending on the solvent used). These reductions occur only when enzymes with reductase are involved, and thus conversion is also used as a measure of viable cells [40].

5.7.4.4 Drug Loading and Drug Release

The absorbance of this colored solution is then quantified at a given wavelength (depending on the solvent used) using a spectrophotometer. These reductions only occur when reductase enzymes are involved, and hence conversion is often used as a measure of viable cells. Classical analytical methods such as UV spectroscopy or high-performance liquid chromatography (HPLC) after ultracentrifugation, ultrafiltration, gel filtration, or centrifugal ultrafiltration are the techniques used for this research. The efficiency of encapsulation refers to the ratio of the amount of encapsulated/absorbed drug to the total (theoretical) amount of drug used in respect of the final nanoparticles dispersal drug delivery system. Quantification is achieved with the UV or HPLC spectroscopy. Drug release assays are also close to the drug loading test that is tested for a period of time to analyze drug release mechanisms.

5.8 Applications of Biopolymers

Biopolymers may be used to enhance the efficiency of other biologically active molecules in a product, due to its biocompatible and biodegradable nature [41]. They may also be changed to accommodate different potential applications, like the following:

5.8.1 Biomedical Applications

Due to their biomedical applications, such as those in tissue engineering, pharmaceutical carriers, and medical devices, biopolymer materials have created great interest in recent years. A natural biopolymer, gelatin, has been widely used as an adhesive in medicine for dressing wounds Porous gelatin scaffolds and films were developed as simple porogens with the aid of solvents or gases that allow the scaffolds to carry drugs or nutrients to the wound for healing. Electrospun poly (lactic-*co*-glycolic acid) (PLGA)-based scaffolds have been extensively implemented in biomedical engineering, such as tissue and drug delivery systems. In several areas of tissue engineering, multi walled carbon nanotubes (MWCNT)-incorporated electrospun nanofibers with a high area of volume ratio and porous characteristics have demonstrated possible applications. Biomaterials made from proteins, polysaccharides, and synthetic biopolymers are preferred but lack the desired mechanical properties and stability for medical applications in aqueous environments. Cross-linking enhances the biomaterial properties, but most cross-linkers either cause undesirable changes to the biopolymer structure or contribute to cytotoxicity. Glutaraldehyde, the most commonly used interlinking agent, is difficult to manage and conflicting views on the cytotoxicity of glutaraldehyde-cross-linked materials have been provided [4, 37, 42]

5.8.2 Food Industry

Replacing oil-based packaging materials with bio-based films and containers will offer not only a competitive advantage due to a more sustainable and greener appearance but also some improved technological properties. Biopolymers are commonly used for usable foodstuffs in food coatings, food packaging products, and encapsulation matrices. They have innovative ways to increase product shelf-life while also reducing the total footprint of food packaging-associated carbon. Those bio-based materials are especially useful in food-related applications in three main areas: food packaging, food processing, and food and encapsulation edible films. Other biodegradable polyesters and thermoplastics such as starch, poly-lactic acid (PLA), polyhydroxyalkonates (PHA), and so on are the most commercially viable products in food packaging which can be handled by traditional equipment. These products are already used in the food packaging industry for a variety of monolayer and multilayer applications. The biopolymers of starch and PLA are probably the most desirable forms of biodegradable material. This is due to the quality of their properties and the fact they have become available commercially. Due to its excellent clarity and relatively good water resistance, PLA is of particular interest in food packaging. The challenge for these particular biomaterials is to strengthen their barriers and thermal properties in order to work like polyethylene terephthalate (PET). Other materials derived from agricultural products such as proteins (e.g. zein), polysaccharides (e.g. chitosan), and lipids (e.g. waxes) also have excellent potential as barriers to gas and flavoring. The key disadvantages of these kinds of materials are the inherent high rigidity and the complexity of processing them in traditional equipment. Most biopolymers have a hydrophilic quality that affects their use as high-end products. Moisture absorption allows these materials to be plasticized thereby degrading the barrier properties of such materials [43].

It also used recycled polymers for encapsulation purposes. Encapsulation has been previously identified as a technology designed to protect sensitive substances against adverse environmental influences. The term "microencapsulation" refers to a given method of wrapping solids, liquids, or gasses in small capsules which may release their contents under specific circumstances. These developments are of great concern for the pharmaceutical industry. Owing to their ability to integrate a variety of functional ingredients, the growing interest in edible films and coatings using biopolymers. Plasticizers such as glycerol, acetylated monoglycerides, and polyethylene glycol that are used to alter the mechanical properties of the film or coating make substantial changes to the film's barrier properties. However, the main advantage of the coatings is that they can be used as a method for incorporating natural or chemical active ingredients such as antioxidants and antimicrobials, enzymes, or ingredients that function like probiotics, minerals, and vitamins. With the food these ingredients can be eaten, thus increasing safety, nutritional, and sensory attributes. Edible films may be used as carriers of flavor or aroma besides providing a barrier to the loss of aroma [44].

As an antimicrobial packaging agent, chitosan has shown great potential to preserve food against a large range of microorganisms. The introduction of antimicrobial compounds into edible films or coatings offers an innovative way of enhancing ready-to-eat food quality and shelf life. Lysozyme is one of the most commonly used antimicrobial enzymes in packaging products, as it is an enzyme that naturally occurs. When mixed with plasticizers, biopolymers like amylose have an excellent capacity for forming thin films for various food and packaging applications. Starch has a high sensitivity to relative humidity (RH) due to its hydrophilic nature, and this can be minimized by implementing plasticizers that improve matrix flexibility. But this strategy also has some drawbacks, due to the complex interactions between hydrophilic plasticizers and starch. If the structure of the plasticizer molecule and the polymer matrix is not compatible, an "antiplasticization" phase takes place with increased rigidity of the matrix [43].

5.8.3 Packaging Applications

Currently, some biodegradable polyesters are the most economically viable materials in food packaging and can be handled by traditional equipment. These products are already used in the food packaging industry for a variety of monolayer and multilayer applications. Sustainable biopolymers used in monolayer packaging include starch, PHA, and PLA amongst the most commonly studied thermoplastics. The biopolymers of starch and PLA are probably the most desirable forms of biodegradable material. This is because of the quality of their properties and because they have become available commercially. The challenge for these particular biomaterials is to strengthen their barriers and thermal properties in order to work like PET. Other materials derived from agricultural products such as proteins (e.g. zein), polysaccharides (e.g. chitosan), and lipids (e.g. waxes) also have excellent potential as barriers to gas and flavoring. The key disadvantages of these types of materials are their naturally high rigidity and the difficulty of traditional equipment to process them [45].

The most important parameter to consider for bio-based food-packaging applications is its barrier properties. Hydrophilic polymers generally have poor resistance to moisture which causes water vapor transmission through packaging and thus affects food quality. This results in shorter shelf life, higher prices, and consequently more waste. Another technique to strengthen biopolymers' barrier properties is to incorporate various nanofillers, such as nanoclays, metal oxide nanoparticles, and so on. Polyglycolic acid (PGA) has excellent barrier properties between bioplastics and is thus one of the most promising new barrier polymers commercially available. Now it can produce its counterpart, glycolic acid, via the glyoxylate cycle, a natural metabolic pathway [46].

5.8.4 Water Purification

Healthy drinking water is an effective but basic evolutionary predictor. Nanotechnology has shown promising advances in providing healthy drinking water by means of efficient purification mechanisms. Various nanomaterials have already demonstrated antibacterial and antifungal properties. Developing inexpensive materials that can continuously release these antibacterial materials such as silver nanoparticles into water is an efficient way to provide healthy drinking water for all microbial content. Developing various nanocomposites with functional materials that can scavenge various toxic metals such as arsenic, lead, etc. from water along with the antibacterial agents will result in inexpensive water purifiers that can operate without electricity. The key challenge in this technology is the development of stable materials that can release nanoparticles that continuously resolve the scaling of nanomaterials induced by different complex species present in water.

Chitosan, a relatively new biopolymer, demonstrates superior efficiency where many traditional polymers fail. It is a flexible polymer with applications in the water treatment fields, pharmaceutical, and dietary supplements. Chitosan is used as a flocculant in processes of water treatment and can biodegrade in the atmosphere for weeks or months rather than years. Many aggressive and cheap synthetic flocculants are available compared to chitosan but leave a residual environmental effect. By forming chelates, chitosan extracts the metals from water [47].

Chelation may be a process by which many binding sites within the polymer chain join along with the metal to create a metal cage sort of a frame, to get rid of it from an answer. This property of chitosan and its biodegradability make it an eminent candidate for the treatment of difficult industrial stormwater and wastewater, where traditional methods have failed to reduce contaminant levels. Porous GO-biopolymer gels can effectively extract cationic dyes from wastewater and heavy metal ions. Nanocrystalline metal oxyhydroxide-chitosan granular composite materials prepared by an aqueous route at near room temperature were also effective in water purification. Nanofibre membranes can enhance water filtration without harming the environment. The combination of different nanomaterials with biopolymers will effectively limit the formation of biofilms on the surface of polymers [48].

5.9 The Role of Biopolymers for Sustainable Development

Sustainable development can be defined as that "which meets the needs of the present generation without compromising the ability of future generations to provide their own needs" [49]. The economic and social goals of sustainable development are not difficult to achieve if one can manage natural resources properly. To meet these goals, it is important to maintain a list of issues (or conditions) such as efficient use/consumption of resources, effective recycling, ecological mentality, development (or innovation) of efficient technologies for renewable resources, ecosystem preservation, and social participation [50]. Concerns related to conventional polymers have been increasing because of their nonrenewability. Such polymers also are known to be responsible for releasing carbon into the atmosphere. The carbon footprint of plastic is around 6 kg CO_2 per kg plastic throughout the life cycle (i.e., production to incineration) [51]. Furthermore, the nondegradability of polymers creates environmental pollution that seriously affects ecosystems, wild and marine life, and/or humans. As biopolymers are derived from renewable sources, they can be considered as sustainable alternatives for conventional plastics. Their neutral carbon lifecycle also helps decrease the release of carbon dioxide into the atmosphere, reducing the "human footprint" on the environment. Biopolymers have low carbon footprint, with less cradle-to-plant-gate greenhouse gas emissions than fossil fuel. As for example, cradle-to-plant-gate greenhouse gas emissions of Polyamide 410 are nearly zero [52]. As the amount of carbon dioxide generated during its manufacture is offset by the amount of carbon dioxide absorbed during the growth of its source castor bean plant. According to the Intergovernmental Panel on Climate Change (IPCC), biopolymers could make significant contribution by removing up to 20% of the CO_2 to limit global warming [53].

Apart from the known sources of biopolymers, scientists around the world are researching, developing, and even in some cases commercializing new products from a variety of renewable sources and waste materials. Newlight Technology, USA collected concentrated methane and air from landfills, farms, and industrial sources for use in the production of AirCarbon. Newlight's biocatalyst separates carbon and hydrogen from methane and oxygen from air in the conversion reactor [54].

The isolated carbon, hydrogen, and oxygen molecules are then linked together to make a long chain thermopolymer known as AirCarbon. AirCarbon is compatible with traditional polyethylene so that it can be used in cast film, blown film, fiber spinning, thermoforming, etc. Unorthodox raw materials from animal sources (e.g. wool, feathers, silk, and horns) also are used as fillers for polymers. Poultry feathers consisting of keratin protein with several amino acids can be used to produce biopolymer. Such polymers can be used to make plastic plates and cups. Biodegradable thermoplastic has been produced from bakery waste (bread crust, stale bread, and sponge cake) by adding suitable additives like natural antioxidants and plasticizers [55].

5.10 Conclusion

Increasing awareness of sustainable development has led researchers to think about natural and biodegradable polymers for different applications to replace synthetic polymers. Nanotechnology has already emerged as the technology of the future and can allow major improvements to the current technology. Under restricted conditions, the bacteria generate various biopolymers for storing carbon and energy resources. Incorporating different nanofillers into these biopolymers will improve their mechanical and barrier properties which will enhance their different applications. Along with the biodegradability of these materials, the biocompatibility of these materials promotes the use of them in everyday applications.

References

1 Mohan Bhagyaraj, S., Oluwafemi, O., Kalarikkal, N. et al. (2016). Biopolymers – application in nanoscience and nanotechnology. *Recent Advances in Biopolymers*.

2 Hernández, N., Williams, R.C., and Cochran, E.W. (2014). The battle for the 'green' polymer. Different approaches for biopolymer synthesis: bioadvantaged vs. bioreplacement. *Organic & Biomolecular Chemistry* 12 (18): 2834–2849.

3 Mohan, S., Oluwafemi, O.S., Kalarikkal, N. et al. (2016). Biopolymers – application in nanoscience and nanotechnology. *Recent Advances in Biopolymers*.

4 Sundar, S., Kundu, J., and Kundu, S.C. (2010). Biopolymeric nanoparticles. *Science and Technology of Advanced Materials* 11 (1): 14104.

5 Snell, K.D., Singh, V., and Brumbley, S.M. (2015). Production of novel biopolymers in plants: recent technological advances and future prospects. *Current Opinion in Biotechnology* 32: 68–75.

6 Scheffel, U., Rhodes, B.A., Natarajan, T.K., and Wagner, H.N.J. (1972). Albumin microspheres for study of the reticuloendothelial system. *Journal of Nuclear Medicine* 13 (7): 498–503.

7 Couvreur, P., Kante, B., Roland, M. et al. (1979). Polycyanoacrylate nanocapsules as potential lysosomotropic carriers: preparation, morphological and sorptive properties. *The Journal of Pharmacy and Pharmacology* 31 (5): 331–332.

8 Kreuter, J. (1995). Nanoparticles as adjuvants for vaccines BT vaccine design: the subunit and adjuvant approach. In: *Pharmaceutical Biotechnology*, vol. 6 (eds. M.F. Powell and M.J. Newman), 463–472. Boston, MA: Springer US.

9 Landfester, K. (2009). Miniemulsion polymerization and the structure of polymer and hybrid nanoparticles. *Angewandte Chemie, International Edition* 48 (25): 4488–4507.

10 Kamaly, N., Yameen, B., Wu, J., and Farokhzad, O.C. (2016). Degradable controlled-release polymers and polymeric nanoparticles: mechanisms of controlling drug release. *Chemical Reviews* 116 (4): 2602–2663.

11 Pinto Reis, C., Neufeld, R.J., Ribeiro, A.J., and Veiga, F. (2006). Nanoencapsulation I. Methods for preparation of drug-loaded polymeric nanoparticles. *Nanomedicine: Nanotechnology, Biology and Medicine* 2 (1): 8–21.

12 Jahanshahi, M. and Babaei, Z. (2008). Protein nanoparticle: a unique system as drug delivery vehicles. *African Journal of Biotechnology* 7: 4926–4934.

13 Rao, J.P. and Geckeler, K.E. (2011). Polymer nanoparticles: preparation techniques and size-control parameters. *Progress in Polymer Science* 36 (7): 887–913.

14 Rubino, O.P., Kowalsky, R., and Swarbrick, J. (1993). Albumin microspheres as a drug delivery system: relation among turbidity ratio, degree of cross-linking, and drug release. *Pharmaceutical Research* 10 (7): 1059–1065.

15 Lynn, A.K., Yannas, I.V., and Bonfield, W. (2004). Antigenicity and immunogenicity of collagen. *Journal of Biomedical Materials Research Part B: Applied Biomaterials* 71B (2): 343–354.

16 Kasaai, M.R. (2020). Chapter 7 - Biopolymer-based nanomaterials for food, nutrition, and healthcare sectors: an overview on their properties, functions, and applications. In: *Handbook of functionalized Nanomaterials for Industrial Applications* (ed. I.A.M. Hussain), 167–184. Elsevier.

17 Bajpai, A.K. and Choubey, J. (2006). Design of gelatin nanoparticles as swelling controlled delivery system for chloroquine phosphate. *Journal of Materials Science. Materials in Medicine* 17 (4): 345–358.

18 Nguyen, T.P., Nguyen, Q.V., Nyugen, V.-H. et al. (2019). Silk fibroin-based biomaterials for biomedical applications: a review. *Polymers (Basel)* 11 (12): 1933.

19 Nayak, S. and Kundu, S. (2016). Silk protein sericin: promising biopolymer for biological and biomedical applications. In: *Biomaterials from Nature for Advanced Devices and Therapies*, 142–158. Wiley-society for Biomaterials.

20 Saha, S., Arshad, M., Zubair, M., and Ullah, A. (2019). *Keratin as a Biopolymer*, 163–185. Cham: Springer.

21 Liu, Z., Jiao, Y., Wang, Y. et al. (2008). Polysaccharides-based nanoparticles as drug delivery systems. *Advanced Drug Delivery Reviews* 60 (15): 1650–1662.

22 Hu, K. and Mcclements, D. (2015). Fabrication of biopolymer nanoparticles by antisolvent precipitation and electrostatic deposition: Zein-alginate core/shell nanoparticles. *Food Hydrocolloids* 44: 101–108.

23 Zhao, D., Yu, S., Sun, B. et al. (2018). Biomedical applications of chitosan and its derivative nanoparticles. *Polymers (Basel)* 10 (4): 1.

24 Casariego, A., Souza, B.W.S., Cerqueria, M.A. et al. (2009). Chitosan/clay films' properties as affected by biopolymer and clay micro/nanoparticles' concentrations. *Food Hydrocolloids* 23 (7): 1895–1902.

25 Negm, N., Hefni, H., and Abd-Elaal, A.A. (2017). *Assessment of Agricultural Wastes as Biosorbents for Heavy Metal Ions Removal from Wastewater*, 465–491. CRC Press.

26 Bouchemal, K., Briançon, S., Perrier, E., and Fessi, H. (2004). Nano-emulsion formulation using spontaneous emulsification: solvent, oil and surfactant optimisation. *International Journal of Pharmaceutics* 280 (1–2): 241–251.

27 Weber, C., Kreuter, J., and Langer, K. (2000). Desolvation process and surface characteristics of HSA-nanoparticles. *International Journal of Pharmaceutics* 196 (2): 197–200.

28 Birrenbach, G. and Speiser, P.P. (1976). Polymerized micelles and their use as adjuvants in immunology. *Journal of Pharmaceutical Sciences* 65 (12): 1763–1766.

29 Joshi, S. Can nanotechnology improve the sustainability of biobased products ? The case of layered silicate biopolymer. *Journal of Environmental Management* 12 (3).

30 Sinha, S., Yamada, K., Okamoto, M. et al. (2003). New polylactide / layered silicate nanocomposites. 5. Designing of materials with desired properties. 44: 6633–6646.

31 Potts, J.R., Dreyer, D.R., Bielawski, C.W., and Ruoff, R.S. (2011). Graphene-based polymer nanocomposites. *Polymer (Guildf)* 52 (1): 5–25.

32 Paul, D.R. and Robeson, L.M. (2008). Polymer nanotechnology: nanocomposites. *Polymer (Guildf)* 49 (15): 3187–3204.

33 Sheregii, E. (2015). *High Resolution Infrared Spectroscopy of Phonons in the II-VI Alloys — The Temperature Dependencies Study*, 63–86. INTECH.

34 Redhead, H.M., Davis, S.S., and Illum, L. (2001). Drug delivery in poly(lactide-co-glycolide) nanoparticles surface modified with poloxamer 407 and poloxamine 908: in vitro characterisation and in vivo evaluation. *Journal of Controlled Release* 70 (3): 353–363.

35 Tomoda, B.T., Yassue, P.H., Lopes, J.V. et al. (2020). Chapter 3 - Characterization of biopolymer membranes and films: physicochemical, mechanical, barrier, and biological properties. In: *Biopolymer Membranes and Films* (eds. M.A. de Moraes, C.F. da Silva and F. Vieira), 67–95. Elsevier.

36 Al, G., Aydemir, D., Kaygin, B. et al. (2018). Preparation and characterization of biopolymer nanocomposites from cellulose nanofibrils and nanoclays. *Journal of Composite Materials* 52: 689–700.

37 Bhagyaraj, S.M., Oluwafemi, O.S., Kalarikkal, N., and Songca, S. (2018). *Biopolymers – Application in Nanoscience and World's largest Science, Technology & Medicine*. Open Access Books no. March 2016.

38 Wokadala, O.C., Ray, S.S., Bandyopadhyay, J. et al. (2015). Morphology, thermal properties and crystallization kinetics of ternary blends of the polylactide and starch biopolymers and nanoclay: the role of nanoclay hydrophobicity. *Polymer (Guildf)* 71: 82–92.

39 Dillen, K., Bozdag, S., Vandervoort, J., and Ludwig, A. (2007). Evaluation of the physicochemical characteristics and activity of various kinds of ciprofloxacin HCl-loaded cationic nanoparticles. *Journal of Drug Delivery Science and Technology* 17 (1): 49–56.

40 Mosmann, T. (1983). Rapid colorimetric assay for cellular growth and survival: application to proliferation and cytotoxicity assays. *Journal of Immunological Methods* 65 (1–2): 55–63.

41 Rydzkowski, T. and Thakur, V.K. Biopolymers for biomedical and pharmaceutical applications : recent advances and overview of alginate electrospinning. *Nanomaterials*: 7–9.

42 Reddy, N., Reddy, R., and Jiang, Q. (2015). Crosslinking biopolymers for biomedical applications. *Trends in Biotechnology* 33 (6): 362–369.

43 Vroman, I. and Tighzert, L. (2009). Biodegradable polymers. *Materials (Basel)* 2 (2): 307–344.

44 Shit, S.C. and Shah, P.M. (2014). Edible polymers: challenges and opportunities. *Journal of Polymers* 2014.

45 Majeed, K. et al. (2013). Potential materials for food packaging from nanoclay / natural fibres filled hybrid composites. *Materials & Design* 46: 391–410.

46 Siracusa, V., Blanco, I., Romani, S. et al. (2012). Poly(lactic acid)-modified films for food packaging application: physical, mechanical, and barrier behavior. *Journal of Applied Polymer Science* 125 (S2): E390–E401.

47 Cheng, C., Lei, B., He, A. et al. (2013). Toward 3D graphene oxide gels based adsorbents for high-efficient water treatment via the promotion of biopolymers. *Journal of Hazardous Materials* 263: 467–478.

48 Sankar, M.U. et al. (2013). Biopolymer-reinforced synthetic granular nanocomposites for affordable point-of-use water purification. *Proceedings of the National Academy of Sciences of the United States of America* 110 (21): 8459–8464.

49 Udayakumar, M., Kollar, M., Nemeth, Z. et al. (2020). Temperature and time dependence of the solvent-induced crystallization of poly(l-lactide). *Polymers (Basel)* 12 (5): 1065.

50 Höglund, A., Odelius, K., and Albertsson, A.-C. (2012). Crucial differences in the hydrolytic degradation between industrial polylactide and laboratory-scale poly(l-lactide). *ACS Applied Materials & Interfaces* 4 (5): 2788–2793.

51 Dormer, A., Finn, D.P., Ward, P., and Cullen, J. (2013). Carbon footprint analysis in plastics manufacturing. *Journal of Cleaner Production* 51: 133–141.

52 Ashby, M.F. (2016). Chapter 3 - Assessing sustainable developments: the steps. In: *Materials and Sustainable Development* (ed. S.D. Ashby), 39–54. Boston: Butterworth-Heinemann.

53 Ashby, M.F. (2016). Chapter 6 - Corporate sustainability and materials. In: *Materials and Sustainable Development* (ed. S.D. Ashby), 101–110. Boston: Butterworth-Heinemann.

54 Popescu, P.A., Popa, M.E., Mitelut, A. et al. (2017). Screening of different methods to establish the biodegradability of packaging materials - a useful tool in environmental risk assessment approach. *AES Bioflux* 9 (1): 30–36.

55 Brodard, P., Dabros, M., Marti, R. et al. (2019). Materials science at Swiss Universities of applied sciences. *CHIMIA International Journal for Chemistry* 73: 645–655.

6

Toward Eco-friendly Nanotechnology-based Polymers for Drug Delivery Applications

Prachi Pandey[1], Jayvadan Patel[2], and Samarth Kumar[3]

[1] Babaria Institute of Pharmacy, BITS Edu Campus, Vadodara, Gujarat, India
[2] Nootan Pharmacy College, Sankalchand Patel University, Visnagar, Gujarat, India
[3] Formulation R & D, Sun Pharmaceutical Industries Ltd, Vadodara, Gujarat, India

6.1 Introduction

In recent years, pharmaceutical research has been focused on the development of nanotechnology-based drug delivery systems applicable in the field of drug delivery. Currently, the utilization of biopolymers as nanoparticles (NPs) represents a viable alternative with huge potential for the targeted distribution of drugs or biological macromolecules in the body [1]. In recent years, it was realized by the scientific community that there is a need to utilize ecofriendly nanotechnology-based polymers for formulating drug delivery systems. Their promising role and applications in the pharmaceutical research may be attributable not only to the fact that they are biodegradable and toxicologically harmless materials of low cost and relative abundance compared to their and synthetic counter parts [1, 2] but also because natural resources are renewable, and if cultivated or harvested in a sustainable manner, they can provide a constant supply of raw materials.

The use of ecofriendly biodegradable polymers has been witnessed in several studies in the past. Biopolymer-based NPs can be used efficaciously to provide bioactive molecules for *in vivo* and *in vitro* applications. Usually, nanotechnology-based drug delivery systems are an extended field of research characterized by the use of materials with sizes ranging between 1 and 1000 nm [3]. Biopolymers in pharmaceuticals are currently being used as polymeric drugs, polymer–drug conjugates, polymer–protein conjugates, polymeric micelles, and polyplexes [4]. These systems allow active principles, peptides, and proteins as well as genes to be delivered through localized release in the targeted tissues [4, 5]. The use of nanometric systems allows them to easily permeate through the cells and, therefore, leads to the targeted distribution of the agent to be delivered. The therapeutic agents after administration are usually distributed in the body by virtue of its physicochemical characteristics, but mostly through systemic circulation. To reach an effective drug concentration at the site of action, administration in high and repeated doses is necessary with some possible side effects. The utilization of biodegradable NPs is toward drug targeting, that is, the selective delivery of a therapeutic agent to its site of action independently of the mode of administration or body compartment [6]. The active therapeutic agents can be dispersed, encapsulated, or adsorbed on the surface of the NPs. Nanobiopolymers are also being utilized in the field of enzyme replacement therapy (ERT). The prospects of using NPs constituted by biocompatible and biodegradable polymers to deliver enzymes in tissues represent an enormous advantage by combating a series of ERT problems.

The use of biodegradable NPs as molecule transporters is one of the most promising strategies for controlled-release systems (CRS). The fundamental requirement for a biomaterial to be used in this sense is its biocompatibility, that is, the ability to be metabolized without any harmful effects. Systems of this type must be able to ensure characteristics such as their ability to cross the body's anatomical barriers, typically the blood–brain barrier (BBB) or the ophthalmic barrier, controlling the concentration of the drug over time and the release of the active molecule at the site of action [7].

The release of the drug from nanoparticles occurs through diffusion or with the dissociation of the NPs into monomeric molecules. This dissociation can be triggered

Sustainable Nanotechnology: Strategies, Products, and Applications, First Edition.
Edited by Yashwant V. Pathak, Govindan Parayil, and Jayvadan K. Patel.

by local variations (particularly pH or temperature) or external stimuli (e.g. radiation and ultrasound). During the process, the loaded NPs prevent the untimely leakage of the drug, thus release it into the targeted cellular compartments, and then the produced products of degradation (the monomeric units) can be easily eliminated from the organism. pH-switchable NPs have been developed in some studies to match the pH conditions that the NPs can encounter in the human body. The drug-loaded aggregates must be stable during transportation in the blood plasma at a nearly neutral pH. It also needs to turn on the drug release inside the lysosomal and endosomal compartments of a cell or within the interstitial space of solid tumors, where the pH is close to 5 [8, 9].

These environmentally responsive NPs have therefore earned the epithet of "smart" drug delivery systems, owing to their ability to modulate and control drug release inside the targeted cells [10–12].

The nanoscale dimensions of nanoparticles make them extremely promising, as they can be administered in various ways and in different regions of the body from which they can enter into the target cells and tissues. However, by means of functionalization processes, it is possible to evade the endosome–exosome system and to deliver the drug directly to the site of action [13]. There are numerous advantages derived from the use of such systems. It is possible to enhance the effect of the drug in the affected site and also minimize the side effects associated with the classic pharmacological treatment, such as the systemic toxicity caused by the drug. Furthermore, hydrophobic molecules within this type of carrier system are increasingly stabilized [14, 15]. The drug or molecule can be encapsulated in a nanoparticle system obtaining, in this case, nanocapsules or dispersion in the polymer matrix to form nanospheres [16]. It is possible to create NPs of different sizes, but in general those with a diameter between 1 and 100 nm have better pharmacokinetic properties, while smaller NPs are eliminated through the renal filtration system and larger NPs are absorbed by the mononuclear phagocytic system (MPS) present mainly in the liver and spleen [17].

In recent research, the development of nanoparticles based on biocompatible and biodegradable polymers such as polylactic acid (PLA) polyesters and their copolymers with glycolic acid (PLGA, poly lactic-*co*-glycolic acid) has increased. Despite being synthetic, these polymers can be degraded in the body into oligomers and monomers that are further eliminated through the normal metabolic pathways, such as the Krebs cycle [18–21].

When polymeric nanoparticles are administered intravenously, they are often subjected to the opsonization reaction, which may result in their phagocytosis by the monocytes–macrophages. To overcome this problem, the particles can be made stealth with the use of hydrophilic polymers such as polyethylene glycol (PEG), which prevents the recognition of the NPs by the reticuloendothelial system.

6.2 Eco-friendly Biodegradable Polymers and Their Nanotechnology-based Drug Delivery Applications

Biodegradable polymers are of utmost importance in the field of drug delivery and nanomedicine. The therapeutic biodegradable polymer must meet some vital criteria such as: (i) must be nontoxic in order to eliminate foreign body response; (ii) the time for the degradation of polymer should be proportional to the time required for therapy; (iii) the products obtained after biodegradation must not be cytotoxic and should be readily eliminated from the body; (iv) the material must undergo easy processing in order to alter the mechanical properties for the development of desired drug delivery system; (v) can be easily sterilized; and (vi) must have acceptable shelf life. The major advantage of a biodegradable drug delivery system is its ability as the drug carrier to target the drug release to a specific site in the body and then degrade into nontoxic end products that can be easily eliminated from the body through natural metabolic pathways [20–24]. Such biodegradable polymers slowly degrade into smaller components, and there is controlled ability to release a drug. The drugs slowly get released as polymer degrades. For example, PLA, PLGA, and poly(caprolactone), all of which are biodegradable, have been earlier utilized to carry anticancer drugs [22–25]. Encapsulating the therapeutic moiety in a polymer shell and adding targeting agents can decrease the toxicity of the drug to healthy cells and nontarget organs. Because of their excellent safety profile, good biocompatibility, low levels of immunogenicity and toxicity, and the predictable rate of biodegradation *in vivo*, these polymers have been approved by the US Food and Drug Administration (FDA) and European Medicines Agency (EMA) as effective carriers for drug delivery in humans [26–28].

Besides that, there are several ecofriendly biodegradable polymers of natural origin from the proteins category such as gelatin, collagen, and silk, and from the polysaccharides category such as hyaluronic acid, alginates, agarose, and chitosan. In this chapter, the utilization of eco-friendly biodegradable polymers and their nanotechnology-based drug delivery applications have been discussed.

6.2.1 Polylactic Acid (PLA)

PLA is a thermoplastic-biodegradable polymer produced synthetically by polymerization of lactic acid monomers or cyclic lactide dimmers. Lactic acid is produced by fermentation of natural carbohydrates, for example maize and wheat, or waste products from the agricultural or food industry [27] (Figure 6.1).

The final crystallinity and mechanical properties of the polymer depend on the stereochemistry of the polymer backbone. PLA is degraded by hydrolysis of the backbone esters of the polymer [28]. The ester bonds are broken randomly, so that the PLA chains in the material become shorter and shorter until the monomers of lactic acid start to become loose and the material gets dissolved. This process is called bulk degradation. PLA does not degrade by microbial attack. PLA has a number of biomedical applications, such as sutures, stents, dialysis media, and drug delivery devices [29].

6.2.2 Polyglycolic acid (PGA)

PGA is commonly obtained by ring-opening polymerization of the cyclic diester of glycolic acid, glycolide. PGA is a hard and crystalline polymer with a melting point of 225 °C and a glass transition temperature, T_g, of 36 °C. As compared to the closely related polyesters such as PLA, PGA is insoluble in most common polymer solvents [28, 29]. PGA has excellent fiber-forming properties and was commercially introduced in 1970 as the first synthetic absorbable suture under the trade name Dexon™. Due to the properties such as low solubility and high melting point, the use of PGA is limited for drug delivery applications, since it cannot be made into films, rods, capsules, or microspheres using solvent or melt techniques [29, 30] (Figure 6.2).

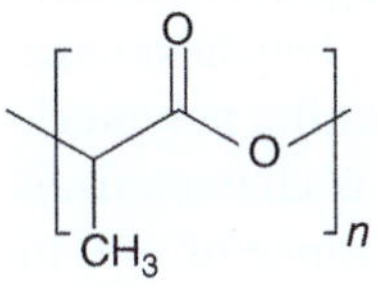

Figure 6.1 Structure of polylactic acid.

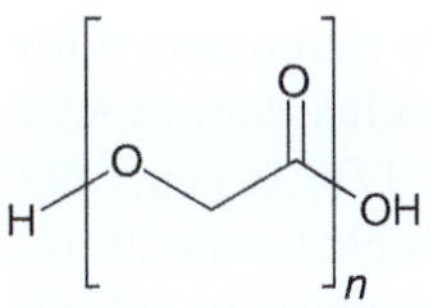

Figure 6.2 Structure of polyglycolic acid.

6.2.3 Polyhydroxybutyrate (PHB)

PHB is a biodegradable polymer, which is mostly present in all living organisms. Several bacteria produce PHB in large quantities as their storage material. It is nontoxic and totally biodegradable. This polymer is basically a carbon assimilation product from glucose or starch and is utilized by microorganisms as a form of energy molecule which needs to be metabolized when other common energy sources are not available [28–30]. PHB and its copolymers have drawn much attention in research as they are produced biosynthetically from renewable resources. Microcapsules from PHB have been prepared using various techniques and investigated for the release characteristics of bovine serum albumin. PHB has also been tested in earlier studies as a suitable matrix for drug delivery of veterinary medicine [29, 30] (Figure 6.3).

6.2.4 Poly Lactic-*co*-glycolic Acid (PLGA)

Since the past two decades, PLGA has been considered as the most promising polymeric candidate used to develop drug delivery devices. PLGA is biocompatible and biodegradable with a broad range of erosion periods; thus, it has modifiable mechanical properties, and most importantly, it is an FDA approved polymer for therapeutic applications. Particularly, PLGA has been extensively researched for the development of devices for controlled delivery of anticancer drug and for small molecule drugs such as proteins and peptides [30] (Figure 6.4).

In PLGA, the ratio of glycolide to lactide at different compositions allows control of the degree of crystallinity of the polymers. When the crystalline PGA is copolymerized with PLA, the degree of crystallinity is reduced, and as a result, this leads to increases in the rates of hydration and

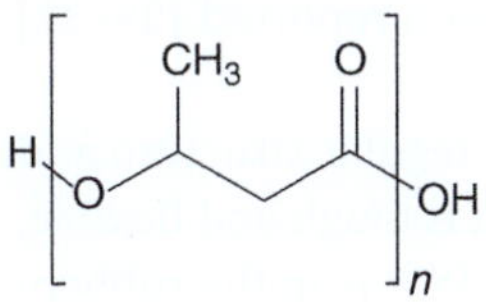

Figure 6.3 Structure of polyhydroxybutyrate.

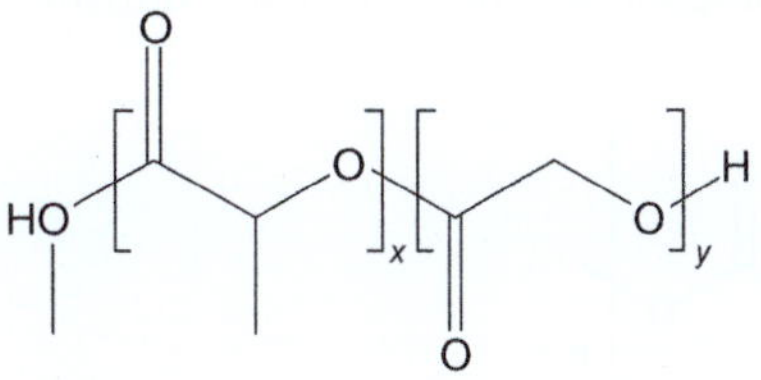

Figure 6.4 Structure of poly lactide-*co*-glycolide.

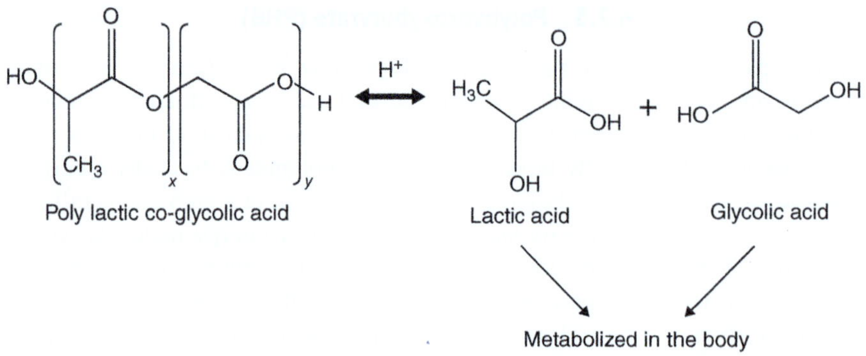

Figure 6.5 Hydrolytic degradation of PLGA.

hydrolysis [30, 31]. It can therefore be concluded that the degradation time of the copolymer is related to the ratio of monomers used in the synthesis. It is observed that, as the content of glycolide becomes higher, the rate of degradation also becomes faster. However, an exception is the 50 : 50 ratio of PGA:PLA, which exhibits the fastest degradation of PLGA, making it suitable for drug delivery applications [30–32] (Figure 6.5).

In water, PLGA biodegrades by hydrolysis of its ester linkages, resulting in biodegradable components. The biodegradability of PLGA is based on the hydrolytic degradation through the process of de-esterification of the polymers to generate the lactic and glycolic acid monomeric components, which are metabolized and then removed by the body by natural pathways (such as the Krebs cycle) [31, 32].

6.2.5 Poly-e-caprolactone (PCL)

PCL is obtained by the ring-opening polymerization of the six-membered lactone, e-caprolactone (e-CL). Anionic, cationic, coordination, or radical polymerization routes are all applicable. Recently, enzymatic catalyzed polymerization of e-CL has been reported [28–32] (Figure 6.6).

PCL crystallizes readily due to the regular structure and has a melting temperature of 61 °C. It is tough and flexible. The T_g of PCL is low (−60 °C). Thus, PCL is in the rubbery state and exhibits high permeability to low molecular species at body temperature. These properties, combined with documented biocompatibility, make PCL a promising candidate for controlled-release applications [32]. PCL degradation proceeds through hydrolysis of backbone ester bonds as well as by enzymatic attack. Hence, PCL degrades under a range of conditions, biotically in soil, lake waters, sewage sludge, *in vivo*, and in compost, and abiotically in phosphate buffer solutions. Hydrolysis of PCL yields 6-hydroxycaproic acid, an intermediate of the w-oxidation, which enters the citric acid cycle and is completely metabolized [30–32].

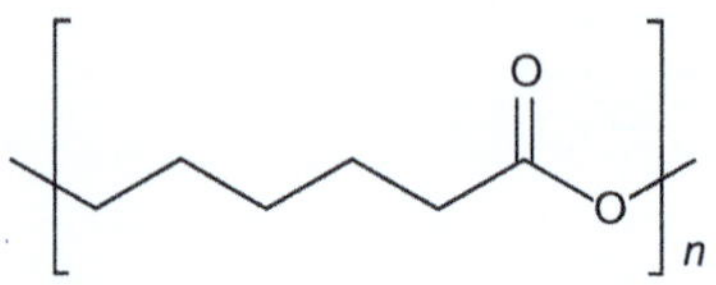

Figure 6.6 Structure of poly-e-caprolactone.

Hydrolysis, however, proceeds by homogeneous erosion at a much slower rate than PLA and PLGA. Hydrolysis of PCL is faster at basic pH and higher temperatures. PCL hydrolyzes slowly compared to APLA and PLGA; it is most suitable for long-term drug delivery. Capronor®, a one-year contraceptive, represents such a system [29–32].

6.2.6 Polydioxanone (PDS)

Although biodegradable polylactides and glycolides have been used to develop versatile resorbable multifilament structures, there is growing research involving the development of materials that form monofilament sutures. Multifilament sutures have a higher risk of infection associated with their use and cause a greater amount of friction when penetrating tissues. Polydioxanone (referred to as PDS) is made by a ring-opening polymerization of the *p*-dioxanone monomer. It is characterized by a glass transition temperature in the range of −10 to 0 °C and a degree of crystallinity of about 55%. Materials prepared with PDS show enhanced flexibility due to the presence of ether oxygen within the backbone of the polymer chain [29–33].

When used *in vivo*, it degrades into monomers with low toxicity and also has a lower modulus than PLA or PGA. For the production of sutures, PDS is generally extruded into fibers at the lowest possible temperature, in order to avoid its spontaneous depolymerization back to the monomer [30–34].

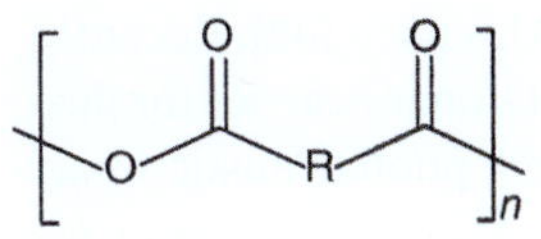

Figure 6.7 Structure of polyanhydride.

6.2.7 Polyanhydrides

Polyanhydrides are a class of biodegradable polymers characterized by anhydride bonds that connect the repeat units of the polymer backbone chain. Polyanhydride esters are polymeric compounds consisting of salicylic acid moieties bridged by linker structures [29–33] (Figure 6.7).

In the presence of water, both bonds may degrade hydrolytically, releasing salicylic acid (B) and sebacic acid (C). Salicylic acid (B) is the active form of aspirin, an anti-inflammatory agent, and sebacic acid is currently used in drug delivery system. *In vivo* mice studies have indicated that this polymer assists in wound healing and that it promotes bone growth. The release of salicylic acid (B) via bond hydrolysis opens up a variety of possibilities for creating drug delivery systems. Potential applications include treatment of inflammatory bowel disease, dental implants, and tissue scaffolding [33, 34].

6.2.8 Applications in PLGA-based Polymers in Drug Delivery Systems

Due to the biodegradation of PLGA due to hydrolysis, several studies have been conducted utilizing the potential of PLGA in altering the drug release and drug targeting and witnessed the therapeutic potential as drug carriers.

Nonsteroidal anti-inflammatory drugs, e.g. diflunisal and diclofenac sodium, have been incorporated into PLGA microspheres and investigated for the treatment of rheumatoid arthritis (RA), osteoarthritis, and related diseases. The encapsulation of biomacromolecules, e.g. proteins and vaccines, into polymeric microspheres presents a formidable problem because of the delicacy of these agents; bioactivity might be lost during preparation, and the release may be poor due to adsorption and/or aggregation. For instance, the release of recombinant human interferon-g from PLGA microspheres was incomplete, and the instability of the system limited its use to seven days or less. Similarly, incomplete release of lysozyme, recombinant human growth hormone, and a nerve growth factor from PLGA microspheres was reported. Hence, much effort has been spent in evaluating PLGA delivery systems, with special regard to microsphere preparation, protein stability, and release characteristics. Proteins on which studies have been done include bovine serum albumin, lysozyme, transferrin, and trypsin. Several peptides, including vapreotide and rismorelin porcine, have been successfully incorporated and released from PLGA microspheres. Systems for the controlled release of antigens have a great potential as vaccine adjuvants. Recently, several studies of controlled-release systems for DNA have been presented. DNA of different sizes has successfully been incorporated into PLGA microspheres, but the loss of DNA integrity and activity still remains an important issue to be solved for these systems.

Their physicochemical and mechanical properties can be modified through the selection of the polymer molecular weight, capability of copolymerization, and their functional properties. PEG is the most popular hydrophilic polymer for surface modification of both hydrophobic polymers PLA and PLGA to form an amphiphilic block copolymer [22, 25]. Their applications in formulating drug delivery systems are mainly focused on involving nanoparticles, micelles, and hydrogels. Poly(ethylene glycol)–poly(lactic-*co*-glycolide) (PEG–*b*-PLGA) diblock copolymer micelles represent one of the potential platforms for drug delivery, where the hydrophobic PLGA core can efficiently encapsulate many therapeutic agents, while the hydrophilic PEG shell prevents the adsorption of proteins and phagocytes, thus extending the circulation time in the blood [23]. In recent research works, PEG–PLGA diblock copolymer micelles have been tested extensively in humans for the incorporation and controlled drug delivery of small molecules and many hydrophobic anticancer compounds [23–25]. Recent researches also present evidence for the development of PLGA nanocarriers for the delivery of therapeutic biomacromolecules which can maintain their colloidal stability and can maximize their loading efficiency even in the harsh physiological environment condition of the diseased tissues [24–26].

The process of chemical conjugation of the PEG–PLGA copolymer facilitates a high drug loading, characterized by an induced localization of the drug in the inner hydrophobic chains.

In a recent study, doxorubicin (DOX)-conjugated PLGA–PEG micellar nanocarriers with a higher DOX loading showed a more sustained drug release behavior compared to physically incorporated DOX in PEG–PLGA micelles [35]. However, up to 50% drug release from conjugated DOX–PLGA–PEG micelles was obtained over two weeks, whereas the total drug release from physically entrapped PEG–PLGA micelles took only three days.

The PEG–PLGA nanocarriers have been reported to encapsulate proteins and peptide drugs, such as insulin, calcitonin, and DNA, in several studies [34]. Also, the suitable combination of the concept of imaging and the use of functional nanoparticles has enabled concurrent diagnosis and therapy of disease-affected tissue through the development of theranostic nanocarriers [36].

Recently, a PLGA–PEG–folate (FA) theranostic system was combined with dual imaging tracers, namely, near-infrared and ^{19}F magnetic resonance imaging with the chemotherapeutic agent doxorubicin DOX [36]. The observations of *in vitro* cytotoxicity assay revealed that FA-targeted PLGA–PEG nanoparticles can kill cancer cells more efficiently than non-FA-conjugated particles [35].

Various preliminary animal studies have showed the great potential of these PLA and PLGA-based nanocarriers in the treatment of various diseases including diabetes, cancer, cardiac disorder, bacterial and viral infection, autoimmune diseases, and cartilage damage [20–24].

In a recent study, a pH-dependent aptamer-based functionalized PLGA nanocarrier system has been reported to enhance anticancer activity of doxorubicin in human lung cancer cells, with reduced toxicity to healthy cells. Also, a superparamagnetic iron oxide-encapsulated nanocarrier for docetaxel has demonstrated favorable pharmacokinetic properties and a higher degree of drug uptake in breast cancer cells.

Since Allison and Honjo were awarded the 2018 Nobel Prize in Physiology and Medicine "for their discovery of cancer therapy by inhibition of negative immune regulation," the field of nanomedical research into cancer immunotherapy has received significant attention. This approach addresses the utilization of pharmaceutical agents to activate a patient's immune system to fight cancers in contrast to traditional chemotherapy which involves directly affecting the cancer cells.

More recently, a sustained and controlled-release PLGA-based nanosystem was formulated to activate the antitumor immune response in mice with melanoma and colon cancer [37]. A PLGA system containing an immune adjuvant together with an enzyme was found to enhance the efficacy of radiation therapy, thus enabling the feasibility of combination of immunotherapy and targeted radiotherapy in BALB mice [38, 39]. Polylactide-*co*-glycolide has applications for tumor diagnostics also as it is able to deliver imaging agents to cancer cells with specificity and controlled biodistribution. Advances in nanotheranostics, which is the incorporation of imaging and therapeutic agents in one nanocarrier, have shown prospects for real-time imaging throughout a patient's treatment regimen [40]. A novel theranostic PLGA nanocarrier with a near-infrared imaging agent, further attached with gold nanoparticles, has been prepared that showed increased activity and photodynamic properties in tumor-grafted BALB mice [41]. In another study, targeted PLGA-based nanobubble was designed with an ultrasound contrast agent which demonstrated specificity and imaging capabilities for breast cancer in BALB mice [42]. Recently, an image-guided photothermal PLGA nanocarrier for doxorubicin showed scope for real-time photoacoustic imaging in tumor-bearing mice [43]. Also, a near infra-red dye-loaded PEGylated PLGA nanocarrier was capable of providing information on the circulation and distribution of the nanoparticles in mice [44].

The delivery of antiretro viral drugs suffers from limitations of conventional drug delivery, and therefore, biomaterials with low toxicity such as PLGA-based nanocarriers have enormous potential for applications in formulations to treat HIV. In a research, mannosylated PLGA nanoparticle carriers have shown hope for targeted delivery of antiretro viral drugs to the brain [45]. It was observed that the use of microfluidics technology can enable the novel synthesis of efiravine-loaded PLGA nanoparticles [46]. A recent study reported on PLGA-based nanocarrier for the combination of the antiretro virals griffithsin and dapivirine displayed a long-acting treatment profile [47], and in a proof-of-concept study, long-acting bictegravir-encapsulated PLGA nanocarrier was found to exhibit further scope [48]. The specific delivery of anti-inflammatory agents to the target site of action could potentially increase the therapeutic concentration of drugs in the inflamed tissue with lesser deposit in nontarget organs [49]. The use of PLGA is particularly suitable for this application because of its biodegradability and nonimmunogenicity [50]. In another study, the authors described a carrier system using intestinal organoids to effectively deliver 5-ASA-encapsulated PLGA nanoparticle to treat inflammatory bowel disease [51]. Another research reported the development of a metformin-loaded nanoformulation that showed efficacy against periodontal inflammation in diabetic rats [52]. Some researchers formulated a dactolisib–PLGA nanoparticle that showed activity against inflamed endothelial cells [53], and recently in 2019, the synthesis of crocetin-loaded nanoparticles was performed that reduced the level of pro-inflammatory cytokines in renal tissue and therefore displayed prospects for the treatment of diabetes-induced nephropathy [54].

A hyaluronic acid-functionalized PLGA-based nanocarrier for methotrexate has been developed in a study for targeted treatment of RA [55]. A PLGA nanoparticle with protease inhibitor minimized the gastrointestinal limitations of oral insulin delivery in rats [56]. Also, a PLGA-chitosan-based nanocarrier has been formulated, which was found to be specific for human antigen-presenting cells [57]. A potential DNA vaccine delivery system has been developed using PLGA-based nanocarriers [58]. The PLGA-based delivery system encapsulating vitamin D has recently displayed activity against various markers for Alzheimer's disease in mice [59]. An optimized fluorometholone-PLGA nanoparticle has been developed that showed enhanced efficacy in treating ocular inflammation as compared to the commercial formula [60].

The biocompatibility, biodegradability, and versatility of PLGA have made it suitable for a wide range of clinical applications. PLGA-containing chemotherapeutic formulations approved for clinical use include Lupron Depot R, for sustained release of leuprolide, which has application in the management of prostate cancer [61]; Trelstar R, a triptorelin-containing suspension for the treatment of prostate cancer; and Zoladex, a goserelin-containing implant used in the treatment of breast and prostate cancer and endometriosis. Formulations approved for other applications include Risperdal R Consta R (risperidone), Vivitrol R (naltrexone), and Arestin R (minocycline) for the treatment of schizophrenia, opioid dependence, and periodontal disease, respectively [62]. A promising approach for clinical development is the development of PLGA-based systems with imaging agents to monitor disease state and relapse patterns using magnetic resonance imaging (MRI). Studies have supported the fact that these structures are noninvasive and cost effective with excellent safety profiles [63]. Further, there are numerous PLGA-based systems that are being used in ongoing clinical trials.

The studies on the rate of degradation of polyanhydride polymer and the rate of release of salicylic acid as a function of pH have shown that this polymer takes about three months to degrade in acidic and neutral environment, but at basic pH it degrades in approximately 19–40 hours. Since the upper gastrointestinal tract is acidic or neutral, this polymeric drug can reach the intestines undamaged and release salicylic acid directly to the lower intestinal tract to treat the disease. Aspirin is also used in dentistry, for example in cases of tooth breakage when there cannot be an immediate operation. In this procedure, the fast influx of salicylic acid irritates the surrounding tissue. Slower release of the medication, provided by the poly(anhydride-esters), would be a gentler solution [64].

6.2.9 Chitosan

Chitosan (CS) is a cationic biological polysaccharide, considered as a promising nanomaterial with extensive medical applications [65, 66]. CS is one of the most abundant biopolymers derived from natural chitin that is commonly present in the exoskeletons of arthropods, crustacean shells, insects, and fungal cell walls [67]. Structurally, CS is a representative polysaccharide with native positively charged amine groups (Figure 6.8).

It is a biodegradable and biocompatible polymer that can be easily modified to perform specific functions for a wide range of drug delivery applications [66]. Chitosan-based nanoparticles have been investigated in various drug delivery applications including parenteral and nonparenteral routes of administration and also for treatment of dermatologic and gastrointestinal diseases, pulmonary diseases, and drug delivery to the brain and ocular areas [67]. Polymeric micelle nanoparticles based on amphiphilic chitosan derivatives have been recently prepared via self-aggregation in water [68]. The self-assembled amphiphilic micelles based on CS and polycaprolactone (PCL) were produced and used as carriers of paclitaxel (PTX) to improve its intestinal pharmacokinetic profile [67]. Some research works indicated that chemical modification of chitosan nanoparticles can improve their drug targeting and bioavailability profile. Recent advances also indicate the use of chitosan nanoparticles for tumor targeting [68], imaging and therapeutic applications [69], and development of targeted drug delivery systems. A chitosan-based nasal formulation of morphine (Rylomine TM) is currently being studied in phase 3 clinical trials in the United States and phase 2 clinical trials in the United Kingdom and the European Union.

The remarkable fact about chitosan is that it possesses unique properties such as nontoxicity, biocompatibility, biodegradability, bioactivity, and mucoadhesion [69, 70].

It can be degraded by enzymes, such as lysozymes and chitosanases to oligosaccharides and monosaccharides, which are subsequently absorbed by the body tissues [71]. Despite its unique physicochemical and

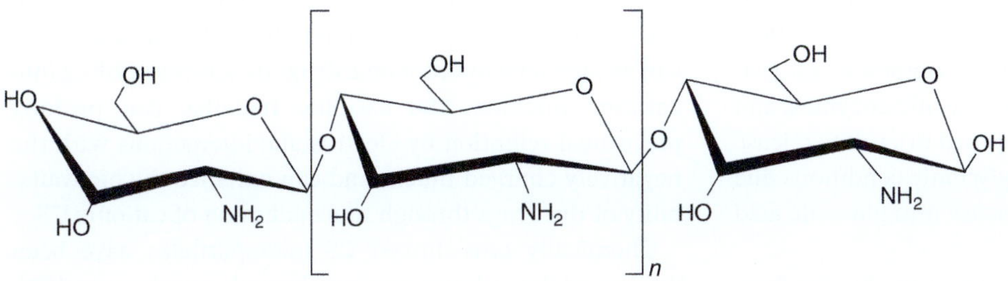

Figure 6.8 Structure of chitosan.

biological characteristics, CS has not been commonly utilized in the clinical studies due to its low solubility and poor mechanical properties [72, 73]. A wide range of CS derivatives with improved solubility have been developed utilizing free amino and hydroxyl groups, and such derivatives displayed high affinity with functional proteins and the capability of self-assembly [74, 75]. Thus, CS has been widely employed in various biomedical and pharmaceutical applications such as drug, gene and vaccine delivery, tissue engineering, wound healing, and manufacture of cosmetic products [74, 75]. Over the past few years, novel nanomaterials successfully developed from CS have been increasing tremendously with potential applications in targeted drug delivery based on the characteristic features such as biocompatibility and ability to serve as reaction sites with other bioactive compounds, protecting unstable drug molecules from strong gastric acids, ability to adhere to mucosal tissues to improve the absorption of specific drugs, ease in combining with anionic biomacromolecules such as DNA by electrostatic action, and colon-targeted administration. Some studies about chitosan-based nanoparticle systems for disease treatment have explained about biological applications of chitosan [76, 77].

It is convenient to take insulin orally, although this route has a number of obstacles. Insulin permeability across membranes is generally poor due to its hydrophilicity and high molecular weight [78]. Also, insulin is readily inactivated in the stomach by gastric acids. Therefore, scientists are working over the development of oral delivery system of insulin for better treatment of diabetes. Insulin is an anionic protein that is convenient to combine with chitosan via ionic interactions. A uniform core–shell nanoparticle has been developed via electrostatic complexation to deliver insulin through oral route. *N*-(2-Hydroxypropyl)-3-trimethylammonium chloride was modified with chitosan (HTCC) and insulin as positively charged core, whereas thiolated hyaluronic acid (HA-SH) acted as anionic shell of the nanoparticle [79]. The particles had an average size of 100 nm and with satisfactory encapsulation efficiency and loading capacity, thereby rendering it as a promising approach for the delivery of insulin in diabetes. The drug delivery system is also well supported by *in vitro* and *ex vivo* experimentation data. Anderson et al. designed closed-loop insulin-injectable delivery through nanocapsules [80]. The developed nanocapsules were composed of pH-responsive chitosan matrix, glucose-specific enzymes, and insulin. The microgel system is supposed to swell to release insulin at higher rate under hyperglycemic conditions due to the enzymatic conversion of glucose into gluconic acid and chitosan-based protonation.

Gastric cancer exhibits malignant tumor originating from the gastric mucosa. A polyphenol-epigallocatechin-3-gallate (EGCG), which is extracted from green tea, has been proved to inhibit tumor growth by antiangiogenesis, proliferation inhibition, and apoptosis induction. Due to low mucosal permeability and instability in the gastrointestinal environment, it is difficult to apply the EGCG to cure gastric cancer. To overcome this problem, Feng et al. developed targeted NPs encapsulated with EGCG [81]. These nanomaterials are composed of three components including the fucose-conjugated chitosan (fucose-chitosan, FCS), PEG-conjugated chitosan (PEG-chitosan, PCS) complexes, and gelatin (Gel), respectively. Fructose is a deoxyhexose sugar that assists in drug targeting due to its ability to modify various types of molecules physiologically. Chitosan enhances the gastrointestinal absorption via electrostatic interaction with abundant *N*-acetylneuraminic acid in gastric mucosa, whereas PEG functions as a cross-linker that prolongs protein circulation in blood. It was observed that the EGCG-loaded NPs were more effective against gastric tumors through *in vivo* and *in vitro* experiments.

The chitosan nanodrugs are also administrated by the parenteral route to achieve therapeutic effects. A stepwise pH-responsive nanoparticle has been developed to release doxorubicin on the stimuli [82]. Yang et al. reported a curcumin-loaded lipid carrier for ophthalmic delivery [73]. The reaction between the primary amino groups of CS and the carboxylic acid groups of *N*-acetylcysteine (NAC) resulted in covalently modified molecules. The modification improved the solubility and performance of the CS vehicle. Cysteine is a thiol group containing compound that may form disulfide bonds between molecules and body tissues, and the interaction can prolong the retention time of drugs. There are several sulfhydryl residues in proteins and mucus on the surface of body tissues that can interact with thiol groups present in the modified CS vehicle. CS plays an important role in mucoadhesion, and the modified thiol group enhances the effect of bioadhesion and precorneal retention. It is proved to be a smart approach that can evidently prolong the residence time between the drug and the cornea. Shi et al. prepared a novel block polymer composed of cationic CS and methoxy polyethylene glycol–polycaprolactone (MPEG–PCL) to deliver the drug diclofenac in ocular diseases [83]. They conjugated the MPEG–PCL block polymer onto chitosan (MPEG–PCL–CS) by covalent bonding. The block polymer can encapsulate hydrophobic drugs by self-assembling into cationic micelle. The cationic micelles can prolong precorneal retention by electrostatic interactions with the negatively charged mucin and can enhance the bioavailability of the drugs through mucoadhesion of cationic CS.

Chemically cross-linked CS microparticles have been developed for administration through nasal route [84]. The versatile particles were formulated by covalently cross-linking CS with different quantities of glutaraldehyde for which the mucin adsorption ability was compared. All the particles displayed high affinity for the mucin *in vitro* and substantially extended the drug release time. The microspheres with mucoadhesive chitosan is a promising novel delivery system for nasal administration. CS is also an ideal molecule for the formulation of targeted colon drug delivery as it is a high-affinity substrate to be degraded by intestinal flora. Opanasopit et al. formulated a pH-sensitive CS polymer to encapsulate curcumin (CUR) for colon-targeted drug delivery [71].

The release of CUR was found to be significantly increased at GI tract pH levels. Situ et al. formulated CS-based particles via ionic cross-linking [85]. They encapsulated proteins in the granules by cross-linking to maintain protein vitality during storage and administration through oral route. The particle morphology, protein loading, and release characteristics under different conditions indicated that the particles can efficiently target the colon.

Carcinoma involves the uncontrolled proliferation of cells. Surgery and chemotherapy usually are combined as a comprehensive strategy to conquer the tumor. Chemotherapy is a systemic treatment with the advantage of eliminating the remaining potential metastatic lesions after surgery. It has the advantage to treat multiple lesions simultaneously, but the disadvantage is also apparent due to the systemic side effects that may affect healthy tissue [86]. Various optimized chemotherapy strategies have been developed to solve the abovementioned issues. Nanomaterials have been designed to target specific tissues or respond to particular environmental conditions. CS is generally utilized to develop antitumor nanovehicles for the treatment of carcinoma because of its unique properties such as mucoadhesiveness and structural variability.

Yin et al. developed hydrophilic NPs with cationic TMC codelivered of DOX and interleukin-2 (IL-2) for enhanced antitumor efficacy [87]. *cis*-Aconitic anhydride was used to covalently graft DOX onto TMC to form nanocomplex with the release of DOX in a pH-sensitive pattern. IL-2 was combined with NPs via electrostatic adsorption without compromising its bioactivity. FA has also been loaded on nanocarrier for targeted delivery. The subspherical nanocomplexes have an average size of about 200 nm and are positively charged. The optimized combination therapy strategy showed improved antitumor efficacy and reduced *in vivo* size effects.

Murugan et al. designed a combinatorial nanocarrier for amalgamation of antitumor agents in breast cancer cells and introduced the drug topotecan (TPT) and quercetin (QT) into a single system [88]. They prepared the mesoporous silica nanoparticles (MSNs), and the MSN core was then loaded with TPT and covered with poly(acrylic acid) (PAA) and CS as shell and conjugated with QT on the surface of TPT loaded in the MSN core. Finally, the arginine–glycine–aspartic acid (cRGD) was attached to the integrated drug-containing NPs, which targeted the cancer cells via integrin receptor-mediated endocytosis. Drug release was triggered by intertissue/intracellular pH change and CS degradation. This strategy has been shown to be effective in breast cancer treatment.

There are many publications about chitosan-based nanomaterials for antitumor drug delivery in innovative ways. Effective drug delivery systems are developed for anticancer therapy based on environmental response and targeting principles to deliver drugs, vaccines, etc.

Genes-encoded specific proteins are essential for various physiological processes of the body, and their mutation often results in disease. Gene therapy is a promising strategy to treat genetic diseases. Gene therapy refers to curing genetic diseases via the introduction of a foreign gene into a target cell to correct or repair a damaged gene. However, DNA and RNA molecules can be destroyed by harsh acids and enzymes that are produced in the body. In addition, DNA and RNA are anionic polymers that have good affinity with cationic polymers such as CS [89].

Small interfering RNA (siRNA) is a double-stranded RNA of 20–25 nucleotides in length and has many important functions in biology. It primarily plays a unique role in the RNA interference events that regulate gene expression in a specific manner. siRNA technology has been widely used in various disease areas such as tumors and inflammation.

Stride et al. developed a CS-deoxycholic acid nanodroplets loading on the magnetic nanoparticle for siRNA delivery [90]. They covalently grafted deoxycholic acid on the CS and sequentially had perfluoropentane and iron oxide modified on the surface. siRNA was combined on the surface of the particle via electrostatic interactions with a cationic CS matrix. The integrated siRNA and nanoparticles displayed stability in serum at 37 °C for up to 4 hours. Interestingly, the nanodroplets would undergo a phase change under the assistance of ultrasound from a liquid state to a microbubble state with a larger volume and higher energy and release the contained siRNA, and the iron oxide-containing NP may be directed to the correct site while applying an external magnetic force. Importantly, treatment efficacy was positively correlated with the energy of these emissions that were produced by ultrasound. Tumor protein p53, an unequal protein, has been shown to respond to cellular stresses, thereby acting as a tumor suppressor. Activating the proliferation suppressor gene may thus be a promising strategy for tumor therapy.

Feron et al. developed noncovalent PEGylated CS nanoparticles for siRNA-targeted delivery [91]. siRNA was loaded via electrostatic interaction with cationic CS. Targeting function was achieved by RGD peptides analog, and PEG would detach from the nanocarriers when necessary to facilitate cellular entry. Notch1 is a signaling receptor that plays a critical role in RA.

Vaccine refers to biological products applied to prevent or control the occurrence and spread of infectious diseases. The nature of the vaccine may be microbes or their toxins, enzymes, human or animal serum, and cells. The emergence of vaccines has made a significant contribution to the prevention and control of diseases. However, there are many problems affecting the quality of the vaccine during the preparation, storage, and administration. It is greatly necessary to take high quality control during the preparation of vaccines, because the inferior quality vaccines pose a great threat to human health. Nanomaterial-encapsulated vaccines are thus promising vaccine-transport vehicles. CS is a particularly attractive choice for vaccine delivery because of its low immunogenicity, low toxicity, biocompatibility, and biodegradability [92].

Tumor cell lysates can act as antigens to stimulate dendritic cells to produce immunity. Kong et al. developed tumor cell lysate-loaded (TCL) CS nanovaccine to enhance antitumor immunity by targeting dendritic cells (DCs) [93]. They modified chitosan with mannose (Man) for specific DCs targeting and then loaded TCL (Man-CTS-TCL NPs) to trigger an immune response. For the vaccine viability experiment, the efficacy of Man-CTS-TCL NPs as cancer vaccine was evaluated *in vitro* and *in vivo*. Specifically, the activation of DCs by Man-CTS-TCL NPs was studied at the cellular level, and the mice were challenged with B16 melanoma cells after they were vaccinated with the prepared vaccine. The experiment results showed that the Man-CTS-TCL NPs are effective antitumor vaccines.

Tirelli et al. developed a dual-targeting vaccine vector. CS showed an electrolyte interaction with HA, which potentially played an important role to load the siRNA [94]. Simultaneously, the mannosylated modification on HA further improved the targeting effect. The interaction of mannose with lectin promoted the internalization of NPs when hyaluronic acid was compatible with CD44, a surface receptor that has been reported to affect DC maturation and adhesion. The construction of targeting nanocarriers based on CS matrix is an effective approach for gene delivery with enhancement of immunomodulation effects.

6.2.10 Gelatin

Gelatin is a natural water-soluble polymer which is a protein in nature and normally produced by denaturing collagen. It is widely used in pharmaceutical and medical applications due to its excellent properties such as biodegradability, biocompatibility, and low antigenicity. In addition, gelatin can be easily altered due to its isoelectric point that allows it to change from negative to positive charge in an appropriate physiological environment. Gelatin as a polymer is used for gene delivery, cell culture, and recently in tissue engineering. Gelatin-based drug delivery systems have the properties to control the release of bioactive agents such as drugs and protein. It is also reported that liposome-loaded bioactive compounds can be incorporated into PEG-gelatin gel which can function as porous scaffold gelatin-based temporary depots for controlled drug release over prolonged time periods. However, some drawbacks have been identified with the use of gelatin-based systems in pharmaceutical applications. These drawbacks include poor mechanical strength and ineffectiveness in the management of drug release at infected sites (Figure 6.9).

Gelatin micro- and nanoparticles have been extensively studied as carrier systems for many applications. Microparticles have the advantage of a large surface area, which enables sufficient exchange of nutrients and metabolic wastes and allows rapid cell development [95]. They can therefore serve as vehicles for cell amplification and can simplify the delivery of these expanded cells or other large bioactive molecules to the desired site. The size of

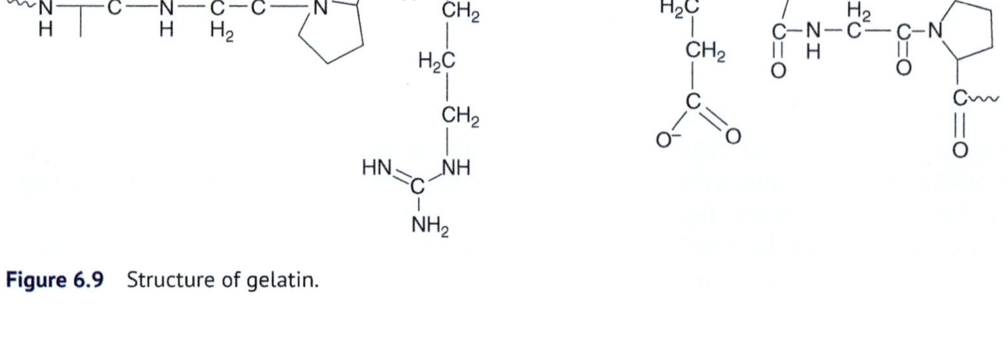

Figure 6.9 Structure of gelatin.

nanoparticles offers other advantages, including a relatively higher intracellular uptake by various areas of the body. For example, nanoparticles are better suited for intravenous delivery [96] or for delivery of drugs to the brain, due to their ability to better accumulate in macrophage-rich organs as they can easily cross the blood–brain barrier [97]. Due to their unique design, liposomes have the ability to incorporate both hydrophilic and -phobic drugs, protect them from degradation, target them to the desired site, and reduce the toxicity or side effects of those molecules [98, 99]. Unfortunately, liposomes have low encapsulation efficiency and poor storage stability, and water-soluble drugs can rapidly leak in the presence of body fluids [68]. However, some researchers showed that embedding liposomes into a gelatin-based system resulted in an improvement in their stability, viscosity, and the half-life of the loaded drug and the liposome [100]. Nanoparticles have good *in vivo* stability, can be easily sterilized and scaled up, and their manufacture can be free of contamination with pyrogens. These advantages can be used in the fields of biomaterials and medicine. Nanoparticles can be used to deliver hydrophilic and hydrophobic drugs, proteins, vaccines, and other types of bioactive molecules to various areas in the body, such as the lymphatic system, brain, arterial walls, lungs, liver, spleen, or even for long-term systemic circulation. A successful nanoparticle system should have high loading capacity to reduce the amount of carrier needed for administration. Depending on the drug and carrier properties, the drug can be dissolved, entrapped, encapsulated, or attached to the nanoparticle. The loading of the drug can be performed at the time of the nanoparticle production, during the polymerization process, or after the formation of the nanoparticles by incubating them with the drug solution. The amount of bound drug and the type of interaction between the drug and the nanoparticles depend on the chemical structure of the drug and the polymer and the conditions of the drug-loading procedure. Gelatin nanoparticles have been widely used for encapsulating many bioactive molecules. Li et al. encapsulated bovine serum albumin (BSA) as a model protein drug, using gelatin-based nanoparticles. BSA release was followed by a diffusion-controlled mechanism and was enhanced by the nanoparticles' water uptake capacity of 51–72% [101]. Nanoparticles can better accumulate in macrophage-rich organs which can easily cross the blood--brain barrier and reach the brain, compared to other carrier systems.

In order to reduce the toxic side effects of chloroquine phosphate, a well-known antimalarial drug, it was encapsulated in gelatin nanoparticles [102]. The results showed that chloroquine phosphate release was controlled via diffusion, increased with increasing temperature, but decreased in the physiological fluids and was found to be optimal near the physiological pH (7.4). The addition of the cross-linker glutaraldehyde to the nanoparticles decreased the drug release rate.

Ibuprofen sodium, an anti-inflammatory drug, was loaded into PEGylated gelatin nanoparticles. The results demonstrated an improved plasma half-life of ibuprofen sodium when encapsulated within the nanoparticles. These results might allow lowering the rate of ibuprofen sodium needed for treatment [103]. Targeting nano- and microvesicles to enhance drug uptake only in tumor cells, while sparing healthy cells, is a unique strategy which is currently being investigated. It is known that the extracellular pH of tumor tissue is lower than that of the normal tissue. The pH-responsive carriers would therefore have an advantage in accelerating the local drug release in tumor tissue. Doxorubicin, an anticancer drug, was encapsulated in an amphiphilic gelatin–iron oxide core/calcium phosphate shell nanoparticle. The addition of a calcium phosphate shell acts as a drug reservoir and turned the nanoparticles into highly pH-responsive drug-release carriers [104].

Gelatin microparticles have been extensively studied as a carrier system for many applications. They can be produced by several methods such as emulsion polymerization, solvent evaporation, coacervation, and spray-drying. However, gelatin microparticles have poor mechanical properties and rapid dissolution rates in aqueous environments, which accelerate drug release at body temperature. Gelatin microparticles are typically cross-linked with formaldehyde, glutaraldehyde, genipin, and/or carbodiimides in order to solve these problems. However, methacrylation of gelatin has been reported to be less cytotoxic and to enable a larger range of cross-linking densities compared to the traditional chemical cross-linking methods, thus offering an alternative method to better control the extent of hydrogel cross-linking [71]. It was shown that fewer methacrylated microparticles had decreased elastic moduli and larger mesh sizes but could be better for an action, ensuring sustained drug delivery to the lungs and as a result improving patient compliance. Therefore, a new polymeric microparticle system based on gelatin covalently bound to isoniazid and containing rifampicin was developed. The results showed that the microparticles demonstrated low cytotoxicity, were able to encapsulate both rifampicin ($51 \pm 6\%$) and isoniazid ($22 \pm 1\%$), and had a good nebulization efficiency, which is important in pulmonary antitubercular drug delivery systems [105]. Polymeric microparticle system based on gelatin covalently bound to isoniazid and containing rifampicin was developed. The results showed that the microparticles demonstrated low cytotoxicity, were able to encapsulate both rifampicin ($51 \pm 6\%$) and isoniazid (22 ± 1%), and had a good nebulization efficiency, which is important in pulmonary antitubercular drug delivery systems [105].

Gelatin microparticles showed several advantages for pulmonary delivery, such as a large distribution to lung epithelial cells and higher availability of bioactive molecules to the infected cells. Rifampicin and isoniazid are part of the therapeutic treatment against tuberculosis. However, these drugs are relatively toxic, can cause several side effects, and might cause drug resistance when not followed to completion. The use of microcarriers loaded with these drugs may allow reducing the therapeutic dose, extending the duration of action, ensuring sustained drug delivery to the lungs, and as a result improving patient compliance.

The utilization of gelatin-based liposomes has been reported in several studies on nanotechnology-based drug delivery systems. Liposomes are small spheres of an aqueous core entrapped by one or more phospholipids that form closed bilayered systems. Liposomes are widely used as an advanced technology for delivering bioactive molecules due to their high biocompatibility and their ability to incorporate hydrophilic and hydrophobic drugs and deliver bioactive molecules directly to the desired site. Hydrophobic drugs are usually encapsulated in the lipid bilayers of liposomes, whereas hydrophilic drugs may either be encapsulated inside the aqueous cores of liposomes or be located in the external aqueous phase. A liposomal hydrogel system consisting of a PEG–gelatin hydrogel loaded with liposomes containing the antibiotic ciprofloxacin was developed in order to reduce bacterial adhesion to silicone catheter material. Liposomal hydrogel-coated catheters were shown to have an antimicrobial efficacy against *Pseudomonas aeruginosa* [106]. Another study demonstrated a controlled release of liposomes loaded with calcein fluorescence dye or calcein labeled with rhodamine from gelatin carboxymethyl cellulose films. The release rate of the loaded liposomes depended mainly on the amount of liposomes entrapped inside the films, the swelling degree and the network density of the film, and the glutaraldehyde cross-linking degree [107]. Embedding liposomal drug delivery systems into a polymer-based system improves the liposome stability, viscosity, half-life of the loaded drug, and the embedded liposome. It also allows a sustained and efficient drug release over prolonged periods of time.

6.2.11 Alginate

Alginate is also a naturally occurring linear polysaccharide. It is extracted from seaweed, algae, and bacteria. The basic chemical structure of alginate consists of (1–4)-b-D-mannuronic acid (M) and (1–4)-a-L-guluronic acid (G) units. This is a nontoxic, biologically safe, biocompatible, and biodegradable polysaccharide in nature (Figure 6.10).

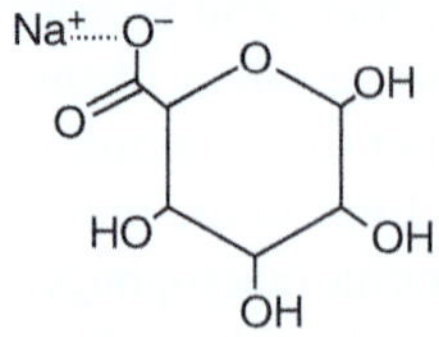

Figure 6.10 Structure of sodium alginate.

Alginate and their derivatives are widely used by many pharmaceutical scientists for drug delivery and tissue engineering applications due to their unique characteristics such as biocompatibility, biodegradability, low toxicity, nonimmunogenicity, solubility in water, relatively economical, gelation ability, stabilizing properties, and high viscosity in aqueous solutions. Since alginate is anionic, fabrication of alginate hydrogels has successively been achieved through a reaction with cross-linking agents such as divalent or trivalent cations mainly calcium ions, water-soluble carbodiimide, and/or glutaraldehyde. The cross-linking was performed at room temperature and physiological pH conditions. The success in fabricating highly porous 3D alginate scaffolds has been through lyophilization. Thus far, alginate-based systems have been successfully used as a matrix for the encapsulation of stem cells and for controlled release of proteins, genes, and drugs. Alginate-based drug delivery systems have also been used as depots for bioactive agent-loaded liposomes for sustained drug release. Highly increased efficacy has been reported from these integrated delivery systems when compared to polymeric- or liposome-based systems alone [108].

Alginates have been widely used as tablet disintegrant, binding agent, viscosity-modifying agent, as a stabilizer in disperse system in the production of suspension and emulsion, and also as thickening agent in pharmaceutical industries. The most important advantage of using alginate as a matrix for controlled-release (CR) formulations is its biodegradability, because it is degraded and is absorbed by the body during and/or after drug release without any toxic effects. This allows bypass of surgical removal of the device. Hence, alginates can be a suitable matrix-forming component for sustained release of various drugs. As the drug delivery can be controlled primarily because of the properties of polymer devices, controlled release is possible for conventional low molecular weight drugs as well as macromolecular drugs including peptide hormones (e.g. insulin and growth hormone), polysaccharides (e.g. heparin), antibiotics, antigens, and enzymes. The drug release from alginate beads occurs basically by diffusion through matrix and due to erosion mechanism at certain pH. Drug release can be controlled by coating of matrix beads with sodium alginate. Sodium alginate has also been evaluated as release-controlling diluent in CR capsules. Several drugs have been incorporated into alginate matrices in a variety of forms (e.g. beads, micro spheres, films, and tablets) for CR therapies [109].

Alginate can play a significant role in the design of a controlled-release product because the alginate molecule can undergo immediate hydration to create a hydrocolloidal layer of high viscosity. This phenomenon makes up a diffusion barrier, thereby decreasing the migration of small molecules and drugs. So far, alginate has mainly been applied in systems based on diffusion. Diffusion systems based on alginate can be divided into two main categories. In the polymeric membrane system, the drug formulation is encapsulated in a drug reservoir compartment. The drug formulation getting encapsulated may exist as a solid or suspension or in solution form. The drug release is controlled by an encapsulating polymeric membrane having a specific permeability. The encapsulation of drug is accomplished by various techniques, e.g. spray coating and microencapsulation [108, 109]. In one study, the compound theophylline was encapsulated, and the drug release rate was significantly reduced compared to the matrix-type alginate gel beads. The release rate became lower as the coat thickness increased. The release followed zero-order kinetics as expected [109]. A further decrease in release rate can be obtained by incorporating additives such as carnauba wax into the drug reservoir. This is demonstrated for indomethacin, a nonsteroidal anti-inflammatory drug which is highly irritating to the mucosa in the upper gastrointestinal (GI) tract. In the polymeric matrix system, the drug is homogeneously dispersed in the rate-controlling polymer matrix. The developed product may be in the form of swellable microspheres or conventional tablets. When such drug delivery systems are subjected to dissolution, drug release is controlled by diffusion through matrix swelling and dissolution/erosion at the matrix periphery. The "swelling–dissolution–erosion" process is highly complex. Sodium alginate cross-linked with calcium chloride forms a system where the osmotic pressure gradient existing between the alginate gel and the environment plays an important role in the swelling process [110]. Under acidic conditions (e.g. in the stomach), swelling of the calcium alginate beads scarcely occurs. A drug is likely to be released by diffusion through the insoluble matrix. Under neutral conditions present in the intestine, the beads undergo swelling, and the drug release depends on the swelling and erosion process. The swelling behavior of calcium alginate has been thoroughly exploited for the development of a multiple-unit, controlled-release drug delivery system [111]. Alginate gel beads seem to be most effective in retarding drugs at higher alginate concentrations [110, 111] and when the alginates are rich in guluronic acid [111].

The molecular weight and viscosity of the alginates did not affect the drug release of nicardipine HCl in neutral medium [112]. Interestingly, the release of the basic drug pindolol was, however, demonstrated to be dependent on the alginate molecular size [113]. The slowest *in vitro* release rate of pindolol (at neutral pH) was observed for the beads prepared by alginate of low molecular weight, although this showed the fastest *in vivo* absorption rate. The drug:alginate ratio and calcium chloride concentration affect the drug release. The release of nicardipine from alginate particles prepared in a ratio of 1 : 1 was delayed more than that from 1 : 2 particles [112]. Calcium alginate beads displayed prolonged release profiles when compared to alginate beads prepared from other cross-link agents.

In nasal drug delivery systems, nasal mucociliary clearance is one of the most important limiting factors. It severely limits the time allowed for drug absorption to occur and effectively prevents sustained nasal drug administration. Thus, to retain the drug in the nasal cavity, a particle formulation would be preferable to solution. In addition, mucoadhesive polymers have been introduced to prepare microparticles, which could further overcome the problems of poor bioavailability by increasing the residence time at the applied site. Mucoadhesion requires a highly expanded and hydrated polymer network, which could promote an intimate contact between microspheres and the mucus layer. Thus, mucoadhesive microspheres have been developed to decrease the effect of mucociliary clearance. The microparticles form a gel-like layer, which is cleared slowly from the nasal cavity, resulting in a prolonged residence time of the drug formulation. Mucoadhesive microspheres significantly increase the systemic absorption of conventional drugs as well as polypeptides across the nasal membrane without the use of absorption-enhancing agents that have the potential for irritation or damage [114]. Alginate is an anionic mucoadhesive polymer, which is known for its ability to create hydrogen bonds with mucin-type glycoproteins through carboxyl–hydroxyl interactions. This anionic biopolymer is used in many pharmaceutical and biotechnological applications [115]. Sodium alginate is a natural water-soluble linear polysaccharide molecule that is widely used to produce polymer matrix due to its nontoxicity, biocompatibility, and gel-forming ability. Therefore, in several studies it has been used as the matrix to prepare microparticles.

In a study, a formulation has been developed in which magnetic iron oxide nanoparticles were encapsulated into alginate beads with controlled size and activated by magnetic external stimuli [116].

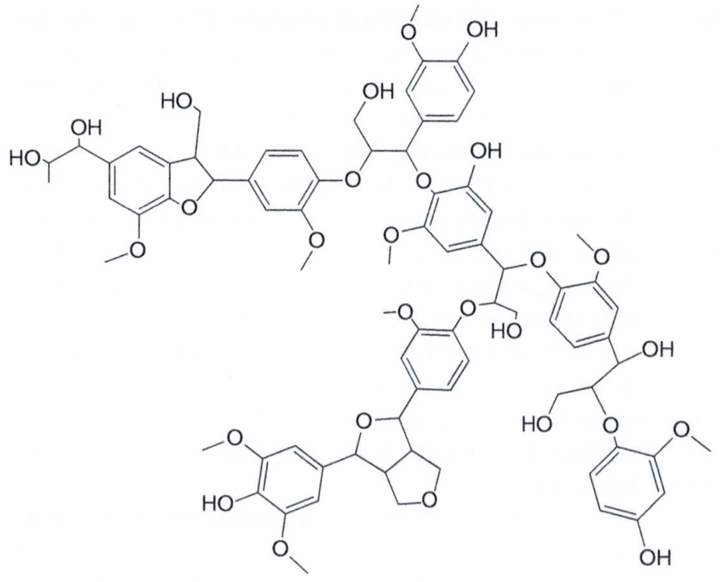

Figure 6.11 Structure of lignin.

6.2.12 Lignin

6.2.12.1 Introduction

Lignin is present in the cell wall of various plants. Next to cellulose, lignin is the most abundant biopolymer. Presence of phenolic and aliphatic hydroxyl groups in the structure of lignin makes it suitable to form a new material by structural modification [9] (Figure 6.11).

Lignin has the potential to produce nanoparticles for encapsulation of different compounds for different pharmaceutical applications such as anticancer therapy [117]. For the first time, Frangville et al. explored the use of nontoxic LNPs for encapsulation of hydrophilic compounds, i.e. Rhodamine 6G with high loading capacity [118]. Thereafter, Tortora et al. prepared the hydrophobic compound Coumarin-6 loaded lignin microcapsule (LMC) and showed that products had low cytotoxicity and could be effectively [119] internalized into Chinese hamster ovary cells. The Coumarin-6 release in the presence of SDS solution 5% w/v showed that almost 100% of entrapped Coumarin-6 was released in 60 min. There may be two concomitant reasons associated with the release of entrapped Coumarin-6 in the presence of SDS such as deterioration on lignin hydrophobic interactions and the high affinity of the molecule for the new dispersion fluid. The SDS molecule interferes with the lignin's stabilizing characteristics at the core shell boundaries which induce the release of Coumarin-6 because of the high solubility of the hydrophobic molecule in SDS solution. From these results, it can be understood that lignin micro/nanocapsules can be used as the topical applications since the pool of antioxidants and essential oils may get released if they contact with the lipid tissue or dermatological diseases. One step further, Lievonen et al. proposed an easy and practical method for the preparation of pH-sensitive lignin-based complex spherical micelles in green solvents using pure AL for the encapsulation of Ibuprofen (IBU) with the help of hydrophobic interaction. The *in vitro* release characteristics of IBU were observed to be pH dependent and demonstrated controlled-release properties. It is possible to preserve 75% of IBU in simulated gastric fluid, whereas in simulated intestinal fluid, 90% could be released smoothly. This work offers a unique technique for the formulation and fabrication of oral drug delivery carrier and possesses significant importance in the value-added use of lignin [120]. Yiamsawas et al. generated lignin-based polyurethane hollow nanocapsules at the interface of water cyclohexane which possess 2,4-toluene diisocyanate (TDI) and a surfactant. The encapsulating capacity of the capsules was calculated with the help of hydrophilic fluorescent dye sulforhodamine, and the long-time stability was observed over several months in both aqueous and organic phases. The release of the dye could be obtained by an enzymatic degradation of the lignin shell [121]. Interestingly, Chen et al. developed LNP through self-assembly process using renewable and nontoxic aqueous sodium *p*-toluenesulfonate (*p*TsONa) solution. A number of water-soluble or even -insoluble drugs can be dissolved in *p*TsONa and undergo 90% encapsulation and possess sustained drug-releasing capability. In addition to the above, it is easy to recycle the nonloaded drugs as well as free pTsONa for multiple times to achieve environmental sustainability [122]. This synthesis approach with broad processing window could realize the industrial scale-up production of LNPs and have wide potential applications.

6.2.13 Cellulose Derivatives

Cellulose is the most abundant naturally occurring biopolymer. It is the main constituent of various natural fibers. It is composed of long chains of anhydro-D-glucopyranose units (AGU) with each cellulose molecule with three hydroxyl groups per AGU, with the exception of the terminal ends (Figure 6.12).

Cellulose is insoluble in water and most common solvents; the poor solubility is attributed primarily to the strong intra- and intermolecular hydrogen bonding between the individual chains. Despite its poor water solubility, cellulose is used for a wide range of pharmaceutical applications. The chemically modified cellulose has improved processing capability, and the cellulose derivatives can be tailored for specific pharmaceutical applications. Large-scale commercial cellulose ethers include carboxymethyl cellulose (CMC), methyl cellulose (MC), hydroxyethylcellulose (HEC), hydroxypropyl methyl cellulose (HPMC), hydroxylpropyl cellulose (HFC), ethyl hydroxyethyl cellulose (EHEC), and methyl hydroxyethyl cellulose (MHEC) [123].

Cellulose and its derivatives are extensively utilized in the drug delivery systems basically for modification of the solubility and gelation of the drugs that resulted in the control of the release profile of the same [124]. Elseoud et al. [125] investigated the utilization of cellulose nanocrystals and chitosan nanoparticles for the oral releasing of repaglinide (an anti-hyperglycemic—RPG). The chitosan nanoparticles demonstrated a mean size distribution of 197 nm, whereas the oxidized cellulose nanocrystals containing RPG had a mean diameter of 251–310 nm. The presence of the hydrogen bonds between the cellulose nanocrystals and the drug resulted in sustained release of the same, and subsequently, the nanoparticles made with oxidized cellulose nanocrystals presented lower release

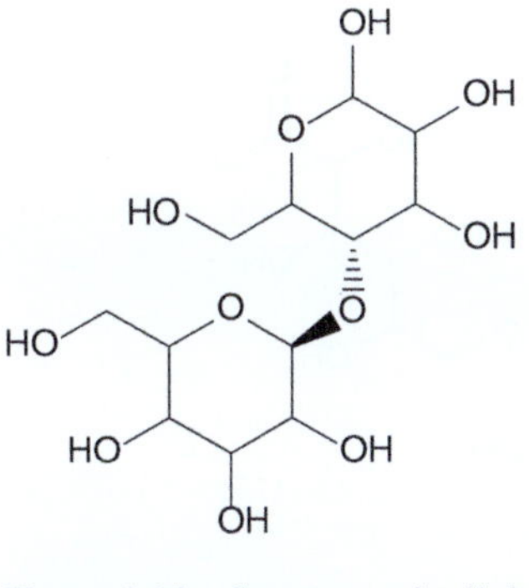

Figure 6.12 Structure of cellulose derivatives.

when compared to the nanoparticles produced with native cellulose nanocrystals.

Agarwal et al. [126] have developed a drug-targeting mechanism, which is based on the conjugation of calcium alginate beads with carboxymethylcellulose (CMC)-loaded 5-fluoroacyl (5-FU) and is targeted to the colon. The beads with lower CMC proportions presented greater swelling and mucoadhesiveness in the simulated colonic environment. In the presence of colonic enzymes, 90% release has been observed from 5-FU encapsulated in the beads. Hansen et al. [127] investigated four cellulose derivatives, including methylcellulose, hydroxypropyl methylcellulose, sodium carboxymethylcellulose, and cationic hydroxyethyl cellulose, for application in drug release into the nasal mucosa. The association of these cellulose derivatives with an additional excipient was also evaluated. The drug model employed in this process was acyclovir. The viability of the polymers as excipients for nasal release applications was also scrutinized for its ciliary beat frequency (CBF) and its infusion through the tissue system of the nostril cavity. An increase in thermally induced viscosity was observed when the cellulose derivatives were mixed with polymer graft copolymer. Also, the permeation of acyclovir into the nasal mucosa was found to be increased, when it was combined with cationic hydroxyethylcellulose. None of the cellulose derivatives caused negative effects on tissues and cells of the nasal mucosa, as assessed by CBF.

6.2.14 Albumin

Albumin can be used to produce various formulations, such as conjugates, nanoparticles, and/or complexes loaded with drugs, peptides, and/or antibodies. Albumin-based drug delivery systems have been widely studied as versatile carriers for anticancer agents, and three important formulations have been marketed or evaluated in a clinically advanced stage: Abraxane, PTX albumin nanoparticles, which are manufactured using the proprietary nab technology from American Bioscience, resulting in nanoparticles with a diameter of approximately 130 nm; MTX albumin conjugate, where MTX is linked to albumin by a lysine (amide bond with an amino group of lysine in albumin); and Aldoxorubicin, the albumin-binding prodrug of doxorubicin, which is an acidic-sensitive prodrug (thiol-binding prodrug) that binds covalently to the Cys-34 position of circulating albumin after intravenous administration [128].

Albumin and the respective formulations can be considered ideal drug carriers for targeting malignant and also inflamed tissue sources, due to the described capacity of albumin to accumulate either in malignant or in inflamed tissue [129, 130]. Several mechanisms are common features in inflammation and in tumors, such as angiogenesis and its humoral regulation by growth factors, for example VEGF. Moreover, the extravasation of macromolecules into the inflamed and in malignant tissues is upregulated in a similar way. Activated cells, such as tumor or inflammatory cells, metabolize albumin to cover their increased need for amino acids and energy, and the increased extravasation helps to satisfy this need [129]. The uptake of albumin by synovial fibroblasts of RA patients and the efficacy of MTX and MTX covalently coupled to HSA (MTX-HSA) in arthritic mice were studied, and the results demonstrated the favorable properties of albumin as a drug carrier to RA, increasing the therapeutic efficacy comparatively to drug alone and reducing the side effects. In addition to these applications, there are other market-approved albumin-based products, such as fatty acid derivatives of human insulin and the glucagon-like-1 peptide (Levemir®, Tresiba®, and Victoza®) for treating diabetes [131, 132].

The structural features of albumin, such as the presence of various specific sites on protein (Cys-34 position, lysine side-chain amines, and active sulfhydryl groups), led to the possibility of binding various peptides and synthetically modified drugs to protein for development of drug–protein conjugates (Figure 6.13).

For the preparation of drug–albumin conjugates, many methods, including direct coupling through lysine residues, chemical coupling methods such as carbodiimide coupling method and thiol/maleimide coupling method, and several other methodologies such as albumin fusion technology, have been used. Thus, albumin is considered as a good candidate for conjugation with different drugs as it provides good pharmacokinetics properties [133]. The drug–albumin conjugates, the most extensively studied conjugates of drugs up to now, can be posteriorly used for

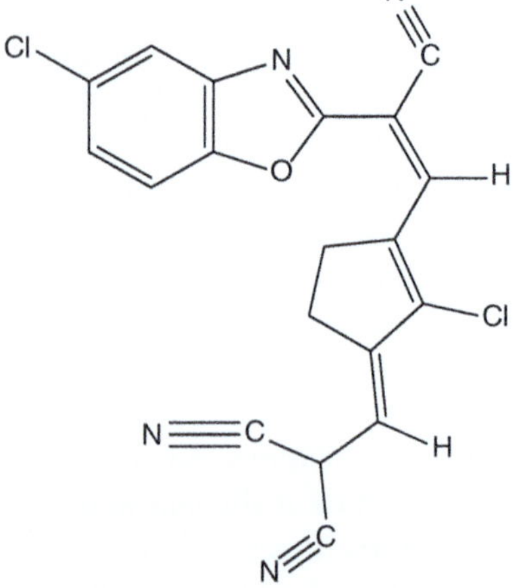

Figure 6.13 Structure of albumin.

preparation of albumin-based nanoparticles, with several benefits in the delivery of cytotoxic agents as previously described [134]. In our recent study, albumin–drug conjugates were produced with either MTX, a potent anticancer agent, or vancomycin (VCM), a potent antibiotic, and those conjugates were then used to produce functionalized albumin nanoemulsions [135]. The albumin–MTX conjugate was prepared via carbodiimide linkage, and the albumin–VCM conjugate was synthesized via thiol-maleimide addition. The anticancer effect of the conjugate albumin–MTX and the nanoemulsions prepared using this conjugate was compared with the effect of MTX alone in human colonic epithelial cell line (Caco-2 cell line). Nanoemulsions loaded with MTX, albumin–MTX conjugate, and MTX alone demonstrated very similar results; however, these developed conjugates and nanoemulsions present important characteristics to overcome the problems associated to this drug, namely, their side effects. The determination of minimum inhibitory concentration (MIC) of VCM, albumin–VCM conjugate solution, and albumin nanoemulsions loaded with VCM was performed using *Staphylococcus aureus* strain Newman. The results demonstrated that albumin–VCM conjugate and nanoemulsions loaded with VCM potentiate the antimicrobial activity of VCM against this strain, with such nanoemulsions presenting advantages for intravenous administration (small and stealth nanoparticles) and protection of the active principle. Thus, the development of these albumin–drugs conjugates and their incorporation in the albumin nanoparticles allowed the production of functionalized and effective nanoparticles, which can be suitable for drug delivery application in a wide range of clinical settings [135].

6.2.15 Dextran

Dextran is a natural linear polymer of glucose linked by a 1–6-linked glucopyranoside and some branching of 1,3-linked side chains (Figure 6.14).

Dextran is prepared from sucrose through lactic-acid bacteria, mostly by *Leuconostoc mesenteroides* and

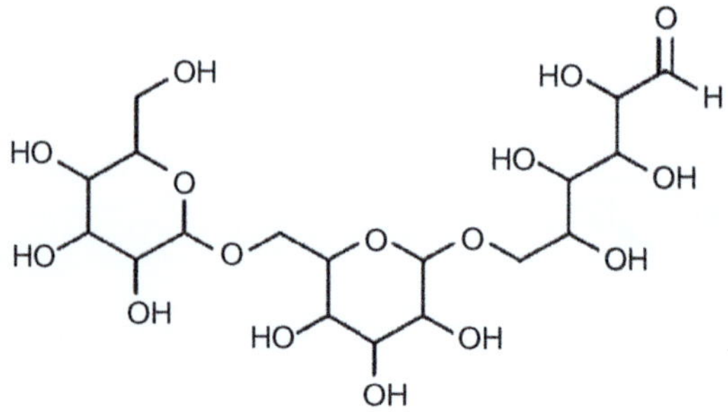

Figure 6.14 Structure of dextran.

Streptococcus mutans. There are two commercial preparations available, namely dextran 40 kilodaltons (kDa) (Rheomacrodex) and dextran 70 kilodaltons (kDa) (Macrodex). In pharmaceutics, dextran has been used as a model of drug delivery due to its unique characteristics that differentiate it from other types of polysaccharides, which include water solubility, biocompatibility, and biodegradability. In recent studies, dextran has been regarded as a potential polysaccharide polymer that can sustain the delivery of proteins, vaccines, and drugs. In a study, interleukin-2, which is a highly effective anticancer drug, has been delivered through liposomes and an injectable dextran hydrogel. A study conducted by Stenekes and coworkers observed the successful encapsulation of a drug in liposome and further formulation of a depot system into a dextran polymer-based material. The polymer-based materials were prepared using a two-phase system; the first phase was water and poly(ethylene glycol), and the second one water-methacrylated dextran. The slower degradation of dextran polymeric material resulted in sustained release of liposomes for a duration of 100 days. Liposomes released from the depot were found to be intact, and there was no significant change observed in the liposomal size. In a gene therapy study by Liptay and coworkers, it was reported that recombinant DNA (which contains chloramphenicol acetyltransferase) was successively encapsulated in cationic liposomes and then integrated within the dextran [136].

DX is neutrally charged, and systems with no surface charge have been shown to experience reduced plasma protein adsorption and an increased rate of nonspecific cellular uptake [137]. Accordingly, it has been used as a colloidal protectant, preventing particles from aggregation through steric hindrance [138]. Owing to the presence of high amount of hydroxyl groups and inspired by the classical example of DX–bleomycin complex and its efficiency in targeting the lymphatics [139], DX has been functionalized with various pharmaceutical agents, such as naproxen [140], daunorubicin [141], mitomycin C [142], doxorubicin [143, 144], and cisplatin [145] as efficient prodrugs aiming at achieving better drug and/or improving of targeting efficiency.

The drug delivery systems that have improved tumor penetration properties are valuable tools for delivery of anticancer agents.

Dextran is water soluble, and hydrophobically modified dextran forms micelles which can be used to either encapsulate drugs or be functionalized with drugs via its numerous hydroxyl groups. Drug-loaded nanocarrier formulations including those containing dextran can exhibit improved pharmacokinetics and reduced systemic toxicity as compared to their parent drug as a result of passive targeting in which circulating drug nanocarriers deposit preferentially into tumor tissue due to the leaky vasculature often associated with these tumors (EPR effect). The dextran–doxorubicin nanocarriers are capable of full penetration into neuroblastoma tumor spheroids with an average diameter of ~400–500 μm, while free doxorubicin can only penetrate ~50 μm into these structures. Due to this enhanced penetration, these nanocarriers display enhanced efficacy over free doxorubicin in 3D tumor spheroids [146].

6.2.16 Collagen

Collagen is a natural protein component found in mammals. Outside the body, collagen is prepared from glycine–proline–(hydroxy) proline repeat units to form a triple-helix molecular structure. So far, 19 types of collagen molecules have been isolated, characterized, and reported in both medical and pharmaceutical applications (Figure 6.15).

Collagen is extensively utilized in pharmaceutical field because it fulfills the requirements of a safe drug delivery system due to properties such as good biocompatibility, low antigenicity, and biodegradability upon implantation in cells. Collagen gels are among the natural polymers that can be used as a promising matrix system for drug delivery and tissue engineering applications. The biodegradable collagen-based systems have been used as 3D scaffold for cell culture and gene therapy.

The combination of liposomes and collagen-based system has been utilized for drug delivery applications since the 1980s. In several studies, drugs and other bioactive agents were first encapsulated in the liposomes and then embedded inside a depot composed of collagen-based

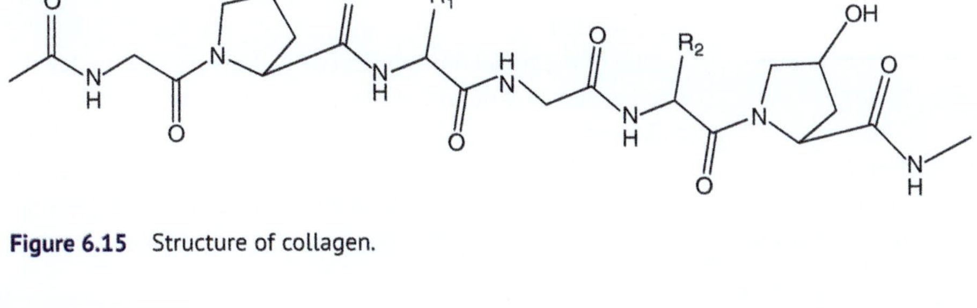

Figure 6.15 Structure of collagen.

systems, including scaffolds and gels. The combination of liposomes and collagen-based system has improved the stability on storage, prolonged drug release, and increased therapeutic efficacy [147]. Due to biocompatibility and low antigenicity, collagen-based nanoparticles have been used in some studies for the delivery of pharmaceuticals such as theophylline, retinol, tretinoin, and lidocaine [148, 149].

It is observed that collagen shows similarity to the microenvironment of some tumor cells, thereby allowing collagen nanoparticles to effectively infiltrate the tumor areas and deliver anticancer drugs [150]. The physical properties of collagen nanoparticles such as size, surface area, and absorption capacity are easy to configure [151]. Such properties make collagen nanoparticles a suitable candidate for controlled drug release strategies.

The advanced drug delivery formulations usually include formulations such as nanosuspension, nanoparticles, liposomes, niosomes, iontophoresis, and ophthalmic delivery system. Among these formulations, collagen finds its application mainly in the preparation of collagen shields for ophthalmic application, and in other formulations, it is being used as an additional substance due to its biocompatibility and easy availability [151].

Controlled and sustained release of drugs has shown its importance over conventional ophthalmic preparations due to its reduced side effects, sustained and controlled drug delivery, increased ocular bioavailability, and targeted drug delivery. This improved and controlled way of drug delivery effects provides an efficient therapeutic activity and better sustainability of drug within the body when compared to the other forms of dosages since the frequency of drug administration is reduced when a controlled drug delivery is maintained [152].

Collagen in the form of gel acts as a drug delivery matrix due to its characteristic properties such as flowability, injectability, and biocompatibility. These essential properties help in achieving sustained release action of therapeutic molecules and to be an attractive biomaterial in tissue engineering application. Collagen gel with suitable pore size alone can act as an effective biomaterial when compared to the other commercially available forms of collagen. Collagen–synthetic polymer composites and collagen-based diffusion membranes are generally used for controlled drug delivery and prolonged drug release treatment. These forms of dosages have shown an improved prophylactic activity when loaded with suitable antibiotics to treat infections of bone and soft tissues [147, 152].

6.2.17 Hyaluronic acid

Hyaluronic acid (also called as hyaluronan and hyaluronate (HA) and sodium hyaluronate (SA), the sodium salt form of hyaluronic acid) is a biodegradable, biocompatible,

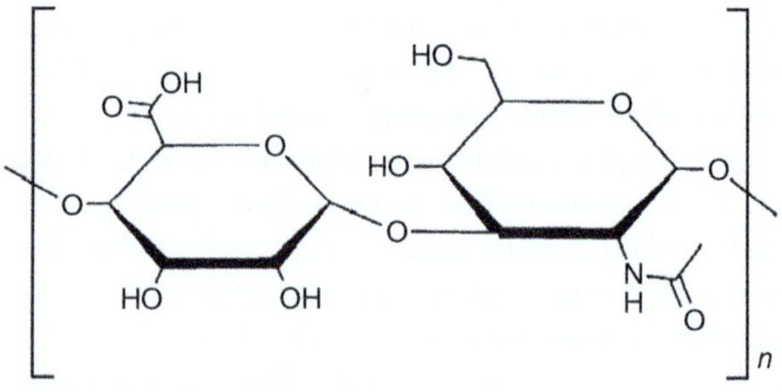

Figure 6.16 Structure of hyaluronic acid.

and viscoelastic linear polysaccharide of a wide molecular weight range (1000–10 000 000 Da) (Figure 6.16).

It is a naturally occurring biopolymer, which serves important biological functions in bacteria and higher animals including humans. Naturally occurring hyaluronic acid may be found in the tissue of higher animals, particularly as intercellular space filler. It is found in greatest concentrations in the vitreous humor of the eye and in the synovial fluid of articular joints. Hyaluronic acid is composed of linear, unbranched, and polyanionic disaccharide units consisting of glucuronic acid (GlcUA) and *N*-acetyl glucosamine (GlcNAc) joined alternately by β-1-3 and β-1-4 glycosidic bonds. Hyaluronic acid solutions are characteristically viscoelastic and pseudoplastic. The viscoelastic property of hyaluronic acid solutions that is important in its use as a biomaterial is controlled by the concentration and molecular weight of the hyaluronic acid chains. As a microcapsule, it can be used for targeted drug delivery [152].

Coating nanoparticles with HA could form an HA surface layer, which could improve biocompatibility, blood circulation time, and most importantly, active targeting capacity mediated by CD44 overexpressed on many cancer cells. The HA coating or modification could be used to target tumor stem cells because of the overexpression of CD44 on the tumor stem cells [153]. Shen et al. [154] coated solid lipid nanoparticles (SLNs) with HA (HA-SLNs) for targeting delivery of PTX to melanoma stem-like cells. It showed that the melanoma stem-like cells expressed high level of CD44, while normal cancer cells expressed low level. *In vitro*, the PTX-loaded HA-SLNs showed higher cellular uptake in stem-like cells and induced higher percentage of apoptosis cells than the PTX-loaded SLNs.

One of the novel nanotopical drug delivery system developed using HA-modified ethosomes targets CD44 in the inflamed epidermis [155]. Ethosomes are considered as deformable liposomes prepared by dispersing liposomes in small-chain biocompatible alcohols, and they are demonstrated to be better performing than classic liposomes with significantly increased skin retention of drugs [156]. The incorporation of curcumin in HA-modified ethosomes targets CD44 in the inflamed epidermis [155]. Recent studies have found that CD44 protein is highly expressed in the epidermis of psoriatic inflamed skin, suggesting that CD44 can serve as a potential target of novel active-targeting nanocarriers for topical administration to increase skin drug retention and enhance drug efficacy [157].

6.2.18 Starch

Starch is the most abundant natural polymer. Starch is obtained from various classes of plants such as roots, cereals or grains, tubers, legumes, and fruits. It has found wide use in drug delivery systems as binder, disintegrant, and filler. The bioavailability of poor-soluble drugs has been improved using starch as a polymer (Figure 6.17).

Starch has been proposed as a carrier for a number of drugs due to its properties such as biodegradability, non-toxicity, good biocompatibility, and simple production combined with low cost and ready availability, which makes it suitable for use in biomedical and pharmaceutical formulations [158].

A superparamagnetic iron oxide nanoparticle is coated by starch for injectable drug delivery. The starch-coated SPION shows a biocompatibility and possibility of being transported in the extracellular space as well as being internalized in the nerve cells [159].

An eco-friendly nanoprecipitation technique was utilized to synthesize native starch nanoparticles loaded with insoluble drugs such as indomethacin (IND) and acyclovir (ACV). The results obtained could be attributed to the dependence of the *in vitro* release rate of IND and ACV on the viscosity of starch when native starch has higher holding capacity and entrapment efficiency for loading the drugs under investigation [160].

In the study of the medical application of the cost-effective cross-linked starch nanoparticles, for the transdermal delivery using diclofenac sodium (DS), the overall results indicate that the starch nanoparticles could be considered as a good carrier for DS drug regarding the enhancement in its controlled release and successful permeation, thus offering a promising nanoparticulate system for the transdermal delivery of nonsteroidal anti-inflammatory drug (NSAID) [161].

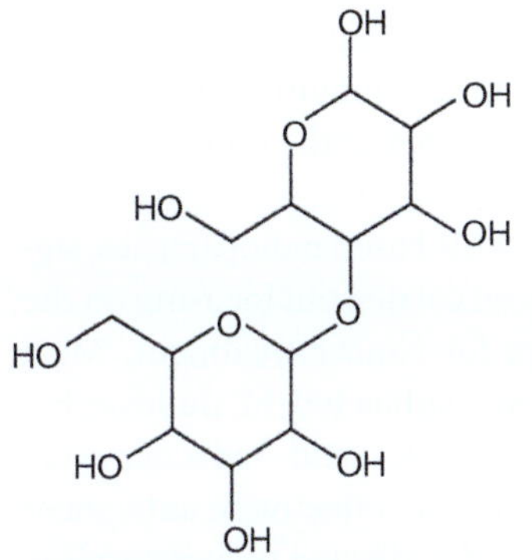

Figure 6.17 Structure of starch.

6.2.19 Guar Gum

Guar gum is a hydrocolloid that is a naturally occurring plant polysaccharide. It is a water-soluble polysaccharide composed of sugars such as mannose and galactose. Guar gum is found in dairy products, condiments, and backed goods. It is a natural nontoxic, biodegradable, mucoadhesive, and cost-effective polymer. It has been mainly used for intestinal drug delivery [162] (Figure 6.18).

Guar gum has been used in studies as nanoparticulate drug delivery systems in the form of nanospheres and nanocapsules based on the use of cross-linked guar gum, capable of incorporating a higher amount of tamoxifen citrate. Guar gum nanospheres containing tamoxifen citrate were prepared and characterized for using it as a carrier for targeted drug delivery for treatment of breast cancer [163].

A mesoporous silica-based colonic enzyme-responsive oral drug delivery system has been developed through guar gum capping. They had also demonstrated *in vitro* that the

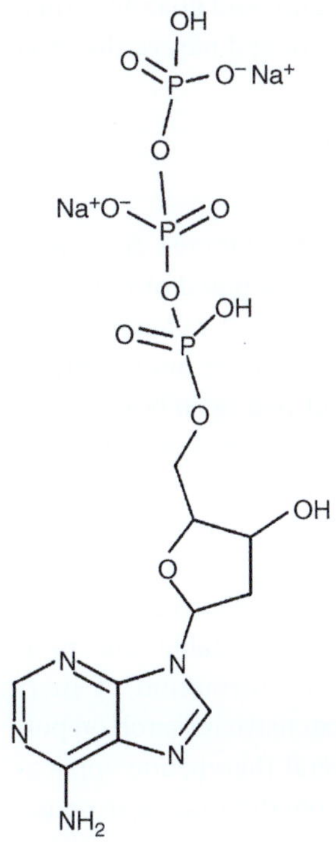

Figure 6.18 Structure of guar gum.

drug-loaded guar gum–mesoporous silica nanoparticles did not exhibit any undesired drug leakage in different pH conditions of GIT [164].

6.2.20 Xanthan Gum

Xanthan gum is obtained from the microorganism *Xanthomonas campestris*. It is used in pharmaceutical industries because of its nontoxic, biodegradable, and hydrophilic characteristics. It is soluble in hot water as well as cold water. Enzymatic degradation is prevented, and it is stable over a wide range of pH and temperature conditions. Xanthan gum shows synergistic effect and improved viscosity with guar gum. With mannan, it forms elastic and thermally reversible gel. In pharmaceutical industries, it is mostly used as a control release agent in the case of solid dosage form and used as a thickening agent, suspending agent, and emulsion stabilizer in the case of liquid dosage form [165].

In a study by Pooja et al., the authors investigated the use of xanthan gum as a reducing agent in the synthesis of gold nanoparticles. These nanoparticles were found to be effective in drug delivery because of their size and efficient targeted drug release in cancer cells. The gold nanoparticles synthesized using xanthan gum were nontoxic and biocompatible as observed in the hemolysis study. The formulation showed high drug loading, stability, and enhanced cytotoxicity in lung cancer cells [166]. In another study, it was also reported that the viscoelastic gel produced by the synergistic interaction of xanthan gum and guar gum mixtures led to the stabilization of micro- and nanoscaled iron particles [167].

6.2.21 Agarose

Agarose is a natural polysaccharide obtained from red-purple seaweeds [18]. It is basically composed of repeating units of D-galactose and 3,6-anhydro-L-galactose [34]. Agarose is generally insoluble in cold water but soluble in boiling water [18]. The formation of hydrogen bonds takes place between the agarose chains as the agarose solution is cooled below 45 °C. The gelling temperature is dependent on the monomer composition and the concentration of the solution [168] (Figure 6.19).

Agarose is commonly used in different biotechnological applications due to its natural, biodegradable, and nontoxic properties, which lead to the formation of inert nanoparticles. In recent studies, extensive research on polysaccharide nanoparticles for several therapeutic applications has been performed. The monodisperse agarose gel nanoparticles were developed using W/O microemulsion system to control the size of the particles.

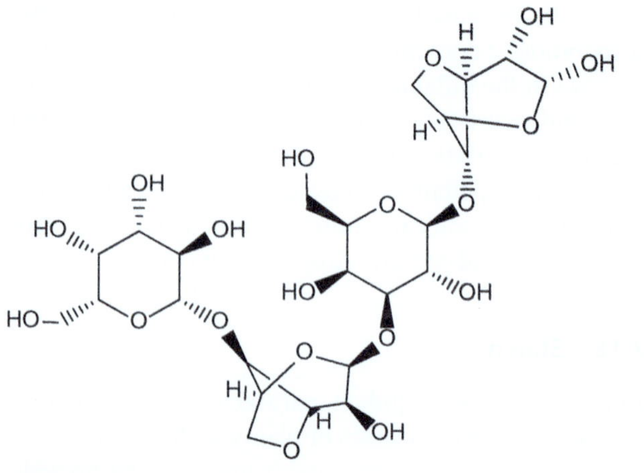

Figure 6.19 Structure or agarose.

The hydrogel particles using agarose for the encapsulation and release of a synthetic insecticide azamethiphos was tested in a study, where the results witnessed that insecticide molecules encapsulated in the fabricated nanoparticles were released following a diffusion-controlled mechanism [168].

Further, the use of agarose for the fabrication of nanoparticles has been widely extended [169, 170]. Wang and Wu [170] described the formulation of agarose hydrogel nanoparticles using an emulsion to suspension conversion method and also explained the applications of the obtained particles for encapsulating proteins and peptides molecules.

6.2.22 Silk

Silk fibroin (SF) is obtained in nature from the silkworm, *Bombyx mori*. It is a natural protein in nature. According to reports, it is used as a biomaterial to deliver the drug in different dosage forms. Silk is also used in burn-wound dressing, enzyme immobilization matrices, nets, vascular prostheses, and implants. Silk degradation is difficult because it has a compact structure along with special orientation and crystallinity. Many articles prove that silk is a biodegradable material, but it takes time in degradation. Two types of proteins are present in silk fiber, fibroin and serine. Both are present in different amounts. Silk has good strength and toughness as compared to the other biomaterial [171] (Figure 6.20).

For small drug delivery from SF-based nanoparticles, significant research work has been carried out focusing on the delivery of anticancer drugs for cancer treatment. Most current anticancer agents are subjected to undesirable biodistribution, systemic toxicity, and adverse side effects [172]. In order to treat cancer effectively, anticancer drugs should be delivered into the affected tissues surpassing the hindrances in the body with minimum loss of their

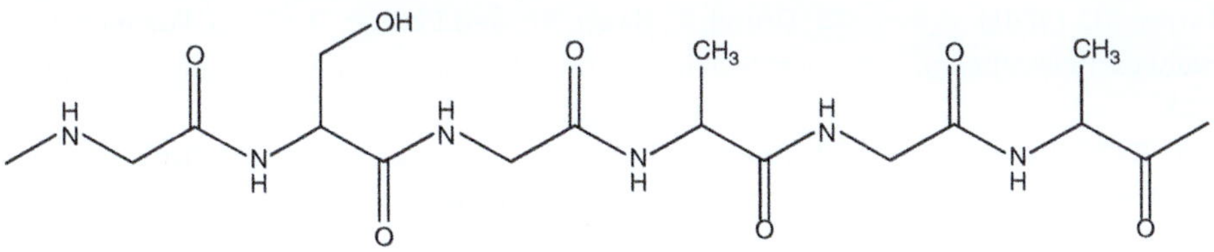

Figure 6.20 Structure of silk.

therapeutic potential in the systemic circulation. Upon reaching the area of tumor tissues, the anticancer drug should selectively kill the targeted tumor cells without affecting the healthy cells by passive and active targeting [173, 174]. Besides, the drugs should be released in a controlled manner in order to have the desired therapeutic effect.

Recently, an anticancer drug-loaded SF nanoparticle has been developed, which has shown significant potential for cancer treatment. Notably, the incorporation of the anticancer drugs such as PTX, doxorubicin (DOX), floxuridine, methotrexate, curcumin, emodin, and cis-dichlorodiaminoplatinum into SF nanoparticles has generated lots of interest in this field [175–181]. Chen et al. [175] formulated PTX-loaded SF nanoparticles ranging in size from 270 to 520 nm. The preparation was done by addition of PTX–ethanol solution into regenerated SF solution under gentle stirring conditions. According to reports, the release duration of PTX-SF nanoparticles can be extended for up to two weeks when the drug is loaded to about 3.0%. Using similar methods, Wu et al. also prepared the PTX–SF nanoparticles of diameter 130 nm. PTX maintained its pharmacological activity after incorporation into PTX–SF nanoparticles. The *in vivo* antitumor studies of PTX–SF nanoparticles on gastric cancer nude mice xenograft model indicated that due to regional delivery, PTX–SF nanoparticles showed better antitumor efficacy by retarding the growth of tumor cells and reducing the weight of the tumor compared with systemic administration [176]. Yu et al. [177] also reported the development of hydrophilic anticancer drug floxuridine-loaded SF nanoparticles with a particle size of 200–500 nm using a similar method of preparation. The maximum drug loading was found to be about 6.8%, and the release time of floxuridine was observed to be more than two days. The floxuridine-loaded SF nanoparticles also demonstrated the similar therapeutic effect to inhibit HeLa cell growth as compared to the free floxuridine. After 24 hours, floxuridine-loaded SF nanoparticles demonstrated inhibition of more than 80% of HeLa cells. The results together suggest that SF-based nanocarriers for anticancer drug have future scope and potential for chemotherapy and in clinical applications [182].

6.3 Future Prospects of Eco-friendly Polymers for Drug Delivery Applications

Eco-friendly biodegradable polymers are a *potential* carrier material for specific *drug delivery* because of their nontoxic and biocompatible nature. Advances in biodegradable polymer science have led to the development of several novel nanotechnology-based drug delivery system. With the consistent development of newer polymer compositions, it is expected that increasingly specific and versatile drug carriers are going to be introduced in future. Drug delivery systems developed using the eco-friendly polymers will lead to the efficacious and safe delivery of drugs that have serious possible biopharmaceutical and toxicity constraints. In spite of several advancements in the field of biodegradable polymers, there are still many challenges and opportunities existing together to transform the responsive polymers from laboratory to clinic. Advanced treatment of diseases with the use of ecofriendly biodegradable polymers requires carriers that can deliver their payload in a highly site-specific way to achieve therapeutically relevant concentrations in affected tissues.

References

1 Mishra, R.K., Ha, S.K., Verma, K. et al. (2018). Recent progress in selected bio-nanomaterials and their engineering applications: an overview. *J. Sci. Adv. Mater. Dev.* 3: 263–288.

2 Joshi, J.R. and Patel, R. Role of biodegradable polymers in drug delivery. *Int. J. Curr. Pharm. Res.* 4 (4): 74–81.

3 Sahoo, S.K. and Labhasetwar, V. (2003). Nanotech approaches to drug delivery and imaging. *Drug Discov.* 8: 1112–1120.

4 Duncan, R. (2003). The dawning era of polymer therapeutics. *Nat. Rev. Drug Discov.* 2: 347–360.

5 Moghimi, S.M., Hunter, A.C., and Murray, J.C. (2001). Long-circulating and target-specific nanoparticles: theory to practice. *Pharmacol. Rev.* 53: 283–318.

6 Ulery, B.D., Nair, L.S., and Laurencin, C.T. (2011). Biomedical applications of biodegradable polymers. *J. Polym. Sci. B Polym. Phys.* 49: 832–864.

7 Honey, P.J., Rijo, J., Anju, A., and Anoop, K.R. (2014). Smart polymers for the controlled delivery of drugs-a concise overview. *Acta Pharm. Sin. B* 4: 120–127.

8 Fleige, E., Quadir, M.A., and Haag, R. (2012). Stimuli-responsive polymeric nanocarriers for the controlled transport of active compounds: concepts and applications. *Adv. Drug Deliv. Rev.* 64: 866–884.

9 Chen, W., Meng, F., Cheng, R., and Zhong, Z. (2010). pH-Sensitive degradable polymersomes for triggered release of anticancer drugs: a comparative study with micelles. *J. Control. Release* 142: 40–46.

10 Khanal, S., Adhikari, U., Rijal, N.P. et al. (2016). pH-responsive PLGA nanoparticle for controlled payload delivery of diclofenac sodium. *J. Funct. Biomater.* 7: 21.

11 Nishiyama, N., Bae, Y., Miyata, K. et al. (2005). Smart polymeric micelles for gene and drug delivery. *Drug Discov. Today* 2: 21–26.

12 Sawant, R.M., Hurley, J.P., Salmaso, S. et al. (2006). SMART drug delivery systems: double-targeted pH-responsive pharmaceutical nanocarriers. *Bioconjugate Chem.* 17: 943–949.

13 Torchilin, V.P. (2001). Structure and design of polymeric surfactant-based drug delivery systems. *J. Control. Release* 73: 137–172.

14 Emerich, D.F. and Thanos, C.G. (2006). The pinpoint promise of nanoparticle-based drug delivery and molecular diagnosis. *Biomol. Eng.* 23: 171–184.

15 Farokhzad, O.C. and Langer, R. (2006). Nanomedicine: developing smarter therapeutic and diagnostic modalities. *Adv. Drug Deliv. Rev.* 58: 1456–1459.

16 Beck, R.C.R., Chaves, P.S., Goyanes, A. et al. (2017). 3D printed tablets loaded with polymeric nanocapsules: An innovative approach to produce customized drug delivery systems. *Int. J. Pharm.* 528: 268–279.

17 Gaumet, M., Vargas, A., Gurny, R. et al. (2008). Nanoparticles for drug delivery: the need for precision in reporting particle size parameters. *Eur. J. Pharm. Biopharm.* 69: 1–9.

18 Wissing, S.A., Kayser, O., and Müller, R.H. (2004). Solid lipid nanoparticles for parenteral drug delivery. *Adv. Drug Deliv. Rev.* 56: 1257–1272.

19 Duncan, R. (2006). Polymer conjugates as anticancer nanomedicines. *Nat. Rev. Cancer* 6: 688–701.

20 Lammers, T., Subr, V., Ulbrich, K. et al. (2010). Polymeric nanomedicines for image-guided drug delivery and tumor-targeted combination therapy. *Nano Today* 5: 197–212.

21 Grund, S., Bauer, M., and Fischer, D. (2011). Polymers in drug delivery-state of the art and future trends. *Adv. Eng. Mater.* 13: B61–B87.

22 Tyler, B., Gullotti, D., Mangraviti, A. et al. (2016). Polylactic acid (PLA) controlled delivery carriers for biomedical applications. *Adv. Drug Deliv. Rev.* 107: 163–175.

23 Wang, J., Li, S., Han, Y. et al. (2018). Poly (ethylene glycol)–polylactide micelles for cancer therapy. *Front. Pharmacol.* 9: 202.

24 Ding, D. and Zhu, Q. (2018). Recent advances of PLGA micro/nanoparticles for the delivery of biomacromolecular therapeutics. *Mater. Sci. Eng.* 92: 1041–1060.

25 Zhang, K., Tang, X., Zhang, J. et al. (2014). PEG-PLGA copolymers: their structure and structure-influenced drug delivery applications. *J. Control. Release* 183 (10): 77–86.

26 Goel, A.S. (2012). A drug delivery: special emphasis given on biodegradable polymers. *Adv. Polym. Sci. Technol. Int. J.* 2 (1): 1–15.

27 Chandra, R. and Rustgi, R. (1998). Biodegradable polymers. *Prog. Polym. Sci.* 23: 1273–1335.

28 Vyas, S.P. and Khar, R.K. (2002). *Controlled Drug Delivery: Concepts and Advances*, 1e, 156–189. New Delhi: Vallabh Prakashan.

29 Uhrich, K.E., Cannizzaro, S.M., and Langer, R.S. (1999). Polymeric systems for controlled drug release. *Chem. Rev.* 99: 3181–3198.

30 Mody, V.V. (2010). Introduction to Polymeric Drug Delivery. *Internet J. Med. Update* 5 (2): 1–2.

31 Rasmesh, O.P. (2001). Polymers in drug delivery. *Curr. Opin. Chem. Biol.* 5 (4): 447–451.

32 Clochard, M., Dinand, E., Rankin, S. et al. (2001). New strategies for polymer development in pharmaceutical science short review. *J. Pharm. Pharmacol.* 53 (9): 1175–1184.

33 Mohanty, A.K., Misra, M., and Hinrichsen, G. (2000). Biodegradable polymers and biocomposites: an overview. *Macromol. Mater. Eng.*: 276–277.

34 Uhrich, K.E., Cannizzaro, S.M., and Langer, R.S. (1999). Polymeric system for controlled drug release. *Chem. Rev.* 99: 3181–3198.

35 Din, F.U., Aman, W., Ullah, I. et al. (2017). Effective use of nanocarriers as drug delivery systems for the treatment of selected tumors. *Int. J. Nanomed.* 12: 7291–7309.

36 Vu-Quang, H., Vinding, M.S., Nielsen, T. et al. (2016). Theranostic tumor targeted nanoparticles combining drug delivery with dual near infrared and 19F magnetic resonance imaging modalities, Nanomedicine: Nanotechnology. *Biol. Med.* 12 (7): 1873–1884.

37 Yin, J., Chen, Y., Zhang, Z.H., and Han, X. (2016). Stimuli-responsive block copolymer-based assemblies for cargo delivery and theranostic applications. *Polymers* 8 (7): 268.

38 Chen, J., Wu, Q., Luo, L. et al. (2017). Dual tumor-targeted poly(lactic-co-glycolic acid)-polyethylene glycol-folic acid nanoparticles: a novel biodegradable nanocarrier for secure and efficient antitumor drug delivery. *Int. J. Nanomed.* 12: 5745–5760.

39 Chen, Q., Chen, J., Yang, Z. et al. (2019). Nanoparticle enhanced radiotherapy to trigger robust cancer immunotherapy. *Adv. Mater.* 31: 180–228.

40 Chapman, S., Dobrovolskaia, M., Farahani, K. et al. (2013). Nanoparticles for cancer imaging: the good, the bad, and the promise. *Nano Today* 8: 454–460.

41 Xi, J., Wang, W., Da, L. et al. (2018). Au-PLGA hybrid nanoparticles with catalase-mimicking and near-infrared photothermal activities for photoacoustic imaging-guided cancer therapy. *ACS Biomater. Sci. Eng.* 4: 1083–1091.

42 Du, J., Li, X.-Y., Hu, H. et al. (2018). Preparation and imaging investigation of dual-targeted C 3 F 8-filled PLGA nanobubbles as a novel ultrasound contrast agent for breast cancer. *Sci. Rep.* 8: 3887.

43 Shen, X., Li, T., Chen, Z. et al. (2019). NIR-light-triggered anticancer strategy for dual-modality imaging-guided combination therapy via a bioinspired hybrid PLGA nanoplatform. *Mol. Pharm.* 16: 1367–1384.

44 Saravanakumar, K., Hu, X., Shanmugam, S. et al. (2019). Enhanced cancer therapy with pH-dependent and aptamer functionalized doxorubicin loaded polymeric (poly d,l-lactic-co-glycolic acid) nanoparticles. *Arch. Biochem. Biophys.* 671: 143–151.

45 Patel, B.K., Parikh, R.H., and Patel, N. (2018). Targeted delivery of mannosylated-PLGA nanoparticles of antiretroviral drug to brain. *Int. J. Nanomed.* 13: 97–100.

46 Martins, C., Araújo, F., Gomes, M.J. et al. (2019). Using microfluidic platforms to develop CNS-targeted polymeric nanoparticles for HIV therapy. *Eur. J. Pharm. Biopharm.* 138: 111–124.

47 Yang, H., Li, J., Patel, S.K. et al. (2019). Design of poly (lactic-co-glycolic acid)(PLGA) nanoparticles for vaginal co-delivery of griffithsin and dapivirine and their synergistic effect for HIV prophylaxis. *Pharmaceutics* 11: E184.

48 Mandal, S., Prathipati, P.K., Belshan, M. et al. (2019). A potential long-acting bictegravir loaded nano-drug delivery system for HIV-1 infection: a proof-of-concept study. *Antiviral Res.* 167: 83–88.

49 Gendelman, H.E., Anantharam, V., Bronich, T. et al. (2015). Nanoneuromedicines for degenerative, inflammatory, and infectious nervous system diseases. *Nanomedicine* 11: 751–767.

50 Lamprecht, A., Ubrich, N., Yamamoto, H. et al. (2001). Biodegradable nanoparticles for targeted drug delivery in treatment of inflammatory bowel disease. *J. Pharmacol. Exp. Ther.* 299: 775–781.

51 Davoudi, Z., Peroutka-Bigus, N., Bellaire, B. et al. (2018). Intestinal organoids containing poly(lactic-co-glycolic acid) nanoparticles for the treatment of inflammatory bowel diseases. *J. Biomed. Mater. Res. A* 106: 876–886.

52 Pereira, A.D.S.B.F., Brito, G.A.D.C., Lima, M.F.S. et al. (2018). Metformin hydrochloride-loaded PLGA nanoparticle in periodontal disease experimental model using diabetic rats. *Int. J. Mol. Sci.* 19: 3488.

53 Gholizadeh, S., Kamps, J.A.A.M., Hennink, W.E. et al. (2018). PLGA-PEG nanoparticles for targeted delivery of the mTOR/PI3kinase inhibitor dactolisib to inflamed endothelium. *Int. J. Pharm.* 548: 747–758.

54 Yang, X. (2019). Design and optimization of crocetin loaded PLGA nanoparticles against diabetic nephropathy via suppression of inflammatory biomarkers: a formulation approach to preclinical study. *Drug Deliv.* 26: 849–859.

55 Trujillo-Nolasco, R.M., Morales-Avila, E., Ocampo-García, B.E. et al. (2019). Preparation and in vitro evaluation of radiolabeled HA-PLGA nanoparticles as novel MTX delivery system for local treatment of rheumatoid arthritis. *Mater. Sci. Eng. C Mater. Biol. Appl.* 103: 109766.

56 Faheem, A.M., Abdeåkader, D., Osman, M.A. et al. (2019). Oral insulin delivery in diabetic rats by PLGA nanoparticles combined with a protease inhibitor (*N*-ethylmaleimide). *Br. J. Pharm.* 4: S2–S3.

57 Duran, V., Yasar, H., Becker, J., Thiyagarajan, D., et al. Preferential uptake of chitosan-coated PLGA nanoparticles by primary human antigen presenting cells. Nanomedicine. 2019; 21:102073

58 Besumbes, E.S., Fornaguera, C., Monge, M. et al. (2019). PLGA cationic nanoparticles, obtained from nanoemulsion templating, as potential DNA vaccines. *Eur. Polym. J.* 120: 109229.

59 Jeon, S.G., Cha, M.-Y., Kim, J.-I. et al. (2019). Vitamin D-binding protein-loaded PLGA nanoparticles suppress Alzheimer's disease-related pathology in 5XFAD mice. *Nanomedicine* 17: 297–307.

60 Gonzalez-Pizarro, R., Silva-Abreu, M., Calpena, A.C. et al. (2018). Development of fluorometholone-loaded PLGA nanoparticles for treatment of inflammatory disorders of anterior and posterior segments of the eye. *Int. J. Pharm.* 547: 338–346.

61 Swider, E., Koshkina, O., Tel, J. et al. (2018). Customizing poly(lactic-co-glycolic acid) particles for biomedical applications. *Acta Biomater.* 73: 38–51.

62 Jain, A., Kunduru, K.R., Basu, A. et al. (2016). Injectable formulations of poly(lactic acid) and its copolymers in clinical use. *Adv. Drug Deliv. Rev.* 107: 213–227.

63 Strohbehn, G., Coman, D., Han, L. et al. (2015). Imaging the delivery of brain-penetrating PLGA nanoparticles in the brain using magnetic resonance. *J. Neurooncol.* 121: 441–449.

64 Erdmann, L. and Uhrich, K.E. (2000). Synthesis and degradation characteristics of salicylic acid-derived poly(anhydride-esters). *Biomaterials* 21: 1941–1946.

65 Sheng, J., Han, L., Qin, J. et al. (2015). N-trimethyl chitosan chloride-coated PLGA nanoparticles overcoming multiple barriers to oral insulin absorption. *ACS Appl. Mater. Interfaces* 7: 15430–15441.

66 Senel, S. and McClure, S.J. (2004). Potential applications of chitosan in veterinary medicine. *Adv. Drug Deliv. Rev.* 56: 1467–1480.

67 Motiei, M., Kashanian, S., Lucia, L.A. et al. (2017). Intrinsic parameters for the synthesis and tuned properties of amphiphilic chitosan drug delivery nanocarriers. *J. Control. Release* 260: 213–225.

68 Prabaharan, M. (2008). Review paper: chitosan derivatives as promising materials for controlled drug delivery. *J. Biomater. Appl.* 23: 5–36.

69 Kean, T. and Thanou, M. (2010). Biodegradation, biodistribution and toxicity of chitosan. *Adv. Drug Deliv. Rev.* 62: 3–11.

70 Diebold, Y., Jarrin, M., Saez, V. et al. (2007). Ocular drug delivery by liposome-chitosan nanoparticle complexes (LCS-NP). *Biomaterials* 28: 1553–1564.

71 Woraphatphadung, T., Sajomsang, W., Rojanarata, T. et al. (2018). Development of chitosan-based pH-sensitive polymeric micelles containing curcumin for colon-targeted drug delivery. *AAPS Pharm. Sci. Tech.* 19: 991–1000.

72 Wang, Y., Li, B., Xu, F. et al. (2018). Tough magnetic chitosan hydrogel nanocomposites for remotely stimulated drug release. *Biomacromolecules* 19: 3351–3360.

73 Liu, D., Li, J., Pan, H. et al. (2016). Potential advantages of a novel chitosan-N-acetylcysteine surface modified nanostructured lipid carrier on the performance of ophthalmic delivery of curcumin. *Sci. Rep.* 6: 28796–28809.

74 Ho, D.K., Frisch, S., Biehl, A. et al. (2018). Farnesylated glycol chitosan as a platform for drug delivery: synthesis, characterization, and investigation of mucus-particle interactions. *Biomacromolecules* 19: 3489–3501.

75 Ali, A. and Ahmed, S. (2018). A review on chitosan and its nanocomposites in drug delivery. *Int. J. Biol. Macromol.* 109: 273–286.

76 Ahsan, S.M., Thomas, M., Reddy, K.K. et al. (2018). Chitosan as biomaterial in drug delivery and tissue engineering. *Int. J. Biol. Macromol.* 110: 97–109.

77 Dobhal, A.; Bangde, P.; Dey, A.; et al. Chitosan-based nanoparticulate systems: implication towards therapeutics application. In Particulate Technology for Delivery of Therapeutics; Jana, S., Ed.; Springer: Singapore, 2017; 167–225.

78 Liu, M., Zhang, J., Zhu, X. et al. (2016). Efficient mucus permeation and tight junction opening by dissociable "mucus-inert" agent coated trimethyl chitosan nanoparticles for oral insulin delivery. *J. Control. Release* 222: 67–77.

79 Tian, H., He, Z., Sun, C. et al. (2018). Uniform core-shell nanoparticles with thiolated hyaluronic acid coating to enhance oral delivery of insulin. *Adv. Healthc. Mater.* 7: 1800285–1800296.

80 Gu, Z., Dang, T., Ma, M. et al. (2013). Glucose-responsive microgels integrated with enzyme nanocapsules for closed-loop insulin delivery. *ACS Nano* 7: 6758–6766.

81 Lin, Y.H., Chen, Z.R., Lai, C.H. et al. (2015). Active targeted nanoparticles for oral administration of gastric cancer therapy. *Biomacromolecules* 16: 3021–3032.

82 Wei-liang Chen, Fang Li, Yan T., et al. Stepwise pH-responsive nanoparticles for enhanced cellular uptake and on-demand intracellular release of doxorubicin, *Int. J. Nanomed.* 2017; 12: 4241–4256

83 Shi, S., Zhang, Z., Luo, Z. et al. (2015). Chitosan grafted methoxy poly(ethylene glycol)-poly(epsilon-caprolactone) nanosuspension for ocular delivery of hydrophobic diclofenac. *Sci. Rep.* 5: 11337–11348.

84 Katsarov, P., Pilicheva, B., Uzunova, Y. et al. (2018). Chemical cross-linking: a feasible approach to prolong doxylamine/pyridoxine release from spray-dried chitosan microspheres. *Eur. J. Pharm.Sci.* 123: 387–394.

85 Situ, W., Xiang, T., and Liang, Y. (2018). Chitosan-based particles for protection of proteins during storage and oral administration. *Int. J. Biol. Macromol.* 117: 308–314.

86 Ye, Z., Wu, W.R., Qin, Y.F. et al. (2018). An integrated therapeutic delivery system for enhanced treatment of hepatocellular carcinoma. *Adv. Funct. Mater.* 28: 1706600.

87 Wu, J., Tang, C., and Yin, C. (2017). Co-delivery of doxorubicin and interleukin-2 via chitosan based nanoparticles for enhanced antitumor efficacy. *Acta Biomater.* 47: 81–90.

88 Murugan, C., Rayappan, K., Thangam, R. et al. (2016). Combinatorial nanocarrier based drug delivery approach for amalgamation of anti-tumor agents in breast cancer cells: An improved nanomedicine strategy. *Sci. Rep.* 6: 34053–34064.

89 Shen, J.W., Li, J., Zhao, Z. et al. (2017). Molecular dynamics study on the mechanism of polynucleotide encapsulation by chitosan. *Sci. Rep.* 7: 5050–5058.

90 Lee, J.Y., Crake, C., Teo, B. et al. (2017). Ultrasound-enhanced siRNA delivery using magnetic nanoparticle-loaded chitosan-deoxycholic acid nanodroplets. *Adv. Healthc. Mater.* 6: 1601246–1601254.

91 Corbet, C., Ragelle, H., Pourcelle, V. et al. (2016). Delivery of siRNA targeting tumor metabolism using non-covalent PEGylated chitosan nanoparticles: identification of an optimal combination of ligand structure, linker and grafting method. *J. Control. Release* 223: 53–63.

92 Seidi, F., Jenjob, R., Phakkeeree, T., and Crespy, D. (2018). Saccharides, oligosaccharides, and polysaccharides nanoparticles for biomedical applications. *J. Control. Release* 284: 188–212.

93 Shi, G.N., Zhang, C.N., Xu, R. et al. (2017). Enhanced antitumor immunity by targeting dendritic cells with tumor cell lysate-loaded chitosan nanoparticles vaccine. *Biomaterials* 113: 191–202.

94 Gennari, A., Pelliccia, M., Donno, R. et al. (2016). Mannosylation allows for synergic (CD44/C-type lectin) uptake of hyaluronic acid nanoparticles in dendritic cells, but only upon correct ligand presentation. *Adv. Healthc. Mater.* 5: 966–976.

95 Foox, M. and Zilberman, M. (2015). Drug delivery from gelatin-based systems. *Exp. Opin. Drug Deliv.* 12 (9): 1547–1563.

96 Hans, M.L. and Lowman, A.M. (2002). Biodegradable nanoparticles for drug delivery and targeting. *Curr. Opin. Solid State Mater. Sci.* 6: 319–327.

97 Kreuter, J. (2001). Nanoparticulate systems for brain delivery of drugs. *Adv. Drug Deliv. Rev.* 47: 65–81.

98 Immordino, M.L., Dosio, F., and Cattel, L. (2006). Stealth liposomes: review of the basic science, rationale, and clinical applications, existing and potential. *Int. J. Nanomed.* 1: 297.

99 Chen, C., Han, D., Cai, C., and Tang, X. (2010). An overview of liposome lyophilization and its future potential. *J. Control. Release* 142: 299–311.

100 Mufamadi, M.S., Pillay, V., Choonara, Y.E. et al. (2011). A review on composite liposomal technologies for specialized drug delivery. *J. Drug Deliv.* 2011: 939851.

101 Li, J.K., Wang, N., and Wu, X.S. (1998). Gelatin nanoencapsulation of protein/peptide drugs using an emulsifier-free emulsion method. *J. Microencapsul.* 15: 163–172.

102 Bajpai, A.K. and Choubey, J. (2006). Design of gelatin nanoparticles as swelling controlled delivery system for chloroquine phosphate. *J. Mater. Sci.* 17: 345–358.

103 Narayanan, D., Koyakutty, M., Nair, S., and Menon, D. (2013). Poly-(ethylene glycol) modified gelatin nanoparticles for sustained delivery of the anti-inflammatory drug Ibuprofen-Sodium: an in vitro and in vivo analysis. *Nanomedicine* 9: 818–828.

104 Lee, E.S., Gao, Z., and Bae, Y.H. (2008). Recent progress in tumor pH targeting nanotechnology. *J. Control. Release* 132: 164–170.

105 Manca, M.L., Cassano, R., Valenti, D. et al. (2013). Isoniazid-gelatin conjugate microparticles containing rifampicin for the treatment of tuberculosis. *J. Pharm. Pharmacol.* 65: 1302–1311.

106 DiTizio, V., Ferguson, G.W., Mittelman, M.W. et al. (1998). A liposomal hydrogel for the prevention of bacterial adhesion to catheters. *Biomaterials* 19: 1877–1884.

107 Peptu, C., Popa, M., and Antimisiaris, S.G. (2008). Release of liposome-encapsulated calcein from liposome entrapping gelatincarboxymethylcellulose films: a presentation of different possibilities. *J. Nanosci. Nanotechnol.* 8: 2249–2258.

108 Tonnesen, H.H. and Karlsen, J. (2002). Alginate in drug delivery systems. *Drug Dev. Ind. Pharm.* 28 (6): 621–630.

109 Lee, K.Y. and Mooney Alginate, D.J. (2012). Properties andbiomedical applications. *Progr. Polym. Sci.* 37 (1): 106–126.

110 Joseph, I. and Venkataram, S. (1995). Indomethacin sustained release from alginate-gelatin or pectin-gelatin coacervates. *Int. J. Pharm.* 126 (1): 161–168.

111 Draget, K.I., Gaserod, O. et al. (2001). Effects of molecular weight and elastic segment flexibility on syneresis in Ca-alginate gels. *Food Hydrocolloids* 15 (4): 485–490.

112 Takka, S., Ocak, O.H. et al. (1998). Formulation and investigation of nicardipine HCl–alginate gel beads with factorial design-based studies. *Eur. J. Pharm. Sci.* 6 (3): 241–246.

113 Imai, T., Kawasaki, C. et al. (2000). Comparison of the pharmaceutical properties of sustained-release gel beads prepared by alginate having different molecular size with commercial sustained release tablet. *Die Pharm.* 55 (3): 218.

114 Illum, L. (2002). Nasal drug delivery: new developments and strategies. *Drug Discov. Today* 7 (23): 1184–1189.

115 Stokke, B.T., Draget, K.I. et al. (2000). Small-angle X-ray scattering and rheological characterization of alginate gels. 1. Caalginate gels. *Macromolecules* 33 (5): 1853–1863.

116 Morales, M.A., Finotelli, P.V., Coaquira, J.A.H. et al. (2008). In situ synthesis and magnetic studies of iron oxide nanoparticles in calcium-alginate matrix for biomedical applications. *Mater. Sci. Eng.* 28: 253–257.

117 Figueiredo, P., Lintinen, K., Hirvonen, J.T. et al. (2008). Properties and chemical modifications of lignin: towards lignin-based nanomaterials for biomedical applications. *Progr. Mater. Sci.* 93: 233–269.

118 Frangville, C., Rutkevicius, M., Richter, A.P. et al. (2012). Fabrication of environmentally biodegradable lignin nanoparticles. *Chem. Phys. Chem.* 13: 4235–4243.

119 Tortora, M., Cavalieri, F., Mosesso, P., et.al.Ultrasound driven assembly of lignin into microcapsules for storage and delivery of hydrophobic molecules. *Biomacromolecules*2014; 15: 1634–1643.

120 Lievonen, M., Valle-Delgado, J.J., Mattinen, M.L. et al. (2016). A simple process for lignin nanoparticle preparation. *Green Chem.* 18: 1416–1422.

121 Yiamsawas, D., Baier, G., Thines, E. et al. (2014). Biodegradable lignin nanocontainers. *RSC Adv.* 4: 11661–11663.

122 Chen, L., Zhou, X., Shi, Y. et al. (2018). Green synthesis of lignin nanoparticle in aqueous hydrotropic solution toward broadening the window for its processing and application. *Chem. Eng. J.* 346: 217–225.

123 Ikoni, J.O., Elijah, I.N., and Jennifer, D.A. (2011). Advances in natural polymers as pharmaceutical excipients. *Pharm. Anal. Acta* 3 (1): 146.

124 Sun, B., Zhang, M., Shen, J. et al. (2017). Applications of cellulose based materials in sustained drug delivery systems. *Curr. Med. Chem.* https://doi.org/10.2174/0929867324666170705143308.

125 Elseoud, W.S.A., Hassan, M.L., Sabaa, M.W. et al. (2018). Chitosan nanoparticles/cellulose nanocrystals nanocomposites as a carrier system for the controlled release of repaglinide. *Int. J. Biol. Macromol.* 111: 604–613.

126 Agarwal, T., Narayana, S.G.H., Pal, K. et al. (2015). Calcium alginate-carboxymethyl cellulose beads for colon-targeted drug delivery. *Int. J. Biol. Macromol.* 75: 409–417.

127 Hansen, K., Kim, G., Desai, K.G. et al. (2015). Feasibility investigation of cellulose polymers for mucoadhesive nasal drug delivery applications. *Mol. Pharm.* 12: 2732–2741.

128 Kratz, F. (2014). A clinical update of using albumin as a drug vehicle—a commentary. *J. Control. Release* 190: 331–336.

129 Hudson, D. and Margaritis, A. (2014). Biopolymer nanoparticle production for controlled release of biopharmaceuticals. *Crit. Rev. Biotechnol.* 34 (2): 161–179.

130 Neumann, E., Frei, E., Funk, D. et al. (2010). Native albumin for targeted drug delivery. *Exp. Opin. Drug Deliv.* 7 (8): 915–925.

131 Sethi, A., Sher, M., Akram, M.R. et al. (2013). Albumin as a drug delivery and diagnostic tool and its market approved products. *Acta Pol. Pharm.* 70 (4): 597–600.

132 Loureiro, A., Azoia, N.G., Gomes, A.C. et al. (2016). Albumin-based nanodevices as drug carriers. *Curr. Pharm. Des.* 2: 000–0001.

133 Yewale, C., Baradia, D., Vhora, I., and Misra, A. (2013). Proteins: emerging carrier for delivery of cancer therapeutics. *Exp. Opin. Drug Deliv.* 10 (10): 1429–1448.

134 Taheri, A., Atyabi, F., Salman Nouri, F. et al. (2011). Nanoparticles of conjugated methotrexate-human serum albumin: preparation and cytotoxicity evaluations. *J. Nanomater.* 2011.

135 Loureiro, A., Abreu, A.S., Sarria, M.P. et al. (2015). Functionalized protein nanoemulsions by incorporation of chemically modified BSA. *RSC Adv.* 5 (7): 4976–4983.

136 Huang, G. and Huang, H. (2018). Application of dextran as nanoscale drug carriers. *Nanomedicine* 13 (24): 3149–3158.

137 Salmaso, S. and Caliceti, P. (2013). Stealth properties to improve therapeutic efficacy of drug nanocarriers. *J. Drug Deliv.* 2013: 374252.

138 Yoshikawa, H., Sezaki, H., and Muranishi, S. (1983). Mechanism for selective transfer of bleomycin into lymphatics by a bifunctional delivery system via the lumen of the large intestine. *Int. J. Pharm.* 13: 321–332.

139 Harboe, E., Larsen, C., Johansen, M., and Olesen, H. (1989). Macromolecular prodrugs. XV. Colon-targeted delivery--bioavailability of naproxen from orally administered dextran-naproxen ester prodrugs varying in molecular size in the pig. *Pharm. Res.* 6: 919–923.

140 Levi-Schaffer, F., Bernstein, A., Meshorer, A., and Arnon, R. (1982). Reduced toxicity of daunorubicin by conjugation to dextran. *Cancer Treat. Rep.* 66: 107–114.

141 Takakura, Y., Matsumoto, S., Hashida, M., and Sezaki, H. (1984). Enhanced lymphatic delivery of mitomycin C conjugated with dextran. *Cancer Res.* 44: 2505–2510.

142 Aboud-Pirak, E., Hurwitz, E., Bellot, F. et al. (1989). Inhibition of human tumor growth in nude mice by a conjugate of doxorubicin with monoclonal antibodies to epidermal growth factor receptor (growth factor receptors/cytotoxicity). *Proc. Natl. Acad. Sci. USA* 86: 3778–3781.

143 Lam, W., Leung, C., Chan, H., and Fong, W. (2000). Toxicity and DNA binding of dextran-doxorubicin conjugates in multidrug-resistant KB-V1 cells: optimization of dextran size. *Anticancer Drugs* 11: 377–384.

144 Mitra, S., Gaur, U., Ghosh, P., and Maitra, A. (2001). Tumour targeted delivery of encapsulated dextran-doxorubicin conjugate using chitosan nanoparticles as carrier. *J. Control. Release* 74: 317–323.

145 Ohya, Y., Oue, H., Nagatomi, K., and Ouchi, T. (2001). Design of macromolecular prodrug of cisplatin using dextran with branched galactose units as targeting moieties to hepatoma cells. *Biomacromolecules* 2: 927–933.

146 Sagnella, S.M., Duong, H. et al. (2014). Dextran-based doxorubicin nanocarriers with improved tumor penetration. *Biomacromolecules* 15: 262–275.

147 Chak, V., Kumar, D., and Visht, S. (2013). A review on collagen based drug delivery systems. *Int. J. Pharm. Teach. Pract.* 4 (4): 811–820.

148 Nagarajan, U., Kawakami, K., Zhang, S. et al. (2014). Fabrication of solid collagennanoparticles using electrospray deposition. *Chem. Pharm. Bull.* 62: 422–428.

149 Rossler, B., Kreuter, J., and Scherer, D. (1995). Collagen microparticles: preparation and properties. *J. Microencapsul.* 12: 49–57.

150 Le, V.M., Lang, M.D., Shi, W.B. et al. (2016). A collagen-based multicellular tumor spheroid model for evaluation of the efficiency of nanoparticle drug delivery. *Artif. Cells Nanomed. Biotechnol.* 44: 540–544.

151 Nitta, S.K. and Numata, K. (2013). Biopolymer-based nanoparticles for drug/gene delivery and tissue engineering. *Int. J. Mol. Sci.* 14: 1629–1654.

152 DeFrates, K., Markiewicz, T., and Gallo, P. Protein polymer-based nanoparticles: fabrication and medical applications. *Int. J. Mol. Sci.* 2018, 19: 1717.

153 Kim, H., Jeong, H., Han, S. et al. (2017). Hyaluronate and its derivatives for customized biomedical applications. *Biomaterials* 123: 155e71.

154 Shen, H., Shi, S., Zhang, Z. et al. (2015). Coating solid lipid nanoparticles with hyaluronic acid enhances antitumor activity against melanoma stem-like cells. *Theranostics* 5: 755e71.

155 Zhang, Y., Xia, Q., Li, Y. et al. (2019). CD44 assists the topical anti-psoriatic efficacy of curcumin-loaded hyaluronan-modified ethosomes: A new strategy for clustering drug in inflammatory skin. *Theranostics* 9 (1): 48.

156 Touitou, E., Dayan, N., Bergelson, L. et al. (2000). Ethosomes - novel vesicular carriers for enhanced delivery: characterization and skin penetration properties. *J. Control. Release* 65 (3): 403–418.

157 Lindqvist, U., Pihl-Lundin, I., and Engström-Laurent, A. (2012). Dermal distribution of hyaluronan in psoriatic arthritis: coexistence of CD44, MMP3 and MMP9. *Acta Derm. Venereol.* 92 (4): 372–377.

158 Aggarwal, S. (2012). Goel A drug delivery: special emphasis given on biodegradable polymers. *Adv. Polym. Sci. Technol. Int. J.* 2 (1): 1–15.

159 Kim, D.K., Mikhaylova, M., Wang, F.H. et al. (2003). Starch-coated superparamagnetic nanoparticles as MR contrast agents. *Chem. Mater.* 15: 4343–4351.

160 El-Feky, G.S., El-Rafie, M.H., and El-Sheikh, M.A. (2015). Utilization of crosslinked starch nanoparticles as a carrier for indomethacin and acyclovir drugs. *Nanomed. Nanotechnol.* 6: 1.

161 El-Naggar, M.E., El-Rafie, M.H., El-sheikh, M.A. et al. (2015). Synthesis, characterization, release kinetics and toxicity profile of drug-loaded starch nanoparticles. *Int. J. Biol. Macromol.* 81: 718–729.

162 Prabaharan, M. (2011). Prospective of guar gum and its derivatives as controlled drug delivery systems. *Int. J. Biol. Macromol.* 49 (2): 117–124.

163 Sarmah, J.K., Bhattacharjee, S.K., Mahanta, R. et al. (2009). Preparation of cross- linked guar gum nanospheres containing tamoxifen citrate by single step emulsion in situ polymer cross-linking method. *J. Incl. Phenom. Macrocycl. Chem.* 65: 329.

164 Balmiki, K., Senthilguru, K., Animesh, M. et al. (2017). Mesoporous silica nanoparticle based enzyme responsive system for colon specific drug delivery through guar gum capping. *Colloids Surf. B: Biointerf.* 150: 352–361.

165 Abdulsalam, A., Nafaa, A., Marwan, A. et al. (2020). Xanthan gum, its biopharmaceutical applications: an overview. *World J. Pharm. Pharmaceut. Sci.* 7 (1): 1536–1548.

166 Pooja, D., Panyaram, S., Kulhari, H. et al. (2014). Xanthan gum stabilized gold nanoparticles: Characterization, biocompatibility, stability and cytotoxicity. *Carbohydr. Polym.* 110: 1–9.

167 Xue, D. and Sethi, R. (2012). Viscoelastic gels of guar and xanthan gum mixtures provide longterm stabilization of iron micro and nanoparticles. *J. Nanoparticle Res.* 14 (11): 1–14.

168 Argudo, P.G., Eduardo, G., and Alejandro, L. (2018). Nanoparticles preparation and application in drug storage and delivery of agarose. *Int. J. Polym. Sci.* 9: 7823587.

169 Satar, R., Iizhar, S.A., Rasool, M. et al. (2016). Investigating the antibacterial potential of agarose nanoparticles synthesized by nanoprecipitation technology. *Pol. J. Chem. Technol.* 18 (2): 9–12.

170 Wang, N. and Wu, X.S. (1997). Preparation and characterization of agarose hydrogel nanoparticles for protein and peptide drug delivery. *Pharmaceut. Dev. Technol.* 2 (2): 135–142.

171 Zheng, Z., Li, Y., and Xie, M.-B. (2015). Silk fibroin-based nanoparticles for drug delivery. *Int. J. Mol. Sci.* 16: 4880–4903.

172 Egusquiaguirre, S.P., Igartua, M., Hernández, R.M. et al. (2012). Nanoparticle delivery systems for cancer therapy: advances in clinical and preclinical research. *Clin. Transl. Oncol.* 14: 83–93.

173 Alexis, F., Pridgen, E.M., Langer, R. et al. (2010). Nanoparticle technologies for cancer therapy. *Handb. Exp. Pharmacol.* 197: 55–86.

174 Zhang, J., Lan, C.Q., Post, M. et al. (2006). Design of nanoparticles as drug carriers for cancer therapy. *Cancer Genom. Proteom.* 3: 147–157.

175 Chen, M., Shao, Z., and Chen, X. (2012). Paclitaxel-loaded silk fibroin nanospheres. *J. Biomed. Mater. Res. A* 100: 203–210.

176 Wu, P., Liu, Q., Li, R. et al. (2013). Facile preparation of paclitaxel loaded silk fibroin nanoparticles for enhanced antitumor efficacy by locoregional drug delivery. *ACS Appl. Mater. Interfaces* 5: 12638–12645.

177 Yu, S., Yang, W., Chen, S. et al. (2014). Floxuridine-loaded silk fibroin nanospheres. *RSC Adv.* 4: 18171–18177.

178 Gobin, A.S., Rhea, R., Newman, R.A. et al. (2006). Silk-fibroin-coated liposomes for long-term and targeted drug delivery. *Int. J. Nanomed.* 1: 81–87.

179 Cheema, S.K., Gobin, A.S., Rhea, R. et al. (2007). Silk fibroin mediated delivery of liposomal emodin to breast cancer cells. *Int. J. Pharm.* 341: 221–229.

180 Subia, B. and Kundu, S.C. (2013). Drug loading and release on tumor cells using silk fibroin-albumin nanoparticles as carriers. *Nanotechnology* 24: 035103.

181 Seib, F.P., Jones, G.T., Rnjak-Kovacina, J. et al. (2013). pH-dependent anticancer drug release from silk nanoparticles. *Adv. Healthc. Mater.* 2: 1606–1611.

182 Tian, Y., Jiang, X., Chen, X. et al. (2014). Doxorubicin-loaded magnetic Silk fibroin nanoparticles for targeted therapy of multidrug-resistant cancer. *Adv. Mater.* 26: 7393–7398.

7

Green-Nanotechnology-Driven Drug Delivery Systems

Manish Patel[1], Jayvadan Patel[2], and Richa Dayaramani[3]

[1] *L M College of Pharmacy, Ahmedabad, India*
[2] *Nootan Pharmacy College, Sankalchand Patel University, Visnagar, Gujarat, India*
[3] *Khyati College of Pharmacy, Palodiya, Ahmedabad, India*

7.1 Introduction

Nanoparticles (NPs) have drawn considerable interest due to their peculiar and interesting properties and applications that are beneficial over their bulk counterparts, with one or more dimensions on the order of 100 nm or less [1]. To synthesize various types of NPs, there are a large number of physical, chemical, biological, and hybrid methods available. While physical and chemical methods are more common in the synthesis of NPs, their biomedical applications, especially in clinical fields, are greatly restricted by the use of toxic chemicals. Therefore, to expand their biomedical applications, the development of reliable, nontoxic, and eco-friendly methods for the synthesis of NPs is of utmost importance. One of the options is to use microorganisms to synthesize NPs to achieve this objective [2]. In several respects, NPs produced by a biogenic enzymatic process are far superior to those particles produced by chemical methods. Although the latter methods are capable of producing large amounts of NPs with a given size and shape in a relatively short period of time, they are complex, obsolete, expensive, and inefficient, producing hazardous toxic waste that is harmful not only to the environment but also to human health. The use of expensive chemicals is eliminated through an enzymatic process, and the more acceptable "green" route is not as energy intensive as the chemical method and is also environmentally friendly [3].

Furthermore, the "biogenic" approach is reinforced by the fact that most bacteria live under atmospheric conditions of varying temperature, pH, and pressure. Because of the bacterial carrier matrix, the particles generated by these processes have higher catalytic reactivity, greater specific surface area, and improved contact between the enzyme and the metal salt in question. When microorganisms pick up target ions from their environment, NPs are biosynthesized and then transform the metal ions into the element metal through enzymes generated by cell activities [3]. In a variety of applications, biosynthesized NPs have been used, including drug carriers for targeted delivery, cancer treatment, gene therapy and DNA analysis, antibacterial agents, biosensors, reaction rate improvement, separation science, and magnetic resonance imaging (MRI). The use of the principles of green chemistry for the improved synthesis and modification of nanodrug delivery systems is discussed in this review. This is followed by a brief overview of the current research activities focusing on the biological synthesis of metallic NPs, oxide NPs, sulfide NPs, and other types of NPs, among the many papers on the synthesis and implementation of nanodrug delivery systems. The paper concludes with discussions on greener and more sustainable remediation (nano) solutions [4–6].

7.2 Unique Properties of Nanoparticles

Attention due to its unique size-dependent physical and chemical properties, the past decade has witnessed a major emphasis on NPs and -materials. Compared to bulk material, NPs show a variety of unique properties and also have unique visible properties since they are small enough to confine their electrons and create quantum effects. Because of these special properties, NPs such as gold are commonly used in diverse fields such as photonics, catalysis, electronics, and biomedicine [7]. NPs have been used to bioremediate

Sustainable Nanotechnology: Strategies, Products, and Applications, First Edition.
Edited by Yashwant V. Pathak, Govindan Parayil, and Jayvadan K. Patel.

radioactive waste from nuclear power plants and to manufacture nuclear weapons, such as uranium. *Bacillus sphaericus* JG-A12 cells and S-layer proteins have been shown to have special capabilities for cleaning up uranium-contaminated waste water [8]. There is a unique potential for biological systems to be self-organized and to synthesize molecules with highly selective properties. A specific plant study indicates that certain nanomaterials can inhibit germination of seeds and root growth [9].

7.3 Greener Synthetic Strategies

Our specific approach is to establish "greener" synthetic strategies using benign reagents in the matrix in which they are to be used to prepare these nanomaterials through many chemical pathways, thus minimizing the risk of contamination or removing the use and generation of the hazardous substances usually used [10].

7.3.1 Plant Polyphenols and Agricultural Residues

The "greener" synthesis of nanometal/nanometal oxide/nanostructured polymers and the subsequent stabilization (using dispersants, biodegradable polymers, etc.) includes the use of naturally renewable resources such as tea and coffee plant extracts and polyphenol antioxidants [11], biodegradable polymers such as carboxymethyl cellulose (CMC), and reducing sugar and agricultural residual waste (red grape pomace from winery waste) [12].

7.3.2 Vitamins

Our sustainable synthetic activity includes benign alternatives for the preparation of NPs that minimize or eliminate the use and production of hazardous substances [13]. Vitamins B1, B2, C, and tea [11] and wine polyphenols, both reducing and capping agents, provide extremely simple green synthetic methods, without the need for large quantities of insoluble templates, for bulk quantities of nanospheres, nanorods, nanowires, and nanoballs of aligned nanobelts and nanoplates of water-based metals [14].

7.3.3 Microwave Heating

Microwave (MW) technology is emerging as a sufficiently efficient alternative source of energy to achieve chemical transformations in minutes, rather than hours or even days. It is more important when the realization of the material properties is focused solely on the size and shape in the sense of nanomaterial synthesis; control over the synthetic methodologies is crucial. This is because nanoscale material growth is largely dependent on the thermodynamic and kinetic barriers in the reaction, as described by the trajectory of the reaction, and is affected by vacancies, defects, and reconstructions of the surface. In order to drive the reaction in which the reaction vessel serves as an intermediary for energy transfer from the heating mantle to the solvent and finally to the reactant molecules, traditional thermal techniques are based on the conduction of blackbody radiation [15]. This can cause sharp thermal gradients and inefficient, nonuniform reaction conditions in the bulk solution. In the synthesis of nanomaterials, this has been a troublesome issue where uniform nucleation and growth rates are important for material quality. In traditional thermal techniques, MW heating methods can resolve the problems of heating inhomogeneity as their use provides increased reaction kinetics, rapid initial heating, and thus increased reaction speeds, resulting in clean reaction products with rapid starting material consumption and higher yields. The technique is also applicable to enzymatic and biological systems under a number of conditions [16].

Through MW-assisted spontaneous reduction of noble metal salts using aqueous solution of a-D-glucose, sucrose, and maltose, a bulk and shape-controlled synthesis of noble nanostructures with various shapes such as prisms, cubes, and hexagons occurs. By varying the concentration of the sugars, the size of the resulting NPs can simply be controlled; higher concentration provides uniformly smaller particles that increase with a decrease in sugar concentration. For the cross-linking reaction of poly(vinyl alcohol) (PVA) with metallic systems such as Pt, Cu, and In; bimetallic systems such as Pt–In, Ag–Pt, Pt–Fe, Cu–Pd, Pt–Pd, and Pd–Fe [13]; and single-wall carbon nanotubes (SWCNTs), multiwall carbon nanotubes (MWCNTs), and Buckminsterfullerenes (C-60), a general method has been developed. The strategy applies to the development of biodegradable CMC composite films with noble nanometals [8]; it is possible to decorate and align carbon nanotubes (CNT) in CMC by means of an MW-assisted approach, which allows for the shape-controlled bulk synthesis of poly(ethylene glycol) (PEG) Ag and Fe nanorods. Using an ethyl glycol-mediated route, a cleaner approach to the formation of tantalum oxide NPs is optimized [17].

A newer type of carbon-doped porous titanium has been prepared using a benign natural polymer, dextrose, which can be useful for visible-light-induced photodegradation of pollutants. Spontaneous heating of the solvent, water and its subsequent evaporation, and combustion of the combustible sugar, dextrose, are the reasons behind the fluffy existence of TiO_2. To build a spongy porous structure, this general and eco-friendly protocol utilizes dextrose and can be applied to other transition metal oxides such as ZrO_2, Al_2O_3, and SiO_2. To produce noble nanocomposites, the noble nanocrystals undergo catalytic oxidation with monomers such as pyrrole, which have potential roles in catalysis, biosensors, energy storage systems, and nanodevices. Without using any surfactant/capping agent and/or prototype, wet synthesis of Ag cables wrapped with polypyrrole was demonstrated at room temperature [18].

The MW hydrothermal process produces magnetic nanoferrites and structured micropine catalysts, and 3D nanostructured metal oxides are obtainable from readily available metal salts. Without using any reducing or capping reagent, these materials were readily prepared from cheap starting materials in water. Ultimately, this economically and environmentally viable, synthetic concept will allow material responses to magnetic, electrical, optical, and mechanical stimuli to be fine-tuned. Several well-defined morphologies with particles in the size range of 100–500 nm, including octahedron, sphere, triangular rod, pine, and hexagonal snowflake, were obtained [19].

7.3.4 Magnetic Nanocatalysts

Sustainable NP routes and their numerous eco-friendly catalysis applications via magnetically recoverable and recyclable nanocatalysts for reducing, oxidizing, and condensing reactions have had a tremendous impact on green chemical pathway growth [10]. A new magnetically recoverable and NP-supported organocatalyst has been produced to catalyze the high-yield Paal–Knorr reaction in a pure aqueous medium that prevents the use of toxic organic solvents, even during the workup phase. The postsynthetic modification of dopamine nanoferrites and then the anchoring of metal particles offer broader possibilities for the very successful deployment of these nanocatalysts. As an example, ruthenium hydroxide with hydration of benzonitrile on magnetic nanoferrites transforms it into benzamide in water [20]. The reaction mixture became apparent after the completion of the reaction, as the stirring ended, and the catalyst was deposited on the magnetic bar because of the paramagnetic nature of the nano-$Ru(OH)_x$. Using an external magnet, the catalyst was easily removed, so a filtration phase was avoided. After catalyst separation, the clear reaction mixture was slowly cooled, and benzamide crystals of appropriate purity were precipitated [21].

The whole process was conducted in pure aqueous medium, and no organic solvents were used during the reaction or during the workup phase. The significant aspect of these protocols is that high turnover numbers (due to the use of a nanocatalyst) and high turnover frequencies (due to the use of MWs) were involved in the response. The catalyst was able to be used many times successfully and left no metal remnants in the finished product. Numerous applications in different asymmetric C—C bond-formation reactions, asymmetric hydrogenations, and asymmetric cycloaddition reactions, among others, will be found in the novel design of magnetically recoverable heterogeneous asymmetric catalysts assisted on [Fe_3O_4] NP systems in the coming years [22].

7.3.5 Nanometal Compounds

Green chemistry for the synthesis of metal nanodrugs in several fields has been implemented in recent years [3]. Significant amounts of toxic as well as unnecessary and hazardous substances are often created by conventional methods for the synthesis of metal NPs. Such metallic NPs, however, have shown great promise in medicine [23]. For example, in drug delivery, treatment, and imaging of integrins on tumor cells as well as therapeutic outcomes, maghemite NPs (γ-Fe_2O_3-NPs) and magnetite NPs (Fe_3O_4-NPs) were extremely helpful [24]. As an example, biocompatible nontoxic superparamagnetic and magnetic NPs have been attached to therapeutic compounds and injected into the body, whereas magnetic fields are produced outside the body to concentrate their accumulation in vivo on specific targets. Due to their super magnetic properties, high surface area, biocompatibility, and safe and fast degradability, iron oxide NPs have attracted great interest. The use of iron oxide magnetic NPs has already been investigated for inductive hyperthermia, chemotherapy, gene carriers for gene therapy, in vitro diagnostic magnetic sensors, vaccines, antibody agents, and therapeutic agents for the treatment of hyperthermia-based cancer [2–4]. However, the field has learned about some of the harmful consequences of using magnetic NPs through such studies, especially in vivo [25]. For instance, due to drugs and NPs breaching the blood–brain barrier (BBB), the successful treatment of brain disorders requires careful consideration. The medicinal properties of NP anions of iron oxide and the adequacy of lysophosphatidic acid (LPA) to temporarily interrupt tight junctions and allow NP anions of iron oxide to reach brain cells have created enthusiasm for the treatment of neural diseases and caution for the unintended accumulation of iron in the brain [26].

Under normal conditions, NP anions of iron oxide have a plasma half-life of 6 minutes, with the main deposition organs being the liver and spleen. No signs of peripheral immune cell infiltration in the brain and no substantial activation of microglia or astrocytes were discovered by the treatment with LPA modified with iron oxide NP anions in the brain and spleen in mice. Following LPA administration, the analysis showed improved delivery efficiency of iron oxide NP anions. These results indicate a temporary disruption of the BBB, which could be safe and successful in increasing the delivery of NP anion iron oxide to the brain [26]. Magnesium (Mg) NPs have recently been shown to have exceptional merit in terms of altering the exclusivity of therapy and thermal ablation effects on tumors. In addition, it is worth noting the low toxicity and side effects of Mg-based chemicals that remain in the body after surgery. Some of the key advantages of magnetic nanocompounds for hyperthermia-based therapy and controlled drug delivery. Using folic acid (FA), which was chosen as the targeting agent combined with PEG by Zhang et al., biocompatible and monodispersed iron oxide superparamagnetic NPs were updated. The biological study showed that altering NPs with PEG-FA greatly enhanced the target cells' intracellular uptake of NPs [23].

Another example of active locomotive intestinal capsules is drug delivery vehicles based on two ring-type soft magnets and a basic plastic hinge. The current drug delivery vehicle has a strong drug-loading capability and can conquer unfavorable body-wide distribution, which is typical for far larger vehicles of other drug delivery vehicles [27]. Often, to minimize damaging effects, the alteration of NP properties is used. For instance, biodegradable iron stents, such as chronic inflammation and premature recoil, have great potential to minimize limitations. In porcine coronary arteries randomly located in the coronary arteries of juvenile domestic pigs, White et al. studied the protection and efficacy of biodegradable iron and cobalt–chromium stents. No impressive variations between iron and cobalt–chromium stents were seen in their performance. Some of the most important examples of a healthy nanomaterial are these biocorrodable iron stents, helping to decrease the side effects. Without using any other organic target for cancer therapy, the production of biodegradable drug delivery compounds with strong targeted recognition and controlled release has led to the manufacture of polyphenol-coated porous nanomaterials (e.g. silica NPs). For the immobilization of targeting moieties, such as aptamers, assisted polyphenol with colloidal stability and low premature drug leakage provided a scaffold [28].

A PEG shielding and tumor microenvironment were designed to efficiently deliver antitumor drugs to cancer cells and reduce their toxic side effects on normal cells, activating a cascade of a pH-sensitive hollow mesoporous silica NP (HMSN) drug delivery system. The 3-(3,4-dihydroxyphenyl) propionic acid-functionalized beta-cyclodextrin (β-CD) was loaded onto the surfaces of HMSNs using boronic acid–catechol ester bonds. Then, PEG-conjugated adamantine (Ada) was hosted on an HMSNs-β-CD nanocarrier by the host–guest interaction. In vivo results confirmed that the drug-loaded HMSNs prevented tumor growth considerably with minimal toxic side effects. This method introduced new insights into the extension of the new production of green chemistry-derived drug delivery carriers by the tumor microenvironment [29].

Based on polyacrylate/nanosilica, other positive green polymers have been prepared. However, polyacrylate coatings have allowed pleural effusion, pericardial effusion, and granuloma and pulmonary fibrosis, which are thought to be due to the elevated loading stage of the NPs. NPs have been used more and more for various industrial purposes. Concerns have been raised, however, about their harmful and destructive effects on humans and the environment. Some studies have determined the presence of NPs in patient biopsies and identified possible harmful effects on human lungs. Silica NPs have been detected using electron microscopy and energy-dispersive X-ray analysis as an example in macrophages, pulmonary microvessels, vascular endothelial cells, microlymphatic vessels and pleural effusions, and a few in alveolar epithelial cells and interstitial pulmonary tissues and have been shown to damage alveolar epithelial cells, macrophages, and pulmonary interstitial tissue. Based on the well-documented toxicity of the silica nanocompound, it is likely that some of the diseases identified by researchers could be associated with these silica NPs [30].

As drug carriers, certain nanopolymer matrices prepared by the chemical reduction of metal ions in an aqueous medium have been used. A functional polymer matrix of poly(1-vinyl-1,2,4-triazole-*co*-*N*-vinylpyrrolidone)(poly[VT-co-VP]) prepared by free radical-initiated polymerization was composed of these novel hydrophilic healthy nanocomposites containing silver nanoparticles (AgNPs). AgNPs do not precipitate and/or alter in size even after storage in an aqueous medium due to their adequate stabilization by functional classes. In white mice, the toxicity of the initial poly(VT-*co*-VP) copolymer and nanocomposite containing AgNPs was recognized, with a value of $5000\,mg\,kg^{-1}$. There was a marked antimicrobial activity of the nanocomposite against various Gram-negative and -positive bacterial strains. At concentrations ranging from 0.5 to $8\,\mu g\,ml^{-1}$, the minimum inhibitory concentration (MIC) decreased microorganism growth, and the minimum bactericidal concentration (MBC) ranged from 0.5 to $16\,\mu g\,ml^{-1}$. Thus, in a poly(VT-co-VP) matrix, these new silver (Ag) nanocomposites are attractive carriers for hydrophilic antiseptics and antimicrobial drugs for medical applications to treat many infectious diseases, including those acquired surgically [31].

For tumor therapy, a molybdenum disulfide (MoS_2) nanosheet has been developed as a photothermal agent. During a simple approach to the preparation of soybean phospholipid-encapsulated MoS_2 (SP-MoS2) nanosheets with excellent colloidal stability, surface modification of

PEG was carried out. The SP-MoS_2 nanosheets demonstrate good photothermal conversion efficiency and photothermal stability during breast tumor photothermal therapy by not observing clear in vitro and in vivo hemolysis, coagulation, and cyto/histotoxicity. Low cost, easy processing, and strong in vivo hemo/histocompatibility were seen in the prepared SP-MoS_2 nanosheets, supporting a promising capacity for cancer treatment. Bovine serum albumin (BSA) has often been used to modify upconversion NPs (UCNPs) based on NaGdY4. The acquired UCNP@BSA NPs demonstrated excellent water solubility as well as physiological solutions. Two separate dye molecules, including a photosensitizer, Rose Bengal (RB), and near-infrared absorbing dyes, IR825, can be effectively loaded onto UCNP@BSA NPs using the hydrophobic domains in the BSA protein. On the one hand, the dual-dye-loaded NPs obtained could serve as a dual-modal magnetic resonance (MR) and upconversion optical imaging probe, while on the other hand, both photodynamic and photothermal therapies are successful, which could result in excellent in vitro and in vivo synergetic cancer-killing properties if combined together [31].

7.3.6 Polymer Nanocomposite (PNC)

One way to decrease toxicity and improve protection is to provide a polymer nanocomposite (PNC) containing a polymer or copolymer with NPs distributed in the polymer matrix. Of course, these nanocomposites can have different shapes (e.g. platelets, fibers, and spheroids), but in the nanometer range, they should have at least one dimension. For decades, polymeric NPs have been prepared for use in a number of high-performance materials, such as nanodrug delivery or drug-functionalized delivery, demonstrating their safety if proper chemistry is used [32].

A new polymeric nanogene delivery vehicle consisting of polyethylenimine-grafted chitosan oligosaccharide (CSO-PEI) with hyaluronic acid (HA) and small interfering RNA (siRNA) was prepared to introduce a novel and healthy drug candidate for the treatment of endometriosis. There was no clear difference in the size observed between (CSO-PEI/siRNA)HA and CSO-PEI/siRNA, but fluorescence accumulation in the endometriotic lesion was more important for (CSO-PEI/siRNA)HA than for CSO-PEI/siRNA because of the unique binding of HA to CD44. Furthermore, the (CSO-PEI/siRNA)HA NP gene therapy greatly decreased the size of the endometriotic lesion with ectopic endometrium atrophy and degeneration. Ectopic endometrium epithelial cells from rat endometriosis models displayed significantly lower CD44 expression than controls after treatment with (CSO-PEI/siRNA)HA. In addition, studies under an electron microscope showed no obvious toxic effects on reproductive organs and verified that the HA gene delivery system (CSO-PEI/siRNA) can be used as a safe and efficient way to treat endometriosis. Further research on molybdenum NPs involves phosphate glasses that contain vanadium and molybdenum oxides as useful compounds for the material. The modified vancomycin molecules exhibit greater bonding of hydrogen with vanadium-doped glasses and thus slower release of drugs over 14 days, indicating better surface modification with drug molecules. The extreme coherence of drug components to the glass surface compared to the free molybdenum will explain this. Hydrogen bonding between the amino functional groups of vancomycin and the hydrated P–O–H groups in the glass network is responsible for the tight connection [33].

Another compound to improve NP performance is a prepared nanogel from dextran and poly(lactide) by in situ cross-linking with a homobifunctional cross-linker via a traditional radical polymerization technique. The biodegradability of the nanogel was verified by degradation experiments using hen egg lysozyme (Ly). In vitro studies of cytocompatibility with human mesenchymal stem cells (hMSCs) showed that native nanogel had nontoxic effects on cancer cells, whereas nanogels loaded with doxorubicin (DOX) showed high toxicity for cancer cells. The above-mentioned nanogel, injectable, multifunctional, and biodegradable poly lactic-*co*-glycolic acid (PLGA), was synthesized and tested for breast cancer cells. These particles were filled with a prodigiosin (PG) anticancer drug obtained from *Serratia marcescens* subsp. bacteria [34].

With hydrophilic PEG shells and a hydrophobic PLGA heart, a novel reactive oxygen species-cleavable diblock polymer was prepared. This latest nanocompound has been used for cancer chemotherapy, with high stability, effective drug delivery, excellent sensitivity, and good compatibility [35]. High antimicrobial activity against a wide variety of bacteria and yeast strains, such as *Staphylococcus aureus*, *Escherichia coli*, *Pseudomonas aeruginosa*, and *Candida albicans*, was shown by different size ranges of poly(tannic acid) particles prepared by lecithin/gasoline microemulated media. The use of safe materials (e.g. solvents and separation agents) was based on green chemistry principles. The polyester dendron-polymer's hydrophilic–hydrophobic character contributes to the aqueous solution formation of flower-like micelles. Through the increased permeability and conservation effect, all nanocopolymer compounds were effective for passive tumor targeting [36].

Cardiovascular disorders are responsible for ~25% of all deaths worldwide, including ischemic heart disease and stroke. The easy availability of nanobiomaterials facilitates their use as drug delivery vehicles or defensive nanoshells to improve the biocompatibility of imaging agents. Due to their biodegradability and local drug delivery, poly-L-lactic acid (PLLA) stents are safer than metallic stents. In addition, the efficacy and safety of PLLA composite coronary biodegradable polymer stents in humans has been demonstrated. No major clinical incidents related to PLLA stent implantation in up to 15 patients after 1 month of implantation were recorded in the results. After 6 months, both patients exhibited appropriate rates of restenosis and target lesion revascularization without death, myocardial infarction, or grafting of coronary artery bypass [37].

As a long-circulating, passively targeted delivery system that responds to intracellular glutathione (GSH) concentrations to improve DNA delivery and transfection, PEG-modified thiolated gelatin (PEG-SHGel) NPs have been developed. The NPs were encapsulated with the Reporter plasmid expressing enhanced green fluorescent protein (EGFP-N1). The PEG-SHGel NPs have released encapsulated plasmid DNA in response to differing GSH concentrations. The results of this study showed that PEG-SHGel NPs could serve as a very effective nanoparticulate vector for the delivery of systemic DNA to solid tumors where it is understood that the cells have significantly higher levels of intracellular GSH [38].

To overcome the relatively high rates of restenosis and target lesion revascularization with bare metal stents, new drug-eluting stents (DES) have been developed. For its ability to hold drugs more effectively within the intimate region rather than concentrating the drug around the stent struts and to balance supporting erosion with drug release, a new and absorbable sirolimus-eluting stent (AC-SES) coating was prepared. The supporting struts were composed of absorbable PLGA and a dry-powder electrostatic process deposited by crystalline sirolimus. In a porcine coronary implant model, the AC-SES findings showed reasonable drug stability under simulated conditions and reliable drug delivery. A large number of research efforts have also been carried out to assess if corrodible substances can be applied as low-toxic biodegradable cardiovascular implants [39]. Corrodible iron stents with a stent configuration similar to a commercially available stent used were prepared from pure iron and laser cut. The stents were inserted into the native descending aorta of white rabbits from New Zealand. During a 6–18-month follow-up, there were no thromboembolic complications and no substantial neointimal proliferation with any significant inflammatory effect or systemic toxicity. Therefore, this critical in vivo research showed that degradable iron stents can be safely implanted without substantial stent vessel obstruction caused by inflammation, neointimal proliferation, or thrombotic events [40]. The toxicity of a corrodible stent prepared from pure iron in a peripheral stent has been determined in this area. Without technical issues, the stent was inserted into the descending aorta of minipigs with an overstretch injury. No symptoms of iron overloaded and local toxicity of iron due to corrosion products were confirmed by histopathological review of heart, lung, spleen, liver, kidney, and paraaortic lymphatic nodes. Healthy and nontoxic Mg stents are one of the safest ways to administer drugs that have been tested in comprehensive clinical trials in patients with acute coronary syndrome, myocardial infarction, and diabetic patients [41].

Using a reaction of polycondensation and ring-opening polymerization of ε-caprolactone in the presence of poly(propylene adipate), a new biodegradable and safe copolymer for NP incorporation was prepared. The prepared nanocomposite exhibited low cytotoxicity and was used to deliver an iron-chelating compound, desferrioxamine. In addition, based on the poorly water-soluble fenofibrate, a novel and very soluble electrosprayed nanospherule was prepared. Several electrosprayed nanospherules filled with fenofibrate have been synthesized with polyvinylpyrrolidone (PVP) and implemented using the electrospray technique in drug delivery. Results verified that, relative to the free drug, the electrosprayed nanospherule approach substantially improved aqueous solubility and dissolution [42]. For the synthesis of core–shell micelle delivery vehicles for cobalt-based pharmaceuticals, similar results have been published by Withey's group. The cobalt complex was synthesized during the self-assembly of poly(trimethylsilyl propargyl methacrylate)-*b*-poly(PEG methyl ether methacrylate). Biodegradable nanocomposites based on poly(ester amine) (PEA) and poly(amido amine) (PAA) were recently prepared for the successful delivery of siRNA to human umbilical endothelial vein cells (HUVECs). The effects of the nanocomposites were calculated and compared with research using DNA-containing particles on small molecule delivery to cells. Results showed that using green fluorescent protein-encoding plasmid DNA (up to 50–60% transfection efficiency), PEAs and linear nondendrimeric PAAs were very successful for DNA transmission to HUVECs [43].

However, the deficiencies in the delivery of small molecular antitumor drugs have been greatly improved by the entry of such small molecules into nanovehicles as well as the regulation of drug release and tumor responsiveness of drug nanocarriers. It has been shown to use the central cross-linking method to synthesize a micelle with improved drug encapsulation capacity and drug release sensitivity in tumors. A micelle made up of PEG and FA-PEG as hydrophilic units, pyridyl disulfide as cross-linkable and hydrophobic units, and disulfide bond as cross-linker, for instance showed considerable promise for cellular uptake, improved in vitro smart tumor responsiveness, and improved cytotoxicity against HeLa cells of cross-linked micelle-curcumin core. Different forms of drug-delivering vehicles are shown in between non-cross-linked micelles (NCM) and FA-core cross-linked micelles [44]. The pH-sensitive DOX encapsulated in the hydrophobic cores of the amphiphilic triblock copolymers is another synthesized nanodrug delivery vehicle focused on green chemistry. Using a lysosome tracker system, the copolymer micelles entered into lysosomes. It is possible to use these healthy and efficient micelles. NP formulations with chitosan are another instance of a green medicinal compound. By mixing aqueous chitosan solutions and GO in dilute acetic acid as a solvent, chitosan-graphene oxide (GO) green nanocomposite films were prepared. Quaternized chitosan (QC)/BSA/rectories were used in the nanocomposite (REC-DOX was used to investigate the efficiency of entrapment and release patterns in NPs). Results have shown that the use of REC may improve the capacity for drug encapsulation and loading ability [45]. Pancreatic cancer is a lethal malignancy whose development is highly dependent on the microenvironments of the nervous system. Recently, pancreatic cancer progression has also been regulated by neural drug-supported ferritin NPs (Ft NPs), which regulate the nervous microenvironment. Through passive targeting of improved permeability and retention effects in tumors and active targeting through transferrin receptor 1 (TfR1) binding on tumor cells, the drug-supported Ft NPs can target pancreatic tumors with a triggered drug release in an acidic tumor medium. Two drugs were introduced into the Ft NPs, one that stimulates neural activity (carbachol) and the other that impairs neural activity (atropine), to form two kinds of nanodrugs, Nano-Cab NPs and Nano-Ato NPs, respectively. Nano-Cab NPs activation of the nervous micromedia dramatically prevented pancreatic tumor progression, while Nano-Ato NPs blockage of the neural niche generally destroys neurogenesis in tumors and pancreatic cancer growth. Therefore, for new anticancer therapies, the Ft-based NPs provide a good and stable way of carrying neural drugs [46].

In order to minimize NP toxicity, alginate also reflects a green nanomedicine solution. For instance, sodium alginate NPs assisted by quinapyramine sulfate (QS) have been prepared to reduce the unpleasant toxic effects of QS against the *Trypanosoma evansi* parasite, a trypanosomosis causative factor. The biocompatibility of QS-NPs was tested using Vero, HeLa cell lines, and horse erythrocytes in a dose-dependent manner to identify the toxicity of the novel NPs and showed no toxicity at effective trypanocidal doses and even at doses several times higher than the effective dose. Another example of safe, biocompatible, and bioresorbable heart delivery vehicles includes bioactive drug delivery vehicles. Without major toxicity or interfering with any functional properties, this new nanodrug delivery vehicle was effectively internalized into cardiomyocytes [47].

Another kind of nanocompound with high potential for applications in drug delivery, biodetection, and diagnosis are two-dimensional silicon NPs (Si-NPs). Modified or functionalized mesoporous silica showed high biocompatibility, protection, and controlled drug release in comparison to amorphous colloidal and porous silica [42]. Biocompatible Si-NPs with amine functional groups synthesized using a one-step green protocol open the way for more biomolecular conjugation, which is an interesting parameter for biomedical applications of NPs, such as biolabels, contrast agents, and vector delivery, and may appear as a bionanostructure for good clinical translation [48].

Injectable polymethylmethacrylate (PMMA) bone cement in cemented arthroplasty, vertebroplasty, and osteoporosis fractures is a widely used bone replacement. However, aseptic loosening of implanted cement at the bone–cement interface could still be observed in a high rate of patients due to improper stiffness, low bioactivity, and high polymerization temperature of PMMA. Artificial extracellular matrices, such as chitosan-glycerophosphate (CS-GP) thermosensitive hydrogel, have been introduced into PMMA to boost the efficiency of PMMA, acting as a pore-forming agent and as a carrier of osteoconductive nanosized hydroxyapatite (nano-HA)/antibiotic gentamicin (GM). Results showed that the CS-GP thermosensitive hydrogel can effectively create open pores on the surface of the PMMA cement, which is believed to facilitate the ingrowth of bone tissue and enhance the anchorage of cement in future clinical applications at the bone–cement interface. More importantly, without affecting cell survival, nano-HA and GM enriched the CS-GP thermosensitive hydrogel effectively to enhance the overall efficiency of the PMMA cement, indicating that the injectable p-PMMA/CS-GP/nano-HA/GM cement will hold strong promise for future applications for bone reconstruction [44].

7.3.7 Biosynthesis of Nanodrug Delivery Vehicles

New advances in nanocarrier biosynthesis, specifically nanocompounds synthesis with complex sizes and shapes, are contributing to the advancement of novel NPs in the treatment and diagnosis of cancer (e.g. drug and gene delivery, imaging, phototherapy, and enhancement of radiotherapy treatments). In addition, in various medical and drug delivery applications, certain significant properties, such as high inherent protection, high surface area, and tunable stability, of NPs are very useful [49].

An eco-friendly and cost-effective approach for the preparation of applicable nanomaterials using biomaterials is to use a new solvent for the assay of drug release from a polysaccharide-based nanocompound for a colon-specific drug delivery device. For this, in the colonic area, probiotic cultures of *Bacteroides*, *Bifidobacterium*, *Lactobacillus* species, *Eubacterium*, and *Streptococcus* were produced and placed in the solvent media and compared with those obtained from the fermentation model based on rat cercal and human fecal. The results obtained with the probiotic system suggested that for the drug release of any polysaccharide-based oral formulation intended for colonic delivery, a probiotic dissolution process can be used [50].

Due to their intriguing nanostructures and properties, gold nanoparticles (AuNPs) and other noble metal NPs have a fascinating and progressive presence in various fields of research, for instance as a reducing/capping agent, the green synthesis of AuNPs for drug delivery based on chitosan. Single oil-in-water (O/W) emulsions were used to synthesize a biocompatible controlled release of rifampicin in the body. Using Ag nitrate ($AgNO_3$) and freshly extracted egg whites, another healthy, cost-effective, and environmentally friendly method was shown to prepare AgNPs in aqueous media. The findings showed that biomolecule-loaded AgNPs were largely spherical in shape and had a mean size of 20 nm. With distinct functional groups, the proteins of egg whites posed great tasks in reducing Ag^+ to generate a stable product. In vitro cytotoxicity studies showed that during cancer radiation therapy, Ag-protein nanocompounds showed reasonable biocompatibility with mouse fibroblast cells. Due to the growing need for pharmacological products, such as antibacterial, -fungal, -viral, and -tumor properties, the important role of seaweed as a marine origin for the biomimetic preparation of nanometal particles has increased. In order to prepare AuNPs in aurum solutions with different broth concentrations of brown seaweed leaf extracts, the bioreduction of bulk gold chloride was synthesized, and AuNPs were used in the lung and liver tumor cells without any side effects [51].

As a potential liver cancer-targeted drug carrier with high efficacy and low side effects, liver-specific biopolymers were chosen. In order to generate new targeted drug delivery and imaging of cancer, pullulan-supported AuNPs were linked with 5-fluorouracil and FA. The in vitro cytotoxicity on tumor cells of free pullulan-AuNPs showed that the amount of pullulan-AuNPs required to achieve 50% of inhibition growth was much smaller [52].

Biodegradable polymers are an important community of macromolecular networks that can retain within their structures a large quantity of an aqueous solvent. A new form of amphiphilic cetylated polyethyleneimine (PEI) was prepared, and polylactic-*co*-glycolic acid cetylated polyethyleneimine/HA NPs (PCPH NPs) was then synthesized as a new type of gene-delivering device by self-assembly. The delivery ability and cytotoxicity of PCPH NPs were calculated as transfecting plasmid DNA vectors (liver hepatocellular carcinoma in vitro) and showed much lower cytotoxicity and higher gene delivery efficiency than PEI and the commercial transfection reagents [53]. Strawberry extract polyphenols (negative groups) associated with positively protonated chitosan amino groups supported. Curcumin and emu oil extracted from emu birds (*Dromaius novaehollandiae*) have shown good results against inflammation. This approach will increase the bioavailability and sustained release of phytochemicals with lower bioavailability. However, because of low solubility and poor transmission, the distribution of curcumin decreases the health issue. Nanoemulsions were synthesized as the oil process, surfactant, and cosurfactant using emu oil, Cremophor, and Labrafil. Major increases in anti-inflammatory arthritic scoring, paw length, and biochemical, genetic, radiological, and histological studies have been shown in anti-inflammatory activity with curcumin-containing formulations compared to pure curcumin with emu oil [54]. Environmentally friendly synthesis of biodegradable semi-interpenetrating hydrogel networks using *N,N′*-methylenebis(acrylamide) (MBA). AgNPs were then assisted on hydrogel networks as nanoreactors used green chemistry methods to extract $AgNO_3$ and *Azadirachta indica* (Neem) plant extracts under ambient conditions. Furthermore, the dispersion between polymeric nanocomposites and the drug has been shown to influence the degree of matrix swelling, its porosity, and the release phase of diffusion. Another effective, inexpensive, and sustainable strategy for the preparation of AgNP was carried out using light emitted as the catalyst in aqueous media. NP seeds or toxic reactants are independent of this technique. The preprepared nanocomposite demonstrated high antimicrobial activity against the bacterial strains studied [55]. *Solanum tricobatum*, *Syzygium cumini*, *Centella asiatica*, and *Citrus sinensis* extracts are other plants for the green synthesis of AgNPs as antimicrobials. AgNPs demonstrated increased bactericidal activity against various drug-resistant human pathogens in these healthy green chemistry AgNPs. The research presented showed the potential role of plantlet extracts extracted in vitro for the functional synthesis of antimicrobial AgNPs, which can be used as nanodrugs in many treatments. The cell-free supernatant from the *Delftia* species is used for the control of microbial infections in another inexpensive and green technique for the preparation of AgNPs, with results showing that the preparation of microbial infections uses the cell-free supernatant from the *Delftia* species [56].

7.3.8 Carbon Nanotubes – CNTs

In cancer diagnosis and treatment, CNTs, with individual physical and chemical properties, have become popular products. The most promising functions of these compounds are, for example the identification of cancer cells and the transport of drugs or small therapeutic molecules to tumor cells. CNTs, however, show not only a high capacity for drug loading but also a potential for cell penetration. Some problems appear with low solubility, clumping or agglomeration, and long half-lives, in addition to many applications of CNTs in drug delivery systems. Solubility and tumor targeting/drug delivery have altered the modification of single-walled nanotubes (SWNTs) with various compounds. This decreased their cytotoxicity and altered the function of immune cells, such as PEG and PEI, cationic glycopolymers, vinblastine, and lentinan [57].

Using mice-bearing syngeneic murine squamous cell tumors, Huang et al. published good findings. They injected SWCNTs intratemporally and irradiated the cancer cells for 10 minutes with a low-power near-infrared-reflectance (NIR) laser. A thermal ablation threshold was observed at a maximum tumor temperature of ~55 °C, and a decrease in the growth of cancer cells was observed [58]. In the same study, Liu et al. systematically analyzed the dependence between the duration and density of PEGylation on the SWCNT surface and the biodistribution, aggregation of tumors, and ablation ability of PEG-functionalized ultraviolet syringed SWCNTs. The findings showed long half-lives of blood circulation for heavily PEGylated SWCNTs and strong uptake of the tumor. This study suggested that in order to increase their unknown tumor-targeting capacity, the surface properties of CNTs are significant [57].

7.4 Toxicity Aspects

As has been observed for FeNP and AgNPs, plant polyphenols from tea or winery waste and GSH-capped NPs may help design new strategies to extend the existence of such entities with reduced toxicity [33]. Choosing environmentally friendly stabilizing agents and functionalization techniques, such as biodegradable polymers and enzymes that have greater biocompatibility, is crucial [59].

7.5 Bioinspired Green Nanomaterial Synthesis

Different processes have been developed by Nature for the synthesis of nano- and micro-length-sized inorganic materials that have contributed to the creation of a relatively new and largely unexplored research field focused on nanomaterial biosynthesis [5, 11]. NP biosynthesis is a kind of bottom-up method where reduction/oxidation is the main reaction. Microbial enzymes or plant phytochemicals with antioxidants or antioxidant properties are generally responsible for reducing the occurrence of nanoparticular metal compounds [60, 61].

7.5.1 Metallic Nanoparticles

Some typical metal NPs produced by plants are listed in Table 7.1, and the metallic NPs produced by microorganisms are described in the content.

Table 7.1 Some typical metal nanoparticles produced by plants [9].

Sr no.	Nanoparticles	Plants
1.	Silicon–germanium (Si–Ge) nanoparticles	• *Freshwater diatom Stauroneis* sp.
2.	Gold and silver nanoparticles	• *Citrus sinensis* • *Diospyros kaki (Persimmon)* • *Pelargonium graveolens* • *Hibiscus rosa sinensis* • *Coriandrum sativum* • *Emblica officinalis* • *Phyllanthium Mushroom extract*
3.	Silver nanoparticles	• *Elettaria cardamomum* • *Parthenium hysterophorus* • *Ocimum* sp. • *Euphorbia hirta,* • *Nerium indicum* • *Azadirachta indica* • *Brassica juncea* • *Pongamia pinnata* • *Clerodendrum inerme* • *Gliricidia sepium* • *Desmodium triflorum* • *Opuntia ficus indica* • *Coriandrum sativum* • *Carica papaya (fruit)* • *Pelargonium graveolens* • *Aloe vera extract* • *Capsicum annum* • *Avicennia marina* • *Rhizophora mucronata* • *Ceriops tagal* • *Rumex hymenosepalus* • *Pterocarpus santalinus* • *Sonchus asper*

(Continued)

Table 7.1 (Continued)

Sr no.	Nanoparticles	Plants
4.	Gold nanoparticles	• *Terminalia catappa* • *Banana peel* • *Mucuna pruriens* • *Cinnamomum zeylanicum* • *Medicago sativa* • *Magnolia kobus and Diospyros kaki* • *Allium cepa* L. • *Azadirachta indica A. Juss.* • *Camellia sinensis* L. • *Chenopodium album* L. • *Justicia gendarussa* L. • *Macrotyloma uniflorum (Lam) Verde* • *Mentha piperita* L. • *Mirabilis jalapa* L. • *Syzygium aromaticum* (L.) • *Terminalia catappa* L. • *Amaranthus spinosus*
5.	Lead nanoparticles	• *Vitus vinifera* L. • *Jatropha curcas* L.
6.	Magnetic nanoparticles	• *Aloe vera*
7.	Palladium nanoparticles	• *Cinnamomum zeylanicum Blume.* • *Cinnamomum camphora* L. • *Gardenia jasminoides Ellis.* • *Soybean (Glycine Max)* L.
8.	Platinum nanoparticles	• *Diospyros kaki* • *Ocimum sanctum* L.

7.5.1.1 Gold Nanoparticles

AuNPs have a long chemical history, dating back to ancient Roman times, where they were used for decorative purposes to stain glasses. AuNPs were already used centuries earlier to treat different diseases. More than 150 years ago, with the work of Michael Faraday, who was probably the first to observe that colloidal gold solutions have properties that differ from bulk gold, the modern age of AuNP synthesis began. Due to an increasing need to improve environmentally sustainable technologies in material synthesis, biosynthesis of NPs as evolving bionanotechnology (the intersection of nanotechnology and biotechnology) has received significant attention. Extracellular synthesis of AuNPs by the fungus *Fusarium oxysporum* and actinomycete *Thermomonospora* sp. has been reported by Sastry and coworkers [62]. The intracellular synthesis of *Verticillium* sp. fungus AuNPs has been documented as well. Studies have shown that by incubating cells with Au^{3+} ions, gold particles of nanoscale dimensions can readily be precipitated within bacterial cells. NPs of monodisperse gold were synthesized using the alkalotolerant *Rhodococcus* sp. in extreme biological conditions, such as alkaline and slightly elevated conditions of temperature. Production of gold nanostructures in various shapes (spherical, cubic, and octahedral) by filamentous Au(I)-thiosulfate and Au(III)-chloride complex cyanobacteria and study of their mechanisms of formation. The growth of nanocrystals and nanoalloys using *Lactobacillus* was reported [63, 64].

7.5.1.2 Silver Nanoparticles

Similar to their bulk equivalents, AgNPs demonstrate effective antimicrobial activity against Gram-positive and -negative bacteria, including extremely multiresistant strains such as *S. aureus*, resistant to methicillin. The secrets discovered by Nature have led to biomimetic approaches to advanced nanomaterial growth being developed. Recently, researchers have made efforts to make use of microorganisms as potential eco-friendly nanofactories for the production of AgNPs. In order to form AgNPs, various microbes are known to reduce Ag^+ ions, most of which are found to be spherical particles [46]. Studies have shown that the bacterium *Pseudomonas stutzeri* AG259, isolated from a silver mine, played a major role in the reduction of Ag^+ ions and the formation of well-defined AgNPs and distinct topographies within the bacteria's periplasmic space when placed in a concentrated aqueous solution of silver nitrate. When the fungi *Verticillium*, *F. oxysporum*, and *Aspergillus flavus* were used, AgNPs were synthesized in the form of a film or formed in a solution or accumulated on the surface of their cell [65].

7.5.1.3 Alloy Nanoparticles

Because of their applications in catalysis, electronics, optical materials, and coatings, alloy NPs are of great interest. In the bimetallic Au–Ag alloy synthesis of *F. Oxysporum* and deciding the composition of Au–Ag alloy NPs, the secreted cofactor nicotinamide adenine dinucleotide (NADH) plays a significant role. Au–Ag alloy NPs biosynthesized by yeast cells were studied. Microscopic fluorescence and transmission electron microscopic characteristics suggested that the NPs of the Au–Ag alloy were primarily synthesized using an extracellular approach and typically resided in irregular polygonal NPs. Electrochemical studies have shown that the glass carbon electrode-modified vanillin sensor based on Au–Ag alloy NPs has been able to increase the electrochemical response of vanillin by at least five times. The synthesis of core–shell Au–Ag alloy NPs from the fungal strain *Fusarium semitectum* has shown that the suspensions of the NP have been very stable for many weeks [66].

7.5.1.4 Other Metallic Nanoparticles

Heavy metals are believed to be harmful to microorganisms. In nature, microbial resistance to most toxic heavy metals is due to their chemical detoxification as well as membrane proteins that act either as ATPase or as chemiosmotic cation or proton antitransporters due to energy-dependent ion efflux from the cell. In microbial resistance, alteration in solubility also plays a role. With the help of metal ion, reducing bacterium can achieve resting S phase. At room temperature and neutral pH within 60 min when lactate was provided as the electron donor, algae were able to reduce aqueous PtCl6 2 ions into elemental platinum. Located in the periplasm were platinum NPs of around 5 nm. Studies reported that *Enterobacter* sp. cells would synthesize mercury NPs. The synthesis of uniform-sized 2–5 nm, spherical, and monodispersed intracellular mercury NPs is facilitated by culture conditions (pH 8.0 and lower mercury concentration). An anaerobic hyperthermophilic microorganism, *Pyrobaculum islandicum*, has been reported to reduce many heavy metals with hydrogen as the electron donor, including U(VI), Tc(VII), Cr(VI), Co(III), and Mn(IV). The sulfate-reducing bacterium, *Desulfovibrio desulfuricans*, and the metal ion-reducing bacterium, *S. algae*, could synthesize palladium NPs [63, 67].

7.5.2 Oxide Nanoparticles

A significant compound NP synthesized by microbes is the oxide NP. In this section, we review the two aspects of biosynthesized oxide NPs: NPs of magnetic oxide and NPs of nonmagnetic oxide. Mostly the magnetotactic bacteria were used in development of NP by magnetic oxide and biological systems [63].

7.5.2.1 Magnetic Nanoparticles

Due to their unusual microconfiguration and properties such as super paramagnetic and high coercive force, and their prospect for broad applications in the fields of biological separation and biomedicine, magnetic NPs are new materials recently created. It is understood that magnetic NPs such as Fe_3O_4 (magnetite) and Fe_2O_3 (maghemite) are biocompatible. Targeted treatment of cancer (magnetic hyperthermia), stem cell sorting and modification, guided drug delivery, gene therapy, DNA analysis, and MRI have been actively studied [63].

Intracellular magnetic particles containing iron oxide, iron sulfides, or both are synthesized by magnetotactic bacteria. They are referred to as bacterial magnetic particles (BacMPs) to differentiate these particles from artificially synthesized magnetic particles (AMPs). BacMPs, aligned within the bacterium in chains, are postulated to function as biological compass needles that enable the bacterium, under the influence of the geomagnetic field of the Earth, to migrate along oxygen gradients in aquatic environments. In aqueous solutions, BacMPs can quickly spread because they are enveloped by organic membranes that consist primarily of phospholipids and proteins. In addition, there is a single magnetic domain or magnetite in an individual BacMP that yields superior magnetic properties [68].

Various morphological forms, including cocci, spirilla, vibrios, ovoid bacteria, rod-shaped bacteria, and multicellular bacteria with specific characteristics, have been described and observed in different aquatic environments since the first report of magnetotactic bacteria in 1975. For instance, magnetotactic cocci have shown high diversity and distribution and have been found frequently at the surface of aquatic sediments. The discovery of this type of bacteria, including the only cultivated strain of magnetotactic coccus MC-1, has indicated that they are microaerophilic. Three optional anaerobic marine vibrios, strains MV-1, MV-2, and MV-4, were isolated from estuarine salt marshes in the case of the vibrio bacterium. These bacteria were described as members of alpha-proteobacteria, probably belonging to the family of Rhodospirillaceae, and were found to synthesize truncated hexa-octahedron-shaped BacMPs and to expand both chemoorganoheterotrophically and chemolithoautotrophically [68]. On the other side, the Magnetospirillaceae family members can be found in fresh water sediments. A large number of the magnetotactic bacteria isolated to date have been found to be members of this family with the use of the growth medium and magnetic isolation techniques developed. The first member of the family to be isolated was the *Magnetospirillum magnetotacticum* strain MS-1, while the *Magnetospirillum gryphiswaldense* strain MSR-1 is also well studied in terms of both its physiological and genetic properties. *Magnetospirillum magneticum*, the facultatively anaerobic magnetotactic spirilla, was AMB-1 isolated. Since 2000, a variety of new magnetotactic bacteria have been found in different aquatic environments. Numerous ecosystems have seen uncultivated magnetotactic bacteria. Most recognized magnetotactic cultured bacteria are mesophilic and do not appear to grow much above 30 °C. In most cases, uncultured magnetotactic bacteria were at 30 °C and below. Thermophilic magnetotactic bacteria are only described in a few studies. Studies reported that in samples from springs whose temperatures ranged from 32 to 63 °C, one of the magnetotactic bacteria known as HSMV-1 was found.

A single polar flagellum and a single chain of bullet-shaped magnetosomes were seen in the transmission electron microscopy (TEM) images of HSMV-1 unstained cells. The average number per cell of magnetosome crystals is 12 ± 6 with a mean size of 113 ± 34 nm by 40 ± 5 nm. The paper's findings have clearly shown that some magnetotactic bacteria can be considered at least moderately thermophilic. In conditions where magnetotactic bacteria occur and are likely to expand (63 °C) and where magnetosome magnetite is deposited, the upper temperature limit has been extended. Studies stated that mesoporous structure magnetic Fe_3O_4 materials were synthesized using yeast cells as a template by the coprecipitation method [66, 67].

7.5.2.2 Nonmagnetic Oxide Nanoparticles

Other oxide NPs, including TiO_2, Sb_2O_3, SiO_2, $BaTiO_3$, and ZrO_2 NPs, have also been studied in addition to magnetic oxide NPs. A low-cost and reproducible green *Saccharomyces cerevisiae*-mediated biosynthesis of Sb_2O_3 NPs was discovered by Jha and colleagues. The synthesis was carried out in compliance with room temperature. Analysis showed that the unit of Sb_2O_3 NPs was a spherical aggregate of 2–10 nm. *F. Ooxysporum* (Fungus) is used for the processing of SiO2 and TiO2 NPnanoparticles from SiF6 2 and TiF6 2 aqueous anionic complexes, respectively. They also prepared NPs of tetragonal $BaTiO_3$ and quasi-spherical ZrO_2 from *F. oxysporum* with a 4–5 and 3–11 nm range of scale, respectively [66, 67].

7.5.3 Sulfide Nanoparticles

In addition to oxide NPs, due to their fascinating and novel electronic and optical properties, sulfide NPs have attracted considerable attention in both basic research and technological applications as quantum-dot fluorescent biomarkers and cell labeling agents. One common form of sulfide NP is the CdS nanocrystal and has been synthesized by microorganisms. Cunningham and Lundie found that *Clostridium thermoaceticum*, in the presence of cysteine hydrochloride in the growth medium where cysteine most likely acts as the source of sulfide, could precipitate CdS on the cell surface as well as in the medium of CdCl2. *Klebsiella pneumoniae* was found to form 20–200 nm CdS on the cell surface when exposed to Cd^{2+} ions in the growth medium [69]. When *E. coli* is incubated with $CdCl_2$ and Na_2SO_4, intracellular CdS nanocrystals, composed of a wurtzite crystal phase, are formed. Depending on the growth stage of the cells, nanocrystal formation varies significantly and increases in *E. coli* by about 20-fold. Compared to that cultivated in the late logarithmic period, *E. coli* grew in the stationary phase. Glabrata (yeasts) for the processing of intracellular CdS NPs with the solution of cadmium salt. Biological systems have successfully synthesized ZnS and PbS NPs. *Rhodobacter sphaeroides* and Desulfobacteraceae were used to obtain intracellular ZnS NPs with a mean diameter of 8 and 2–5 nm [70], respectively. *R. sphaeroides* were also used to synthesize PbS NPs whose diameters were regulated by the culture period. Eukaryotic species such as fungi have been found to be a strong candidate for extracellular synthesis of metal sulfide NPs. The fungus *F. oxysporum* can generate some stable metal sulfide NPs, such as CdS, ZnS, PbS, and MoS_2, extracellularly when exposed to metal sulfate in an aqueous solution. The quantum dots were produced by the sulfide ion reaction of Cd^{2+} ions created by the enzymatic reduction of sulfate ions to sulfide ions [63].

Magnetic Fe_3S_4, or FeS NP, is another type of sulfide NP. The development of Fe_3S_4 by uncultivated magnetotactic bacteria has been documented. A sediment sample containing approximately 105 magnetotactic bacteria per cm^3 was tested, and approximately 105 cells were acquired by the racetrack method after purification. In the uncultivated cells, magnetosomes showed an elongated rectangular shape. The average number of magnetosomes per cell was around 40, and they were predominantly distributed inside the cell as a large cluster. Besides the large cluster, aligned magnetosomes forming a chain-like structure were also observed. Sulfate-reducing bacteria were able to generate NPs of magnetic FeS [71].

7.5.4 Other Nanoparticles

In biological systems, through the use of biopolymers such as protein and microbe cells, a wide variety of species form organic/inorganic composites with ordered structures. $PbCO_3$, $CdCO_3$, $SrCO_3$, PHB, $Zn_3(PO_4)_2$, and CdSe NPs have been reported to be synthesized by microbes. $SrCO_3$ crystals were obtained when aqueous Sr^{2+} ions were incubated with the demanding fungi. The protein secretion is responsible for modulating the morphology of strontianite crystals during the growth of the fungus *F. oxysporum* and directing their hierarchical assembly into higher order superstructures. Yeasts were synthesized with zinc phosphate nanopowders as biotemplates. The $Zn_3(PO_4)_2$ powder was synthesized having butterfly-like microstructure with a size range of 10–80 nm in width and 80–200 nm in length. *F. oxysporum* can synthesize extremely luminescent CdSe quantum dots at room temperature [72, 73].

7.6 Greener and Sustainable (Nano) Solutions for Remediation

Although several engineered nanomaterials, ENMs, published in the literature show outstanding results, the potential safety drawbacks associated with their use in habitats, associated with possible bioaccumulation due to ingestion, dermal contact, and inhalation, are still controversial in terms of water and soil decontamination effectiveness. A number of studies have failed to reveal a nanodimensional risk of materials per se, as ENM effects are difficult to distinguish from those of bulk materials. Nevertheless, national and international legislation frequently take a cautious stance under this ambiguity, preventing the field use of ENMs. This indicates the need to design new technologies capable of taking these crucial factors into account.

In this context, the simple idea of moving from nanosized materials to nanostructured structures, transferring the advantages of nanotechnology to macrodimensioned systems, is a valuable alternative strategy to address ecotoxicological and regulatory issues related to the use of ENMs for environmental remediation. If ENMs, such as NPs and nanofibers, are not used directly in the remediation process but become the building blocks of stable micro- and nanoporosity-enhanced nanostructured systems, a new class of high-surface sorbent units, capable of extracting organic and inorganic contaminants from polluted water, air, and soil, can be given. An optimized system should maintain the benefits extracted from ENMs and prevent their release into the environment to achieve this objective. Moreover, if the new ENMs are obtained starting from the simple and scalable processing of renewable sources, this method could be considered even more useful. For this cause, it is becoming an essential goal to choose biopolymers as starting materials. Polysaccharides suit most of the ENM design specifications well, as they combine strong chemical reactivity with high biodegradability and negligible toxicity for further nanostructuring processes, due to the existence of many hydroxyl functional groups on the polymer backbone. Cellulose is an abundant, renewable, and low-cost natural source of polysaccharides for the production of materials for water remediation, especially when extracted from agricultural and industrial by-products [74]. For the removal of heavy metal ions from polluted water, sugarcane bagasse, fruit peel, biomass, and rice husks were proposed as cellulose-based matrices. In addition, waste paper would also represent an alternative source of cellulose, even cheaper, indicating the virtuous "recycling to remediate" method [75]. In addition, waste paper would also be an alternative, even cheaper source of cellulose.

However, its intrinsic hierarchical structure is what makes cellulose so appealing as a source for the production of advanced materials. The cellulose fiber composite is made of cellulose, hemicellulose, and lignin macrofibers. Macrofibers are composed of microfibrils that, in turn, are formed from cellulose nanofibrils. The ability to cleave the original native cellulose structure and generate cellulose nanofibers (CNF) opens up exciting possibilities for a wide variety of applications, including the treatment of waste water. With the 2,2,6,6-tetramethylpiperidinyloxyl (TEMPO)-mediated method, cellulose can be preliminarily oxidized following the simplest CNF processing procedure, selectively converting the primary C6-hydroxyl groups of the glucose units to the corresponding carboxylic groups. Defibrillation of TEMPO-oxidized cellulose nanofibers (TOCNF) can be accomplished by increasing the solution's pH according to this method. In fact, the deprotonation of carboxyl groups favors the electrostatic repulsion of single fibrils that are negatively charged, resulting in the physical separation of single fibrils. As effective and reusable heavy metal ion adsorbents, hydrogels obtained from TOCNF have been reported [70]. However, TOCNF, taking advantage of the modern carboxylic moieties introduced on the polymer backbone, can also be used for further cross-linking. Since this method would lead to macrodimensioned nanostructured structures, with all the advantages discussed previously, it would be possible to incorporate additional properties and functional groups to select the ideal cross-linker, enhancing the flexibility of the structures. In this context, a thermal route for the manufacture of a new class of aerogels was recently published, beginning with TOCNF and following a simple thermal protocol in the presence of branched-polyethyleneimine (bPEI) [76]. High reticulation into sponge-like, water-stable structures, which demonstrate high efficiency in extracting heavy metals and phenolic derivatives from waste water, has favored the formation of amide bonds between the carboxylic and the amine moieties. The possibility of selectively functionalizing the cross-linker amino groups and using these devices as templates for more organic and inorganic coating indicates the promise of this new ENM, the properties of which can be modulated in order to selectively absorb and degrade target contaminants. In addition, their safe use for environmental remediation is implemented by the introduction of these systems for biomedical applications in the area of drug delivery.

The basic application of hydrogels obtained from TOCNF and evaluated for their ecosafety will seek to establish new eco-friendly nanotechnologies for sludge and dredged sediment remediation within the context of the NANOBOND project (nanomaterials for remediation of environmental matrices associated with dewatering). The NANOBOND project, funded under the POR CReO FESR Tuscany 2014–2020 scheme, seeks to build a groundbreaking method for the treatment of polluted sludge and dredged sediments by incorporating the use of eco-friendly nanostructured materials with classical geotexile dewatering tubes. This new approach would allow, in terms of transport volumes and costs, the reduction of polluted sludge and sediments and also the conversion of the resulting solid and liquid waste into a sustainable clean resource to be used, for example in riverbank settlements and any other applications. Through the production of dewatering-related nanoremediation techniques, NANOBOND aims to explore new solutions to the management of dredging and sludge linked to hydrogeological destruction and the maintenance of harbor areas, emerging problems that are becoming increasingly global. This groundbreaking approach aims to become a successful method for dramatically reducing sludge and sediment pollution by nanoremediation, as it is also easily scalable for competitively priced large-scale in situ applications.

The NANOBOND consortium, made up of 70% of industrial partnerships directly with companies involved in the disposal of sludge and dredged sediments as well as in their risk assessment and 30% of academic and research institutes for the synthesis, ecosafety, and life cycle assessment of nanostructured materials, fulfilled the required technology transfer and business development requirements; both are still very less developed in European countries in terms of industrial productivity and innovation.

Other examples include the INTERREG EUROPE TANIA project (TREATING Pollution THROUGH NANOREMEDIATION) with the objective of improving EU regional policies on the treatment of pollution through nanoremediation in European countries and of introducing regional development policies in the field of pollutant safety and environmental prevention. Specifically, TANIA addresses creative and low-cost technical (nano) remediation methods for polluted soil and water [74].

Green nanotechnology applies to the use of nanotechnology to increase the sustainability of processes that create harmful externalities to the environment. It also refers to the use of materials in nanotechnology to increase sustainability. It involves the development of green nanoproducts and the use of nanoproducts to encourage sustainability.

7.7 Conclusions

The advantages over conventionally used processes of the new green-synthetic protocols for NPs generation include: first, only naturally occurring nonhazardous materials are used; second, no hazardous waste is produced; third, reduced processing effort is needed; fourth, the materials are more durable, easier to store, and easier to transport; and fifth, the materials can be more readily produced around the world using common and often biorenewable materials. At this early stage of nanotechnology development, the application of the principles of green chemistry to the progress of new nanomaterials and applications is all the more important and could lead to new design rules that are eco-friendly and benign in the sense of environmental and human health protection.

References

1 Luechinger, N.A., Grass, R.N., Athanassiou, E.K., and Stark, W.J. (2010). Bottom-up fabrication of metal/metal nanocomposites from nanoparticles of immiscible metals. *Chemistry of Materials* 22 (1): 155–160.

2 Amutha, M., Lalitha, P., and Firdhouse, M.J. (2014). Biosynthesis of silver nanoparticles using *Kedrostis foetidissima (Jacq.) Cogn. Journal of Nanotechnology* 2014: 860875.

3 Liu, J., Qiao, S.Z., Hu, Q.H., and Lu, G.Q.M. (2011). Magnetic nanocomposites with mesoporous structures: synthesis and applications. *Small* 7 (4): 425–443.

4 Tiwari, D.K., Behari, J., and Sen, P. (2008). Time and dose-dependent antimicrobial potential of Ag nanoparticles synthesized by top-down approach. *Current Science* 95: 647–655.

5 Zhang, X., Yan, S., Tyagi, R.D., and Surampalli, R.Y. (2011). Synthesis of nanoparticles by microorganisms and their application in enhancing microbiological reaction rates. *Chemosphere* 82 (4): 489–494.

6 Odularu, A.T. (2018). Metal nanoparticles: thermal decomposition, biomedicinal applications to cancer treatment, and future perspectives. *Bioinorganic Chemistry and Applications* 2018: 9354708.

7 Durán, N., Marcato, P.D., De Souza, G.I.H. et al. (2007). Antibacterial effect of silver nanoparticles produced by fungal process on textile fabrics and their effluent treatment. *Journal of Biomedical Nanotechnology* 3 (2): 203–208.

8 Nadagouda, M.N. and Varma, R.S. (2007). Preparation of novel metallic and bimetallic cross-linked poly(vinyl alcohol) nanocomposites under microwave irradiation. *Macromolecular Rapid Communications* 28 (4): 465–472.

9 Yadav, K.K., Singh, J.K., Gupta, N., and Kumar, V. (2017). A review of nanobioremediation technologies for environmental cleanup: a novel biological approach. *Journal of Materials and Environmental Science* 8 (2): 740–757.

10 Varma, R.S. (2012). Greener approach to nanomaterials and their sustainable applications. *Current Opinion in Chemical Engineering* 1 (2): 123–128.

11 Nadagouda, M.N. and Varma, R.S. (2008). Green synthesis of silver and palladium nanoparticles at room temperature using coffee and tea extract. *Green Chemistry* 10 (8): 859–862.

12 Nadagouda, M.N. and Varma, R.S. (2007). Synthesis of thermally stable carboxymethyl cellulose/metal biodegradable nanocomposites for potential biological applications. *Biomacromolecules* 8 (9): 2762–2767.

13 Nadagouda, M.N. and Varma, R.S. (2007). A greener synthesis of core (Fe, Cu)-shell (Au, Pt, Pd, and Ag) nanocrystals using aqueous vitamin C. *Crystal Growth & Design* 7 (12): 2582–2587.

14 Baruwati, B. and Varma, R.S. (2009). High value products from waste: grape pomace extract—a three-in-one package for the synthesis of metal nanoparticles. *ChemSusChem* 2 (11): 1041–1044.

15 Nadagouda, M.N., Speth, T.F., and Varma, R.S. (2011). Microwave-assisted green synthesis of silver nanostructures. *Accounts of Chemical Research* 44 (7): 469–478.

16 Nadagouda, M.N. and Varma, R.S. (2007). Microwave-assisted synthesis of crosslinked poly(vinyl alcohol) nanocomposites comprising single-walled carbon nanotubes, multi-walled carbon nanotubes, and buckminsterfullerene. *Macromolecular Rapid Communications* 28 (7): 842–847.

17 Zhu, P., Zhang, J., Wu, Z., and Zhang, Z. (2008). Microwave-assisted synthesis of various ZnO hierarchical nanostructures: effects of heating parameters of microwave oven. *Crystal Growth & Design* 8 (9): 3148–3153.

18 Baruwati, B., Nadagouda, M.N., and Varma, R.S. (2008). Bulk synthesis of monodisperse ferrite nanoparticles at water–organic interfaces under conventional and microwave hydrothermal treatment and their surface functionalization. *Journal of Physical Chemistry C* 112 (47): 18399–18404.

19 Nadagouda, M.N. and Varma, R.S. (2007). Room temperature bulk synthesis of silver nanocables wrapped with polypyrrole. *Macromolecular Rapid Communications* 28 (21): 2106–2111.

20 Polshettiwar, V. and Varma, R.S. (2010). Green chemistry by nano-catalysis. *Green Chemistry* 12 (5): 743–754.

21 Polshettiwar, V., Baruwati, B., and Varma, R.S. (2009). Self-assembly of metal oxides into three-dimensional nanostructures: synthesis and application in catalysis. *ACS Nano* 3 (3): 728–736.

22 Luque, R., Baruwati, B., and Varma, R.S. (2010). Magnetically separable nanoferrite-anchored glutathione: aqueous homocoupling of arylboronic acids under microwave irradiation. *Green Chemistry* 12 (9): 1540–1543.

23 Jahangirian, H., Lemraski, E.G., Webster, T.J. et al. (2017). A review of drug delivery systems based on nanotechnology and green chemistry: green nanomedicine. *International Journal of Nanomedicine* 12: 2957–2978.

24 Arruebo, M., Fernández-Pacheco, R., Ibarra, M.R., and Santamaría, J. (2007). Magnetic nanoparticles for drug delivery. *Nano Today* 2 (3): 22–32.

25 Sun, Z., Matthew, W., James, A. et al. (2016). Biodistribution of negatively charged iron oxide nanoparticles (IONPs) in mice and enhanced brain delivery using lysophosphatidic acid (LPA). *Nanomedicine* 12 (7): 1775–1784.

26 Kharissova, O.V., Kharisov, B.I., Oliva González, C.M. et al. (2019). Greener synthesis of chemical compounds and materials. *Royal Society Open Science* 6 (11): 191378.

27 Kumar, C.S.S.R. and Mohammad, F. (2011). Magnetic nanomaterials for hyperthermia-based therapy and controlled drug delivery. *Advanced Drug Delivery Reviews* 63 (9): 789–808.

28 Li, J., Wu, S., Qui, C. et al. (2016). Versatile surface engineering of porous nanomaterials with bioinspired polyphenol coatings for targeted and controlled drug delivery. *Nanoscale* 8 (16): 8600–8606.

29 Zhang, Y., Sun, C., Kohler, N., and Zhang, M. (2004). Self-assembled coatings on individual monodisperse magnetite nanoparticles for efficient intracellular uptake. *Biomedical Microdevices* 6 (1): 33–40.

30 Zhu, X., Cao, W., Chaang, B. et al. (2016). Polyacrylate/nanosilica causes pleural and pericardial effusion, and pulmonary fibrosis and granuloma in rats similar to those observed in exposed workers. *International Journal of Nanomedicine* 11: 1593–1605.

31 Chen, Q., Wang, C., Cheng, L. et al. (2014). Protein modified upconversion nanoparticles for imaging-guided combined photothermal and photodynamic therapy. *Biomaterials* 35 (9): 2915–2923.

32 Li, X., Yun, G., and Jun, L. (2016). Facile synthesis of soybean phospholipid-encapsulated MoS_2 nanosheets for efficient in vitro and in vivo photothermal regression of breast tumor. *International Journal of Nanomedicine* 11: 1819–1833.

33 Pozdnyakov, A.S., Emel'yanov, A.I., Kuznetsova, N.P. et al. (2016). Nontoxic hydrophilic polymeric nanocomposites containing silver nanoparticles with strong antimicrobial activity. *International Journal of Nanomedicine* 11: 1295–1304.

34 Li, Q., Wen, Y., Wen, J. et al. (2016). A new biosafe reactive oxygen species (ROS)-responsive nanoplatform for drug delivery. *RSC Advances* 6 (45): 38984–38989.

35 Das, D., Patra, P., Ghosh, P. et al. (2016). Dextrin and poly(lactide)-based biocompatible and biodegradable nanogel for cancer targeted delivery of doxorubicin hydrochloride. *Polymer Chemistry* 7 (17): 2965–2975.

36 Obayemi, J.D., Danyuo, Y., Dozie-Nwachuku, S. et al. (2016). PLGA-based microparticles loaded with bacterial-synthesized prodigiosin for anticancer drug release: effects of particle size on drug release kinetics and cell viability. *Materials Science & Engineering. C, Materials for Biological Applications* 66: 51–65.

37 Sahiner, N., Sagbas, S., and Aktas, N. (2016). Preparation of macro-, micro-, and nano-sized poly(tannic acid) particles with controllable degradability and multiple biomedical uses. *Polymer Degradation and Stability* 129 (C): 96–105.

38 Kommareddy, S. and Amiji, M. (2007). Poly(ethylene glycol)-modified thiolated gelatin nanoparticles for glutathione-responsive intracellular DNA delivery. *Nanomedicine* 3 (1): 32–42.

39 Hermawan, H., Dubé, D., and Mantovani, D. (2010). Developments in metallic biodegradable stents. *Acta Biomaterialia* 6 (5): 1693–1697.

40 Carlyle, W.C., James, M.B., Abraham, T.R. et al. (2012). Enhanced drug delivery capabilities from stents coated with absorbable polymer and crystalline drug. *Journal of Controlled Release* 162 (3): 561–567.

41 Nanaki, S.G., Pantopoulos, K., and Bikiaris, D.N. (2011). Synthesis of biocompatible poly(ε-caprolactone)-block-poly(propylene adipate) copolymers appropriate for drug nanoencapsulation in the form of core-shell nanoparticles. *International Journal of Nanomedicine* 6: 2981–2995.

42 Yousaf, A.M., Mustapha, O., Kim, D.W. et al. (2016). Novel electrosprayed nanospherules for enhanced aqueous solubility and oral bioavailability of poorly water-soluble fenofibrate. *International Journal of Nanomedicine* 11: 213–221.

43 Peeters, P., Bosiers, M., Verbist, J. et al. (2005). Preliminary results after application of absorbable metal stents in patients with critical limb ischemia. *Journal of Endovascular Therapy* 12 (1): 1–5.

44 Tzeng, S.Y., Yang, P.H., Grayson, W.L., and Green, J.J. (2011). Synthetic poly(ester amine) and poly(amido amine) nanoparticles for efficient DNA and siRNA delivery to human endothelial cells. *International Journal of Nanomedicine* 6: 3309–3322.

45 Tu, H., Lu, Y., Wu, Y. et al. (2015). Fabrication of rectorite-contained nanoparticles for drug delivery with a green and one-step synthesis method. *International Journal of Pharmaceutics* 493 (1–2): 426–433.

46 Di Mauro, V., Iafisco, M., Vacchiano, N. et al. (2016). Bioinspired negatively charged calcium phosphate nanocarriers for cardiac delivery of microRNAs. *Nanomedicine (London, England)* 11 (8): 891–906.

47 Mohammed, M., Abdel-Gawad, E., Awwad, S. et al. (2016). Therapeutic role of a synthesized calcium phosphate nanocomposite material on hepatocarcinogenesis in rats. *Biochemistry and Cell Biology* 94 (3): 279–288.

48 Zhang, Y., Zhou, J., Yang, C. et al. (2016). Folic acid-targeted disulfide-based cross-linking micelle for enhanced drug encapsulation stability and site-specific drug delivery against tumors. *International Journal of Nanomedicine* 11: 1119–1130.

49 Sampath, M., Balakumaran, M.D., Kalaichelvan, P., and Balasubramanian, S. (2013). Green synthesis of gold nanoparticles for controlled delivery. *Advanced Materials Letters* 4: 933–940.

50 Kotla, N.G., Singh, S., Maddiboyina, B. et al. (2016). A novel dissolution media for testing drug release from a nanostructured polysaccharide-based colon specific drug delivery system: an approach to alternative colon media. *International Journal of Nanomedicine* 11: 1089–1095.

51 Paciotti, G.F., Myer, L., Weinreich, D. et al. (2004). Colloidal gold: a novel nanoparticle vector for tumor directed drug delivery. *Drug Delivery* 11 (3): 169–183.

52 Boisselier, E. and Astruc, D. (2009). Gold nanoparticles in nanomedicine: preparations, imaging, diagnostics, therapies and toxicity. *Chemical Society Reviews* 38 (6): 1759–1782.

53 Ma, N., Zhang, B., Liu, J. et al. (2015). Green fabricated reduced graphene oxide: evaluation of its application as nano-carrier for pH-sensitive drug delivery. *International Journal of Pharmaceutics* 496 (2): 984–992.

54 Jeengar, M.K., Romipicharia, S.V.K., Shrivastava, S. et al. (2016). Emu oil based nano-emulgel for topical delivery of curcumin. *International Journal of Pharmaceutics* 506 (1–2): 222–236.

55 Logeswari, P., Silambarasan, S., and Abraham, J. (2013). Sharif University of Technology Ecofriendly synthesis of silver nanoparticles from commercially available plant powders and their antibacterial properties. *Scientia Iranica* 20 (3): 1049–1054.

56 Kumar, C.G. and Poornachandra, Y. (2015). Biodirected synthesis of miconazole-conjugated bacterial silver nanoparticles and their application as antifungal agents and drug delivery vehicles. *Colloids and Surfaces B: Biointerfaces* 125: 110–119.

57 Liu, X., Tao, H., Yang, K. et al. (2011). Optimization of surface chemistry on single-walled carbon nanotubes for in vivo photothermal ablation of tumors. *Biomaterials* 32 (1): 144–151.

58 Minhee, K., Kim, D., Kim, K.S. et al. (2018). Applications of cancer cell-specific aptamers in targeted delivery of anticancer therapeutic agents. *Molecules* 23 (4): 830.

59 Singh, M., Kumar, M., Subramanian, M. et al. (2014). Drug delivery system for controlled cancer therapy using physico-chemically stabilized bioconjugated gold nanoparticles synthesized from marine macroalgae, *Padina gymnospora*. *Journal of Nanomedicine & Nanotechnology* 5: 1–7.

60 Durán, N., Marcato, P.D., Alves, O.L. et al. (2005). Mechanistic aspects of biosynthesis of silver nanoparticles by several *Fusarium oxysporum* strains. *Journal of Nanbiotechnology* 3 (1): 8.

61 Mohanpuria, P., Rana, N., and Yadav, S.K. (2008). Biosynthesis of nanoparticles: technological concepts and future applications. *Journal of Nanoparticle Research* 10 (3): 507–517.

62 Mukherjee, P., Ahmad, A., Mandal, D. et al. (2001). Bioreduction of $AuCl_4^-$ ions by the fungus, *Verticillium* sp. and surface trapping of the gold nanoparticles formed. *Angewandte Chemie, International Edition* 40: 3585–3588.

63 Li, X., Xu, H., Chen, Z.-S., and Chen, G. (2011). Biosynthesis of nanoparticles by microorganisms and their applications. *Journal of Nanomaterials* 2011: 270974.

64 Ahmad, A., Senapati, S., Khan, M.I. et al. (2003). Extracellular biosynthesis of monodisperse gold nanoparticles by a novel extremophilic actinomycete, *Thermomonospora* sp. *Langmuir* 19 (8): 3550–3553.

65 Fayaz, A.M., Balaji, K., Girilal, M. et al. (2010). Biogenic synthesis of silver nanoparticles and their synergistic effect with antibiotics: a study against gram-positive and gram-negative bacteria. *Nanomedicine: Nanotechnology, Biology and Medicine* 6 (1): 103–109.

66 Kalimuthu, K., Suresh Babu, R., Venkataraman, D. et al. (2008). Biosynthesis of silver nanocrystals by *Bacillus licheniformis*. *Colloids and Surfaces B: Biointerfaces* 65 (1): 150–153.

67 De Windt, W., Aelterman, P., and Verstraete, W. (2005). Bioreductive deposition of palladium (0) nanoparticles on Shewanella oneidensis with catalytic activity towards reductive dechlorination of polychlorinated biphenyls. *Environmental Microbiology* 7 (3): 314–325.

68 Jain, N., Bhargava, A., Majumdar, S. et al. (2011). Extracellular biosynthesis and characterization of silver nanoparticles using *Aspergillus flavus* NJP08: a mechanism perspective. *Nanoscale* 3 (2): 635–641.

69 Alsaggaf, M.S., Elbaz, A.F., El Badawy, S., and Moussa, S.H. (2020). Anticancer and antibacterial activity of cadmium sulfide nanoparticles by *Aspergillus niger*. *Advances in Polymer Technology* 2020: 4909054.

70 Isobe, F., Hiroaki, O., Hitoshi, S., and Tetsuo, I. (2013). Nanolayered diamond sintered compact obtained by direct conversion from highly oriented graphite under high pressure and high temperature. *Journal of Nanomaterials* 2013: 1–7.

71 Sana, P., Hashmi, L., and Malik, M.M. (2012). Luminescence and morphological kinetics of functionalized ZnS colloidal nanocrystals. *ISRN Optics* 2012: 621908.

72 Narayanan, K.B. and Sakthivel, N. (2010). Biological synthesis of metal nanoparticles by microbes. *Advances in Colloid and Interface Science* 156 (1–2): 1–13.

73 Ram Kumar Pandian, S., Deepak, V., Kalishwaralal, K. et al. (2009). Synthesis of PHB nanoparticles from optimized medium utilizing dairy industrial waste using *Brevibacterium casei* SRKP2: a green chemistry approach. *Colloids and Surfaces B: Biointerfaces* 74 (1): 266–273.

74 Krishnani, K.K. and Ayyappan, S. (2006). Heavy metals remediation of water using plants and lignocellulosic agrowastes. *Reviews of Environmental Contamination and Toxicology* 188: 59–84.

75 Setyono, D. and Valiyaveettil, S. (2016). Functionalized paper—a readily accessible adsorbent for removal of dissolved heavy metal salts and nanoparticles from water. *Journal of Hazardous Materials* 302: 120–128.

76 Corsi, I., Fiorati, A., Grassi, G. et al. (2018). Environmentally sustainable and ecosafe polysaccharide-based materials for water nano-treatment: an eco-design study. *Materials (Basel)* 11 (7): 1228.

8

Green Synthesis of Titanium Dioxide Nanoparticles and Their Applications

Tabassum Siddiqui, Nida Jamil Khan, and Tasneem Fatma

Department of Biosciences, Jamia Millia Islamia, New Delhi, India

8.1 Introduction

Nanotechnology refers to the use of different methods and technologies aimed at obtaining materials with unique functions and properties [1]. Nanoparticles are the particles that possess size less than 100 nm and have greater surface-to-volume ratio (per unit mass) as compared to their non–nanoscale counterparts [2]. Increasing use of nanoparticles has led to continuously growing industrial, academic, and consumer interest in them. Unique surface chemistry and smaller size of nanoparticles are being utilized in the field of medicine, nutrition, and energy. Nanomaterials are referred as a wonder of modern medicine because they can kill upto 650 different disease-causing organisms [3].

Titanium belongs to the family of transition metal oxides. It is a strong, lustrous, corrosion-resistant metal. There are four commonly known polymorphs of TiO_2: anatase (tetragonal), brookite (orthorhombic), rutile (tetragonal), and TiO_2 (B) (monoclinic) [4]. Various polymorphs of TiO_2 differ from each other with respect to their respective band-gap energies, i.e. 3.2, 3.02, and 2.96 eV for the anatase, rutile, and brookite phases, respectively [5].

Synthesis of nanoparticles involves two general approaches, namely: (a) bottom-up and (b) top-down approach (Figure 8.1). Bottom-up approach involves the scaling down the material in question up to atomic level and then assembling it with each other leading to the synthesis of nanosized particles. The process of assemblage of basic structures to form compact stable nanoparticles is assisted by physical forces acting at nanoscale dimension. This methodology is also referred as building-up approach. Green synthesis of nanoparticles uses this approach [7]. Top-down approach makes use of precursor structures of macroscopic size and transforms them into nanoparticles by controlling the external parameters [8]. It employs destructive methodology beginning from larger molecule that is broken down into smaller molecules which are eventually transformed into nanoparticles. Examples include mechanical milling, chemical etching, etc.

Various physical and chemical methods that are used to synthesize titanium dioxide nanoparticles (TiO_2 NPs) includes chemical vapor deposition, microemulsion, chemical precipitation, hydrothermal crystallization, and sol–gel methods but nanoparticles produced via these techniques does not possess much potential to be used in medical field as these methods incorporate high temperature, pressure, and toxic chemicals, and hence they can jeopardize human health and environment. Therefore, there is an increasing demand for the environment-friendly outlook for the synthesis of nanomaterials on commercial scale. This could only be accomplished by exploiting biological entities as the source for nanoparticles synthesis [9].

TiO_2 NPs are known to possess novel optical and dielectric properties. It is widely employed as a self-cleaning and self-disinfecting material. It is further used in environmental purification applications because of its nontoxicity, photo-induced super-hydrophobicity, and anti-fogging nature [10]. TiO_2NPs are among the one of the most widely produced nanoparticles due to their unique characteristics which include photoactivity, whitening ability, and transparency in the nanoparticulate form. These properties are being used in manufacture of paints, sunscreens, cosmetics, and solar technologies [11]. It is also exploited in the synthesis of tints, textiles, papers, and foodstuffs. Colloidal TiO_2 NPs are utilized in the process which employs degradation of toxins in water. They can be found in commodities of daily use such as coatings, plastics, soaps, nanofibers and nanowires, bandages, alloys, and textiles. Its use as a color additive in food, drugs, cosmetics, and contact lenses

Sustainable Nanotechnology: Strategies, Products, and Applications, First Edition.
Edited by Yashwant V. Pathak, Govindan Parayil, and Jayvadan K. Patel.

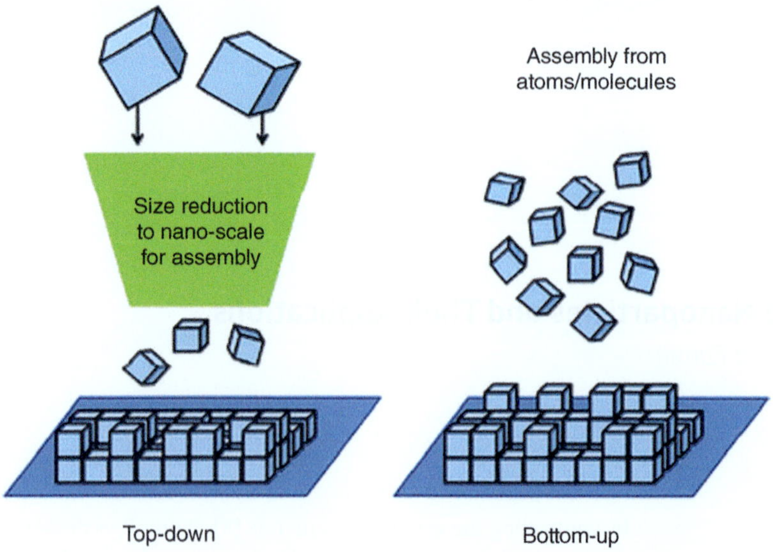

Figure 8.1 Top-down and bottom-up approaches of nanoparticle synthesis. *Source:* Birol et al. [6].

has already been approved by the US Food and Drug Administration. TiO_2 NPs have potential of removing inorganic and organic contaminants in wastewater treatment plants and in the remediation of polluted subsurface environments [12]. Presently, TiO_2 NPs are being used for antimicrobial, antibiotic, and antifungal treatments. Besides this, TiO_2 NPs are also used as ultraviolet (UV) blockers, antiscratch additives, and catalysts. Biogenic TiO_2 NPs synthesized from *Propionibacterium jensenii* are biocompatible and nontoxic nature. Cell viability and hemolytic assays suggested their wide range of biological applications [13].

8.2 Green Synthesis of TiO_2 Nanoparticles From Various Biological Sources

Green nanotechnology involves the synthesis of nanoparticles from various eco-friendly sources including enzymes, vitamins, polysaccharides, plant extracts, biodegradable polymers, and microorganisms (Figure 8.2). Biological sources for the synthesis of nanoparticles have always been popular because they incorporate the use of mild reaction parameters without using harmful chemical agents. For the biosynthesis of TiO_2 NPs, the most common salt used is titanium oxyhydrate; however, other salts [titanium tetraisoproxide, titanium dioxide, titanium(IV) bis(ammonium lactato)-dihydroxide, titanyl sulfate, titanium tetrachloride, etc.] are also used. Basic methodology used for the biosynthesis of TiO_2 NPs requires preparation of salt solution which is then mixed with the culture/extract of the biological source (plant, bacteria, fungi, etc.). The reaction mixture is then maintained at specific pH, temperature, and time under constant stirring conditions. Completion of the reaction is indicated by color change which in most cases is light green. Nanoparticles are then collected via centrifugation or filtration and dried via calcination, oven, or lyophilization which are then characterized by Fourier transform infrared spectroscopy (FTIR), X-ray diffraction (XRD), Scanning electron microscopy-Energy dispersive X-ray (SEM-EDX), Transmission electron microscopy (TEM), etc.

8.3 TiO_2 NPs Synthesis Through Bacteria

Chemical mode of synthesis of nanoparticles utilizes noxious and expensive agents that are detrimental to the environment, hence an eco-friendly and inexpensive approach is needed. Since microbial cells are easy to grow and maintain, the process is low cost and highly efficient so they turned out to be the promising candidate for the green synthesis of nanoparticles. Numerous bacterial species have been exploited for the synthesis of various nanoparticles including TiO_2 NPs [14].

TiO_2 NPs synthesized via *Planomicrobium* sp. isolated from melted ice showed maximum absorbance peak at 400 nm by UV–Visible spectroscopy. FTIR indicated that proteins acted as the reducing agents and XRD revealed the crystalline nature of nanoparticles that possessed the mean size of 8.89 nm. Microorganisms like *Bacillus subtilis* have been reported to synthesize TiO_2 NPs via green approach [15]. The nanostructures gave the onset

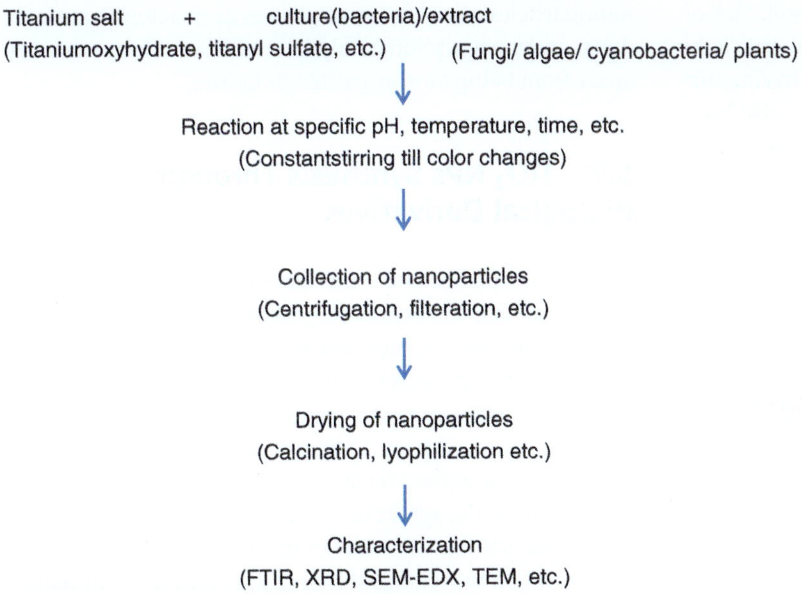

Figure 8.2 Biosynthesis of titanium dioxide nanoparticles.

wavelength of the optical absorption at 366 nm when characterized by UV–Vis spectroscopy. The significant FTIR peaks were found at 3430, 1578, 1451, 1123 cm^{-1} due to various alcohols, lipids, and proteins. Atomic force microscopy (AFM) analysis further revealed the formation of anatase and rutile form of TiO_2 NPs. SEM images showed that nanoparticles were spherical and oval having size range of 66–77 nm [16]. Jayaseelan et al. [17] synthesized TiO_2 NPs using *Aeromonas hydrophila*. Their characterization via FTIR indicated prominent peaks at 3430, 2937, 1643, and 1079 suggesting that alcohols, phenols, primary amines, lactones, and aliphatic amines were responsible for the synthesis of the nanoparticles. XRD and Field emission scanning electron microscopy (FE-SEM) revealed that the mean size of biosynthesized nanoparticles was 40.50 nm. Babitha and Korrapati [13] synthesized TiO_2NPs using *P. jensenii* [KC545833] isolated from the coal fly ash effluent. Their FE-SEM analysis revealed that the nanoparticles were smooth and spherical having size ranging from 15 to 80 nm. The EDAX spectrum revealed the atomic percentage of Ti and O was 54.73 and 45.27, respectively. Priyaragini et al. [18] isolated actinobacteria PSV3 from marine sediments of Chennai and synthesized TiO_2NPs which showed absorbance peak in the range of 400–420 nm. Their FTIR analysis revealed that the functional groups like aldehyde, alkenes, nitro, and alkynes were involved in the synthesis process. FTIR revealed the average size of 58.3 nm of the biosynthesized nanoparticles. Biosynthesis of TiO_2 NPs using *Bacillus amyloliquefaciens* by Khan and Fulekar [14] showed that the nanoparticles were spherical and their size ranges from 22.11 to 97.28 nm. Prasad et al. [19] did XRD and TEM of TiO_2NPs and reported that *Lactobacillus* sp. assisted TiO_2 NPs had the mean size of 40 nm. Similarly, *Bacillus mycoides* was also used for the synthesis of TiO_2 NPs having cut-off wavelength near 380 nm observed via UV–Vis spectroscopy as described by Órdenes-Aenishanslins et al. [20]. Biosynthesized TiO_2 NPs displayed spherical shape and size range of 40–60 nm by TEM while FTIR gave prominent bands or peaks at 3431, 2985, 1646, 1554, 1462 cm^{-1} and 1400, 1246, 1047 cm^{-1} attributed to biomolecules like peptides or carbohydrates bound to the TiO_2NPs synthesized by *B. mycoides*.

8.4 TiO_2 NPs Synthesis Through Fungi

The exploitation of fungi for the synthesis of nanoparticles is a comparatively new addition to the repertoire of microorganisms capable of synthesizing nanoparticles. The use of fungi to produce nanoparticles grabs special attention because they are able to secrete large amounts of enzymes. A novel protocol was developed for the synthesis of TiO_2 NPs by using fungus *Aspergillus flavus*. The nanoparticles were characterized via various techniques in order to know their physicochemical properties. X-ray diffraction analysis showed that the synthesized nanoparticles were of rutile form while SEM revealed spherical and oval nanoparticles with size range of 62–74 nm [21]. Raliya and Tarafdar [22] reported the synthesis of TiO_2 NPs synthesized using fungi (*Aspergillus flavus*, *Aspergillus terreus*, *Aspergillus tubingensis*, *Aspergillus niger*, *Rhizoctonia bataticola*, *Aspergillus fumigatus*, and *Aspergillus oryzae*) isolated from soil. Size of the synthesized nanoparticles determined via Dynamic light scattering (DLS) was found to be lower than 6 nm having surface charge of −6.25 mV. *Fusarium oxysporum* has also been exploited for the synthesis of TiO_2 NPs [23]. TEM images revealed that the average size of spherical particles was 10.2 nm, Selected area electron diffraction (SAED) and XRD indicated that the crystalline particles were of brookite polymorph of TiO_2 and FTIR showed that the proteins in the fungus are responsible for the biosynthesis of nanoparticles.

8.5 TiO_2 NPs Synthesis Through Algae and Cyanobacteria

Both algae and cyanobacteria are rich source of biologically active compounds which can successfully act as the reducing and capping agents in the biosynthesis of various metallic nanoparticles. Cyanobacteria and eukaryotic algae have been exploited as a low-cost source for synthesizing silver, gold, selenium nanoparticles etc. [24–27]. To the best of our knowledge, so far algae and cyanobacteria are not utilized as a raw material for the synthesis of TiO_2 NPs. However, there are few reports related to the algal and cyanobacterial studies in relation to titanium metal salt with respect to its toxicity and population dynamics [11, 28, 29]. So, we have carried out the screening of 37 cyanobacterial strains for the synthesis of TiO_2 NPs using titanium dioxide salt and obtained spherical and oval TiO_2 NPs at ambient temperature (30 °C) with mean particle size ranging from 82.7 to 162.8 nm. Determination of their bioactivity is in the process for their future applications (accepted).

8.6 TiO_2 NPs Synthesis Through Plants

Many plants like *Justicia gendarussa* [30], *Calotropis gigantea* [31], *Psidium guajava* [32], *Annona squamosal* [33], *Echinacea purpurea* [34], *Eclipta prostrata* [35], *Vigna radiata* [36], *Solanum trilobatum* [37], *Nyctanthes* [38], *Cicer arietinum* L. [39], *Ageratina alttissima* (L.) [40] have been exploited for the synthesis of TiO_2 NPs. Their FTIR analysis revealed that hydroxyl alcohol group, alkenes/aldehydes, alkenes, carboxylic acids, primary amines and amides, alkynes, alkanes, aliphatic amines, nitro compounds, vinyl ethers, beta lactones, phenol, ammonium ions, vinyl, azo compounds act as reducing and capping agents for the nanoparticle biosynthesis. Synthesized nanoparticles were spherical and oval in morphology and their mean size ranged from 10 to 120 nm. XRD evinced that the crystalline nanoparticles were either in rutile or anatase phase. The biosynthesized nanoparticles were known to possess antibacterial, antioxidant, anticancer, photocatalytic, and acaricidal properties apart from being biocompatible in nature.

8.7 TiO_2 NPs Synthesis Through Biological Derivatives

Apart from plants and microorganisms, many biological derivatives like R5 peptide, alpha-amylase, soluble starch, poly(allylamine), etc. have also been used as source of nanoparticle synthesis. Sewell and Wright [41] synthesized TiO_2 NPs of average size 50 nm using R5 peptide, a bioinspired analogue from diatom *Cylindrotheca fusiformis*. Ahmad et al. [42] used alpha-amylase as the sole reducing and capping agent for the synthesis of TiO_2 NPs and found that the average particle size decreased from 50 to 25 nm when enzyme concentration was changed from 2 to 15 mg ml^{-1}. Muniandy et al. [43] synthesized TiO_2 NPs using soluble starch and concluded that tetragonal anatase TiO_2 NPs were formed having mean size of 99.2 nm at pH 5.0 and 64.19 nm at pH 9.0. Cole et al. [44] reported the biosynthesis of amorphous TiO_2 NPs using poly(allylamine) and the R5 peptide, in the presence of phosphate having particles ranging from 2 nm to 5 μm. Kyuá and Hyeoká [45] reported titanium–tungsten oxide composites by lysozyme-mediated biomineralization. Lysozyme was obtained from chicken egg white. After calcination, the average size reduced from 100 to 50 nm.

8.8 Applications

TiO_2 NPs are one of the top five nanoparticles used in the consumer products [46]. Following are some important applications of TiO_2 NPs:

8.8.1 Biomedical Field

TiO_2NPs synthesized from biological source showed increased collagen stabilization and thereby enabled the preparation of collagen-based biological wound dressing [13]. Titanium nanostructures are one of the most sought biomaterials for biomedical devices due to their potential for cellular interaction and protein adhesion. Commonly, TiO_2NPs are used in orthopedic and dental implants, bladder stents, and blood-contacting or drug delivery devices [47].

8.8.2 Antibacterial Activity

Due to increasing microbial resistance against metal ions, antibiotics, and development of resistant bacterial strains, researchers are continuously looking for alternatives. Since

TiO_2 NPs possess formidable antibacterial activity so they can be regarded as future antibacterial agents. Studies of TiO_2 NPs synthesized via *P. guajava* showed better antibacterial activity than standard tetracycline disk against *Staphylococcus aureus* and *Escherichia coli* [32]. TiO_2 NPs synthesized via alpha-amylase have the minimum inhibitory concentration (MIC) 62.50 $\mu g\,ml^{-1}$ for *S. aureus* and *E. coli* [42].

8.8.3 Antioxidant Activity

Antioxidant activity plays an important part in eliminating the oxidative rancidity of food by scavenging the free radicals generated during oxidation [48, 49]. NanoTiO_2 has been reported to increase the activity of catalase and stimulate glutathione transferase in *Mytilus galloprovincialis* [50].

8.8.4 Acaricidal Activity

TiO_2 NPs synthesized from flower aqueous extract of *C. gigantea* possess prominent antiparasitic activity against the tick larvae (*Rhipicephalus (Boophilus) microplus*) and the adult of flies (*Haemaphysalis bispinosa*) [31]. (*R. (B.) microplus*) is the common parasite of cattle, buffalo, horses, donkeys, goats, sheep, deer, pigs, dogs, and some wild animals [51] whereas (*H. bispinosa*) parasitize various wild and domestic mammals such as domestic cattle, goats, and sheep and various bird species and it is a potential vector of Kyasanur Forest disease virus and *Bartonella bovis* which causes Bartonellosis [52].

8.8.5 Anticancer Activity

TiO_2 NPs can act as effective anticancer agents as their exposure is injurious to proteins, lipids, and DNA which eventually leads to subcellular organelles damage and causes cell death. They cause apoptosis by inducing chromatin condensation, nuclear blebbing, and caspase activation [53]. TiO_2 NPs exhibited dose-dependent cytotoxicity on treated breast cancer MCF-7 cells. IC_{50} against MCF-7 cells by TiO_2 NPs was 60 $\mu g\,ml^{-1}$. TiO_2NPs exposure to the cells resulted in observable morphological changes such as cytoplasmic condensation, cell shrinkage, formation of several cell surface protuberances at the plasma membrane, and the aggregation of the nuclear chromatin into dense masses under the nuclear membrane. AO/EtBr for fluorescence microscopic study revealed the induction of apoptosis following treatment with IC_{50}concentration of TiO_2 NPs. Nuclear shrinkage and blebbing created orange-colored apoptotic bodies and apoptotic cells and red-colored necrotic cells. 4′,6-diamidino-2-phenylindole (DAPI) staining revealed that TiO_2NPs induce nuclear fragmentation in the cells. Number of apoptotic cells increased with increasing concentration of TiO_2 NPs. Breast cancer MCF-7 cells when treated with TiO_2 NPs activated cytochrome c expression profiles hinting that apoptosis is initiated via mitochondria-mediated pathway [54].

8.8.6 Photocatalytic Activity in Pollution Control

Several studies have shown that TiO_2 NPs have properties that can be used for the reduction of dyes and pharmaceutical drugs. Ability of TiO_2 NPs to degrade dyes in natural environment greatly contributes toward curbing water pollution. TiO_2 NPs have been reported to degrade many dyes such as Reactive Red 31 [14], Acid Red 79, Acid Red 80 [18], methylene blue, alizarin red, crystal violet, and methyl orange [40]. According to Khan and Fulekar [14], doping of TiO_2 NPs with metals like Ag, La, Zn, and Pt enhanced their photocatalytic activity. Doping is the process of adding impurities to intrinsic semiconductors to alter their properties. Surface properties of TiO_2 are changed by metal ion doping, it also obstructs the recombination of photogenerated electron-hole pairs, and numbers of active sites are increased [55, 56]. Dhandapani et al. [57] showed that biogenic TiO_2NPs suppress the growth of the aquatic biofilm formed by stagnant pond water by generation of free radicals.

8.8.7 As Mosquitocides

TiO_2 NPs have great potential to be used as future mosquitocides and chemotherapeutic agents. Larvicidal and pupicidal studies revealed that LC_{50} values of NPs were 4.02 ppm (larva I), 4.962 ppm (larva II), 5.671 ppm (larva III), 6.485 ppm (larva IV), and 7.527 ppm (pupa) against *Aedes aegypti*, the chief dengue causing mosquito [54].

8.8.8 In Agriculture

TiO_2 NP enhances metabolic activity of the plants thereby leading to increased biomass production. They also promote microbial activities which help in efficient utilization of local nutrients. Foliar application of TiO_2 NPs (10 $mg\,l^{-1}$ for 14 days) on mung bean plant promoted growth as evinced by increase in shoot length by 17%, root length by 49.6%, root area by 43%, and root nodule by 67.5%. TiO_2 NPs at 10 $mg\,l^{-1}$concentration increased chlorophyll content by 46.4 % and the total soluble leaf protein content by 94%. Its application also led to the increase in population of rhizospheric bacteria, fungi, and actinomycete between 21.4% and 48.1% when applied at critical growth stage of mung bean crop. Explosion of microbial population also led to the increased activity of dehydrogenase (108.7%), phytase (64%), acid phosphatase (67.3%), and alkaline phosphatase (72%) in the rhizosphere [58]. Native phosphorous mobilization in rhizosphere may increase due to the activity of phytase and phosphatase [59].

8.8.9 Reproduction of Silkworm

Low concentrations of TiO_2 NPs affect the reproduction rate of *Bombyx mori*. Speedy development of sex organs and up-regulation of reproduction-related genes was observed when *B. mori* was fed TiO_2 NPs [60]. TiO_2 NPs can increase reproduction of silkworms and thus silk production.

Acknowledgments

The authors are highly thankful to the Department of Biosciences, Jamia Millia Islamia and University Grants Commission (UGC) for the financial assistance.

References

1 Lusvardi, G., Barani, C., Giubertoni, F., and Paganelli, G. (2017). Synthesis and characterization of TiO_2 nanoparticles for the reduction of water pollutants. *Materials (Basel)* 10 (10): 1208. https://doi.org/10.3390/ma10101208.

2 Borzabadi-Farahani, A., Borzabadi, E., and Lynch, E. (2014). Nanoparticles in orthodontics, a review of antimicrobial and anti-caries applications. *Acta Odontologica Scandinavica* 72 (6): 413–417. https://doi.org/10.3109/00016357.2013.859728.

3 Shukla, A.K. and Iravani, S. (2017). Metallic nanoparticles: green synthesis and spectroscopic characterization. *Environment Chemical Letters* 15: 223. https://doi.org/10.1007/s10311-017-0618-2.

4 Carp, O., Huisman, C.L., and Reller, A. (2004). Photoinduced reactivity of titanium dioxide. *Progress in Solid State Chemistry* 32: 33–117.

5 Wunderlich, W., Oekermann, T., and Miao, L. (2004). Electronic properties of nano-porous TiO_2 and ZnO thin films comparison of simulations and experiments. *Journal of Ceramic Processing Research* 5: 343–354.

6 Birol, H., Rambo, C.R., Guiotokuc, M., and Hotza, D. (2013). Preparation of ceramic nanoparticles via cellulose-assisted glycine nitrate process: a review. *RSC Advances* 3: 2873–2884.

7 Iravani, S. (2011). Green synthesis of metal nanoparticles using plants. *Green Chemistry* 13: 2638. https://doi.org/10.1039/c1gc15386b.

8 Soni, M., Mehta, P., Soni, A., and Goswami, G.K. (2018). Nanoparticles: synthesis and applications. *Journal of Biotechnology and Biochemistry* 4: 78–83.

9 Nadeem, M., Tungmunnithum, D., Hano, C. et al. (2018). The current trends in the green syntheses of titanium oxide nanoparticles and their applications. *Green Chemistry Letters and Reviews* 11 (4): 492–502. https://doi.org/10.1080/17518253.2018.1538430.

10 Haghi, M., Hekmatafshar, M., Janipour, M.B. et al. (2012). Antibacterial effect of TiO_2 nanoparticles on pathogenic strain of *E. coli*. *International Journal of Advanced Biotechnology and Research* 3 (3): 621–624.

11 Kulacki, K.J. and Cardinale, B.J. (2012). Effects of nano-titanium dioxide on freshwater algal population dynamics. *PLoS One* 7 (10): e47130. https://doi.org/10.1371/journal.pone.0047130.

12 Rebecca, F., Jacobson, A., Bojeongkim, I.S. et al. (2009). Influence of ionic strength, pH, and cation valence on aggregation kinetics of titanium dioxide nanoparticles. *Environmental Science and Technology* 43: 1354–1359.

13 Babitha, S. and Korrapati, P.S. (2013). Biosynthesis of titanium dioxide nanoparticles using a probiotic from coal fly ash effluent. *Materials Research Bulletin* 48: 4738–4742.

14 Khan, R. and Fulekar, M.H. (2016). Biosynthesis of titanium dioxide nanoparticles using *Bacillus amyloliquefaciens* culture and enhancement of its photocatalytic activity for the degradation of a sulfonated textile dye Reactive Red 31. *Journal of Colloid and Interface Science* 475: 184–191.

15 Kirthi, A.V., Rahuman, A.A., Rajakumar, G. et al. (2011). Biosynthesis of titanium dioxide nanoparticles using bacterium *Bacillus subtilis*. *Materials Letters* 65: 2745–2747.

16 Malarkodi, C., Chitra, K., Rajeshkumar, S. et al. (2013). Novel eco-friendly synthesis of titanium oxide nanoparticles by using *Planomicrobium* sp. and its antimicrobial evaluation. *Pelagia Research Library Der Pharmacia Sinica* 4 (3): 59–66.

17 Jayaseelan, C., Rahuman, A.A., Roopan, S.M. et al. (2013). Biological approach to synthesize TiO_2 nanoparticles using *Aeromonas hydrophila* and its antibacterial activity. *Spectrochimica Acta Part A: Molecular and Biomolecular Spectroscopy* 107: 82–89.

18 Priyaragini, S., Veena, S., Swetha, D. et al. (2014). Evaluating the effectiveness of marine actinobacterial extract and its mediated titanium dioxide nanoparticles in the degradation of azo dyes. *Journal of Environmental Sciences* 26: 775–782.

19 Prasad, K., Jha, A.K., and Kulkarni (2007). *Lactobacillus* assisted synthesis of titanium nanoparticles. *Nanoscale Research Letters* 2 (5): 248.

20 Órdenes-Aenishanslins, N.A., Saona, L.A., Durán-Toro, V.M. et al. (2014). Use of titanium dioxide nanoparticles biosynthesized by *Bacillus mycoides* in quantum dot sensitized solar cells. *Microbial Cell Factories* 13 (1): 90.

21 Rajakumar, G., Rahumana, A.A., Roopan, S.M. et al. (2012). Fungus-mediated biosynthesis and characterization of TiO_2 nanoparticles and their activity against pathogenic bacteria. *Spectrochimica Acta Part A* 91: 23–29.

22 Raliya, R. and Tarafdar, J. (2014). Biosynthesis and characterization of zinc, magnesium and titanium nanoparticles: an eco-friendly approach. *International Nano Letters* 4 (1): 93.

23 Bansal, V., Rautaray, D., Bharde, A. et al. (2005). Fungus-mediated biosynthesis of silica and titania particles. *Journal of Materials Chemistry* 15 (26): 2583–2589.

24 Aziz, N., Faraz, M., Pandey, R. et al. (2015). Facile algae–derived route to biogenic silver nanoparticles: synthesis, antibacterial and photocatalytic, properties. *Langmuir* 31: 11605–11612. https://doi.org/10.1021/acs.langmuir.5b03081.

25 Hussain, S., Afreen, S., Hemlata, Y.D. et al. (2019). Cyanobacteria as a bioreactor for the synthesis of silver nanoparticles- an effect of different reaction conditions on size of nanoparticles and their dye decolorization ability. *Journal of Microbial Methods 162*: 77–82. https://doi.org/10.1016/j.mimet.2019.05.011.

26 Hussain, S., Sardar, M., and Fatma, T. (2015). Screening of cyanobacterial extracts for synthesis of silver nanoparticle. *World Journal of Microbiology and Biotechnology* 31: 1279–1283. https://doi.org/10.10007/s11274-015-1869-3.

27 Shah, M., Fawcett, D., Sharma, S. et al. (2015). Green synthesis of metallic nanoparticles via biological entities. *Materials (Basel)* 8 (11): 7278–7308.

28 Cherchi, C. and Gu, A.Z. (2010). Impact of titanium dioxide nanomaterials on nitrogen fixation rate and intracellular nitrogen storage in *Anabaena variabilis*. *Environmental Science and Technology 44*: 8302–8307.

29 Ozkaleli, M. and Erdem, A. (2018). Biotoxicity of TiO_2 nanoparticles on *Raphidocelis subcapitata* microalgae exemplified by membrane deformation. *International Journal of Environmental Research and Public Health* 15: 416.

30 Senthilkumar, S. and Rajendran, A. (2018). Biosynthesis of TiO_2 nanoparticles using *Justicia gendarussa* leaves for photocatalytic and toxicity studies. *Research on Chemical Intermediates 44*: 5923. https://doi.org/10.1007/s11164-018-3464-3.

31 Marimuthu, S., Rahuman, A.A., Jayaseelan, C. et al. (2013). Acaricidal activity of synthesized titanium dioxide nanoparticles using *Calotropisgigantea* against *Rhipicephalusmicroplus* and *Haemaphysalisbispinosa*. *Asian Pacific Journal of Tropical Medicine* 6: 682–688.

32 Santhoshkumar, T., Rahuman, A.A., Jayaseelan, C. et al. (2014). Green synthesis of titanium dioxide nanoparticles using *Psidium guajava* extract and its antibacterial and antioxidant properties. *Asian Pacific Journal of Tropical Medicine* 6: 968–976.

33 Roopan, S.M., Bharathi, A., Prabhakarn, A. et al. (2012). Efficient phyto-synthesis and structural characterization of rutile TiO_2 nanoparticles using *Annona squamosa* peel extract. *Spectrochimica Acta Part A* 98: 86–90.

34 Dobrucka, R. (2017). Synthesis of titanium dioxide nanoparticles using *Echinacea purpurea* Herba. *Iranian Journal of Pharmaceutical Research* 16 (2): 756–762.

35 Rajakumar, G., Rahuman, A.A., Priyamvada, B. et al. (2012). *Ecliptaprostrata* leaf aqueous extract mediated synthesis of titanium dioxide nanoparticles. *Materials Letters* 68: 115–117.

36 Chatterjee, A., Nishanthini, D., Sandhiya, N., and Abraham, J. (2016). Biosynthesis of titanium dioxide nanoparticles using *Vigna radiata*. *Asian Journal of Pharmaceutical and Clinical Research* 9 (4): 1–4.

37 Rajakumar, G., Rahuman, A.A., Jayaseelan, C. et al. (2014). *Solanum trilobatum* extract-mediated synthesis of titanium dioxide nanoparticles to control *Pediculus humanus capitis, Hyalomma anatolicum anatolicum and Anopheles subpictus*. *Parasitology Research* 113 (2): 469–479.

38 Sundrarajan, M. and Gowri, S. (2011). Green synthesis of titanium dioxide nanoparticles by *Nyctanthes arbor-tristis* leaves extract. *Chalcogenide Letters* 8 (8): 447–451.

39 Kashale, A.A., Gattu, K.P., Ghule, K. et al. (2016). Biomediated green synthesis of TiO_2 nanoparticles for lithium ion battery application. *Composites Part B Engineering* 99: 297–304.

40 Ganesan, S., Babu, I.G., Mahendran, D. et al. (2016). Green engineering of titanium dioxide nanoparticles using *Ageratina altissima* (L.) King and H.E. Robines. Medicinal plant aqueous leaf extracts for enhanced photocatalytic activity. *Annals of Phytomedicine* 5 (2): 69–75.

41 Sewell, S.L. and Wright, D.W. (2006). Biomimetic synthesis of titanium dioxide utilizing the R5 peptide derived from *Cylindrotheca fusiformis*. *Chemistry of Materials* 18: 3108–3113.

42 Ahmad, R., Mohsin, M., Ahmad, T., and Sardar, M. (2015). Alpha amylase assisted synthesis of TiO_2 nanoparticles: structural characterization and application as antibacterial agents. *Journal of Hazardous Materials* 283: 171–177.

43 Muniandy, S.S., Kaus, N.H.M., Jiang, Z.-T. et al. (2017). Green synthesis of mesoporous anatase TiO_2 nanoparticles and their photocatalytic activities. *RSC Advances* 7 (76): 48083–48094.

44 Cole, K.E., Ortiz, A.N., Schoonen, M.A., and Valentine, A.M. (2006). Peptide-and long-chain polyamine-induced synthesis of micro-and nanostructured titanium phosphate and protein encapsulation. *Chemistry of Materials* 18 (19): 4592–4599.

45 Kyuá, K.J. and Hyeoká, P.J. (2014). Lysozyme-mediated biomineralization of titanium–tungsten oxide hybrid nanoparticles with high photocatalytic activity. *Chemical Communications* 50 (82): 12392–12395.

46 Shukla, R.K., Sharma, V., Pandey, A.K. et al. (2011). ROS-mediated genotoxicity induced by titanium dioxide nanoparticles in human epidermal cells. *Toxicology in vitro* 25: 231–241. https://doi.org/10.1016/j.tiv.2010.11.008.

47 Kulkarni, M., Mazare, A., Gongadze, E. et al. (2015). Titanium nanostructures for biomedical applications. *Nanotechnology* 26: 062002. (18 pp).

48 Beltran, E., Pla, R., Yuste, J., and Mor-Mur, M. (2004). Use of antioxidants to minimize rancidity in pressurized and cooked chicken slurries. *Meat Science* 66 (3): 719–725.

49 Coppin, E.A. and Pike, O.A. (2001). Oil stability index correlated with sensory determination of oxidative stability in light-exposed soybean oil. *Journal of American Oil Chemist's Society* 78: 13–18.

50 Canesi, L., Fabbri, R., Gallo, G. et al. (2010). Biomarkers in *Mytilus galloprovincialis* exposed to suspensions of selected nanoparticles (nano carbon black, C60 fullerene, nano-TiO_2, nano-SiO_2). *Aquatic Toxicology* 100 (2): 168–177.

51 Harry, H., Boo-Liat, L., and George, A. (1969). *Haemaphysalis (Kaiseriana) bispinosa* Neumann (*Ixodoidea: Ixodidae*): evidence for consideration as an introduced species in the Malay Peninsula and Borneo. *The Journal of Parasitology* 55 (5): 1075–1077. https://doi.org/10.2307/3277178.

52 Kho, K., KohFui-Xian, J.T., Nizam, Q.N.H., and Tay, S. (2015). Prevalence and molecular heterogeneity of *Bartonella bovis* in cattle and *Haemaphysalis bispinosa* ticks in Peninsular Malaysia. *BMC Veterinary Research* 11: 153. https://doi.org/10.1186/s12917-015-0470-1.

53 Park, E.J., Yi, J., Chung, K.H. et al. (2008). Oxidative stress and apoptosis induced by titanium dioxide nanoparticles in cultured BEAS-2B cells. *Toxicology Letters* 180: 222–229.

54 Murugan, K., Dinesh, D., Kavithaa, K. et al. (2016). Hydrothermal synthesis of titanium dioxide nanoparticles: mosquitocidal potential and anticancer activity on human breast cancer cells (MCF-7). *Parasitology Research* 115: 1085–1096. https://doi.org/10.1007/s00436-015-4838-8.

55 Hsieh, S.H., Chen, W.J., and Wu, C.T. (2015). Pt-TiO_2/graphene photocatalysts for degradation of AO7 dye under visible light. *Applied Surface Science* 340: 9.

56 Khalid, N.R., Ahmed, E., Hong, Z. et al. (2013). Graphene modified Nd/TiO_2 photocatalyst for methyl orange degradation under visible light irradiation. *Ceramics International* 39: 3569.

57 Dhandapani, P., Maruthamuthu, S., and Rajagopal, G. (2012). Bio-mediated synthesis of TiO_2 nanoparticles and its photocatalytic effect on aquatic biofilm. *Journal of Photochemistry and Photobiology B: Biology* 110: 43–49.

58 Raliya, R., Biswas, P., and Tarafdar, J.C. (2015). TiO_2 nanoparticles biosynthesis and its physiological effect on mung bean (*Vignaradiata* L.). *Biotechnology Reports* 5: 22–26.

59 Richardson, A.E. and Simpson, R.J. (2011). Soil microorganisms mediating phosphorus availability update on microbial phosphorus. *Plant Physiology* 156: 989–996.

60 Ni, M., Li, F., Wang, B. et al. (2015). Effect of TiO_2 nanoparticles on the reproduction of silkworm. *Biological Trace Element Research* 164: 106–113.

9

Sustainable and Eco-safe Nanocellulose-based Materials for Water *Nano*-treatment

Carlo Punta[1], Andrea Fiorati[1], Laura Riva[1], Giacomo Grassi[2], Giulia Liberatori[2], and Ilaria Corsi[2]

[1] *Department of Chemistry, Materials, and Chemical Engineering "G. Natta", Politecnico di Milano and INSTM Local Unit, Milano, Italy*
[2] *Department of Physical, Earth and Environmental Sciences, University of Siena and INSTM Local Unit, Siena, Italy*

9.1 Eco-safe Materials: Behind a Story

Circular Economy and Sustainability. In the last decade, these two keywords have become more and more crucial in the development of new products and processes. From one side, an increasing demand for virtuous waste management is pushing toward not only simple recycling and reusing of discharged materials but rather toward their valorization by the design of up-scaled secondary products. On the other side, the whole circular process has to be sustainable from cradle-to-gate (or to grave), which could be declined in three key steps: (i) sustainability of the sources of choice, (ii) sustainability of the production processes, and (iii) sustainability of the new products in the final environment of use, which means eco-safety of the new devices.

Engineered nanomaterials (ENMs) do not represent an exception in this approach. On the contrary, the high concerns related to the use of ENMs, associated to the potential risks for environment and human health, require an even more careful attention in the development of new nanotechnologies since the very beginning of their design [1].

In this chapter, we want to tell the story of a recently developed nanocellulose-based material, a nanosponge that has been successfully applied for fresh and marine water remediation, and whose eco-design has a lot of to do with the key concepts of circular economy and sustainability herein introduced.

First described in 2015 in their primitive design [2], these cellulose nanosponges (CNS) have been further developed within the project NANOBOND (Nanomaterials for Remediation of Environmental Matrices associated with Dewatering, Nanomateriali per la Bonificaassociata a Dewatering di matriciambientali), following a safe-by-design strategy in order to guarantee an environmentally sustainable solution for water treatment.

The polysaccharide building block chosen for their construction represents a first important indication in this direction [3, 4].

9.1.1 Nanocellulose-based Materials

Cellulose is the most abundant biopolymer on Earth. It is formed by about 10 000–15 000 β-D-glucopyranose monomers linked together by β-1,4-glycosidic bonds. It is a renewable, biodegradable, and nontoxic raw starting material that can be extracted and recovered from a wide range of biomass sources, including agricultural, wood, and paper-making waste.

The cleaving of the hierarchical structure of cellulose fibers by mechanical, enzymatic, or chemical approaches allows producing nano-sized cellulose (nanocellulose (NC)), characterized by at least one dimension under 100 nm. NC can be obtained in the form of either cellulose nanocrystals (CNC) or cellulose nanofibers (CNF), depending on the production process and the biopolymer source [5–7]. (Figure 9.1)

The unique characteristics of NC include thermal resistance, tensile and mechanical strength, and a high and highly reactive surface area, which makes it suitable for potential applications in different production and industrial fields [7]. These properties also well combine with a negligible cytological and inhalation toxicity [8].

Sustainable Nanotechnology: Strategies, Products, and Applications, First Edition.
Edited by Yashwant V. Pathak, Govindan Parayil, and Jayvadan K. Patel.

Figure 9.1 Nanocellulose from agricultural, wood, and paper making waste sources.

Moreover, in spite the issue of being nano-sized building blocks, both CNC and CNF have resulted to be eco-safe, showing low toxicity toward developing zebrafishes, chosen as model organisms [9]. These findings enforce the use of NC especially for water and wastewater treatment processes [10–14], in line with a choice of sustainability in the design of new nanostructured materials.

Finally, yet importantly, the researched nano-dimension of the fibers opens the way for the valorization of that cellulose waste which does not find further market as raw material in products such as paper or textiles, due to the excessive deterioration of the fibers. For this reason, waste paper has become our preferred biomass source [15] for developing CNS within the NanoBonD project.

If for a long period CNF production has found a low commercial interest, due to the energy intensive processes required [16], in the last decade, it has attracted a renewed interest thanks to the introduction of chemical and enzymatic pretreatments, which have significantly reduced the energy demand. Nowadays, different companies worldwide produce CNF in a scale ranging from 100 to 2000 kg per day (http://www.tappinano.org, 2019).

Among the new chemical approaches proposed for CNF production, the TEMPO (2,2,6,6-tetramethylpiperidine 1-oxyl free radical)-mediated NaClO/NaBr oxidation of cellulose fibers is attracting more and more attention [17, 18]. It consists into the selective conversion of C6 alcoholic groups of the glucopyranose units to the corresponding carboxylic acids. Moving at basic pH, it is possible to induce the deprotonation of the carboxylic moieties, favoring the nano-defibrillation of TEMPO-oxidized cellulose nanofibers (TOCNF) via sonication or refining/homogenization, by taking advantage of the electrostatic repulsion among the negatively charged cellulose chains.

It has been demonstrated by means of life cycle assessment how this synthetic route would have the lowest environmental impact in NC production if compared with other processes, especially in terms of energy and global warming potential [19]. Moreover, it can be efficiently scaled up with a low energy consumption [20].

By considering the mechanical and strength properties of TOCNF, in the last years, they were suggested as ideal building blocks for the design of cellular nanostructured materials. In most cases, this result is achieved by undergoing TOCNF hydrogels to simple freeze-drying treatments [21]. Following this approach, we proposed TOCNF aerogels as ideal templates for further ceramic and organic coatings [22, 23]. More recently, we also envisioned the possibility of taking advantage of the introduction of carboxylic moieties on TOCNF backbone to promote the formation of nanostructured materials by the addition of a proper cross-linker, possibly avoiding the use of additional chemicals in order to make the overall process as sustainable as possible. This approach has opened the way to the development of a new class of CNS to be used as adsorbent units for water decontamination from organic and inorganic contaminants.

9.2 Synthesis of CNS: Original Formulations

9.2.1 The First Approach

The first synthetic approach for CNS production involved the use of two fundamental building blocks: TOCNF and a branched polyamine. The cellulose nanofibers were obtained by the oxidation of cotton cellulose through the previously described TEMPO/NaClO/KBr system [22]. This simple reaction, carried out in water at room temperature, allowed to obtain cellulose nanofibers with a partial conversion of the hydroxyl groups on the C6 carbon position into the corresponding carboxyl groups, as previously described. The average degree of oxidation used for the synthesis of the nanosponges was about 1.1 mmol g^{-1}, evaluated through a conductimetric titration [24].

The polyamine used as second building block was the branched polyethyleneimine 25 kDa (bPEI). This polymer, completely characterized by ^{13}C NMR studies in D_2O [2, 25], has primary (7.43 mmol g^{-1}), secondary (9.42 mmol g^{-1}), and tertiary (6.36 mmol g^{-1}) amines. The synthesis of the bPEI-TOCNF co-polymeric material is schematically shown in Figure 9.2. A variable amount of bPEI was added to an aqueous suspension of TOCNF, subsequently increasing the volume of water until reaching a TOCNF concentration of about 3.5% w/v (concentration that allows having a handy suspension and guaranteeing a mechanical resistance to the final material). The mixture was then placed in molds and freeze-dried obtaining a solid that was then thermally treated in oven, by carrying out a thermal ramp from 60 to 102 °C. The final CNS were then sequentially washed with HCl, NH_4OH, and acetone, and finally dried in air for 48 hours and in oven at 70 °C for 2 hours.

We have demonstrated how the freeze-drying step is crucial to promote micro-pores formation with a two-dimensional sheet-like morphology, due to the ice crystals acting as templates (Figure 9.3) (see next paragraph for details) [21]. Instead, the final physical and chemical resistance of CNS should be ascribed to the subsequent thermal treatment, which favors the formation of amidic bonds between carboxylic moieties of TOCNF and primary amino groups of bPEI, as evidenced by NMR analysis.

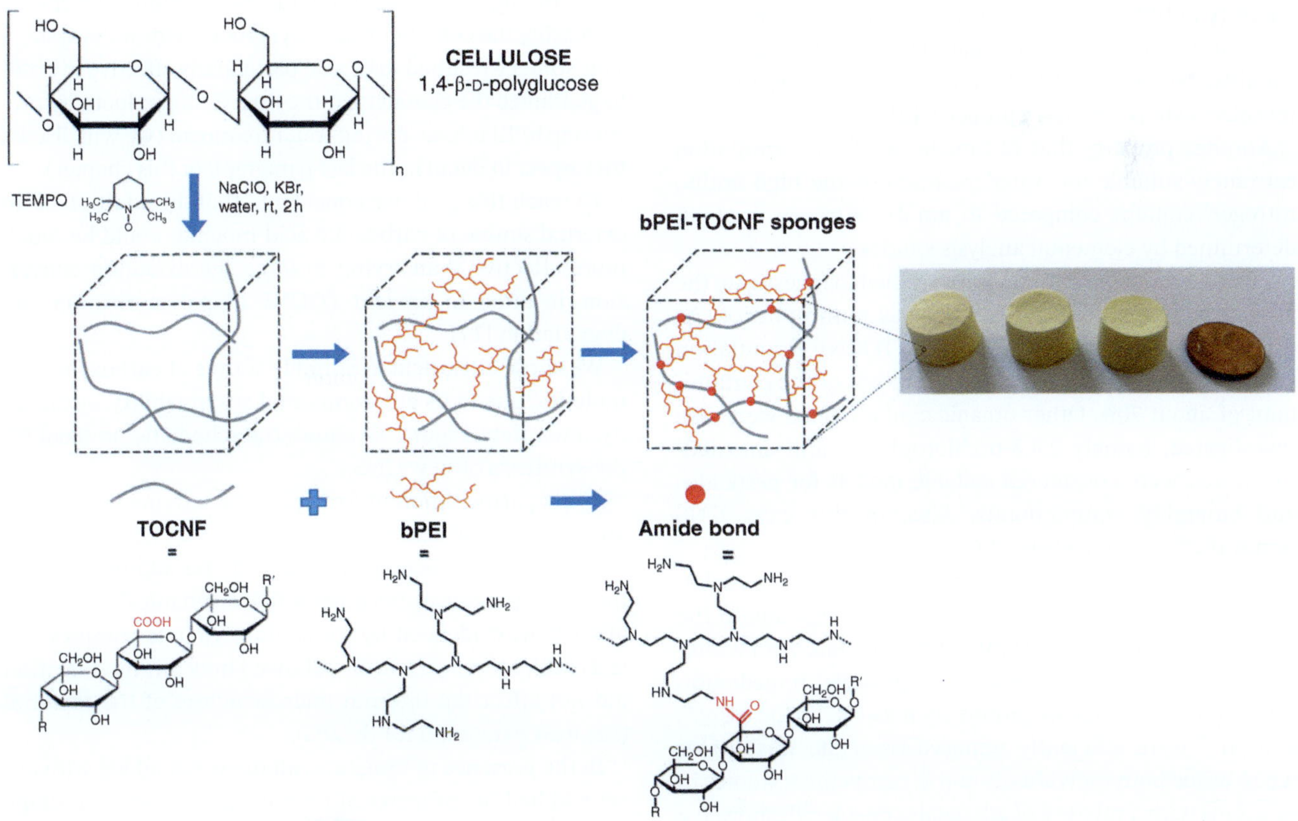

Figure 9.2 Synthesis of the CNS, first synthetic approach. The steps for the preparation of the material are the addition of a bPEI solution to TOCNF hydrogel, freeze-drying at −80 °C, 24 hours and finally heating up to 102 °C for 10 hours.

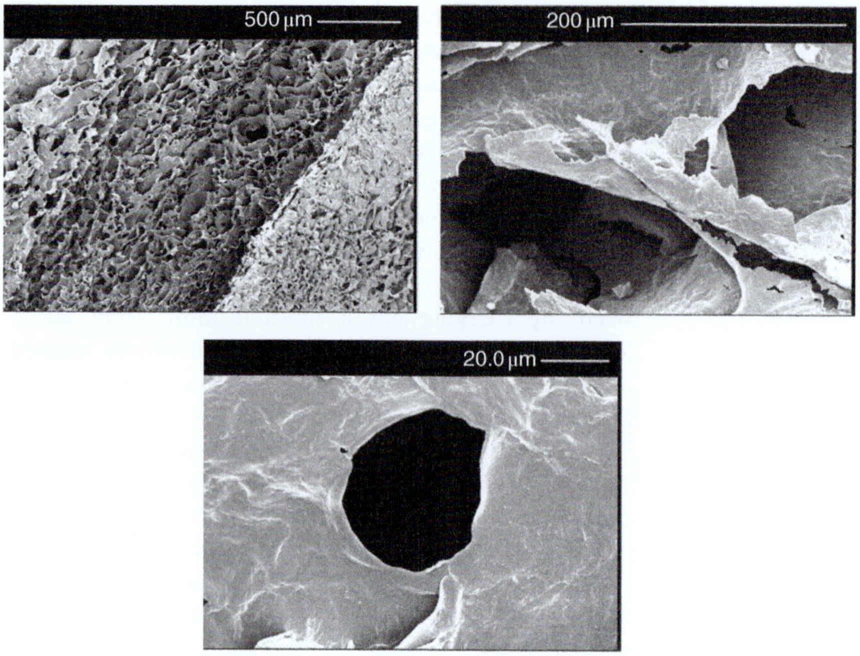

Figure 9.3 SEM images of bPEI-TOCNF (2 : 1) nanosponges at different magnifications.

Three different bPEI/TOCNF ratios were tested (0.2 : 1, 0.5 : 1, and 2 : 1), showing a correlation between the amount of bPEI and the final morphological/mechanical properties of CNS. In fact, the progressive increase in the quantity of bPEI has led to a constant improvement in the structural strength of the material, guaranteeing good stability. Moreover, CNS with a bPEI/TOCNF ratio 2 : 1 revealed a shape-recovery property [2].

Another property that has made the 2 : 1 formulation extremely suitable for water treatment is the high amino nitrogen content compared to amidic nitrogen content, determined by elemental analysis studies.

The first model of contaminant studied to investigate the absorbent properties of this CNS was *p*-nitrophenol. As expected, the CNS with a 2 : 1 bPEI/TOCNF formulation gave the best results, promoting a removal of the contaminant of about 70%. Other organic contaminants were also investigated, namely 2,4,5-trichlorophenol and amoxicillin, which were considered suitable models for pesticides and emerging contaminants. Also, in this case, CNS removal efficiency was about 65–75%.

Finally, by considering the high concentration of chelating amino groups in the cellulose nanosponge due to the presence of bPEI, the removal of heavy metals from wastewater was also considered as a possible remediation action [26, 27]. Bivalent cations such as Cd^{2+}, Cu^{2+}, Co^{2+}, and Ni^{2+} were efficiently removed from the wastewater when tested both individually and in competitive studies in solutions with a mixture of all metals considered, showed a higher efficiency of the sponge toward copper, which was absorbed preferentially than cadmium, cobalt, and nickel.

9.2.2 An Evolution in the Formulation

In order to obtain CNS with a higher chemical and mechanical stability, we considered the possibility to increasing the cross-linking degree between the two building blocks by increasing the content of carboxyl groups within the system. This strategy resulted to be also particularly effective in order to guarantee the ecosafety of the final CNS products, by preventing bPEI release during water treatment (we will discuss this aspect in detail in the last paragraph of this chapter).

To reach this goal, we considered that the addition of an external source of carboxylic acid moieties could be much more effective than trying to force the oxidation conversion, in order to prevent TOCNF de-polymerization and degradation [17].

We chose citric acid as suitable source of carboxylic acid moieties, due to its environmental sustainability, availability, and safety. Figure 9.4 shows the schematic protocol for the synthesis of new CNS.

By preparing different formulations, varying the relative amount of citric acid respect to TOCNF/bPEI content, we could verify the beneficial effect of the addition of this third component in improving the mechanical properties of CNS, as evidenced by static and dynamic compression tests [28]. Moreover, we also showed how this modification did not affect the internal microstructure of the material (see next paragraph for details).

In the presence of moisture, all materials added with citric acid had the property of recovering their original shape without excessive loss of mechanical strength. However, the best formulation resulted to be the one with 18% equivalents of citric acid respect to the NH_2 groups of bPEI in the final aqueous suspension, with a 2 : 1 bPEI/TOCNF ratio. In fact, only this formulation in the absence of moisture showed no plastic deformation, maintaining high mechanical strength for ten test cycles. On the contrary, all other formulations showed, under the same conditions, a decrease in mechanical resistance as the number of test cycles increased [28].

9.3 Structure of CNS: Nano-dimensioned Fibers to Build a Macroscopic Nano-structured Material

In this paragraph, we will describe the structural behavior of CNS. We will start looking the sponges' morphology by naked eye, and gradually zooming by means of different analytical approaches up to reach the nanoscale dimension.

9.3.1 Macrostructure of the Sponges

Thanks to their adaptable and scalable synthetic protocol, it is possible to easily control the final shape of CNS. Indeed, since the mixture of sponges components (TOCNFs, bPEI, and citric acid) possess the behavior of viscous liquid and this is placed in a mold before the freeze-drying process, by carefully choosing the shape of the container it is possible to confer to the material any desired shape. As an example, the sponges reported in Figures 9.2 and 9.4 look like cylinders of about 1.3 cm of diameter because in this case the mixture was frozen in cylindrical molds. By the way, it is possible to obtain the same materials as plates, cubes, or parallelepipeds. Although the macroscopic dimension (in the range of centimeters) and the shape of the sponges do not affect their chemical behavior, the morphology of the final material strongly influences some physical properties, in particular, the mechanical ones and the diffusion of solutions and solutes through the sponges [29]. The issue associated to the mass transport phenomena through the material can be easily overcome by exploiting the micro- and nano-structure of the sponges. Indeed, by grinding them to a powder, which still possesses micro and nano porosity, it is possible to obtain a material that maintains and even increases the adsorption properties.

9.3.2 Microstructure of the Sponges

The microstructure (on micrometer scales) of CNS in function of the amount of citric acid as co-cross-linker has been recently investigated [28]. In this work, a scanning electron microscopy (SEM) micrographs inspection on sections of different sponges qualitatively demonstrated that the addition of different amounts of citric acid in the formulation does not lead to significant differences in terms of micro-porosity.

The same result was confirmed by data obtained from micro-computed tomography (μ-CT). Four groups of sponges containing variable amounts of citric acid were

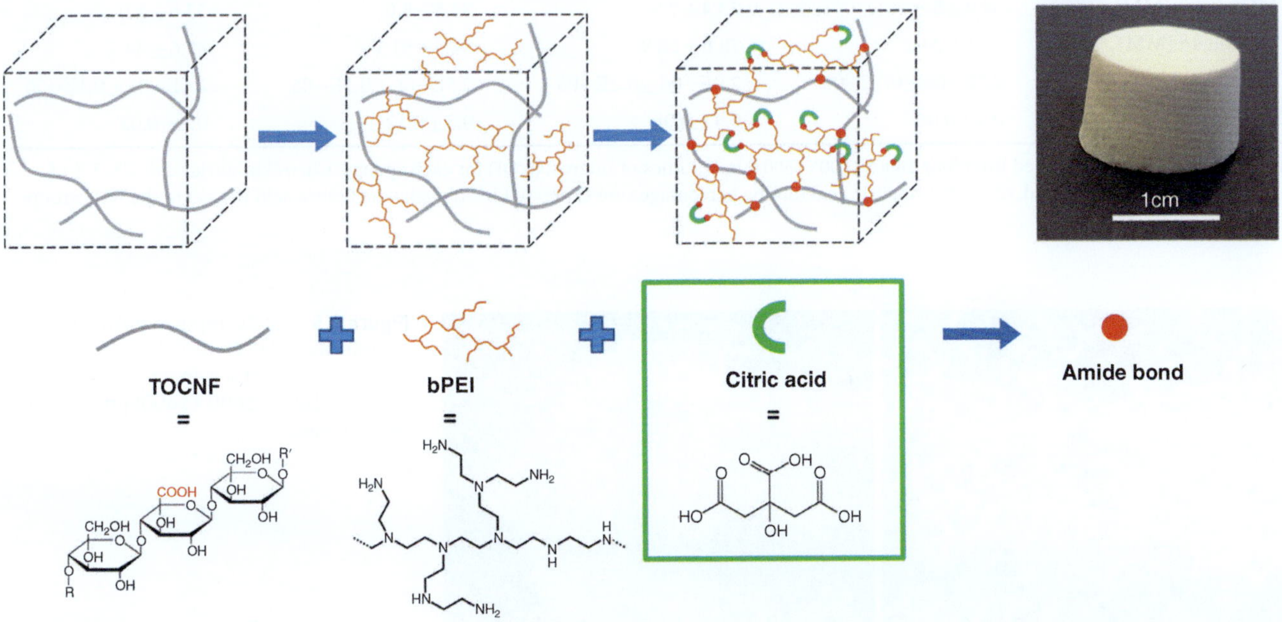

Figure 9.4 Synthesis of the CNS with the addition of citric acid. For the synthesis of the material different steps are required. Firstly, the addition of a bPEI solution and a citric acid solution to TOCNF hydrogel, then freeze-drying at −80 °C for 24 hours and finally heating up to 102 °C for 10 hours.

investigated, measuring different parameters like porosity, trabecular thickness (Tb.Th), mean trabecular separation (Tb.Sp), connectivity density (Conn.D), and degree of anisotropy (DA). None of these factors show statistically significant differences in function of the amount of citric acid added (Table 9.1), confirming that the microstructure of the CNS is only slightly affected by the material formulation.

In Figure 9.5a an internal slice of a typical CNS sponge is reported and Figure 9.5b shows a 3D reconstruction of the inner part of a CNS volume.

9.3.3 Nanostructure of the Sponges

The nanostructure of the sponges was systematically assessed by small-angle neutron scattering (SANS) as a function of both the amount of cross-linker and the hydration degree [30]. Experimental data were analyzed applying a correlation length model (CLM) [31] and the role of the cross-linker (bPEI) and the hydration degree in determining the density of CNFs at high (larger than hundreds of Å) and low (in the range 10–100 Å) spatial scale was determined (Figure 9.6) [32, 33]. The computed Lorentzian (m) exponent and the correlation length (ξ) were used to figure out the mesh size and the dimensions of the pores of the sponge-like material (Figure 9.6) in the range of 10–100 Å. The power law exponent (n) was used for describing inhomogeneities at scales of hundreds of Å. The inhomogeneity in CNFs density was associated to the presence of nano dimensioned pores. Another interesting information retrieved from SANS analysis is related to the random distribution of TOCNFs inside the sponges, as expected according to the synthetic procedure.

In particular, the dimension of these pores' results is affected by the hydration level and by the cross-linking degree. Generally, the mesh size (ξ) is in the range of 2–3 nm. Interestingly, when the cross-link is low (bPEI-TOCNF sponges without citric acid, circle and triangle symbols in Figure 9.7) the increase of hydration level (h) led to a decrease of the mesh size (ξ) and to an increase of the dimensions of inhomogeneities at low spatial scales (m). On the contrary, when the cross-linking is enhanced by the adding citric acid, the dimensions of these inhomogeneities result constant at all hydration levels (rhombus

Table 9.1 Quantitative 3D characterization of the samples' microstructure.

	C0	C3	C9	C18
Porosity	70% ± 0%	71% ± 0%	74% ± 0%	73% ± 0%
Tb.Th mean (MM)	34.0 ± 8.8	33.4 ± 7.9	33.5 ± 8.0	33.6 ± 8.0
Tb.Sp mean (MM)	72.1 ± 24.1	78.6 ± 28.9	84.5 ± 51.1	85.6 ± 44.2
Conn.D (MM^{-3})	2.7E−06 ± 0.1E−06	2.9E−06 ± 0.2E−06	2.5E−06 ± 0.1E−06	2.6E−06 ± 0.1E−06
DA	0.4 ± 0.06	0.4 ± 0.08	0.3 ± 0.12	0.3 ± 0.05

Each value is the mean obtained from four 1600V1600V1600 mm volumes of interest (VOI) for each sponge. C0, 0 % of citric acid; C3, 3 % of citric acid; C9, 9 % of citric acid; C18, 18 % of citric acid (all the percentages are expressed in equivalents of citric acid respect to the NH_2 groups of bPEI.

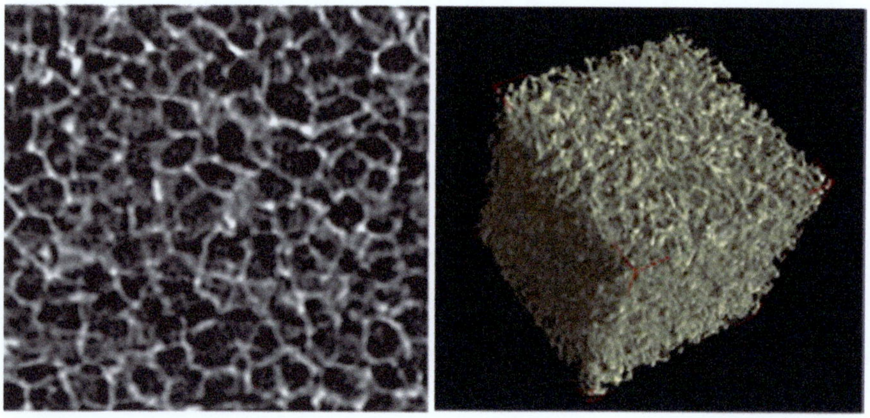

Figure 9.5 (a) 2D representation of sample slice (1600 × 1600) μm^2, (b) 3D representation of a sample volume of interest (1600 × 1600 × 1600) μm^3, and (c) SEM images.

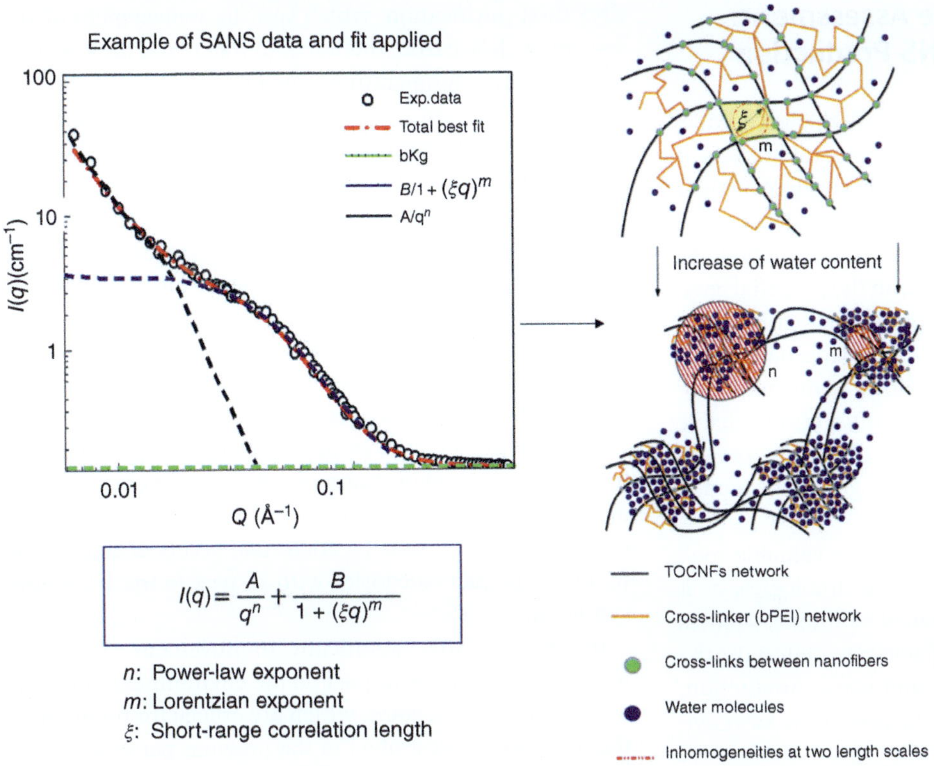

Figure 9.6 Example of SANS data fitting with correlation length model (CLM), and schematic and not in scale representation of the nanostructures of the CNS network.

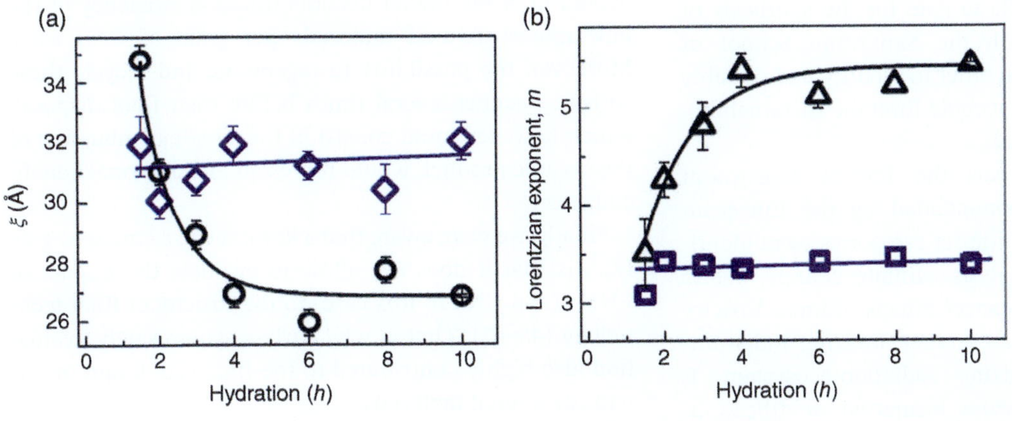

Figure 9.7 (a) Hydration dependence of a short-range correlation length ξ and (b) Lorentzian exponent m for CNS (2 : 1) (circle and triangle symbols) and CNS (2 : 1)–CA (rhombus and square symbols) [32].

and square symbols in Figure 9.7). This behavior at nanoscale reflects the increase of stiffness observable at macroscopic scale.

In conclusion, CNS sponges can be considered as nanostructured macro-materials. Indeed, the cellulose nanofibers are used as "brick" linked together with bPEI and, eventually, citric acid forming a macroscopic (centimeter scale) assembly which possesses a porosity on different scales ranging from nano (2–3 nm) to micro dimensions (30–100 μm). This means that the sponges possess features similar to nanomaterials, but they can be easily handled, possibly overcoming the main concerns related to the impact of ENMs on environment.

9.4 Early-stage Life Cycle Assessment: A Tool for a Sustainable CNS Production

Life cycle assessment (LCA) has become, in the last decade, a key tool in the evaluation of the environmental impacts of innovative technologies since the very beginning of their design. The correct use of this analysis should favor the choice and development of a sustainable production, by helping to identify, already at lab scale, all the potential environmental effects of the new process and their significance.

From one side, the early stage LCA approach (also defined as "prospective LCA" [34]) shows many methodological issues, mainly due to the difficulty to find reliable and quality data, to envision the impact of scaling-up, and to compare results with existing technologies developed at industrial scale [35, 36].

Nevertheless, prospective LCA remains a valuable tool, especially for the development of nanotechnologies, as it can drive the adoption of eco-designed solutions, such as the correct choice of safe and available raw materials, the limitation of energy consumption and waste production, and so on [37–39], leading to the reduction of the key environmental impacts.

For example, it has been reported how usually nanotechnologies have a higher impact in terms of energy consumption than their conventional technological alternatives [40, 41].

For these reasons, we considered the opportunity of an early-stage LCA from cradle to gate for the synthesis of CNS, in collaboration with the Sant'Anna School of Advanced Studies of Pisa, in order to identify and possibly improve all those aspects that could limit the sustainability of the synthetic protocol [42].

ILCD 2011 Midpoint+was the impact assessment method of choice, as recommended by the European Commission [43], while 16 impact categories were identified as crucial for our analysis: climate change, ozone depletion, human toxicity-cancer effects, human toxicity noncancer effects, particulate matter, ionizing radiation-human health HH, ionizing radiation-ecosystem E (interim), photochemical ozone formation, acidification, terrestrial eutrophication, freshwater eutrophication, marine eutrophication, freshwater ecotoxicity, land use, water resource depletion, and mineral, fossil, and renewable resource depletion.

Since the very first lab-scale analysis conducted in Pisa, we could verify how the energy-consuming steps of the synthetic protocol, namely TOCNF sonication, CNS freeze-drying and heating in oven, and purification protocol at high temperatures, were those which mainly influenced the overall environmental impact. We could immediately overcome the latter limit at the lab-scale by modifying the CNS final purification, which saw the replacement of the solvent of choice from methanol to water, and the lowering of the washing temperature from 60 to 30 °C. According to a second optimized LCA, the first intervention reduced the impact of the materials on climate change of 88%, while the second one reduced the effect of energy-consuming processes on the same impact category of 11%. However, the overall environmental impact remained within the same order of magnitude of that determined at the first stage.

In order to further optimize the energy-consuming processes, we developed an additional scale-up model of the CNS production, by adapting the equations reported by Piccinno et al. [44], and assuming a recycling of water used for purification, while energy recovery was not considered at this stage. This third analysis provided a reduction of the environmental impacts by about two orders of magnitude in all the impact categories with respect to the laboratory synthesis.

In spite of this significant improvement, LCA still revealed the negative impact on the environment of freeze-drying and heating steps, which are crucial for modulating the morphology described in the previous paragraph. This result could push to consider the CNS production less competitive if compared with other adsorbent materials used for water remediation.

However, it is worth to outline how this study should be now extended from cradle to gate, by considering in the overall LCA the higher decontamination efficiency of the new nanostructured materials per gram of CNS used. Moreover, the possibility to regenerate and recycle these sorbent systems several times before their final disposal, which however could consist in the simple combustion of the organic product, would represent an additional beneficial effect.

Finally, we were aware that a key issue in a cradle-to-gate LCA is that it does not allow to evaluate the impact of ENMs (and CNS in this case) to the toxicity of final technology [45–47]. That is why early-stage ecosafety evaluation also highly contributed to the final eco-design of our nanostructured materials.

9.5 Environmental Safety of Nanoscale Materials

In the last 20 years, great attention has been devoted to nanotechnologies for soil and groundwater/wastewater remediation by the use of engineered nanoscale materials (ENMs) and engineered nanoscale particles (ENPs) from which the term *nanoremediation* [48, 49]. Based on the EU consensus agreement, a nanomaterial can be defined as "a material with one or more dimensions in the size range 1–100 nm" [43] or "any organic, inorganic, or organometallic materials presenting specific chemical, physical, and/or electrical properties that vary as a function of the size and shape of the material" [50]. By virtue of such small dimension, nanoscale materials or particles possess unique physicochemical properties compared to their larger-scale (bulk) counterparts such as, for instance, high surface-to-volume ratio, greater strength, lighter weight, increased control within the visible spectrum, and greater chemical reactivity [51]. Reduction, oxidation, sorption, or a combination of these processes associated to nanoscale dimension have been used to develop ENM/Ps with tunable properties for in situ treatment of contaminated groundwater and soil [52]. While nanoremediation emerged as a promising technology able to successfully overcome the limits of current existing processes [53, 54], uncertainties due to the lack of knowledge on potential environmental implications as, for instance, toxicological impact on natural ecosystems have recently increased [1, 55–58]. Those unique features, which make ENM/Ps suitable for pollutant removal from soil and water matrices, can in fact rebound on the safety of their in situ applications in natural environmental compartments [56, 59, 60]. High mobility and overall reactivity, both enhanced by small dimensions, will favor their exchanges within water and soil compartments hence becoming potentially harmful toward unexpected biological targets [61]. In the past twenty years, the impact of ENM/Ps on terrestrial and aquatic species has been largely documented [62] raising regulatory concerns and public awareness in terms of risk-benefit of nanotechnologies including *nanoremediation* [55, 56, 63, 64]. Consensus agreement has been more recently reached on the need to adopt environmentally realistic exposure scenarios for ENM/Ps risk assessment in order to support policy-makers, legislators, and stakeholders in limiting any risk associated with their use and application including in situ remediation [65–68].

Synergistic as well as antagonistic effects on aquatic biota have been reported on selected ENMs for environmental application as, for instance, titanium dioxide nanoparticles (n-TiO_2) and contaminants to be remediated as cadmium and dioxin (2,3,7,8-TCDD) [69–74]. Moreover, transformation occurring due to interaction with abiotic and biotic factors (ionic strength, NOM, biota) of the receiving environmental compartments have been identified as major drivers of the observed toxicity and able to affect their ability to interact with contaminants to be remediated [75–78]. ENM/Ps agglomeration and aggregation by Brownian motion and/or interaction with other organic and inorganic particles in the absence of repulsive energy barriers can reduce their reactivity but also mobility in soil and water media and sequestration in selected environmental compartments [79]. However, in water bodies secondary processes such as resuspension and disaggregation cannot be excluded and make a consistent fraction of ENM/Ps still bioavailable to aquatic organisms both pelagic (surface) and benthonic [80].

Therefore, the pursuit of more ecologically safe and sustainable nanotechnologies for water and soil remediation represents a major breakthrough [81]. Current scientific gaps call for a thorough evaluation of ENM/Ps potential "side effects" for in situ nanoremediation and a case-by-case analysis must be undertaken to prove their environmental safety (ecosafety) [1].

Besides that, together with predictive fate and transport models able to quantify ENM/Ps transformations in realistic exposure scenarios [82, 83], the introduction of a more ecologically safe predictive assessment approach (ecosafety) will prevent against potential hazards natural communities and ecosystems services [1, 80]. It is hence wise to foresee possible scenarios of ENMs interactions with natural compartments (soil and water) and to screen for their potential ecotoxicity toward different levels of biological organization [84, 85].

9.5.1 A New Eco-design Paradigm for CNS Production

The "safety by design" paradigm and the more recently adopted definition of *eco-design* are equally recognized as a successful strategy to overcome undesirable properties of materials, including those at nanoscale, which are hazardous for humans and the environment [65, 86–88]. Only those properties which will guarantee environmental safety while maintaining ENM/Ps efficacy toward decontamination processes should be incorporated in the material design process (re-design) during product development. The final aim will be to not only to *re-design* already existing ENM/Ps which have been shown clear toxicity toward soil and aquatic organisms but redefine the way they should be manipulated from the beginning up to the end of their life cycle [42]. Therefore, both old and new ENM/Ps could be tested and redesign in an eco-safer way finally solving those uncertainties regarding human and environmental risks of in situ application.

In line with these considerations, we have recently outlined the fruitful contribution of an ecotoxicological approach to aid the synthesis of *ecosafe* CNS [29]. In particular, a trial-and-error like process in which ecotoxicological assessment aids the formulation of materials in a stepwise fashion (i.e. by testing single components and synthesized materials at the final stage) along the synthetic procedure has been proven to be useful in the production of an intrinsically *ecosafe* material [29]. By mimic environmental realistic aquatic scenarios in which both planktonic and filter-feeders benthic species could potentially become targets of nanoscale/nanostructured ENM/Ps used for pollutants removal, an *eco-design* strategy has been developed with the aim to *re-design* the material based on those *ecologically safer* properties identified by assessing possible negative outcomes of their application.

The case study investigates different batches of CNS, all of them able to remove heavy metals from seawater, by using ecotoxicity bioassays with marine microalgae and bivalve mollusks as representative of planktonic and benthic marine species which could be encountered by CNS once released in seawater and/or during the remediation process [29].

An initial batch of bulk, nonsoluble solid CNS (10×3 mm) was added to algal medium and tested with a standardized ecotoxicity bioassay using marine microalgae *Dunaliella tertiolecta* (OECD 202). At the end of the test, microalgae resulted retained by the external surface of the bPEI-TOCNF CNS regardless of their different formulations and microalgae cells count showed high inhibition of their growth which is considered a clear sign of toxicity (Figure 9.8). In order to overcome such limitation, we went back to the lab and grind the bPEI-TOCNF CNS (*1°re-design*) to a powder and further tested using the microalgae growth inhibition assay. In order to envisage a possible application scenario in which bPEI-TOCNF CNS itself is not released directly in situ, we decided to incubate the CNS powder in artificial seawater (ASW) for 24 hours and then upon its removal the resulting seawater used for further run the microalgae growth inhibition test. Despite being in powder, all bPEI-TOCNF CNS batches still resulted toxic for microalgae with an average of growth inhibition higher than 70% (Figure 9.8). Such findings suggest to us that again the CNS powder at this stage was unsuitable for an environmental application and we decided to target their synthetic process in order to overcome the observed microalgal toxicity. Among chemicals used for their synthesis, bPEI was identified as a constitutive ingredient of CNS formulation and despite its wide usage for the reticulation of cellulose nanofibers and the formation of the final nanostructured material, its cell toxicity is highly documented (DNA damages, apoptosis, and necrosis) with an extent that is directly proportional to the molecular branching and charge density [89, 90]. Upon testing the inherent toxicity of bPEI toward the microalgae *D. tertiolecta*, a concentration-dependent inhibition of cell growth was observed with an EC_{50} of 2.75 $\mu g\,ml^{-1}$ (Figure 9.8).

Such findings dictated our *2° re-design* step which was to progressively substitute bPEI with citric acid which not only reduced the release of bPEI in water but also maintained the macroscopic assembly and a micro-nanoporosity of the new CNS formulations (Figure 9.8) (see Section 9.3.3 for further details). In order to verify the efficacy of our new CA CNS batches, we carried out a new algal (*D. tertiolecta*) growth inhibition assays. Overall, we observed very mild effects on *D. tertiolecta* growth with a maximum growth inhibition of an average of 10% which is considered nontoxic for microalgae. These encouraging results highlighted how the new CA CNS formulation was nontoxic yet showing outstanding heavy metal removal efficiencies (refer to [29]). To further support the results of the microalgal tests, we performed an in vivo exposure tests with marine mussels *Mytilus galloprovincialis* by using the same testing conditions by incubating CNS with ASW and upon CNS removal the obtained ASW was tested on mussels. Lysosomal membrane stability (LMS) and DNA damage (micronucleus test) of mussel's circulating immune cells were measured as potential targets of CNS toxicity. Both mussel's immune cells' responses are known to be sensitive to exposure to different classes of pollutants including heavy metals and hydrocarbons [91–94]. Upon exposure to ASW incubated with CNS, both stability of lysosomal membranes and micronucleus frequencies were similar to those recorded in controls suggesting the complete absence of any cellular damage upon exposure to CNS. Overall findings obtained from microalgal tests and biological responses in mussels confirmed our new CACNS batch as *ecologically safer* compared to the initial batches and the best compromise between ecosafety and decontamination efficiency of CNS in seawater.

In our study, we demonstrated that a combination of cellular responses in both microalgae and filter-feeders in standardized test conditions (in vitro and in vivo) was successful in recognizing those CNS batches which were lacking the ecosafety properties. We could identify bPEI as the main toxic agent in their formulation and to target it and redirect the synthesis efforts to significantly reduce it. Indeed, by the application of an ecologically based approach in the *eco-design* process, the resulting final formulations of CNS were proven to be *ecosafe* for two different marine trophic levels, photoautotrophs phytoplankton and filter-feeders benthic species [29]. Marine phytoplanktonic organisms play a key ecological role in coastal and oceanic ecosystems, being primary producers at the basis of the food web able to transfer energy at higher trophic levels [95, 96]. Therefore, toxic effect sustained by these organisms, either of anthropogenic or natural origin, can rebound upon higher levels of biological organization deeply affecting overall ecosystem health. Moreover, owing to their unicellular nature and ubiquity in natural waters, photoautotrophic phytoplankton are likely to be important targets of any kind of substance and/or material which would come in contact with including engineered or

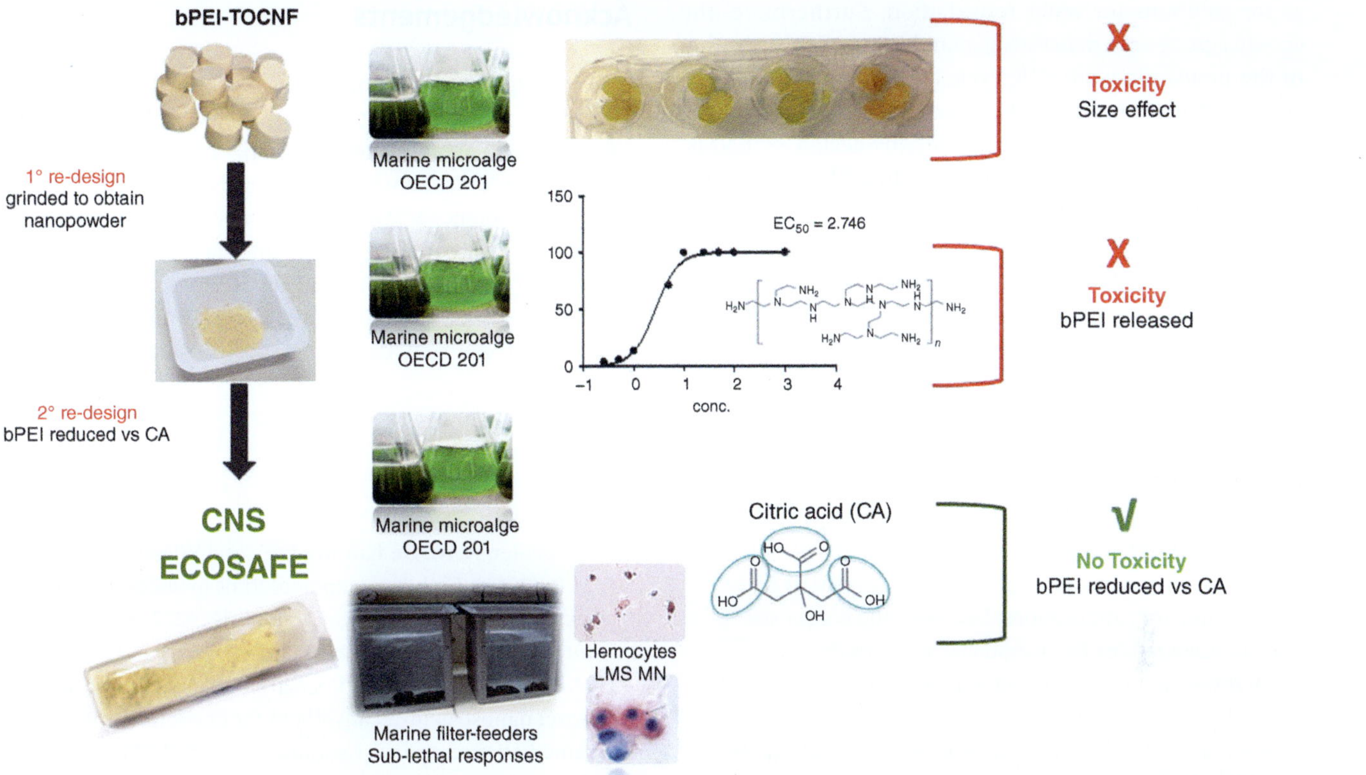

Figure 9.8 Schematic representation of the eco-design concept study developed within Fiorati et al. [29] in which ecotoxicological bioassays with marine microalgae and bivalve filter-feeders have been used to test environmental safety of CNS for heavy metal removals from seawater. For further details on the experimental plan refers to Fiorati et al. [29]. (a) *Source:* Fiorati, A. et.al, (2020), Fig 01, Pg 15 / with permission of Elsevier.

natural nanoscale particles and colloids. The use of phytoplanktonic organisms in standardized toxicity tests is a staple requirement to elucidate the environmental hazard caused by the exposure to a certain compound, also providing scientific evidence for the classification and eventual regulation of potentially toxic substances. Marine and freshwater microalgae, such as the diatom *Phaeodactylum tricornutum* or the green unicellular alga *D. tertiolecta*, are included in OECD guideline for the testing of chemicals [97] and are nowadays a common and useful tool employed not only for routine testing and regulatory purposes but also in ecotoxicology and nanoecotoxicology studies concerning compounds of emerging environmental concern as ENM/Ps [98]. To date, the harmful effects of ENM/Ps applied for environmental remediation had been largely documented on this marine phytoplanktonic microalgae as well as ecotoxicity mechanism elucidated [99–101]. As benthonicfilter-feeders widely distributed on the seafloor or on rocky shores along the coasts, marine bivalve mollusks possess great ability to respond to a large variety of environmental stressors including chemical pollutants and to bioconcentrate them in soft tissues over time, all peculiar abilities which have justified their widespread use in marine pollution monitoring as sentinel species since the 1980s [102]. For the same reasons, they are widely recognized as a suitable model in ecotoxicology and more recently in nanoecotoxicology by showing ENM/Ps cell-tissue specific effects as well as bioaccumulation and translocation within the body [75, 103–105].

Thanks to our proof of concept *eco-design* study, we underscored the importance of including an ecologically based testing strategy in material design and support ecotoxicology as a promising tool for environmental risk assessment of nanoscale/nanostructured materials, whose intended use will be in natural compartments. The proposed *eco-design* will ameliorate material properties with the aim to better fit within environmental protection and potentially act as a further driver for market development of the nanoremediation sector [106].

9.6 Final Remarks

We recommend that the concept of sustainability be introduced since the early stage of materials design, especially when their final use is in direct contact with environment, as for solutions for water remediation. Furthermore, the eco-design of new materials is mandatory when operating in the nanoscale word. Life cycle assessment and ecotoxicology evaluation are the two needful tools to reach the goal of producing "sustainable-by-design" ENMs. Moreover, the choice of starting building blocks, possibly deriving from renewed sources, allows combining the concepts of sustainability and circular economy in an ecologically oriented approach. The case study herein described on CNS enforces our persuasion that nanocellulose is in the running to become an important resource for the design of ENMs for the present and the future.

Acknowledgements

The corresponding authors want to acknowledge Regione Toscana for funding the project NANOBOND (Nanomaterials for Remediation of Environmental Matrices associated with Dewatering, Nanomateriali per la Bonificaassociata a Dewatering di matriciambientali) within the European Regional Funding POR CReO FESR Toscana 2014-2020 (30/07/2014- LA 1.1.5 CUP 3389.30072014.067000007), in which they have been involved as PIs and developed the ecosafe strategy for CNS design.

References

1 Corsi, I., Winther-Nielsen, M., Sethi, R. et al. (2018a). Ecofriendly nanotechnologies and nanomaterials for environmental applications: key issue and consensus recommendations for sustainable and ecosafe nanoremediation. *Ecotoxicol. Environ. Saf.* 154: 237–244. https://doi.org/10.1016/j.ecoenv.2018.02.037.

2 Melone, L., Rossi, B., Pastori, N. et al. (2015). TEMPO-oxidized cellulose cross-linked with branched polyethyleneimine: nanostructured adsorbent sponges for water remediation. *Chempluschem* 80: 1408. https://doi.org/10.1002/cplu.201500145.

3 Corsi, I., Fiorati, A., Grassi, G. et al. (2018). Environmentally sustainable and ecosafe polysaccharide-based materials for water nano-treatment: an eco-design study. *Materials (Basel)* 11: 1228. https://doi.org/10.3390/ma11071228.

4 Crini, G. (2005). Recent developments in polysaccharide-based materials used as adsorbents in wastewater treatment. *Prog. Polym. Sci.* 30 (1): 38–70. https://doi.org/10.1016/j.progpolymsci.2004.11.002.

5 Kargarzadeh, H., Ahmad, I., Thomas, S., and Dufresne, A. (2017). *Handbook of Nanocellulose and Cellulose Nanocomposites*. Wiley.

6 Lee, K.-Y. (2018). *Nanocellulose and Sustainability: Production, Properties, Applications, and Case Studies*. CRC Press.

7 Thomas, B., Raj, M.C., Athira, K.B. et al. (2018). Nanocellulose, a versatile green platform: from biosources to materials and their applications. *Chem. Rev.* 118: 11575–11625. https://doi.org/10.1021/acs.chemrev.7b00627.

8 Roman, M. (2015). Toxicity of cellulose nanocrystals: a review. *Ind. Biotechnol.* 11: 25–33.

9 Harper, B.J., Clendaniel, A., Sinche, F. et al. (2016). Impacts of chemical modification on the toxicity of diverse nanocellulose materials to developing zebrafish. *Cellulose* 23: 1763–1775. https://doi.org/10.1007/s10570-016-0947-5.

10 Carpenter, A.W., De Lannoy, C.F., and Wiesner, M.R. (2015). Cellulose nanomaterials in water treatment technologies. *Environ. Sci. Technol.* 49: 5277–5287. https://doi.org/10.1021/es506351r.

11 Mahfoudhi, N. and Boufi, S. (2017). Nanocellulose as a novel nanostructured adsorbent for environmental remediation: a review. *Cellulose* 24: 1171–1197. https://doi.org/10.1007/s10570-017-1194-0.

12 Mohammed, N., Grishkewich, N., and Tam, K.C. (2018). Cellulose nanomaterials: promising sustainable nanomaterials for application in water/wastewater treatment processes. *Environ. Sci. Nano* 5: 623–658. https://doi.org/10.1039/c7en01029j.

13 Ngwabebhoh, F.A. and Yildiz, U. (2019). Nature-derived fibrous nanomaterial toward biomedicine and environmental remediation: today's state and future prospects. *J. Appl. Polym. Sci.* 136: 1–21. https://doi.org/10.1002/app.47878.

14 Wang, D. (2019). A critical review of cellulose-based nanomaterials for water purification in industrial processes. *Cellulose* 26: 687–701. https://doi.org/10.1007/s10570-018-2143-2.

15 Kumar, V., Pathak, P., and Bhardwaj, N.K. (2020). Waste paper: an underutilized but promising source for nanocellulose mining. *Waste Manag.* 102: 281–303. https://doi.org/10.1016/j.wasman.2019.10.041.

16 Spence, K.L., Venditti, R.A., Rojas, O.J. et al. (2011). A comparative study of energy consumption and physical properties of microfibrillated cellulose produced by different processing methods. *Cellulose* 18 (4): 1097–1111. https://doi.org/10.1007/s10570-01-9533-z.

17 Isogai, A., Saito, T., and Fukuzumi, H. (2011). TEMPO-oxidized cellulose nanofibers. *Nanoscale* 3: 71–85. https://doi.org/10.1039/c0nr00583e.

18 Pierre, G., Punta, C., Delattre, C. et al. (2017). TEMPO-mediated oxidation of polysaccharides: an ongoing story.

Carbohydr. Polym. 165: 71–85. https://doi.org/10.1016/j.carbpol.2017.02.028.

19 Li, Q., McGinnis, S., Sydnor, C. et al. (2013). Nanocellulose life cycle assessment. *ACS Sustain. Chem. Eng.* 1: 919–928. https://doi.org/10.1021/sc4000225.

20 Myja, D., Loranger, E., and Lanouette, R. (2019). 4-acetamido-TEMPO-mediated oxidation of wood chips and thermomechanical pulp in large scale. *Ind. Eng. Chem. Res.* 58: 12723. https://doi.org/10.1021/acs.iecr.9b02049.

21 Chen, W., Yu, H., Li, Q. et al. (2011). Ultralight and highly flexible aerogels with long cellulose I nanofibers. *Soft Matter* 7: 10360–10368. https://doi.org/10.1039/c1sm06179h.

22 Melone, L., Altomare, L., Alfieri, I. et al. (2013). Ceramic aerogels from TEMPO-oxidized cellulose nanofibre templates: synthesis, characterization, and photocatalytic properties. *J. Photochem. Photobiol. A Chem.* 261: 53–60. https://doi.org/10.1016/j.jphotochem.2013.04.004.

23 Panzella, L., Melone, L., Pezzella, A. et al. (2016). Surface-functionalization of nanostructured cellulose aerogels by solid state eumelanin coating. *Biomacromolecules* 17: 564–571. https://doi.org/10.1021/acs.biomac.5b01497.

24 da Silva Perez, D., Montanari, S., and Vignon, M.R. (2003). TEMPO-mediated oxidation of cellulose III. *Biomacromolecules* 4 (5): 1417–1425. https://doi.org/10.1021/bm034144s.

25 Von Harpe, A., Petersen, H., Li, Y., and Kissel, T. (2000). Characterization of commercially available and synthesized polyethylenimines for gene delivery. *J. Control. Release* 69 (2): 309–322. https://doi.org/10.1016/S0168-3659(00)00317-5.

26 Kislenko, V.N. and Oliynyk, L.P. (2002). Complex formation of polyethyleneimine with copper(II), nickel(II), and cobalt(II) ions. *J. Polym. Sci. Part A Polym. Chem.* 40: 914–922. https://doi.org/10.1002/pola.10157.

27 Kobayashi, S., Hiroishi, K., Tokunoh, M., and Saegusa, T. (1987). Chelating properties of linear and branched poly(ethylenimines). *Macromolecules* 20 (7): 1496–1500. https://doi.org/10.1021/ma00173a009.

28 Fiorati, A., Turco, G., Travan, A. et al. (2017). Mechanical and drug release properties of sponges from cross-linked cellulose nanofibers. *Chempluschem* 82: 848. https://doi.org/10.1002/cplu.201700185.

29 Fiorati, A., Grassi, G., Graziano, A. et al. (2020). Eco-design of nanostructured cellulose sponges for sea-water decontamination from heavy metal ions. *J. Clean. Prod.* 246: 119009. https://doi.org/10.1016/j.jclepro.2019.119009.

30 Paladini, G., Venuti, V., Almásy, L. et al. (2019). Cross-linked cellulose nano-sponges: a small angle neutron scattering (SANS) study. *Cellulose* 26: 9005–9019.

31 Rossi, B., Paciaroni, A., Venuti, V. et al. (2017). SANS investigation of water adsorption in tunable cyclodextrin-based polymeric hydrogels. *Phys. Chem. Chem. Phys.* 19: 6022–6029.

32 Hammouda, B. (2006). Solvation characteristics of a model water-soluble polymer. *J. Polym. Sci. Part B Polym. Phys.* 44 (22): 3195–3199.

33 Hule, R.A., Nagarkar, R.P., Hammouda, B. et al. (2009). Dependence of self-assembled peptide hydrogel network structure on local fibril nanostructure. *Macromolecules* 42 (18): 7137–7145.

34 Arvidsson, R.A., Tillman, M., Sandén, B.A. et al. (2018). Environmental assessment of emerging technologies: recommendations for prospective LCA. *J. Ind. Ecol.* 22 (6): 1286–1294.

35 Cucurachi, S., van der Giesen, C., and Guinée, J. (2018). Ex-ante LCA of emerging technologies. *Procedia CIRP* 69: 463–468.

36 Hetherington, A.C., Borrion, A.L., Griffiths, O.G., and McManus, M.C. (2014). Use of LCA as a development tool within early research: challenges and issues across different sectors. *Int. J. Life Cycle Assess.* 19: 130–143. https://doi.org/10.1007/s11367-013-0627-8.

37 Bauer, C., Buchgeister, J., Hischier, R. et al. (2008). Towards a framework for life cycle thinking in the assessment of nanotechnology. *J. Clean. Prod.* 16: 910–926. https://doi.org/10.1016/j.jclepro.2007.04.022.

38 Tsuzuki, T. (2014). Life cycle thinking and green nanotechnology. *Austin J. Nanomed. Nanotech.* 2: 1010.

39 Wardak, A., Gorman, M.E., Swami, N., and Deshpande, S. (2008). Identification of risks in the life cycle of nanotechnology-based products. *J. Ind. Ecol.* 12 (3): 435–448.

40 Grubb, G.F. and Bakshi, B.R. (2011). Life cycle of titanium dioxide nanoparticle production: impact of emissions and use of resources. *J. Ind. Ecol.* 15 (1): 81–95.

41 Kim, H.C. and Fthenakis, V. (2013). Life cycle energy and climate change implications of nanotechnologies. *J. Ind. Ecol.* 17 (4): 528–541.

42 Bartolozzi, I., Daddi, T., Punta, C. et al. (2020). Life cycle assessment of emerging environmental technologies in the early stage of development. *J. Ind. Ecol.* 24: 101–115. https://doi.org/10.1111/jiec.12959.

43 European Commission (2011). 2011/696/EU Commission Recommendation of 18 October 2011 on the definition of nanomaterial. *OJEU* L275 (54): 38.2.

44 Piccinno, F., Hischier, R., Seeger, S., and Som, C. (2016). From laboratory to industrial scale: a scale-up framework for chemical processes in life cycle assessment studies. *J. Clean. Prod.* 135: 1085–1097.

45 Gallagher, M.J., Allen, C., Buchman, J.T. et al. (2017). Research highlights: applications of life-cycle assessment as a tool for characterizing environmental impacts of engineered nanomaterials. *Environ. Sci. Nano* 4: 276–281.

46 Hischier, R. and Walser, T. (2012). Life cycle assessment of engineered nanomaterials: state of the art and strategies to overcome existing gaps. *Sci. Total Environ.* 425: 271–282.

47 Miseljic, M. and Olsen, S.I. (2014). Life-cycle assessment of engineered nanomaterials: aliterature review of assessment status. *J. Nanopart. Res.* 16 (6): 1–33.

48 Joint Research Centre (JRC) (2007). Report from the Workshop on Nanotechnologies for Environmental Remediation. JRC Ispra, 16–17 April 2007. David Rickerby and Mark Morrison. www.nanowerk.com/nanotechnology/reports/reportpdf/report101.pdf (accessed 21 September 2021).

49 Nel, A., Xia, T., Madler, L., and Li, N. (2006). Toxic potential of materials at the nanolevel. *Science* 311: 622. https://doi.org/10.1126/science.1114397.

50 Hochella, M.F. Jr., Mogk, D.W., Ranville, J. et al. (2019). Natural, incidental, and engineered nanomaterials and their impacts on the Earth system. *Science* 363: 6434–8299. https://doi.org/10.1126/science.aau8299.

51 Nowack, B. and Bucheli, T.D. (2007). Occurrence, behaviour and effects of nanoparticles in the environment. *Environ. Pollut.* 150: 5–22.

52 Grieger, K.D., Hjorth, R., Rice, J., Kumar, N. and Bang, J. (2015). Nano-remediation: tiny particles cleaning up big environmental problems. IUCN. https://www.iucn.org/sites/dev/files/import/downloads/nanoremediation.pdf (accessed 21 September 2021).

53 Karn, B., Kuiken, T., and Otto, M. (2009). Nanotechnology and in situ remediation: a review of the benefits and potential risks. *Environ. Health Perspect.* 117: 1823–1831. https://doi.org/10.1289/ehp.0900793.

54 Otto, M., Floyd, M., and Bajpai, S. (2008). Nanotechnology for site remediation. *Remediation* 19: 99–108. https://doi.org/10.1002/rem.20194.

55 Grieger, K.D., Fjordbøge, A., Hartmann, N.B. et al. (2010). Environmental benefits and risks of zero-valent iron nanoparticles (nZVI) for in situ remediation: risk mitigation or trade-off? *J. Contam. Hydrol.* 118: 165–183. https://doi.org/10.1016/j.jconhyd.2010.07.011.

56 Patil, S.S., Shedbalkar, U.U., Truskewycz, A. et al. (2016). Nanoparticles for environmental clean up: a review of potential risks and emerging solutions. *Environ. Technol. Innov.* 5: 10–21. https://doi.org/10.1016/j.eti.2015.11.001.

57 Sánchez, A., Recillas, S., Font, X. et al. (2011). Ecotoxicity of, and remediation with, engineered inorganic nanoparticles in the environment. *TrAC-Trend. Anal. Chem.* 30: 507–516. https://doi.org/10.1016/j.trac.2010.11.011.

58 Trujillo-Reyes, J., Peralta-Videa, J.R., and Gardea-Torresdey, J.L. (2014). Supported and unsupported nanomaterials for water and soil remediation: are they a useful solution for worldwide pollution? *J. Hazard. Mater.* 280: 487–503. https://doi.org/10.1016/j.jhazmat.2014.08.029.

59 Bardos, P., Merly, C., Kvapil, P., and Koschitzky, H. (2018). Status of nanoremediation and its potential for future deployment: risk-benefit and benchmarking appraisals. *Remediation* 28: 43–56. https://doi.org/10.1002/rem.21559.

60 Lofrano, G., Libralato, G., and Brown, J. (eds.) (2017). *Nanotechnologies for Environmental Remediation – Applications and Implications*. Springer.

61 Garner, K.L. and Keller, A.A. (2014). Emerging patterns for engineered nanomaterials in the environment: a review of fate and toxicity studies. *J. Nanopart. Res.* 16: 2503. https://doi.org/10.1007/s11051-014-2503-2.

62 Blasco, J. and Corsi, I. (eds.) (2019). *Ecotoxicity of Nanoparticles in the Aquatic Ecosystems*. Boca Raton, FL: CRC Press, Taylor & Francis Group https://doi.org/10.1201/9781315158761.

63 Corsi, I. and Grassi, G. (2019). The role of ecotoxicology in the eco-design of nanomaterials for water remediation. In: *Ecotoxicology of Nanoparticles in Aquatic Systems* (eds. J. Blasco and I. Corsi), 219–229. CRC Taylor & Francis Group.

64 Grieger, K., Wickson, F., Andersen, H.B., and Renn, O. (2012). Improving risk governance of emerging technologies through public engagement: the neglected case of nano-remediation? *Int. J. Emerging Techn. Soc.* 10: 61–78.

65 Hjorth, R., Coutris, C., Nguyen, N.H.A. et al. (2017a). Ecotoxicity testing and environmental risk assessment of iron nanomaterials for sub-surface remediation – recommendations from the FP7 project NanoRem. *Chemosphere* 182: 525–531. https://doi.org/10.1016/j.chemosphere.2017.05.060.

66 Holden, P.A., Gardea-Torresdey, J.L., Klaessig, F. et al. (2016). Considerations of environmentally relevant test conditions for improved evaluation of ecological hazards of engineered nanomaterials. *Environ. Sci. Technol.* 50: 6124–6145. https://doi.org/10.1021/acs.est.6b00608.

67 Oomen, A.G., Steinhäuser, K.G., Bleeker, E.A.J. et al. (2018). Risk assessment frameworks for nanomaterials: scope, link to regulations, applicability, and outline for future directions in view of needed increase in efficiency. *NanoImpact* 9: 1–13. https://doi.org/10.1016/j.impact.2017.09.001.

68 Petersen, E.J., Diamond, S.A., Kennedy, A.J. et al. (2015). Adapting OECD aquatic toxicity tests for use with manufactured nanomaterials: key issues and consensus recommendations. *Environ. Sci. Technol.* 49: 9532–9547. https://doi.org/10.1021/acs.est.5b00997.

69 Canesi, L., Frenzilli, G., Balbi, T. et al. (2014). Interactive effects on n-TiO_2 and 2,3,7,8-TCDD on the marine bivalve *Mytilus galloprovincialis*. *Aquat. Toxicol.* 153: 53–65. https://doi.org/10.1016/j.aquatox.2013.11.002.

70 Della Torre, C., Buonocore, F., Frenzilli, G. et al. (2015). Influence of titanium dioxide nanoparticles on 2,3,7,8-tetrachlorodibenzo-*p*-dioxin bioconcentration and toxicity in the marine fish European sea bass (Dicentrarchus labrax). *Environ. Pollut.* 196: 185–193. https://doi.org/10.1016/j.envpol.2014.09.020.

71 Della Torre, C., Balbi, T., Grassi, G. et al. (2015a). Titanium dioxide nanoparticles modulate the toxicological response to cadmium in the gills of *Mytilus galloprovincialis*. *J. Hazard. Mater.* 297: 92–100. https://doi.org/10.1016/j.jhazmat.2015.04.072.

72 Nigro, M., Berbardeschi, M., Costagliola, D. et al. (2015). Co-exposure to titanium dioxide nanoparticles and cadmium: genomic, DNA and chromosomal damage evaluation in the marine fish European sea bass *Dicentrarchus labrax*. *Aquat. Toxicol.* 168: 72–77. https://doi.org/10.1016/j.aquatox.2015.09.013.

73 Rocco, L., Santonastasio, M., Nigro, M. et al. (2015). Genomic and chromosomal damage in the marine mussel *Mytilus galloprovincialis*: effects of the combined exposure titanium dioxide nanoparticles and cadmium chloride. *Mar. Environ. Res.* 111: 144–148. https://doi.org/10.1016/j.marenvres.2015.09.004.

74 Vannuccini, M.L., Grassi, G., Leaver, M.J., and Corsi, I. (2015). Combination effects of nano-TiO_2 and 2,3,7,8-tetrachlorodibenzo-p-dioxin (TCDD) on biotransformation genes expression in the liver of European sea bass *Dicentrarchus labrax*. *Comp. Biochem. Physiol. C* 176–177: 71–78. https://doi.org/10.1016/j.cbpc.2015.07.009.

75 Canesi, L. and Corsi, I. (2016). Effects of nanomaterials in marine invertebrates. *Sci. Total Environ.* 565: 393–340. https://doi.org/10.1016/j.scitotenv.2016.01.085.

76 Klaine, S.J., Alvarez, P.J.J., Batley, G.E. et al. (2008). Nanomaterials in the environment: behavior, fate, bioavailability, and effects. *Environ. Toxicol. Chem.* 27: 1825–1851. https://doi.org/10.1897/08-090.1.

77 Lowry, G.V., Gregory, K.B., Apte, S.C., and Lead, J.R. (2012). Transformations of nanomaterials in the environment. *Environ. Sci. Technol.* 46: 6893–6899. https://doi.org/10.1021/es300839e.

78 Micić, V., Bossa, N., Schmid, D. et al. (2019). Groundwater chemistry has a greater influence on the mobility of nanoparticles used for remediation than the chemical heterogeneity of aquifer media. *Environ. Sci. Technol.* 54 (2): 1250–1257. https://doi.org/10.1021/acs.est.9b06135.

79 Petosa, A.R., Jaisi, D.P., Quevedo, I.R. et al. (2010). Aggregation and deposition of engineered nanomaterials in aquatic environments: role of physicochemical interactions. *Environ. Sci. Technol.* 44: 6532–6549. https://doi.org/10.1021/es100598h.

80 Corsi, I., Cherr, G.N., Lenihan, H.S. et al. (2014). Common strategies and technologies for the ecosafety assessment and design of nanomaterials entering the marine environment. *ACS Nano* 8: 9694–9709. https://doi.org/10.1021/nn504684k.

81 Holland, K.S. (2011). A framework for sustainable remediation. *Environ. Sci. Technol.* 45: 7116–7117. https://doi.org/10.1021/es202595w.

82 Garner, K.L., Suh, S., and Keller, A.A. (2017). Assessing the risk of engineered nanomaterials in the environment: development and application of the nanofate model. *Environ. Sci. Technol.* 51: 5541–5551. https://doi.org/10.1021/acs.est.6b05279.

83 Nowack, B., Ranville, J.F., Diamond, S. et al. (2012). Potential scenarios for nanomaterial release and subsequent alteration in the environment. *Environ. Toxicol. Chem.* 31: 50–59. https://doi.org/10.1002/etc.726.

84 Bour, A., Mouchet, F., Silvestre, J. et al. (2015). Environmentally relevant approaches to assess nanoparticles ecotoxicity: a review. *J. Hazard. Mater.* 283: 764–777. https://doi.org/10.1016/j.jhazmat.2014.10.021.

85 Scott-Fordsmand, J., Peijnenburg, W., Semenzin, E. et al. (2017). Environmental risk assessment strategy for nanomaterials. *Int. J. Environ. Res. Public Health* 14: 1251. https://doi.org/10.3390/ijerph14101251.

86 Bastus, N.G. and Puntes, V. (2018). Nanosafety: towards safer nanoparticles by design. *Curr. Med. Chem.* 25: 4587–4601. https://doi.org/10.217 4/0929867324666170413124915.

87 Hjorth, R., van Hove, L., and Wickson, F. (2017). What can nanosafety learn from drug development? The feasibility of "safety by design". *Nanotoxicology* 11: 305–312. https://doi.org/10.1080/17435390.2017.1299891.

88 Singh, A.V., Laux, P., Luch, A. et al. (2019). Review of emerging concepts in nanotoxicology: opportunities and challenges for safer nanomaterial design. *Toxicol. Mech. Methods* 29 (5): 378–387. https://doi.org/10.1080/1537651 6.2019.1566425.

89 Fischer, D., Li, Y., Ahlemeyer, B. et al. (2003). in vitro cytotoxicity testing of polycations: influence of polymer structure on cell viability and hemolysis. *Biomaterials* 24: 1121–1131.

90 Kafil, V. and Omidi, Y. (2011). Cytotoxic impacts of linear and branched polyethylenimine nanostructures in A431 cells. *Bioimpacts* 1: 23–30.

91 Bolognesi, C. and Fenech, M. (2012). Mussel micronucleus cytome assay. *Nat. Protoc.* 7: 1125–1137.

92 Davies, I.M., Gubbins, M., Hylland, K. et al. (2012, 2012). Technical annex: assessment criteria for biological effects measurements, 209–212. In: *Integrated Monitoring of Chemicals and Their Effects. ICES Cooperative Research Report, No. 315* (eds. I.M. Davies and A.D. Vethaak), 277. ICES.

93 Izaguirre, U., Ruiz, P., and Marigómez, I. (2009). Time-course study of the early lysosomal responses to pollutants in mussel digestive cells using acid phosphatase as lysosomal marker enzyme. *Comp. Biochem. Physiol. C Toxicol. Pharmacol.* 149: 587–597.

94 Marigómez, I., Izagirre, U., and Lekube, X. (2005). Lysosomal enlargement in digestive cells of mussels exposed to cadmium, benzo[a]pyrene and their combination. *Comp. Biochem. Physiol. C Toxicol. Pharmacol.* 141: 188–193.

95 Behrenfeld, M.J., O'Malley, R.T., Siegel, D.A. et al. (2006). Climate-driven trends in contemporary ocean productivity. *Nature* 444: 752–755. https://doi.org/10.1038/nature05317.

96 Field, C.B., Field, C.B., Behrenfeld, M.J., and Randerson, J.T. (2009). Primary production of the biosphere: integrating terrestrial and oceanic components. *Science* 281: 237–240. https://doi.org/10.1126/science.281.5374.237.

97 OECD (1994). *OECD guidelines for the testing of chemicals (123).* OECD https://doi.org/10.1787/9789264203785-en.

98 Bondarenko, O., Juganson, K., Ivask, A. et al. (2013). Toxicity of Ag, CuO and ZnO nanoparticles to selected environmentally relevant test organisms and mammalian cells in vitro: a critical review. *Arch. Toxicol.* 87: 1181–1200. https://doi.org/10.1007/s00204-013-1079-4.

99 Manzo, S., Buono, S., Rametta, G. et al. (2015). The diverse toxic effect of SiO_2 and TiO_2 nanoparticles toward the marine microalgae *Dunaliella tertiolecta. Environ. Sci. Pollut. Res.* 22: 15941–15951. https://doi.org/10.1007/s11356-015-4790-2.

100 Morelli, E., Gabellieri, E., Bonomini, A. et al. (2018). TiO_2 nanoparticles in seawater: aggregation and interactions with the green alga *Dunaliella tertiolecta. Ecotoxicol. Environ. Saf.* 148: 184–193. https://doi.org/10.1016/j.ecoenv.2017.10.024.

101 Sendra, M., Moreno-Garrido, I., Yeste, M.P. et al. (2017). Toxicity of TiO_2, in nanoparticle or bulk form to freshwater and marine microalgae under visible light and UV-A radiation. *Environ. Pollut.* 227: 39–48. https://doi.org/10.1016/j.envpol.2017.04.053.

102 Goldberg, E.D., Bowen, V.T., Farrington, J.W. et al. (1978). The mussel watch. *Environ. Conserv.* 5 (2): 101–125. https://doi.org/10.1017/S0376892900005555.

103 Beyer, J., Green, N.W., Brooks, S. et al. (2017). Blue mussels (*Mytilus edulis* spp.) as sentinel organisms in coastal pollution monitoring: a review. *Mar. Environ. Res.* 130: 338–365. https://doi.org/10.1016/j.marenvres.2017.07.024.

104 Canesi, L., Ciacci, C., Fabbri, R. et al. (2012). Bivalve molluscs as a unique target group for nanoparticle toxicity. *Mar. Environ. Res.* 76: 16–21. https://doi.org/10.1016/j.marenvres.2011.06.005.

105 Rocha, T.L., Gomes, T., Serrão Sousa, V. et al. (2015). Ecotoxicological impact of engineered nanomaterials in bivalve molluscs: an overview. *Mar. Environ. Res.* 111: 74–88. https://doi.org/10.1016/j.marenvres.2015.06.013.

106 Bartke, S., Hagemann, N., Harries, N. et al. (2018). Market potential of nanoremediation in Europe: market drivers and interventions identified in a deliberative scenario approach. *Sci. Total Environ.* 619–620: 1040–1048. https://doi.org/10.1016/j.scitotenv.2017.11.215.

10

Nanotechnology Applications in Natural Nanoclays Production and Application for Better Sustainability

Manjir Sarma Kataki, Bibhuti Bhusan Kakoti, Kangkan Deka, and Ananya Rajkumari

Department of Pharmaceutical Sciences, Dibrugarh University, Dibrugarh, Assam, India

10.1 Introduction

Nanoclays are fine-grained crystalline materials composed of layered mineral silicates. Nanoclays are compounds based on silicon, oxygen, and other elements. A layer is the basic structural unit of nanoclays, and these layers are likely to position themselves over one another like pages of a book. Individual layers are composed of the tetrahedral and/or octahedral sheets, and this arrangement of sheets plays a crucial role in defining and differentiating these clay minerals. In tetrahedral sheet, the silicon–oxygen tetrahedra are linked to the neighboring tetrahedra by sharing three corners, while the fourth corner of each tetrahedron forms a part to the adjacent octahedral sheet. The octahedral sheet is typically composed of aluminum or magnesium in a sixfold coordination with oxygen from the tetrahedral sheet and with hydroxyl. Tetrahedral sheets consist of silicon–oxygen tetrahedra linked to the neighboring tetrahedra, sharing three corners, while the fourth corner of each tetrahedron sheet is linked to an adjacent octahedral sheet via a covalent bond. The arrangements of these sheets impact various defining and distinguishing aspects of nanoclays. Depending on their mineralogical composition, there exist about 30 different types of nanoclays, which depending on their properties are used in different applications. Three major 1 : 1, 2 : 1, and 2 : 1 : 1 sheet arrangements are observed in common nanoclay materials as shown in Table 10.1.

Depending on the nanoparticle morphology and chemical composition, nanoclays are classified into a few classes such as montmorillonite (MMT), bentonite, kaolinite, hectorite, and halloysite. Owing to the widespread accessibility, relatively low cost, and low environmental impact, nanoclays have been explored and developed for numerous applications [1]. With the tremendous growth of nanotechnology, clay minerals are progressively utilized as natural nanomaterials [2].

10.2 Occurrence and Production of Nanoclays

Nanoclays are primarily originated from the naturally occurring sedimentary rock "bentonite," which comprises 60–80% MMT. Bentonite has been called after a clay deposit in the United States in the state of Wyoming near Fort Benton. There are other places known for important deposits in Europe, e.g. on the Greek island Milos, in Turkey, and in Germany. In Bavaria, the largest mining areas for the mineral can be spotted. Bentonite is one of the well-known clay minerals that have interlayer of silica and alumina. It is naturally designed from volcanic ash weathering in the presence of water. Bentonite can be used as such without any modification or can be modified via physiochemical treatments, depending on their applications [3]. It is primarily used in ceramic bodies, structural polymers, drilling fluids [4], and as catalyst [5]. MMT is the most common plate-like nanoclay (smectite) that consists of ~1-nm-thick aluminosilicate layers surface substituted with metal cations and stacked in ~10 μm-sized multilayer stacks. MMT is named after a deposit near Montmorillon in southern France. Based on the superficial alteration of the clay layers, MMT can be disseminated in a polymer matrix to form polymer–clay nanocomposite [6]. Halloysite is a naturally occurring nanotube composed of aluminosilicate and chemically similar to kaoline with an average dimension of

Sustainable Nanotechnology: Strategies, Products, and Applications, First Edition.
Edited by Yashwant V. Pathak, Govindan Parayil, and Jayvadan K. Patel.

Table 10.1 Classification of clay minerals.

Type of clay layer	Clay group	Clay species
1:1	Kaolinite, Halloysite, Rectorite, Chyrsotile	Kellyte, Nacrite, Lizardite, Cronstedtite, Berthierine, Fraipontite, Brindleyite, Dickite
2:1	Pyrophyllite talc, Mica, Brittle Mica, Vermiculite, Smectite	Montmorillonite, Bentonite, Hectorite, Talc, Sepiolite, Vermiculite, Laponite, Mucovite, Pyrophyllite, Clintonite, Paragonite, Bityite
2:1:1	Chlorite	Amesite, Cookeite

15×1000 nm, which is comparable to carbon nanotubes. Due to the specialized hollow tube structure, halloysite can be employed for controlled delivery and controlled release of drugs. In addition, these nanostructures find their use in various rheology modification applications. There are organically modified nanoclays (organoclays) with potential applications in polymer nanocomposites, as rheological modifiers, gas absorbents, and drug delivery carriers. There are various methods for nanoclay production from clay minerals, such as freeze drying, centrifugation, cross-flow filtration, energetic stirring, and through ultracentrifugation [7, 8].

10.3 Nanoclays for Biomedical Applications

10.3.1 Biocompatibility of Nanoclays

Clay materials are explored well and regarded to be safe or environmentally friendly [9, 10], therefore making them attractive for a variety of biomedical applications [9]. Owing to the rise in engineering of nanoclay-containing products, including a widespread range of industrial applications, data on the toxicological and health effects of nanoclay exposure is utmost important. The various conventional as well as novel methods of nanotoxicity assessment can be explored or adapted for toxicity evaluation of nanoclay materials also. There are many developed *in vitro* and *in vivo* techniques including biochemical and cell culture methods for evaluation of toxicity of nanoclay materials. Most studies designate that nanoclays are comparatively safe, and toxicity is frequently related to nanoclay functionalized with organic modifiers [11]. Studies have demonstrated the safety of the nanosilicate platelets (NSPs) derived from natural MMT clay for likely usages in biomedical fields [9]. Li et al. [9] demonstrated not any mutagenic effect in three different genetic toxicity test systems without any apparent changes in cell morphology upon accumulation and adherence onto cell surfaces.

Another report revealed that halloysite nanoclay is nontoxic up to concentrations of 75 $\mu g\ ml^{-1}$ in two different cell lines, although parallel laser confocal visualization of cell uptake of fluorescently labeled halloysite demonstrated its location within the cells in the nuclear vicinity. This research suggested that halloysite nanoclays are not toxic for cells [12]. Two more zeolite nanoclays, namely, clinoptilolite and sepiolite, proved to be well tolerated in extremely phagocytic cultures [13], with results equivalent to asbestos-free talc powder. The results suggested that both the zeolite nanoclays could be highly biocompatible. Nevertheless, when compared, clinoptilolite seems to be less toxic than sepiolite [13]. However, unmodified nanoclay, Cloisite $Na^{+®}$, and the organically modified nanoclay, Cloisite 93A®, were reported to be highly cytotoxic in the human hepatic HepG2 cell line, indicating a possible risk to human health. Overall, nanoclays are mostly biocompatible and safe for development of clay nanoparticles toward bio and medical applications [14]. Most studies indicate that nanoclays are relatively safe, and toxicity is often linked to nanoclay functionalization with organic modifiers [11].

10.3.2 Nanoclay in Bone Cement

Bones are complex composite materials composed of inorganic calcium phosphates that provide strength and organic collagen that provides flexibility [15, 16]. Polymethyl methacrylate (PMMA) is commonly and widely used as a bone cement with the goal of fixing hip and knee replacement implants into adjacent bones [17]; though, PMMA has a few disadvantages including poor fatigue strength, overall inadequate mechanical properties [15], and possibility of necrosis and loosening of implants over time [18]. In this context, nanoclay materials have been studied as reinforcements for PMMA composites for bone cement applications [8, 19] or bone implants [15] with improved activity and mechanical properties. For example, novel PMMA/HA/nanoclay cement

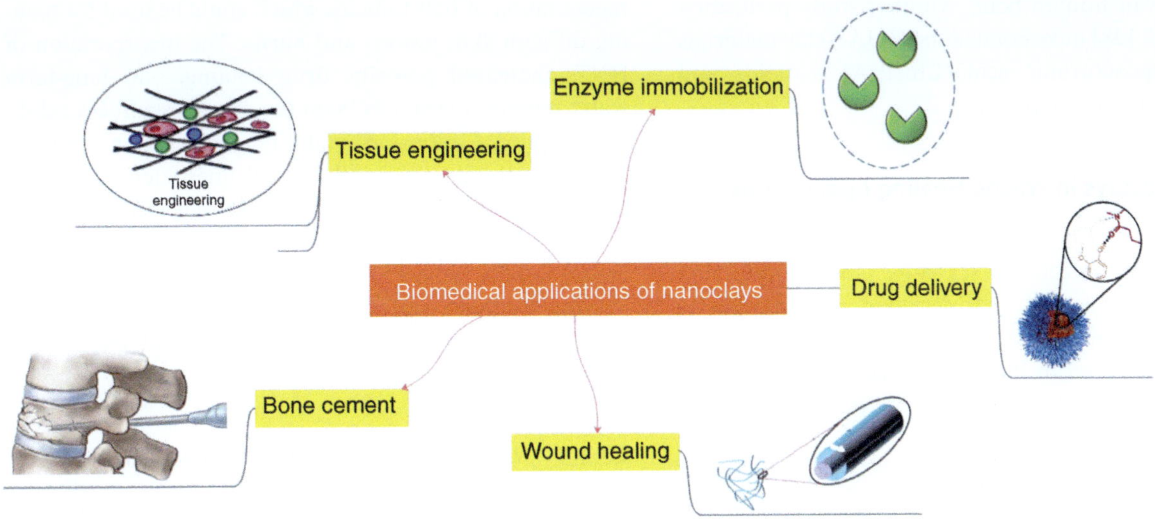

Figure 10.1 Biomedical applications of nanoclays.

was developed by dry blending of PMMA with HA (hyaluronic acid) and nanoclay and subjected to Brazilian disk tests that demonstrated that the addition of 1.0 wt.% nanoclay into the nanocomposites improves the fracture resistance and strain energy release rate of the bone–cement interface [19]. Another study [18] arranged PMMA bone cement/layered silicate nanohybrids with nanoclays based on MMT in concentrations of 0.5, 1.0, 1.5, and 2.0 wt.% and subjected to mechanical testing, cell culture studies, and *in vivo* studies. The results showed a significant decrease in the exothermic polymerization temperature, which could actually reduce cell necrosis, and loosening of implants. Young's modulus, toughness, and strength were also found to be improved by the nanoclay fillers because of their suppression of crack growth with enhanced biocompatibility [14, 18] (Figure 10.1).

10.3.3 Nanoclay in Tissue Engineering

The main goal of tissue engineering is to advance, maintain, and restore tissue functions significantly [20]. The addition and incorporation of nanoclays renders various polymers or polysaccharides tunable in terms of physical and mechanical properties according to the desired application [20, 21]. A grown body of publications indicated a widespread fabrication of nanoclay materials for biomedical application. De Silva et al. [22] showed that the addition of natural nanotubes of halloysite (HNTs) as an environmentally friendly nanofiller to chitosan (CS) biopolymer can increase the possible applications of this biopolymer, particularly when the mechanical properties are concerned. The addition fabrication enhanced mechanical properties as well as thermal stability. A study by Bonifacio et al. [7] successfully fabricated a hydrogel of gellan gum (GG) and glycerol, where HNTs were incorporated for soft tissue engineering applications such as pancreas, liver, and skin regeneration. The addition of the HNTs reduced the water uptake by 30–35%. Nitya et al. [23] prepared a fibrous polycaprolactone/HNT composite scaffold by electrospinning and evaluated *in vitro* for bone tissue engineering applications. These scaffolds were very advantageous in terms of enhanced mineralization, greater protein adsorption, and faster proliferation of human mesenchymal stem cells (hMSCs) seeded on these scaffolds. Zhou et al. [24] exhibited that adding HNTs to poly(vinyl alcohol) (PVA) bionanocomposite films resulted in nanotopographical changes and surface chemistry of PVA. These modifications significantly raise the level of cell adhesion. Furthermore, the mechanical properties of the films were significantly enhanced. Katti et al. [25] synthesized a biopolymer with an intercalated structure consisting of CS mixed with HAP, and MMT resulted, improving the thermal stability and nanomechanical properties. Aliabadi et al. [26] produced a biocompatible CS ammonium salt *N*-(2-hydroxy) propyl-3-trimethylammonium CS chloride (HTCC)-modified MMT with antibacterial properties. The samples were found to be effective in both Gram-negative and -positive bacteria. The antibacterial efficiency provided by MMT was due to the entrapment of bacteria between the intercalated structures of HTCC in MMT. Due to their outstanding results, these composites were proposed for tissue engineering applications. Further, varied state-of-the-art applications of clay nanomaterials for tissue engineering have been explored. For example, Payne et al. [27] modified sodium-MMT (Na-MMT) by an amino acid in order to mineralize synthetic HAP, resembling biogenic HAP in human bone. Another study performed by Ambre et al. [28] incorporated these HAP-clay materials into CS/polygalacturonic acid (Chi/PgA) scaffolds and films for bone tissue engineering.

10.3.4 Nanoclays in Wound Healing Applications

Nanoclays have also been greatly explored in wound healing to prevent infection, scarring, and minimizing pain [29–33]. Sabaa et al. [34] designed a biodegradable PVA composite with carboxymethyl chitosan (CMCh) and MMT, which resulted in improved swelling behavior and good antimicrobial efficacy compared to standard drugs such as penicillin G.

Modified collagen/*N*-isopropylamide hydrogels were prepared by incorporating MMT nanoparticles with the purpose of adjusting their stimuli response as scaffold with enhanced healing and regenerative properties [35]. The incorporation of MMT nanoparticles allowed for the formation of new bonds and formed a 3D network of interconnected pores in the modified hydrogels. Nanoclays improved the mechanical properties and chemical stability, making the host polymer/material suitable for wound healing applications with significantly improved properties. Yang et al. [33] evaluated cross-linked nanoclays, such as semi-IPN sericin/poly(NIPAm/LMSH) (HSP) nanocomposite hydrogels, for wound dressing. The wound healing area treated with the nanocomposites increased threefold over the area covered by gauze after 6 days and exhibited nearly complete recovery by the 13th day. In another study by Pacelli et al. [36], laponite was combined in a wound dressing material composed of gellan gum methacrylate (GG-MA) for efficient drug delivery to the wound site. The laponite nanoclay modulated the swelling behavior of the GG-MA hydrogel network and was able to lower the amount of antibiotic released during the first 8 hour, compared to unfilled hydrogels.

CS nanocomposites were also widely fabricated and evaluated as a potential modality in wound healing applications. Sandri et al. [30] fabricated CS oligosaccharide nanocomposites modulated with HNTs and further evaluated for wound healing applications. These nanoclay-fabricated nanocomposites demonstrated to be biocompatible with normal human fibroblasts and exhibited an enhancement of cell proliferation in an *in vitro* wound healing test. Cross-linked collagen and thermoresponsive poly(*N*-isopropylacrylamide) with embedded MMT was evaluated for wound healing and demonstrated good cytotoxicity and biocompatibility [35]. Demirci et al. [37] fabricated and evaluated HA cryogels as hydrogels with enhanced porosity and mechanical strength and demonstrated a progressive degree of revascularization and regeneration of hair follicles which could be used for treating difficult skin lesions and burns. The incorporation of HNTs increased porosity, drug loading, and long-term water retention [38]. HNTs on the cryogel scaffolds exhibited improved adhesion, viability, proliferation, and the growth of the diverse kinds of cell lines such as cervical carcinoma cells (HeLa), colon cancer cells (HCT116), and mesenchymal stem cells (MSCs). HNT cryogel composites are repeatedly proved as a capable material as tissue engineering scaffolds with their improved macroporous structure, thermal and mechanical stability, and good blood compatibility.

10.3.5 Nanoclays in Enzyme Immobilization

Immobilization is the key to optimizing the functional and operational capabilities of an enzyme in numerous industrial processes. Immobilized enzymes are robust and resistant to environmental changes compared to free enzymes in solution. The heterogeneity of the immobilized enzyme systems permits an easy retrieval of both enzymes and products, continuous operation of enzymatic processes, multiple reuse of enzymes, and rapid termination of reactions [39]. Nanoclays have also been explored for their capability to immobilize enzymes [40]. Tzialla et al. [40] characterized two types of MMTs in the immobilization of lipase B from *Candida antarctica* on Laponite, SWy-2 and Kunipia (smectite group nanoclays), and showed enhanced activity and stability in low-aqueous media. MMT was also used for immobilization of microbial phytases from *Escherichia coli* and *Aspergillus niger* where γ-aminopropyltriethoxysilane had been utilized for functionalization of HNTs for the purpose of enzyme immobilization [41]. Bugatti et al. [42] fabricated lysozyme-encapsulated HNTs for incorporation into PLA nanocomposite, which demonstrated improved barrier as well as mechanical properties. In a nutshell, nanoclays are reported to be very efficient in enzyme immobilization. Acevedo et al. [43] reported nanoclay as a suitable support material for MnP (manganese peroxidase from *Anthracophyllum discolor*) immobilization. Nanoclay-immobilized MnP exhibited an augmented stability at higher pH, temperature, and storage time.

10.3.6 Polymer Nanocomposites: Nanoclays in Drug Delivery

Polymer nanocomposites (PNCs) are the amalgamation of polymer continuous phase and nanoparticles as discontinuous phase, which demonstrate many advantages in terms of electrical, mechanical, and optical properties as compared with individual components. These exceptional and nonpareil attractive properties of PNCs have fascinated the scientists around the globe in recent years [44]. Nanoclays and polymer nanocomposites find their way to the numerous drug delivery applications ranging from antibiotics, antihistamines, anti-inflammatories, antimicrobial, antifungal, antialgal, and anticancer treatments [45]. Roozbahani et al. [45] investigated laponite nanoplatelets (LAPs) as a possible podium for effective sustained release of anionic dexamethasone (DEX). The study encapsulated the anionic dexamethasone into the interlayer spacing of LAP nanodisks with a very high efficiency. The results revealed a pH-dependent release of the drug with faster rates in acidic environments.

Chitosan/organic rectorite (chitosan/OREC) nanocomposite films loaded with drug were effectively fabricated by a casting/solvent evaporation method [46]. Bovine serum albumin (BSA) was the model drug used to study the drug delivery behavior which was dependent on the amount and interlayer spacing of OREC [46]. Zeynabad et al. [47] utilized the same nanoclay delivery system for studying a pH-dependent dual drug delivery for cancer therapy. This study explored the encapsulation of dual drugs (the anticancer drug methotrexate and an antibacterial agent ciprofloxacin) in an organo-modified laponite–polymer composite system [47]. Mura at al. [48] designed a "drug – in cyclodextrin – in nanoclays" hybrid single delivery system using sepiolite, attapulgite, and bentonite, respectively, and demonstrated an improved dissolution of oxaprozin (an anti-inflammatory drug) with sepiolite. Nanoclays for the entrapment of the poorly soluble drugs are suggested to be an improvised tool for augmenting the therapeutic effectiveness of poorly soluble drugs such as oxaprozin by reducing the quantity of cyclodextrin required for gaining the needed drug solubility.

Nanocomposite hydrogels composed of CS and MMT (2 wt.% MMT) were fabricated and investigated for drug release behavior following electrostimulation [49]. This study revealed a mechanically reliable and efficient pulsatile release profile of the drug (vitamin B_{12}) alongside an excellent antifatigue behavior, compared with that of the pure CS. The exfoliated MMT sheets were able to increase the cross-linking density, increase the fatigue behavior of CS, and improve the long-term drug release performance.

Likewise, Campbell et al. [50] developed paracetamol-loaded nanocomposites of poly(ε-caprolactone) filled with MMT without affecting the mechanical properties of the polymer matrix.

Furthermore, Campbell et al. [51], in a different study, successfully entrapped paracetamol molecules in a polyethylene glycol and nanoclay system without affecting the structure of paracetamol during the hot-melt process. The entrapment significantly reduces the rate of diffusion, producing a controlled release of the drug as required.

HNTs are perhaps the most widely studied nanoclay materials for development of various drug delivery systems attributable to their microtubule shape [52, 53]. There is a pH-dependent binding of drugs to HNTs, suggesting sustainability in the release of drugs [54]. Cationic drugs (antibiotics and antihistamines) are the most suitable candidates for delivery using HNTs as evident via growing body of research in the field [55]. Tu et al. [56] designed and developed a biodegradable HNT nanocomposite hydrogel with a polymer matrix comprising oligo(trimethylene carbonate)–poly(ethylene glycol)–oligo(trimethylene carbonate) diacrylate (TPT) and alginate sodium (AG) with improved mechanical properties and drug release behavior. These HNTs are highly biocompatible and efficient in various terms. Research also identified higher level of cytocompatibility of certain composite nanofibers of HNTs in poly(lactic-*co*-glycolic acid) (PLGA) [57]. The composite nanofibers were reported to release the drug in a sustained manner for up to 42 days. Occurrence of postsurgical infection due to an early failure of orthopedic and orthodontic implants is very likely and possesses significant risk as well as problem issues. Hence, to address this issue, calcium alginate and calcium phosphate bone cement with HNTs loaded with antibiotic gentamicin were designed and developed by Karnik et al. [58]. These nanoclay/ hydrogel nanocomposites exhibited a sustained and extended drug release with a boosted antimicrobial activity. Control release of anticancer agents in HNTs has also been attained by capping their tube ends [8]. Stimuli-triggered drug delivery vehicles of HNTs had also been well studied and developed for targeted intracellular drug delivery in cancer chemotherapy [59].

Dzamukova et al. developed a novel strategy for anticancer drug loading into brilliant green HNTs coated with dextrin stoppers for controlled release. The buildup and enzymatically induced release of drug only happened for malignant cells; therefore, noncancerous cells remained unaffected [59].

Demirci et al. [37] fabricated a HA hydrogel named as cryogel with HNTs loaded with sodium diclofenac as a model drug. The porosity as well as the pore volume augmented 5% and 6.2%, respectively, and the drug release was found to be 31%. In another study, polymer-grafted magnetic HNTs were developed which were loaded with the antibiotic norfloxacin (a cationic agent) for controlled release and indicated a higher drug loading capacity as compared to pristine HNTs [60].

In a nutshell, the HNTs possess various attractive properties including high adsorption capacity, magnetic properties, and sustainability in the drug release, which further

Table 10.2 Some drugs in nanoclay drug delivery systems.

Drug	Nanoclay type	References
5-Fluorouracil (anticancer)	Halloysite nanotubes	[62]
Ibuprofen (nonsteroidal anti-inflammatory)	Poly(ε-caprolactone)-layered silicate nanocomposites	[63]
Amino acids	Biomineralized hydroxyapatite amino acid-modified nanoclay	[64]
BSA (model protein)	Poly(ethersulfone) nanocomposite membrane filled with nanoclay	[65]
Plasmid DNA (gene delivery)	Halloysite nanotubes (HNTs)	[66]
Donepezil (Alzheimer)	Smectite clays (laponite, saponite, and montmorillonite)	[67]
Paclitaxel (anticancer drug)	Rectorite-immobilized cellulose acetate nanofibers	[68]
Lidocaine (local anesthetic drug)	Thermosensitive hybrid system based on pluronic F127 and nanoclay laponite	[69]
Doxorubicin (anticancer drug)	Laponite–doxorubicin nanohybrids	[70]
Vitamin B1	Nanoclays with montmorillonite structure	[70]
Tamoxifen (anticancer drug)	Montmorillonite system	[71]
Timolol maleate (β-adrenergic blocking agent)	Thiolated chitosan-immobilized amino acid-intercalated montmorillonite	[72]
Procainamide hydrochloride (antiarrhythmic drug)	Montmorillonite poly-L-lactide microcomposites	[73]
Ranitidine hydrochloride (antacid)	Montmorillonite (MMT)	[74]

regard it as a potential candidate drug delivery system or vehicle for targeted drug delivery in diverse forms of cancer. Rao et al. [61] designed and synthesized hydrogels for colon cancer drug delivery comprising sodium hyaluronate in a poly(hydroxyethyl methacrylate) matrix and HNTs loaded with an anticancer drug. *In vitro* experimental findings in a simulated gastric fluid and intestinal gastric fluid demonstrated a pH-dependent controlled release of the encapsulated drug with less than 10% release in the gastric region and higher release in a controlled manner in the intestinal fluid. This release behavior makes this kind of HNT delivery system very efficient for colon cancer drug delivery (Table 10.2).

10.3.7 Nanoclays in Food and Beverage Packaging

Diverse forms of nanoclays are incorporated into the polymers to improve the characteristics of the nanoclays, viz. MMT (MMT and MMT-Na^+) and organophilic MMT (organically modified MMT, OMMT). These nanoclays have gained utmost consideration in the packaging field, as evident from both academic and industrial researchers. The observed popularity of these nanoclays is due to the high surface area with a large aspect ratio (50–1000) and good compatibility with most of the organic thermoplastics [75]. The use of nanoclays in food packaging has started since 1990s for the improvement of the mechanical and barrier properties of the packaging [76]. Polymer nanocomposite, a resultant hybrid material from the incorporation of an inorganic nanoclay into an organic polymer matrix, possesses lower weight with superior properties than conventional microcomposites. The concentration of nanoclay in the fabrication of polymer nanocomposite plays a significant role in the properties of the polymer nanocomposite to be used in the production of packaging materials with improved properties. Generally, a nanocomposite can be fabricated via four different methods: solution intercalation, *in situ* intercalative polymerization, *in situ* direct synthesis, and melt intercalation. However, due to the cost-effectiveness, simplicity, feasibility, and being the environmentally benign process, *in situ* direct synthesis and melt intercalation techniques have gained more popularity among the industrial and academic researchers [75, 76] (Table 10.3).

Polymer-based packaging materials is the evolving field in food packaging where the utilization of nanoclays in polymer-based packaging materials is becoming widely prevalent owing to its certified functional property reinforcement and barrier enhancement.

Table 10.3 A few commercial nanoclays (organically modified clay minerals) used in food-packaging industry.

Name	Clay	Chemical structure of organic modifier	Manufacturer/Company
Cloisite® Na^+	MMT	Unmodified clay	Southern Clay Products Inc. (USA)
Nanocor® I.24TL	MMT	12-Aminododecanoic acid	Nanocor Inc. (USA)
Nanolin® DK1	MMT	Octadecyltrimethylammonium	Zhejiang Fenghong Clay Chemicals Co. Ltd. (China)
Dellite® LVF	MMT	Unmodified clay	Laviosa Chimica Mineraria SpA (Italy)
Shelsite Na^+	MMT	Unmodified clay	Nanoshel LLC (USA)

10.3.8 Biocompatible Functionalization of Nanoclays for Improved Environmental Remediation

MMT, kaolinite, halloysite, and palygorskite or attapulgite are the most common nanoclay minerals used in environmental applications [77, 78]. These mineral clays have been explored very well in clay science and well publicized [79]. Owing to the natural occurrence and abundance, clay minerals are largely observed as inexpensive environmentally friendly materials [79]. Because of the growing awareness of the negative effect of chemical pollution in an overpopulated world [80], there is a pressing necessity to develop more environmentally compatible as well as functionally modified nanoclays for advancing remediation efforts. Natural nanoclays are highly biocompatible and safe; however, as remediating agents, natural nanoclays often need some forms of physicochemical modification to obtain engineered nanoclays or nanocomposites which are appropriate to perform a definite functional chore in relation to the sorption toxic substances or dispersion of toxic chemicals [81]. In the process, various chemical agents are utilized to achieve the desired changes to the nanoclays, which in turn renders the natural nanoclays questionable in terms of their environment friendliness. In this regard, there is a necessity to assess the benefits against the risks of using remediating products before their application in combating the effects of anthropogenic pollution [81]. In addition, owing to the use of various conventional surfactants in organically modified clay minerals (organoclays), organoclays may possess poor biocompatibility with a potential to be toxic to environmental flora and fauna [82]. However, the growing research in biomedical, food and packaging industries suggested several research activities in biocompatibility evaluations as well as compliance testing of nanoclay products [83–85]. Recently, potential toxicity and compatibility of halloysite nanotubes and their modified materials were reviewed by Lazzara et al. [86] for agricultural uses.

10.3.8.1 Nanorobots Constructed from Nanoclay

Micro/nanomotors are the new generation of nanotechnology that permits matters to thrust and perform autonomous motion [87]. They have displayed potential applications in environmental remediation [10, 77, 81, 83] and in drug delivery [8, 12, 17]. Maric et al. [88] used Halloysite nanoclay to fabricate functional nanomotors with the incorporation of platinum and further evaluated their potential application in the elimination of Zn^{2+} and Cd^{2+} in aqueous solution. These nanotubes of halloysite nanoclay can be excavated in large quantities and can be used as a base for the fabrication of nanomachines. Also, it is well documented that halloysite nanoclay surface groups can bind sturdily with heavy metals, which renders it potentially valuable in environmental remediation. It was also observed that the heavy metals in aqueous environment get easily adsorbed onto the fabricated nanomotors. The results suggested an efficient removal efficiency of 86% for both Zn^{2+} and Cd^{2+}. The equilibrium adsorption of both the studied heavy metals was also achieved very fast within 5 minutes. The fabricated platinum–halloysite nanoclay nanomotors possess several benefits including low cost, ease of fabrication, and potential for large-scale manufacturing of the nanomotors. In addition, the nanomotors have demonstrated appropriate adsorption performance and capabilities for the removal of heavy metals.

10.3.8.2 Wastewater Treatment

Wastewater treatment is crucial and needs urgent attention. With technological advancement, the methods of wastewater treatment have been evolving continuously. The application of organoclays in wastewater treatment has become a common modality in industry today. The application of organoclays in wastewater treatment works in line with the existing unit operational processes in a synergistic way. The preexisting unit processes include reverse osmosis, granular-activated charcoal, and air strippers, which were further augmented by the use of organoclays.

Granular-activated carbon is predominantly operative and efficient in terms of removing a large range of organic molecules from water. However, the same technique is not as efficient in eliminating large molecules such as humic acid as well as emulsified oil and grease. In this context, organoclays have recognized to be the technology of choice for treating oily wastewaters [89, 90]. Conventional techniques of wastewater treatment failed to remove humic acid which is one of the most common contaminants in potable water. Therefore, the use of organoclays has proven to be very worthwhile, superior as well as efficient for many water treatment applications. The mechanism of organoclays as a wastewater treatment modality depends on a partitioning phenomenon and found to possess a synergistic effect with activated carbon and other unit processes such as reverse osmosis. Another important application of organoclays includes the trihalomethane control in drinking water. With the growing concern about the carcinogenic effects of trihalomethanes, the organoclays find it way to industrial commercialization [90].

10.3.9 Nanoclays and Agricultural Sustainable Development

Nanotechnology renders numerous likely benefits in agriculture sector such as reduction of agricultural inputs, augmentation of food quality and safety, and enrichment of absorbing nanoscale nutrients from the soil. Agriculture, natural resources, and food are a part of those challenges such as susceptibility, sustainability, healthy life, and human health. Water quality and pesticide residues or limits are very important for sustainable development of agriculture [91], which can be monitored and subsequently controlled using nanotechnology applications [92]. Nanomaterials are special and have diverse assets for which it is not possible yet to evaluate every possibility of these nanomaterials including the possible health risks as well as environmental risk [91]. Nanotoxicity assessment is crucial and becomes a growing area of research as the nanomaterials or nanoparticles possess very different properties. In the agrifood sector, there are growing applications of biosensors, nanotubes, fullerenes, controlled delivery systems, nanofiltration, etc. [93].

Application of nanotechnology finds its way in drug delivery in plants and soil fertility maintenance and further being evaluated progressively in the use of biomass and agricultural waste as well as in food processing and food packaging system [94]. Nanodetection technology as nanosensors further pushed the agricultural sector a step forward, and the electrochemical nanosensors, electronic nose, electronic tongue, optical nanosensors, and nanobarcode technology and wireless nanosensors have revolutionized the sensing in food and agriculture sectors with multiplex and real-time sensing capabilities [93, 95]. Nanoclays help in the degradation of waste and toxic materials by acting as a catalyst. Further, it can aid the soil microbes to act faster and efficiently in degradation of waste and toxic products. This whole concept of utilization and enhancement of soil microorganism for benefit is generally discussed as bioremediation (beneficial microbes). Similarly, phytoremediation (beneficial plants) and mycoremediation (beneficial fungi and mushrooms) terms are used accordingly. With bioremediation, microorganism can be utilized proficiently to remove heavy metals from soil and water in an environmentally friendly manner [96]. Therefore, the agricultural bioremediation aids in sustainable remediation skills to resolve and restore the natural condition of the soil [93, 96].

10.3.9.1 Nanofertilizers

Nanofertilizers are developing and appearing in the fields in recent decade. These may contain nanozinc, silica, iron and titanium dioxide, InP/ZnS core–shell QDs (quantum dots), ZnCdSe/ZnS core–shell QDs, Mn/ZnSe QDs, gold nanorods, core–shell QDs, etc. Many studies were performed and data reported regarding the uptake, biological fate, and toxicity of several metal oxide NPs, viz. Al_2O_3, TiO_2, CeO_2, FeO, and ZnONPs [97, 98]. A plethora of research indicated the deficiency of zinc as one of the key problem issues in limiting agricultural productivity in the alkaline nature of soils [99]. Using the nanofertilizers, this issue has been addressed effortlessly along with many other problem issues of similar kind.

10.3.10 Nanoclays for Reducing the Radioactive Contamination

Radioactive contamination of the environment is a potential threat seriously affecting the health and life on earth. The possibility of contamination may arise due to nuclear reactor leakage or processing of uranium or other radioactive isotopes. The probable mechanism leading to widespread contamination involves leaching of the radioactive material into the ground water, which can further lead to the contamination of the drinking water bodies of large populated areas. Hence, it is of utmost importance to fabricate materials or technologies which can efficiently remove the radioactive ions by selectively and irreversibly trapping the radioactive ions from environment, waste waters, or other contaminated sources. Natural inorganic cation-exchange materials, such as clays and zeolites, have been broadly studied and used in the removal of radioactive ions from water efficiently. These nanominerals remove the ions by ion-exchange mechanisms and are afterward disposed of in a safe way [100].

In a recent research, researchers also described the use of titanate nanofibers as intelligent absorbents for the removal of radioactive ions from waste material or water. Furthermore, the titanate nanofiber can selectively remove the radioactive ions in the presence of plenteous competitive ions. These nanofibers are reported to be very efficient and worthy from all the possible angles as required. The unique properties of these titanate nanofibers include their ability to trap radioactive ion efficiently, selectively in the presence of other ions and irreversibly.

Research also demonstrated the low fabrication cost of the nanofibers as compared to other candidates. Moreover, these fibers can be easily dispersed into the solutions and can be separated from any liquid after the absorption simply by filtration, sedimentation, or centrifugation due to their fibril morphology [100].

10.4 Conclusions and Prospects

The evolution and fast development of nanotechnology enthused modulation of properties and potential applications of natural nanoclays including MMT, kaolinite, halloysite, rectorite, and allophane. Organoclays further extended the applicability of these nanoclays in numerous industrial areas including cosmetics, grease, paint, and paper and as adsorbents of organic contaminants. However, the single most significant application of organoclays over the recent years is in the formulation and synthesis of polymer–clay nanocomposites, which further find its usage in various fields including biomedical applications such as drug delivery, tissue engineering, bone cementing, wound healing/dressing, enzyme immobilizations, food packaging industry, in waste water management, and radioactive decontamination. In addition, the applicability of nanoclays or nanominerals in agriculture sector is enormous and growing.

This chapter reviewed and suggested that natural nanoclays or modified nanoclays have a great potential for industrial and environmental applications. However, the number of researches in terms of published literature were found to be scanty. Another important aspect which needs to be addressed includes the nanotoxicity assessment of the nanoclays to extensively evaluate the potential biological as well as environmental risks. More research is required to improve our understanding of the surface properties, structural alignments, and reactivity of these nanoclays or nanominerals and the mechanisms underlying their interactions with external solutes.

References

1 Bordes, P., Pollet, E., and Avérous, L. (2009). Nano-biocomposites: biodegradable polyester/nanoclay systems. *Progress in Polymer Science* **34** (2): 125–155.

2 Jahromi, S.G. and Khodaii, A. (2009). Effects of nanoclay on rheological properties of bitumen binder. *Construction Building Materials* **23** (8): 2894–2904.

3 Koosha, M. and Hamedi, S. (2019). Intelligent Chitosan/PVA nanocomposite films containing black carrot anthocyanin and bentonite nanoclays with improved mechanical, thermal and antibacterial properties. *Progress in Organic Coatings* **127**: 338–347.

4 Vipulanandan, C. and Mohammed, A. (2015). Effect of nanoclay on the electrical resistivity and rheological properties of smart and sensing bentonite drilling muds. *Journal of Petroleum Science Engineering* **130**: 86–95.

5 Derakhshani, E. and Naghizadeh, A. (2018). Optimization of humic acid removal by adsorption onto bentonite and montmorillonite nanoparticles. *Journal of Molecular Liquids* **259**: 76–81.

6 Ma, J., Xu, J., Ren, J.-H. et al. (2003). A new approach to polymer/montmorillonite nanocomposites. *Polymer* **44** (16): 4619–4624.

7 Bonifacio, M.A., Gentile, P., Ferreira, A.M. et al. (2017). Insight into halloysite nanotubes-loaded gellan gum hydrogels for soft tissue engineering applications. *Carbohydrate Polymers* **163**: 280–291.

8 Lvov, Y.M., DeVilliers, M.M., and Fakhrullin, R.F. (2016). The application of halloysite tubule nanoclay in drug delivery. *Expert Opinion on Drug Delivery* **13** (7): 977–986. https://doi.org/10.1517/17425247.2016.1169271.

9 Li, P.-R., Wei, J.-C., Chiu, Y.-F. et al. (2010). Evaluation on cytotoxicity and genotoxicity of the exfoliated silicate nanoclay. *ACS Applied Materials & Interfaces* **2** (6): 1608–1613. https://doi.org/10.1021/am1001162.

10 Hetzer, M. and De Kee, D. (2008). Wood/polymer/nanoclay composites, environmentally friendly sustainable technology: a review. *Chemical Engineering Research and Design* **86** (10): 1083–1093. https://doi.org/10.1016/j.cherd.2008.05.003.

11 Brandelli, A. (2018). Toxicity and Safety Evaluation of Nanoclays. In: *Nanomaterials: Ecotoxicity, Safety, and Public Perception* (eds. M. Rai and J.K. Biswas), 57–76. Cham: Springer International Publishing.

12 Vergaro, V., Abdullayev, E., Lvov, Y.M. et al. (2010). Cytocompatibility and uptake of halloysite clay

nanotubes. *Biomacromolecules* **11** (3): 820–826. https://doi.org/10.1021/bm9014446.

13 Toledano-Magaña, Y., Flores-Santos, L., Montes de Oca, G. et al. (2015). Effect of clinoptilolite and sepiolite nanoclays on human and parasitic highly phagocytic cells. *BioMed Research International* 164980 (2015) https://doi.org/10.1155/2015/164980.

14 Lordan, S., Kennedy, J.E., and Higginbotham, C.L. (2011). Cytotoxic effects induced by unmodified and organically modified nanoclays in the human hepatic HepG2 cell line. *Journal of Applied Toxicology* **31** (1): 27–35. https://doi.org/10.1002/jat.1564.

15 Michael, F.M., Khalid, M., Walvekar, R. et al. (2016). Effect of nanofillers on the physico-mechanical properties of load bearing bone implants. *Materials Science Engineering: C* **67**: 792–806.

16 Bankoff, A.D.P. (2012). Biomechanical characteristics of the bone. *Human Musculoskeletal Biomechanics* **61**: 86.

17 Al Thaher, Y., Perni, S., and Prokopovich, P. (2017). Nano-carrier based drug delivery systems for sustained antimicrobial agent release from orthopaedic cementous material. *Advances in Colloid Interface Science* **249**: 234–247.

18 Kapusetti, G., Misra, N., Singh, V. et al. (2014). Bone cement based nanohybrid as a super biomaterial for bone healing. *Journal of Materials Chemistry B* **2** (25): 3984–3997.

19 Wang, C.X. and Tong, J. (2008). Interfacial strength of novel PMMA/HA/nanoclay bone cement. *Bio-Medical Materials and Engineering* **18**: 367–375. https://doi.org/10.3233/BME-2008-0553.

20 Ambre, A.H., Katti, K.S., and Katti, D.R. (2010). Nanoclay based composite scaffolds for bone tissue engineering applications. *Journal of Nanotechnology in Engineering Medicine* **1** (3): 031013.

21 Olad, A. and Azhar, F.F. (2014). The synergetic effect of bioactive ceramic and nanoclay on the properties of chitosan–gelatin/nanohydroxyapatite–montmorillonite scaffold for bone tissue engineering. *Ceramics International* **40** (7): 10061–10072.

22 De Silva, R.T., Pasbakhsh, P., Goh, K.-L. et al. (2013). Physico-chemical characterisation of chitosan/halloysite composite membranes. *Polymer Testing* **32** (2): 265–271.

23 Nitya, G., Nair, G.T., Mony, U. et al. (2012). in vitro evaluation of electrospun PCL/nanoclay composite scaffold for bone tissue engineering. *Journal of Materials Science: Materials in Medicine* **23** (7): 1749–1761.

24 Zhou, W.Y., Guo, B., Liu, M. et al. (2010). Poly (vinyl alcohol)/halloysite nanotubes bionanocomposite films: Properties and in vitro osteoblasts and fibroblasts response. *Journal of Biomedical Materials Research Part A* **93** (4): 1574–1587.

25 Katti, K.S., Katti, D.R., and Dash, R. (2008). Synthesis and characterization of a novel chitosan/montmorillonite/hydroxyapatite nanocomposite for bone tissue engineering. *Biomedical Materials* **3** (3): 034122.

26 Aliabadi, M., Dastjerdi, R., and Kabiri, K. (2013). HTCC-modified nanoclay for tissue engineering applications: a synergistic cell growth and antibacterial efficiency. *BioMed Research International* 2013.

27 Payne, S.A., Katti, D.R., and Katti, K.S. (2016). Probing electronic structure of biomineralized hydroxyapatite inside nanoclay galleries. *Micron* **90**: 78–86.

28 Ambre, A.H., Katti, D.R., and Katti, K.S. (2013). Nanoclays mediate stem cell differentiation and mineralized ECM formation on biopolymer scaffolds. *Journal of Biomedical Materials Research Part A* **101** (9): 2644–2660.

29 Noori, S., Kokabi, M., and Hassan, Z. (2015). Nanoclay enhanced the mechanical properties of poly (vinyl alcohol)/chitosan/montmorillonite nanocomposite hydrogel as wound dressing. *Procedia Materials Science* **11**: 152–156.

30 Sandri, G., Aguzzi, C., Rossi, S. et al. (2017). Halloysite and chitosan oligosaccharide nanocomposite for wound healing. *Acta Biomaterialia* **57**: 216–224.

31 Liu, M., Shen, Y., Ao, P. et al. (2014). The improvement of hemostatic and wound healing property of chitosan by halloysite nanotubes. *RSC Advances* **4** (45): 23540–23553.

32 Heydary, H.A., Karamian, E., Poorazizi, E. et al. (2015). A novel nano-fiber of Iranian gum tragacanth-polyvinyl alcohol/nanoclay composite for wound healing applications. *Procedia Materials Science* **11**: 176–182.

33 Yang, C., Xue, R., Zhang, Q. et al. (2017). Nanoclay cross-linked semi-IPN silk sericin/poly (NIPAm/LMSH) nanocomposite hydrogel: an outstanding antibacterial wound dressing. *Materials Science Engineering: C* **81**: 303–313.

34 Sabaa, M.W., Abdallah, H.M., Mohamed, N.A., and Mohamed, R.R. (2015). Synthesis, characterization and application of biodegradable crosslinked carboxymethyl chitosan/poly (vinyl alcohol) clay nanocomposites. *Materials Science Engineering: C* **56**: 363–373.

35 Nistor, M.T., Vasile, C., and Chiriac, A.P. (2015). Hybrid collagen-based hydrogels with embedded montmorillonite nanoparticles. *Materials Science Engineering: C* **53**: 212–221.

36 Pacelli, S., Paolicelli, P., Moretti, G. et al. (2016). Gellan gum methacrylate and laponite as an innovative nanocomposite hydrogel for biomedical applications. *European Polymer Journal* **77**: 114–123.

37 Demirci, S., Suner, S.S., Sahiner, M., and Sahiner, N. (2017). Superporous hyaluronic acid cryogel composites embedding synthetic polyethyleneimine microgels and Halloysite Nanotubes as natural clay. *European Polymer Journal* **93**: 775–784. https://doi.org/10.1016/j.eurpolymj.2017.04.022.

38 Suner, S.S., Demirci, S., Yetiskin, B. et al. (2019). Cryogel composites based on hyaluronic acid and halloysite nanotubes as scaffold for tissue engineering. *International Journal of Biological Macromolecules* **130**: 627–635. https://doi.org/10.1016/j.ijbiomac.2019.03.025.

39 Gomes-Ruffi, C.R., Henrique da Cunha, R., Almeida, E.L. et al. (2012). Effect of the emulsifier sodium stearoyl lactylate and of the enzyme maltogenic amylase on the quality of pan bread during storage. *LWT – Food Science and Technology* **49**: 96–101.

40 Tzialla, A.A., Pavlidis, I.V., Felicissimo, M.P. et al. (2010). Lipase immobilization on smectite nanoclays: Characterization and application to the epoxidation of α-pinene. *Bioresource Technology* **101** (6): 1587–1594. https://doi.org/10.1016/j.biortech.2009.10.023.

41 Menezes-Blackburn, D., Jorquera, M., Gianfreda, L. et al. (2011). Activity stabilization of Aspergillus niger and Escherichia coli phytases immobilized on allophanic synthetic compounds and montmorillonite nanoclays. *Bioresource Technology* **102** (20): 9360–9367.

42 Bugatti, V., Sorrentino, A., and Gorrasi, G. (2017). Encapsulation of lysozyme into halloysite nanotubes and dispersion in PLA: structural and physical properties and controlled release analysis. *European Polymer Journal* **93**: 495–506.

43 Acevedo, F., Pizzul, L., Castillo, M.D. et al. (2010). Degradation of polycyclic aromatic hydrocarbons by free and nanoclay-immobilized manganese peroxidase from Anthracophyllum discolor. *Chemosphere* **80** (3): 271–278. https://doi.org/10.1016/j.chemosphere.2010.04.022.

44 Das, R., Pattanayak, A.J., and Swain, S.K. (2018). 7 - Polymer nanocomposites for sensor devices. In: *Polymer-based Nanocomposites for Energy and Environmental Applications* (eds. M. Jawaid and M.M. Khan), 205–218. Woodhead Publishing.

45 Roozbahani, M., Kharaziha, M., and Emadi, R. (2017). pH sensitive dexamethasone encapsulated laponite nanoplatelets: Release mechanism and cytotoxicity. *International Journal of Pharmaceutics* **518** (1–2): 312–319.

46 Wang, X., Du, Y., Luo, J. et al. (2007). Chitosan/organic rectorite nanocomposite films: Structure, characteristic and drug delivery behaviour. *Carbohydrate Polymers* **69** (1): 41–49.

47 Zeynabad, F.B., Salehi, R., and Mahkam, M. (2017). Design of pH-responsive antimicrobial nanocomposite as dual drug delivery system for tumor therapy. *Applied Clay Science* **141**: 23–35.

48 Mura, P., Maestrelli, F., Aguzzi, C., and Viseras, C. (2016). Hybrid systems based on "drug–in cyclodextrin–in nanoclays" for improving oxaprozin dissolution properties. *International Journal of Pharmaceutics* **509** (1-2): 8–15.

49 Liu, K.-H., Liu, T.-Y., Chen, S.-Y., and Liu, D.-M. (2008). Drug release behavior of chitosan–montmorillonite nanocomposite hydrogels following electrostimulation. *Acta Biomaterialia* **4** (4): 1038–1045.

50 Campbell, K., Qi, S., Craig, D.Q., and McNally, T. (2009). Paracetamol-loaded poly (ε-caprolactone) layered silicate nanocomposites prepared using hot-melt extrusion. *Journal of Pharmaceutical Sciences* **98** (12): 4831–4843.

51 Campbell, K., Craig, D.Q., and McNally, T. (2008). Poly (ethylene glycol) layered silicate nanocomposites for retarded drug release prepared by hot-melt extrusion. *International Journal of Pharmaceutics* **363** (1-2): 126–131.

52 Lvov, Y. and Abdullayev, E. (2013). Functional polymer–clay nanotube composites with sustained release of chemical agents. *Progress in Polymer Science* **38** (10): 1690–1719. https://doi.org/10.1016/j.progpolymsci.2013.05.009.

53 Fakhrullina, G.I., Akhatova, F.S., Lvov, Y.M., and Fakhrullin, R.F. (2015). Toxicity of halloysite clay nanotubes in vivo: a *Caenorhabditis elegans* study. *Environmental Science: Nano* **2** (1): 54–59. https://doi.org/10.1039/C4EN00135D.

54 Levis, S.R. and Deasy, P.B. (2002). Characterisation of halloysite for use as a microtubular drug delivery system. *International Journal of Pharmaceutics* **243** (1): 125–134. https://doi.org/10.1016/S0378-5173(02)00274-0.

55 Jiang, W.-T., Chang, P.-H., Tsai, Y., and Li, Z. (2016). Halloysite nanotubes as a carrier for the uptake of selected pharmaceuticals. *Microporous and Mesoporous Materials* **220**: 298–307. https://doi.org/10.1016/j.micromeso.2015.09.011.

56 Tu, J., Cao, Z., Jing, Y. et al. (2013). Halloysite nanotube nanocomposite hydrogels with tunable mechanical properties and drug release behavior. *Composites Science and Technology* **85**: 126–130. https://doi.org/10.1016/j.compscitech.2013.06.011.

57 Qi, R., Guo, R., Zheng, F. et al. (2013). Controlled release and antibacterial activity of antibiotic-loaded electrospun halloysite/poly(lactic-co-glycolic acid) composite nanofibers. *Colloids and Surfaces B: Biointerfaces* **110**: 148–155. https://doi.org/10.1016/j.colsurfb.2013.04.036.

58 Karnik, S., Jammalamadaka, U.M., Tappa, K.K. et al. (2016). Performance evaluation of nanoclay enriched anti-microbial hydrogels for biomedical applications. *Heliyon* **2** (2): e00072. https://doi.org/10.1016/j.heliyon.2016.e00072.

59 Dzamukova, M.R., Naumenko, E.A., Lvov, Y.M., and Fakhrullin, R.F. (2015). Enzyme-activated intracellular drug delivery with tubule clay nanoformulation. *Scientific Reports* **5** (1): 10560. https://doi.org/10.1038/srep10560.

60 Fizir, M., Dramou, P., Zhang, K. et al. (2017). Polymer grafted-magnetic halloysite nanotube for controlled and sustained release of cationic drug. *Journal of Colloid and Interface Science* **505**: 476–488. https://doi.org/10.1016/j.jcis.2017.04.011.

61 Rao, K.M., Nagappan, S., Seo, D.J., and Ha, C.-S. (2014). pH sensitive halloysite-sodium hyaluronate/poly(hydroxyethyl methacrylate) nanocomposites for colon cancer drug delivery. *Applied Clay Science* **97-98**: 33–42. https://doi.org/10.1016/j.clay.2014.06.002.

62 Harikrishnan, S., Sedev, R., Beh, C.C. et al. (2020). Loading of 5-fluorouracil onto Halloysite nanotubes for targeted drug delivery using a subcritical gas antisolvent process (GAS). *The Journal of Supercritical Fluids* **159**: 104756. https://doi.org/10.1016/j.supflu.2020.104756.

63 Campbell, K.T., Craig, D.Q.M., and McNally, T. (2010). Ibuprofen-loaded poly(ε-caprolactone) layered silicate nanocomposites prepared by hot melt extrusion. *Journal of Materials Science: Materials in Medicine* **21** (8): 2307–2316. https://doi.org/10.1007/s10856-009-3963-2.

64 Katti, D.R., Sharma, A., Ambre, A.H., and Katti, K.S. (2015). Molecular interactions in biomineralized hydroxyapatite amino acid modified nanoclay: in silico design of bone biomaterials. *Materials Science and Engineering: C* **46**: 207–217. https://doi.org/10.1016/j.msec.2014.07.057.

65 Abedini, R. (2019). Enhanced antifouling properties of poly(ethersulfone) nano-composite membrane filled with nano-clay particles. *Polymer Bulletin* **76** (4): 1737–1753. https://doi.org/10.1007/s00289-018-2464-1.

66 Batasheva, S., Kryuchkova, M., Fakhrullin, R. et al. (2020). Facile fabrication of natural polyelectrolyte-nanoclay composites: halloysite nanotubes, nucleotides and DNA study. *Molecules* **25** (15): 3557.

67 Park, J.K., Choy, Y.B., Oh, J.-M. et al. (2008). Controlled release of donepezil intercalated in smectite clays. *International Journal of Pharmaceutics* **359** (1): 198–204. https://doi.org/10.1016/j.ijpharm.2008.04.012.

68 Liu, Y., Wang, Q., Lu, Y. et al. (2020). Synergistic enhancement of cytotoxicity against cancer cells by incorporation of rectorite into the paclitaxel immobilized cellulose acetate nanofibers. *International Journal of Biological Macromolecules* **152**: 672–680. https://doi.org/10.1016/j.ijbiomac.2020.02.184.

69 Hou, G., Gao, Y., Xie, Y. et al. (2018). Thermosensitive hybrid system based on pluronic F127 and nanoclay laponite for sustained local release of lidocaine hydrochloride. *Journal of Materials and Applications* **7** (1): 41–46.

70 Xiao, S., Castro, R., Maciel, D. et al. (2016). Fine tuning of the pH-sensitivity of laponite–doxorubicin nanohybrids by polyelectrolyte multilayer coating. *Materials Science and Engineering: C* **60**: 348–356. https://doi.org/10.1016/j.msec.2015.11.051.

71 Silva, D.T.C., Arruda, I.E.S., França, L.M. et al. (2019). Tamoxifen/montmorillonite system – effect of the experimental conditions. *Applied Clay Science* **180**: 105142. https://doi.org/10.1016/j.clay.2019.105142.

72 Anirudhan, T.S. and Parvathy, J. (2015). Preparation and characterization of thiolated chitosan immobilized-amino acid intercalated montmorillonite for the controlled release of timolol maleate, a β-adrenergic blocking agent. *International Journal of Polymeric Materials and Polymeric Biomaterials* **64** (1): 25–31. https://doi.org/10.1080/00914037.2014.886240.

73 Kevadiya, B.D., Patel, T.K., Patel, P.B. et al. (2013). Montmorillonite poly-l-lactide microcomposites of procainamide for controlled drug delivery: in vitro and in vivo evaluation. *Indian Journal of Pharmaceutical Sciences* **75** (6): 722–726.

74 Joshi, G.V., Kevadiya, B.D., and Bajaj, H.C. (2010). Controlled release formulation of ranitidine-containing montmorillonite and Eudragit® E-100. *Drug Development and Industrial Pharmacy* **36** (9): 1046–1053. https://doi.org/10.3109/03639041003642073.

75 Bumbudsanpharoke, N. and Ko, S. (2019). Nanoclays in food and beverage packaging. *Journal of Nanomaterials* 2019: 8927167. https://doi.org/10.1155/2019/8927167.

76 Brody, A.L., Bugusu, B., Han, J.H. et al. (2008). Innovative food packaging solutions. Scientific status summary. *Journal of Food Science* 73 (8): 1750–3841.

77 Biswas, B., Warr, L.N., Hilder, E.F. et al. (2019). Biocompatible functionalisation of nanoclays for improved environmental remediation. *Chemical Society Reviews* **48** (14): 3740–3770. https://doi.org/10.1039/C8CS01019F.

78 Wu, Q., Que, Z., Li, Z. et al. (2018). Photodegradation of ciprofloxacin adsorbed in the intracrystalline space of montmorillonite. *Journal of Hazardous Materials* **359**: 414–420. https://doi.org/10.1016/j.jhazmat.2018.07.041.

79 Brigatti, M.F., Galan, E., and Theng, B. (2013). Structure and mineralogy of clay minerals. In: *Developments in Clay Science*, vol. 5, 21–81. Elsevier.

80 Biswas, B., Qi, F., Biswas, J.K. et al. (2018). The fate of chemical pollutants with soil properties and processes in the climate change paradigm—a review. *Soil Systems* **2** (3): 51.

81 Sarkar, B., Rusmin, R., Ugochukwu, U.C., Mukhopadhyay, R., Manjaiah, K.M.: Modified clay minerals for environmental applications. In: *Modified*

Clay and Zeolite Nanocomposite Materials. pp. 113–127. Elsevier, (2019)

82 Olkowska, E., Polkowska, Z., and Namiesnik, J. (2011). Analytics of surfactants in the environment: problems and challenges. *Chemical Reviews* **111** (9): 5667–5700.

83 Biswas, J.K. and Rai, M. (2018). *Nanomaterials: Ecotoxicity, Safety, and Public Perception*. Springer International Publishing: Imprint: Springer.

84 Laridon, Y., Touchaleaume, F., Gontard, N., and Peyron, S. (2020). Food-grade PE recycling: effect of nanoclays on the decontamination efficacy. *Polymers (Basel)* **12** (4) https://doi.org/10.3390/polym12040822.

85 Sumantha, A., Larroche, C., and Pandey, A. (2006). Microbiology and industrial biotechnology of food-grade proteases: a perspective. *Food Technology and Biotechnology* **44**: 211–220.

86 Lazzara, G., Massaro, M., and Riela, S. (2018). Current status of nanoclay phytotoxicity. In: *Phytotoxicity of Nanoparticles*, 151–174. Springer.

87 Wang, J. (2013). *Nanomachines: Fundamentals and Applications*. Wiley.

88 Maric, T., Mayorga-Martinez, C.C., Khezri, B. et al. (2018). Nanorobots constructed from nanoclay: using nature to create self-propelled autonomous nanomachines. *Advanced Functional Materials* **28** (40): 1802762.

89 Kumar, P., Jasra, R.V., and Bhat, T.S.G. (1995). Evolution of porosity and surface acidity in montmorillonite clay on acid activation. *Industrial & Engineering Chemistry Research* **34** (4): 1440–1448. https://doi.org/10.1021/ie00043a053.

90 Beall, G.W. (2003). The use of organo-clays in water treatment. *Applied Clay Science* **24** (1): 11–20. https://doi.org/10.1016/j.clay.2003.07.006.

91 Prasad, R., Kumar, V., and Prasad, K.S. (2014). Nanotechnology in sustainable agriculture: present concerns and future aspects. *African Journal of Biotechnology* **13** (6): 705–713.

92 Mukhopadhyay, S.S. (2014). Nanotechnology in agriculture: prospects and constraints. *Nanotechnology, Science and Applications* **7**: 63–71. https://doi.org/10.2147/NSA.S39409.

93 Ion, A.C., Ion, I., and Culetu, A. (2010). Carbon-based nanomaterials: environmental applications. *University Politehnica of Bucharest* **38**: 129–132.

94 Floros, J.D., Newsome, R., Fisher, W. et al. (2010). Feeding the world today and tomorrow: the importance of food science and technology. *Comprehensive Reviews in Food Science and Food Safety* **9**: 572–599. https://doi.org/10.1111/j.1541-4337.2010.00127.x.

95 Kaushal, M. and Wani, S. (2017). Nanosensors: frontiers in precision agriculture. *Nanotechnology*: 279–291. https://doi.org/10.1007/978-981-10-4573-8_13.

96 Dixit, R., Malaviya, D., Pandiyan, K. et al. (2015). Bioremediation of heavy metals from soil and aquatic environment: an overview of principles and criteria of fundamental processes. *Sustainability* **7** (2): 2189–2212.

97 Dimkpa, C.O. (2014). Can nanotechnology deliver the promised benefits without negatively impacting soil microbial life? *Journal of Basic Microbiology* **54**: 889–904. https://doi.org/10.1002/jobm.201400298.

98 Nayak, A.P., Yuan, Z., Cao, B. et al. (2015). Pressure-modulated conductivity, carrier density, and mobility of multilayered tungsten disulfide. *ACS Nano* **9** (9): 9117–9223.

99 Sadeghzadeh, B. (2013). A review of zinc nutrition and plant breeding. *Journal of Soil Science and Plant Nutrition* **13**: 905–927. https://doi.org/10.4067/S0718-95162013005000072.

100 Yang, D., Zheng, Z., Liu, H. et al. (2008). Layered titanate nanofibers as efficient adsorbents for removal of toxic radioactive and heavy metal ions from water. *The Journal of Physical Chemistry C* **112** (42): 16275–16280. https://doi.org/10.1021/jp803826g.

11

Eco-friendly, Biodegradable, and Biocompatible Electrospun Nanofiber Membranes and Applications

Sylvia Thomas, Bianca Seufert, William Serrano-Garcia, Manopriya Devisetty, Ridita Khan, Kavyashree Puttananjegowda, and Norma Alcantar

University of South Florida (USF) College of Engineering, USF, Tampa, FL, USA

11.1 Introduction

Materials are imperative in sustaining the environment, the economy, human life, and society, and as new materials are explored for nanotechnology, there is a direct correlation to sustainability. Sustainable materials can be classified as eco-friendly, "green," biodegradable, biocompatible, or biobased, ranging from inorganic to organic materials, and are used to produce a vast array of nanostructures (fibers, tubes, spheres, wires, particles, ribbons, coaxials/multiaxials, membranes/meshes, etc.) for the development of transformative technology. Over the past several years, breakthroughs in polymeric nanofiber membrane (PNM) technology have gained substantial interest by industry, government, and academia for the need of clean energy, energy harvesting, point-of-care testing platforms, clean water, disease prevention, electronic miniaturization, and the Internet-of-Things (IoT). PNM technology is leading the way for sustainable nanotechnology that is environmentally conscious, recyclable, renewable, biodegradable, biocompatible, and cost effective.

Electrospinning has emerged as the process of choice for the fabrication of PNM's technology due to versatility in producing 2D–3D patterns and manipulative morphologies; cost-effectiveness; enhanced functionality from large surface area and ultrasmall size; structural interconnectivity; and ability to use different materials and composites [1–5]. This technique offers some sustainable features, but more importantly, the electrospinning technique can utilize sustainable materials, particularly eco-friendly and biobased (biodegradable and biocompatible) polymers, to fabricate nanofiber membrane structures/devices that do not adversely affect the environment or human life.

This chapter provides an overview of eco- and biobased polymeric membrane materials and highlights the versatility of the electrospinning fabrication of nanofiber membranes for different applications. In addition, the chapter discusses different aspects of how nanofiber membrane technology has transformed solar cells and photovoltaics, batteries, the water-degenerative disease nexus, cancer treatment, tissue engineering, glucose monitoring, and the integration of miniaturized-flexible-wearable electronics, offering a future perspective on the technology transition to the IoT, inclusive of data mining and nanocomputing.

11.2 Fabrication of Electrospun Nanofiber-based Membranes

Nanofiber-based membranes have a variety of applications from basic filtration to the complex drug delivery systems and flexible electronic devices. *Electrospinning is a versatile approach to fabricating nanofiber-based membranes. This unique technique utilizes an electric force to overcome the surface tension of charged threads of various types of polymer solutions.* This allows for the use of sustainable, eco-friendly, and biobased polymers to create these nanofiber-based membranes for the fabrication of structures and devices. Moreover, electrospinning of inorganic polymers such as silica nanofibers can be prepared for benign applications such as drug delivery and wound dressing. In addition to the variety of compatible materials, electrospinning provides versatility in producing 2D and 3D structures for wound dressing and bone tissue engineering, respectively [6, 7]. The ability to create innovative structures with fine-tuned morphologies has many benefits. For instance,

Sustainable Nanotechnology: Strategies, Products, and Applications, First Edition.
Edited by Yashwant V. Pathak, Govindan Parayil, and Jayvadan K. Patel.

electrospinning has been widely used for the fabrication of tissue scaffolding; this technique offers structures that are flexible and promote cell migration. While electrospinning opens many avenues into exploring new medical applications such as completely organic-based drug delivery systems and biosensors, it also has nonmedical applications. Applications such as miniaturized, flexible, and wearable electronics, batteries, and solar cells have benefited by the electrospinning of sustainable materials. Furthermore, the simplicity, reliability, and overwhelming enhancements this technique can offer for the fabrication of nanofiber structures with sustainable materials are too great to disregard. Advancements in this technique could allow for development in cutting-edge medical devices and new age wearable technology. In the following sections, more details will be provided on electrospinning processes, advantages of nanofiber morphology, and the types of sustainable compatible materials electrospinning uses for a variety of applications.

11.2.1 Electrospinning of PNMs

Electrospinning is a fiber production process that has many possibilities to advance products in the sustainability realm. Historically, the process is modeled after the behavior of a water droplet and soap bubbles [8] under the influence of electrostatic forces, and eventually the behavior was explained by an electric charge mechanism of electrodynamic excitation on a dielectric, leading to the electrospinning technique [9–13]. The working principle is comparable to the electrospraying technique but diverging in the outcome by forming particles via atomization instead of fibers.

The electrospinning technique has the ability to produce a vast array of nanostructures such as nanofibers, nanotubes, nanoribbons, and coaxials/multiaxials, forming membranes and meshes. Advances in this area have led to increased industry interest and further application of this technology for manufacturing purposes.

11.2.1.1 Electrospinning Processes

The potential of nanotechnology can be appreciated extensively with the economical and efficient manufacturing of nanomaterials. Electrospinning is a cost-effective and adaptable technique that uses high voltage for the fabrication of fibers, mainly from a polymeric solution, with diameters ranging from tens of nanometers to several microns [4, 5]. The technique is scalable and can be tuned, offering excellent control over the dimension of the fabricated structure.

Two common classifications for electrospinning comprise solution electrospinning and melt electrospinning.

Solution electrospinning (Figure 11.1) uses polymeric solvent-based solutions. Here, the viscosity of the solution can be mainly controlled by the amount of solvent added to form the solution. These solutions are formed from a mixture of a solvent and polymer (powder, fibrils, pellet, or flake), and when the solution is enclosed in a syringe with a needle spinneret, fibers of diameters down to the nanometer range are produced. A high voltage is applied, which overcomes the surface tension of the polymeric solution, promoting the formation of the well-known Taylor cone. This occurs when the critical voltage for the formation of the Taylor cone, V_{tc} (kV), forms the conical-shape stream and helps for the elongation and outburst of a polymer droplet, causing electrostatic repulsive whipping to generate a fiber that is deposited onto a grounded collector plate [13–15]. Historically, the Taylor cone voltage, V_{tc}, was found using Eq. (1) [14]:

$$V_{tc}^2 = 4\frac{L^2}{h^2}\left(\ln\left(\frac{2h}{R}\right)-1.5\right)(0.117\pi R\gamma) \tag{1}$$

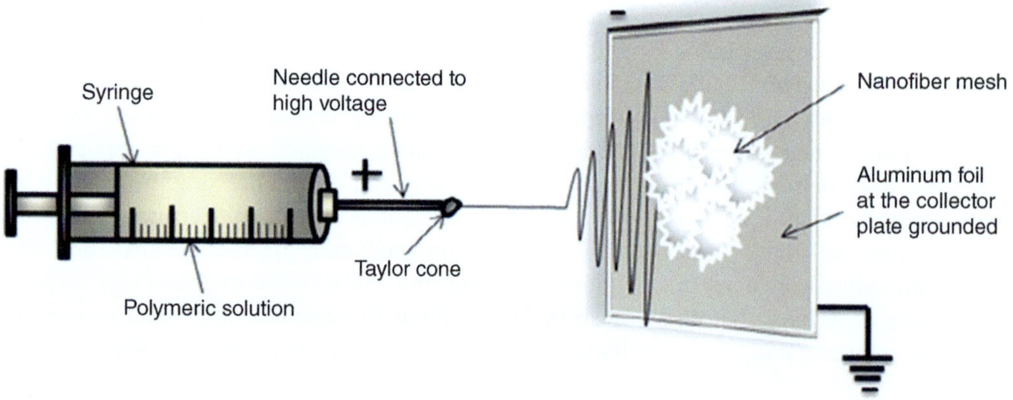

Figure 11.1 Schematic of horizontal solution electrospinning used for nanofiber fabrication.

where L is the distance (in centimeters) between the needle tip and the collector plate, h is the length (in centimeters) of the cylindrical needle, R is the needle inner radius (in centimeters), and γ is the surface tension of the solution (in dyn/cm, where $1000\,\text{dyn}\,\text{cm}^{-1} = 1\,\text{N}\,\text{m}^{-1}$).

In solution electrospinning, depending on the solvent of choice, there could be low solvent evaporation (removal) rates and elevated toxicity, which increases the cost of processing, requiring additional steps of solvent recycling or removal. In contrast, there are biobased polymers that can be dissolved using solvents that are eco-friendly (water, ethanol, D-limonene, etc.) and can be used to fabricate biobased nanofibers.

Melt electrospinning uses temperature to control polymer viscosity [16]. It can be viewed as a combination of additive manufacturing (e.g. 3D printing) and solution electrospinning. Melt electrospinning uses a polymeric melt solution to typically produce microfibers of an average diameter of 5–50 μm down to the submicron diameter [16–20].

Melt electrospinning is advantageous for 3D regenerative tissue scaffolds, wearable electronics, and sensing surfaces, where controlled porosity, alignment, and higher resolution are a priority. Melt electrospinning can be classified as a greener process, than solution electrospinning, as toxic solvents are not needed in the melt electrospinning process and solution whipping is not a dominant interaction. In this process, the melt electrospinning technique heats the spinneret containing the polymer (powder, pellet, flake, or filament) to the minimum temperature needed for molten polymer, and a fiber is formed using air pressure to direct the flow through a syringe. There is also the capability of direct writing a melt electrospun pattern using the technique of melt direct-write electrospinning, where a 3D pattern is formed using a movable *xyz* stage [20].

Both electrospinning techniques have been widely used, solution more than melt, and PNMs fabricated using these techniques will be discussed in this chapter.

11.2.1.2 Nanofiber Morphology

The electrospinning process can produce versatile nanofiber membranes in 2D and 3D patterns for manipulative morphology [6, 7, 18, 20–22]. This technique brings new opportunities to the assembly process of nanoscale structures by increasing the cost-effectiveness and structural interconnectivity. In addition, electrospun nanofiber structures can be layered to "print" in 2D and 3D patterns, scaffolds, or composites [6, 7, 23]. These patterns could be simple 1D membranes or 3D meshes with several layers. These patterns play key roles in filters and biosensors. Moving forward, the electrospinning nanostructure configurations and morphology has the potential to go beyond 2D or 3D patterns to the transformative 4D patterns which could revolutionize the way products are fabricated by industry today [23]. Further advancement in the creation of 4D transformative structures is theoretically possible. It allows for structures to be printed in their 3D form and then change form once placed in its intended environment.

Electrospun nanostructures are adaptable and used in various ways throughout industry. For example, nanocoaxial/multiaxial fibers are helping to achieve advanced drug delivery systems for patients suffering from cancer [24–27]. The uniqueness of multiaxial electrospinning is the spinning of more than one type of material to be used to generate a multilayered fiber. These fibers can then be woven into a patch that can be placed in a specific biological area where the drug will be delivered. Therefore, if each layer of the multiaxial fiber contained a cancer treatment drug woven into a patch, then doctors could administer the patch upon resection of patient's tissue. This technique will allow for targeted cancer treatment in one biological area instead of administering delivery into the bloodstream.

Furthermore, electrospinning allows for nanofiber morphologies to occur after printing. For instance, electrospun nanofibers made of one type of material can be sintered after spinning to decrease the pore size and enhance fiber-to-fiber bonding (fiber reinforced composite) [28]. A nanofiber matrix-like morphology is well suited for filtering since pores can be controlled, targeting specific dimensions for precise filtration. Another example is when the nanofibers are fabricated using multiple materials in a blended composition or a multiaxial array. In this case, specific molecules of polymers can be removed using orthogonal solvents, producing fibers with rougher topology or thinner fibers, respectively. Such morphologies can be fine-tuned to promote cell migration in tissue engineering.

Overall electrospinning is quite versatile with its vast array of compatible materials and wide range of nanostructure configurations. Using eco-friendly materials to spin different nanostructures such as nanotubes can push the boundaries of this technology for more sustainable applications.

Figure 11.2 offers various nanostructures of diverse morphologies and geometries including single, multiaxial, honeycomb, aligned, yarned, and patterned electrospun fibers and membranes [20, 29–41]. A direct write electrospinning process [16, 17, 19] offers another technique to create electrospun fibers with superior patterned morphological features.

Next-generation materials will be fabricated with the electrospinning technology to be used across many industries, not just medicine or computing hardware. This technology will continue to advance the future of bio and advanced manufacturing to meet grand challenges.

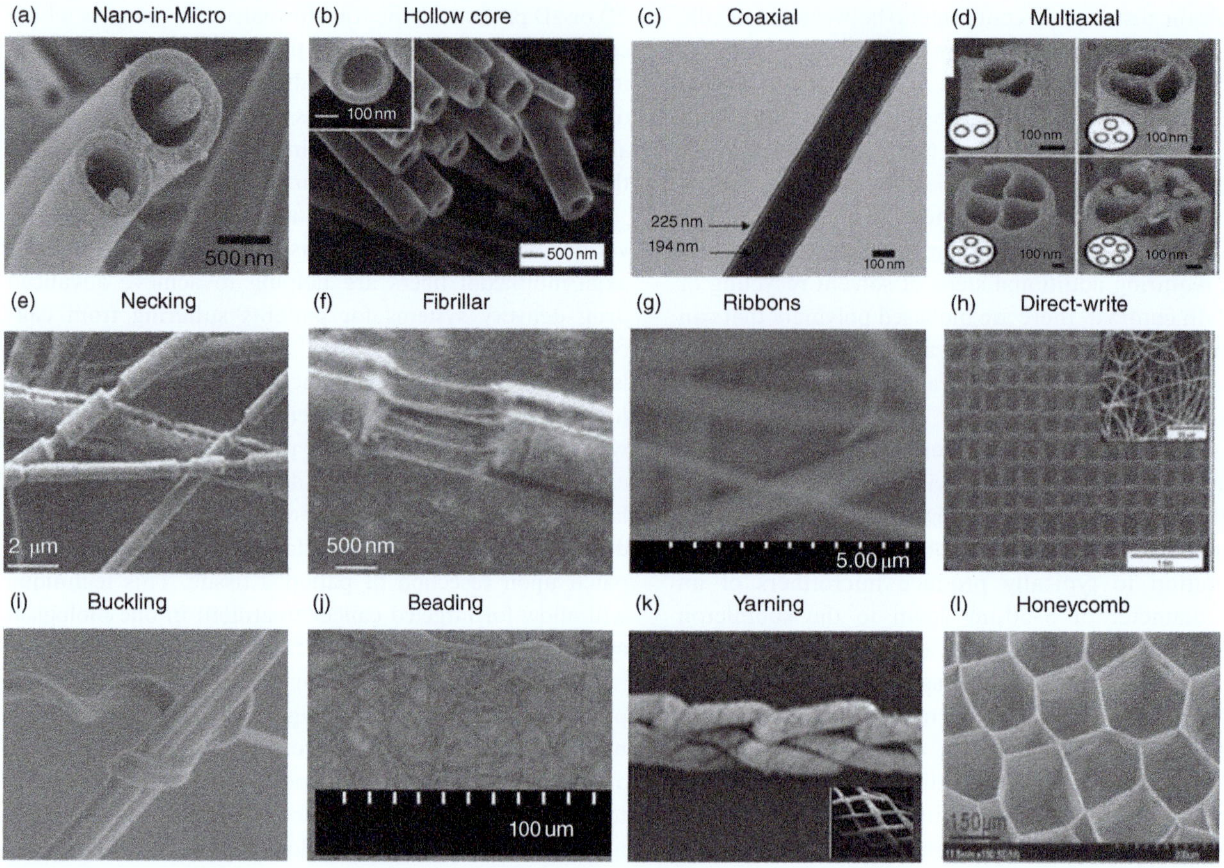

Figure 11.2 Scanning electron microscopy (SEM) of diverse nanostructures: (a) nanowire in microtube with hollow core [29]. *Source:* Chen et al. [29], figure 11.02(a) (p.008)/with permission of American Chemical Society; (b) hollow core/shell fibers [30]. *Source:* Li and Xia [30], figure 11.02(b) (p.008)/with permission ofAmerican Chemical Society; (c) coaxial core/shell nanofiber of P3HT/BBL (poly(benz imidazobenzophenanthroline)); (d) multiaxial core/shell nanofiber structures [31]. *Source:* Zhao et al. [31], figure 11.02(d) (p.008)/with permission ofAmerican Chemical Society; (e) fiber necking produced by extended stretching on electrospinning collector drum [32]. *Source:* Zussman et al. [32], figure 11.02(e) (p.008)/with permission of AIP Publishing LLC; (f) fibrillar section of stretched fiber after necking [32]. *Source:* Zussman et al. [32], figure 11.02(f) (p.008)/with permission of AIP Publishing LLC; (g) ribbon-like fibers; (h) patterned nanofibers from direct-write electrospinning [33]. *Source:* Lee et al. [33], figure 11.02(h) (p.008)/with permission of American Chemical Society; (i) buckling due to coiling of fibers [34]. *Source:* Tao and Shivkumar [34], figure 11.02(i) (p.008)/with permission of Elsevier; (j) thread-beading of nanofibers; (k) twisted fibers from electrospun yarning apparatus (inset of braided fiber) [35]. *Source:* Joseph et al. [35], figure 11.02(k) (p.008)/with permission ofAmerican Chemical Society; (l) nanofibers patterned in honeycomb-like structure [36]. *Source:* Yan et al. [36], figure 11.02(l) (p.008) / with permission of American Chemical Society.

11.2.2 Eco-friendly, Biocompatible, and Biodegradable Materials for Nanofiber Membranes

Sustainability is a critical feature when it comes to creating new products. The sustainability of PNM technology is driven primarily by the selection of materials used to synthesize and fabricate these membranes. The critical stipulation is that the technology must meet "the needs of the present, without compromising the ability of future generations to meet their needs or endangering natural systems (atmosphere, waters, soils, and living things) that support life on Earth" – the definition of sustainable development [42].

Sustainable materials, otherwise known as eco-friendly, "green," biodegradable, biocompatible, or biobased materials, have become increasingly important to many in industry and academia. Classes of materials for PNM technology span environmentally or eco-friendly polymers, biodegradable polymers, biocompatible polymers, and a diverse combination of each. Biodegradable polymers completely degrade or deteriorate, physically and chemically upon exposure to anaerobic, aerobic, and microorganism processes [43]. Biocompatible polymers maintain material integrity when in contact with a living system with no adverse effects for *in vitro* and *in vivo* applications [43].

To enhance the functional characteristics of the PNM, nanofillers, such as nanoparticles, nanofibers, nanotubes, and nanorods, can be embedded into or dispersed onto the fiber. More specifically, these diverse combinations capture blends of biodegradable and biocompatible polymers, copolymerization of biodegradable and biocompatible monomers, and/or sheath-core polymeric formations. By ensuring that a product can be created with sustainable materials and not pollute the environment, we are establishing a safer future for ourselves and society. Using local sustainably sourced materials, a secure and reliable supply chain of these materials is made possible. Exploration of these kinds of materials often leads to the discovery of common characteristics among them. For example, eco-friendly materials are often biodegradable, and materials that are biodegradable can also be biocompatible. These traits are not mutually exclusive.

Environmentally or eco-friendly polymers are often defined as materials that are nontoxic, biodegradable, renewable, and will not enhance CO_2 emissions. Polymers that are naturally biodegradable include chitosan/chitin, soy, proteins from plants and animals, starch, collagen, polysaccharides, and cellulose. These source materials are less toxic, cost effective, and abundant as compared to the commonly used synthetic biodegradable polymers, e.g. poly(glycolic acid) (PGA), poly(e-caprolactone) (PCL), poly(lactic acid) (PLA), and polyvinyl alcohol (PVA). To meet the enhanced properties of lower glass transition temperatures and selective permeability and higher mechanical strength, natural biobased polymers can be embedded with nanofillers into the polymeric solution (*in situ*) or on the surface (*ex situ*) of the nanofiber membrane. Strengthening these properties will enhance the sustainability of PNM for specific applications. For example, a PNM having a hydrophilic nature will expand or dissolve, depending on the humidity of the environment; this same PNM, when filled with a plant-based mucilage (polysaccharide), can act as a chemical filter for arsenic or *E. coli* [44]. Another example: some eco-friendly plant and animal proteins are only biodegradable and have great potential to be used in various health treatments, while other eco-friendly materials, such as aloe, are biodegradable and biocompatible [45]. Volatile organic compounds, such as chloroform and benzene, that adsorb by flexible polyacrylonitrile (PAN) composite membranes are biodegradable but not biocompatible [46]. Furthermore, eco-friendly materials often lead to products that are biodegradable, i.e. not harmful to the earth. The research advancements of these materials have shed light on the possibility of using these biodegradable materials for electronic devices. There is concern that electronic waste (E-waste) from electronic devices can cause serious environmental problems, such as the release of hazardous chemicals and toxic heavy metals. With these biodegradable materials, the creation of biodegradable insulators and conductors can drastically reduce the amount of E-waste generated from discarded electronics. It has been estimated that the world has 50 million metric tons of e-waste every year, and only 15–20% is recyclable, with 80–85% going to landfill or burned, where toxins and hazardous materials may leach into the soil and/or ground water [43].

Moreover, materials such as cellulose, shellac, PLA, and PCL are biodegradable and biocompatible, which means these materials can be used in electronic devices as well as in implantable medical devices. Because these materials are biodegradable, it enhances the application of electronic devices that can get absorbed by the surrounding tissue without an increased toxicity. Biodegradable and biocompatible materials are ideal for health treatments, such as treating burns, covering wounds, and delivering drugs. Due to the unique and often small-scale applications of these materials, the technique of electrospinning can provide many opportunities for product assembly.

Electrospinning, as a fiber production method, is compatible with eco-friendly, biocompatible, and biodegradable materials to create nanofiber membranes and various other nanofiber membrane technologies. Nanofiber membranes can be used in implantable medical devices such as biosensors used to monitor patient vitals or cancer drug delivery systems. Solar cells, photovoltaics, and batteries have all been impacted by the advancements of nanofiber membrane technology. Eco-friendly, biodegradable, and biocompatible materials highlight the versatility of the electrospinning process for the fabrication of nanofiber membranes.

With further advancement in this field, mass-scale manufacturing of electrospun products is possible. The use of eco-friendly, biodegradable, and biocompatible materials for large-scale manufacturing using electrospinning could be advantageous for large-scale bio and advanced additive manufacturing. Electrospinning brings many opportunities for large-scale development for a variety of different bio- or eco-based products that could revolutionize the health care industry, further securing the safety of future generations through the use of less-hazardous products consumed by the society today.

The electrospinning process uses sustainable, eco-friendly, and biobased polymers to develop sustainable nanotechnology for energy harvesting; filtration; health care sensing and detecting; flexible, wearable, and edible electronics; and drug delivery [47–54]. Eco-friendly and biobased materials are critical for the fabrication of nanofiber membrane structures/devices that can be placed in the human body or used in any environment, without

Table 11.1 Biobased polymers and materials that can be electrospun.

Material	Biocompatible	Biodegradable	Ecofriendly	Electrospun
Cellulose	√	√	√	√
Silk	√	√	√	√
Chitosan	√	√	√	√
Shellac	√	√	√	√
Poly(vinyl alcohol) PVA	√		√	√
Polyvinylpyrrolidone (PVP)	√	√	√	√
Poly(lactic acid) (PLA)	√	√	√	√
Polycaprolactone (PCL)	√	√	√	√
Poly(ethylene glycol) (PEG)	√		√	√
Poly(lactic-co-glycolic acid) (PLGA)	√	√	√	√
Polydimethylsiloxane (PDMS)	√		√	√
Albumen	√	√	√	√
Polyaniline (PANI)	√		√	√
Polypyrrole (PPy)	√		√	√
Poly(3,4-ethylenedioxythiophene) (PEDOT)	√		√	√
Poly(3,4-ethylenedioxythiophene):poly-(styrene sulfonate) (PEDOT:PSS)	√			√
Polyvinylidene fluoride (PVDF)	√			√

causing harm to the body or society. Table 11.1 presents a list of biobased polymers and materials that have been electrospun.

One example of one of the most popular polymers, which is biocompatible, biodegradable, and compostable, is PLA. PLA is a well-known thermoplastic polymer extracted from plant starches (i.e. sugar cane and corn) from its monomer and can also be found on the human body due to anaerobic respiration. While PLA has been extensively used in the food-packaging industry, it is also the most recognized polymer in 3D printing technologies, in conjunction with acrylonitrile butadiene styrene (ABS).

In conclusion, sustainable materials, also referred to as eco-friendly materials, are gaining a larger role in today's products because of their many benefits to the environment. These materials showcase the versatility of the electrospinning process. Using sustainable materials to form electrospun nanofibrous membranes has great potential to create a variety of new products. The benefits of having biodegradable and/or biocompatible characteristics have led to many discoveries for nanofiber membrane technology. From biosensors to biomanufacturing, nanofiber membrane technology is pushing the boundaries of current technology, ensuring a healthier and safer future.

11.3 Applications of Electrospun Nanofiber Membranes

Electrospun nanofiber membranes have become viable for various arenas, such as environmental water/air filtering; military warfare detection of chemical and biological contaminants; protective coatings for blocking/repelling chemical, bio, or electrical inputs; flexible and wearable electronics; alternative energy, regenerative processes; health care sensing; and drug delivery [55–60].

11.3.1 Nanoenergy Harvesting

PNMs have made contributions to the sustainability of renewable energy systems as battery separators, anti-reflective coatings (ARCs), cathodes, and anodes. Being able to provide a structure that has high mechanical strength, good electrolyte durability, and thermal stability for eco-efficient and environmentally friendly energy, devices can be accomplished using electrospun nanofiber membranes [6, 61].

Separators play an important role in the performance, safety, and sustainability of energy-storage systems. At elevated temperatures, separators with weak mechanical strength, insufficient electrolyte uptake, and poor thermal stability characteristics can result in battery failures, even explosion when the cathode and anode make physical contact. Therefore, continuous efforts have been realized for nanofiber separators in batteries, particularly Li-ion battery performance [62–67].

Chitin nanofiber membranes (~25 μm thick) used as battery separators in $LiFePO_4$/Li half-cells and $Na_3V2(PO4)3$/Na cells had comparable electrochemical performance to a commercial polypropylene (PP) separator, as shown in Figure 11.3. As the porosity of the chitin nanofiber membranes increased from 40 to 60%, the electrolyte uptake increased from 192 to 252%, while the PP separator had a porosity of 41% and uptake of only 160% [68].

Using nanofibers allows for adjustment of pore size in the separator membrane, leading to improvements in the electrochemical response. In addition, multiple-layered nanofiber membranes have increased wettability and electrolyte adsorption due to the large surface area and high porosity. Electrolyte uptake (stored inside the membrane) can be measured using the wet and dry weights of the nanofiber membrane. The weight difference of the nanofiber membrane separator, pre- and postelectrolyte soak, will provide the uptake based on the following equation:

$$\text{Uptake\%} = \frac{W_w - W_d}{W_d} \times 100\% \tag{2}$$

where W_w is the postsoak (wet) weight, and W_d is the presoak (dry) weight [69].

ARCs of nanofiber membranes, as an active layer in a solar cell, have the potential to maximize the intensity of light waves increasing the converted energy [70–72]. The material and nanostructure choice of an ARC can provide the characteristics of low reflectivity, high transmittance, and low absorption, which promotes the avoidance of loss of photons. Antimony tin oxide (ATO) layers have been shown to improve the thermal stability of ARCs in dye-sensitized (DSC) and cadmium telluride (CdTe) solar cells [73, 74]. ATO proves to be a viable material because of its electrical conductivity, crystalline structure, visible light transparency, ultraviolet light absorption, and near and mid-infrared reflection capabilities [7, 21, 22]. When cobalt II oxide (CoO) is added to ATO and electrospun into a nanofiber membrane (Figure 11.4), the magnetic properties as well as the infrared reflectivity of a solar cell can be enhanced [75–78]. Despite the successful fabrication and utilization of CoO ATO nanofiber membranes, there is still a concern regarding the eco-friendly nature of the material as cobalt is lethal in its pure elemental form. However, oxides are not hazardous and tend to be more chemically stable.

In summary, electrospinning fabrication of nanofiber separators and ARCs of eco-friendly materials can contribute significantly to sustainable renewable energy systems.

11.3.2 Nanofiltration

Nanofiltration spans areas of health care to water sustainability, and the use of nanofiber membranes in filtration systems has become viable for separation technology applications in various arenas, such as water, air, and biological filtering. Transformative products continue to emerge in the use of nanofiber filtration membranes [78–90]. Scientists, engineers, and health care providers developed semitransparent masks (a class 1 medical device) using electrospun polymeric nanofiber membranes [91]. These semitransparent masks are promoted as biodegradable, recyclable, and breathable and should have an impact on bacteria filtration efficiency, airborne trajectory, and

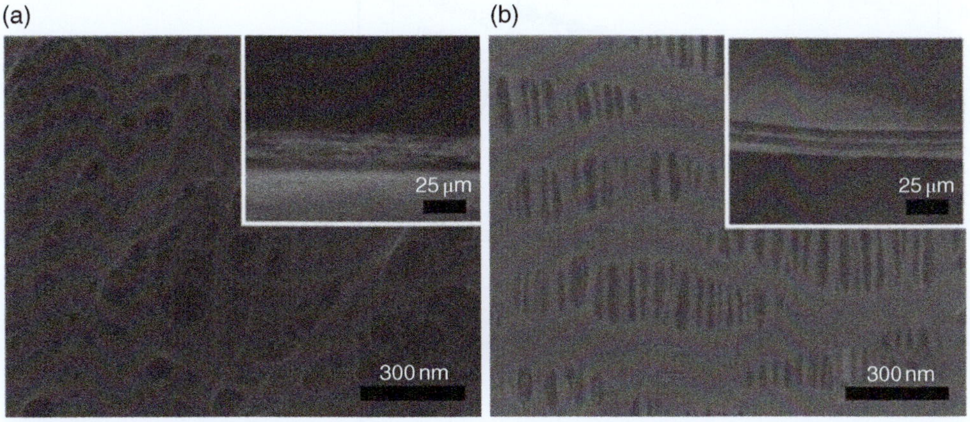

Figure 11.3 Pore size of chitin nanofiber membrane separator (a) versus commercialized polypropylene (PP) separator (b). *Source:* Reproduced with permission from Ref. [68].

air permeability reduction. This nanofiltration technology has the potential to impact the mitigation of disease pandemics such as the coronavirus disease (COVID-19), an airborne virus of the severe acute respiratory syndrome (SARS-CoV-2) [92].

Polyvinylidene fluoride (PVDF) nanofiber membranes treated by corona discharge (electrostatic charges from 15 kV for 60 s) were studied to filter airborne viruses, such as COVID-19, and nanoaerosols (60–140 nm). Figure 11.5 illustrates the increased filtration efficiency, 39–62%, for a charged one-layered (1L) nanofiber membrane with decreasing fiber diameter, 525 84 nm, respectively. This increased efficiency was for sodium chloride aerosol size, Dp, range from 50 to 500, filtered through 0.191 g per square meter (gsm) of PVDF membrane.

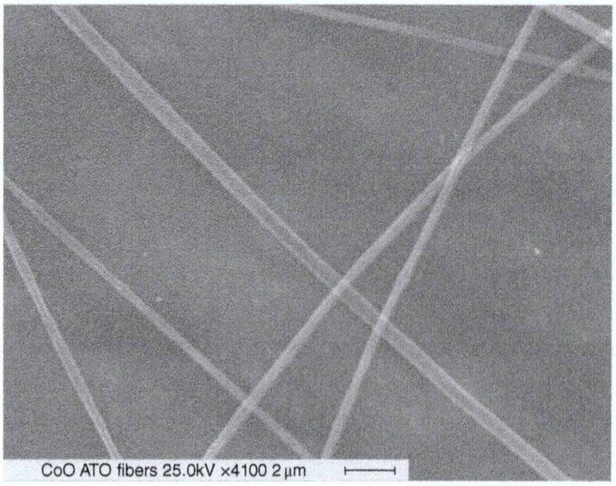

Figure 11.4 CoO-ATO nanofiber membrane.

Higher filtration efficiency was achieved using multiple layers of small-diameter nanofiber membranes as in Figure 11.6, where four layers (4L) of PVDF nanofibers, having an average diameter of 84 nm and weight of 0.77 gsm, demonstrated an efficiency as high as 96.5%.

Nanofiber membranes are also critical in sustaining water systems and mitigating the interdependence of poor water quality and degenerative diseases, e.g. ingestion of elevated amounts of arsenic in water, leading to cancer, heart disease, skin lesions, diabetes, etc. [94–96]. Natural materials, such as plant-based nontoxic polysaccharides, such as *Opuntia ficus indica* (Ofi) cactus mucilage, can be combined with polymers to produce biocompatible and biodegradable nanofiber filtering membranes [97–100]. Mucilage is an inexpensive and abundant material containing organic species that interact with biological substances, metals, and ionized species to promote the removal of particulates, arsenic, *E. coli*, and bacteria from drinking water [101–104].

Figure 11.7 indicates a nanofiber membrane with a 50 : 50 volume-to-volume ratio of polystyrene to mucilage that has an arsenic removal of 18.93% comparable to sand removal at 18.33% [1, 99, 105]. These plant-based PNMs can be used as a sustainable method for water filtration and contaminant sensing.

Nanofiber membrane functionality has been used for decades in applications to remove, detect, or block solids and bacteria. The size of the filtering pore is important to determine the size of particles that can be filtered out. In addition, the filtering efficiency is typically dependent on the structural properties, such as pore size, thickness, fiber diameter, weight of the fiber membrane, and, if layered, the layer matrix.

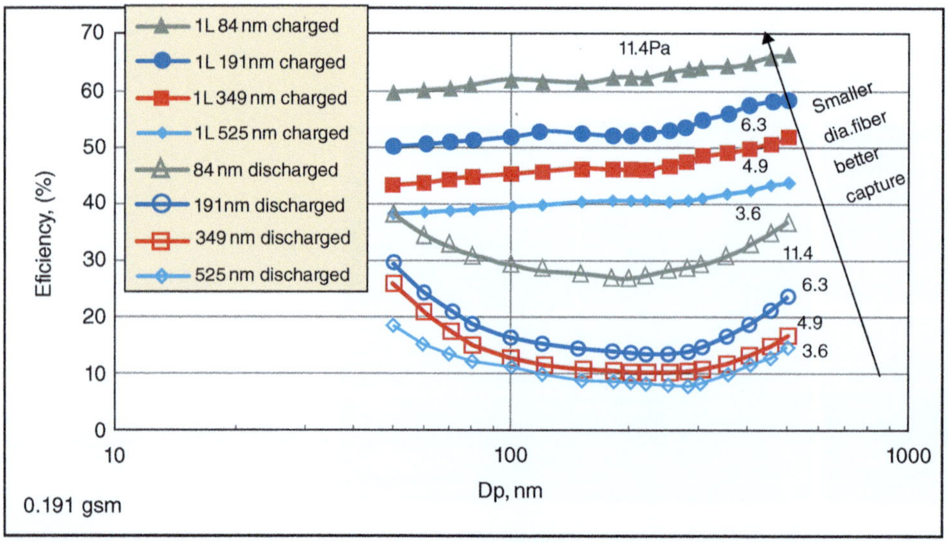

Figure 11.5 Filtration efficiency of charged and discharged one-layered (1L) PVDF nanofiber membrane. *Source:* Reproduced with permission from Ref. [93].

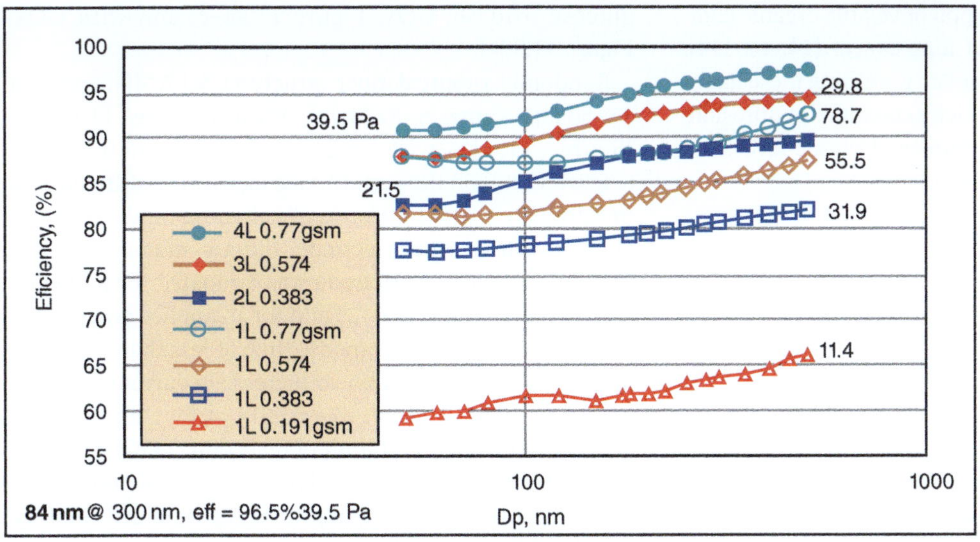

Figure 11.6 Filtration efficiency of multilayered (1L-4L) charged PVDF nanofiber membranes. *Source:* Reproduced with permission from Ref. [93].

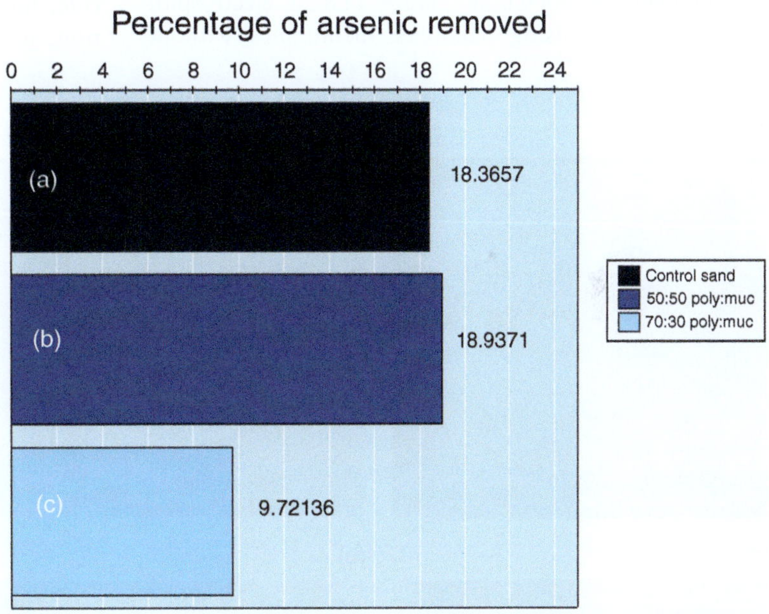

Figure 11.7 Comparison of the amount of arsenic absorbed in (a) sand (control), (b) 50 : 50 PS : mucilage PNM, and (c) 70 : 30 PS : mucilage PNM.

11.3.3 Nanosensing Health Care

Nanosensing devices in health care are imparting enhanced capabilities for predictive diagnostics, targeted disease detection, and point-of-care/telemedicine. PNM technology is impacting all levels of the sensing platform in health care from *in vivo* and *in vitro* sensing, sensing data integration and analysis through artificial intelligence (AI) and machine learning (ML), to the future elevation of PNM technology, bridging the gap between these areas to revolutionize telemedicine platforms and patient point-of-care engagement.

Through PNM technology, which is primarily biocompatible, biodegradable, and eco-friendly, health care sensing has opportunities for more precise response, better analytic performance, and enhanced surface immobilization [24].

Of particular interest to 18.1 million cancer patients diagnosed across 36 different kinds of cancers is the advancement of nanosensing for predictive diagnosis with biomarkers, high selectivity, and sensitivity with modified surface area [106–116]. Polymer nanofibers of PMMA, PCL, PANI/PEG, and PDMS have been used in PNM biomarker sensing structures to detect down to the ppb of volatile organic compounds (VOCs) that can predict lung cancer [24, 46, 114]. The fundamental sensing mechanism governing these nanosensing PNM structures follows Fick's equations of diffusion, where solute molecules diffuse onto an electrode formed or covered by a nanofiber membrane,

$$1^{st}\text{ Fick's law}: J_D = -D\frac{\partial n}{\partial x} \tag{3}$$

$$2^{nd}\text{ Fick's law}: \frac{\partial n(x,t)}{\partial t} = D\frac{\partial^2 n}{\partial x^2} \tag{4}$$

where J is the diffusion flux, D is the diffusion coefficient, and n is the concentration across time (t) and space (x).

A biocompatible intrinsically conductive polymer, PEDOT : PSS, as a PNM electrode demonstrates selective sensing capabilities for glucose sensing and stretchability characteristics for flexible health care sensing [25, 26, 117]. Scanning electron, transmission electron, and atomic force microscopies are shown in Figure 11.8 of electrospun PEDOT : PSS sensing nanofiber membranes without glucose oxidase, GOx, Figure 11.8a–c, and with GOx, Figure 11.8d–f.

Randomly oriented fiber structures with diameters of 25–50 nm show a SEM of PNM (Figure 11.8a vs. Figure 11.8d) with higher GOx immobilization (Figure 11.8d) on the nanofibrous matrix; this difference in morphology is also indicative of enhanced binding of GOx enzyme within the electrode fibrous membrane. The PEDOT : PSS PNM electrode was evaluated for a glucose concentration range of 0–25 mM for an applied potential of 0.5 V and resulted in a response time of less than 4 seconds for glucose detection. A linear increase in current occurred from $0.23 \pm 0.0014\,\mu A$ to $19.7 \pm 0.0015\,\mu A$ for a sensitivity of $5.11\,\mu A\,mM^{-1}\,cm^{-2}$ and limit of detection (LOD) of 2.3 μM as shown in Figure 11.9. The PNM electrode demonstrates improved sustainability qualities of repeatability, stability, and durability [117–120].

Selectivity (Figure 11.10) of the PNM glucose sensor was validated through CA analysis at +0.5 V under the interference of various ingredients, such as sodium chloride, and potassium ferricyanide with a 2 mM concentration, as shown in Figure 11.10. A significant increase in

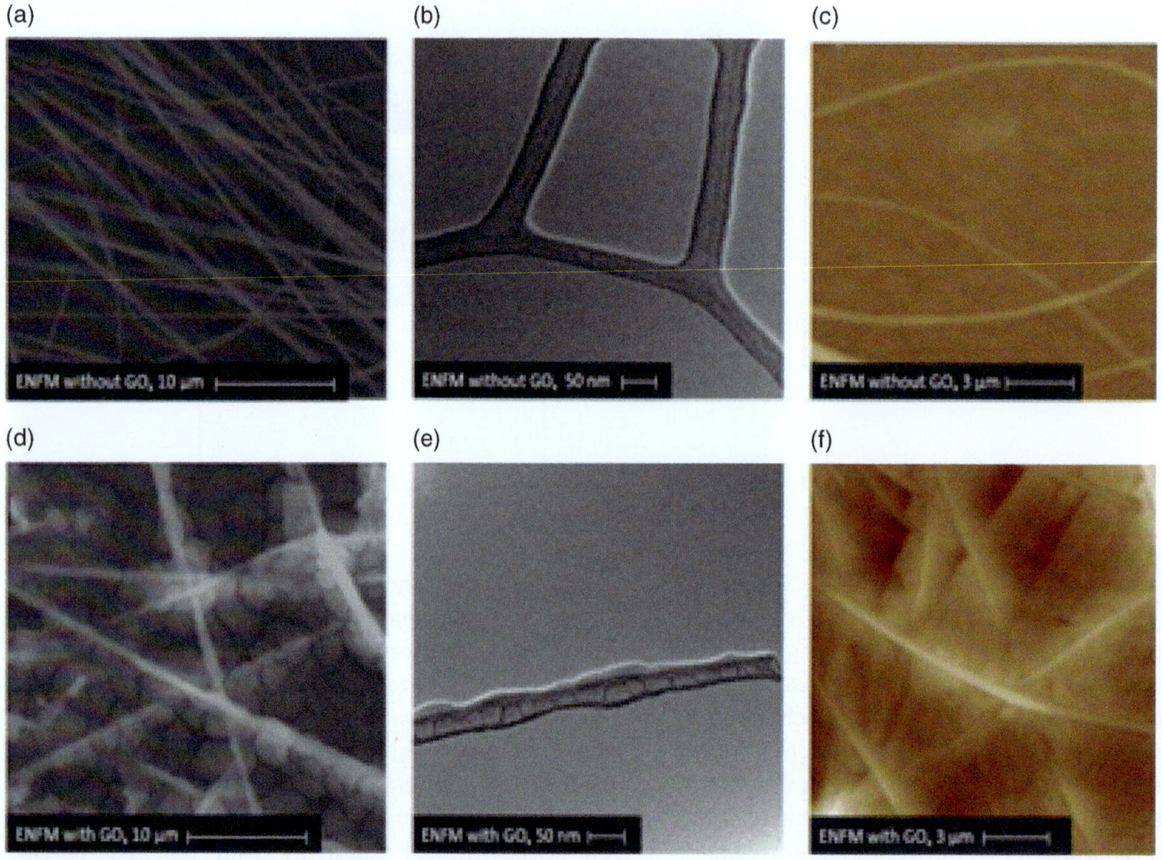

Figure 11.8 (a*) and (d**) SEM, (b*) and (e**) TEM, and (c*) and (f**) AFM images of a PNM glucose sensing electrode. *a, b, and c are the images of an PNM without GOx. **d, e, and f are the images of a PNM with GOx.

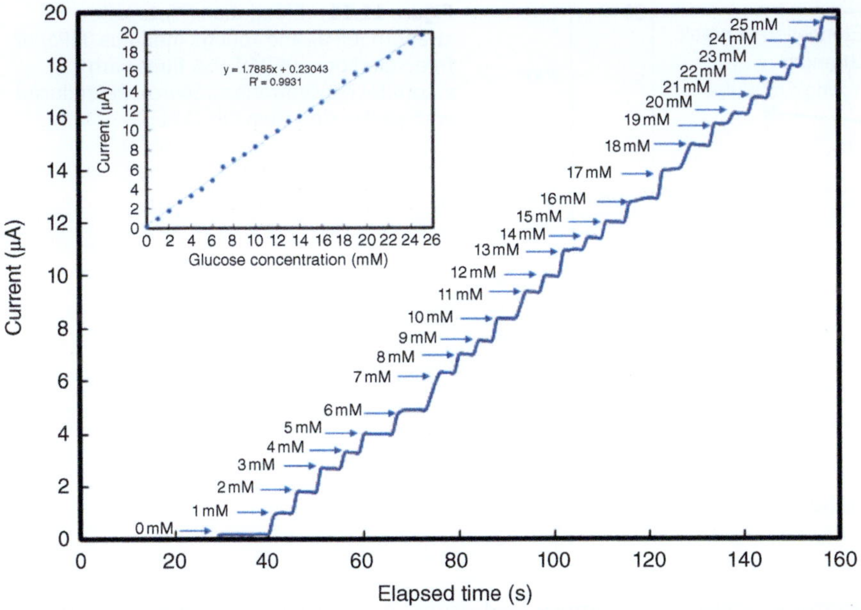

Figure 11.9 Chronoamperometric analysis of PNM glucose sensor for successive increase in glucose concentration at +0.5 V potential.

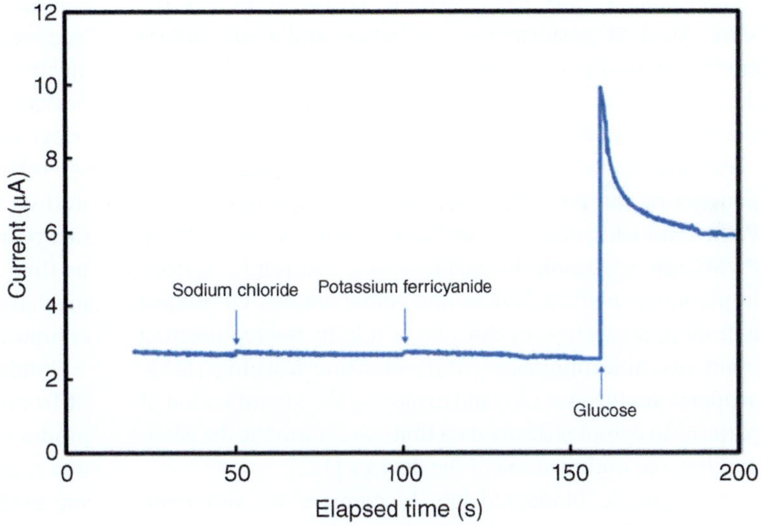

Figure 11.10 Selectivity test for the PNM-based glucose sensor.

current was observed when the 2 mM concentration of glucose was injected, which demonstrates that the PNM electrode acts as a selective glucose sensor. These electrochemical analysis results are indicative of good repeatability, durability, and selectivity through strong electrochemical responses when using PNM electrodes, encouraging the application of nanofiber structures for the fabrication of biosensors.

By slightly doping PEDOT : PSS with biocompatible D-sorbitol, the stretchability and conductivity are improved, indicative of a decrease in tensile strength (23.6–1.23 MPa) and Young's modulus (307–1.87 MPa) and conductivity (>1000 S cm^{-1}) [121], as shown in Figure 11.11.

This demonstrates the flexibility and ability to tune these PNMs to become more sustainable, and there are still intensive efforts to further extend the biobased and eco-friendly polymer nanofiber membranes for nanosensing.

11.3.4 Nanoelectronics

Nanoelectronics are capable of impacting almost every aspect of emerging technology challenges from edible, wearable, and flexible sensors to smart electronics for AI. However, there is tremendous growth in electronic waste surplus, which can be a detrimental challenge to the

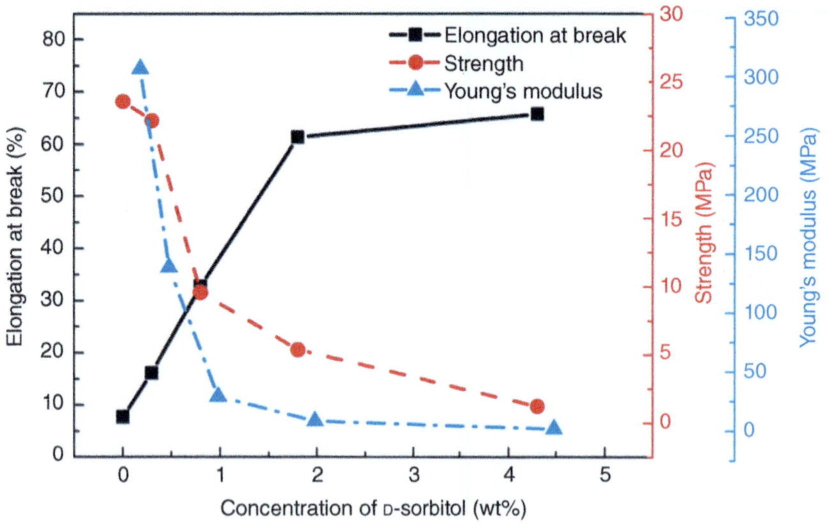

Figure 11.11 Elongation (%), tensile strength (MPa), and Young's modulus (MPa) of free-standing s-PEDOT:PSS films with D-sorbitol concentrations. *Source:* Reproduced with permission from Ref. [121].

environment. Many components of the electronic devices are not recyclable, and when disposed of or burned, there is leaching of toxins or harmful chemical by-products. An alternative is fabricating electronic devices using biodegradable and biocompatible polymers to get the same level of performance, or better, and manufacture environmentally safe, low-cost, and eco-friendly electronic devices, which will expand the use of nanoelectronics in the biomedical field.

Polymer nanofiber membranes have been used for components of the building blocks for electronic devices, i.e. PNM conductor, dielectric, and semiconductor layers. These PNMs are applicable to biodegradable electronic devices, implantable medical devices, and edible sensors. Of interest is how nanoelectronics can play a role in interconnecting brain machine interfaces (BMI)–machine learning (ML)–artificial intelligence (AI) and exploring the identification of patterns in complex device data that can inform the decision-making and optimization of the devices [122].

One specific biodegradable, biocompatible, and compostable polymer is PLA, which has been found to be advantageous for use in nanodevices. The macro capabilities of being able to be molded (injection molding technology) and extruded (3D printing) are not the only advantages. PLA can also be spun into fibers and yarns in the micro-scale, and under the right circumstances, using an electrostatic method such as the electrospinning technique, fibers down to the nanoscale can be produced. Such advantages in the nanoscale include the large surface area-to-volume ratio, which is excellent for filter and drug delivery applications. Another facet this polymer can have is the ability to be a mechanical support to promote fiber formations of other materials that cannot be formed into fibers by their own such as low molecular weight molecules, particles, ceramics, and drugs. A remarkable example is the use of PLA for the fabrication on electronic devices.

PLA, in conjunction with other polymers that are electrically conductive, has opened a new scope of research with this appraised material. By a blend of PLA and a semiconductive polymer such as poly(3-hexylthiophene) (P3HT), electroactive nanofibers can be fabricated, leading to the design of organic diodes. Through electrospinning, nanodevices as shown in Figure 11.12 can be prepared. Due to the inherent characteristics of the semiconductor polymer, in this case P3HT, a depletion region is formed in the contact point of the PLA/P3HT and the other semiconductor, in this case silicon. In this structure, PLA provides the mechanical stability for the P3HT, making an electroactive composite able to function as rectifier and a sensor.

Conductive polymers such as polyaniline can also benefit from the usage of PLA. In this case, PLA enhances the production of fibers with conductive behavior. Polyaniline was successfully used to fabricate a single fiber VOC's sensor. as shown in Figure 11.13.

Huge amounts of PLA can be recycled mechanically and chemically for further usage. While the quality of the PLA may decrease after the recycling process, the outcome can be directed for more general usages that do not require high purity or, as preferred, composted for the generation of new crops that can lead to new PLA.

11.3.5 Nanodelivery

Multiaxial nanofiber membrane drug delivery systems for the preclusion of lung cancer recurrence continue to be on the forefront of drug delivery systems. Cancer is the foremost cause of death globally, accountable for approximately 9.6 million deaths in 2018. According to the World

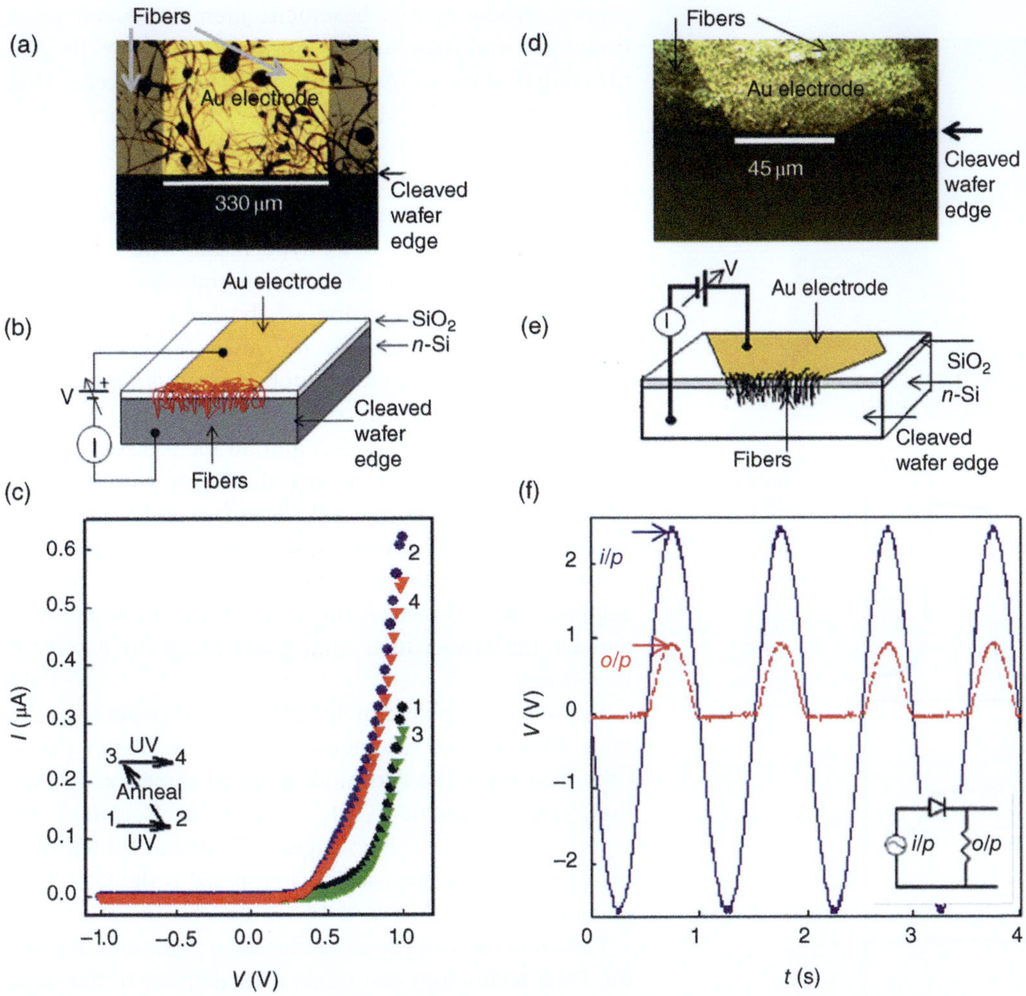

Figure 11.12 Electroactive PLA composite used as mechanical support for fiber fabrication. When blended with a semiconductive polymer such as P3HT and electrospun into fibers, it can be deposited over gold electrodes in order to make an electronic device. Optical microscope images, (a) and (d), show the fibers over the gold electrode. Electrical sketches, (b) and (e), show the fibers making electrical contact with the silicon over a cleaved surface with an electrode-silicon distance of 200 nm. The electrical characterizations show a functional and a UV radiation tunable rectifier (c) and signal rectification (f) [123, 124]. *Source:* (d) Serrano et al. [124], Fig 05, Pg 05/with permission of Elsevier.

Health Organization, among all cancers diagnosed in 2018, lung cancer was the most common (2.09 million cases) and the leading cause of death (1.76 million deaths) [94]. The circulation of anticancer drugs throughout the body, effecting both cancerous and healthy cells, and the cancerous cells developing a defense mechanism to specific drugs over time are challenging aspects of cancer research. To address these concerns, researchers have used nanoengineered structures (nanoparticles, nanofibers, nanotubes, etc.) to improve the release profiles for drugs, local chemotherapy, and combinational drug delivery. The fabrication of tri- and multiaxial nanofiber membrane delivery systems can be used for the delivery of combinatorial chemotherapeutics targeted at the pathological site after surgical resection.

The new regime of anticancer drug delivery using triaxial nanofibers allows for gaining an understanding of drug release kinetics from a tri/multiaxial nanoengineered structure. The delivery system using the fabricated nanofiber is opted to enhance the efficacy of the drug delivery process through the use of this innovative concept, which has the potential to reduce the major circulation of chemotherapeutics to nonmalignant cells. While tri/multiaxial membrane models suggest enhancing release kinetics [126, 127], there is still a need to explore multicore loading and release kinetics in a tri/multiaxial system for specific types of cancers, e.g. non-small cell lung cancer (NSCLC). With drug-loaded multicores and an antibody-loaded shell, nanoengineered multiaxial fibers can deliver anticancer drugs over various release times.

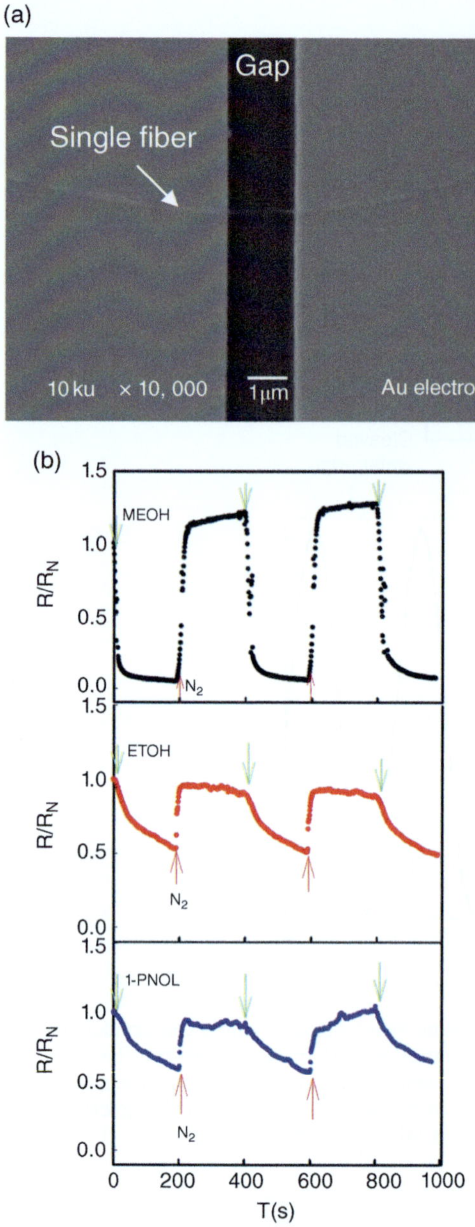

Figure 11.13 (a) PLA/polyaniline composite single fiber as the active material for a gas sensor. (b) Sensing response (one cycle = 400 s) of PLA/Polyaniline nanofibers to methanol, ethanol, and 1-propanol, respectively [125]. *Source:* (a) Serrano et al. [125], Fig 02, Pg 02 / with permission of IEEE.

Cancer is an abnormal accretion of cell growth within a specific region. There are distinct phases through which the disease advances. It can start developing from a benign stage, which may further extend to a precancerous lesion that can aggressively go through metastatic stage spreading cancer throughout the body. Different stages of cancer are explained in Figure 11.14.

A healthy tissue in general contains normal cells supported by a basement membrane. Extracellular matrix protein produces this basement membrane and gives functional and structural properties to healthy cells. The functional area contains stem/progenitor cells that replenish dead cells. However, there are instances when the progenitor cells go through alterations in their genetic sequence, which causes mutation, and the mutated cell develops abnormal proliferation in that region. This is classified as hyperplasia where the cells are identical but in excessive numbers. Additional genetic sequence alteration in the hyperplasia region forms aggressive clusters of cells, forming benign tumors. Benign tumors are not cancerous and can be treated for a complete cure.

The benign tumor can alter and, in the process, employ their blood supply from the surrounding network of blood vessels. Employment of a new interface of blood vessels around the benign tumor allows the source of nourishment for the tumor to survive and continue its growth aggressively. At this stage, the tumor enhances its ability to damage the extracellular matrix and break the basement membrane.

The uncontrollable mutated cells can move out from the main site once the basement membrane is broken. These cells then enter the surrounding blood vessels and make their way to a secondary site through the circulatory systems. In the process, the cancer cells spread through the body producing a metastasis. This spread is the fatal form of cancer.

Therefore, we can depict cancer as an ingenious disease, the PNM technology can target drug delivery to the stage that becomes resistant to the drug, and potentially mitigate a reoccurrence after a treatment [24, 113–115, 126–129]. Even though some specific forms of cancers can be addressed with surgeries and well-established medicine, some can reoccur. This is possible by surpassing the precancerous lesion and succeeding to go through the metastatic spread without responding to the medication.

Engineering multiaxial PNMs with loaded combinational chemotherapeutics, antibodies, and inhibitors, at the nanoscale, provides emerging ways to fight destructive forms of cancer [126–129]. Electrospun nanofibers have been found to be effective for several areas of cancer research, including treatment of cancer and cell capturing and behavior. A method for this treatment can involve the usage of nanofiber membranes carrying the chemotherapy drugs for localized delivery, and therefore reducing the damage of healthy cells. Indeed, a model for a multiaxial drug delivery system that has the potential to administer several chemotherapy drugs and test the nanoengineered system to find out its efficacy in delivering chemotherapeutics, antibodies, and inhibitors to the pathological site could be of great value for this area.

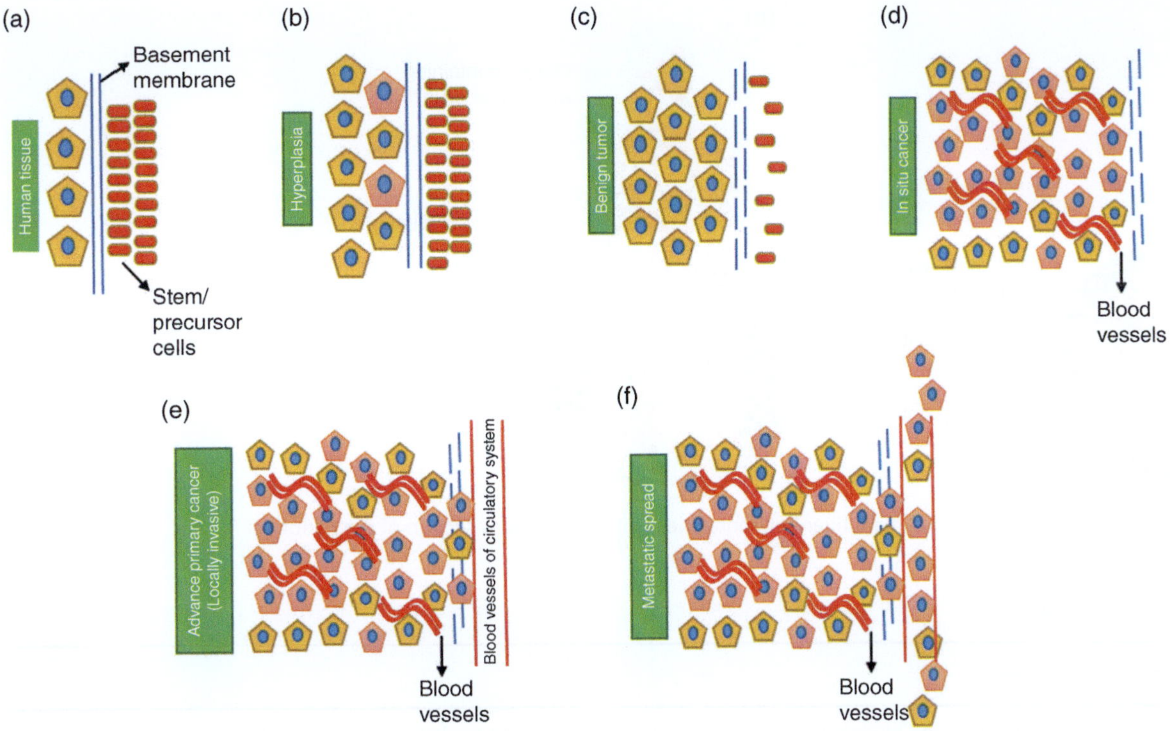

Figure 11.14 Cancer multiaxial nanofiber drug delivery as (a) through (f) shows development of cancer.

The tri/multiaxial nanofibrous delivery system, as shown in Figure 11.15, offers a combinatorial drug delivery regimen of three phases composed of a central core (core1 drug), tubular core (core2 drug), and the shell (shell antibody) configuration, where the antibody will detect the overexpressed antigens while threating the specific areas with the loaded drugs. The loading of multiple drugs for a time-released treatment of cancer, e.g. NSCLC, after surgery in the process will complement chemotherapy and radiation therapy. The targeted delivery can be used for delivery of a combination of chemotherapeutics placed at the pathological site after surgical resection; thus, an innovative concept of reducing the major systemic circulation of chemotherapeutics to healthy tissue is one challenge of nanoparticle drug delivery systems [130].

While fabricating a PNM, both material properties and experimental variables are considered very carefully. Parameters such as chemical composition of the polymeric solution, concentration of the solution, viscosity, surface tension, molecular weight, solution conductivity, and molecular weight distribution affect the dimension of the fiber (length and size) and its porosity. Considering both material property and experimental variables can increase the drug solubility and modulate control release of drugs to a specific target. Also, process variables such as voltage, the flow rate of the solution in the syringe pump, and distance from the collector plate affect the formation of specific fiber.

Research has shown a PCL/PVP (polyvinylpyrrolidone) homogeneous, coaxial, and triaxial fiber used to simulate drug release profiles using two color dyes (KAB, keyacid blue and KAU, keyacid uranine) [126]. The multiaxial configuration incorporated a barrier (unloaded) layer between the core and the shell, which was used to circumvent core leaching [127–129]. Color dye optical absorption was measured to model the drug release mechanism of this triaxial showing a slower core drug release (~24X) and more constant shell release (~80% h^{-1}) for long- and short-term treatments, respectively. Others presented a triaxial model using ethyl cellulose in inner, middle, and outer layers, loaded with different concentrations of the nonsteroidal anti-inflammatory drug, ketoprofen (KET). This triaxial demonstrated that the fibers can have linear release of KET over 24 hours [127]. Each case demonstrates the potential of the triaxial as a drug release system. Challenges still remain with the adaptation of loading specific therapeutic drugs in the triaxial membrane with biocompatible polymers, such as poly(D,L-lactide-co-glycolide) (PLGA), which has both hydrophobic and -philic compound release functionality, Table 11.2. Other potential biocompatible and FDA-approved materials are poly(ethylene oxide) (PEO) and polyethylenimine (PEI), and a drug-loaded membrane

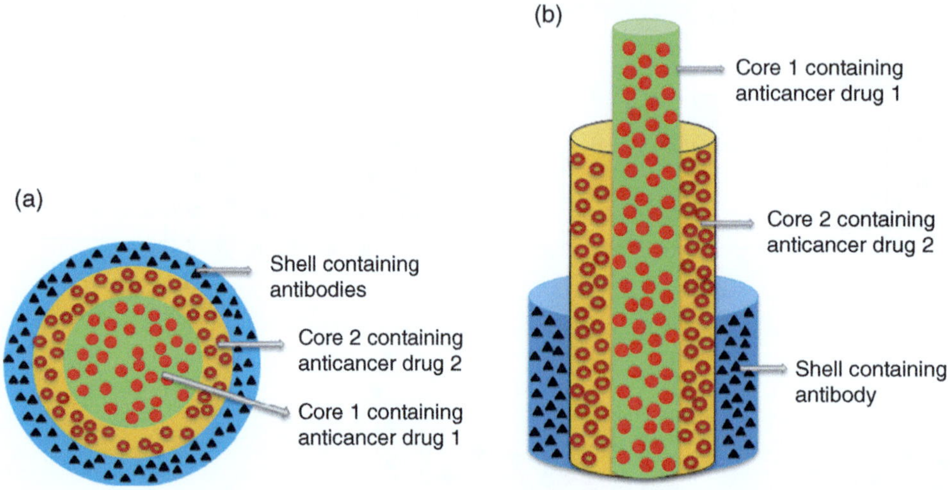

Figure 11.15 (a) Cross section of triaxial nanofiber horizontal view. (b) Vertical cross section of the nanofiber.

Table 11.2 Material and properties.

Material	Type	Properties
PLGA	Polymer	Biocompatible and biodegradable FDA approved for use in cancer drug delivery. Therapeutic agent for cancer. Controlled release.
PEO	Polymer	Biocompatible and can be used for drug delivery system. FDA approved. Hydrophilic and hydrophobic interactions.
PEI	Polymer	Biocompatible and can be used for cancer drug delivery. FDA approved.

fabricated from these polymers could be placed in the pathological site after surgical resection.

When compared to nanoparticle drug delivery applications, particularly for lung cancer, nanofibrous drug-loaded membranes can be targeted to specific areas and size, e.g. wedge resection or segmented resection. PNM drug delivery can have limited burst drug release, foster greater extracellular matrix-like cell interactions, exhibit diverse surface topography, have enhanced encapsulation, and have higher efficacy and better controlled drug release [24, 113–115, 126–129]. Some chemotherapeutic agents are unable to differentiate between malignant cells and normal cells. Since the chemotherapeutic agents cannot critically determine the malignant site, they cause cytotoxicity to nearly every cell they penetrate. The adjuvant application of this nanoscale triaxial material drug delivery system will have a positive postoperative impact on survival rates, recurrence disease, therapeutic efficacy, targeted and controlled drug release, and low toxicity.

Properties of fabricated multiaxial nanofiber membrane for delivering anticancer drug in the pathological site also depend on the orientation of the nanofiber. A smart release system should be engineered to release different drugs and tumor cell defense mechanisms, disabling agents at different rates. Different orientations of PNMs (aligned and random, Figure 11.16) will offer different release rates of the drug to optimize the efficacy of the system.

For both orientations, drug entrapment efficiency to deduce the ratio of the released drug to the target to the total drug loaded/added to the nanofiber can be found using Eq. (5) [126]:

$$\text{Entrapment efficiency} = \frac{\text{Overall mass of drug released from nanofiber}}{\text{Mass of total drug added}} * 100 \tag{5}$$

Also, the study of the degree of swelling (Eq. (6)) and weight loss (Eq. (7)) is important to understand the profile of the release study [107, 108]. The drug release kinetics is directly interrelated to the nanofiber mesh weight loss and the time variants in swelling,

$$\text{Degree of swelling\%}\left(W_s\right) = \frac{W_1 - W_2}{W_2} \times 100 \tag{6}$$

and

$$\text{Weight loss\%}\left(W_L\right) = \frac{W_2 - W_3}{W_2} \times 100 \tag{7}$$

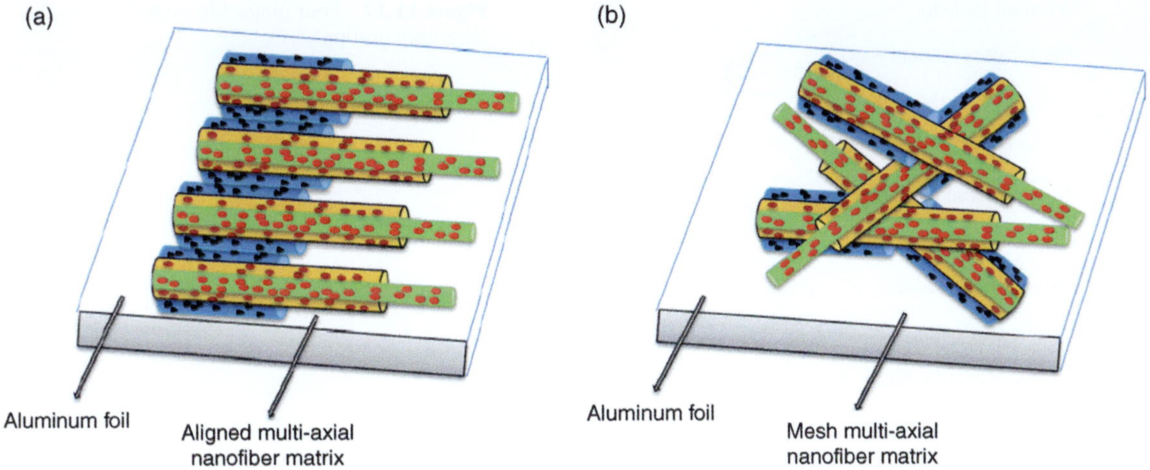

Figure 11.16 (a) Aligned multiaxial PNM and (b) random ordered PNM.

where
W_1 is the swollen nanofiber weight
W_2 is the initial sample weight
W_3 is the weight of the dehydrated sample.

Overall, using PNMs, whether uniaxial or multiaxial, for drug delivery systems for placement on the surface of the surgical site after resection of metastases, will have a global impact.

11.3.6 Summary

In summary, one sustainability challenge for eco-friendly and biobased materials and PNM processing is the lack of environmental impact data for manufacturing and end-of-life disposal. The electrospinning of nanofiber membranes must, therefore, consist of a robust environmentally friendly process for a variety of natural materials, thermally sensitive additives, and a wide range of polymers, including biodegradable and biocompatible materials. A holistic processing approach for PNM manufacturing evaluates membranes by performance and environmental impact and is adaptive to a variety of material systems.

Environmental impacts of fabrication or manufacturing processes can be evaluated using modeling tools, or software are used. For example, SimaPro® uses a life cycle impact (LCI) to analyze the life cycle assessment (LCA) of materials or processes. Consider PNMs of cobalt oxide-antimony-doped tin oxide (CoO-ATO) with polystyrene electrospun using two different solvents. To evaluate this LCA, a functional unit is established of 1 g of the final product, produced on one electrospinning system. Comparative studies can also be done using different manufacturing processes and/or usage of several machines; this would take into consideration if one of the products is more energy efficient than the other. When an end-of-life scenario, such as reuse, recycle, or dispose, is known, adding this data to the LCA gives a complete environmental impacts assessment. However, it should be noted that an environmental impact is assessed for one process or an entire manufacturing process. The data required for a manufacturing process LCA will be the list of all materials used and the amount of energy consumed during the process.

LCA software has a vast database, and when materials are not listed, the processes and material stages can be created and added to the database. In the case study for CoO-ATO PNMs, a process stage for antimony-doped tin oxide (ATO) and cobalt oxide (CoO) is created in SimaPro®. ATO is made by substituting tin (Sn) atoms in tin oxide (SnO_2) with antimony (Sb) by hydrolysis of tin chloride with antimony chloride. Cobalt oxide is made by electrolyzing a solution of cobalt chloride:

$$CoCl_2 + H_2O \rightarrow CoO + H_2 + Cl_2 \quad (8)$$

While including this data, the electrical energy and/or heat energy is also added. The power consumption during a mixing or sonicating process can be determined using a wattmeter, where the meter reads the voltage and current drawn into the device and the power reading. The electrical energy input is made in watt-hours, or for heat energy, the manufacturer's instruction manual should include energy consumption data. This will be used in LCA data.

In the LCA analysis, when a raw material from the database inventory is added to the process, all the information regarding the manufacturing, sourcing, and the energy that was used in making the materials is included. For the CoO-ATO case study, the process stages were created in SimaPro for ATO, CoO, and the solvent D-limonene, an extraction from citrus peel. The analysis estimated that 1.55 MJ of energy is used to make 0.7 ml of D-limonene.

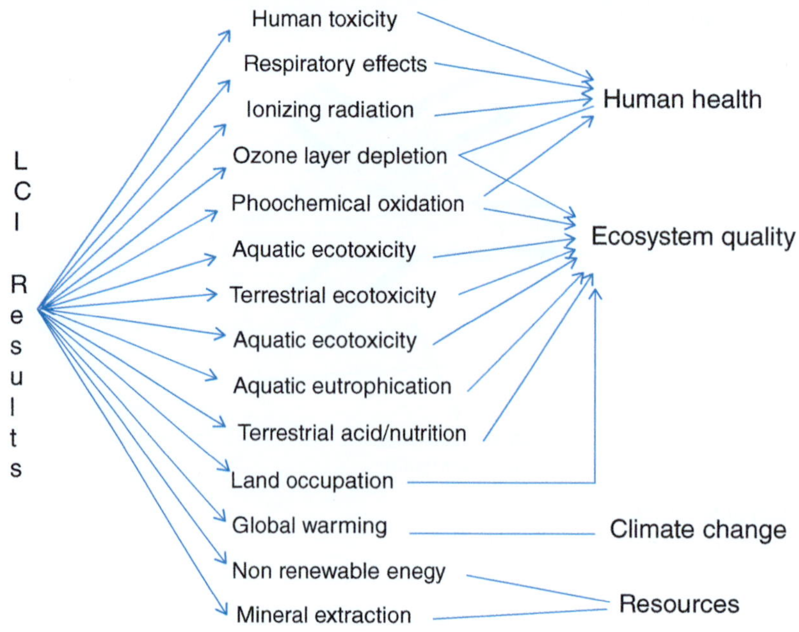

Figure 11.17 Four major life cycle inventory groups.

The SimaPro inventory already consists of parameters for the polymer, polystyrene (PS), and the second solvent, toluene. These stages are used in the environmental impact analysis of PS: D-limonene CoO:ATO and PS:Toluene CoOATO are used to produce CoO-ATO nanofiber membranes. Using the Eco-indicator 99 impact assessment method in SimaPro, impacts can be determined. Impacts are classified into four major groups, human health, ecosystem quality, climate change, and resources, as shown in Figure 11.17.

Figure 11.18 is a sample of a network representation of the energy consumed in each process stage of the production of PNMs of PS:CoO-ATO nanofibers made with toluene. Each of the boxes is a process stage. The thermometer in each process stage is an indication of its contribution to energy consumption. The functional unit for this case study is the quantity of nanofibers made from 1 ml of the precursor solution of polystyrene, cobalt oxide, ATO, and toluene.

For overall sustainability, the electrospinning of PNMs must consist of a holistic evaluation approach (Figure 11.19) adaptive to a variety of material systems to determine the environmental impacts of waste, water use, carbon emission, and energy use as part of a LCA.

11.4 Future Perspective and Nanocomputing/Data Mining/IoT

Beyond the existing applications of PNMs is the emerging potential to integrate and utilize PNM in the building of nanocomputing – chemical nanocomputing, medical (DNA) nanocomputing, mechanical nanocomputing, and quantum nanocomputing technologies. Toward the development of a <50-nm-sized computer, nanofibers of different configurations and compositions will be the building blocks for nanosized systems-on-a-chip of low power, low noise, and low leakage, and 100 to 10 000-fold in circuit functionality [131–134]. The nanosize and large surface area of the PNM technology will provide computing power that will increase the interconnectivity of the IoT and more data accuracy in data mining and machine learning.

Also, nanofibers will be used for digital computing bits, interconnects for molecular switching crossbars or cellular neural networks (CNNs), and 3D bioprinting of embedded nanocomputers with discrete functions. The material composites being used for these potential nanocomputing nanofiber structures will need extensive investigations to move the needle toward sustainable nanotechnology.

Magnetic nanofibers of the synthetic polymer, PAN, containing (Figure 11.20) nickel ferrite (Figure 11.21) nanoparticles, were studied for the characteristics suitable for applications in neuromorphic computing and spintronics [27]. The magnetic PNMs, calcinated and carbonized to result in membranes exhibiting magnetic properties based on the size of the nanoparticles matrix, are necessary to fully understand the magnetic behavior of the electrospun, stabilized, and carbonized magnetic nanofibers. Micromagnetic simulations and design of experiments are needed to further understand magnetization reversal and dynamics of the magnetic nanofiber membranes, which contained some beading which are known to impact the

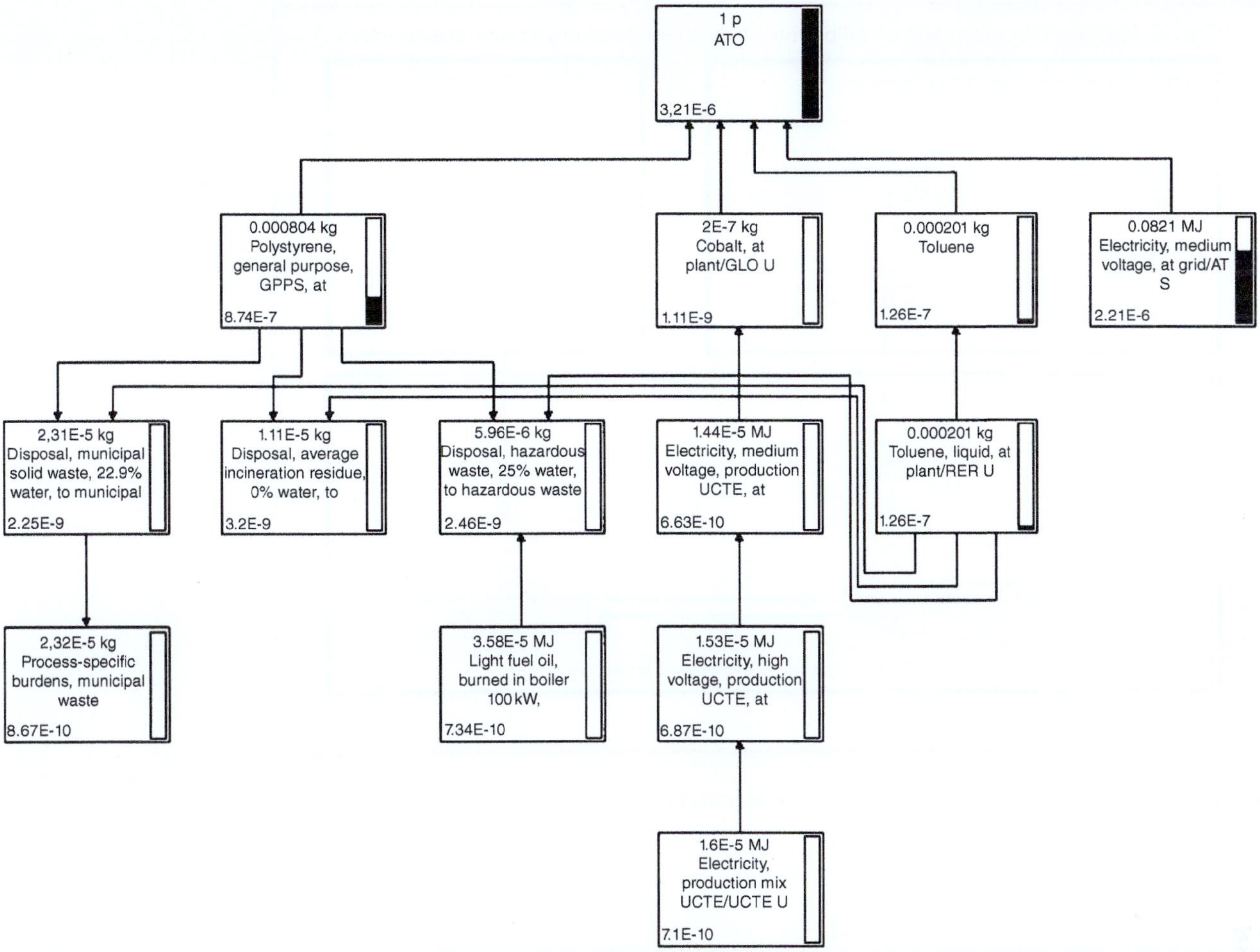

Figure 11.18 Network representation of the energy consumed in each process stage of the production of CoO ATO polystyrene nanofibers.

properties of the fiber. Furthermore, these additional studies will bring clarity to the potential of these structures for switching, transport for neuromorphic computing, or data storage.

Nanofibers can exhibit properties of dielectrics and semiconductors, switches, and diodes. Another illustration of nanofibers for computing uses silica nanofibers to observe the optical transmission and properties for quantum computing. Researchers continue to probe into this area to reproduce ultrahigh transmission optical nanofibers with a diameter of 530 nm and elongation of 84 mm with a single-mode transmission of $99.95 \pm 0.02\%$. Because of the nano-sized features, light waves propagate through the optical membrane and trap atoms, which can then be used as quantum bits (qubits) for quantum computing [132, 135].

To code information, the qubit, the basic unit of quantum information, must be anywhere on the sphere (superposition of states), unlike the classical binary bit that is in one of the two states, Figure 11.22 [135].

These concepts are being investigated for future practical applications of quantum science in the advancement of transformative quantum communications, sensors, information technology, and computing [136].

This capability for the IoT will be critical in the years to come as a network of embedded technology seeks to connect, sense, exchange data, data mine, communicate, or even conduct telemedicine over the Internet.

11.5 Conclusion

In conclusion, the classification of sustainable materials for PNM technology can be very convoluted when considering the raw material sources (plants, biobased monomers, proteins, polysaccharides, and polymers), material waste (cost and disposal of excess raw material), material transformation (low energy processes used to produce the nanofibers), material transport (energy

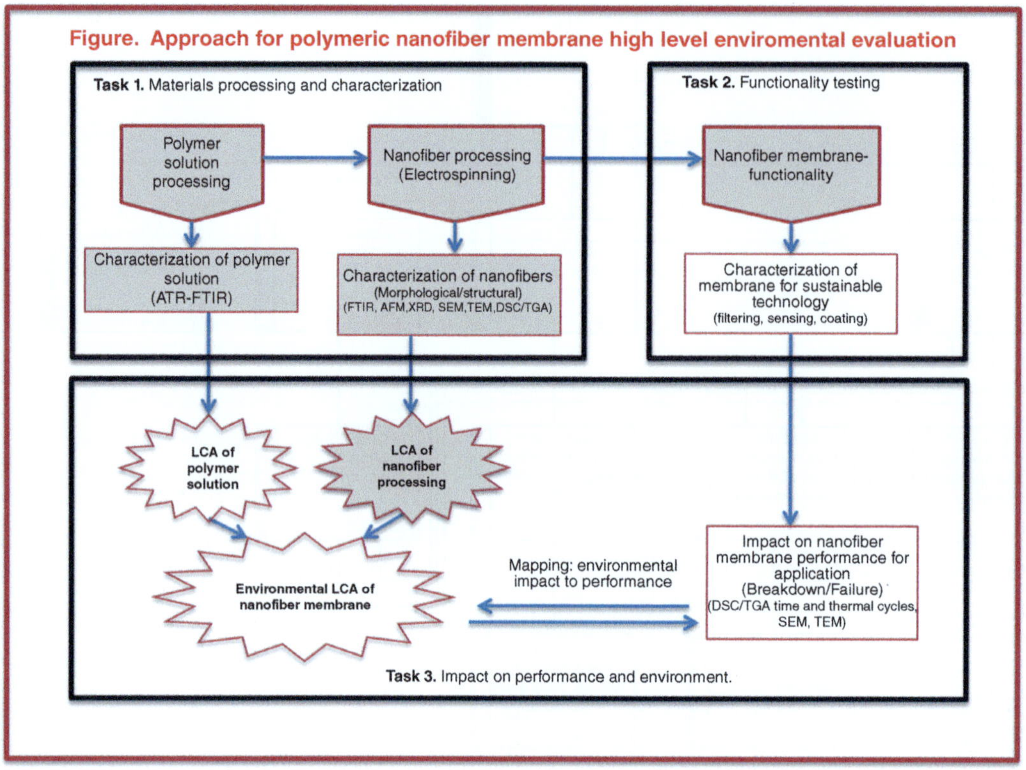

Figure 11.19 Holistic processing approach for nanofiber membrane performance and environmental impact.

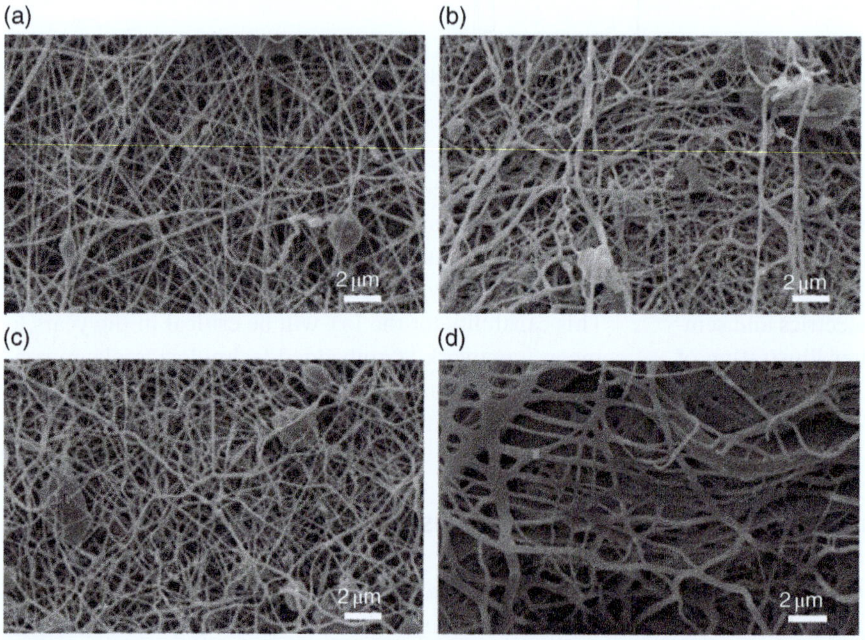

Figure 11.20 Magnetic PAN/magnetite nanofibers after (a) electrospinning, (b) stabilization, (c) 500 °C carbonization, and (d) 800 °C carbonization [27]. *Source:* Fokin et al. [27], Fig 2a, Pg 04 / MDPI / CC BY 4.0.

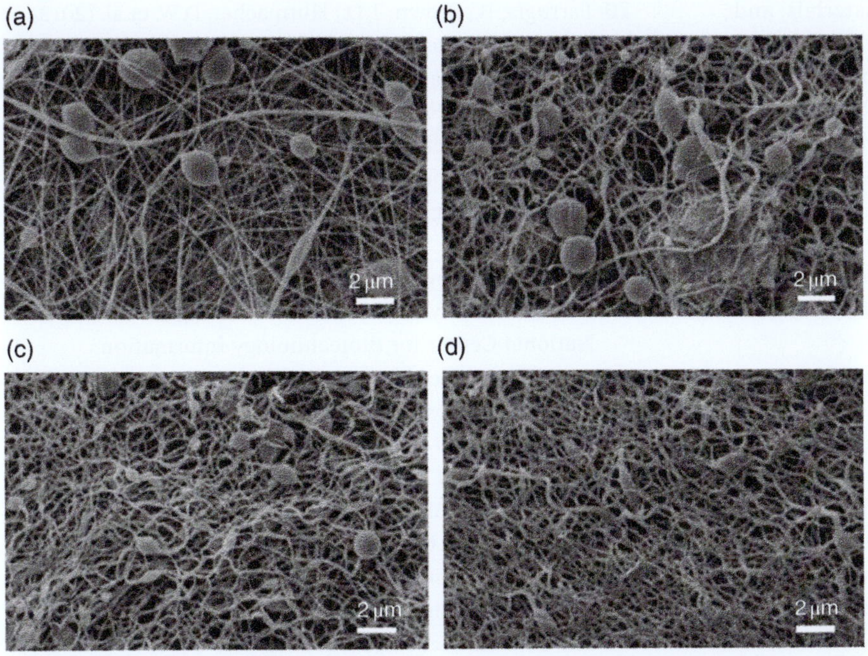

Figure 11.21 Magnetic PAN/nickel-ferrite nanofibers after (a) electrospinning, (b) stabilization, (c) 500 °C carbonization, and (d) 800 °C carbonization [27]. *Source:* Fokin et al. [27], Fig 3a Pg 05 / MDPI / CC BY 4.0.

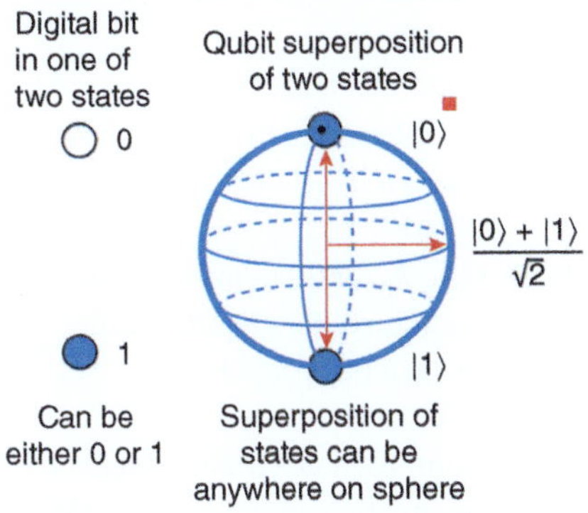

Figure 11.22 Schematic of binary bit and qubit – two-state quantum device [135].

exerted to transfer raw materials), and material end-of-life (decomposition, biodegradability, biocompatibility, and recyclability). Existing processing efforts to fabricate biodegradable, biocompatible, and eco-friendly electrospun multifunctional membranes will address manufacturability, safety, toxicity, and environmental impacts stressed in the national challenges of the Environmental Protection Agency, National Science Foundation, and National Institute of Standards and Technology. The global challenge to establish a green platform for nanofiber industrial production is being addressed, and researchers are continuing to establish protocols for sustainable nanotechnology.

References

1 Thomas, S., Devisetty, M., Katakam, H. et al. (2015). Investigation of novel opuntia ficus-indica mucilage nanofiber membrane filtration for water systems. MRS Proceedings, 1745, Mrsf14-1745-ff04-05. doi:https://doi.org/10.1557/opl.2015.582.

2 Julien, T.C., Subramanyam, M.D., Katakam, H.C. et al. (2019). Ultrasoft polycarbonate polyurethane nanofibers made by electrospinning: fabrication and characterization. *Polym. Eng. Sci.* 59: 838–845. https://doi.org/10.1002/pen.25021.

3 Jamie, A.G., Hamilton, A.O., Archibong, E. et al. (2019). Synthesis and characterization of novel polyethylene oxide–dinuclear Cu(II) complex electrospun nanofibers. *Mater. Lett.* 238: 58–61.

4 Agarwal, S., Greiner, A., and Wendorff, J.H. (2009). Electrospinning of manmade and biopolymer

nanofibers: progress in techniques, materials, and applications. *Adv. Funct. Mater.* 19 (18): 2863.
5 Chronakis, I.S. (2005). Novel nanocomposites and nanoceramics based on polymer nanofibers using electrospinning process-a review. *J. Mater. Process. Technol.* 167: 283–293.
6 Niu, H., Zhou, H., and Wang, H. (2020). Chapter 1 – Electrospinning: an advanced nanofiber production technology. In: *Energy Harvesting Properties of Electrospun Nanofibers* (eds. J. Fang and T. Lin), 1–44. IOP Publishing Ltd. https://doi.org/10.1088/978-0-7503-2005-4.
7 Richard, B., Alcantar, N., and Thomas, S. (2012). Cobalt-doped ATO sol-gels on carbon-silicon layers for modeling sol-gel-carbon fiber interfaces. *Japanese Journal of Applied Physics* 51: 11PG15.
8 Wilson, C. and Taylor, G. (1925). The bursting of soap-bubbles in a uniform electric field. *Math. Proc. Camb. Philos. Soc.* 22 (5): 728–730. https://doi.org/10.1017/S0305004100009609.
9 Teo, W.E. and Ramakrishna, S. (2006). A review on electrospinning design and nanofibre assemblies. *Nanotechnology* 17 (14): R89.
10 Lamor, J. (1898). Note on the complete scheme of electrodynamic equations of a moving material medium and on electrostriction. *Proc. R. Soc.* 63: 365.
11 Cooley, J.F. (1902). Apparatus for electrically dispersing fluids. US Patent 692,631.
12 Morton, W.J. (1902). Method of dispersing fluids. US Patent 705,691.
13 Gray, S. A letter concerning the electricity of water, from Mr. Stephen Gray to Cromwell Mortimer, M.D. Secr. R. S. *Philos. Trans. R. Soc. Lond.* 37 (422): 227–260.
14 Taylor, G. 1969. Electrically driven jets. Proceedings of the Royal Society of London. Series A, Containing Papers of a Mathematical and Physical Character, 453–475.
15 Long, Y.-Z., Xu, Y., Wang, X.-X. et al. (2019). Chapter 2 - Electrospinning: the setup and procedure. In: *Electrospinning: Nanofabrication and Applications*, 21–52. https://doi.org/10.1016/B978-0-323-51270-1.00002-9.
16 Hutmacher, D.W. and Dalton, P.D. (2011). Melt electrospinning. *Chem. Asian J.* 6: 44–56.
17 Dalton, P.D., Vaquette, C., Farrugia, B. et al. Electrospinning and additive manufacturing: converging technologies. *Biomater. Sci.* 1: 171.
18 Brown, T.D., Dalton, P.D., and Hutmacher, D.W. (2011). Direct writing by way of melt electrospinning. *Adv. Mater.* 23: 5651–5657.
19 Brown, T.D., Slotosch, A., Thibaudeau, L. et al. (2012). Design and fabrication of tubular scaffolds by direct writing in a melt electrospinning mode. *Biointerphases* 7: 13. https://doi.org/10.1007/s13758-011-0013-7.
20 Farrugia, B., Brown, T.D., Hutmacher, D.W. et al. (2013). Dermal fibroblast infiltration of poly(ε-caprolactone) scaffolds fabricated by melt electrospinning in a direct writing mode. *Biofabrication* 5: 025001.
21 Richard, B., Alcantar, N., Hoff, A., and Thomas, S. (2012). Optical properties of ATO sol-gel coated carbon fibers. *MRS Proceedings* 1454: 221–226. https://doi.org/10.1557/opl.2012.1253.
22 Thomas, S., Devisetty, M., Khan, R. et al. (2018). Cobalt oxide-antimony tin oxide anti-reflecting coating. National Center for Biotechnology Information. "PubChem Patent Summary for US-2020023407-A1" PubChem. https://pubchem.ncbi.nlm.nih.gov/patent/US-2020023407-A1.
23 Radacsi, N. and Nuansing, W. (2020). Chapter 7 – Fabrication of 3D and 4D polymer micro- and nanostructures based on electrospinning. In: *3D and 4D Printing of Polymer Nanocomposite Materials* (eds. K.K. Sadasivuni, K. Deshmukh and M.A. Almaadeed), 191–229. Elsevier https://doi.org/10.1016/B978-0-12-816805-9.00007-7.
24 Mane, P.P., Ambekar, R.S., and Kandasubramanian, B. (2020). Electrospun nanofiber-based cancer sensors: a review. *Int. J. Pharm.* 583: 119364. https://doi.org/10.1016/j.ijpharm.2020.119364.
25 Puttananjegowda, K., Takshi, A., and Thomas, S. (2020). Electrospun nanofibrous membrane based electrochemical glucose sensor. *IEEE Sensors Lett.* 4: 1–4.
26 Puttananjegowda, K., Takshi, A., and Thomas, S. (2020). Electrospun nanofibrous structures for electrochemical enzymatic glucose biosensing: a perspective. *J. Electrochem. Soc.* 167 (3): 037553.
27 Fokin, N., Grothe, T., Mamun, A. et al. (2020). Magnetic properties of electrospun magnetic nanofiber mats after stabilization and carbonization. *Materials (Basel, Switzerland)* 13 (7): 1552. https://doi.org/10.3390/ma13071552.
28 Agarwal, K., Sahay, R., Budiman, A., and Baji, A. (2021). Chapter 9 – Development of strong and tough electrospun fiber-reinforced composites. In: *Electrospun Polymers and Composites*, Woodhead Publishing Series in Composites Science and Engineering (eds. Y. Dong, A. Baji and S. Ramakrishna), 287–313. Woodhead Publishing https://doi.org/10.1016/B978-0-12-819611-3.00009-1.
29 Chen, H.Y., Wang, N., Di, J.C. et al. (2010). Nanowire-in-microtube structured core/shell fibers via multifluidic coaxial electrospinning. *Langmuir* 26: 11291–11296.
30 Li, D. and Xia, Y.N. (2004). Direct fabrication of Composite and ceramic hollow nanofibers by electrospinning. *Nano Lett.* 4: 933–938.

31 Zhao, Y., Cao, X.Y., and Jiang, L. (2007). Bio-mimic multichannel microtubes by a facile method. *J. Am. Chem. Soc.* 129: 764–765.

32 Zussman, E., Rittel, D., and Yarin, A.L. (2003). Failure modes of electrospun nanofibers. *Appl. Phys. Lett.* 82 (22): 3958–3960.

33 Lee, J., Lee, S.Y., Jang, J. et al. (2012). Fabrication of patterned nanofibrous mats using direct-write electrospinning. *Langmuir* 28: 7267–7275.

34 Tao, J. and Shivkumar, S. (2007). Molecular weight dependent structural regimes during the electrospinning of PVA. *Mater. Lett.* 61: 2325–2328.

35 Joseph, J., Nair, S.V., and Menon, D. (2015). Integrating substrateless electrospinning with textile technology for creating biodegradable three-dimensional structures. *Nano Lett.* 15 (8): 5420–5426. https://doi.org/10.1021/acs.nanolett.5b01815.

36 Yan, G., Yu, J., Qiu, Y. et al. (2011). Self-assembly of electrospun polymer nanofibers: a general phenomenon generating honeycomb-patterned nanofibrous structures. *Langmuir* 27 (8): 4285.

37 Chrisey, D.B. (2000). The power of direct writing. *Science* 289 (5481): 879.

38 Cho, S.J., Kim, B., An, T., and Lim, G. (2010). Replicable multilayered nanofibrous patterns on a flexible film. *Langmuir* 26 (18): 14395.

39 Jao, P.F., Machado, M., Cheng, X. et al. (2011). Fabrication of nanoporous membrane and its nonlithographic patterning using Electrospinning and Stamp-thru-mold (ESTM). *Proceedings of The 24th IEEE International Conference on Micro Electro Mechanical Systems (MEMS 2011)*, Cancun, Mexico, 23–27 January 2011.

40 Carlberg, B., Wang, T., and Liu, J. (2010). Direct photolithographic patterning of electrospun films for defined nanofibrillar microarchitectures. *Langmuir* 26 (4): 2235.

41 Zhang, D. and Chang, J. (2008). Electrospinning of three-dimensional nanofibrous tubes with controllable architectures. *Nano Lett.* 8 (10): 3283.

42 *Our Common Future* (1987). Report of the World Commission on Environment and Development. Published as Annex to General Assembly document A/42/427, Development and International Co-operation: Environment 2 August 1987.

43 Cao, Y. and Uhrich, K.E. (2019). Biodegradable and biocompatible polymers for electronic applications: a review. *J. Bioact. Compat. Polym.* 34 (1): 3–15. https://doi.org/10.1177/0883911518818075.

44 Thomas, S., Devisetty, M., Katakam, H. et al. (2015). Investigation of novel opuntia ficus-indica mucilage nanofiber membrane filtration for water systems. *MRS Proc.* 1745: Mrsf14-1745-ff04-05. https://doi.org/10.1557/opl.2015.582.

45 Ezhilarasu, H., Ramalingam, R., Dhand, C. et al. (2019). Biocompatible aloe vera and tetracycline hydrochloride loaded hybrid nanofibrous scaffolds for skin tissue engineering. *Int. J. Mol. Sci.* 20 (20): 5174. https://doi.org/10.3390/ijms20205174.

46 Yu, Y., Ma, Q., Zhang, J.-b., and Liu, G.-b. (2020). Electrospun SiO2 aerogel/polyacrylonitrile composited nanofibers with enhanced adsorption performance of volatile organic compounds. *Appl. Surf. Sci.* 512: 145697. https://doi.org/10.1016/j.apsusc.2020.145697.

47 Zong, X., Bien, H., Chung, C.Y. et al. (2005). Electrospun fine-textured scaffolds for heart tissue constructs. *Biomaterials* 26 (26): 5330.

48 Bhattarai, S.R., Bhattarai, N., Yi, H.K. et al. (2004). Novel biodegradable electrospun membrane: scaffold for tissue engineering. *Biomaterials* 25 (13): 2595.

49 Wnek, G.E., Carr, M.E., Simpson, D.G., and Bowlin, G.L. (2003). Electrospinning of nanofiber fibrinogen structures. *Nano Lett.* 3 (2): 213.

50 Serrano-Garcia, W., Jayathilaka, W.A.D.M., Chinnappan, A. et al. (2019). Nanocomposites for electronic applications that can be embedded for textiles and wearables. *Sci. China Technol. Sci.* 62: 895–902. https://doi.org/10.1007/s11431-018-9436-6.

51 Jayathilaka, W.A.D.M., Qi, K., Qin, Y. et al. (2019). Significance of nanomaterials in wearables: a review on wearable actuators and sensors. *Adv. Mater.* 31: 1805921. https://doi.org/10.1002/adma.201805921.

52 Langer, R. (1990). New methods of drug delivery. *Science* 249 (4976): 1527.

53 Choi, S.W., Jo, S.M., Lee, W.S., and Kim, Y.R. (2003). An electrospun poly (vinylidene fluoride) nanofibrous membrane and its battery applications. *Adv. Mater.* 15 (23): 2027.

54 Miao, J., Miyauchi, M., Simmons, T.J. et al. (2010). Electrospinning of nanomaterials and applications in electronic components and devices. *J. Nanosci. Nanotechnol.* 10 (9): 5507.

55 Yang, C., Huang, Y., Hassler, B.L. et al. (2009). Amperometric electrochemical microsystem for a miniaturized protein biosensor array. *IEEE Trans. Biomed. Circuits Syst.* 3 (3): 160–168.

56 Heller, A. and Feldman, B. (2008). Electrochemical glucose sensors and their applications in diabetes management. *Chem. Rev.* 108 (7): 2482–2505.

57 Bai, H. and Shi, G.Q. (2007). Gas sensors based on conducting polymers. *Sensors* 7: 267–230.

58 Sangodkar, H., Sukeerthi, S., Srinivasa, R.S. et al. (1996). A biosensor array based on Polyaniline. *Anal. Chem.* 68: 779.

59 Wisitsoraat, A., Pakapongpan, S., Sriprachuabwong, C. et al. (2013). Graphene-PEEDOT:PSS on screen printed carbon electrode for enzymatic biosensing. *J. Electroanal. Chem.* 704: 208–213.

60 Rahman, M.A., Kumar, P., Park, D.S., and Shim, Y.B. (2008). Electrochemical sensors based on organic conjugated polymers. *Sensors* 8: 118–141.

61 Li, X.Y., Chen, W.C., Qian, Q.R. et al. (2020). Electrospinning-based strategies for battery materials. *Adv. Energy Mater.* 11: 2000845. https://doi.org/10.1002/aenm.202000845.

62 Chun, S.-J., Choi, E.-S., Lee, E.-H. et al. (2012). Eco-friendly cellulose nanofiber paper-derived separator membranes featuring tunable nanoporous network channels for lithium-ion batteries. *Mater. Chem.* 22: 16618–16626.

63 Zhang, J., Yue, L., Hu, P. et al. (2014). Taichi-inspired rigid-flexible coupling cellulose-supported solid polymer electrolyte for high-performance lithium batteries. *Sci. Rep.* 4: 6272. https://doi.org/10.1038/srep06272.

64 Hu, L.B., Wu, H., La Mantia, F. et al. (2010, 2010). Thin, flexible secondary Li-ion paper batteries. *ACS Nano* 4 (10): 5843–5848. https://doi.org/10.1021/nn1018158.

65 Xu, Q., Kong, Q.S., Liu, Z.H. et al. (2014). Polydopamine-coated cellulose microfibrillated membrane as high performance lithium-ion battery separator. *RSC Adv.* 4: 7845–7850.

66 Kim, J.-H., Kim, J.-H., Choi, E.-S. et al. (2013). Colloidal silica nanoparticle-assisted structural control of cellulose nanofiber paper separators for lithium-ion batteries. *J. Power Sources* 242: 533–540. https://doi.org/10.1016/j.jpowsour.2013.05.142.

67 Zhang, L.C., Sun, X., Hu, Z. et al. (2012). Rice paper as a separator membrane in lithium-ion batteries. *J. Power Sources* 204: 149–154. https://doi.org/10.1016/j.jpowsour.2011.12.028.

68 Zhang, T.-W., Shen, B., Yao, H.-B. et al. (2017). Prawn shell derived chitin nanofiber membranes as advanced sustainable separators for Li/Na-ion batteries. *Nano Lett.* 17 (8): 4894–4901. https://doi.org/10.1021/acs.nanolett.7b01875.

69 Liu, J., Liu, Y., Yang, W. et al. (2018). Lithium ion battery separator with high performance and high safety enabled by tri-layered SiO2@PI/m-PE/SiO2@PI nanofiber composite membrane. *J. Power Sources* 396: 265–275. https://doi.org/10.1016/j.jpowsour.2018.06.008.

70 Zhurbina, I.A., Tsetlin, O.I., and Timoshenko, V.Y. (2011). Optical generation of free charge carriers in thin films of tin oxide. *Semiconductors* 45 (2): 236–240.

71 Muraoka, Y.N., Takubo, N., and Hiroi, Z. (2009). Photoinduced conductivity in tin dioxide thin films. *J. Appl. Phys.* 105 (10): 103702.

72 Singh, M.K., Mathpal, M.C., and Agarwal, A. (2012). Optical properties of quantum dots synthesized by Laser ablation in liquid. *Chem. Phys. Lett.* 536: 87–91.

73 Ngasinlapasaithian, S., Yoshikazu, S., Thammanoon, S., and Susumu, Y. (2006). Doubled layered ITO/SnO_2 conducting glass for substrate of dye-sensitized solar cells. *Sol. Energy Mater. Sol. cells* 90 (14): 2129–2140.

74 Chambuleyron, I. and Saucedo, E. (1979). Properties of chemically sprayed SnO_2 antireflecting films on Si solar cells. *Sol. Energy Mater.* 1 (3–4): 299–311.

75 Khan, R.R. (2017). Modeling, simulation and characterization of optoelectronic properties of 2D-3D CoO-ATO nano structures. Graduate Theses and Dissertations. https://scholarcommons.usf.edu/etd/7414.

76 Shanthi, E. (1982). Electrical and optical-properties of tin oxide-films doped with F and (SB+F). *J. Appl. Phys.* 53 (3): 1615–1612.

77 Stjerna, B., Olsson, E., and Granqvist, C.G. (1994). Optical and electrical properties of radio frequency sputtered tin oxide films doped with oxygen vacancies, F, Sb, or Mo. *J. Appl. Phys.* 76 (6): 1397.

78 Van der Bruggen, B. and Vandecasteele, C. (2003). Removal of pollutants from surface water and groundwater by nanofiltration: overview of possible applications in the drinking water industry. *Environ. Pollut.* 122: 435–445.

79 Gopal, R., Kaur, S., Ma, Z. et al. (2006). Electrospun nanofibrous filtration membrane. *J. Membr. Sci.* 281: 581–586.

80 Thavasi, V., Singh, G., and Ramakrishna, S. (2008). Electrospun nanofibers in energy and environmental applications. *Energy Environ. Sci.* 1 (2): 205–221.

81 Sereshti, H., Amini, F., and Najarzadekan, H. (2015). Electrospun polyethylene terephthalate (PET) nanofibers as a new adsorbent for micro-solid phase extraction of chromium(VI) in environmental water samples. *RSC Adv.* 5: 89195–89203. https://doi.org/10.1039/C5RA14788C.

82 Kang, Y.-L., Zhang, J., Wu, G. et al. (2018). Full-biobased nanofiber membranes toward decontamination of wastewater containing multiple pollutants. *ACS Sustain. Chem. Eng.* 6 (9): 11783–11792. https://doi.org/10.1021/acssuschemeng.8b01996.

83 Martín, D.M., Ahmed, M.M., Rodríguez, M. et al. (2017). Aminated polyethylene terephthalate (PET) nanofibers for the selective removal of Pb(II) from polluted water. *Materials* 10: 1352. https://doi.org/10.3390/ma10121352.

84 Wu, J., Wang, N., Zhao, Y., and Jiang, L. (2013). Electrospinning of multilevel structured functional micro-/nanofibers and their applications. *J. Mater. Chem. A* 1: 7290–7305.

85 Wang, X., Yu, J., Sun, G., and Ding, B. (2016). Electrospun nanofibrous materials: a versatile medium for effective oil/water separation. *Mater. Today* 19 (7): 403–414.

86 Ganesh, V.A., Ranganath, A.S., Baji, A. et al. (2016). Electrospun differential wetting membranes for efficient oil–water separation. *Macromol. Mater. Eng.* 301 (7): 812–817.

87 Tai, M.H., Gao, P., Tan, B.Y.L. et al. (2014). Highly efficient and flexible electrospun carbon–silica nanofibrous membrane for ultrafast gravity-driven oil–water separation. *ACS Appl. Mater. Interfaces* 6 (12): 9393–9401.

88 Ge, J., Zong, D., Jin, Q. et al. (2018). Biomimetic and superwettable nanofibrous skins for highly efficient separation of oil-in-water emulsions. *Adv. Funct. Mater.* 28 (10): 1705051.

89 Jiang, S., Hou, H., Agarwal, S., and Greiner, A. (2016). Polyimide nanofibers by "Green" electrospinning via aqueous solution for filtration applications. *ACS Sustain. Chem. Eng.* 4 (9): 4797–4804.

90 Schäfer, A.I., Stelzl, K., Faghih, M. et al. (2018). Poly(ether sulfone) nanofibers impregnated with β-cyclodextrin for increased micropollutant removal from water. *ACS Sustain. Chem. Eng.* 6 (3): 2942–2953.

91 SWI swissinfo.ch (2020). Swiss develop first see-through surgical mask. https://www.swissinfo.ch/eng/swiss-develop-first-see-through-mask/45824518 (accessed 9 June 2020).

92 Yuen, K.S., Ye, Z.W., Fung, S.Y. et al. (2020). SARS-CoV-2 and COVID-19: the most important research questions. *Cell Biosci.* 10: 40. https://doi.org/10.1186/s13578-020-00404-4; https://www.healthline.com/health/coronavirus-vs-sars.

93 Leung, W.W.F. and Sun, Q. (2020). Electrostatic charged nanofiber filter for filtering airborne novel coronavirus (COVID-19) and nano-aerosols. *Sep. Purif. Technol.* 250: 116886. https://doi.org/10.1016/j.seppur.2020.116886.

94 World Health Organization (2018). WHO/L. Chitsulo. https://www.who.int/news-room/fact-sheets/detail/arsenic (accessed 15 February 2018).

95 Argos, M., Kalra, T., Rathouz, P.J. et al. (2010). Arsenic exposure from drinking water, and all-cause and chronic-disease mortalities in Bangladesh (HEALS): a prospective cohort study. *Lancet* 376 (9737): 252–258.

96 Flanagan, S.V., Johnston, R.B., and Zheng, Y. (2012). Arsenic in tube well water in Bangladesh: health and economic impacts and implications for arsenic mitigation. *Bull. World Health Organ.* 90: 839–846.

97 Thomas, S.W., Pais, Y., and Alcantar, N.A. (2012). Electrospun cactus mucilage nanofibers. US Patent 9,555,392.

98 Eppili, V. (2016). Electrospinning of polymeric solutions using opuntia ficus-indica mucilage and iron oxide for nanofiber membranes for treating arsenic contaminated water. Graduate thesis, University of South Florida, Tampa, FL.

99 Muppaneni, R. (2015). Investigation of opuntia ficus-indica mucilage nanofiber membrane filtration for water systems. https://scholarcommons.usf.edu/etd/5541 (accessed 2 December 2020).

100 Thomas, S.W. and Alcantar, N.A. (2017). System integration of functionalized natural materials. *MRS Bull.* 42 (05): 343.

101 Stebbins, D., Buttice, A.L., Fox, D. et al. (2013). Cactus Mucilage as an emergency response biomaterial to provide clean drinking water. In: *Monitoring Water Quality: Pollution Assessment, Analysis, and Remediation* (ed. S. Ahuja), 249–260. Oxford, UK: Elsevier.

102 Alcantar, N.A., Joseph, B., and Young, K. (2007). Water purification method using plant molecules for removal of arsenic. US Patent US7,943,049 B1, University of South Florida, Tampa, FL.

103 Buttice, A.L. and Alcantar, N.A. (2014). Sediment removal with the opuntia fi cusIndica cactus: a water purification method for communities in Latin America. In: *Comprehensive Water Quality and Purification*, vol. 1 (ed. S. Ahuja), 98. New York: Elsevier.

104 Alcantar, N.A., Fox D.I., Thomas, S., and Toomey, R.G. (2012). Use of cactus Mucilage as a dispersant and absorbant for oil in oil-water mixtures. US Patent 9,163,374, University of South Florida, Tampa, FL, 56.

105 Pais, Y. (2011). Fabrication and characterization of electrospun cactus Mucilage nanofibers. Graduate theses and dissertations. http://scholarcommons.usf.edu/etd/3279.

106 Balasubramanian, K., Yadav, R., and Prajith, P. (2015). Antibacterial nanofibers of polyoxymethylene/gold for pro-hygiene applications. *Int. J. Plast. Technol.* **19**: 363–367. https://doi.org/10.1007/s12588-015-9127-y.

107 Prajapati, D.G. and Kandasubramanian, B. (2019). Progress in the development of intrinsically conducting polymer composites as biosensors. *Macromol. Chem. Phys.* 220: 1800561. https://doi.org/10.1002/macp.201800561.

108 Zhou, H., Shi, Z., Wan, X. et al. (2019). The relationships between process parameters and polymeric nanofibers fabricated using a modified coaxial electrospinning. *Nanomaterials* 9: 843.

109 Kandasubramanian, B. and Govindaraj, P. (2014). Peeling model for cell adhesion on electrospun polymer nanofibres. *J. Adhes. Sci. Technol.* 28 (2): 171–185. https://doi.org/10.1080/01694243.2013.833402.

110 Zhao, K., Wang, W., Yang, Y. et al. (2019). From taylor cone to solid nanofiber in tri-axial electrospinning: size relationships. *Results Phys.* 15: 102770. https://doi.org/10.1016/j.rinp.2019.102770.

111 Gore, P., Khraisheh, M., and Kandasubramanian, B. (2018). Nanofibers of resorcinol–formaldehyde for effective adsorption of As (III) ions from mimicked effluents. *Environ. Sci. Pollut. Res.* **25**: 11729–11745. https://doi.org/10.1007/s11356-018-1304-z.

112 Hui, N., Sun, X., Niu, S., and Luo, X. (2017). PEGylated polyaniline nanofibers: antifouling and conducting biomaterial for electrochemical DNA sensing. *ACS Appl. Mater. Interfaces* 9 (3): 2914–2923. https://doi.org/10.1021/acsami.6b11682.

113 Nagarajan, S., Bechelany, M., Kalkura, N.S. et al. (2019). Chapter 20 – Electrospun nanofibers for drug delivery in regenerative medicine. In: *Micro and Nano Technologies,Applications of Targeted Nano Drugs and Delivery Systems* (eds. S.S. Mohapatra, S. Ranjan, N. Dasgupta, et al.), 595–625. Elsevier https://doi.org/10.1016/B978-0-12-814029-1.00020-X.

114 Chen, Z., Chen, Z., Zhang, A. et al. (2016). Electrospun nanofibers for cancer diagnosis and therapy. *Biomater. Sci.* 4: 922–932.

115 Kosaki, Y., Izawa, H., Ishihara, S. et al. (2013). Nanoporous carbon sensor with cage-in-fiber structure: highly selective aniline adsorbent toward cancer risk management. *ACS Appl. Mater. Interfaces* 5: 2930–2934. https://doi.org/10.1021/am400940q.

116 Puttananjegowda, K. and Thomas, S. (2020). A low-power low-noise multi-stage transimpedance amplifier for ampherometric based blood glucose monitoring systems. *Analog Integr. Circuits Signal Process. J.* 102: 659–666.

117 Thomas, S.W., Khan, R.R., Puttananjegowda, K., and Serrano-Garcia, W. (2020). Chapter 9 – Conductive polymers and metal oxide polymeric composites for nanostructures and nanodevices. In: *Advances in Nanostructured Materials and Nanopatterning Technologies, Applications for Healthcare, Environmental and Energy, Advanced Nanomaterials* (eds. V. Guarino, M.L. Focarete and D. Pisignano), 243–271. Elsevier https://doi.org/10.1016/B978-0-12-816865-3.00009-3.

118 Puttananjegowda, K. and Thomas, S. (2018). A CNTFET based multi-stage transimpedance amplifier for blood glucose monitoring system. IEEE IEMCON, November 2018.

119 Puttananjegowda, K. and Thomas, S. (2019). Cascode common source transimpedance amplifiers for analyte monitoring systems. US Patent App. 16/008,864, April 2019.

120 Puttananjegowda, K. (2020). Electrospun nanofibrous membrane based glucose sensor with integration of potentiostat circuit. Graduate theses and dissertations.

121 He, H., Zhang, L., Guan, X. et al. (2019). Biocompatible conductive polymers with high conductivity and high stretchability. *ACS Appl. Mater. Interfaces* 11 (29): 26185–26193. https://doi.org/10.1021/acsami.9b07325.

122 Silva, G.A. (2018) https://doi.org/10.3389/fnins.2018.00843). A new frontier: the convergence of nanotechnology, brain machine interfaces, and artificial intelligence. *Front. Neurosci.* 12: 843.

123 Pinto, N.J. and Serrano, W. (2012). Composite nanofibers of electroactive polymers prepared via electrospinning. *15th European Conference on Composite Materials (ECCM15)*, Venice, Italy, 24–28 June 2012.

124 Serrano, W., Meléndez, A., Ramos, I., and Pinto, N.J. (2014). Electrospun composite poly(lactic acid)/polyaniline nanofibers from low concentrations in CHCl3: making a biocompatible polyester electro-active. *Polymer* 55: 5727–5733.

125 Serrano, W., Meléndez, A., Ramos, I. and Pinto, N.J. (2014). Sensor response of electrospun poly(lactic acid)/polyaniline nanofibers to aliphatic alcohol vapors of varying sizes. *2014 IEEE 9th IberoAmerican Congress on Sensors*, Bogota, pp. 1–4, doi: 10.1109/IBERSENSOR.2014.6995537.

126 Han, D. and Steckl, A.J. (2013). Triaxial electrospun nanofiber membranes for controlled dual release of functional molecules. *ACS Appl. Mater. Interfaces* 5 (16): 8241–8245. https://doi.org/10.1021/am402376c.

127 Yu, D.-G., Li, X.Y., Wang, X. et al. (2015). Nanofibers fabricated using triaxial electrospinning as zero order drug delivery systems. *ACS Appl. Mater. Interfaces* 7: 18891–18897.

128 Li, J., Xu, W., Li, D. et al. (2018). Locally deployable nanofiber patch for sequential drug delivery in treatment of primary and advanced orthotopic hepatomas. *ACS Nano* 12 (7): 6685–6699. https://doi.org/10.1021/acsnano.8b01729.

129 Pant, B., Park, M., and Park, S.-J. (2019). Drug delivery applications of core-sheath nanofibers prepared by coaxial electrospinning: a review. *Pharmaceutics* 11: 305. https://doi.org/10.3390/pharmaceutics11070305.

130 Rizvi, S.A.A. and Saleh, A.M. (2018). Applications of nanoparticle systems in drug delivery technology. *Saudi Pharm. J.* 26 (1): 64–70. https://doi.org/10.1016/j.jsps.2017.10.012.

131 Nanocomputing. Computer Sciences. Retrieved 22 September 2021 from Encyclopedia.com. https://www.encyclopedia.com/computing/news-wires-white-papers-and-books/nanocomputing.

132 Hoffman, J.E., Ravets, S., Grover, J.A. et al. (2014). Ultrahigh transmission optical nanofibers. *AIP Adv.* 4: 067124. https://doi.org/10.1063/1.4879799.

133 Hecht, J. (2020). Quantum science: the quest for quantum information technology expands Hope for a new generation of information technology based on quantum science led the U.S. to launch a National Quantum Initiative; it's a big challenge with a potentially huge payoff, 14 July 2020.

134 US National Science and Technology Council Committee on Technology (2011). National nanotechnology initiative strategic plan, subcommittee on nanoscale science, engineering, and technology. February 2011. http://www.nano.gov/sites/default/files/pub_resource/2011_strategic_plan.pdf (accessed 2 November 2020).

135 Reich, D.M., Gualdi, G., and Koch, C.P. (2014). Optimal qudit operator bases for efficient characterization of quantum gates. *J. Phys. A Math. Theor.* 47 (38): 385305.

136 Laroche, G., Marois, Y., Guidoin, R. et al. (1995). Polyvinylidene fluoride (PVDF) as a biomaterial: from polymeric raw material to monofilament vascular suture. *J. Biomed. Mater. Res.* 29 (12): 1525–1536. https://doi.org/10.1002/jbm.820291209.

137 Hearon, K., Nash, L.D., Rodriguez, J.N. et al. (2014). A high-performance recycling solution for polystyrene achieved by the synthesis of renewable poly(thioether) networks derived from d-Limonene. *Adv. Mater.* 26: 1552–1558. https://doi.org/10.1002/adma.201304370.

138 Bosworth, L., Alam, N., Wong, J., and Downes, S. (2013). Investigation of 2D and 3D electrospun scaffolds intended for tendon repair. *J. Mater. Sci.: Mater. Med.* 24: 1605–1614. https://doi.org/10.1007/s10856-013-4911-8.

139 Keirouz, A., Chung, M., Kwon, J. et al. (2020). 2D and 3D electrospinning technologies for the fabrication of nanofibrous scaffolds for skin tissue engineering: a review. *WIREs Nanomed. Nanobiotechnol.* 12: e01626. https://doi.org/10.1002/wnan.1626.

140 Feig, V.R., Tran, H., and Bao, Z. (2018). Biodegradable polymeric materials in degradable electronic devices. *ACS Cent. Sci.* 4 (3): 337–348. https://doi.org/10.1021/acscentsci.7b00595.

12

Plants for Nanomaterial: Improving the Environmental Sustainability

Debjani Nath, Baishakhi Bairagi, Pratyusha Banerjee, Anugrah Ray, and Puspendu Roy

Department of Zoology, University of Kalyani, Kalyani Nadia, West Bengal, India

12.1 Nanotechnology: Small Material With Large Potential

According to the US National Nanotechnology Initiative, "nanotechnology is the understanding and control of matter at dimensions of roughly 1 to 100 nanometers, where unique phenomena enable novel applications. Nanotechnology involves imaging, measuring, modeling, and manipulating matter at this length scale and encompasses nanoscale science, engineering, and technology." The ASTM (American Society for Testing and Materials), International Committee E56 standard (E 2456-06) recently defined nanotechnology as "a term referring to a wide range of technologies that measure, manipulate, or incorporate materials and/or features with at least one dimension between approximately 1 and 100 nanometers (nm). Such applications exploit the properties, distinct from bulk/macroscopic systems, of nanoscale components." The parameters of nanotechnology are clear that it deals with materials having a dimension about 1–100 nm having some unique properties [1]. While nanomaterials are defined by their size range and properties, their composition does not. It permeates all industries as it is the most powerful technology due to its potentiality. Nanomaterials could be nanorods, nanotubes, nanofiber, nanopolymers in shape composed of pure metal atoms, metal–metal compounds, or metal oxides, carbon nanotubes, or soccer ball-shaped fullerenes of carbon or other atoms, or nanoscaled branched polymers (dendrimers), or nanometer diameter wires, or combinations and derivatives of any of these. The main theme of nanotechnology is that when a matter comes to its nanoscale levels, its physical, electrical, magnetic, and optical properties become changed [2]. Quantum mechanics can explain it clearly because at nanoscale level classical physics cannot explain the properties of matter. Gold is chemically inert, but at nanoscale level it becomes a potential chemical catalyst. Generally, two approaches are used in nanotechnology such as the bottom up and top down methods (Figure 12.1). In the case of bottom-up approach, small molecular components are assembled together chemically and they can recognize themselves by molecular construction. In the case of top-down method, a nanoscale material is constructed by larger entities.

12.1.1 Scale of Nanoparticles at Which Much of the Material Properties Occur

The important parameters of these materials are their shape (including aspect ratios), size, and the substructural morphology of the substance. Nanoparticles may be present as an emulsion (two liquid phases) or aerosol (mostly solid or liquid phase in air) or a suspension (mostly solid in liquids). The surface and interfacial properties of nanoformulation may be modified in the presence of chemical agents (surfactants). Indirectly such agents can modify the outmost layer of the particle stabilizing against coagulation or aggregation by conserving particle charge. Very complex compositions, possibly with complex mixtures of adsorbents, have to be expected depending on the growth history and the lifetime of a nanoparticle. Different sunscreens are prepared by nanoparticles such as zinc oxide and titanium oxide. Photocatalytic glasses and UV protective clothes are also prepared by the nanomaterials. Aluminum silicate is helpful in producing scratch-resistant coatings. This is an important application of nanoparticles. Silver nanoparticle is very much important as an antimicrobial agent. These are also useful in preparing disinfectants. Nanosized oil droplets are mainly helpful in enhancing bacterial activities. Drug delivery should be possible by the help of different nanomaterials. In the case of cancer chemotherapy

Sustainable Nanotechnology: Strategies, Products, and Applications, First Edition.
Edited by Yashwant V. Pathak, Govindan Parayil, and Jayvadan K. Patel.

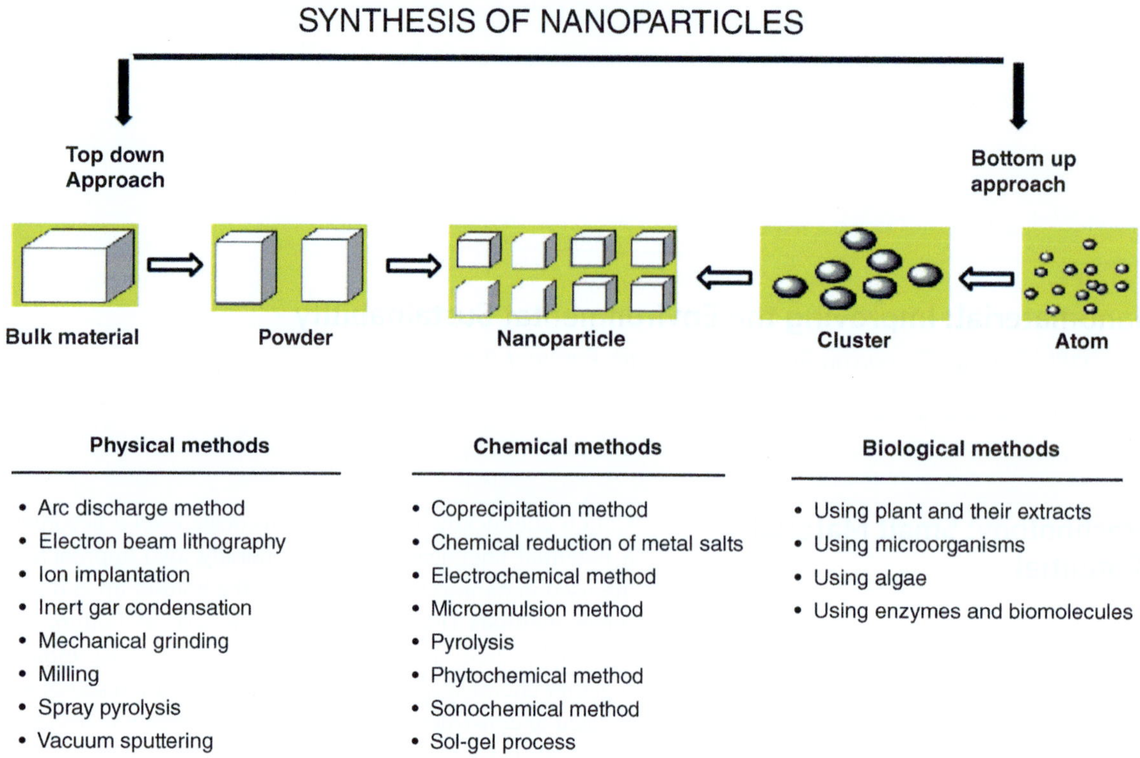

Figure 12.1 Different approach of synthesis of nanoparticle.

targeted drug is so much important, otherwise healthy cells may be destroyed. In this targeted drug delivery system, nanomaterials play very much important role. Nanomaterials are also helpful in preparing liquid crystal devices (LCDs). Now, the tennis rackets are modified by clay nanomaterials. Deep-penetrating cosmetic creams contain nanomaterials that will act as wrinkle resistant. Nanocantilever arrays and nanocontainers are helpful in targeting different biological diseases. Carbon nanotube, small cylindrical structure, is the allotrope of carbon. It is so much important in material science due to its high electrical conductivity. In biological and biomedical application nanotubes are used. In preparing electrical nanosensor, nanotubes play a vital role. Nanocomposites are also important because of their thermal stability, optical clarity, etc. For detecting phosgene, polyaniline nanofiber composites with amines are used. Nanopolymers such as dendrimer and fullerene are so much important in drug delivery. Nanofiber also has different applications such as in preparing nanocomposites and in other medical applications. Nanoshells are of spherical structure, and it is a new approach for optical applications of nanomaterials. Nanohorns are of very small structure, and when becomes united, they act as electrodes. Nanowires also have unique optical and electrical properties and act as light-emitting diodes. Nanosprings are modified nanowires that are helpful in preparing highly effective magnetic field detector. Nanocrystals are with specific shape, and these are much harder than other nanomaterials. They are mainly helpful in preparing different electronic parts. Nanomaterials have important role in preparing DVDs, chip fabrication, batteries, nanocantilever arrays, nanocontainers, nanovector, etc.

12.2 Environmental Sustainability

Environmental sustainability is the ability to maintain the qualities of values in the physical environment. For example, if people want to maintain human life, different conditions are essential such as clean water, clean air, and responsible interaction with the environment. Sustenance also relates to improve quality of life, the livability, and beauty of the environment, avoiding depletion or degradation of natural resources. It has three informal pillars such as people, planet, and profits. Traditionally, when environmental problems arise, people think how to reduce the damage. But it is not always easy to work out exactly when and where threats will have their effects, and often, the impacts are not easy to reverse. So, increasingly environmental managers adopt strategies and aim to prevent damages that are related to environmental hazards. A fully sustainable program needs to include actions to prevent the environment from threats and damage and restore or reverse the damage already done. Environmentally sustainable programs include actions to reduce more the use of physical resources, the adoption of "recycle everything/buy the recycled" approach, the use of renewable resources, and the design of production processes and products to eliminate the toxic materials, protecting and restoring the natural habitats and environments valued for their livability or beauty. Perhaps, tackling environmental sustainability is the most pressing global public policy challenge. Some of the issues that pose major environmental sustainability problems include:

- Destruction of the living environments (habitats) of native species.
- Discharge of polluting chemicals and other materials into the environment.
- Emission of greenhouses gases into the atmosphere that can cause climate change.
- Depletion of low cost oil and other fossil fuels.

Human depend on the physical environment in different ways sustaining desired environmental conditions, which contributes to the sustaining human societies, that is to social sustainability. The viability of the economy clearly depends on environmental resources and service flows, so economic sustainability depends on environmental sustainability. More generally, it can be seen that sustainability in one domain can be necessary for sustainability in another.

The 1980 World Conservation Strategy produced by the International Union for the Conservation of Nature and Natural Resources (IUCN or World Conservation Union) put forward the concept of "sustainable development," meaning development that would allow ecosystem services and biodiversity to be sustained. Then, the 1992 UN Conference on Environment and Development (UNCED) in Rio set in train processes such as Agenda 21 and Local Agenda 21 that resulted in many people coming to the view that sustainability equals the integration or balancing of environmental, social, and economic issues or simultaneous progress in the environmental, social, and economic domains, often in the context of strong programs of consultation and participation. The 1987 Brundtland Report shifted the meaning of sustainable development to mean "development that meets the needs of the present without compromising the ability of future generations to meet their own needs" [3].

Sustainability requires a significant change in today's practice, including the minimization of the manufacturing footprint of a material, and also the sustainable gains of its use during the life cycle and clever reuse of the material or its components. Integral sustainability must become a driver for new energy technologies, to produce durable systems to convert, produce, and store clean energy. Resources for energy (fossil origin) and raw materials (rare elements) are depleting, and this requires a transition to sustainable energy production as well as the reduction, replacement, or recycling of rare elements and the further development of bio-based materials. The transition to a sustainable society will likely have a tremendous impact. While initial efforts are aimed at reducing the footprint by making existing technologies more efficient, the final goal is a (circular) society based on truly sustainable resources for energy and materials. In this transition to a sustainable society, advanced materials will play a crucial role. These materials will have in common less nonrenewable energy use and less greenhouse gas emission during the synthesis, construction, processing, packaging, transportation usage, recycling, and reuse.

12.3 Natural Phytoconstituents: A Protective Package for Sustainable Materials

Phytochemicals found in different parts of plants have the potential to protect the plants from attack of viruses, bacteria, and fungi. These phytochemicals also have medicinal applications as in diabetes, heart diseases, hypertension, depression, etc.

12.3.1 Chemical Nature of Most Phytochemicals

Alkaloids: Alkaloids are the largest group of secondary metabolites within the plant parts containing nitrogen bases that is obtained from amino acids. Alkaloids are mainly found in seeds and roots. They have important role in the central nervous system and as an anesthetic. Alkaloids also find their usefulness as pain relievers. Atropine is one type of alkaloid that is used as an antidote to organophosphate poisoning. Caffeine is one of the phytochemicals that stimulates the CNS as well as the respiratory systems. Caffeine also serves as an antidote to barbiturate and morphine that are poisonous, and emetine (from *Cephaelis ipecacuanha*) root is very much important for protozoan infections, e.g. amebic dysentery [4].

Tannins: These are mostly found in fruits containing phenolic group. Due to the presence of this group they have antiseptic activity. They are also helpful in gastrointestinal problems such as diarrhea and dysentery. So, they have very important antibacterial activity that can destroy gastrointestinal diseases. They are also treated as useful antioxidants [5].

Saponins: These are another group of phytochemicals that contains amphipathic glycoside group. Saponins are therapeutically important because they can be used in curing malignant tumors due to their anticancer activities. Saponins are also helpful in producing inhibitory effect on inflammation. They have the capability in lowering cholesterol level. These activities of saponins help to reduce the probabilities of cardiovascular diseases as well as stroke [6].

Glycosides: These are one type of carbohydrates stored in plant parts. They help in regulating the cell growth and metabolism. Glycosides function in reducing the thyroxin level. They have the antiprotozoal activities that cause killing of different pathogenic protozoan species. They are also useful in cancer therapy, skin disease, heart diseases, etc. [7].

Anthraquinones: These are phenolic and glycosidic compounds and mainly used for industrial purposes [8].

Cardiac glycosides: Strophanthus is an important phytochemical that is known as heart drug. It is actually helpful in cardiac failure and also helps in strengthening of weakened heart.

Cyanogenic glycosides: These glycosides can produce HCl on hydrolysis. No specific use of the phytochemicals in pharmaceutical can be seen due to their poisonous activities.

Flavonoids: These are an important group of polyphenols. Flavonoids function to reduce the risk of coronary heart diseases and have anticoagulant, anti-inflammatory, and aphrodisiac properties [9].

Phenols: These are also phytochemicals that can be found in different plants. Phenolics are the natural antioxidants that can be found in different fruits also. They can fight against cancer cells, act as anti-inflammatory agents, and are also helpful in heart diseases [10].

Steroids: These are mostly known as cardiac or steroid glycosides that can work on cardiac muscle directly. They are so much helpful in osteoporosis. Steroids have the ability to exhibit activities such as antifungal, antiviral, antileukemia, hypnotic, antipyretic, and muscle-relaxant and found in large quantities in many plants [11].

Carotenoids: These are one of the coloring agents that can be found in different fruits and vegetables. They act as good antioxidant that is so important against the free radicals [12].

12.3.2 Different Phytochemicals and Their Mode of Action

Carvacrol: This is one of the most important phytochemical found in different herbs and other medicinal plants. It can inhibit COX-mediated inflammation and regulates body temperature by controlling ion channels. It also has dopamine-controlling activities in the brain and helpful in depression ([13]) (Figure 12.2).

Curcumin: This phenolic compound also has medicinal application as antidepressant. Different chronic stress and shocks causing behavioral changes can be improved by curcumin. Curcumin action involves 5-HT receptors, specifically 5-HT1A/1B and 5-HT2C subtypes. In the case of

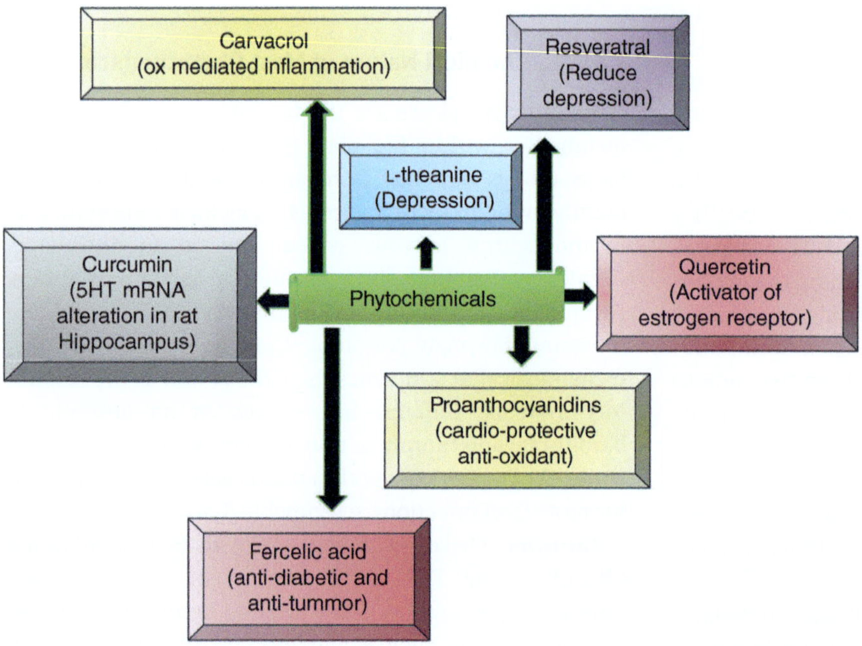

Figure 12.2 Different important phytochemicals and their mode of action.

rat hippocampus, 5HT mRNA alteration can be seen due to the effect of curcumin [14].

Ferulic acid: This phytochemical mainly acts as an antioxidant. Different cell-signaling pathways are also modified by the actions of ferulic acid as it enhances phosphorylation of CREB and different neurotrophic factor mRNAs. It also has antidiabetic and antitumor activities [15].

L-theanine: It is an amino acid that can bind to receptors acting as antagonist of AMPA and kinate receptors. It also has serotonin and dopamine reuptake activities. It also improves sleep disturbance, anxiety, depression, etc. [16].

Proanthocyanidins: These can be found in different plants and contain polymeric flavan-3-ols group that has pharmaceutical application. They have cardioprotective and antioxidant effect and antinociceptive application ([17]).

Quercetin: This is most important antioxidant that can be found in different fruits, vegetables, herbs, etc. It can act as protein kinase inhibitor as well as activator of estrogen receptor. It also reduces the hyperactivation of hypothalamic-pituitary-adrenal axis [18].

Resveratrol: This phytochemical is found in different fruits, and it has anti-inflammatory and neuroprotective activities. According to experimental reports, it inhibits 5-HT/noradrenaline uptake and MAO activity in rats. It also reduces depression that is caused by stress. It can enhance the levels of 5-HT/dopamine/noradrenaline as well as decreases MAO activity that proves the interaction of resveratrol with the monoaminergic system for other activities ([19]).

12.3.3 Mechanism of Action of Phytochemicals

Phytochemicals work as a modulator of metabolic processes, activator of gene expression, and control different signal transduction processes. Phytochemicals can show their functions in different ways (Figure 12.3):

Antioxidant: Reactive oxygen species are produced in the body and removed by the antioxidants. Actually, the free radical can be converted to a stable form by the help of different phytochemicals. Plant parts contain lots of free radicals that scavenge superoxide anions and hydroxyl radical. Thus, the phytochemicals reduce the oxidative damages and diseased condition of cell organelles as mitochondria, cell membrane, DNA, etc. [20].

Anticarcinogenesis: Polyphenols as in plant phytochemicals have the potential for inhibiting the carcinogenesis [21]. Phenolics have the capability to minimize the formation of the specific cancer-promoting nitrosamines formed by the dietary nitrites and nitrates. Different vegetables are the source of good anticarcinogens. Glucosinolates from various vegetables are very effective for colon cancer. Phytochemicals can increase the

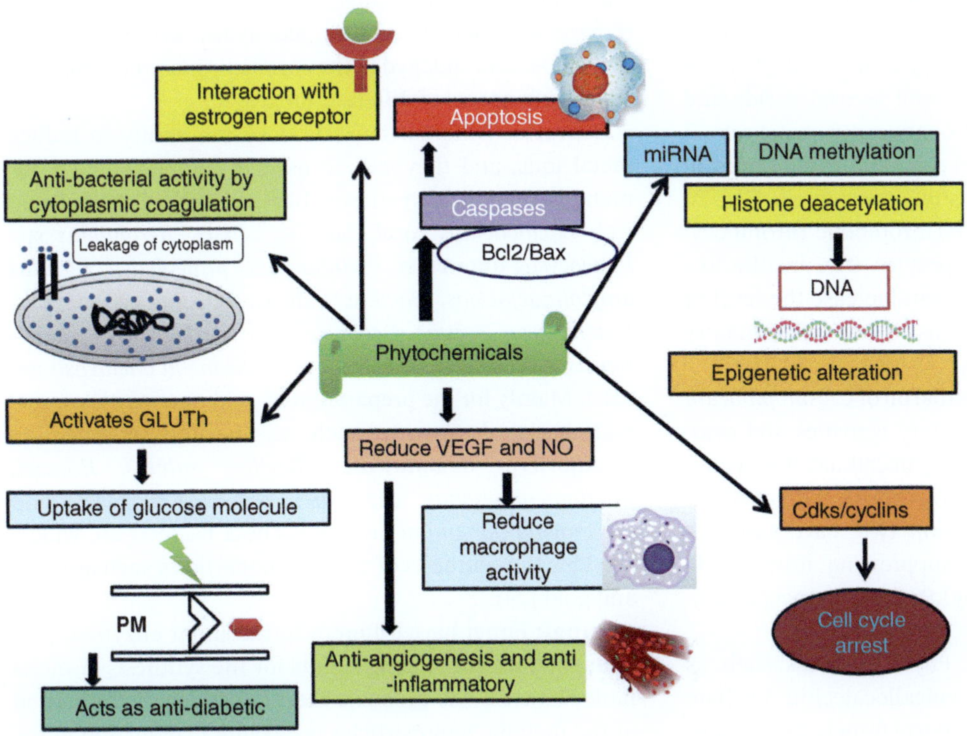

Figure 12.3 Mechanism of action of phytochemicals in biological system.

glutathione-S-transferase and also reduce the urinary purine metabolite, an indicator of cancerous DNA degradation. Isothiocyanates and the indole-3-carbinols detoxify the metabolites and are responsible to change the structure of DNA by activating the phase ii enzyme NAD (P) H quinone reductase or glutathione. Sulforaphane also has the capability to arrest the cell cycle and induce the apoptosis of cancerous cells. Indole-3-carbinol can also reduce urinary cancer as well as breast cancer by blocking the estrogen receptor. It upregulates p21 and p27 as well as downregulates CDK 6 in prostate cancer. It works in different ways for destructing the tumors by regulating the Akt, NF-kappa B, MAPK, and Bel-2-signaling pathways.

Antimicrobial activity: Mainly, the phenolic compound of the phytochemicals reduces the adherence capability of microorganism to the bladder and teeth and also reduces UTIs. Different volatile oils of plant origin in combination with high carbon dioxide level can resist microorganisms.

Antiulcer: It is reported that plant extracts have the capability to inhibit both the growth of *H. pylori* in vitro and its urease activity. The capability to reduce microorganisms of some plant extracts in liquid medium and at low pH levels can be increased in the human stomach. The plant extract also has the capability to inhibit the effect of ulcers on the intestine, to reduce kidney Na+/K+ ATPase activity, and also has effect on alanine transport in rat jejunum.

Antidiabetic: Different phytoconstituents are responsible for reducing diabetes. Cinnamaldehyde, a phytoconstituent, has significant antihyperglycemic effect that can result in lowering cholesterol as well as triglyceride and increases HDL cholesterol in STZ-induced diabetic rats. It is reported that polyphenols with procyanidin polymer can increase the amount of IR (insulin resistance), GLUT4 (glucose transporter-4), and TPP (thrombotic thrombocytopenic purpura) in 3T3-L1 adipocytes. The insulin-like activity of *Cinnamon* polyphenol can increase the level of TTP, IRβ, and GLUT4. It also acts as an anti-inflammatory and antiangiogenesis agent [22].

Anti-inflammatory: The essential oil of *C. osmophloeum* twigs has excellent anti-inflammatory activities and cytotoxicity against HepG2 (human hepatocellular liver carcinoma cell line) cells. According to reports it is indicated that the constituents of *C. osmophloeum* twig have excellent anti-inflammatory activities by suppressing nitric oxide production that can be increased by LPS (lipopolysaccharide)-stimulated macrophages [23].

Multifunctional activities: Phytochemicals such as curcumin from turmeric, epigallocatechin-3-gallate (EGCG) from green tea, and resveratrol from red wine have different functions, but their proper functions are not exactly known. They can affect cell cycle proteins, cell adhesion molecules, different protein kinases, etc. Phytochemicals can also interfere with the transcription and a translation process of bacterial cell as well as it has effect in modulating transcription factors at molecular level. There are different molecular functions of phytoconstituents, but the proper signaling mechanisms are not known [20].

12.4 Green Synthesis of Nanoparticles

In materials science, "green" synthesis of a wide range of materials/nanomaterials including metal/metal oxides nanomaterials, hybrid materials, and bioinspired materials has gained extensive attention as a reliable, sustainable, and eco-friendly protocol for synthesis. Green synthesis actually means a controlling process that will be helpful in environmental friendliness as it should be less toxic to the environment. "Green" syntheses are required to avoid the production of hazardous by-products by applying reliable, sustainable, and eco-friendly synthesis procedures. Nowadays, the green synthesis of nanoparticles can be achieved by different biological materials such as bacteria, fungi, algae, and plant extract (Figure 12.4). Green synthesis of nanoparticles also requires controlling of different parameters such as pH, temperature, time of incubation, pressure, and different solvent systems. Actually, the plant parts contain different phytochemicals such as flavonoids, terpenoids, phenols, amines, aldehydes, ketones, carboxylic acids, and aldehydes that are acting as reducing and stabilizing agents of different metals.

Bacteria: Different bacteria have the ability to reduce metal ions, and they can be used in nanotechnology for metal nanoparticle synthesis. Bacteria such as *Escherichia coli*, *Lactobacillus casei*, *Bacillus cereus*, *Aeromonas* sp., *Phaeocystis antarctica*, *Pseudomonas proteolytica*, *Bacillus amyloliquefaciens*, *Bacillus indicus*, *Bacillus cecembensis*, *Enterobacter cloacae*, *Geobacter* spp., *Arthrobacter gangotriensis*, *Corynebacterium* sp., and *Shewanella oneidensis* are used. Mainly for the preparation of gold nanoparticles, several bacterial species such as *Bacillus megaterium*, *Desulfovibrio desulfuricans*, *Bacillus subtilis*, *E. coli*, *Rhodopseudomonas capsulate*, *Shewanella alga*, and *Plectonema boryanum* are used. These bacteria are widely used for the synthesis of metal nanoparticles such as silver and gold [24].

Fungi: Fungi have different intracellular enzymes, proteins that act as reducing agents for the synthesis of metal nanoparticles. The probable mechanism for the formation of the metallic nanoparticles is enzymatic reduction in the cell wall or inside the fungal cell. Fungi such as *Rhizopus*

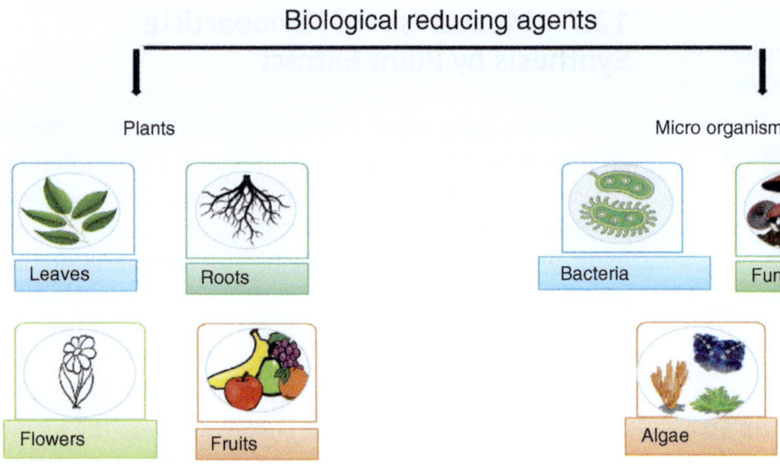

Figure 12.4 Biological agents used for nanoparticle synthesis.

nigricans, Aspergillus fumigates, Phanerochaete chrysosporium, Fusarium semitectum, Cladosporium cladosporioides, Penicillium brevicompactum, Phoma glomerata, Alternaria alternata, Trichoderma viride, Verticillium luteoalbum, Fusarium oxysporum, and *Aspergillus flavus* are helpful in the synthesis of metal nanoparticles [25].

Yeast: Yeasts are unicellular, eukaryotic microorganisms. A total of 1500 yeast species have been identified. Numerous research groups reported successful synthesis of nanoparticles/nanomaterials by yeast. The biosynthesis of silver and gold nanoparticles by a silver-tolerant yeast strain and *Saccharomyces cerevisiae* broth has been reported. MKY3 and *Saccharomyces cerevisiae* are helpful in synthesizing nanoparticles [26].

Plants: Different phytochemicals of different parts (leaves, root, bark, fruit, and seeds) of plants help in synthesizing nanoparticles that are to be reduced by specific agents. Plants have the potentiality to accumulate certain amounts of heavy metals in their body parts. By the help of proper methods plant phytochemicals are used to synthesize the metal or metal oxide nanoparticles. Plants such as *Cinnamomum zeylanicum, Cinnamomum camphora, Desmodium triflorum, Parthenium hysterophorus, Solanum xanthocarpum, Trianthema decandra, Zingiber officinale, Hibiscus cannabinus,* and *Lantana camara* help in the synthesis of nanoparticles. The major advantages of using plant extracts for the synthesis of AgNPs are that they are easily available, safe, and nontoxic in most of the cases; they have a broad variety of metabolites enabling them to reduce the silver ions; and this process is quicker than microbes during the synthesis. Due to the presence of different phytochemicals, the plant extract has the ability to reduce the metal ions. This can be considered as the major mechanism of green synthesis of AgNPs. The main phytochemicals responsible for the immediate reduction of the metal ions include water-soluble phytochemicals such as terpenoids, flavones, ketones, aldehydes, amides, carboxylic acids, organic acids, and quinines. Therefore, Jha et al. [27] suggested that the phytochemicals are involved directly in the reduction of the metal ions and formation of AgNPs. Other nanoparticles are also synthesized by the same procedure.

Others: Nanoparticles are also produced with the help of DNA, proteins, immunoglobulin, serum albumins, etc.

12.5 Solvent System-Based Green Synthesis of Nanoparticles

Proper and ideal solvent systems are so important for the synthesis of nanoparticles. According to reports, different solvent systems are used, but water is the ideal and suitable solvent system for ever. It is said that "the best solvent is no solvent, and if a solvent is desirable then water is ideal." Gold nanoparticles were prepared by the laser ablation technique in an aqueous solution. The oxygen of the water is used for partial oxidation of the synthesized gold nanoparticles that finally enhanced its chemical reactivity and also had a great impact on its growth. In the case of green synthesis, renewable resources are used, and these resources are natural. Ionic and supercritical liquids are nowadays one of the best examples. Ionic liquids (ILs) are composed of ions that have melting points below 100 °C. Ionic liquids, acknowledged as "room temperature ionic liquids," have been used in synthesizing several metal nanoparticles (e.g. Au, Ag, Al, Te, Ru, Ir, and Pt). Such types of ionic liquids act as both reducing and protective agents that are needed for the synthesis of proper stable nanoparticles. The advantages of using ILs are that it can dissolve polar organic compounds, solubility of it depends on anions and cations, it is not volatile, etc. But it also has disadvantages that it is not suitable for the synthesis of metallic nanoparticles due to presence of both anions and cations. So, according to requirements, different types of solvent systems are required for the synthesis of nanoparticles.

12.6 Stability and Toxicity of Nanoparticles

The stability of the nanoparticles can be evaluated by estimating their aggregation, determining their size, and estimating their affinity toward other surrounding media. The aim of the green synthesis is to synthesize less-toxic and stable nanoparticles. Aggregation of nanoparticles is a time-dependent phenomenon that must be overcome using metabolites of plants. By manipulating the size, shape, and capping agents, aggregation of nanoparticles can be controlled (Figure 12.5). As evidenced, nanoparticles synthesized by plant parts are less toxic due to the presence of phytochemicals that are also less toxic and environmentally friendly.

12.7 Mechanism of Nanoparticle Synthesis by Plant Extract

For synthesizing metal nanoparticles different parameters must be taken such as temperature, pH, different types of phytochemicals that are present, concentration of phytochemicals, and metal salt concentration. The main phytochemicals present in different parts of the plants are flavones, terpenoids, sugars, ketones, aldehydes, carboxylic acids, and amides, which are responsible for the bioreduction of nanoparticles (Table 12.1). These phytochemicals are responsible for the reduction of metals and metal oxides as well as stabilize the metal ions and thus facilitate the synthesis of nanoparticles (Figure 12.1). Flavonoids contain various functional groups, which have an enhanced ability to reduce metal ions. In the case of oxidation–reduction process, one molecule is reduced here as another molecule becomes oxidized. During nanoparticle synthesis phytochemicals become oxidized by converting the enol form to the keto form. According to reports, the reactive hydrogen atom is released due to tautomeric transformations in flavonoids by which the enol form is converted into the keto form. This process is carried out by the reduction of

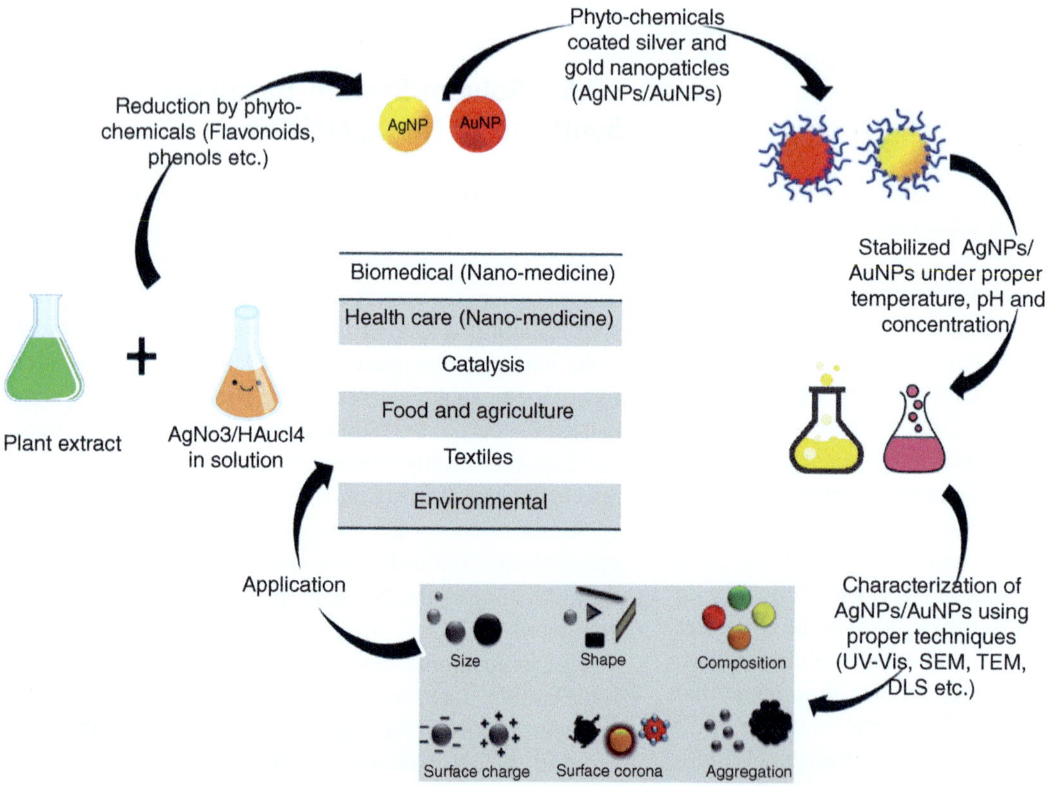

Figure 12.5 Manipulation of different characteristic features of nanoparticles.

Table 12.1 Name of few plant-reduced nanoparticles.

Sl No.	Source	Biomolecule involved	Nanoparticles size	References
1	*Allium cepa*	Vitamin C	Au ~100 nm	[28]
2	*Allium sativum*	Sucrose and fructose	Ag 4.4 ± 1.5 nm	[29]
3	*Achyranthus aspera*	Polyols	Ag, 2030 nm	[30]
4	*Anacardium occidentale*	Polyols and proteins	Au, Ag, Au–Ag alloy and Au core-Ag shell, 20–60 nm	[31]
5	*Andrographis paniculata*	Hydroxyflavones Catechins	Ag, 28 nm	[32]
6	*Astragalus gummifer*	Proteins	Ag, 13.1 ± 1.0 nm	[33]
7	*Azadirachta indica*	Salannin, nimbin, azadirone, and azadirachtins	Au, Ag, 2–100 nm	[34]
8	*Camellia sinensis*	Polyphenolic compounds	Au, 25 nm	[35]
9	*Carica papaya*	Hydroxyflavones and catechins.	Ag, 15 nm	
10	*Centella asiatica*	Terpenoid and flavonoid	Ag, 20–45 nm	[36]
11	*Chenopodium album*	Oxalic acid	Ag-12 nm Au-10 nm	[37]
12	*Coleus aromaticus*	Flavonoids	Ag-40–50 nm	[38]
13	*Cinnamomum zeylanicum*	Terpenoids	Pd, 15–20 nm	[39]
14	*Cinnamomum camphora*	Polyols and heterocyclic Components	Pd, 3.2–6.0 nm	[40]
15	*Citrullus colocynthis*	Polyphenols with aromatic ring and bound amide region	Ag, 31 nm	[41]
16	*Datura metel*	Plastohydroquinone or plastrocohydroquinol	Ag, 16–40 nm	[42]
17	*Desmodium triflorum*	Water-soluble antioxidative agents such as ascorbic acids	Ag, 5–20 nm	[43]
18	*Dioscorea bulbifera*	Polyphenols or flavonoids	Ag, 8–20 nm	[44]
19	*Dioscorea oppositifolia*	Polyphenols with aromatic ring and bound amide region	Ag, 14 nm	[45]
20	*Elettaria cardamomum*	Alcohols, carboxylic acids, ethers, esters, and aliphatic amines	Ag, 31 nm	[46]
21	*Gardenia jasminoides*	Geniposide, chlorogenic acid, crocins, and crocetin	P3–5 nm	[27]
22	*Glycyrrhiza glabra*	Flavonoids and terpenoids	Ag, 20 nm	[47]
23	*Hibiscus cannabinus*	Ascorbic acid	Ag, 9 nm	[48]
24	*Hydrilla verticilata*	Proteins	Ag, 65.55 nm	[49]
25	*Jatropha curcas*	Curcacycline (a nonapeptide) curcain	ZnS–0 nm Pb 10–12.5 nm	[50]
26	*Justicia gendarussa*	Polyphenol and lavonoid	Au-27 nm	[51] [53]
27	*Lantana camara*	Carbohydrates, glycosides, and flavonoids	Ag-12.55	[52]
28	*Leonuri herba*	Polyphenols and hydroxyl groups	Ag 9.9–13.0 nm	[53]
29	*Mentha piperita*	Menthol	Ag-90 nm Au-150 nm	[54] [54]
30	*Mirabilis jalapa*	Polyols	Au-100 nm	[55]
31	*Morinda pubescens*	Hydroxyflavones, catechins	Ag 25–50 nm	[56]
32	*Ocimum sanctum*	Phenolic and flavanoid compounds, proteins, ascorbic acid, gallic acid, and terpenoids	Ag~10 nm Ag, 4–30 nm Pt- 23 nm	 [57] [57]

(*Continued*)

Table 12.1 (Continued)

Sl No.	Source	Biomolecule involved	Nanoparticles size	References
33	*Parthenium hysterophorus*	Hydroxyflavones and catechins	Ag-10 nm	[58]
34	*Pedilanthus tithymaloides*	Proteins and enzymes	Ag15–30 nm	[59]
35	*Piper betle*	Proteins	Ag-37 nm	[60]
36	*Piper nigrum*	Proteins	Ag-5–50 nm	[61]
37	*Plumeria rubra*	Proteins	Ag-32 nm	[62]
38	*Sesuvium portulacastrum*	Proteins, flavones, and terpenids	Ag-5–20 nm	[63]
39	*Solanum xanthocarpum*	Phenolics, alkaloids, and sugars	Ag-10 nm	[64]
40	*Glycine max*	Proteins and amino acids	Pd~15 nm	[65]
41	*Swietenia mahogany*	Polyhydroxy limonoids	Ag, Au	[66]
42	*Syzygium aromaticum*	Flavonoids	Au-5-100 nm	[67]
43	*Terminalia catappa*	Hydrolysable tannins	Au10–35 nm	[68]
44	*Trianthema decandra*	Hydroxyflavones and catechins.	Ag10–50 nm	[69]
45	*Tridax procumbens*	Water-soluble carbohydrates	Cuo2	[70]
46	*Vitus vinifera*	Flavone and anthocyanins	Pb-61 nm	[71]
47	*Zingiber officinale*	Alkanoids, flavonoids	Ag-10 nm	[72]

metal ions into metal nanoparticles. Glucose and fructose are also helpful for the synthesis of metal nanoparticles as these molecules have the capability to reduce metal ions.

Amino acids also have the ability to reduce the metal ions. Clem [73] observed that amino acids such as cysteine, arginine, lysine, and methionine are efficient in binding with silver ions. Tan et al. [74] tested all of the 20 natural α-amino acids to establish their potentiality for the reduction of gold metal ions. According to different reports, Fourier transform infrared spectroscopy (FTIR) analysis confirms that different functional groups are present in phytochemicals such as —C—O—C—, —C—O—, —C=C—, and —C=O—. According to Huang et al., the absorption peaks of FTIR spectra at (i) 1042 and 1077, (ii) 1606 and 1622, and (iii) 1700–1800 cm^{-1} imply the stretching of (i) —C—O—C— or —C—O—, (ii) —C=C—, and (iii) —C=O, respectively. Based on the FTIR analysis, they confirmed that functional groups such as —C—O—C—, —C—O—, —C=C—, and —C=O are the capping agents of the nanoparticles. The main role of the capping agents is to reduce the agglomeration of nanoparticles. Kesharwani et al. [42] covered photographic films using an emulsion of silver bromide. The silver bromide was sensitized when the light hit the film that was placed into a solution of hydroquinone. The hydroquinone was further oxidized to quinone by the action of sensitized silver ion. The silver ion was reduced to silver metal, which was in the emulsified form. Hydroquinone or plastohydroquinone or quinol serves as a main reducing agent responsible for the reduction of Au and Ag ions through noncyclic photophosphorylation. Sangaru et al. [75] confirmed that the terpenoids of geranium leaf extract actively take part in the conversion of silver ions into nanoparticles. Eugenol is a main terpenoid component of *Cinnamomum zeylanisum* (cinnamon) extracts, and this compound plays a critical role in the bioreduction of HAuCl4 and AgNO3 metal salts into their respective metal nanoparticles. FTIR data showed that during the formation of Au and Ag nanoparticles, —OH groups originating from eugenol disappear. After the formation of Au nanoparticles, carbonyl, alkenes, and chloride functional groups appeared. Before and after the production of Au nanoparticles, several other groups [e.g. R—CH and —OH (aqueous)] were also found. Thus, the formation of nanoparticles requires three phases: activation phase in which the bioreduction occurs, the growth phase, and the termination phase where the nanoparticles are finally synthesized.

12.8 Environmentally Friendly Activity of Green-Synthesized Nanoparticles

12.8.1 Antimicrobial Activity

According to a literature survey, the antimicrobial potential of silver nanoparticles can be described properly. Nanoparticles can denature the bacterial outer membrane, generate the gaps in the bacterial cell membrane, and that can cause fragmentation of the cell membrane. Different metabolic processes of bacterial cell can be destroyed by the interactions between Ag NPs and disulfide or sulfhydryl groups of enzymes of the bacteria. This event leads to bacterial cell death. The shape-dependent antimicrobial activity was also examined. Because the high-atom-density surfaces have enhanced antimicrobial activity, truncated triangular nanoparticles are highly reactive in nature [76]. Green-synthesized nanoparticles are more efficient than chemically synthesized nanoparticles because green-synthesized nanoparticles are stabilized or capped by plant extracts that have different medicinal application. According to different reports, inhibition zone on agar plates is large enough in the case of green-synthesized nanoparticles than the chemically synthesized nanoparticles. The antimicrobial capability of different nanoparticles is different. One type of nanoparticles can destroy different types of bacteria, and the other can destroy very small number of bacteria.

12.8.2 Catalytic Activity

Nanoparticles have an important feature such as high surface area-to-volume ratio, and hence, these are with higher adsorption capacity. This is for their catalytic potentiality. As, for example 4-nitrophenol is used in the synthesis of different pesticides, fungicides, herbicides, insecticides, etc. and also is considered as an organic pollutant that must be reduced, or the ecosystem will be disturbed. So, reduction of this organic matter is necessary. Reduced products can also be used in different fields such as in preparing sulfur dyes, and film developer. The simplest and most effective way to reduce 4-nitrophenol is to introduce $NaBH_4$ as a reductant and a metal catalyst such as Au NPs, Ag NPs, CuONPs, and Pd NPs. By increasing the adsorption of reactants on their surface, and by reducing the activation of energy barriers, metallic NPs can promote the rate of reaction. If activation energy remains higher, then the reaction rate will be slowed down.

12.8.3 Removal of Pollutant Dyes

Paper mills, plastic factories, leather, and food industries are dealing with different dyes that are discharged in the nearby water. These types of dyes are the major pollutants of water and are also responsible for the death of different aquatic lives as well as ecological balance. So, it is a big challenge for us to control the pollutant dyes. Now, by applying metal and metal ion semiconductor nanomaterials, the unnecessary dyes can be controlled. Now, the nanoparticles, prepared by the help of plant extracts, are used in this field. Due to the presence of high surface area-to-volume ratio, nanoparticles can remove a high rate of contaminant at a time at low concentration. Silver nanoparticles, prepared by different plant extracts, also have the capability to degrade the pollutants photocatalytically.

12.8.4 Heavy Metal Ion Sensing

Heavy metals (such as Ni, Cu, Fe, Cr, Zn, Co, Cd, Pb, Cr, Hg, and Mn) are well known for acting as pollutants in air, soil, and water. There are different sources of heavy metal pollution such as mining waste, vehicle emissions, natural gas, paper, plastic, coal, and dye industries. Heavy metal is an important topic to be discussed because due to increasing level of heavy metal, aquatic life and total biological system is going to be destroyed. Specific size and optical sensitivity of different nanoparticles can enhance them to fight with the metal ions by detecting them in biological systems. The advantages of using metal NPs as colorimetric sensors for heavy metal ions in environmental systems include their simplicity, cost-effectiveness, and high sensitivity at ppm levels. Different nanoparticles are differently sensitive for the different types of heavy metals. Heavy metals such as zinc, lead, and mercury are easily detected by green-synthesized nanoparticles.

12.9 Challenges and Future Scope

One important step toward the goal of designing more environmentally friendly nanoparticles is to figure out how different nanoparticles interact with different organisms in the environment. In order to do this, we should investigate the interactions at a molecular level. The goal is to be able to predict interactions and any potential toxicity of nanoparticles with organisms and then use that knowledge to design nanoparticles in a way that will mitigate environmental risk. The use of plants in sustainable green nanotechnology is developing rapidly. Plant-derived nanostructures have applications in diverse fields including solar cell devices, photocatalysis, biosensors, sensing, photonics, catalysis, biomedicine, electronics, environmental clean-up, imaging, biolabeling, and drug and gene delivery. The properties of nontoxicity and biocompatibility of these nanostructures enable their applications in biomedical sciences such as tissue engineering as well as in pharmaceutical industries. But the important challenges and technical problems need to be addressed as in phytoformulation studies developing nanodosage forms including liposomes, solid lipid NPs, nanoemulsions, and protein- and lipid-based drug delivery systems. A detailed study on the bioavailability, enhancement of solubility, prevention of toxicity, enhancement of pharmacological activity and stability, sustained delivery, and avoidance of physical and chemical degradation are some of the important prerequisites.

12.10 Conclusion

Most of us are familiar with the concept of "side effects." This is when something that is designed to be helpful ends up having some harm that goes along with it. Nanoparticles are one example; they have many amazing uses for consumer products, but they can sometimes have harmful impacts on environmental organisms. One goal of the sustainable nanotechnology is to try to find options for designing nanoparticles so that those harmful impacts can be reduced. Greener synthetic strategies avoid the elimination of the use or exposure to or generation of the hazardous substances normally used in the preparation of these nanomaterials. The development of plant-biomimetic and phytofabrication approaches to nanostructures is one of the most promising technological challenges in the coming years for the development bioinspired materials and systems, hierarchically structured materials, adaptive materials, and 3D composites that are compatible with ecological requirements emphasizing the sustainable use of renewable resources. Thus, it can be concluded that the progressive research agenda should aim at reviewing and assessing critically to the objectives related to the activities in global economic law and governance affecting sustainability transformations of nanotechnology research.

Conflict of Interest Statement

The author declares that the research was conducted in the absence of any commercial or financial relationships that could be construed as a potential conflict of interest.

Acknowledgment

The authors gratefully acknowledge the University of Kalyani to support the study.

References

1 Lyddy, R. (2009). Nanotechnology. In: *Information Resources in Toxicology*, 4e (eds. P. Wexler and S.G. Gilbert), 321–328. Academic press.

2 Booth, T.J. and Baker, M.A.B. (2017). Nanotechnology: building and observing at the nanometer scale. In: *Pharmacognosy Fundamentals, Applications and Strategies* (eds. S. Badal and R. Delgoda), 633–643. Academic Press.

3 Commission on the Environment and Development (WCED) (1987). *Our Common Future; Brundtland Report*. Oxford, UK: Oxford University Press.

4 Kokate, C.K., Purohit, A.P., and Gokhale, S.B. (2001). Carbohydrate and derived products, drugs containing glycosides, drugs containing tannins, lipids and protein alkaloids. In: *Textbook of Pharmacognosy*, 7e, 133–166.

5 Scalbert, A. (1991). Antimicrobial properties of tannins. *Phytochemistry* 30: 3875–3883. https://doi.org/10.1016/0031-9422 (91)83426-L.

6 Francis, G., Kerem, Z., Makkar, H.P., and Becker, K. (2002). The biological action of saponins in animal systems: a review. *The British Journal of Nutrition* 88 (6): 587–605.

7 Sarker, S.D. and Nahar, L. (2007). *Chemistry for Pharmacy Students General, Organic and Natural Product Chemistry*, 283–359. England: Wiley.

8 Odugbemi, T.O., Akinsulire, O.R., Aibinu, I.E., and Fabeku, P.O. (2006). Medicinal plants useful for malaria therapy in Okeigbo, Ondo State, Southwest Nigeria. *African Journal of Traditional, Complementary, and Alternative Medicines* 4 (2): 191–198.

9 Rice-Evans, C.A., Miller, N.J., and Paganga, G. (1996). Structure-antioxidant activity relationships of flavonoids and phenolic acids. *Free Radical Biology and Medicine* 20 (7): 933–956.

10 Minatel, I.O., Borges, C.V., Ferreira, M.I. et al. (2017). Phenolic compounds: functional properties, impact of processing and bioavailability. In: *Phenolic Compounds-Biological Activity* (eds. M. Soto-Hernandez, M. Palma-Tenango and M. del Rosario Garcia-Mateos). Intech Open https://doi.org/10.5772/66368.

11 Patel, S.S. and Savjani, J.K. (2015). Systematic review of plant steroids as potential antiinflammatory agents:

current status and future perspectives. *The Journal of Phytopharmacology* 4 (2): 121–125.

12 Dos Reis, L.C.R., Facco, E.M.P., Salvador, M. et al. (2018). Antioxidant potential and physicochemical characterization of yellow, purple and orange passion fruit. *Journal of Food Science and Technology* 55 (7): 2679–2691.

13 Han, X. and Parker, T.L. (2017). Anti-inflammatory, tissue remodeling, immunomodulatory, and anticancer activities of oregano (*Origanum vulgare*) essential oil in a human skin disease model. *Biochimie Open* 4: 73–77.

14 Faheem, N.M. and Askary, A.E. (2017). Neuroprotective role of curcumin on the hippocampus against the structural and serological alterations of streptozotocin-induced diabetes in Sprague Dawley rats. *Iranian Journal of Basic Medical Sciences* 20 (6): 690–699.

15 Sompong, W., Cheng, H., and Adisakwattana, S. (2015). Protective effects of ferulic acid on high glucose-induced protein glycation, lipid peroxidation, and membrane ion pump activity in human erythrocytes. *PLoS One* 10 (6): e0129495.

16 Ceremuga, T.E., Martinson, S., Washington, J. et al. (2014). Effects of L-theanine on posttraumatic stress disorder induced changes in rat brain gene expression. *Scientific World Journal* 2014: 419032. https://doi.org/10.1155/2014/419032.

17 Martín-Fernández, B., de las Heras, N., Valero-Muñoz, M. et al. (2014). Beneficial effects of proanthocyanidins in the cardiac alterations induced by aldosterone in rat heart through mineralocorticoid receptor blockade. *PLoS One* 9 (10): e111104.

18 Ożarowski, M., Mikołajczak, P.Ł., Kujawski, R. et al. (2018). Pharmacological effect of Quercetin in hypertension and its potential application in pregnancy-induced hypertension: review of in vitro, in vivo, and clinical studies. *Evidence-based Complementary and Alternative Medicine* 2018: 7421489. https://doi.org/10.1155/2018/7421489.

19 Lee, G. and Bae, H. (2017). Therapeutic effects of phytochemicals and medicinal herbs on depression. *BioMed Research International* 2017: 6596241. https://doi.org/10.1155/2017/6596241.

20 Andreadi, C.K., Howells, L.M., Atherfold, P.A., and Manson, M.M. (2006). Involvement of Nrf2, p38, B-Raf, and nuclear factor-kappaB, but not phosphatidylinositol 3-kinase, in induction of hemeoxygenase-1 by dietary polyphenols. *Molecular Pharmacology* 69 (3): 1033–1040.

21 Liu, R.H. (2004). Potential synergy of phytochemicals in cancer prevention: mechanism of action. *The Journal of Nutrition* 134 (12 Suppl): 3479S–3485S. https://doi.org/10.1093/jn/134.12.3479S.

22 Cao, H., Polansky, M.M., and Anderson, R.A. (2007). Cinnamon extract and polyphenols affect the expression of tristetraprolin, insulin receptor, and glucose transporter 4 in mouse 3T3-L1 adipocytes. *Archives of Biochemistry and Biophysics* 459 (2): 214–222.

23 Chouhan, S., Sharma, K., and Guleria, S. (2017). Antimicrobial activity of some essential oils—present status and future perspectives. *Medicines (Basel)* 4 (3): 58.

24 Singh, J., Dutta, T., Kim, K. et al. (2018). Green' synthesis of metals and their oxide nanoparticles: applications for environmental remediation. *Journal of Nanobiotechnology* 16: 84. https://doi.org/10.1186/s12951-018-0408-4.

25 Moghaddam, A.B., Namvar, F., Moniri, M. et al. (2015). Nanoparticles biosynthesized by fungi and yeast: a review of their preparation, properties, and medical applications. *Molecules* 20: 16540–16565.

26 Kasemets, K., Käosaar, S., Vija, H. et al. (2019). Toxicity of differently sized and charged silver nanoparticles to yeast *Saccharomyces cerevisiae* BY 4741: a nano-biointeraction perspective. *Nanotoxicology* 13 (8): 1041–1059.

27 Jha, A.K., Prasad, K., and Prasad, D. (2009). A green low-cost biosynthesis of Sb20 nanoparticles. *Biochemical Engineering Journal* 43: 303–306.

28 Parida, U.K., Bindhani, B.K., and Nayak, P. (2011). Green synthesis and characterization of gold nanoparticles using onion (*Allium cepa*) extract. *World Journal of Nano Science and Engineering* 7: 93–98.

29 White, R.J., Budarin, V.L., Moir, J.W.B., and Clark, J.H. (2011). A sweet killer: mesoporous polysaccharide confined silver nanoparticles for antibacterial applications. *International Journal of Molecular Sciences* 12: 5782–5796.

30 Daniel, S.C., Ayyappan, S., Philiphan, N.J.P. et al. (2011). Green synthesis and transfer of silver nanoparticles in a food chain through *Chiranamous larva* to zebra fish – a new approach for therapeutics. *International Journal of NanoScience and Nanotechnology* 2: 159–169.

31 Shenya, D.S., Joseph Mathew, J., and Philip, D. (2011). Phytosynthesis of Au, Ag and Au–Ag bimetallic nanoparticles using aqueous extract and dried leaf of *Anacardium occidentale*. *Spectrochimica Acta. Part A, Molecular and Biomolecular Spectroscopy* 79: 254–262.

32 Sulochana, S., Krishnamoorthy, P., and Sivaranjani, K. (2012). Synthesis of silver nanoparticles using leaf extract of *Andrographis paniculata*. *Journal of Pharmacology and Toxicology* 7: 251–258.

33 Kora, A.J., Sashidhar, R.B., and Arunachalam, J. (2012). Aqueous extract of gum olibanum (*Boswellia serrata*): a reductant and stabilizer for the biosynthesis of antibacterial silver nanoparticles. *Process Biochemistry* 47: 1516–1520.

34 Thirumurugan, A., Jiflin, G.J., Rajagomathi, G. et al. (2010). Biotechnological synthesis of gold nanoparticles of *Azadirachta indica* leaf extract. *International Journal of Systems Biology and Technologies* 1: 75–77.

35 Boruah, S.K., Boruah, P.K., Sarma, P. et al. (2012). Green synthesis of gold nanoparticles using *Camellia sinensis* and kinetics of the reaction. *Advanced Materials Letters* 3: 481–486.

36 Palaniselvam, K., Velanganni, A.A.J., Govindan, S.N., and Karthi (2012). Leaf assisted bioreduction of silver ions using leaves of *Centella asiatica* L. and its bioactivity. *Journal of Life Science* 1: 46–49.

37 Dwivedi, A.D. and Gopal, K. (2011). Plant mediated biosynthesis of silver and gold nanoparticles. *Journal of Biomedical Nanotechnology* 7: 163–164.

38 Vanaja, M. and Annadurai, G. (2012). *Coleus aromaticus* leaf extract mediated synthesis of silver nanoparticles and its bactericidal activity. *Applied Nanoscience* 3: 217–223.

39 Santhoshkumar, T., Rahuman, A.A., Bagavan, A. et al. (2012). Evaluation of stem aqueous extract and synthesized silver nanoparticles using *Cissus quadrangularis* against *Hippobosca maculata* and *Rhipicephalus microplus*. *Experimental Parasitology* 132: 156–165.

40 Xin, Y., Qingbiao, L., Huixuan, W. et al. (2010). Green synthesis of palladium nanoparticles using broth of *Cinnamomum camphora* leaf. *Journal of Nanoparticle Research* 12: 1589–1598.

41 Satyavani, K., Ramanathan, T., and Gurudeeban, S. (2011). Plant mediated synthesis of biomedical silver nanoparticles by using leaf extract of *Citrullus colocynthis*. *Journal of Nanoscience and Nanotechnology* 1: 95–101.

42 Kesharwani, J., Yoon, K.Y., Hwang, J., and Rai, M. (2009). Phytofabrication of silver nanoparticles by leaf extract of *Datura metel*: hypothetical mechanism involved in synthesis. *Journal of Bionanoscience* 3: 39–44.

43 Ahamed, M., Khan, M.A., Siddiqui, M.K.J. et al. (2011). Green synthesis, characterization and evaluation of biocompatibility of silver nanoparticles. *Physica E* 43: 1266–1271.

44 Ghosh, S., Patil, S., Ahire, M. et al. (2012). Synthesis of silver nanoparticles using *Dioscorea bulbifera* tuber extract and evaluation of its synergistic potential in combination with antimicrobial agents. *International Journal of Nanomedicine* 7: 483–496.

45 Maheswari, R.U., Prabha, A.L., Nandagopalan, V., and Anburaja, V. (2012). Green synthesis of silver nanoparticles by using rhizome extract of *Dioscorea oppositifolia* L. and their antimicrobial activity against human pathogens. *IOSR Journal of Pharmacy and Biological Sciences* 1: 38–42.

46 Gnanajobitha, G., Annadurai, G., and Kannan, C. (2012). Green synthesis of silver nanoparticles using *Elettaria cardamomum* and assessment of its antimicrobial activity. *International Journal of Pharmaceutical Sciences and Research* 3: 323–330.

47 Dinesh, S., Karthikeyan, S., and Arumugam, P. (2012). Biosynthesis of silver nanoparticles from *Glycyrrhiza glabra* root extract. *Archives of Applied Science Research* 4: 178–187.

48 Bindhu, M.R. and Umadevi, M. (2013). Synthesis of monodispersed silver nanoparticles using *Hibiscus cannabinus* leaf extract and its antimicrobial activity. *Spectrochimica Acta Part A: Molecular and Biomolecular Spectroscopy* 101: 184–190.

49 Sable, N., Gaikwad, S., Bonde, S. et al. (2012). Phytofabrication of silver nanoparticles by using aquatic plant *Hydrilla verticilata*. *Nusantara Bioscience* 4: 45–49.

50 Hudlikar, M., Joglekar, S., Dhaygude, M., and Kodam, K. (2012). Latex mediated synthesis of ZnS nanoparticles: green synthesis approach. *Journal of Nano Research* 14: 1–6.

51 Fazaludeena, M.F., Manickamb, C., Ashankytyc, M.A.I. et al. (2012). Synthesis and characterizations of gold nanoparticles by *Justicia gendarussa* Burm F. leaf extract. *Journal of Microbiology and Biotechnology Research* 2: 23–34.

52 Sivakumar, P., Nethradevi, C., and Renganathan, S. (2012). Synthesis of silver nanoparticles using *Lantana camara* fruit extract and its effect on pathogens. *Asian Journal of Pharmaceutical and Clinical Research* 5: 97–101.

53 Rang, A., Han, L., Kim, E.R. et al. (2012). Enhanced antibacterial activities of *Leonuri herba* extracts containing silver nanoparticles. *Phytotherapy Research* 26: 1249–1255.

54 Ali, D.M., Thajuddin, N., Jeganathan, K., and Gunasekaran, M. (2011). Plant extract mediated synthesis of silver and gold nanoparticles and its antibacterial activity against clinically isolated pathogens. *Colloids and Surfaces, B: Biointerfaces* 85: 360–365.

55 Vankar, P.S. and Bajpai, D. (2010). Preparation of gold nanoparticles from *Mirabilis jalapa* flowers. *Indian Journal of Biochemistry and Biophysics* 47: 157–160.

56 Mary, E.J. and Inbathamizh, L. (2012). Green synthesis and characterization of nano silver using leaf extract of *Morinda pubescens*. *Asian Journal of Pharmaceutical and Clinical Research* 5: 159–162.

57 Ahmad, N., Sharma, S., Alam, M.K. et al. (2010). Rapid synthesis of silver nanoparticles using dried medicinal plant of basil. *Colloids and Surfaces, B: Biointerfaces* 81: 81–86.

58 Ashok Kumar, D. (2012). Rapid and green synthesis of silver nanoparticles using the leaf extracts of *Parthenium hysterophorus*: a novel biological approach. *International Research Journal of Pharmacy* 3: 169–170.

59 Sundaravadivelan, C. and Nalini, M. (2012). Biolarvicidal effect of phytosynthesized silver nanoparticles using *Pedilanthus tithymaloides* (L.) Poit stem extract against the dengue vector *Aedes aegypti* L. (Diptera; Culicidae). *Asian Pacific Journal of Tropical Biomedicine* 2: 1–8.

60 Mallikarjuna, K., Dillip, G.R., Narashima, G. et al. (2012). Phytofabrication and characterization of silver nanoparticles from *Piper betle* broth. *Research Journal of Nanoscience and Nanotechnology* 2: 17–23.

61 Garg, S. (2012). Rapid biogenic synthesis of silver nanoparticles using black pepper (*Piper nigrum*) corn extract. *International Journal of Innovations in Biological and Chemical Sciences* 3: 5–10.

62 Patil, D.C., Patil, V.S., Borase, P.H. et al. (2012). Larvicidal activity of silver nanoparticles synthesized using *Plumeria rubra* plant latex against *Aedes aegypti* and *Anopheles stephensi*. *Parasitology Research* 110: 1815–1822.

63 Nabikhan, A., Kandasamy, K., Raj, A., and Alikunh, N.M. (2010). Synthesis of antimicrobial silver nanoparticles by callus and leaf extracts from saltmarsh plant, *Sesuvium portulacastrum*. *Colloids and Surfaces, B: Biointerfaces* 79: 488–493.

64 Amin, M., Anwar, F., Janjua, M.R.S.A. et al. (2012). Green synthesis of silver nanoparticles through reduction with *Solanum xanthocarpum* L. berry extract: characterization, antimicrobial and urease inhibitory activities against *Helicobacter pylori*. *International Journal of Molecular Sciences* 13: 9923–9941.

65 Petla, R.K., Vivekanandhan, S., Misra, M. et al. (2012). Soybean (*Glycine Max*) leaf extract based green synthesis of palladium nanoparticles. *Journal of Biomaterials and Nanobiotechnology* 3: 14–19.

66 Mondal, S., Roy, N., Laskar, R.A. et al. (2011). Biogenic synthesis of Ag, Au and bimetallic Au/Ag alloy nanoparticles using aqueous extract of mahogany (*Swietenia mahogani*) leaves. *Colloids and Surfaces, B: Biointerfaces* 82: 497–504.

67 Deshpande, R., Bedre, D.M., Basavaraja, S. et al. (2010). Rapid biosynthesis of irregular shaped gold nanoparticles from macerated aqueous extracellular dried clove buds (*Syzygium aromaticum*) solution. *Colloids and Surfaces, B: Biointerfaces* 79: 235–240.

68 Ankamwar, B. (2010). Biosynthesis of gold nanoparticles (Green gold) using leaf extract of *Terminalia catappa*. *European Journal of Organic Chemistry* 7: 1334–1339.

69 Geethalakshmi, R. and Sarada, D.V.L. (2010). Synthesis of plant mediated silver nanoparticles using *Trianthema decandra* extract and evaluation of their antimicrobial activities. *International Journal of Engineering, Science and Technology* 2 (5): 970–975.

70 Gopalakrishnan, K., Ramesh, C., Ragunathan, V., and Thamilselvan, M. (2012). Antibacterial activity of copper oxide nanoparticles on *Escherichia coli* synthesized from *Tridax procumbens* leaf extract and surface coating with poly- aniline. *Digest Journal of Nanomaterials and Biostructures* 7: 833–839.

71 Pavani, K.V., Swati, T., Snehika, V. et al. (2012). Phytofabrication of lead nanoparticles using grape skin extract. *International Journal of Engineering, Science and Technology* 4: 3376–3380.

72 Singh, C., Sharma, V., Naik, K.R.P. et al. (2011). A green biogenic approach for synthesis of gold and silver nanoparticles using *Zingiber officinale*. *Digest Journal of Nanomaterials and Biostructures* 6: 535–542.

73 Clem, G. (1975). Interaction of amino acids with silver(I) ions. *Biochimica et Biophysica Acta (BBA) – Protein Structure* 386 (1): 270–274.

74 Tan, Y.N., Lee, J.Y., and Wang, D. (2010). Uncovering the design rules for peptide synthesis of metal nanoparticles. *Journal of the American Chemical Society* 132 (16): 5677–5686.

75 Sangaru, S.S., Ahmad, A., and Sastry, M. (2003). Geranium leaf assisted biosynthesis of silver nanoparticles. *Biotechnology Progress* 19: 1627–1631. https://doi.org/10.1021/bp034070w.

76 Pal, S., Tak, Y.K., and Myong, J. (2007). Song does the antibacterial activity of silver nanoparticles depend on the shape of the nanoparticle? A study of the gram-negative bacterium *Escherichia coli*. *Applied and Environmental Microbiology* 73 (6): 1712–1720.

13

Sustainable Nanobiocomposites

Jigar Shah[1], Vimal Patel[1], Vishal Chavda[1], and Jayvadan Patel[2]

[1] *Institute of Pharmacy, Nirma University, Ahmedabad, Gujarat, India*
[2] *Nootan Pharmacy College, Sankalchand Patel University, Visnagar, Gujarat, India*

13.1 Nanocomposites and Nanobiocomposites

In simple words, the materials that are metabolic by-products of living organisms with nanometer-scale dimensions are labeled as nanocomposite materials. These newer environment-friendly and natural nanocomposites are bio-based materials with unique structures and characteristic properties, which have recently gained significant importance from scientists worldwide. Biopolymeric-based nanocomposites are developed as unique category of materials, which are also known as "nanobiocomposites (NBCs)." These natural NBCs are potential candidates for further research and development inputs, and their wide scope of applications makes them suitable not only for biomedical science but also for in-depth research in various other fields; they also have importance from industrial perspectives. The green nanocomposites, which can also be referred as NBCs, are the next-generation sustainable materials. The nanocomposites made from renewable nanoparticles, petroleum-based materials, biopolymers, and synthetic nanofillers are considered as NBCs [1]. These types of materials are different in properties, characteristics, preparation methods, and applications. The most advantageous property of these materials is that they are biodegradable in nature. It means that these biocomposites are very safe for the environment because of their natural degradation with the help of organisms, resulting in organic by-products. Besides this, their availability in abundance in nature and also their renewable property make them the ideal candidates to replace petroleum-derived plastics and similar types of materials [2].

Interestingly, the shape of nanoparticulates plays a significant role in the development of NBCs and is also responsible for the properties shown by them. On the basis of its geometric structure, nanoparticulates are classified as nanoparticles, nanoplatelets, nanosomes, nanofibers, nanocapsules, nanotubes, nanorods, etc. The problems associated with the fabrication of nanocomposites have opened the way for the development of NBCs [3–5]. Nanoparticulates enter into the ecosystems, specifically aquatic systems, by various ways, which include industrial waste chemicals and from discharge of wastewater. These materials then undergo various climatic physical and chemical reactions and transform into several different forms. These different forms of nanoparticles can be taken up by higher organisms, and some aquatic and terrestrial organisms can also assemble these nanoparticles through various processes [6]. Finally, all these processes would result in the development of a new class of nanoparticulates known as NBCs. Increasing the usage of carbon-containing fossil resources leads to lack of these resources as their process of generation is very slow. These resources are limited in quantity and exhaustible in near future. Besides this, these types of carbon-containing materials have an impact on global environment, which creates immense problem to all living creatures on the earth [7, 8]. NBCs originated from organisms, and hence they contain a large amount of bio-based materials. They can strongly give competition to petroleum-based materials with an affordable cost performance ratio and also help in balancing strong relation between ecosystem, technology, and economy. Hence, they can be well-thought-out as green sustainable materials.

Sustainable Nanotechnology: Strategies, Products, and Applications, First Edition.
Edited by Yashwant V. Pathak, Govindan Parayil, and Jayvadan K. Patel.

13.2 Sustainable Nanobiocomposites

Sustainable properties of NBCs have increased the interest of scientists in the past few years. These advanced composites are developed from eco-friendly, low cost, renewable, biocompatible, and biodegradable resources having multifunctional properties and reduced pollution impact on the environment [9–11]. In addition to these unique and improved physicochemical properties, these nanocomposites can be prepared through easy and economical procedures. These materials offer innovative platforms and provide a dynamic interface between material science, biomedical science, biology, and nanotechnology. The current NBCs are derived from biopolymers that are from the class of plant and animal polysaccharides, polypeptides, polynucleic acids, and polyesters, and fillers include metal nanoparticles, hydroxyapatite, and clay [12].

The term "green" means a separate category of materials, methods, and products having characteristics such as being environment-friendly and having no harmful effects on the living organisms on Earth as compared with their traditional counterparts. The important property related to NBCs is its biodegradation behavior, in which the materials decompose naturally in the environment either by converting them into other materials or by changing their physical or mechanical properties and/or chemical structure; hence they are disposed of without harming the ecosystem, rather they have some advantages for the environment [13, 14].

These materials are also considered green due to their renewable property. But the prime problem observed is the production of biocomposites from these "green" polymers. Biopolymers are from natural sources such as pectin, cellulose, starch, chitin, polysaccharides, guar gum, polylactic acid (PLA), alginate, chitosan, gelatin, and lignin. The polymers from synthetic categories such as polyesters and polyvinyl alcohol are classified as biopolymers due to their biodegradability property, but they are not renewable in nature. In this chapter, many of the natural biopolymers and degradable materials, which are considered as green biocomposites, are discussed briefly. Though some of the biocomposites can have applications in different disciplines, their usage is inadequate due to the poor mechanical and barrier properties. However, with the help of various approaches of nanotechnology in developing nanosized products or compounds, the nanocomposites can improve their properties and the scope of applications [15].

13.3 Merits and Demerits of Sustainable Biocomposites

Due to their variety of applications, biocomposites are widely used in different pharmaceutical preparations to enhance biodegradability and chemical stability. They are mostly nontoxic in nature and cost friendly. They are renewable and sustainable, and hence they are frequently used in a variety of dosage forms and pharmaceutical preparations. Due to their cell recognition and cell interaction properties, they are widely acceptable in various modern drug delivery systems for combating specific pathological conditions. Their property to degrade to specific pH and enzymes makes them more applicable in targeted drug delivery approaches. They are chemically stable, and also having a property of semblance to the extracellular matrix makes them more useful in modern drug delivery systems. Apart from their variety of merits, there are a few demerits too. A few biocomposites having variety have impurities in them. Such composites need to be purified, which generally is time consuming but with optimum yields. Due to the low stability of specific compounds, the shelf lives of such pharmaceutical preparations have low batch time profile of withdrawal and reproduction [16].

13.4 Classification of Nanobiocomposites

NBCs are grouped based on different parameters, such as source, shape of fortifications, size, and kinds of matrix utilized. The classification of NBCs is as follows.

13.4.1 Based on Sources

Depending on their origin, NBCs could be categorized into two classes: natural and synthetic NBCs.

13.4.1.1 Natural Nanobiocomposites

A large portion of the natural resources is a blend of natural and clay materials. Matrix utilized naturally in biocomposites includes cellulose, collagen, chitin, and proteoglycans and is strengthened generally with silica, $CaCO_3$, and hydroxyapatite. Among numerous others, bone and wood are notable instances of natural biocomposites [17]. Bone is a natural inorganic biocomposite of solid yet delicate fibrils of protein collagen and rigid yet fragile slim discs of hydroxyapatite containing a few arrangements with fluctuating models. Wood is a composite of solid cellulose fiber with the coagulating material identified as lignin [18, 19].

13.4.1.2 Synthetic Nanobiocomposites

Endeavors have been completed in the recent three eras for quick rise in the fabrication of synthetic biocomposites, specifically joining thin fibers in different plastics. Forecasts predict that the interest in biocomposites will keep on expanding consistently, with ceramic- and metallic-based composites producing a progressively huge commitment. A few activities have also been carried out to create preparation techniques in synthetic ceramic production and composites to deliver microstructure-like biomaterials. Regular biomineralization method is pending as it may be difficult to duplicate completely in an artificial counterfeit framework [20, 21].

13.4.2 Based on Proportions of the Filler Materials

Three kinds of NBCs can be recognized, liable upon the measure of dimensions of the dispersed materials that are in nanometer scale [22, 23].

13.4.2.1 Particulate Nanobiocomposites

Such NBCs comprise dimensional particles as fortifications that moderate a short strengthening impact, due to the small angle proportion. Furthermore, the principal reason for utilizing these fortifications is to diminish porousness, make them inflammable, and bring down the expense of composites [24].

13.4.2.2 Elongated Particle Nanobiocomposites

Such NBCs utilize extended particles, for example carbon nanotubes (CNTs) and cellulose nanofibrils as fortification to frame the extended molecule NBCs. This delineates the high-angle proportion of the support and hence offers better biomechanical performance [25].

13.4.2.3 Layered Particle-reinforced Nanobiocomposites

Layered particle is otherwise called layered NBCs. There can be varied subcategories (shed nanocomposites, flocculated nanocomposites, microcomposites, and intercalated) dependent on the scattering proportion in the matrix. There is no section between the layers owing to particle interactions, resulting into flocculated or stage-isolated nanocomposites. The microcomposites are scattered in the polymeric grid. Moreover, these polymer chains are introduced among the plates of the layered nanoparticle structures in nanocomposites; the section of different plates lies into exfoliated nanocomposites [26].

13.4.3 Classification Based on Matrix (Continuous Phase) Used

Depending on the type of matrix, biocomposite materials can be categorized into three groups.

13.4.3.1 Matrix–Ceramic Nanobiocomposites

Matrix–ceramic nanocomposites (MCNs) re-collect an extraordinary variability of materials in which a matrix (polycrystalline ceramic, glass, or their blend) is fortified with the expansion of strands, flacks, and particles. The structures of multicomponent frameworks upgrade their capacity to get prime quality and sturdiness at raised temperatures. Generally, MCNs include a metal as the subsequent segment. Preferably, all segments, the ceramic one and the metallic one, are finely scattered in one another eliciting the specific nanoscale properties [27, 28]. MCNs from blends showed an enhancement in their optical, electrical, and attractive characteristics just as erosion safe, tribological, and other defensive properties [29]. The production methods for MCNs include blending, strengthening and squeezing of powder matrix and vapor deposition method, for e.g. LAS (lithium aluminosilicate)–glass–clay compiosites and LAS-50% silicon carbide composites. MCNs could be employed for biomedical inserts, devices, load-bearing auxiliary sheets, grating shells, and automotive, aerospace, and power generation applications [30].

13.4.3.2 Polymeric Matrix Nanobiocomposites

The utmost utilized lattices for biocomposites are of polymeric nature, and that is the case in around 90% of all biocomposites. These polymeric materials are composed of fiber and/or particles support [31]. Direct encapsulating nanoparticles to a polymeric matrix can improve their appearance, sensational degree, basically by exploiting the nature and characteristic of nanofiller. These methodologies are especially successful in producing superior nanocomposites, once great scattering of the nanofiller is accomplished, and the characteristics of the nanofiller are considerably extraordinary and better than those of polymer matrix, e.g. strengthening a polymer matrix by a lot of rigid clays of nanoparticles, CNTs, or ceramics. Polymeric matrix NBCs are used in different areas such as medical industries, aerospace, domestic appliances, and electronics [32].

13.4.3.3 Metal–Matrix Nanobiocomposites

In these NBCs, metallic blends remain as constant lattices, and support phases are most important for ceramic production. The support as particulates, hairs, or filaments, e.g. alumina, SiC, and TiC nanoparticles, has been utilized in aluminum metal–matrix NBCs, and carbon fibers have been utilized in Mg metal–matrix NBCs [33, 34]. They possess advanced focal points above solid metals, allowing a higher explicit quality and modulus, lower constants of thermal expansion, better thermal characteristics, and massive wear obstruction. [35].

13.4.4 Classification Based on Polymer Biocomposites

From the basic perspective, clay–polymer biocomposites could be commonly arranged into three types of biocomposites, as shown in Figure 13.1.

13.4.4.1 Conventional Nanobiocomposites

The association of clay nanolayers in regular biocomposites (phase isolated) is held at the point where it is blended in with the polymer; however, there is no intercalation observed in polymer into the clay structures. Thus, clay portion in these NBCs challenge their practical role and play for the most part as filler for economic contemplations [36].

13.4.4.2 Intercalated Nanobiocomposites

Solitary broadened polymer chains are introduced among the nanosilicate sheets bringing about a fairly arranged multilayer matrix developed within inorganic layers and revolving polymeric matrix [37].

13.4.4.3 Exfoliated Nanobiocomposites

At the point where the nanosilicate sheets have been consistently dissolved in a continuous phase polymer matrix, a delaminated structure is acquired. Broad polymer entrance, certain delamination of the nanosilicate sheet and continuous agitation creates exfoliated structures comprising of discrete 1 nm thin nanosilicate sheets dispersed in the polymer matrix [38].

13.5 Natures of Nanobiocomposites

NBCs might be described as a type of nanocomposites. In this manner, the mostly acknowledged meaning of nanomaterials can be functional. It represents the dispersed particles in the nanometer range. In any case, the significant qualities of biopolymers from manufactured, petroleum-based polymers does not permit recognizing NBCs by the nanocomposites. They contrast in the dissolvability in water, biodegradability, biocompatibility, and thermal dependability, which decide the strategies for functionalities, planning, and field of uses of materials [39]. On the off chance that the prior is considered, the accompanying definition for NBCs could be recommended. NBCs are composite constituents that comprising of the organic source and particles in the range of 1–100 nm [40]. It ought to be emphasized that the "constituent(s) of natural inception" is referenced in definition yet not biopolymers that are believed to be only in the NBCs. There are various nanocomposites that are produced using variable biomaterials. It can incorporate low-atomic weight substances, for instance, lecithin, which is most important part of the lipidic matrix of biological membrane and living creatures like microorganisms. Those that are additionally mineralized to produce composites are generally termed as

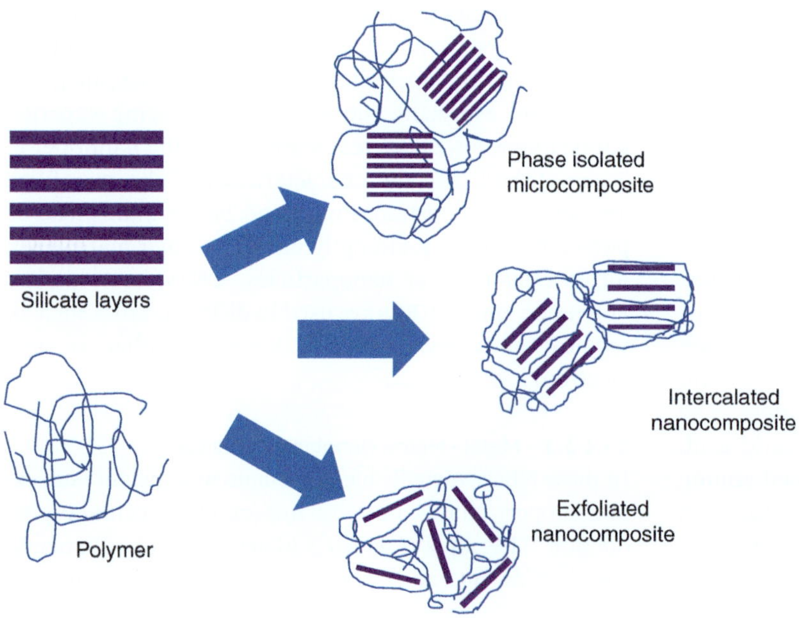

Figure 13.1 Polymer biocomposite classification of NBCs.

"hybrids." This definition does not indicate the nature and properties yet uncovers whether it ought to be called NBCs. This is on the grounds that they incorporate inorganics nanosized and parts of origin, as per the meaning suggested earlier [41].

There are NBCs that are fabricated via joining artificial and common polymers. Biopolymers could be stimulated with synthetic plastic, or fibers are saturated with natural polymer expanding their biodegradability and biocompatibility [32]. Inorganic added substances (e.g. calcium carbonate and clay minerals) are similarly included as a result of their fortification impact, minimal effort, simple accessibility, and nonappearance of negative effect on nature. They are occupied in non-nanoparticulate state [40]. In this manner, these nanofillers strengthen the plastics by measures of a few percent. Composite materials of these types are recognized as bioplastics and biocomposites. These are characterized as "... composites produced using both synthetic plastics and bioplastics saturated with synthetic or natural fibers or both." This term can be utilized to indicate the kind of material [42].

Position of the NBC among other biocomposite constituents as per the above-examined meanings is outlined in Figure 13.2. Composites comprise materials set up by joining microsized inorganic substances and synthetic plastics, for example, layered silicates, calcium carbonate, and carbon black, and are utilized as fillers in the elastic industry from the mid-twentieth century after the fortifying effect was found [21, 43]. Wherever the characteristic rubber is reserved, these composites are among the bioplastics or biocomposites. It is intriguing that bioplastics were most important composite materials prior to 1950s, when petroleum-based compound industry, including polymer formation, began significant production [44, 45]. Nanocomposites vary from the composite with added inorganic substances taken as nanosized state. The natural components are introduced by the petroleum-based polymer(s). In the event that an alternative is required for the biopolymers, the NBC is used. This is not just a conventional alternative. There is a significant change of materials, including the biodegradability, biocompatibility, arrangement techniques, functionalities, and properties [46].

NBCs contrast additionally from biocomposites because they are mostly made from biopolymers; however, they do not possess nanosized materials. As discussed earlier, arrangement of composite materials states their primary types, isolating by each other as per the constituents that decide the properties, functionalities, and structure.

13.5.1 Plant Polysaccharides-based NBCs

Polysaccharides are also referred to as glycans. These are made by monosaccharides that are covalently connected via the glycosidic linkage (see Figure 13.3), shaping a straight or stretched polymer chain. Carbohydrates are the most commonly dispersed and abundant natural combinations on Earth.

13.5.1.1 Nanocellulose

Cellulose is a straight polysaccharide involving D-glucose monomers and comprising glucose-glucose linkages organized into a direct manner, where C-1 of each glucose

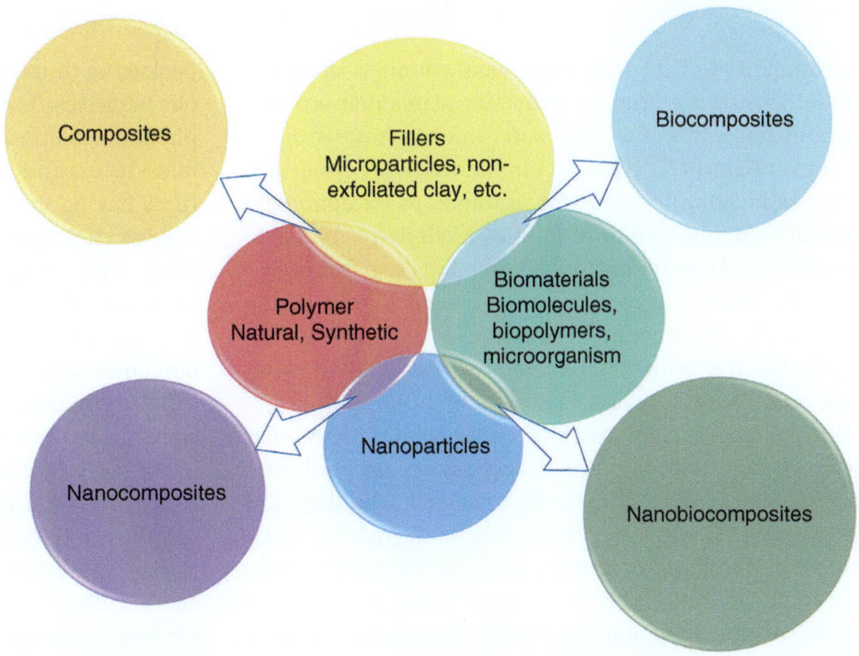

Figure 13.2 Type and composition of composite materials.

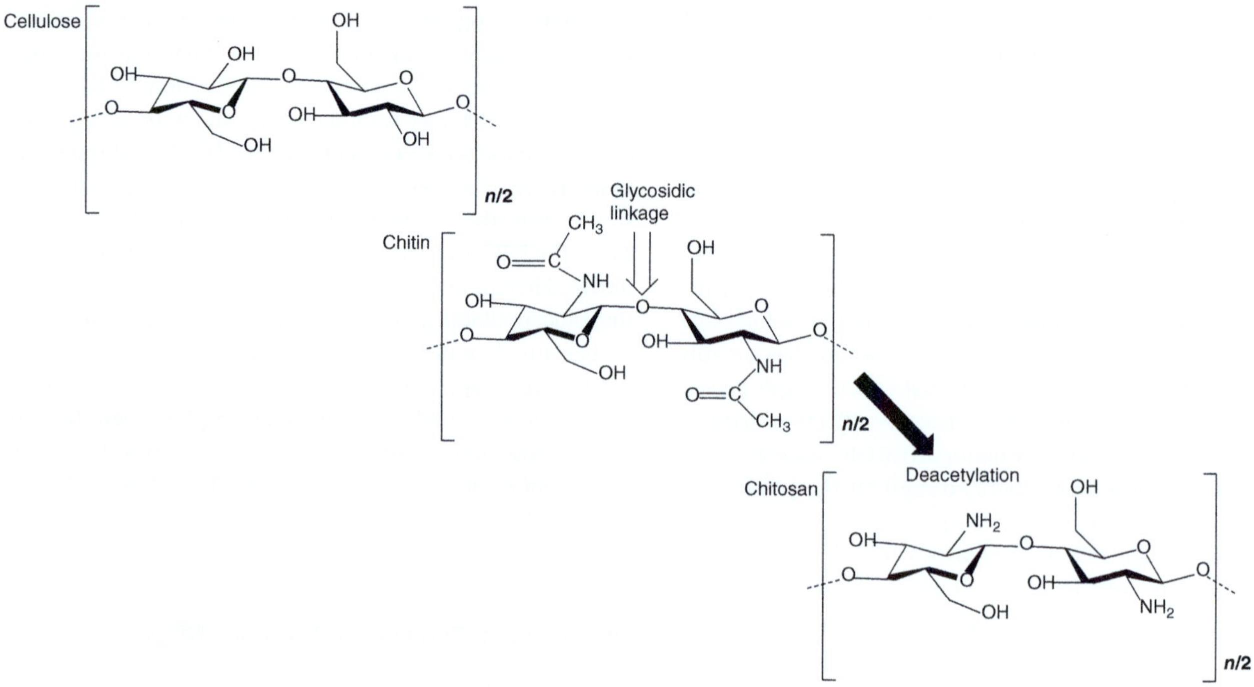

Figure 13.3 Polysaccharide-based biocomposites.

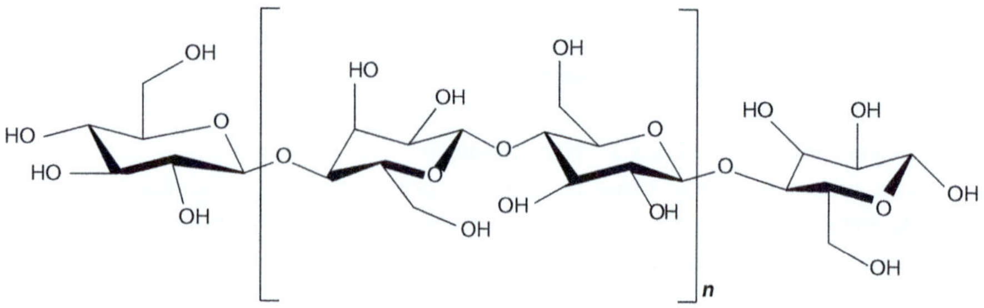

Figure 13.4 Cellulose-based biocomposite preparations.

is clung to C-4 of the following glucose atom as shown in Figure 13.4. It is a primary source for plants that produce carbohydrates from CO_2 and water due to photosynthesis using sunlight [47]. The high atomic weight polymer has other captivating property, for example, its exceptionally crystalline nature and uncertainties. Cellulose is infusible in nature and continuously changing in its two forms, i.e., cellulose ethers and cellulose esters, which kept it highly processable. Additionally, cellulose biocomposites are highly reactive with different biomaterials and showing high modulus and explicit quality [48].

Recent uses of cellulose biocomposites are mentioned here. Green biocomposites dependent on cellulose nanocrystals and kefiran were made by Ghahfarrokhi et al. [49]. Kefiran nanocrystals are a biodegradable NBC, which are most appropriate for packing of food. The elasticity and high modulus of cellulose nanocrystals make them suitable biocomposites [40]. Biopolymer-inferred packaging films could be enhanced by incorporating various additives. The herbal extracts have a huge number of phenolic composites that work as magnificent antimicrobials to expand the food lifespan [50–52]. Newer NBC films having thermal, optical, and mechanical properties used as packaging materials is shown by Miri et al. [53]. A thin film was made by starch polysaccharide--carboxymethyl cellulose (CMC) composite by cellulose nanocrystals as strengthening [54]. Several energizing outcomes were accounted for by Ruan et al. on recovered cellulose/tourmaline nano-stone biocomposite thin films showing antimicrobial activity against *Staphylococcus aureus*. These data revealed outcomes that made NBCs ready for useful packaging applications [55]. Also, nanocellulose is modest, effectively accessible, and revealed numerous advantages, such as catalytic and antimicrobial actions. These make nanocellulose-based NBCs suitable for water purification.

13.5.1.2 Pectin

Plant-derived polysaccharide such as pectin has been used with galacturonic acid and its methyl ester salts as composite of extracellular framework. It possesses more organic similarity and comprises carboxyl functional groups with shaping capacity as per body structure [56]. Skeletal structures of pectin along with different proteins and polysaccharides are more physiochemically stable. Pectin could also take different structures converting liquid solution into thick gels having high atomic weight and polyanionic property subject to the level of the compound response. Because these properties allow pectin molecule to convey signal atoms, it then bolsters different naturally dynamic substances [57]. The structure of pectin in Figure 13.5 is principally made out of rehashing components of galacturonic acid linked through 1 → 4 glycosidic bond, making straight polymer chain [58]. These straight blocks are separated via impartial sugar side chains. Rhamnose has carboxylic galacturonic acid, which remains as free acid either esterified with methanol or neutralized with cations. Also, pectin is basically D-galacturonic homopolymer of (1 → 4)α-D-galactopyranosyluronic acid units with changing degrees of methyl esterified carboxyl groups, and rhamnogalacturonan is a heteropolymer of repeating units of (1 → 2)α-L-rhamnosyl-(1 → L)α-D-galactopyranosyluronic acid with disaccharide, making it a α-D-galacturonan [57].

13.5.1.3 Starch

It is a significant source of energy in plants and is stored in roots, tubers, stems, and seeds for extensive periods [59, 60]. Starch powder is a polysaccharide made by two different polymers: amylase (10–30%) and amylopectin (remaining 70–90%) [61, 62]. Amylose is made by an elongated group of α (1-4)-connected D-glucose units in a straight manner having a high level of polymerization [63]. Amylopectin has an extremely large molecular weight composite with a structure based on amylase saccharides cross-connected by α [1–6] bonds [64]. Different structures of amylopectin and amylose are shown in Figure 13.6. It reveals the amazing characteristics of starch such as biodegradability, biocompatibility, nontoxicity, and attractive strengthening that are essential to produce starch-based NBCs and are appropriate in biomedical applications [65–69]. A few uses include tissue engineering, bone reinforcing treatments, drug delivery matrix, and hydrogels. Previously, CNTs/starch biocomposites have been investigated for medical applications, with their use in protecting and activating bone and in tissue repair mechanisms [70]. Fama et al. showed that a blend of multi-walled carbon nanotubes (MWCNTs) and starch-based NBCs was effective in bone reinforcing treatments [71].

Hydrogels have good antimicrobial activity and are regularly used in tissue designing in the biomedical field. A starch hydrogel has been produced by gamma radiation polymerization strategy [66]. Abdel-Halim and Al-Deyab developed NBC hydrogel together with polyacrylamide, starch, and silver nanoparticles [72]. Antimicrobial starch-based patches were used with silver nanoparticles [73]. The silver nanofibers are made of waxy starch mold by Valodkar et al. delineated phenomenal bactericidal activity against Gram-positive and Gram-negative microbes [69]. Additionally, starch-clay NBCs have also been examined in food packaging applications [74–77].

13.5.1.4 Guar Gum

Guar plant seeds are the main source of guar gum, *Cyamposis tetragonoloba* (Leguminosae); it is an expanded polymer chain of β-D-mannopyranosyl monomers connected [1–4] to single-membered α-D-galactopyranosyl monomers as side chains [78]. Guar gum has been utilized in different areas (suspending agents, development of thicker, and in the ion-exchange resins) [79]. Though, a few data with respect to the presentation of guar gum into superabsorbent have been reported [80].

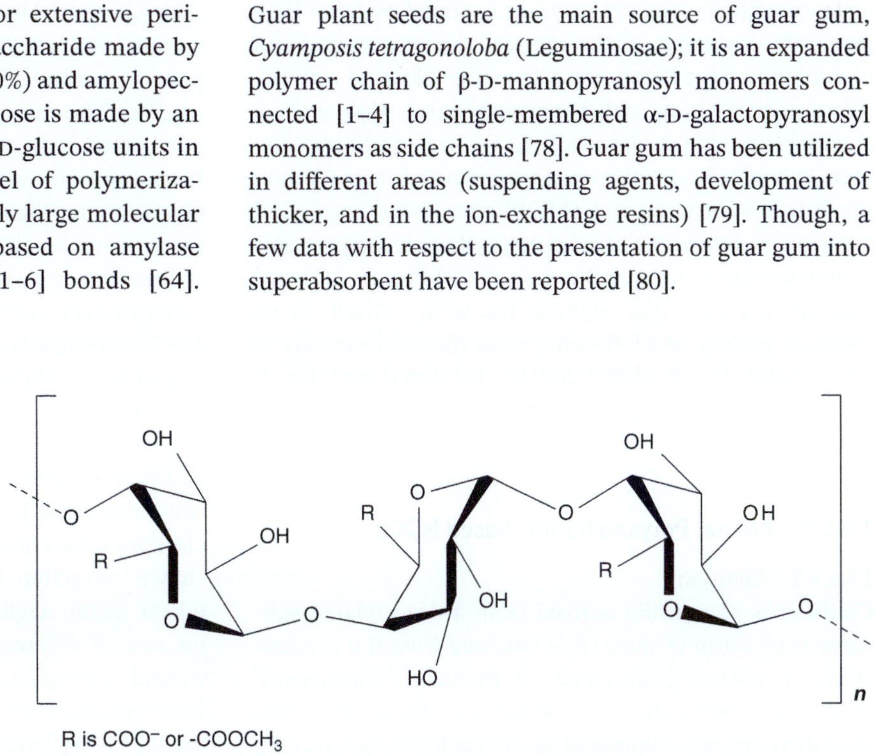

Figure 13.5 Pectin-based biocomposite preparations.

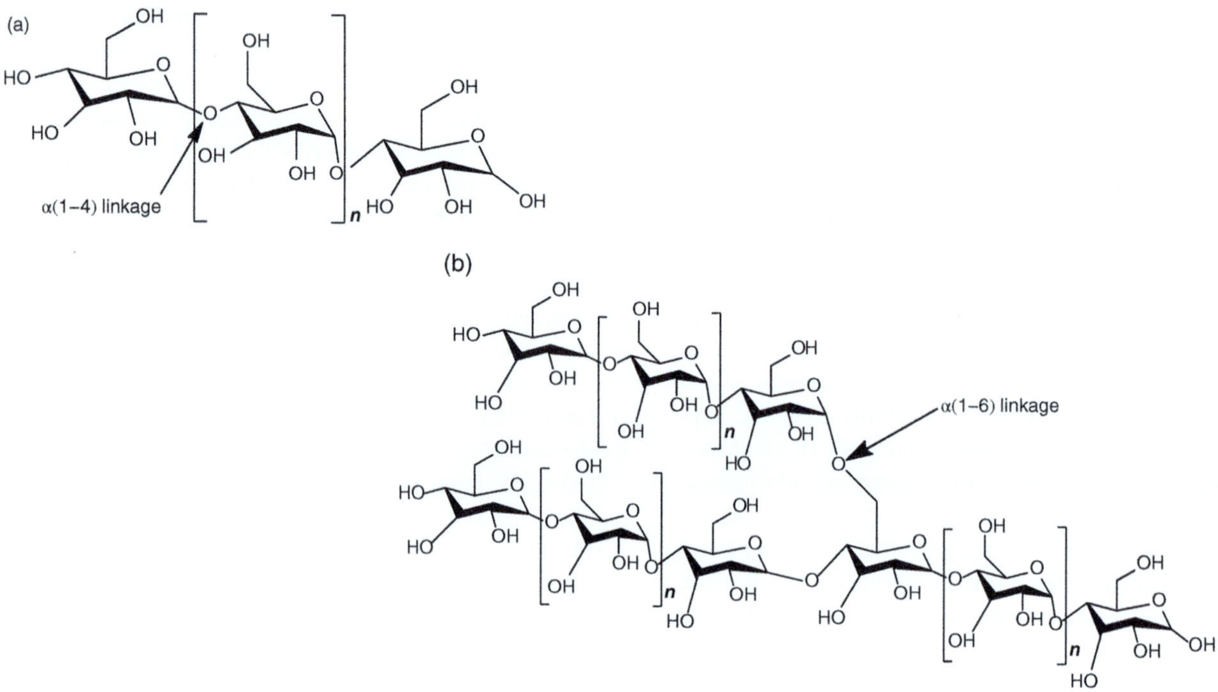

Figure 13.6 Starch-based biocomposite constituents: (a) amylase; (b) amylopectin.

13.5.1.5 Alginates

Wang and Wang [80] describe the primary structure of alginate [81]; it is an anionic polysaccharide produced as basic components of brown marine algae Phaeophyta [82]. Brown marine alga is a plenteous polysaccharide source of alginate with around 40% of the dehydrated material [83], primarily occurring in the intercellular spaces and cell membrane where it comprises insoluble blend of sodium, calcium, potassium, and magnesium salts. Alginate is a cluster block with straight pair of (1→4) glycoside copolymers connected C-5 epimere of α-L-guluronic acid (G) and the α-D-mannuronic acid (M) filtrate [84]. Figure 13.7 explains the image of a characteristic alginate. Its principle action is improving quality and adaptability in seaweed plant [85]. Industrially, alginate has been utilized for its viscosity, gelling, and balancing properties and capacity to retain water [86]. It additionally has perceived potential in expulsion of poisonous substantial metals from industrial trash by biosorption [87].

13.5.2 Animal Polysaccharide-based NBCs

13.5.2.1 Chitosan

Chitosan is chemically derived from a deacetylated substituent of chitin (Figure 13.3) obtained from the exoskeletons of marine species such as mollusks, crustaceans, fungi, and insects [88, 89]. Chitosan is the second most abundant organic compound found on Earth followed by cellulose. Around 100 billion tons of chitin are produced yearly in nature. Interestingly, chitin is yet the most unutilized considering the massive renewal property of its biomass [90]. Concurred that chitin polysaccharide is rare abuse biopolymer of noteworthy flexibility and guarantee, with a potential for applications in numerous fields due to its inimitable antibacterial activity, physicochemical properties, immune-stimulating activities, and wound healing [91]. The entire scope of referenced points of interest is even better than that of the cellulose. Its macromolecule fundamentally comprises of β (1→4) connected D-glucosamine (Figure 13.3). In this manner, NBC development can be performed by means of electrostatic connections of oppositely charged partners. It has an extraordinary latent as a structure block for producing NBCs. Chitosan is subtle to the prevalence of anionic compound in its solutions [92].

In contrast to coal and oil, chitosan is normally recovering strength (e.g. shrimp shells and crab) that could be additionally upgraded by culturing. Chitosan could specifically bind attractive ingredients, for example, metal particles, proteins, cholesterol, and fats. Chitosan has similarly shown attraction to proteins, for example, trypsin and wheat germ agglutinin. Chitosan is extremely helpful because of different properties such as antifungal activity, wound curing, restraint of tumor cells, stimulation of immunomodulator, and increasing speed of plant development [93, 94]. Because of its numerous splendid properties

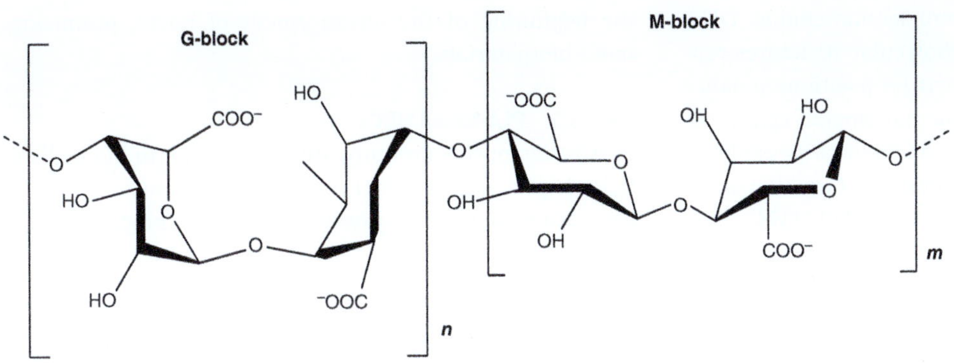

Figure 13.7 Alginate composite consists of β-D-mannuronic (M) and α-L-guluronic (G) acid residues.

such as natural source, reactivity, biodegradability, and abundance, chitosan is frequently used in medical field, biotechnological field, wastewater treatment, and constructions [95, 96].

13.5.2.2 Gelatin

Gelatin is a naturally derived collagen protein and regularly utilized as plasma expander. Gelatin is basically a protein made of an exclusive arrangement of amino acids. Typical highlight of gelatin is that it has essential amino acids, such as hydroxyproline, proline, and glycine. It is a versatile component and provides stability and consistency to the final products because of at least one of the accompanying properties: gel development, film forming, thickening, flexibility, foaming, emulsifying solvency, pH-induced surface charge, biocompatibility, and biodegradability [97]. This versatility has prompted the utilization of gelatin as a profoundly appreciated component in numerous products. It consists of multifunctional groups, $-COOH$ and $-NH_2$, which makes it a reasonable candidate for drug–doxorubicin complex [98] or with polyethylene glycol (PEG) to make reticuloendothelial system escaping complex [99]. Besides, extracellular matrix and cell surface contain fibronectin and also explicit site for imposing cells and a scope for macromolecules through gelatin and collagen [100]. In addition, gelatin could be manufactured as nanospheres and microspheres relying on the method utilized for the development to improve tumoral cell phagocytosis [101]. Other significant characteristic of gelatin is linking by metallic particles, such as sulfhydryl group of cysteine residue, and nitrogen ions in imidazole group of histidine are expected adsorbents for metallic particles [102].

13.5.2.3 Collagen

Collagens are fibrous proteins that contain α-helical state peptide chains. They are the most abundant proteins in warm blooded animals, representing around 20–25% of the total aggregate. They have a connective, defensive, and cooperative character in active bodies [103]. Collagen-I has maximum utilization and is considered a collagen protein, which is available in the ligaments, bones, and skin. Bovine-like Achilles ligaments ordinarily help for its formation. Collagen-I has osteo-inductive, cell adhesive, and mechanical characteristics that makes it appropriate for tissue engineering [104]. Hydroxyapatite containing collagen-I NBCs are mostly used for bone development and tissue biomimicking [105].

13.5.2.4 Silk

Silk proteins, identified as silk fibroins, remain mostly available in a form of fibrillar structure found in nature; *Bombyx mori* silkworms and spiders are the primary source. In silkworm, the fibroin acts as a formational protein containing β-sheets and sericin glycoprotein glue-coated microfilaments bundles, which are linked together to give nanofibrils [106]. Each fiber contains two knotted microfilaments. This structural association, organized with crystalline fibroin, gives fibrillar silk high rigidity and leads to significant strengthening of NBCs [107].

13.5.2.5 Deoxyribonucleic Acid

Deoxyribonucleic acid (DNA) is a biopolymer of incredible organic enormity. Its essential importance includes hereditary data transport and providing synthesis of every material that are organized in whole living structure. Subsequently, DNA is utilized most importantly in genetic engineering; however, it could also help as a matrix structure for NBCs [108, 109]. DNA assembly depends on an anomalous mixture of its features, incorporates: single- and double-helical structure; double- and triple-hybrid macromolecules, and assembles into different 2D and 3D structures. Moreover, width charge of the spine polyphosphate deciding their polyelectrolyte complex arrangement and electrostatic interactions, covalent connection with metallic nanoparticles, and modest conjugation with 3′ as well as 5′ endpoints of the oligonucleotide chains. DNA self-assembly in a form of supramolecular structures renders a countless facility as the matrix for positioning nanosized inorganic substances at the nanometer size [109, 110]. Metallized macromolecules, as often as possible by gold, are very effective in chemotherapy by targeting delivery of anticancer drugs to the tumor cells [111, 112].

Single-walled carbon nanotubes (SWCNTs) could solubilize with single- and double-stranded DNAs, twisting around a dispersed nanotube that permits their distribution in fluids. These connections have high strength and stability and are separated by anion-exchange chromatography [112, 113]. DNA-carbon NBCs have a wide range of applications in materials science such as biomedicine and nanobiotechnology. Negatively charged macromolecules show difficulties in the arrangement of biocomposites with equivalently charged nanoparticles, and with clay minerals it seems to be positively charged, double-layered hydroxides [114]. Fundamental objective of organized DNA NBCs is to develop the methods for diagnosis, design of biochips, biosensors, and nonviral gene delivery [115–117].

13.5.3 Aliphatic Polyester-based NBCs

13.5.3.1 Polyhydroxyalkanoates (PHAs)

Polyhydroxyalkanoates (PHAs) contain two copolymers: the elastomeric poly(3-hydroxyoctanoate) (PHO) and poly(3-hydroxybutyrate-*co*-3-hydroxyvalerate) (PHB/HV); these are biocompatible, biodegradable, and ready to be made through bacterial fermentation from sustainable sources, as they make distinct cytoplasmic combinations within bacterial cells, blended via various bacteria under nutrient control environments in presence of carbon; polyhydroxyesters of 3-, 4-, 5-; and 6-hydroxyalkanoic acids. Nonetheless, they have disadvantages similar to normal thermoplastics such as crystallinity, fragility, and deprived thermal dependability close to their melting point [118]. Clay/PHA NBCs are generally synthesized by melt intercalation and solvent casting techniques [119, 120]. Because of hydrophobic property of PHA, chemical alteration on the clay surface improves scattering in the PHB grid would predicts the type of structure [121].

PHAs have gained the attention of scientists in the past three decades as a promising part of biocomposite materials. Their clear scope lies in their biodegradability into harmless items in varied environments, comprising wetlands and marine environments [118]. Presently, they represent as substitutes of petroleum-based polymers, especially in tissue designing and biomedicine [122]. Hydroxyapatite/PHA nanoparticles-based NBCs have similar mechanical properties as bones [123]. PHAs were merged similarly with layered silicates; however, this is just the beginning of the advancement of novel, promising nano-biomaterials.

13.5.3.2 PLA-based NBCs

PLA is the most commonly utilized polymer in biomedical field. It possesses thermal plasticity, high tensile strength, gas permeability, slow crystallization rate, biocompatibility, biodegradability, and good osteoconductivity [124]. Despite having great mechanical and physical properties, PLA needs some of the fundamental qualities of biocomposites. Those qualities include sustaining physical stability during development of tissues. Besides supporting and stimulating tissue development, PLA ought to keep up the proliferation rate of new tissue even though maintaining its degeneration rate equivalently during tissue growth. This has caught the attention of researchers to investigate more on PLA-based multifunctional NBCs [125, 126].

13.5.3.3 PCL-based NBCs

Poly-ε-caprolactone (PCL) is a hydrophobic, semicrystalline, and completely biodegradable polymer derived from crude oil. PCL could be synthesized by ring-opening polymerization of 2-methylene-1,3-dioxepane free radicals and/or ε-caprolactone with type of coordination catalysts, cationic, and anionic [127]. The barrier, mechanical, and thermal characteristics along great biodegradability of PCL composites could be enhanced by combination with silica, clays, graphene, and CNTs. PCL-based NBCs were synthesized by in situ polymerization, solvent casting, and melt mixing techniques [128]. PCL-chitosan NBCs were prepared by inclusion of polypyrrole, which increased their hydrophobicity and electrical conductivity of the nanofibrous scaffold, which gives them more potential for biomedical applications [129]. Antimicrobial silver-loaded PCL-hydroxyapatite NBC film was developed by chemical precipitation method, which shows good antibacterial activity against *Escherichia coli* and *Staphylococcus epidermis*, and promised to be framed as reconstructive and orthopedic implants [130]. Platelet-rich plasma-treated PCL-tricalcium phosphate NBC porous scaffold increases the expression of marker bone gene and has potential for stability and regeneration of newly prepared bone tissue [131].

13.5.4 Carbon-based NBCs

13.5.4.1 Graphene-based NBCs

Graphene, a unique 2D single layer of honeycomb-like crystal of carbon atom, was serendipitously discovered in 2004 [132]. Graphene is one of the carbon allotropes such as fullerene, CNT, and diamond. In 2010, Novoseloy and Geim received a Nobel Prize for their spectacular experiments on 2D graphene. A single graphene sheet contains the sp^2 hybridized carbon atoms present in hexagonal ring, which includes three in-plane sigma bonds with p^z-orbitals opposite to the planes. Graphene possesses distinct electrical properties, such as electron dexterity of about 250 000 $cm^2\ Vs^{-1}$ at room temperature and uncharacteristic quantum hall effect [133]. Graphene has apparent solubility, chirality, and surface morphology giving rise to its interesting thermal, optical, electrical, magnetic, and chemical properties [134, 135]. Researchers have studied and reported that they have extreme biocompatibility, tunable fluorescence, excellent photostability, cargo carrier loading, low cytotoxicity, and effective targeting features that expose the potential biomedical application in gene and drug delivery, nanomedicine, bioimaging, biosensors, genetic engineering, and as an artificial implant [136–139]. When only 1% graphene was added into polystyrene, a conductivity of the nanocomposite increased about ~0.1 Sm^{-1} and it also improved structural quality, toughness, and tensile strength of NBCs [140].

Graphene-copper oxide nanocomposites have been developed for cost-effective electrochemical nonenzymatic glucose detection [141]. The doxorubicin-loaded graphene quantum dot/CMC nanocomposite hydrogel patches show enhancement in water vapor permeability, swelling, and pH-responsive drug delivery for blood cancer cells K562 without significant toxicity [142]. The theragnostic nanocomposite of graphene oxide with bismuth selenide has been synthesized by single pot solvothermal method in presence of polyvinylpyrrolidone; these nanocomposites have been used as contrast agent for photoacoustic imaging, X-ray CT scanning in vivo, and show cytotoxicity to cancer cell under laser irradiation photothermal therapy [143]. The nanocomposite scaffold of graphene oxide grafted PCL nanofibers was successfully prepared by electrospinning method. These nanocomposites show good biocompatibility and increase MG63 cell proliferation, which make them suitable for application in tissue engineering [144].

13.5.4.2 CNTs-based Nanobiocomposites

CNTs were first discovered in 1991, have been one of the extremely fascinating ingredients with interesting properties for the past 20 years. The CNTs comprise of graphene sheets arranged such that to shape a tube that exists as single, twofold, and multi-walled CNTs, relying on number of graphene layers [145]. The electrical flow transmission limit of CNT is multiple times higher than a copper wire. CNT has high large surface area, toughness, hollow geometry, and strength. CNTs are fortified substances with multifunctional composite matrix as they have surprising optical, electrical, and mechanical properties alongside their ability to conjugate with functional groups and synthetic chemicals [146, 147]. Despite the fact that nanotubes are extremely encouraging in a wide diversity of fields, it has been constrained to larger scale production of CNTs. The principle constraints frustrate its utilization, poor processability, presence of impurities, and struggle in structure control. To improve their useful applications, mass nanotube materials have lately been drawn in expanding considerations, especially through development of composites with polymers [148]. Production and characterization of 3D printed PCL-CNT cardiac tissue scaffolds have been explored, and the cytotoxicity and cell imaging results of H9c2 cell show that composite has increased cell propagation and good biodegradability [149]. CNTs fused gelatin/CMC micro-scaffold was fabricated by cryogelation technique that showed good cell viability, biocompatibility, and conductivity. This customized micro-scaffold has been a potential candidate for drug screening, organ biochip, tissue transplantation, and 3D cell culture modelling [150].

13.6 Sustainable Properties of Nanobiocomposites

In the past five decades, the field of biomedical investigation has progressed and scientists are endeavoring to displace tissues and organs by natural and/or artificial biocomposites with different characteristics, which makes biocomposites mostly realistic, having direct contact with live tissue, and delivered by procedures that frequently utilize or copy natural circumstances. Recently various NBCs have been utilized as biomaterials in various biomedical applications [151].

13.6.1 Biodegradability

Irrespective of whether natural or synthetic, polymers degrade with dynamic losses of properties, e.g. robustness, appearance, and mechanical strength. Currently, biodegradable polymers have gained increased attention of scientists as these materials break down in the body by usual biological processes, taking out the essential step of eliminating a drug delivery matrix after drug release is accomplished [152]. As a biodegradable polymer, it is vital to perceive that decomposition is a biochemical procedure (realized in variables, such as, water, heat, ultrasound, and light); it might happen through bulk hydrolysis, and disintegration happens in static manner that relies on diffusion and dissolution forms [153, 154]. Few biodegradable polymers are prominently the polyorthoesters and polyanhydrides, in which degradation

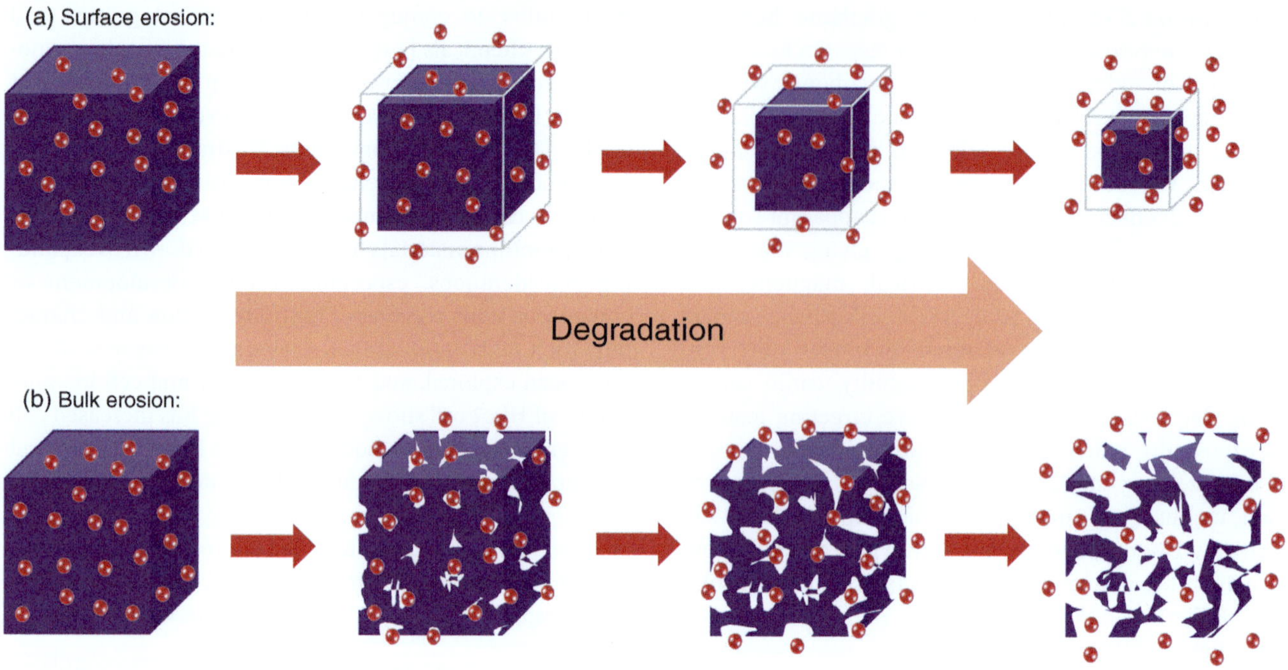

Figure 13.8 Mechanisms of biodegradation: (a) surface erosion; (b) bulk erosion.

happens only at the polymer surface, bringing about a rate of drug release that is equivalent to the surface area of the formulation [155]. Degradation of polymer can be characterized into hydrolytic, biodegradative, oxidative, radiation, thermal, photon, mechanical, and chemical erosion [154]. In degradable polymers, two diverse disintegration processes have been expected: (i) heterogeneous/surface erosion and (ii) homogeneous/bulk erosion (Figure 13.8). On account of surface dissolving biopolymer system, degradation of polymer is quicker than water interruption. Degradation happens, by this way, for the most part in the peripheral polymer layers. Subsequently, disintegration influences just the surface and not the inside of a polymer [156]. Bulk process polymers degrade gradually; water accumulation in the matrix is quicker than degradation of polymer. In this manner, the whole polymer matrix is quickly hydrated and degraded with the formulation. Thus, disintegration is not confined to the polymer surface [157, 158].

13.6.2 Biocompatibility

Biomaterials such as polymers, metals, and ceramics are normally intended to preserve an inactive material, either synthetic or natural substances for biomedical applications, and enhance the personal fulfilment for countless patients consistently. Since they are appropriate for consideration in systems that interface with human body fluid and tissues that replace or expand the volume of injured organ or tissues in a solid, safe, and physiologically worthy means without producing any adverse reactions on another tissue or blood [159, 160], they would show biocompatibility, nontoxicity, immunogenicity, sterility, and practicality. The biocompatibility indicates the capacity of a biomaterial to accomplish through a proper host reaction a particular condition, and it is utilized in a particular application except for any destructive or toxic consequences for organic function [161]. In this specific circumstance, such materials ought to radiopaque, manipulatable, nonabsorbable, nontoxic, and dimensionally robust [162]. An assurance of the biocompatibility of such substances and embedded mechanism includes comprehensive evaluation of the materials and broad examining – first at the in vitro cell culture, second in in vivo animal models, and finally in clinical trials [161]. The impacts of the surface and the structure characteristics of biocompatibility of polymeric biomaterials include (i) the free interfacial energy, (ii) equilibrium among the hydrophobicity and hydrophilicity upon a superficial layer, (iii) organic nature of functional groups, (iv) density and type of surface charges, (v) subatomic mass of the polymer, (vi) surface geography and harshness, and (vii) conformational adaptability of the polymer [163].

13.6.3 Mechanical Properties

Mechanical characteristics of NBC are significant in pharmaceutical applications. For example, sustainability of the drug delivery matrix through the utilization is imperative to acquire. The consent of Food and Drug Administration is not given except if the system is intended to ensure a sensitive drug; for example, protein must keep up its stability till drug is released out of the matrix. Shifting the level of cross-linking agents has been used to accomplish the ideal mechanical properties of the hydrogel. Expanding the level of cross-linking of the matrix would bring about a sturdier hydrogel. However, further extent of cross-linking makes weaker structure. Consequently, the ideal level of cross-linking is to accomplish moderately dense but flexible hydrogel. Copolymerization has similarly been used to accomplish the required mechanical characteristics of hydrogel. Fusing of comonomer that improves the H-holding enhances the quality of hydrogel [57].

13.7 Pharmaceutical Applications of Nanobiocomposites in Drug Delivery Systems

Current advances in drug delivery systems are based on the novel approaches and specific targeted drug delivery systems. The novel approaches are customized with the help of agents that are biodegradable in nature and/or can enhance the property of certain drugs' delivery. Few biodegradable agents are coated to deliver the drugs toward specific targets into human systemic circulation. Such biodegradable agents are called biocomposites. They are classified according to their sources and their applicability in advanced drug delivery systems. Modern class of drug formulations are composed of pharmaceutical ingredient and excipients coated with such biodegradable materials [164]. Excipients supply various functions in drug formulations including binding, lubricating, gelling, bulking, suspending, and flavoring [165]. Biocomposites are usually used in disintegrating, matrix formation, modified drug release, viscosity enhancement, binding, bulking process, flavoring, gel application, lubricating agent, and much more in pharmaceutical approaches. There are various merits and demerits of using biocomposites in various advanced drug delivery systems. The upgrading interest of various biocomposites in pharmaceutical applications is due to its variety of useful applications. Despite great merits, biocomposites also exert various limitations based on its source, nanotoxicity, and time-consuming purification process.

Polysaccharides are widely accepted biocomposites in various biomedical pharmaceutical applications in advanced drug delivery systems. Plant polysaccharide cellulose and synthetic cellulose are widely used in a variety of pharmaceutical applications. Cellulose powder is used as the filler in a variety of pharmaceutical dosage forms. In pharmaceutical industry, microcrystalline cellulose is used as diluents or binder application. It is also used as the viscosity builder in liquid dosage forms. It is also useful as a compressibility enhancer. Another plant polysaccharide, starch, is also widely used in a variety of pharmaceutical applications. It is mostly used as the drug release modifier and film forming agent. It is also useful as binder and bulking agent in various industrial powder formulations. Pectin is a much useful biocomposite and is widely used as bioavailability enhancer and drug release modifier in various sustained or prolonged release drug delivery systems. It is also used as coating agent and taste enhancer in various tablet and capsule formulations. In various gel preparations, it is used as the gel forming agent, e.g. hydrogels. Modified cellulose derivative hydroxy propyl methyl cellulose (HPMC) is widely used as a gelling and coating agent in various gel formulations. It is also used as taste masker in various tablet or capsule formulations to mask its bitter taste to improve its palatability and drug release. methyl cellulose (MC), hydroxy ethyl cellulose (HEC), and hydroxypropyl cellulose (HPC) are also widely used as taste masking agents in various pharmaceutical industrial formulations. Na-CMC is widely used as the binding agent in drug loaded or active pharmaceutical ingredient loaded powder formulations. Various modified organic cellulose esters like cellulose acetate (CA), cellulose acetate phthalate (CAP), cellulose acetate trimellitate (CAT), hydroxypropylmethylcellulose phthalate (HPMCP), and HPMCAP are widely used as coating agents in disease-specific therapeutics and targeted drug delivery systems, e.g. bioadhesive and mucoadhesive drug delivery systems. They are also widely used in semipermeable membrane formulation of osmotic systems, e.g. extended release in gastrointestinal systems. It is also widely used for its delayed release application in enteric-coated solid dosage form for pH-specific targeted drug release systems to enhance its bioavailability and therapeutic potency [e.g. for non-steroidal anti-inflammatory drugs (NSAIDs)]. Cellulose derivatives are widely used as thickening and stabilizing agents in various suspensions and emulsions for specific drugs or active pharmaceutical ingredient in specified drug delivery preparation. The gums and mucilage are repeatedly used biocomposites in pharmaceutical preparations as in thickening, binding, emulsifying, suspending, and stabilizing agents. Guar gum and xanthan gum are natural biocomposites that are widely used as drug release modifier in various extended drug delivery formulations [166]. It is also used as the compressibility enhancer in various tablet formulations. Gellan gum and pullulan gum are widely used in sustained release drug delivery formulations as drug release modifier and in capsules as encapsulating agents. Tragacanth and gum arabic are used in emulsions and suspensions as suspending agent and emulsifying agent. It is also used in advanced drug delivery systems for delayed release as drug release modifier and in capsules as microencapsulating agent. Almond gum and Kyaha gums are used as binding agent and drug release modifier in sustained release drug delivery systems, e.g. transdermal patches. Grewia gum and Okra gum are widely used as binding agent, bioavailability enhancer, drug release modifier (delayed release or extended release systems), hydrophilic polymer matrix agent, bioadhesive, and film forming agent in various modern drug delivery systems. Psyllium gum is popular biocomposite used in laxative preparations and in various delayed release drug delivery systems.

Chitosan-active biocomposite is widely used as the bioabsorption enhancer or bioavailability enhancer and drug release modifier in various targeted drug delivery systems [167]. Due to its nontoxic profile, it is widely used as various central nervous system (CNS) targeted drug delivery systems [168]. Seaweed polysaccharide alginate and agar are frequently used in emulsion and suspensions as stabilizers and suspending agent [169]. They are useful as active biocomposite in drug encapsulation for delayed or targeted drug delivery preparations. Agar is also used as lubricating agent and also in laxative pharmaceutical preparations.

References

1 Darder, M., Aranda, P., and Ruiz-Hitzky, E. (2007). Nanobiocomposites: a new concept of ecological, bioinspired, and functional hybrid materials. *Adv. Mater.* 19: 1309–1319.

2 Liu, X., Sun, W., Wang, H. et al. (2005). Microspheres of corn, zein, for an ivermectin drug delivery system. *Biomaterials* 26: 109–115.

3 Ray, S.S. and Okamoto, M. (2003). Polymer/layered silicate nanocomposites: a review from preparation to processing. *Prog. Polym. Sci.* 28: 1539–1641.

4 Pavlidou, S. and Papaspyrides, C.D. (2008). A review on polymer–layered silicate nanocomposites. *Prog. Polym. Sci.* 33: 1119–1198.

5 Okpala, C.C. (2014). The benefits and applications of nanocomposites. *Int. J. Adv. Eng. Technol.* 5: 12–18.

6 Batley, E.I., Kirby, J.K., and Mclaughlin, M.J. (2013). Fate and risks of nanomaterials in aquatic and terrestrial environments. *Acc. Chem. Res.* 46: 854–862.

7 Sudesh, K. and Iwata, T. (2008). Sustainability of biobased and biodegradable plastics. *CLEAN–Soil, Air, Water.* 36 (5-6): 433–442.

8 Golomb, D.S. and Fay, J.A. (2004). *Energy, Waste, and the Environment: A Geochemical Perspective* (eds. R. Giere and P. Stille), 153–167. London: Geological Society.

9 Thakur, V.K., Thakur, M.K., Raghavan, P., and Kessler, M.R. (2014). Progress in green polymer composites from lignin for multifunctional applications: a review. *ACS Sustain. Chem. Eng.* 2: 1072–1092.

10 Fark, O., Bledzki, A.K., Fink, H.P., and Sain, M. (2014). Progress report on natural fiber reinforced composites. *Macromol. Mater. Eng.* 299: 9–26.

11 Lee, K.Y., Buldum, G., Mantalaris, A., and Bismarck, A. (2014). More than meets the eye in bacterial cellulose: biosynthesis, bioprocessing, and applications in advanced fiber composites. *Macromol. Biosci.* 14: 10–32.

12 Ruiz-Hitzky, E., Darder, M., and Aranda, P. (2005). Functional biopolymer nanocomposites based on layered solids. *J. Mater. Chem.* 15: 3650–3662.

13 Jamshidian, M., Tehrany, E.A., Imran, M. et al. (2010). Poly-lactic acid: production, applications, nanocomposites, and release studies. *Compr. Rev. Food Sci. Food Saf.* 9: 552–571.

14 Pandey, J.K., Chu, W.S., Lee, C.S., and Ahn, S.H. (2007). Preparation characterization and performance evaluation of nanocomposites from natural fiber reinforced biodegradable polymer matrix for automotive applications. Presented at the International Symposium on Polymers and the Environment: Emerging Technology and Science, Bio Environmental Polymer Society (BEPS), Vancouver, WA, 17–20.

15 Adeosun, S.O., Lawal, G.I., Balogun, S.A., and Akpan, E.I. (2012). Review of green polymer nanocomposites. *J. Miner. Mater. Charact. Eng.* 11 (4): 385–416.

16 Vilaplana, F., Strömberg, E., and Karlsson, S. (2010). Environmental and resource aspects of sustainable biocomposites. *Polym. Degrad. Stab.* 95 (11): 2147–2161.

17 Mohanty, A.K., Misra, M., and Drzal, L.T. (eds.) (2005). *Natural Fibers, Biopolymers, and Biocomposites*, 1e, 896. United Kingdom: CRC Press.

18 Sehaqui, H., Allais, M., Zhou, Q., and Berglund, L.A. (2011). Wood cellulose biocomposites with fibrous structures at micro-and nanoscale. *Compos. Sci. Technol.* 71 (3): 382–387.

19 Le Duigou, A., Castro, M., Bevan, R., and Martin, N. (2016). 3D printing of wood fibre biocomposites: from mechanical to actuation functionality. *Mater. Des.* 15 (96): 106–114.

20 Shekar, H.S. and Ramachandra, M. (2018). Green composites: a review. *Mater. Today* 5 (1): 2518–2526.

21 Rogovina, S.Z. (2016). Biodegradable polymer composites based on synthetic and natural polymers of various classes. *Polym. Sci. Ser. C.* 58 (1): 62–73.

22 Alexandre, M. and Dubois, P. (2000). Polymer-layered silicate nanocomposites: preparation, properties and uses of a new class of materials. *Mater. Sci. Eng. R. Rep.* 28 (1-2): 1–63.

23 Pyrz, R. and Bochenek, B. (2007). Discrete-continuum transition at interfaces of nanocomposites. *Bull. Pol. Acad. Sci.: Tech. Sci.* 55 (2): 251–260.

24 Saini, R.K., Bajpai, A.K., and Jain, E. (2017). Advances in nanobiocomposites for biomedical applications. In: *Biodegradable and Biocompatible Polymer Composites: Processing, Properties and Applications*, 1e (ed. N. Shimpi), 351–377. Philadelphia, PA: Elsevier.

25 Wolf, C., Angellier-Coussy, H., Gontard, N. et al. (2018). How the shape of fillers affects the barrier properties of polymer/non-porous particles nanocomposites: a review. *J. Membr. Sci.* 556: 393–418.

26 Thomas, S., Kuruvilla, J., and Malhotra, S.K. (eds.) (2013). Electrical and optical studies of organic light emitting devices using SWCNTs-polymer nanocomposites. In: *Polymer Composites Volume 3, Biocomposite*, 1ste, 1–608. New Jersey, NJ: Wiley.

27 Low, I.M. (ed.) (2018). *Advances in Ceramic Matrix Composites*, 2e, 1–842. Australia: Woodhead Publishing/ Elsevier.

28 Chawla, K.K. (2019). Ceramic matrix composites. In: *Composite Materials*, 3e (ed. K. Chawla), 251–296. Cham: Springer.

29 Suib, S.L. (2017). A review of nanoceramic materials for use in ceramic matrix composites. In: *Sol-Gel Based Nanoceramic Materials: Preparation, Properties and Applications*, 1e (ed. A.K. Mishra), 185–230. Cham: Springer.

30 Low, I.M. (2018). *Advances in Ceramic Matrix Composites: Introduction*, 2e, 1–7. Australia: Woodhead Publishing/ Elsevier.

31 Gowda, T.Y., Sanjay, M.R., Bhat, K.S. et al. (2018). Polymer matrix-natural fiber composites: an overview. *Cogent Eng.* 5 (1): 1446667.

32 Divya, H.V., Naik, L.L., and Yogesha, B. (2016). Processing techniques of polymer matrix composites–a review. *Int. J. Eng. Res. Gen. Sci.* 4 (3): 357–362.

33 Malaki, M., Xu, W., Kasar, A.K. et al. (2019). Advanced metal matrix nanocomposites. *Metals* 9 (3): 330.

34 Kumar, A., Arafath, M.Y., Gupta, P. et al. (2020). Microstructural and mechano-tribological behavior of Al reinforced SiC-TiC hybrid metal matrix composite. *Mater. Today* 21: 1417–1420.

35 Sharma, D.K., Mahant, D., and Upadhyay, G. (2020). Manufacturing of metal matrix composites: a state of review. *Mater. Today* 26 (2): 506–519.

36 Abdelaal, M.Y., Hussein, M.A., and Alosaimi, A.M. (2017). An overview on polysulphone/clay nanocomposites. *Int. J. Biosen. Bioelectron.* 3 (1): 201–205.

37 Xu, T., Xu, X., Gu, Y. et al. (2018). Functional intercalated nanocomposites with chitosan-glutathione-glycylsarcosine and layered double hydroxides for topical ocular drug delivery. *Int. J. Nanomed.* 13: 917.

38 Gohari, D.P., Kalaee, M.R., and Sharif, A. (2019). Interfacial in situ polymerization of layered-silicate/poly (hexamethylene isophthalamide) nanocomposites. *J. Inorg. Organomet. Polym. Mater.* 29 (4): 1243–1251.

39 Mozumder, M.S., Mairpady, A., and Mourad, A.H. (2017). Polymeric nanobiocomposites for biomedical applications. *J. Biomed. Mater. Res. B.* 105 (5): 1241–1259.

40 Cotica, L.F. et al. (2018). Nanobiocomposites: synthesis and environmental applications. In: *Fungal Nanobionics: Principles and Applications*, 1e (eds. R. Prasad, V. Kumar, M. Kumar and S. Wang), 1–19. Singapore: Springer.

41 Patwekar, S.L. (2016). Nanobiocomposite: a new approach to drug delivery system. *Asian J. Pharm.* 10 (04): 646–656.

42 Laadila, M.A., Hegde, K., Rouissi, T. et al. (2017). Green synthesis of novel biocomposites from treated cellulosic fibers and recycled bio-plastic polylactic acid. *J. Clean. Prod.* 164: 575–586.

43 Endres, H.J. (2017). Bioplastics. In: *Biorefineries*, 1e (eds. W. Kurt and T. Nils), 427–468. Cham: Springer.

44 Ashter, S.A. (2016). *Introduction to Bioplastics Engineering*, 1e. William Andrew: New Hampshire.

45 Eg, Ö. and Murat, A.T. (2018). Rethinking sustainability: a research on starch-based bioplastic. *J Sustain. Construct. Mater. Technol.* 3 (3): 249–260.

46 Emadian, S.M., Onay, T.T., and Demirel, B. (2017). Biodegradation of bioplastics in natural environments. *Waste Manag.* 59: 526–536.

47 Feldman, D. (2015). Polyblend nanocomposites. *J. Macromol. Sci. A: Pure Appl. Chem.* 52: 322–329.

48 Nishino, T., Matsuda, I., and Hirao, K. (2004). All-cellulose composite. *Macromolecules* 37: 7683–7687.

49 Ghahfarrokhi, I.S., Khodaiyan, F., and Mousavi, M. (2015). Green bionanocomposite based on kefiran and cellulose nanocrystals produced from beer industrial residues. *Int. J. Biol. Macromol.* 77: 85–91.

50 Gutierrez, M.Q., Echeverria, L., Ihl, M. et al. (2012). Carboxymethylcellulose– montmorillonite nanocomposite films activated with murta (*Ugni molinae* Turcz) leaves extract. *Carbohydr. Polym.* 87: 1495–1502.

51 Alizadeh, A., Khoshkhui, M., Javidnia, K. et al. (2010). Effects of fertilizer on yield, essential oil composition, total phenolic content and antioxidant activity in *Satureja hortensis* L. (Lamiaceae) cultivated in Iran. *J. Med. Plant Res.* 4: 33–40.

52 Krstev, T.M., Radnovi, C.D., Kiti, D. et al. (2009). Chemical composition and antimicrobial activity of *Satureja hortensis* L. essential oil. *Cent. Eur. J. Biol.* 4: 411–416.

53 Miri, N.E., Abdelouahdi, K., Barakat, A. et al. (2015). Bio-nanocomposite films reinforced with cellulose nanocrystals: rheology of film-forming solutions, transparency, water vapor barrier and tensile properties of films. *Carbohydr. Polym.* 129: 156–167.

54 Gonzalez, C.R., Martínez-Hernández, A.L., Castaño, V.M. et al. (2012). Polysaccharide nanocomposites reinforced with graphene oxide and keratin-grafted graphene oxide. *Ind. Eng. Chem. Res.* 51: 3619–3629.

55 Ruan, D., Zhang, L., Zhang, Z., and Xia, X. (2003). Structure and properties of regenerated cellulose/ tourmaline nanocrystal composite films. *J. Polym. Sci. B Polym. Phys.* 42 (3): 367–373.

56 Kokkonen, H.E., Ilvesaro, J.M., Morra, M. et al. (2007). Effect of modified pectin molecules on the growth of bone cells. *Biomacromolecules* 8 (2): 509–515.

57 Saini, R.K., Bajpai, A.K., and Jain, E. (2018). Fundamentals of nanobiocomposites. In: *Biodegradable and Biocompatible Polymer Composites* (ed. N. Shimpi), 351–377. Mumbai, India: Woodhead Publishing.

58 Chen, C.H., Sheu, M.T., Chen, T.F. et al. (2006). Suppression of endotoxin-induced proinflammatory responses by citrus pectin through blocking LPS signaling pathways. *Biochem. Pharmacol.* 72 (8): 1001–1009.

59 Bogracheva, T.Y., Meares, C., and Hedley, C.L. (2006). The effect of heating on the thermodynamic characteristics of potato starch. *Carbohydr. Polym.* 63: 323–330.

60 Blennow, A., Bay-Smidt, A.M., Leonhardt, P. et al. (2003). Starch paste stickiness is a relevant native starch selection criterion for wet-end paper manufacturing. *Starch-Starke* 55 (9): 381–389.

61 Pareta, R. and Edirisinghe, M.J. (2006). A novel method for the preparation of starch films and coatings. *Carbohydr. Polym.* 63: 425–431.

62 Castro, J.V., Dumas, C., Chiou, H. et al. (2005). Mechanistic information from analysis of molecular weight distributions of starch. *Biomacromolecules* 6: 2248–2259.

63 Suortti, T., Gorenstein, M.V., and Roger, P. (1998). Determination of the molecular mass of amylose. *J. Chromatogr. A* 828: 515–521.

64 Xie, F., Pollet, E., Halley, P.J., and Averous, L. (2013). Starch-based nano-biocomposites. *Prog. Polym. Sci.* 38: 1590–1628.

65 Lane, M.E. (2011). Nanoparticles and the skin-applications and limitations. *J. Microencapsul.* 28: 709–716.

66 Eid, M. (2011). Gamma radiation synthesis and characterization of starch-based polyelectrolyte hydrogels loaded silver nanoparticles. *J. Inorg. Organomet. Polym. Mater.* 21: 297–305.

67 Gao, X., Wei, L., Yan, H., and Xu, B. (2011). Green synthesis and characteristic of coreshell structure silver/ starch nanoparticles. *Mater. Lett.* 65: 2963–2965.

68 Schmitt, H., Prashantha, K., Soulestin, J. et al. (2012). Preparation and properties of novel melt-blended halloysite nanotubes/wheat starch nanocomposites. *Carbohydr. Polym.* 89: 920–927.

69 Valodkar, M., Sharma, P., Kanchan, D.K., and Thakore, S. (2010). Conducting and antimicrobial properties of silver nanowire–waxy starch nanocomposites. *Int. J. Green Nanotechnol. Phys. Chem.* 2: 10–19.

70 Harrison, B.S. and Atala, A. (2007). Carbon nanotube applications for tissue engineering. *Biomaterials* 28: 344–353.

71 Fama, L.M., Pettarin, V., Goyanes, S.N., and Bernal, C.R. (2011). Starch/multi-walled carbon nanotubes composites with improved mechanical properties. *Carbohydr. Polym.* 83: 1226–1231.

72 Abdel-Halim, E.S. and Al-Deyab, S.S. (2014). Antimicrobial activity of silver/starch/polyacrylamide nanocomposite. *Int. J. Biol. Macromol.* 68: 33–38.

73 Yoksan, R. and Chirachanchai, S. (2010). Silver nanoparticle-loaded chitosan–starch based films: fabrication and evaluation of tensile, barrier and antimicrobial properties. *Mater. Sci. Eng. C* 30: 891–897.

74 Park, H.M., Li, X., Jin, C.Z. et al. (2002). Preparation and properties of biodegradable thermoplastic starch/clay hybrids. *Macromol. Mater. Eng.* 287 (8): 553–558.

75 Avella, M., De Vlieger, J.J., Errico, M.E. et al. (2005). Biodegradable starch/clay nanocomposite films for food packaging applications. *Food Chem.* 93: 467–474.

76 Yoon, S. and Deng, Y. (2006). Clay–starch composites and their application in papermaking. *J. Appl. Polym. Sci.* 100 (2): 1032–1038.

77 Cyras, V.P., Manfredi, L.B., Ton-That, M.T., and Vazquez, A. (2008). Physical and mechanical properties of thermoplastic starch/montmorillonite nanocomposite films. *Carbohydr. Polym.* 73 (1): 55–63.

78 Mudgil, D., Barak, S., and Khatkar, B.S. (2014). Guar gum: processing, properties and food applications—a review. *Int. J. Food Sci. Technol.* 51 (3): 409–418.

79 Thombare, N., Jha, U., Mishra, S., and Siddiqui, M.Z. (2016). Guar gum as a promising starting material for diverse applications: a review. *Int. J. Biol. Macromol.* 88: 361–372.

80 Wang, W. and Wang, A. (2009). Preparation, characterization and properties of superabsorbent nanocomposites based on natural guar gum and modified rectorite. *Carbohydr. Polym.* 77 (4): 891–897.

81 Draget, K.I., Smidsrød, O., and Skjåk-Bræk, G. (2002). Alginates from algae. Biopolymers. In: *Polysaccharides II: Polysaccharides from Eukaryotes* (eds. A. Steinbüchel and S. Rhee). Weinheim, Germany: WILEY-VCH Verlag GMBH & Co. KGaA.

82 Painter, T.J. (1983). Algal polysaccharides. In: *The Polysaccharides*, 1e (ed. G.O. Aspinall), 196–286. London: Academic Press.

83 Draget, K.I., Smidsrod, O., and Skjak-Braek, G. (2005). Alginates from algae. In: *Polysaccharides and Polyamides in the Food Industry*, 1e (eds. A. Steinbuchel and S.K. Rhee), 1–30. Weinheim: Wiley-VCH Verlag GmbH & Co.

84 Youssouf, L., Lallemand, L., Giraud, P. et al. (2017). Ultrasound-assisted extraction and structural characterization by NMR of alginates and carrageenans from seaweeds. *Carbohydr. Polym.* 166: 55–63.

85 Abd El-Mohdy, H.L. (2017). Radiation-induced degradation of sodium alginate and its plant growth promotion effect. *Arab. J. Chem.* 10: 431–480.

86 Qin, Y., Jiang, J., Zhao, L. et al. (2018). Applications of alginate as a functional food ingredient. In: *Biopolymers for Food Design*, 1e (eds. A.M. Grumezescu and A.M. Holban), 409–429. London, PA: Academic Press/ Elsevier.

87 Soliman, S.M., Ali, A.M., and Sabaa, M.W. (2017). Alginate-based hydrogel for water treatment. *Desalin. Water Treat.* 94: 129–136.

88 Dutta, P.K. (ed.) (2016). *Chitin and Chitosan for Regenerative Medicine*, 1e. Germany: Springer.

89 Jennings, J.A. and Bumgardner, J.D. (eds.) (2016). *Chitosan Based Biomaterials. Volume 1: Fundamentals*, 1e. Cambridge: Woodhead Publishing.

90 Choi, C., Nam, J.P., and Nah, J.W. (2016). Application of chitosan and chitosan derivatives as biomaterials. *J. Ind. Eng. Chem.* 33: 1–20.

91 Sahariah, P. and Masson, M. (2017). Antimicrobial chitosan and chitosan derivatives: a review of the structure–activity relationship. *Biomacromolecules* 18 (11): 3846–3868.

92 Aranaz, I., Harris, R., and Heras, A. (2010). Chitosan amphiphilic derivatives: chemistry and applications. *Curr. Org. Chem.* 14 (3): 308–330.

93 Muzzarelli, R.A. and Muzzarelli, C. (2005). Chitosan chemistry: relevance to the biomedical sciences. In: *Polysaccharides I. Advances in Polymer Science*, vol. 186 (ed. T. Heinze), 151–209. Heidelberg: Springer.

94 Kurita, K. (1998). Chemistry and application of chitin and chitosan. *Polym. Degrad. Stab.* 59 (1–3): 117–120.

95 Shahidi, F. and Abuzaytoun, R. (2005). Chitin, chitosan, and co-products: chemistry, production, applications, and health effects. *Adv. Food Nutr. Res.* 49: 93–137.

96 Dutta, P.K., Dutta, J., and Tripathi, V.S. (2004). Chitin and chitosan: chemistry, properties and applications. *J. Sci. Ind. Res. India* 63: 20–31.

97 Lin, L., Regenstein, J.M., Lv, S. et al. (2017). An overview of gelatin derived from aquatic animals: properties and modification. *Trends Food Sci. Technol.* 68: 102–112.

98 Tsai, L.C., Hsieh, H.Y., Lu, K.Y. et al. (2016). EGCG/ gelatin-doxorubicin gold nanoparticles enhance therapeutic efficacy of doxorubicin for prostate cancer treatment. *Nanomedicine* 11 (1): 9–30.

99 Truong, V.X., Tsang, K.M., and Forsythe, J.S. (2017). Nonswelling click-cross-linked gelatin and PEG hydrogels with tunable properties using pluronic linkers. *Biomacromolecules* 18 (3): 757–766.

100 Maurer, L.M., Ma, W., and Mosher, D.F. (2016). Dynamic structure of plasma fibronectin. *Crit Rev Biochem Mol.* 51 (4): 213–227.

101 Li, Z., Xiao, C., Yong, T. et al. (2020). Influence of nanomedicine mechanical properties on tumor targeting delivery. *Chem. Soc. Rev.* 49 (8): 2273–2290.

102 Thakur, S., Govender, P.P., Mamo, M.A. et al. (2017). Recent progress in gelatin hydrogel nanocomposites for water purification and beyond. *Vacuum* 146: 396–408.

103 Shoulders, M.D. and Raines, R.T. (2009). Collagen structure and stability. *Annu. Rev. Biochem.* 78: 929–958.

104 Taubenberger, A.V., Woodruff, M.A., Bai, H. et al. (2010). The effect of unlocking RGD-motifs in collagen I on pre-osteoblast adhesion and differentiation. *Biomaterials* 31 (10): 2827–2835.

105 Cahyanto, A., Tsuru, K., Ishikawa, K., and Kikuchi, M. (2017). Mechanical strength improvement of apatite cement using hydroxyapatite/collagen nanocomposite. In: *Multifunction Bio-Ceramics for Innovative Therapy. Key Engineering Materials* (ed. A. El-Ghannam), 167–172. Charlotte: Trans Tech Publications Ltd.

106 Badillo-Sanchez, D., Chelazzi, D., Giorgi, R. et al. (2018). Characterization of the secondary structure of degummed *Bombyx mori* silk in modern and historical samples. *Polym. Degrad. Stab.* 157: 53–62.

107 Su, D., Ding, S., Shi, W. et al. (2019). Bombyx mori silk-based materials with implication in skin repair: Sericin versus regenerated silk fibroin. *J. Biomater. Appl.* 34 (1): 36–46.

108 Putnam, D., Zelikin, A.N., Izumrudov, V.A., and Langer, R. (2003). Polyhistidine–PEG: DNA nanocomposites for gene delivery. *Biomaterials* 24 (24): 4425–4433.

109 Patil, A.J., Vickery, J.L., Scott, T.B., and Mann, S. (2009). Aqueous stabilization and self-assembly of graphene sheets into layered bio-nanocomposites using DNA. *Adv. Mater.* 21 (31): 3159–3164.

110 Travers, A. and Muskhelishvili, G. (2015). DNA structure and function. *FEBS J.* 282 (12): 2279–2295.

111 Xiao, Z., Ji, C., Shi, J. et al. (2012). DNA self-assembly of targeted near-infrared-responsive gold nanoparticles for cancer thermo-chemotherapy. *Angew. Chem.* 124 (47): 12023–12027.

112 Alexander, C.M., Hamner, K.L., Maye, M.M., and Dabrowiak, J.C. (2014). Multifunctional DNA-gold nanoparticles for targeted doxorubicin delivery. *Bioconjug. Chem.* 25 (7): 1261–1271.

113 Kupis-Rozmysłowicz, J., Antonucci, A., and Boghossian, A.A. (2016). Engineering the selectivity of the DNA-SWCNT sensor. *ECS J. Solid State Sci.* 5 (8): M3067.

114 Zheng, M. (2018). Structure-defined DNA-carbon nanotube hybrids and their applications. *ECS Trans.* 85 (13): 511.

115 Jodat, Y.A., Lotfi, P., Abadi, P.P. et al. (2018). pH-responsive DNA nanolinker conjugated hybrid materials for electrochemical microactuator and biosensor applications. *ACS Appl. Nano Mater.* 1 (12): 6630–6640.

116 Singh, S., Kaushal, A., Khare, S., and Kumar, A. (2017). DNA chip based sensor for amperometric detection of infectious pathogens. *Int. J. Biol. Macromol.* 103: 355–359.

117 Pötzinger, Y., Rabel, M., Ahrem, H. et al. (2018). Polyelectrolyte layer assembly of bacterial nanocellulose whiskers with plasmid DNA as biocompatible non-viral gene delivery system. *Cellulose* 25 (3): 1939–1960.

118 Roy, I. and Visakh, P.M. (eds.) (2014). *Polyhydroxyalkanoate (PHA) Based Blends, Composites and Nanocomposites*, 1e. London: Royal Society of Chemistry.

119 Bordes, P., Pollet, E., Bourbigot, S., and Averous, L. (2008). Structure and properties of PHA/clay nano-biocomposites prepared by melt intercalation. *Macromol. Chem. Phys.* 209 (14): 1473–1484.

120 Plackett, D. (2012). PHA/clay nano-biocomposites. In: *Environmental Silicate Nano-Biocomposites*, 1e (ed. E. Pollet), 143–163. London: Springer.

121 Anderson, S., Zhang, J., and Wolcott, M.P. (2013). Effect of interfacial modifiers on mechanical and physical properties of the PHB composite with high wood flour content. *J. Polym. Environ.* 21 (3): 631–639.

122 Keskin, G., Kızıl, G., Bechelany, M. et al. (2017). Potential of polyhydroxyalkanoate (PHA) polymers family as substitutes of petroleum-based polymers for packaging applications and solutions brought by their composites to form barrier materials. *Pure Appl. Chem.* 89 (12): 1841–1848.

123 Ivorra-Martinez, J., Quiles-Carrillo, L., Boronat, T. et al. (2020). Assessment of the mechanical and thermal properties of injection-molded poly (3-hydroxybutyrate-co-3-hydroxyhexanoate)/hydroxyapatite nanoparticles parts for use in bone tissue engineering. *Polymers* 12 (6): 1389.

124 Arrigo, R., Bartoli, M., and Malucelli, G. (2020). Poly (lactic acid)–biochar biocomposites: effect of processing and filler content on rheological, thermal, and mechanical properties. *Polymers* 12 (4): 892.

125 Alam, F., Varadarajan, K.M., and Kumar, S. (2020). 3D printed polylactic acid nanocomposite scaffolds for tissue engineering applications. *Polym. Test.* 81: 106203.

126 Liu, S., Qin, S., He, M. et al. (2020). Current applications of poly (lactic acid) composites in tissue engineering and drug delivery. *Compos Part B: Eng.* 199: 108238.

127 Mohamed, R.M. and Yusoh, K. (2016). A review on the recent research of polycaprolactone (PCL). *Adv. Mater. Res.* 1134: 249–255.

128 Goodwin, D.G., Xia, Z., Gordon, T.B. et al. (2016). Biofilm development on carbon nanotube/polymer nanocomposites. *Environ. Sci. Nano* 3 (3): 545–558.

129 Talebi, A., Labbaf, S., and Karimzadeh, F. (2019). Polycaprolactone-chitosan-polypyrrole conductive biocomposite nanofibrous scaffold for biomedical application. *Polym. Compos.* 41 (2): 645–652.

130 Hoş, A., Tunç, K., and Olgun, U. (2020). Antibacterial nano biocomposite poly(ε-caprolactone) films with nano Ag-hydroxyapatite filler particles. *Compos. Interfaces* 27 (5): 479–493.

131 Wang, F., Guo, Y., Xu, W., and Wang, W. (2020). Development of nano-tricalcium phosphate/

polycaprolactone/platelet-rich plasma biocomposite for bone defect regeneration. *Arab. J. Chem.* 13 (9): 7160–7169.

132 Xu, X., Ray, R., Gu, Y. et al. (2004). Electrophoretic analysis and purification of fluorescent single-walled carbon nanotube fragments. *J. Am. Chem. Soc.* 126 (40): 12736–12737.

133 Novoselov, K.S., Geim, A.K., Morozov, S.V. et al. (2005). Two-dimensional gas of massless Dirac fermions in graphene. *Nature* 438 (7065): 197–200.

134 Smith, A.T., LaChance, A.M., Zeng, S. et al. (2019). Synthesis, properties, and applications of graphene oxide/reduced graphene oxide and their nanocomposites. *Nano Mater. Sci.* 1 (1): 31–47.

135 Liu, B. and Zhou, K. (2019). Recent progress on graphene-analogous 2D nanomaterials: properties, modeling and applications. *Prog. Mater. Sci.* 100: 99–169.

136 Lee, J., Kim, J., Kim, S., and Min, D.H. (2016). Biosensors based on graphene oxide and its biomedical application. *Adv. Drug Deliv. Rev.* 105: 275–287.

137 Shen, H., Zhang, L., Liu, M., and Zhang, Z. (2012). Biomedical applications of graphene. *Theranostics.* 2 (3): 283.

138 Chung, C., Kim, Y.K., Shin, D. et al. (2013). Biomedical applications of graphene and graphene oxide. *Acc. Chem. Res.* 46 (10): 2211–2224.

139 Song, S., Shen, H., Wang, Y. et al. (2020). Biomedical application of graphene: from drug delivery, tumor therapy, to theranostics. *Colloid Surf. B.* 185: 110596.

140 Qi, X.Y., Yan, D., Jiang, Z. et al. (2011). Enhanced electrical conductivity in polystyrene nanocomposites at ultra-low graphene content. *ACS Appl. Mater. Interfaces* 3 (8): 3130–3133.

141 Hsu, Y.W., Hsu, T.K., Sun, C.L. et al. (2012). Synthesis of CuO/graphene nanocomposites for nonenzymatic electrochemical glucose biosensor applications. *Electrochim. Acta* 82: 152–157.

142 Javanbakht, S. and Namazi, H. (2018). Doxorubicin loaded carboxymethyl cellulose/graphene quantum dot nanocomposite hydrogel films as s potential anticancer drug delivery system. *Mater. Sci. Eng. C* 87: 50–59.

143 Zhang, Y., Zhang, H., Wang, Y. et al. (2017). Hydrophilic graphene oxide/bismuth selenide nanocomposites for CT imaging, photoacoustic imaging, and photothermal therapy. *J. Mater. Chem. B* 5 (9): 1846–1855.

144 Mohammadi, S., Shafiei, S.S., Asadi-Eydivand, M. et al. (2017). Graphene oxide-enriched poly (ε-caprolactone) electrospun nanocomposite scaffold for bone tissue engineering applications. *J. Bioact. Compat. Polym.* 32 (3): 325–342.

145 Popov, V.N. (2004). Carbon nanotubes: properties and application. *Mater. Sci. Eng. R* 43 (3): 61–102.

146 Musso, S., Porro, S., Rovere, M. et al. (2007). Physical and mechanical properties of thick self-standing layers of multiwall carbon nanotubes. *Diam. Relat. Mater.* 16 (4-7): 1174–1178.

147 Sun, Y.P., Fu, K., Lin, Y., and Huang, W. (2002). Functionalized carbon nanotubes: properties and applications. *Acc. Chem. Res.* 35 (12): 1096–1104.

148 Andrews, R. and Weisenberger, M.C. (2004). Carbon nanotube polymer composites. *Curr. Opin. Solid State Mater. Sci.* 8 (1): 31–37.

149 Ho, C.M., Mishra, A., Lin, P.T. et al. (2017). 3D printed polycaprolactone carbon nanotube composite scaffolds for cardiac tissue engineering. *Macromol. Biosci.* 17 (4): 1600250.

150 Velasco-Mallorquı, F., Fernández-Costa, J.M., Nevesb, L., and Ramón-Azcón, J. (2020). New volumetric CNT-doped gelatin–cellulose scaffolds for skeletal muscle tissue engineering. *Nanoscale Adv.* 2: 2885–2896.

151 Iqbal, N., Khan, A.S., Asif, A. et al. (2019). Recent concepts in biodegradable polymers for tissue engineering paradigms: a critical review. *Int. Mater. Rev.* 64 (2): 91–126.

152 Scaffaro, R., Maio, A., Sutera, F. et al. (2019). Degradation and recycling of films based on biodegradable polymers: a short review. *Polymers* 11 (4): 651.

153 Allen, N.S. and Edge, M. (eds.) (1993). *Fundamentals of Polymer Degradation and Stabilization*, 1e, 201. Philadelphia, PA: Springer Science & Business Media.

154 La Mantia, F.P., Morreale, M., Botta, L. et al. (2017). Degradation of polymer blends: a brief review. *Polym. Degrad. Stab.* 145: 79–92.

155 Deng, Z., Schweigerdt, A., Norow, A., and Lienkamp, K. (2019). Degradation of polymer films on surfaces: a model study with poly (sebacic anhydride). *Macromol. Chem. Phys.* 220 (15): 1900121.

156 Antil, P., Singh, S., Kumar, S. et al. (2019). Erosion analysis of fiber reinforced epoxy composites. *Mater. Res. Express.* 6 (10): 106520.

157 Sevim, K. and Pan, J. (2018). A model for hydrolytic degradation and erosion of biodegradable polymers. *Acta Biomater.* 66: 192–199.

158 Loh, X.J., Chee, P.L., and Owh, C. (2019). Biodegradable thermo-gelling polymers. *Small Methods.* 3 (3): 1800313.

159 Shimpi, N.G. (ed.) (2017). *Biodegradable and Biocompatible Polymer Composites: Processing, Properties and Applications*, 1e, 438. Mumbai: Woodhead Publishing.

160 Balaji, A.B., Pakalapati, H., Khalid, M. et al. (eds.) (2017). Natural and synthetic biocompatible and biodegradable polymers. In: *Biodegradable and Biocompatible Polymer Composites: Processing, Properties and Applications*, 3–32. Duxford: Woodhead Publishing

Series in Composites Science and Engineering, Woodhead Publishing.

161 Asghari, F., Samiei, M., Adibkia, K. et al. (2017). Biodegradable and biocompatible polymers for tissue engineering application: a review. *Artif. Cell Nanomed. Biotechnol.* 45 (2): 185–192.

162 Arif, U., Haider, S., Haider, A. et al. (2019). Biocompatible polymers and their potential biomedical applications: a review. *Curr. Pharm. Des.* 25 (34): 3608–3619.

163 He, W. and Benson, R. (2017). Polymeric biomaterials. In: *Applied Plastics Engineering Handbook*, 2nde (ed. M. Kutz), 145–164. New York, NY: William Andrew Publishing.

164 AL-Oqla, F.M. and Omari, M.A. (2017). Sustainable biocomposites: challenges, potential and barriers for development. In: *Green Biocomposites. Green Energy and Technology*, 1e (ed. O. Alothman), 3–29. Springer: Cham.

165 Yılmaz, N.D. and Yılmaz, K. (2014). Use of natural polysaccharides in pharmaceutical biocomposites. *Poster session presented at 1st International Conference on Sustainable Composite Technologies*; 2014 November; Isparta, TURKEY, 47–48.

166 Sharma, G., Sharma, S., Kumar, A. et al. (2018). Guar gum and its composites as potential materials for diverse applications: a review. *Carbohydr. Polym.* 199: 534–545.

167 HPS, A.K., Saurabh, C.K., Adnan, A.S. et al. (2016). A review on chitosan-cellulose blends and nanocellulose reinforced chitosan biocomposites: properties and their applications. *Carbohydr. Polym.* 150: 216–226.

168 Cheung, R.C., Ng, T.B., Wong, J.H., and Chan, W.Y. (2015). Chitosan: an update on potential biomedical and pharmaceutical applications. *Mar. Drugs* 13 (8): 5156–5186.

169 Wróblewska-Krepsztul, J., Rydzkowski, T., Michalska-Pożoga, I., and Thakur, V.K. (2019). Biopolymers for biomedical and pharmaceutical applications: recent advances and overview of alginate electrospinning. *Nanomaterials* 9 (3): 404.

14

Role of Eco-friendly Nanotechnology for Green and Clean Technology

Bibhuti Bhusan Kakoti, Kangkan Deka, and Manjir Sarma Kataki

Department of Pharmaceutical Sciences, Dibrugarh University, Dibrugarh, Assam, India

14.1 Nanotechnology

Nanotechnology is the science of manipulating matter on molecular and atomic-scale range. The prefix "nano" corresponds to a Greek prefix meaning "dwarf" or something very small and portrays one thousand millionths of a meter (10^{-9} m) [1]. In his talk "There is plenty of room at the bottom" in 1959, Richard Feynman first presented the idea of nanotechnology, in which he outlined the possibility of synthesis by direct atom manipulation, which is considered as the beginning of advanced nanotechnology [1–4]. In 1974, Norio Taniguchi, a Japanese scientist, for the first time coined and defined the term "nanotechnology" as "nanotechnology consists mainly of separation, consolidation, deformation of materials by one atom or one molecule" [5].

The National Nanotechnology Initiative (NNI) in the United States describes nanotechnology as "a science, engineering, and technology conducted at the nanoscale (1–100 nm), where unique phenomena allow novel applications in a broad range of fields, from chemistry, physics and biology, to medicine, engineering and electronics" (NSET, 2004) [6].

In its 2004 report, the Royal Society and the Royal Academy of Engineering gave the following definition referring to as "nanotechnologies": Nanotechnologies is the design, characterization, production, and application of structures, devices, and systems by controlling shape and size at the nanometer scale. The decision to use "nanotechnologies" rather than "nanotechnology" was made to indicate the versatile and cross-disciplinary nature of the technology [2].

The dictionary definition of nanotechnology is "the design, characterization, manufacture and shape and size-controlled application of matters in the nanoscale" [7]. Another definition from the same dictionary defines nanotechnology as "the careful and controlled manipulation, precision placement, modeling, measurement, and production of materials at the nanoscale in order to make matters, systems, and devices by fundamentally novel properties and functions" [7].

The materials commonly classified as nanoparticles or nanomaterials have transformed the technological universe, largely due to their extraordinary physical, chemical, and biological properties, relative to their bulky counterparts. Because of these properties, nanotechnology has managed to obtain tremendous attention in a number of sectors, such as biomedical [8], electronics [9], chemical sector [10], pharmaceutical industries [11], space industry [12], mechanics [13], and environmental health [14].

14.2 Nanotechnology and Its Potential Impact on Human Health and Environment

Given the strong opportunities for nanotechnology in the future, there is growing concern that certain forms of nanoparticles will lead to severe adverse health and environmental pollution, whether accidentally or deliberately [15]. There have been significant effects in multiple environmental compartments, including climate, water, and soil, that have become evident in recent years. These implications are more severe due to the uncertain shape, size, and chemical compositions of some of the nanomaterials. Nanoparticles have a larger surface area than bulk materials and can do more harm to the environment relative to

Sustainable Nanotechnology: Strategies, Products, and Applications, First Edition.
Edited by Yashwant V. Pathak, Govindan Parayil, and Jayvadan K. Patel.

bulk particles. The potential risks of nanoparticles toward environment are as follows:

- Nanoparticle synthesizing process comes with high energy demand.
- Release of harmful, permanent nanosubstances that cause environmental damage.
- Low rates of recovery and recycle.
- Inadequate knowledge regarding environmental implication of other life cycle stages.
- Lack of well-trained engineers and staff, raising more concern [16].

Nanoparticles may reach the human body by the following pathways: penetration by the nodes of the skin, inhalation by the respiratory system, and absorption by the digestive system. Through respiratory and blood circulation processes, these taken-up particles reach the other organs. The mechanism of absorption and degree of nanoparticle toxicity is the most crucial matter to be understood. Nearly all chemicals, from small heavy metals to organic compound macromolecules, are toxic to cells, plants, and animals above a certain threshold level [15].

It has been proven that with decreasing particle size, the toxicity of nanoparticles increases. Literatures have indicated the production of reactive oxygen species by nanostructures due to the involvement of the surface of nanostructures in many catalytic and oxidative reactions. As a consequence, cytotoxicity can get triggered, and owing to the increased surface-to-volume ratio, the effect is more in nanomaterials compared to bulk materials. Metals and compounds used in nanomedicine possess their own toxic effects, and breakdown of these into respective components can produce toxic effects close to the components themselves. Perhaps, the toxic reaction can be more pronounced due to the scaling of nano-level materials [17].

14.2.1 Approaches to Nanomaterial Synthesis/Generation/Formation

Basically, there are two general approaches to nanoparticles synthesis: top-down and bottom-up approach. The top-down approach uses larger (macroscopic) initial structures. Using suitable lithographic methods such as grinding, sputtering, and milling, the required bulk materials are broken down to fine particles. The bottom-up approach involves miniaturization of the components of the materials (up to atomic level) with a further self-assembly process, which leads to nanostructure production [4, 18].

14.2.1.1 Physical Approach

The most widely employed techniques for the physical approach to nanoparticle production include laser ablation [19], high-energy ball milling [20], electrospraying [21], and laser pyrolysis [22]. Arc discharge method [23], metal sputtering [24], atomization [25], and annealing [26] are other physical techniques. The lack of solvent contamination in formulated thin films and the uniformity of the synthesized nanoparticle distribution make these physical approaches more desirable than the chemical ones [4].

14.2.1.2 Chemical Approach

The most widely used chemical processes are the sol–gel method, the microemulsion technique, hydrothermal synthesis, polyol synthesis, and chemical vapor synthesis. Metallic precursors, stabilizers, and reduction agents (inorganic and organic) constitute the principal components of a chemical approach. Examples of the reducing agents used are odium citrate, ascorbate, sodium borohydride ($NaBH_4$), elemental hydrogen, polyol process, tollens reagent, *N*,*N*-dimethylformamide (DMF), and poly(ethylene glycol) block copolymers [27, 28].

14.2.1.3 Biological Synthesis

Biologic approaches include the synthesis of nanoparticles using bacteria, fungi, or plants. Strategies for producing nanoparticles using naturally occurring reagents such as vitamins, sugars, plant extracts, biodegradable polymers, and microorganisms as reductive and capping agents may be deemed desirable for nanotechnology [29].

14.3 Green Approach to Nanoparticle Synthesis

As mentioned previously, nanoparticles may be made in a number of ways (physical, chemical, or biological processes), but the generation of nanoparticles by physical and chemical processes poses both toxicity and environmental concerns. In addition to being very time consuming, the physical pathway requires a huge amount of space and produces a large amount of heat, increasing the temperature around the source material. The main disadvantage of the chemical process of making nanoparticles is its use of harmful solvents and chemicals. Because of the harmful chemicals, various studies have suggested that these approaches will lead to possible exposure to nanomaterials through inhalation, skin, or ingestion routes, thereby preventing them from becoming economically feasible or environmentally sound. The need for a better, economically feasible, and environmentally sound nanoparticle and nanomaterial synthesis method has therefore been felt worldwide, contributing to the introduction of the idea of green nanotechnology [4]. Green nanotechnology is a technique used to create sustainable technology that can reduce environmental risks as well as improve human health [30]. It is related to the application of nanotechnology products and their processing techniques. Green nanotechnology encompasses the implementation of the concepts of green chemistry to develop nanoscale products, the advancement of manufacturing methods for nanomaterials, and the usage of nanomaterials. The approach necessitates an understanding of the properties of nanomaterials, especially those related to toxicity and ecotoxicity, and the development of nanoscale materials that can be integrated into high-performance goods that pose minimum risk to human health or the environment. It seeks to identify strategies of synthesis that remove the need for hazardous reagents and improve the efficacy of these techniques, while delivering the requisite amount of pure material in an economically viable manner. It sets out design schemes to ensure that nanomaterials generated are innately safer by analyzing the biological and ecological risks in parallel with design [31]. Almost all principles of green chemistry can be implemented in the design and application of nanoscale products, nanosynthesis methods, and nanomaterials [15].

Green nanotechnology may generally be interpreted as the use of biological pathways such as bacteria, fungi, or plants for the synthesis of nanomaterials (or nanoparticles) with the help of several biotechnological methodologies. The nanoparticles thus produced are eco-friendly and free of toxic chemicals [4].

In the process of generating highly stable and well-characterized nanoparticles, the significant aspects that could be considered are as follows:

i) *Selection of the best organisms*: In selecting the best candidates, some of the critical properties of organisms require attention, including growth rate, enzyme activity, and biochemical pathways. And moreover, the selection of the organisms largely depends on the desired application of the resultant nanoparticle [32].
ii) *Optimal conditions for cell growth and enzyme activity*: Higher enzyme production can be achieved by more biomass production. Optimization of the conditions for growth is therefore essential. It is necessary to have optimized conditions in terms of nutrient, inoculum size, light, temperature, pH, buffer strength, and mixing speed. The induction of the responsible enzymes is important. Harvesting time is crucial while using whole cell or crude enzyme [33].
iii) *Optimal reaction conditions*: The yield and the production rate are important issues to consider in order to use the organisms for producing nanoparticles at an industrial scale. Optimized bioreduction conditions in the reaction mixture are thus necessary. Concentration of the substrate, the biocatalyst, the electron donor, temperature, pH, buffer strength, light, and mixing speed need to stay optimized. Use of complementary factors, such as visible light and microwave irradiation, and boiling can be an option, as they tend to influence the size, morphology, and the reaction rate. Optimization of these variables seems to be able to monitor the morphologies and other properties of nanoparticles. Studies are therefore focused on determining the optimal reaction condition and cellular mechanisms responsible for the reduction of metal ion and the formation of nanoparticles [32].

14.4 Methods of Green Synthesis of Nanoparticles

14.4.1 From Microorganisms

Microbial-assisted nanomaterial biosynthesis is a promising biotechnological-based nanomanufacturing approach that represents a "green" alternative strategy to physical and chemical methods of nanomaterial synthesis. Microbial-assisted biosynthesis of metallic (also alloy), nonmetallic, or metal oxide nanoparticles has been documented for several microbial strains of bacteria, yeast, fungi, and microalgae [34]. The mechanism of microbial-assisted biosynthesis is principally based on the interactions between the microorganisms and the target ions of the medium or the environment. This interaction causes emission of enzymes from the microbial cells, thus reducing the metal ions to its elemental form. These nanoparticles may pile up either outside or within the cell wall, and depending on the site of deposition, the process of biosynthesis is categorized into intra- and extracellular synthesis. The trapping of metal ions on the surface of the cell wall occurs in extracellular synthesis of nanoparticles and is reduced by the presence of multiple biomolecules such as enzymes, proteins, and peptides; on the other hand, the intracellular synthesis requires transporting ions into the microbial cell to form nanoparticles in the presence of biomolecules. Steps involved in the microbial-assisted biosynthesis of nanoparticles include (i) microorganism culture preparation under ideal conditions for a certain growth period, (ii) mixing of material salt solution with microbial culture, (iii) bioreduction and capping, and (iv) recovery of nanoparticles through different procedures [35].

14.4.1.1 Bacteria

The most suitable microorganism for the biosynthesis of nanoparticles (NPs) is bacteria, a unicellular prokaryotic microorganism found all over the earth, water, and air.

Bacteria are obvious targets for nanoparticle manufacturing, considering their rapid development, affordable culturing costs, and simple monitoring and handling of growth environments. It is also recognized that many bacterial species have unique mechanisms to suppress metallic or heavy-metal toxicity [35]. Owing to the defense mechanism possessed by bacteria, production of nanoparticles by bacteria comes into action. The bacterial cell's resistance to reactive ions in the environment is responsible for the synthesis of nanoparticles. For bacterial cells, high ion concentrations are usually harmful. Their cellular machinery assists in the conversion of reactive ions to stable atoms to avoid cell death. For biosynthesis of nanoparticles, this property of bacteria is exploited [36]. Studies have demonstrated that bacteria can be used either through intra- or extracellular routes to synthesize metal, organic, inorganic, and other nanoparticles. Bacteria are regarded as possible biofactory for nanomaterial synthesis, such as gold, silver, platinum, palladium, titanium, titanium dioxide, zinc oxide, silicon oxides, and cadmium sulfide [35]. The development of silver nanoparticles by *Pseudomonas stutzeri* AG259 (isolated from silver mining) has been documented. Magnetic nanoparticle synthesis has been reported using magnetotactic bacteria. The bacterial cell supernatant of *Pseudomonas aeruginosa* has recently been used to reduce gold ions, resulting in extracellular biosynthesis of gold nanoparticles [37].

14.4.1.2 Algae

Algae are uni- or multicellular species and are found in fresh or salt water or moist soil. Algae by virtue of its pigments, proteins, carbohydrates, fats, nucleic acid, and secondary metabolites breaks down metallic salts into nanoparticles. At certain temperature, the aqueous medium of the algae extract is supplemented with metal solution of the corresponding pH and concentration, which ultimately leads to the synthesis of nanoparticles without the production of any toxic by-products during the process of synthesis [38]. Many research studies have demonstrated the use of algae for nanoparticles synthesis. *Spirulina platensis* (a blue-green algae) has been reported to be used in the protein-mediated synthesis of uniform-sized (approximately 5 nm) gold nanoparticles, which finds their use in antibacterial assays against *Bacillus subtilis* and *Staphylococcus aureus* [39]. Silver nanoparticles synthesis using *Sargassam wightii* and the evaluation for antibacterial potential have also been reported [40].

14.4.1.3 Fungi

Mycosynthesis of nanoparticles is one of the most effective ways of creating nanoparticles because compared to bacteria, fungi grow faster and are easier to manage and fabricate in a laboratory setup. They serve as an effective secretor of extracellular enzymes that contribute to large-scale enzyme production [36]. Most fungi containing essential metabolites with higher bioaccumulation potential and simple downstream processing are easy to develop for efficient, cost-effective, nanoparticle development. Three potential pathways to describe the mycosynthesis of metal nanoparticles have been postulated: nitrate reductase action; electron shuttle quinones; and both [41]. Fungal mycelial mesh is capable of tolerating conditions such as flow pressure and agitation in bioreactors or other reaction compartments [4].

In order to elucidate the mechanism of formation of nanoparticles, an in vitro technique has been implemented to successfully reduce $AuCl^{4-}$ ion to gold nanoparticles using species-specific NADH-dependent reductase released by *Fusarium oxysporum*. This opened a new *in vitro* approach to nanomaterial synthesis based on fungal/enzyme [37]. For nanoparticle synthesis, several fungi have been utilized, such as *Fusarium* sp. and *Colletotrichum* sp. Fungal enzymes, such as reductase enzymes from *Penicillium* and *F. oxysporum*, nitrate reductase, and alpha-NADPH-dependent reductases, have been found to play an important role in nanoparticle synthesis [35].

14.4.1.4 Actinomycetes

Actinomycetes belong to the phylum Actinobacteria and share distinct fungal characteristics [4]. Although the nanoparticles synthesized using actinomycetes-mediated pathways have strong monodispersity and stability and have significant biocidal activity against different pathogens, the approach remains to be thoroughly explored [41]. Actinomycetes have been found to play a significant role in the synthesis of nanoparticles from metals. *Thermomonospora* species (an alkalothermophilic actinomycetes) has been reported to synthesize gold ions (8-nm sized) under intense alkaline conditions and at elevated temperatures. For the synthesis of gold nanoparticles (5–15 nm range), *Rhodococcus* sp. (an alkalitolerant actinomycetes) were used. Studies have shown that nanoparticles are more abundant on the cell wall than on the membrane of the cell. Even after the development of nanoparticles, the cells continued to grow, demonstrating that the metal ions produced were not toxic to the cells. Extracellular gold nanoparticles were generated by the chemical reaction of biomass with chloroaurate ions [42].

14.4.1.5 Yeast

Yeast is a unicellular eukaryotic microorganism. Owing to their capacity to synthesize semiconductor nanoparticles, specifically cadmium sulfide NPs (CdS), they are commonly known as "semiconductor crystals" or quantum semiconductor crystals. Intracellular synthesis of monodispersed, spherical peptide-bound CdS quantum crystals (2 nm size) by the yeast *Candida glabrata* has been documented [35]. Different experiments have demonstrated that yeast can accumulate large amounts of heavy metals. Researchers further studied this feature of yeast for the development of nanoparticles by either intra- or extracellular enzymes. Yeast has an upper hand over bacterial-mediated nanoparticles synthesis due to the ease of handling and possibility of mass production on a laboratory scale. Metallic nanoparticles were synthesized by analyzing multiple strains of the yeast. For extracellular synthesis of silver nanoparticles, the silver-tolerant yeast strain MKY3 was utilized. *Yarrowia lipolytica* was reported to be used for the synthesis of gold nanoparticles [43].

14.4.1.6 Biotemplates-assisted Biogenesis

For the generation of nanoparticles, different biomolecules such as nucleic acids, cell membranes, and viruses are reportedly used as templates [4]. Proteins, upon being used as a biotemplate, provide multiple binding sites that readily attach to a metal ion and further reduce. Compared to other biotemplates, proteins offer excellent scaffolds for the template-driven arrangement of nanostructured materials of unique shapes and sizes. Proteins used as templates are categorized into two types based on the structure/size, firstly as "spherical protein," used for the synthesis of spherical particles, clusters, nanoplates, and microspheres, and secondly as "filamentous protein," used for the synthesis of nanowires or nanotubes. Examples of spherical proteins are ferritin and bovine, and collagen, silk, actins, keratins, myosins, and flagellins fall in the category of filamentous proteins. They are utilized for the synthesis of diverse structured nanomaterials [44].

Deoxyribonucleic acid (DNA) serves as an ideal biomolecular template owing to its strong affinity toward transition metal ions. Experiments demonstrated gold nanoparticles synthesis upon production of DNA hydrogels wherein they are cross-linked followed by the incorporation of gold (Au) ions into DNA macromolecules. Au^{3+} gets bioreduced to Au^0 atoms and metal clusters formed on the DNA chain to gold nanoparticles. A highly stable, wire-like cluster of silver nanoparticles was synthesized utilizing DNA as biotemplate, which was later used as an ultrasensitive surface-enhanced Raman scattering (SERS) substrate [45]. Biologic membranes have also been used as templates for the synthesis and design of nanoparticles, utilizing the ultrafine pores of the membranes. A rubber membrane made from *Hevea brasiliensis* was used to synthesize gold nanoparticles; the membrane serves as a preservative in the reduction of Au^{+3} in a solution at a temperature of 80 °C [46]. Viruses also find their use as templates for the production of nanoparticles with uniform size and morphology, by virtue of the hollow spaces in the center [47]. Tobacco mosaic virus (TMV) has been used as a template for synthesizing iron oxides by oxidative hydrolysis, CdS and lead(II) sulfide (PbS) cocrystallization, and silicon dioxide (SiO_2) synthesis using the condensation of sun-freeze [48].

14.4.2 Plant-mediated Biosynthesis

Phytonanotechnology, a new field of research, has gained attention in recent years. This approach offers possibility of synthesizing stable nanomaterials in simple, fast, eco-friendly, and economically feasible manner [44]. The plant-mediated biosynthesis of valuable nanostructures and NPs is of considerable significance for its several benefits, such as high yields, very mild reaction conditions, clean reaction profiles, easy processing methods, the use of eco-friendly solvents such as water, and the avoidance of expensive, toxic, and unsafe materials and high temperatures and pressures. Plants contain several bioactive phytochemicals such as glycosides, alkaloids, flavonoids, tannins, terpenoids, resins, and saponins owing to the fine distribution of phytochemicals inside plants; almost every part of the plant (leaves, stems, flowers, roots, glands, and fruit peel) can be utilized in nanomaterial biosynthesis.

Phytosynthesis of nanostructures utilizes reducing molecules called antioxidants, such as phenolic compounds, nitrogen compounds, vitamins, reducing sugars, terpenoids, and some other phytochemicals. Consequently, the selection of plants considering its antioxidant content is of considerable significance for the phytosynthesis of nanomaterials. Plant components may also serve as stabilizing and capping agents when adsorbed on the surface of nanoparticles and shield them from deformation and degradation processes [49].

Gardea-Torresdey et al. were the first to report the formation of gold and silver nanoparticles inside living plants. They demonstrated the synthesis of gold and silver nanoparticles within live *Medicago sativa* (alfalfa) plant by gold and silver ion uptake, respectively, from solid media. Tetrachloroaurate ion ($AuCl_4^-$)-rich environment was used to grow the plants. X-ray absorption spectroscopy (XAS) and transmission electron microscopy (TEM) were used to validate the uptake and formation of gold and silver nanoparticles [50].

14.4.3 Characterization of Nanoparticles

After the NPs are synthesized, their conformational data regarding shape, size, dispersity, homogeneity, and also surface morphology are measured using different methods.

The common characterizing techniques are as follows: ultraviolet–vis absorption spectroscopy, X-ray diffraction (XRD), Fourier transmission infrared (FTIR) spectroscopy, dynamic light scattering (DLS), energy-dispersive X-ray examination (EDAX), scanning electron microscopy (SEM), TEM, etc. [28].

14.5 Factors Affecting Green Synthesis of Nanoparticles

Several studies have demonstrated that the stability, size, shape, monodispersity, and physicochemical property of the biosynthesized nanoparticles are largely dependent on factors such as pH of the reaction medium, temperature, and reaction time. Some of the factors are as follows:

i) *Effect of pH of the reaction medium*: The formation of nanoparticle relies strongly on the pH of the reaction medium, as a slight variation in pH may result in nanoparticles of varying sizes and shapes. For example, at pH 2, large rod-shaped Au nanoparticles (25–85 nm) were formed from *Avena sativa*, while at pH 3 and 4, comparatively smaller nanoparticles (5–20 nm) were formed [51]. The synthesis of spherical Au nanoparticles with a size ranging from 10 to 20 nm using the bacterium *Rhodopseudomonas capsulate* occurs at pH 6 with dilute $AuCl_4$, but spheres and triangular nanoplates were obtained at pH 4 [52]. Additionally, Au nanowires were obtained at pH 6 when the salt concentration increased [53].
ii) *Effect of temperature of the reaction medium*: Temperature has a profound impact on the biosynthesis of nanoparticles in terms of the size, shape, and yield of nanoparticles. For example, upon *Cymbopogon flexuosus* leaf extract-mediated biosynthesis of gold nanoparticles it was observed that lower reaction temperatures resulted in the development of nanotriangles, while more spherical nanoparticles were synthesized at higher reaction temperatures along with the nanotriangles [54]. In *Cyanobacteria*-mediated synthesis of silver nanoparticles, the size of the nanoparticles appeared to increase with increase in temperature, and variation in shape was observed with varying temperature [35].
iii) *Effect of reaction time*: A certain amount of time is needed for all forms of chemical/biological reactions to achieve the desired result. Reaction time can thus affect the quality and morphology of the nanoparticles formed. Long reaction time may result in aggregation and shrinkage and ultimately affect the nanoparticle potential. Gericke and Pinches [55, 56] in a study on gold nanoparticle synthesis found that small particle size and good monodispersity were obtained at a lower cell exposure time of 1 hour to ion solution relative to longer cell exposure time of 24 hours.
iv) *Effect of pressure*: In the synthesis of metallic nanoparticles, pressure also plays an important role as it impacts the shape and size of the synthesized nanoparticles. Studies have demonstrated that phytochemical agents reduce metal ions at a faster rate under ambient pressure condition [4].

14.5.1 Application of Green Nanotechnology

Nanomaterials synthesized using green techniques play important role in the application of nanotechnology in different areas. The nanoparticles synthesized by utilizing various biological methods have many applications in biological and medicine field. Several studies have suggested the use of metal and metal oxide nanoparticles for their use as antibacterial, anticancer, antimycotic, and antiviral activity in therapeutic and diagnostic applications.

Pathogenic microbes are developing resistance against the commercially available antibiotic drug due to which antibiotics are not exhibiting their potency against bacteria that are multidrug resistant. This has opened up the door for the development of new antibacterial medicines with greater effectiveness. In this case, researchers have developed synergistic approaches, for example nanosize with biological active compounds to address multi-drug-resistant bacteria. Nanoparticles can readily penetrate into the cells of pathogenic bacterial strains due to their nanosize. The antibacterial property of the nanosized particles can be enhanced by the presence of biologically active compounds on the surface. Biosynthesized metals and metal oxide nanoparticles, including Ag, Au, ZnO, Cu, CuO, Ni, and NiO, have been reported to have shown a strong inhibitory effect against bacterial strains. Fayaz et al. [57] demonstrated an increased bactericidal tendency against both the Gram-positive and -negative bacterial strains by silver nanoparticles (5–40 nm) synthesized using fungus-mediated technique. Khan et al. [58] and Ijaz et al. [59] demonstrated that plant-mediated synthesis of ZnO, Cu-doped ZnO, and CuO nanoparticles possessed significant bactericidal propensity against both Gram-positive (*S. aureus* and *B. subtilis*) and Gram-negative (*Escherichia coli* and *Klebsiella*) bacteriological species. Biologically synthesized nanoparticles have also been extensively explored against numerous pathogenic mycological organisms for their antimycotic ability. The increased antimycotic activity against various mycological species posed by Ag nanoparticles synthesized using fungal strains such as *Alternaria alternate* has been documented by Gajbhiye et al. [60]. The enhanced antimycotic propensity of ZnO and Cu-doped ZnO synthesized using leaf extracts from plants such as *Abutilon indicum*, *Clerodendrum infortunatum*, and

Clerodendrum inerme against three fungal strains such as *Aspergillus flavus*, *Aspergillus niger*, and *Trichoderma harzianum* has been stated by Khan et al. [58]. Biosynthesized metal and metal oxide nanoparticles have also been extensively studied for anticancer diagnosis and application. Biosynthesized nanoparticles such as silver, gold, zinc oxide, and magnesium oxide have demonstrated to possess anticancer potential. The enhanced anticancer activity of plant-mediated synthesis of ZnO and Cu-doped ZnO nanoparticles against breast-cancer cell lines was stated by Khan et al. [58].

Gusseme et al. [61] documented the utilization of biosynthesized silver produced using *Lactobacillus fermentum* for the removal of viruses from drinking water. For home-water purification systems that contain silver as a disinfectant, Aquapure and QSI-Nano are commercially available [62]. Green nanomaterials have a broad variety of uses in the treatment of surface water, sewage, and wastewater polluted with harmful metal ions, organic and inorganic solutions, and microorganisms [63]. Self-cleaning nanoscale surface coatings can remove many of the cleaning chemicals used in daily maintenance routines [64]. Iron NPs are of great significance due to their increasingly emerging uses for water disinfection and remediation of heavy metals from soil [65–67]. NPs are alternatives to pesticides in the prevention and management of plant disease [68–70] and also serve as efficient fertilizers [71] that have become environmentally safe and improve crop yields.

14.5.2 Green Methods for Clean and Green Environment

Green science has managed to significantly lessen the level of toxic waste emitted to the environment. Green approaches play a vital role in renewable energy research and in the development of innovative ways of producing solar cells, fuel cells, and energy-storage batteries. Among the green technologies used to eliminate or reduce environmentally harmful chemicals, it is important to note the development of several next-generation green catalysts and revamped chemical products.

Nanotechnology has the power to boost all industrial sectors and to affect their progress toward sustainability. Advancement in nanotechnology led to the rise of concept of green nanotechnology, i.e. a sustainable approach to nanotechnology, probably preventing potential environmental concerns by developing safe nanostructures and materials that often can address environmental challenges. Green nanotechnology has encouraged the development of more environmentally sustainable nanomaterials and nanoproducts without the use of harmful chemicals and at low temperatures requiring less energy and more natural resources whenever possible, taking into account the life-cycle concerns at all levels of its design and engineering. Green nanotechnology has led to the development of environmentally friendly nanomaterials with a wide variety of uses, such as cleaning toxic waste areas, desalination of water, and treatment or control of environmental pollutants. Furthermore, many uses of nanomaterials can be implicitly related to green nanotechnology use, such as the presence of nanocomposites in transport vehicle systems to minimize their weight, nano-enabled fuel cells and light-emitting diodes, self-cleaning nanoscale surface coatings, and improved battery life; they all experience the economic gains of green nanotechnology in terms of energy savings.

In general, green innovations are part of any environmental invention that includes energy conservation, recycling, safety and health concerns, and sustainable sources. Green technology is engaged both in material processing and use and includes the use of environmental innovations to track and analyze, to deter and manage pollution, and to remediate and restore them. It is essential to track harmful pollution, whether natural or anthropogenic. Further, the production of environmentally harmful pollutants must be minimized or avoided by modifying human actions in a manner that minimizes environmental harm. In addition, the monitoring of toxic and harmful chemicals before they reach the atmosphere and the enhancement of environmental conditions, degraded by natural or anthropogenic effects, reflect a mechanism of remediation linked to green nanotechnology. Green nanotechnology is also an essential part of environmentally friendly technologies [49].

14.5.3 Limitations and Challenges of Green Nanotechnology

Green nanotechnology is an evolving field which has its own limits and problems to overcome. Issues linked with green nanotechnology are:

- Management of toxicity-related concerns of nanomaterial
- Technical and economic barriers
- Regulatory policies for nanomanufacturing processes
- Deployment of scale-up procedures
- Life-cycle analysis.

Green nanotechnology offers materials that minimize pollution and are environmentally sustainable; however, the costs and uncertainties involved with the manufacture of nano-based products are the main restrictions. While progress has been made in green nanotechnology, there is still concern about the degree of sustainability of greener applications. The products synthesized using green nanotechnology are effective, but upstream processing of the products is the main concern [72]. Work is under way to synthesize and apply greener nanoproducts, but there are still few items commercialized to date while talking on the commercial front [73]. It is a common belief that the business potentials of green nanotechnology would take a few years to truly grasp.

14.6 Conclusion

Green nanotechnology offers substantial environmental, societal, and economic benefits, but possible risks and costs are also present. Progress has been made in the implementation of green nanotechnology, but questions exist about the level of sustainability of such green nanotechnologies and the real or potential threats to the environment, health, and safety (EHS). Although applications of green nanotechnology could save energy and reduce the final product's carbon emissions, there are concerns about the amount of energy that could be needed in the upstream processing of nanomaterial components. Continued development in the area is expected to reduce energy and manufacturing costs for nanomaterial production over time, as process techniques are developed and new materials evolve. However, the cost of producing materials for green nanotechnology applications, energy, waste, and resource extraction remain an essential part of this equation to ensure a responsible implementation of green nanotechnology.

References

1 Bayda, S., Adeel, M., Tuccinardi, T. et al. (2020). The history of nanoscience and nanotechnology: from chemical-physical applications to nanomedicine. *Molecules* 25 (1): 112.

2 Maynard, A.D. (2007). Nanotechnology: the next big thing, or much ado about nothing? *The Annals of Occupational Hygiene* 51 (1): 1–12.

3 Rani, K. and Sridevi, V. (2017). An overview on role of nanotechnology in green and clean technology. *Austin Environmental Sciences* 2 (3): 1–6.

4 Pal, G., Rai, P., and Pandey, A. (2019). Green synthesis of nanoparticles: a greener approach for a cleaner future. In: *Green Synthesis, Characterization and Applications of Nanoparticles* (eds. A.K. Shukla and S. Iravani), 1–26. Elsevier Inc.

5 Singh, M., Subramanian, M., and Kumaraguru, A. (2011). Nanoparticles: a new technology with wide applications. *Research Journal of Nanoscience and Nanotechnology* 1: 1–11.

6 What is Nanotechnology? nano.gov. Available from: https://www.nano.gov/nanotech-101/what/definition.

7 Abad, E. (2005). *Nano Dictionary*. Basilea: Basel Coll.

8 Kulkarni, M., Mazare, A., Gongadze, E. et al. (2015). Titanium nanostructures for biomedical applications. *Nanotechnology* 26 (6): 062002.

9 Bubnova, O. (2016). Twists and turns. *Nature Nanotechnology* https://doi.org/10.1038/nnano.2016.250.

10 Bera, P. and Aruna, S.T. (2017). Solution combustion synthesis, characterization, and catalytic properties of oxide materials. In: *Nanotechnology in Catalysis*, 91–118. Wiley.

11 Brown, P.D. and Patel, P.R. (2015). Nanomedicine: a pharma perspective. *Wiley Interdisciplinary Reviews. Nanomedicine and Nanobiotechnology* 7 (2): 125–130.

12 Meyyappan, M., Koehne, J.E., and Han, J.-W. (2015). Nanoelectronics and nanosensors for space exploration. *MRS Bulletin* 40 (10): 822–828.

13 Wang, X., Tabarraei, A., and Spearot, D.E. (2015). Fracture mechanics of monolayer molybdenum disulfide. *Nanotechnology* 26 (17): 175703.

14 Selvaraj, R., Al-Kindy, S.M.Z., Silanpaa, M., and Kim, Y. (2014). Nanotechnology in environmental remediation: degradation of volatile organic compounds (VOCs) over visible-light-active nanostructured materials. *Reviews on Environmental Health* 29 (1–2): 109–112.

15 Maksimović, M. and Omanović-Mikličanin, E. (2017). Towards green nanotechnology: maximizing benefits and minimizing harm. In: *CMBEBIH 2017*, 164–170. Singapore: Springer.

16 Zhang, B., Misak, H., Dhanasekaran, P.S. et al. (2011). Environmental impacts of nanotechnology and its products. *American Society for Engineering Educatio*: 1–9.

17 Kakoti, B.B., Kataki, M.S., and Pathak, Y. (2016). Nanotoxicity of nanobiomaterials in ocular system and its evaluation. In: *Nano-Biomaterials For Ophthalmic Drug Delivery* (eds. Y. Pathak, V. Sutaria and A.A. Hirani), 495–533. Springer International Publishing.

18 Lakhani, P. and Jain, N. (2018). Nanotechnology and green nanotechnology: a road map for sustainable development, cleaner energy and greener world. *International Journal for Innovative Research in Science & Technology* 3 (1): 580–588.

19 Scaramuzza, S., Agnoli, S., and Amendola, V. (2015). Metastable alloy nanoparticles, metal-oxide nanocrescents and nanoshells generated by laser ablation in liquid solution: influence of the chemical environment on structure and composition. *Physical Chemistry Chemical Physics* 17 (42): 28076–28087.

20 Kuang, L., Mitchell, B.S., and Fink, M.J. (2015). Silicon nanoparticles synthesised through reactive high-energy ball milling: enhancement of optical properties from the removal of iron impurities. *Journal of Experimental Nanoscience* 10 (16): 1214–1222.

21 Foroutan, F., Jokerst, J.V., Gambhir, S.S. et al. (2015). Sol–Gel synthesis and electrospraying of biodegradable (P2O5)55–(CaO)30–(Na2O)15 glass nanospheres as a transient contrast agent for ultrasound stem cell imaging. *ACS Nano* 9 (2): 1868–1877.

22 Bouhadoun, S., Guillard, C., Dapozze, F. et al. (2015). One step synthesis of N-doped and Au-loaded TiO2 nanoparticles by laser pyrolysis: application in photocatalysis. *Applied Catalysis B: Environmental* 174–175: 367–375.

23 Tsuji, T., Iryo, K., Watanabe, N., and Tsuji, M. (2002). Preparation of silver nanoparticles by laser ablation in solution: influence of laser wavelength on particle size. *Applied Surface Science* 202 (1–2): 80–85.

24 Tsuji, T., Kakita, T., and Tsuji, M. (2003). Preparation of nano-size particles of silver with femtosecond laser ablation in water. *Applied Surface Science* 206 (1): 314–320.

25 Okuyama, K. and Wuled, L.I. (2003). Preparation of nanoparticles via spray route. *Chemical Engineering Science* 58 (3): 537–547.

26 Nakade, S., Matsuda, M., Kambe, S. et al. (2002). Dependence of TiO_2 nanoparticle preparation methods and annealing temperature on the efficiency of dye-sensitized solar cells. *The Journal of Physical Chemistry. B* 106 (39): 10004–10010.

27 Zhang, X.-F., Liu, Z.-G., Shen, W., and Gurunathan, S. (2016). Silver nanoparticles: synthesis, characterization, properties, applications, and therapeutic approaches. *International Journal of Molecular Sciences* 17 (9): 1534.

28 Gour, A. and Jain, N.K. (2019). Advances in green synthesis of nanoparticles. *Artificial Cells, Nanomedicine, and Biotechnology* 47 (1): 844–851.

29 Kharissova, O.V., Dias, H.V.R., Kharisov, B.I. et al. (2013). The greener synthesis of nanoparticles. *Trends in Biotechnology* 31 (4): 240–248.

30 Alshammari, A., Kalevaru, V.N., and Martin, A. (2016). Metal nanoparticles as emerging green catalysts. In: *Green Nanotechnology [Internet]* (eds. M.L. Larramendy and S. Soloneski), 1–33. Rijeka: IntechOpen.

31 Dahl, J.A., Maddux, B.L.S., and Hutchison, J.E. (2007). Toward greener nanosynthesis. *Chemical Reviews* 107 (6): 2228–2269.

32 Korbekandi, H., Iravani, S., and Abbasi, S. (2009). Production of nanoparticles using organisms production of nanoparticles using organisms. *Critical Reviews in Biotechnology* 29 (4): 279–306.

33 Korbekandi, H. and Iravani, S. (2013). Biological synthesis of nanoparticles using algae. In: *Green Biosynthesis of Nanoparticles: Mechanisms and Applications* (eds. M. Rai and C. Posten), 53–60. CABI.

34 Grasso, G., Zane, D., and Dragone, R. (2019). Microbial nanotechnology: challenges and prospects for green biocatalytic synthesis of nanoscale materials for sensoristic and biomedical applications. *Nanomaterials* 10 (1): 11.

35 Golhani, D.K., Khare, A., Burra, G.K. et al. (2020). Microbes induced biofabrication of nanoparticles: a review. *Inorganic and Nano-Metal Chemistry* 50 (10): 983–999.

36 Devatha, C.P. and Thalla, A.K. (2018). Green synthesis of nanomaterials. In: *Synthesis of Inorganic Nanomaterials* (eds. S. Mohan Bhagyaraj, O.S. Oluwafemi, N. Kalarikkal and S. Thomas), 169–184. Woodhead Publishing.

37 Mohanpuria, P., Rana, N.K., and Yadav, S.K. (2008). Biosynthesis of nanoparticles: technological concepts and future applications. *Journal of Nanoparticle Research* 10 (3): 507–517.

38 Nadaroglu, H., Alayli, A., and Ince, S. (2017). Synthesis of nanoparticles by green synthesis method. *International Journal of Innovative Research and Reviews* 1 (1): 6–9.

39 Suganya, K.S.U., Govindaraju, K., Kumar, V.G. et al. (2015). Blue green alga mediated synthesis of gold nanoparticles and its antibacterial efficacy against Gram positive organisms. *Materials Science & Engineering. C, Materials for Biological Applications* 47: 351–356.

40 Govindaraju, K., Kiruthiga, V., Kumar, V.G., and Singaravelu, G. (2009 Sep). Extracellular synthesis of silver nanoparticles by a marine alga, *Sargassum wightii* Grevilli and their antibacterial effects. *Journal of Nanoscience and Nanotechnology* 9 (9): 5497–5501.

41 Singh, P., Kim, Y.-J., Zhang, D., and Yang, D.-C. (2016). Biological synthesis of nanoparticles from plants and microorganisms. *Trends in Biotechnology* 34 (7): 588–599.

42 Hulkoti, N.I. and Taranath, T.C. (2014). Biosynthesis of nanoparticles using microbes—a review. *Colloids and Surfaces B: Biointerfaces* 121: 474–483.

43 Khan, T., Abbas, S., Fariq, A., and Yasmin, A. (2018). Microbes: nature's cell factories of nanoparticles synthesis. In: *Exploring the Realms of Nature for*

Nanosynthesis (eds. R. Prasad, A.K. Jha and K. Prasad), 25–50. Cham: Springer International Publishing.

44 Arulmani, S., Anandan, S., and Ashokkumar, M. (2018). Introduction to advanced nanomaterials. In: *Nanomaterials for Green Energy* (eds. B.A. Bhanvase, V.B. Pawade, S.J. Dhoble, et al.), 1–53. Elsevier.

45 Majumdar, D., Singha, A., Mondal, P.K., and Kundu, S. (2013). DNA-mediated wirelike clusters of silver nanoparticles: an ultrasensitive SERS substrate. *ACS Applied Materials & Interfaces* 5 (16): 7798–7807.

46 Cabrera, F.C., Mohan, H., Dos Santos, R.J. et al. (2013). Green synthesis of gold nanoparticles with self-sustained natural rubber membranes. *Journal of Nanomaterials* 2013: 1–10.

47 Pokorski, J.K. and Steinmetz, N.F. (2011). The art of engineering viral nanoparticles. *Molecular Pharmaceutics* 8 (1): 29–43.

48 Shenton, W., Douglas, T., Young, M. et al. (1999). Inorganic–organic nanotube composites from template mineralization of tobacco mosaic virus. *Advanced Materials* 11 (3): 253–256.

49 Nasrollahzadeh, M., Sajjadi, M., Sajadi, S.M., and Issaabadi, Z. (2019). Green nanotechnology. In: *Interface Science and Technology* (eds. M. Nasrollahzadeh, S.M. Sajadi, M. Sajjadi, et al.), 145–198. Elsevier.

50 Iravani, S. (2011). Green synthesis of metal nanoparticles using plants. *Green Chemistry* 13 (10): 2638–2650.

51 Sathishkumar, M., Sneha, K., and Yun, Y.-S. (2010). Immobilization of silver nanoparticles synthesized using *Curcuma longa* tuber powder and extract on cotton cloth for bactericidal activity. *Bioresource Technology* 101 (20): 7958–7965.

52 He, S., Guo, Z., Zhang, Y. et al. (2007). Biosynthesis of gold nanoparticles using the bacteria *Rhodopseudomonas capsulata*. *Materials Letters* 61 (18): 3984–3987.

53 He, S., Zhang, Y., Guo, Z., and Gu, N. (2008). Biological synthesis of gold nanowires using extract of *Rhodopseudomonas capsulata*. *Biotechnology Progress* 24 (2): 476–480.

54 Rai, A., Singh, A., Ahmad, A., and Sastry, M. (2006). Role of halide ions and temperature on the morphology of biologically synthesized gold nanotriangles. *Langmuir* 22 (2): 736–741.

55 Gericke, M. and Pinches, A. (2006). Biological synthesis of metal nanoparticles. *Hydrometallurgy* 83 (1): 132–140.

56 Gericke, M. and Pinches, A. (2006). Microbial production of gold nanoparticles. *Gold Bulletin* 39 (1): 22–28.

57 Fayaz, A.M., Balaji, K., Girilal, M. et al. (2010). Biogenic synthesis of silver nanoparticles and their synergistic effect with antibiotics: a study against gram-positive and gram-negative bacteria. *Nanomedicine* 6 (1): 103–109.

58 Khan, S.A., Noreen, F., Kanwal, S. et al. (2018). Green synthesis of ZnO and Cu-doped ZnO nanoparticles from leaf extracts of *Abutilon indicum, Clerodendrum infortunatum, Clerodendrum inerme* and investigation of their biological and photocatalytic activities. *Materials Science and Engineering: C* 82: 46–59.

59 Ijaz, F., Shahid, S., Khan, S.A. et al. (2017). Green synthesis of copper oxide nanoparticles using *Abutilon indicum* leaf extract: antimicrobial, antioxidant and photocatalytic dye degradation activities. *Tropical Journal of Pharmaceutical Research* 16 (4): 743–753.

60 Gajbhiye, M., Kesharwani, J., Ingle, A. et al. (2009). Fungus-mediated synthesis of silver nanoparticles and their activity against pathogenic fungi in combination with fluconazole. *Nanomedicine* 5 (4): 382–386.

61 Gusseme, B.D., Sintubin, L., Baert, L. et al. (2010). Biogenic silver for disinfection of water contaminated with viruses. *Applied and Environmental Microbiology* 76 (4): 1082–1087.

62 Maynard, A.D. (2007). Nanotechnologies: overview and issues. In: *Nanotechnology – Toxicological Issues and Environmental Safety and Environmental Safety*, NATO Science for Peace and Security Series (eds. P.P. Simeonova, N. Opopol and M.I. Luster), 1–14. Dordrecht: Springer Netherlands.

63 Dhandapani, P., Maruthamuthu, S., and Rajagopal, G. (2012). Bio-mediated synthesis of TiO2 nanoparticles and its photocatalytic effect on aquatic biofilm. *Journal of Photochemistry and Photobiology. B* 110: 43–49.

64 Elango, G. and Roopan, S.M. (2015). Green synthesis, spectroscopic investigation and photocatalytic activity of lead nanoparticles. *Spectrochimica Acta. Part A, Molecular and Biomolecular Spectroscopy* 139: 367–373.

65 Wang, T., Jin, X., Chen, Z. et al. (2014). Green synthesis of Fe nanoparticles using eucalyptus leaf extracts for treatment of eutrophic wastewater. *Science of the Total Environment* 466–467: 210–213.

66 Wang, T., Lin, J., Chen, Z. et al. (2014). Green synthesized iron nanoparticles by green tea and eucalyptus leaves extracts used for removal of nitrate in aqueous solution. *Journal of Cleaner Production* 83: 413–419.

67 Mahdavi, M., Namvar, F., Ahmad, M.B., and Mohamad, R. (2013). Green biosynthesis and characterization of magnetic iron oxide (Fe3O4) nanoparticles using seaweed (*Sargassum muticum*) aqueous extract. *Molecules* 18 (5): 5954–5964.

68 Khot, L.R., Sankaran, S., Maja, J.M. et al. (2012). Applications of nanomaterials in agricultural production and crop protection: a review. *Crop Protection* 1 (35): 64–70.

69 Singh, A., Singh, N., Hussain, I. et al. (2015). Plant-nanoparticle interaction: an approach to improve agricultural practices and plant productivity. *International Journal of Pharmaceutical Science Invention* 4 (8): 25–40.

70 Singh, N.B., Singh, A., Hussain, I. et al. (2015). Synthesis, characterization and application of ruthenium oxide nanoparticles on growth and metabolism of *Brassica oleracea* L. *Advanced Science Letters* 21 (8): 2635–2640.

71 Kottegoda, N., Munaweera, I., Madusanka, N., and Karunaratne, V. (2011). A green slow-release fertilizer composition based on urea-modified hydroxyapatite nanoparticles encapsulated wood. *Current Science* 101: 73–78.

72 Clark, J.H. (1999). Green chemistry: challenges and opportunities. *Green Chemistry* 1 (1): 1–8.

73 Constable, D.J.C., Curzons, A.D., and Cunningham, V.L. (2002). Metrics to 'green' chemistry—which are the best? *Green Chemistry* 4 (6): 521–527.

15

Risk Assessment and Management of Occupational Exposure to Nanopesticides in Agriculture

Anand Patel[1], Bhavin Patel[1], Pranav Shah[2], and Jayvadan Patel[3]

[1] Department of Chemistry, Uka Tarsadia University, Bardoli, Gujarat, India
[2] Maliba Pharmacy College, Uka Tarsadia University, Bardoli, Gujarat, India
[3] Nootan Pharmacy College, Sankalchand Patel University, Visnagar, Gujarat, India

15.1 Introduction

15.1.1 Agriculture

The term agriculture is derived from the Latin word ager, which implies soil and cultural means cultivation. Just we can say soil cultivation. It is the science of agriculture, including work of soil cultivation, crop production, and livestock rearing. It has two main branches: Crop and animals.

Agriculture is the stepping stone in the history of the human race, due to which agriculture (farming) man settled at particular place. Agriculture has been one of human being's ancient and foremost activities. Despite the world's increasing urbanization and industrialization, nearly 50% of the work force is still involved in agriculture. Agriculture has become a major source of income in developing nations and has made a significant contribution to the country's economy.

The basic purpose of agriculture is to encourage stronger and more productive plants and crops and to develop them by enhancing the soil and water supply.

Farming is a backbone of the Indian economy. In India, about 64% of the total population depend on agriculture for their food source.

Current agriculture is heavily dependent on science and technology as well as on the physical and biological sciences. Agricultural chemistry deals with the other important agricultural issues, such as fertilizer application, insecticides and fungicides, soil quality, agriculture products analysis, and farm animal nutritional requirements.

Users demand a year-round availability of high-quality agricultural goods, but due to pests and inclement weather, there are sometimes short-term harvest fluctuations. Chemical pest controls are one way of controlling crop loss variability. In addition to reducing crop losses from pests by pesticides, there are other ways to sustain food and fiber intake.

15.2 Pesticides

Pesticides such as rodenticides, insecticides, acaricides, fungicides, herbicides, nematicides, and molluscicides are found naturally and are widely used in agriculture to increase the crop yield and productivity and generally protect plants from harmful effects such as plant diseases, insects, and weeds. There are many groups of pesticides based on application and chemistry: organochlorine pesticides, organophosphorus pesticides, carbamate pesticides, etc.

15.2.1 Understanding Pesticide Benefits

There are several types of advantages that can be related to pesticides, but the public frequently overlook these benefits. The economic advantages for farmers derived from the security of the product quality and yield and the elimination of other expensive inputs such as labor and fuel are the most evident and simplest to quantify. The rate of global pest losses for several crops in some regions found that pesticide-induced losses are much more than 50% of the achievable crop output [1]. Food production will fall without pesticides, and food prices will increase. With lower productivity and higher prices, farmers will be less

Sustainable Nanotechnology: Strategies, Products, and Applications, First Edition.
Edited by Yashwant V. Pathak, Govindan Parayil, and Jayvadan K. Patel.

profitable for large crops on world market. Trying to prevent or reducing agricultural losses to pests using pesticides increases yields and hence ensures stable supplies of agricultural products at prices that are affordable to customers and enhances product quality in terms of cosmetic appeal that is also essential to buyers. Pesticide benefits can thus rack up to a variety of different users, not just to farmers or customers, and also to the community.

The pesticides available in supermarkets are used to control rodents and insects that are drawn to food and food waste. Davis et al. [2] recorded that virtually all households (97%) used pesticides at least once a year, and two-thirds used pesticides more than five times annually.

15.2.2 Understanding Pesticide Risks

It is obvious that the prevalent use of pesticides in modern farming has long become an established part of the business. At the very same time, the use of such chemicals contains evidence of both direct and indirect risks [3–7]. The use of pesticides can cause adverse effect on human and environment because of their high biological activity and in some cases because of their environmental persistence. Improper disposal of certain pesticides can result in severe acute intoxication; in certain cases, harmful effects can also result from long-term exposure at low levels [8]. Because of their wide usage, a large proportion of the population may also be subjected to or occupied with pesticides in the overall environment. Numerous individuals and groups are at risk of toxicity, characterized by different patterns and exposures. Workers in the pesticide manufacturing and formulation, as well as public health users, are exposed to pesticides during work. In agriculture, the farmers and trained pesticide applicators may be exposed.

15.3 Nanopesticides

Some of the pesticides used have been found to be environmentally harmful; in addition, some of them are dangerous or even poisonous to humans and other species and therefore have been prohibited by the state or foreign agencies. Because efficient and useful pesticides are difficult and costly to create, new technologies were tested. Nanoscale science and nanotechnology showed great promise in delivering new and improved solutions [9–16]. Nanotechnology is seen as one of the twenty-first century primary technologies. Nanosized products alter their chemical and physical properties; this has developed and innovated a range of industrial, pharmaceutical, and medical products [17–21]. Application of nanotechnology in agriculture and food industry is comparatively modern in contrast with its use in pharmacy (drug delivery) [15, 22]. Use of nanotechnologies can make a significant contribution to the safe intensification of agricultural production, as they can not only promote plant defense against pesticides but also control plant growth, ensure a secure increase in global food output, promise improved nutritional quality, and minimize waste [23–26]. Agricultural and Food development is one of the most important areas for application of nanotechnology [27–30].

According to Thornton [31], the use of nanotechnology in agriculture would improve not only the conservation and management of inputs in crops but also animal products and fisheries that will eventually be expressed in improved farming practices. The effectivity of active ingredients, such as fungicides, herbicides, and insecticides, can be improved if applied in the form of nanoscale, such as polymeric nanocapsules [32]. Using nanotechnology in agriculture allows targeted delivery of active ingredients followed by their controlled release and monitoring of crop conditions, including the effect of environmental stress.

The most commonly used nanopesticides (nanoherbicides, -fungicides, -insecticides, and -nematicides), as well as frequently used nanoformulations based on encapsulation of biodegradable organic based matrix systems and silica based nano carriers or silver-and copper – based nanoformulations will be discussed further in this chapter.

15.4 Nanoinsecticides

Insect pests, as they destroy crops and infest stored food and food products, are the major issue in agriculture. The loss is caused by a worsening in food quality and plant disease transmission [33, 34]. Synthetic chemical insecticides take predominance in treatment methods. These really toxic chemicals have unjustified toxicity and lethal effect on nontarget organisms, establish target biological resistance, and cause adverse environmental impacts. Nanotechnology would therefore focus on providing green and efficient alternative for control of insect pests in agriculture without damaging nature [35, 36]. Formulations specifically targeting the pests and in turn preventing pollution, such as the "good buster" (i.e. encapsulated product), break open only when they interact with the alkaline environment, such as the intestines [25].

Insecticides based on active ingredients extracted from plants, as well as essential oils derived from specific species as among the botanical insecticides. The use of nanotechnology-related botanical insecticides provides major potential for increasing crop yields while reducing human and environmental health impacts [37]. For example, the formulation of water dispersible amorphous permethrin NPs with a mean particle size of 151 nm showed higher larvicidal activity against *Culex Quinquefasciatus* than bulk permithrin indicates that, permethrin NPs could be a good option for such an insect as a potent and specific larvicide [38]. The formulation on a nanoscale did not reveal toxicity to nontarget organisms.

Bhagat et al. used a low-molecular-mass gelator to verify a nanogel formulated from a pheromone methyl eugenol, such as all-trans-tri(*p*-phenylenevinylene)-bis-aldoxime [39]. It was found that this formulation is appropriate for effective management of *Bactrocera dorsalis*, a pervasive harmful pest for a number of fruit crops, including guava.

Nanoparticles loaded with neem (*Azadirachta indica*) extracts formulated as a colloidal suspension and (spray-dried) powder ensure maximum percent of larval death of *Plutella xylostella*, and their release profile was based on polymer or polymer erosion swelling and relaxation [40].

Technologies for nanoencapsulation allowed not only the development of built-in crop pesticides but also in-built switches to control the release and subsequent availability of pesticides. These nanoparticle pesticides have improved the absorption capacity in plants and cannot be washed off as readily, thus increasing their efficacy [41, 42].

The nanocapsules can be known to release the active ingredient, such as an insect's stomach, into specific environmental conditions or physiological environments. These intelligent pesticides can provide a more accurate, controlled, and efficient use of pesticides and thus possibly decrease the required concentrations of pesticides used [43].

15.5 Nanoherbicides

The development of target-specific herbicide molecule encapsulated in a nanoparticle is intended for a specific receptor in the origin of target weeds, which enters the root system of the weeds and translocates to parts that restrict glycolysis of food supplies in the root system, ultimately causing the specific weed plant to starve for food and get killed [23, 44, 45]. Controlling parasitic weeds with nanoencapsulated herbicides decreases herbicidal phytotoxicity on crops [46].

Polycaprolactone nanoparticle formulations containing herbicide atrazine were shown to be effective in controlling the target species *Brassica*; nanoparticles have decreased the mobility of atrazine in the soil and have reduced the genotoxicity of the herbicide, thus further lowering the impact of harm to the environment and humans [47].

Jiang et al. have formulated an eco-sustainable nanoemulsion system for water-soluble herbicide glyphosate isopropylamine application [48]. The nanoemulsion system with particle size of the diameter of less than 200 nm reveals substantially lower surface tension than a commercial formulation (Roundup), and its ED50 (0.40 kg a.e ha^{-1}) in regulating the weed *Eleusine indica* was lower than those of Roundup (0.48 kg a.e ha^{-1}). This implies that a nanoemulsion system may be used to increase the penetration and absorption of isopropylamine glyphosate.

The formulations of herbicide-containing nanoparticles were efficient in the pre- and postemergence therapy of the target organism *Raphanus raphanistrum*, but they did not influence nontarget organism's plant growth. Guo et al. evaluated the controlled release of acetochloride from the poly(butyl-meth-acrylate-diacetone acrylamide) formulation and found that improved herbicide incorporation and slower release were linked to potential interaction between the herbicide and the polymer [49].

15.6 Nanofungicides

A fungicide is a particular type of pesticide that controls fungal disease by specifically inhibiting or killing the disease-causing fungus. Fungicides cannot adequately control all diseases caused by the fungi. These include fungal vascular diseases Fusarium and Verticillium. During the formation and maintenance of a crop, fungicides are being used to control disease, to increase a crop's production and decrease blemishes, and to enhance the storage life and quality of the harvested plants. Fungicides are therefore widely used in the agricultural system to control soil, airborne, or seed borne fungal pathogens [50]. The ever-increasing types of fungicides and their pervasive use in modern farming have generated tension about their specific interaction with nontarget organisms directly or indirectly engaged in soil fertility maintenance [51]. As in other classes of nanopesticides, therefore, nanofungicides development was also focused on designing and production of environmentally friendly nanoformulations [23].

Polymerization of citric acid on the exterior of the oxidized multiwall carbon nanotubes resulted in poly(citric acid)-grafted hybrid composites that were not only water-soluble but also capable of trapping water-soluble pesticides, such as zineb and mancozeb, and had a superior toxic effect on *Alternaria alternate* fungi compared to large pesticides [52].

Biodegradable chitosan–lactide copolymer was used as a hydrophobic carrier for pyraclostrobin, a new wide-spectrum foliar fungicide in the chemical class of strobilurin. The fungicide loaded in nanoparticles showed better fungicidal activity against *Colletotrichum gossypii* at long incubation time, which also showed sustained release characteristics.

Kumar et al. [53] created a hexaconazole nanoform (with particle size of 100 nm), using polyethyleneglycol-400 as a surface stabilizer. Hexaconazole nanoparticles were more potent in comparison to that of bulk hexaconazole. The impact of nanohexaconazole on soil nitrifiers was tested in vitro, and there were no significant adverse effects in their numbers observed as compared to the conventional registered formulation, providing the safety of nanofungicide [54].

The activity of Cu nanoparticles with particle size in the range of 3–10 nm coated with cetyltrimethylammonium capping agent bromide was proved to be better than that of the commercially available Bavistin fungicide against four plant pathogenic fungi, *Cochliobolus lunata, A. alternata, Fusarium oxysporum, and Phoma destructiva* [55].

It has been reported that the copper-based nanoparticles are more efficient than the fungicides marketed and can be applied at a lower formulated product and active ingredient rate without any harmful effect on the plant.

15.7 Latest Research by Types of Nanopesticides

15.7.1 Nanoemulsion

Pesticidal AI nanoemulsions are often added to improve the uptake of the AIs, but supporting data in plant protection products remain scarce. However, the results of two recent studies confirm the theory of greater adoption. The first of these studies, experiments on a series of neem oil nanoemulsions, had shown that the LC50 (the concentration required to achieve mortality of 50%; Anjali et al. [56]) decreased with droplet size, which was perceived as implying an increased intake of smaller droplets. In the second experiment, the efficiency of a permethrin nanoemulsion (presented in [38]) was considerably higher than pure AI, again interpreted as specifying increased nanoformulated AI uptake [54].

In a series of articles from a research team based in Malaysia, nanoemulsions for a water-soluble AI (i.e. glyphosate) have recently been presented. Their nanoformulations aimed at increasing the herbicide's bioavailability while trying to avoid a number of adjuvants present in the current glyphosate formulations associated with non-target organism toxicity [57].

15.7.2 Polymer-based Nanopesticides

Many recent publications revealing new polymer-based nanoformulations have shared the common objective of using biodegradable polymers and/or naturally occurring AIs to develop less-harmful plant protection products. Biodegradable materials of biological origins such as corn oil, beeswax, and lecithin have increased in use over the past two years [58, 59]. In addition to being seen as more environmentally friendly, it is probable that such matrix material could be regarded for use in organic crop production when correlated with AIs of natural origin. Many natural substances are known to have pesticidal characteristics but are usually rather unstable and necessitate preterm degradation protection. For this purpose, a number of polymer-based nanoformulations in the form of nanospheres, -gels, or -fibers have been recently proposed.

15.7.2.1 Nanospheres

Lansiumamide B, a molecule extracted from C kernels, has the pesticidal properties. Han and coworkers [60] have briefly covered lansium, but the compound is hard to dissolve and unstable in the environment. Yin et al. (2012) described a polymer-based lansiumamide B nanoformulation whose nematicidal activity was increased compared to that of the nonformulated compound [60]. The performance of lansiumamide B (both formulated and nonformulated) exceeded that of a synthetic nematicide (ethoprophos, pure AI, Yin et al., 2012), demonstrating the potential usefulness of lansiumamide B on the basis of disease progression and average number of rot knots.

15.7.2.2 Nanogels

Nanogels were suggested for use in plant protection products over the past two years as a way of meeting organic farming standards with pheromones, essential oil, or copper as AIs. Bhagat et al. proposed the immobilization of pheromones within a nanogel (without using any potentially toxic chemicals such as cross-linkers or antioxidants) [39]. Evaporation of pheromones in the nanogel was significantly reduced compared to the evaporation of pure AI, extending their effectiveness for up to 33 weeks compared to only three weeks for the pure AI.

Bhagat et al. also demonstrated the effectiveness of the nanoformulations during an adverse season in an open orchard [39].

Brunel et al. suggested the use of pure chitosan nanogels to boost the effectiveness of copper-based antifungal treatments. The potential advantages of using nanogels instead of a solution involve easier management, enhanced distribution on the leaves, and long-term release of copper on or into the soil without loss of antimicrobial properties [61].

15.7.2.3 Electrospun Nanofibers

Electrospun technique shows potential for scaling up and meeting industrial production requirements [62], and nanofibers acquired through electrospinning have in recent times been investigated for plant protection applications. The potential benefit of such nanofibers over spheres or capsules comes from its ability to prevent the release bursts that take place when the AI is not distributed homogeneously within the polymer matrix [63].

Hellman et al. showed the feasibility of effectively integrating pheromones into polyamide or cellulose acetate nanofibers (around 30 wt%) and achieving a near linear release over several weeks. The author suggested that the nanofiber webs could be allocated across the fields to be protected (quite similar to spider webs) so that pheromones can be released uniformly [64].

Xiang et al. presented a similar nanofiber network composed of polylactic acid and cellulose nanocrystals. In a glasshouse experiment, the fibers loaded with thiamethoxam were effective against whitefly over a nine-day period, at 50% of the prescribed dose for pure AI [63].

15.7.3 Hybrid Nanoformulation

Nanoformulations formed in the pharmaceutical industry have motivated researchers to explore the potential of more complicated nanoformulations for pesticide delivery. Below are two examples of the use of solid lipid nanoparticles (SLNPs) and coated liposomes.

SLNPs have been increasingly considered as an alternative to polymeric nanoparticles for pharmaceutical formulations [65]. However, applications in the agrochemical sector remain limited, and by 2011, only two formulations had been submitted. Recently, a second generation of lipid nanoparticles has been formed, integrating liquid lipids into the solid matrix of SLNPs to increase the payload and prevent rapid AI leakage. Two papers about the potential of such formulations to prevent deltamethrin from photodegradation have recently been published [58, 59]. For the formulated AI, both direct and indirect photodegradation was reduced, relative to the pure AI.

Bang et al. first described the preparation of coated liposomes for the slow-releasing insecticide [66]. Since then, two studies have tested the insecticidal efficacy of these formulations and demonstrated the prolonged or delayed nanoformulation activity relative to that of pure AIs [67, 68]. For example, pure pyrifluquinazon had its best lethal efficacy two days after treatment, while it was 14 days after treatment for the nanoformulation [68].

15.8 Pesticides Risk in Agriculture

Pesticides play a significant part in reducing crop losses. But their overuse can have serious adverse effects on both human health and environment. Therefore, it is considered that there could be a risk of pesticide use and overuse to reduce crop losses. The risk of plague depends on two things: exposure (how much?) and toxicity (how toxic?). The exposure is the amount that you receive in or on your body, or the amount released into the environment. The toxicity of pesticide is a measure of how toxic it is to human and the environment (http://npic.orst.edu/health/risk.html).

15.9 Risk Assessment: Aim and Importance

- The risk assessment process aims at assessing hazards and then removing that hazard or minimizing its risk level by adding control measures, as necessary. This has created a safer and healthier work place. The objective is to try and answer the following questions:
- What, and under what situations?
- What are the repercussions of that?
- How likely are there to be any repercussions?
- Is risk completely controlled, or is even more action needed?

Risk assessments are quite essential, as they form an essential part of a plan to manage occupational health and safety. They contribute to:

- Build awareness about hazards and risks.
- Identify who may be at risk (e.g. cleaners, employees, visitors, contractors, and the general public).
- Determine if a control program is necessary for a specific hazard.
- Determine whether existing controls are adequate, or whether more should be done.
- Prevent injuries or diseases, particularly during design or planning phase.
- Making hazards and control measures a priority.
- Comply with applicable legal requirements. (https://www.ccohs.ca/oshanswers/hsprograms/risk_assessment.html)

15.10 Nanopesticides Risk Assessment: Toxicity Testing

Nanotechnology use has been very well developed in various fields such as medicine, engineered materials, and cosmetics. The theory of the validity of this method is to adjust the products or substances according to the demands of the business and/or population to a nanometric scale with greater stability and target properties. A need to reduce the use of pesticides has arisen in agriculture to mitigate their effects on the ecosystems. For example, surfactants increase the effectiveness and dispersion of pesticides by enhancing their water solubility, thus facilitating contamination of the river and representing risks to organism and also to human health. The production of nanopesticides including AI in its center could therefore be a major advancement in the agricultural sector by reducing the quantity of surfactants and pesticides applied to the crops, facilitating their dispersal, and preserving the pesticide properties after applications [69]. These formulations are supposed to present less environmental and human health issue, only in relation to AI. The rapid growth and use of nanotechnology has culminated in the research world having a high interest in this area. Around 60 papers involving nanopesticides have been published in the past decade [70]. For example, to assess the effects on the environment and health, plants, animals insects, soil, and water are used to evaluate toxicity and protection of nanopesticides. It is surprising that various materials are used as AI carriers, with uncertain physiological activities [69]. Therefore, their protection and risks need to be measured. In these studies, it is necessary to note that the dispersion of pesticides in the environment affects both the area and nontarget species, because the residue will deposit and accumulate due to the lipophilic characteristics of these particles [71]. Such nanoformulations could be shaped by polymers (organic) or metals (inorganic), as the desired applications require. Inorganic NPs are associated with increased release and output. The formulation type can therefore vector and encourage the incorporation of AI and improve its biological role [72]. For example, when silver NP-chitosan-encapsulated paraquat was applied to soil for a wide period of one month, it retained herbicide activity and did not affect the soil macro and micronutrients and soil microflora [73]. No change in plant growth was observed when phytotoxicity was investigated in the nontarget Vigna mungo (green gramme), whereas free paraquat affected all growth parameters. This shows how NP can minimize the collateral damage caused by application of pesticides; however, no further risk assessments have been published for this NP to date [73]. The use of nanopesticides has been extensive; however, the current state of awareness is not sufficient to accurately determine its environmental risk. Crop plant experiments analyzing genetic, metabolic, and physiological responses showed improvements in spinach (*Spinacia oleracea*) and lettuce (*Lactuca sativa*) antioxidant-protection mechanisms [74–76]. A study with microcrustacean *Daphnia magna* exposed to Kocide 3000 copper(II) hydroxide revealed that genes involved in detoxification and reproduction have changed their expression, depending on the time of exposure to $Cu(OH)_2$ NPs [77]. Zhang et al. showed that $Cu(OH)_2$ NP was able to stimulate soil microbiota alterations that impeded neonicotinoid thiacloprid degradation [78]. Consequently, if one pesticide or other contaminant deposits in the soil affect all the micro- and macrocosms that exist in the atmosphere, that pose hazards to all living organisms [79]. Organic NP is a promising alternative for reducing potential toxicity. For example, SLNPs provide slow release of the material and can therefore minimize toxicological impacts [80]. SLN-loaded γ-cyhalothrin had less toxicity to fish (*Brachydanio rerio*) and thus possible applicability in nanotechnology for agrochemical agents without harming the environment [81]. Atrazine- and simazine-loaded SLNs displayed great stability and slow release, with low cytotoxicity in fibroblast cells (20% lower than commercial formulation) observed by MTT assay. When the same nanopesticides were tested for nontarget species (*Zea mays*), SLNs did not cause plant growth changes, although the herbicide activity against *R. raphanistrum* was 10 time stronger than free pesticides [82]. Since the results in loaded and unloaded NPs were the same, the authors attributed the toxicity to the SLN formulation and not the pesticides [83]. The insecticide emamectin benzoate loaded into the SLN has been tested against pests (moths and aphid) responsible for crop damage. Compared with traditional pesticide formulations, this formulation has shown greater toxicity against these insects. The formulation has also demonstrated strong stability, great dispersibility, and interesting biological activity [81]. However, further toxicity tests are missing for this formulation. Nanoformulations that are coated with organic polymers are of particular interest, given the greater biocompatibility with different organism and therefore the putatively lower ecotoxicological effect. Imaging studies have shown that the NPs can be stored in the intestines of soil organisms, as earthworms [83, 84]. The aggregation and toxic effects of nanopesticides are nevertheless predicted to be lower than the commercial nanoformulations. Research that examined the parameters of nanoencapsulated pesticides (bifenthrin) distribution and bioavailability in soil earthworm (*Eisenia fetida and Lumbricus terrestris*) have confirmed that nanoformulations have accumulated in worms; however, removal is better than free bifenthrin [85]. One of the most effective examples of risk management is the use of atrazine loaded with poly-x-caprolactone (PCL) established to boost the herbicide activity of atrazine NPs showing better efficacy for mustard (*Brassica Juncea*) plants than free atrazine [47]. In particular, plant growth parameters in non-targeted plant *Z. mays*, which is an indication of pesticide vectoring, have not been affected [82]. Risk assessment studies were performed as genotoxicity (using allium cepa and human cells), cytotoxicity (using human cells), and ecotoxicological studies (using algae *Pseudokirchneriella subcapitata* and worm *Enchytraeus crypticus*), suggesting a decreased toxicity of PCL nanocapsules containing atrazine to nontarget species compared to free herbicide [84, 86, 87]. To use a more complex and vertebrate organism, Andrade et al. showed that atrazine nanoencapsulation attenuated biochemical alterations to the fish *Prochilodus lineatus* induced by the herbicide [88]. A perfect example of how risk evaluation can flow is the PCL-laden atrazine series of studies: from simpler to the more complex species. The use of naturally derived polymers is an environmentally sustainable trend in the production of nanoagrochemicals. Some examples are chitosan, derived from chitin and zein, derived from maize (*Z. mays*), used to manufacture biocompatible, nontoxic, and biodegradable biopolymers, characteristics that are consistent with reducing NM hazards. The literature uses zein as a biopolymer to offer incipient studies of botanical pesticides. For example, geraniol, eugenol, R-citronellal, and cinnamaldehyde-loaded zein NPs showed improved efficacy against pests such as *Tetranychus urticae*, which includes Chrysodeixis compared to emulsified compounds. In particular, encapsulation with zein NPs decreased the compounds' normal cytotoxicity, as observed in vitro in two cell lines (V79-4 and 3 T3) [89]. Pascoli et al. utilized *Allium cepa* and *C. elegans* to analyze the protection of neem oil-loaded zein NPs (nanoparticles). The study found that these NPs decreased genotoxicity and also had no impact on soil microbiota [90]. Scientists attribute these results to chitosan, as it has already been shown that this polymer can alter the levels of serotonin in zebrafish (*Danio rerio*) [91]. Zein data is very promising and gives great prospects for healthy and agricultural practices.

15.11 Prevention of Occupational Pesticide Risk

Preventing workplace health hazard from pesticides requires a comprehensive "risk identification and control" throughout their manufacturing; pre- and postpesticides are approved into use. Risk analysis includes "risk assessment," "risk management," and "risk communication," according to the schemes used by international organizations. The fundamental risk assessment leaps are hazard identification, hazard characterization, exposure evaluation, and risk characterization. Managing risk covers risk assessment, evaluation of options, execution of options, review, and monitoring. Risk assessment is a crucial tool for predicting the probability of man-made harmful impacts and identifying the need for prevention measures. The risk assessment allows to determine for each group of the population at risk:

- The amplitude of the hazard posed by the pesticide product, that is the adverse effect that a particular active ingredient may intrinsically produce;
- The dose response assessment, that is the approximation of the dose-to-incidence and severity of the impact;
- The extent of exposure, that is the evaluation or prediction of the doses exposed to human beings; and
- Risk definition, that is the probability of transmission of adverse consequences due to real or expected exposure conditions, and the extent and seriousness of such consequences.

In the end, risk characterization reflects the relative need for prevention interventions and the components of preventive strategies. With emerging research awareness, a better understanding of physiological toxicity pathways could allow for more precise predictions of risks [92].

15.12 Strategies for Prevention in the Premarketing Phase

Prevention during the premarketing process refers to all practices that may be performed before a consumer product is put on the market. In this process, industry is the key player in prevention along with national regulatory authorities involved in the authentication protocols. The production of less-toxic compounds, or the introduction of modern effective formulations (soluble wrapping, microencapsulation, etc.), and the availability of suitable technologies for the applications of pesticides have produced very successful results. The "toxicological monitoring" of the substance under production is a crucial protective factor in the premarketing process. The responsible regulatory authority should mandate the producer to include animal toxicity monitoring and analyses of impact on the habitats and environmental fate in order to approve the sale and the use of a new pesticide within a region. The quality of the technological dossier you need varies in various countries. Directive 91:414 includes a very complicated series of measures in the European Union intended to determine the substance's full toxicological profile. This profile includes acute toxicity (oral, dermal, and inhalation), inflammation of the eyes and skin, sensitization of the skin, short-term toxicity (28 ± 90 days), mutagenicity, long-term toxicity (two Years), tumors, reproductive toxicity, and other special effects (e.g. neurotoxicity). Strict standards must be fulfilled in developing countries before a substance is approved, and the cost of meeting these specifications is very high. Large numbers of pesticides are transported from developed countries to the developing nations, but effective processes for reporting pesticides are not yet in operation. The lack of sufficient regulation and monitoring systems to implement regulations in developed countries also contributes to the use of the most toxic substances and inadequate control of harmful impact on human and environmental consequences.

International agencies have a vital role to play in helping enact laws, encouraging proactive measures, and providing recommendations to improve risk management methodologies. An example of initiatives taken by multinational agencies applies to the creation of recommendations to strengthen risk management methodologies. In the premarketing process, it is generally very crucial to set appropriate exposure assessment measures that would be used to track its effects on consumers and the environment following the launch of the pesticide. Staff exposure management is an important aspect of the risk assessment of pesticides that are used for both regulatory compliance purposes and postregistration tracking. The Workshop on Pesticide Exposure Assessment Approaches, held in 1993 in Ottawa (Canada), made a significant contribution to the advancement of exposure assessment techniques [93]. It was a collaborative initiative between Health Canada and North Atlantic Treaty Organization, formally sponsored by the U.S. EPA and OECD. Workshop delegates settled on a guide paper on strategies for determining pesticide toxicity, as a way of achieving the goal of international harmonization. The paper was sent to the OCED as an OCED support text, for review.

When risk assessment of harmful impacts on farm staff is conducted in the premarketing process, the purpose of the evaluation is to assess the danger faced from exposure to a certain active ingredient under certain condition of use. Exposure evaluation often focuses on exposure-level estimates for standard exposure conditions, relying on the use of common databases. Generic records are sets of exposure data provided by various field experiments, conducted under different situations of use, for the type of agricultural operation (mixing and packing, spraying), the type of preparation, and the nature and mode of use. The accessible generic dataset consists of the exposure dataset for pesticide handlers (PHED), used in North America; the German dataset, used by BBA in Germany; the predictive operator exposure model (POEM), established in the United Kingdom; the Netherlands model; and, more recently, the European Model (EUROPOEM). These sources vary for the form of research and how the exposure measurement is measured. The health risk appraisal for field employees who use pesticides needs to take into consideration both the acute and long-term health consequences. In the context of European Directive 91:414, the "acceptable operator exposure limit" (AOEL) is an important term which is described as the maximal amount of active agent to which the operator can be exposed without any detrimental effect. The AOEL is determined from sufficient toxicological studies beginning with the NOEL (or NOAEL) and adding the so-called safety factors to compensate for the complexities of animal-to-man extrapolation and the heterogeneity of vulnerability of intraspecies.

The operators' health risk assessment, as presently undertaken in Europe, has to deal with several issues relating to the evaluation process. Exposure in the field is typically very constrained in time or can be repeated for several weeks; however, the toxicological studies used to depict such exposures are based on a prolonged scheme of administration, either short or long term, that is quite different from the actual exposure schedule. Furthermore, a route-to-route extrapolation is very often expected due to the lack of studies performed through dermal administration.

Interspecies extrapolation may not be entirely acceptable, and there is little theoretical evidence for the use of a safety margin, though established.

15.13 Strategies for Prevention in the Postmarketing Phase

Postmarketing protection techniques refer to all those protective steps that can be implemented after a pesticide product has been approved and is on the market. The goal should be both to protect the health of employees in the workplace and to avoid undesired consequences in the general public. Residue testing of food and drinking water is an important method for surveying the general population's capacity for exposure to pesticides. The physiological adaptations of acute pesticides toxicity provide the foundation for prevention programs, such as awareness campaigns to prevent excessive and unsafe usage of pesticides and to protect customers and the environment, or instruction on responsible implementation procedures.

Hazards that occur through the implementation of pesticides are mostly attributed to lack of intelligence, expertise and understanding, inadequate monitoring, and the selling of highly dangerous chemicals on the open market. To this end, all efforts aimed at encouraging a healthy use of pesticides for man and the world need to be promoted and supported.

FAO has established a reference model for enhancing pesticide use. FAO describes good agricultural practice (GAP) in use of pesticides as recommended use of a pesticide under reasonable circumstances (at every point of manufacturing, storage, transportation, delivery, and food processing or other agricultural commodities), taking into consideration the minimum amounts required for appropriate regulation, the pesticide being administered in such a way as to leave a residue of the smallest possible quantity and which can be toxicologically accepted.

In addition to regulatory initiatives, postmarketing prevention measures may include implementing effective occupational health risk control to reduce harmful effects on workers using pesticides. Several methods are used to develop the policy to prevent toxic reactions of workers using pesticides. They are extremely crucial for farm users, where the risk of exposure is significantly greater, and the general hygienic level is lower, particularly in the developing countries.

15.14 Health Risk Management

The aim of the health risk management is to reduce risk to nonsignificant degrees, which can only be accomplished by collaboration among health and safety experts, administrators, and personnel. Staff engagement is important in ensuring that an effective implementation is accomplished, enhancing the flow of information between managers, experts, and workers, and introducing preventive strategies focused on intimate experience of the efficient environment and individual roles and activities. Another goal of the partnership is compliance with the regulatory provisions, which is the responsibility of the management and which the workers must satisfy.

The basic stages of on-the-job health risk management are hazard detection, threat evaluation, risk analysis, and health control.

15.15 Hazard Identification

It is important to define the danger that characterizes the work:

- Identify and store the chemical ingredients to which the workers are exposed;
- Get material safety data sheets (MSDS) from manufacturers for any product;
- Group the products according to the active ingredient chemical category and the WHO hazard classification; and
- Identify the carries solvent or other inert MSDS ingredients and note their particular toxicities.

15.16 Exposure Assessment

Monitoring tasks is a systemic and routine procedure intended to contribute to disciplinary steps, which consists of qualitative and quantitative examination of the activity habits of group of workers subject to homogeneous risks. Ambient monitoring measures environmental damage and assesses health hazard by comparing with appropriate reference and estimates levels of exposure by measuring potential dermal exposure, actual dermal exposure, and exposure to inhalation. Biological testing consists of measuring the active ingredients or their metabolism in biological systems to measure the body pressure of the pesticide being examined [94]. Biological surveillance provides some benefits over environmental monitoring, as an individual dosage representing biological parameter is more closely related to the adverse effects than any environmental assessment. Biological testing provides a comprehensive estimation of exposure across all absorption routes in the body and assesses total exposure as the amount of multiple sources of contamination. In certain cases it tells about long-term exposure and helps to determine a particular subject's exposure as a function of his/her behavior within the job environment and specific variables that affect the physiologic substance's pharmacokinetics. However, where exposure pathways are merged or mixed, environmental testing can be helpful in clarifying which pathway is more important or in defining the substances that must be taken into consideration in the biological monitoring process. In comparison, if the environmental testing revealed very low levels of emissions, the need for biological surveillance could be omitted: in this view, the biological and environmental surveillance should also be considered complementarily rather than as alternate activities.

15.17 Risk Characterization

The risk characterization aim is to determine the extent of danger to public health. The preliminary health risk evaluation is achieved by combining information about the identified danger with the predicted or assessed exposure and the employees' health status. These should be taken into account when results of medical surveillance and biological monitoring are available. This process will result in the evaluation of both the qualitative aspects of the risk (reversibility, affected target organ, function alteration, etc.) and the quantitative aspects (high medium, low probability, or harmful impacts). The final step is to assess whether the predicted risk is small, acceptable, or not reasonable, depending on the magnitude of the adverse consequence and the likelihood of occurrence. If it is determined that the risk is insignificant or acceptably low, there would be no extra regulation, although the finding should be checked by health surveillance.

15.18 Health Surveillance

Health screening requires preemployment physical testing, annual review of the employees' health status, and biological monitoring. The preemployment medical test defines the physical health to perform the applicant worker's position, recognizes any medical problems that could be aggravated by the training, and provides a standard score before the position. The aim of periodic health monitoring is to identify, as early as possible, any potential adverse effects that may have been triggered by the current exposure or working circumstances and any substantial improvements in health status that could affect the ability to resume the work or even deteriorate the health status as the work continues. A guide for the health monitoring of workers subjected to pesticides was published jointly by the WHO, ICOH, and ICPS many years ago [95]. When a person returns to work following a major injury, he or she must undergo a medical test or assessment to create a new standard to determine any substantial improvement in health status that may have affected the job's physical capacity. Managing the health risk is a constant and step-by-step operation. Effective risk assessment involves an iterative process with defined feedback loops.

15.19 Record Keeping

The development of instruments for collecting data supported by the occupational health program is important. This would encourage epidemiological labor force research to be performed.

15.20 Information, Instruction, and Training

User education should be aimed at avoiding unneeded pesticide use, adopting secured application practices, and defending the crops, consumers, and the environment. It is recommended that users become more proficient through specific training programs, especially for the use of potentially dangerous pesticides; licensed applicators and/or professional sprayers are essential for the safe use of certain products, such as fumigants and soil pesticides. Training programs should help workers to understand the psychological hazard of pesticides, adopt appropriate work practices, adequately use their safety equipment, practice personal health and hygiene measures, realize initial symptoms of overexposure or poisoning, and get first aid as quickly as possible. A number of employees should be trained to identify initial overexposure and poisoning symptoms, provide first aid, and recommend medical attention. In addition, these staff must be trained in life-saving strategies as well as first-aid measures. The company or the managerial staff of the health surveillance program should coordinate a medical emergency plan to cover emergency situations due to pesticide exposures. Education as indicated by labels and safety data sheets on safe transportation and handling is a measure of prime importance to eliminate accidental exposure of handlers and bystanders. It is also recommended that the empty containers be disposed of properly.

15.21 National Plans for Prevention of Pesticide Risk

A national pesticide risk management plan must be constructed in each country and reviewed periodically by the national authority. Society and workers protection needs to be a foundational mandate for governments and public health officials around the world. A model organizational scheme that recognizes roles and responsibilities in coordinating actions at the central, regional, and local levels is explored here.

15.21.1 Central Level

A nationalized plant-product protection authority, with regulatory and coordinating functions, should be established in each country. This nationalized authority has specific tasks:

- To enhance regional pesticide-related regulations;
- Coordinating the authorization process for the national registration of plant product protection;
  Assessing the outcomes of surveillance activities in the country frequently:
  - Monitoring sales and uses of pesticides;
  - Monitoring the effects of pesticides on humans and ecosystems; and
  - Surveillance of environmental contamination: water, food, soil, and air residues.

Coordination is required between health, agriculture, environment, and labor departments of the government, so that all aspects related to pesticide products used in agriculture and public health can be handled jointly at national level. To this end, it is crucial that all the government departments directly involved in the aforementioned tasks be described for plant product protection in the national authority.

15.21.2 Regional/Province Level

There needs to be a channel of public health services for every country. Such an infrastructure in local health units (LHUs) should be organized at local level (every 50 000–100 000 inhabitants). Each LHU should include a Prevention Department which guarantees the following range of services:

- Public health services;
- Environmental sanitation service;
- Occupational health and safety services; and
- Veterinary medical service.

These services will oversee and manage all competency issues including pesticides. In rural areas, safety-related activities associated with pesticides can be among the most influential for Occupational Health and Environmental Hygiene services. These services should have the major objectives:

- Promoting and coordinating territorial-competence preventive activities;
- To collect data on pesticide sales and/or use, water and food residues of pesticides, and epidemiological data (severe poisoning of pesticides, prolonged effect);
- Promoting sound first-aid organization and medication of poisoned workers;
- Promoting and carrying out training programs for users, physicians, and health and safety personnel and informing and educating the general public.

Public Health Laboratories should support the activity of the LHUs, equipped with technical and human resources for measuring pesticide residues in biological and environmental matrices. Hospitals or specialized ambulatory facilities should provide clinical activities for the diagnosis of occupational diseases and for workers' medical supervision. A poison control center for the prevention, diagnosis, and the treatment of intense pesticide poisoning should be available at regional or national level.

15.22 Encapsulation of Chemical Nanopesticides

Inventive encapsulated nanopesticides are formulations that comprise the active ingredient that is encircled by wall material, forming a capsule in a core-shell or matrix. Before being used, the CS product that contains the capsule of nanopesticides is usually diluted with water. The active ingredient is discharged from the capsule after application as a spray. Typically these capsule particulates are in a submicron or reduced micron-size scale (around 0.05–20 μm in diameter), which significantly increases the specific surface area (i.e. total surface area per unit particle volume) of these nanopesticides and provides superior characteristics compared to the same active ingredient in the conventional form. This small scale size enhances the rate of dissolution of the active ingredient and improves its distribution, leading to better biological activity. Moreover, encapsulated pesticides may have several beneficial effects, which should be noted as follows:

- Minimize the impact on nontarget organisms, since the active ingredient can only be delivered to the target organism or released under specific customized conditions. This can limit exposure to the poisonous active ingredient of nontarget organisms [96], minimizing the detrimental effects on certain organisms including crop phytotoxicity.
- Improve compatibility and stability within the formulation system as well as enriched compatibility of the tank mix, since the active ingredients may not interact with the external shell components. This feature enables mixed formulations with other active ingredients that are incompatible [97].
- Preserve the integrity of bio-based ingredients (microbes and biomolecules) to avoid any contact with harmful product components.
- Protect unstable active ingredients, from degradation such as photolysis and hydrolysis [98, 99].
- Improve product rain fastness, i.e. granules can adhere to the surface of leaves to prevent the active ingredients being wiped off by rain and watering [100–102].
- Improve pesticide residual activity, extending the effectiveness of the ingredient that may postpone or even minimize consequent pesticide application [103, 104].
- Reduce the volatility rate of high vapor pressure active ingredients such as clomazone, dicamba, and pesticides based on essential oils extracted from botanical extracts and lower the impact of drift and loss of ingredients [105, 106].

Subsequently, active ingredient encapsulation can provide a sustained or delayed release, combining quickly with slow release rates and transferring liquid ingredients into solid particles. Thus, a capsule formulation technology leads to a relatively safe formulation for handling and environment, in addition to boosting the efficacy of active ingredients reducing the required quantity of the conventional analogue product applied to crops. New pesticide products have been inaugurated, which include combo formulation of not only chemical ingredients but also microorganisms, primarily for seed treatment applications. Indeed, there is current pressure to substitute chemicals as active ingredients for natural and harmless components with microorganisms and other bio-based molecules. Encapsulation technology shows some downsides and difficulties which should be recognized and addressed at an industrial scale when designing new commercial products. The major obstacles are the high cost of production and the complexity of processes. It is necessary to consider not only the manufacturing of the encapsulated product in a way that ensures reliable quality at a competitive cost but also the supply chain for industrialization and marketing. One of the challenges is to find an asset with a sufficiently large scale and capability to manufacture the laboratory-developed encapsulation technology. One way to offset investments in assets and new equipment is to use the developed encapsulation technologies in various applications or with multiple active ingredients, keeping industrial production at maximum capacity as much as possible. Companies often manufacture third-party-encapsulated product. It is not uncommon when businesses buy another business that has expertise and know-how to devise their expertise on encapsulation method. Besides this, the regulatory question for the registration of the commercial product in authority agencies should be addressed, in particular for nanosized capsules of the dimension of 100 nm or less where the impacts of nanosized matter are not yet well understood [107, 108]. Apart from these factors, the durability of the formulated substance containing capsules can also pose a long-term storage challenge and should avoid the accumulation of capsules and/or the differentiation of phases. Distribution of the particle size of the capsule can also cause clogging of the nozzle during spray operation. Such normal problems are often encountered during traditional manufacturing, and the consistency of the substance should be assessed.

15.23 Encapsulated Pesticides: Market Value

Encapsulation technology, in addition to regulatory pressure against chemical pesticides, is a very significant demand for crop protection sector with rising acceptance by farmers looking for advanced pest control. The global demand for encapsulated pesticides is projected to cross US$ 800 million in the forecast timeframe from 2019 to 2025 at a CAGR of 11.8% by 2025. Among the various types of pesticides (insecticides, herbicides, fungicides, and others), the category of encapsulated insecticides was the most significant with a share of sales in 2017 of 42.3%. Provided that insecticides typically contain more toxic active ingredients, the need for safer and innocuous goods with superior efficiency propelled the production of encapsulated insecticides and their marketing. Following these rising demands, over the projected period, herbicides are predicted to rise at an accelerated CAGR of 13.9%. Because of the methods utilized, encapsulation also has the potential to take any market share of emulsifiable concentrate (EC) formulations, which in 2017 was about 31% in volume. Europe stood out in the encapsulated pesticides industry in 2017 with a share of 44.9% in terms of area. ADAMA, Arysta LifeScience Corporation, BASF, Bayer, BotanoCap, Belchim Crop Safety, Syngenta, FMC, GAT Microencapsulation, reed pacific, Ecosafe Natural Products Inc., and Monsanto Group, among others, are the major players in the global microencapsulated pesticides industry. Such companies have shown a relentless commitment in the implementation of technologies, product development, and creativity to produce more successful goods in the market. Some companies have introduced strategies for mergers, acquisitions, amalgamations, alliances, and advertising campaigns to speed up product launch [109].

References

1 Oerke, E.-C., Dehne, H.W., Schönbeck, F., and Weber, A. (2012). *Crop Production and Crop Protection: Estimated Losses in Major Food and Cash Crops*. Elsevier.

2 Davis, J.R., Brownson, R.C., and Garcia, R. (1992). Family pesticide use in the home, garden, orchard, and yard. *Archives of Environmental Contamination and Toxicology* **22** (3): 260–266.

3 Koh, D. and Jeyaratnam, J. (1996). Pesticides hazards in developing countries. *Science of the Total Environment* **188** (1): S78.

4 Metcalf, R.L. (1987). Benefit/risk considerations in the use of pesticides. *Agriculture and Human Values* **4** (4): 15–25.

5 Pimentel, D. (2005). Environmental and economic costs of the application of pesticides primarily in the United States. *Environment, Development and Sustainability* **7** (2): 229–252.

6 van der Werf, H.M. (1996). Assessing the impact of pesticides on the environment. *Agriculture, Ecosystems & Environment* **60** (2-3): 81–96.

7 Woodruff, T.J., Kyle, A.D., and Bois, F.Y. (1994). Evaluating health risks from occupational exposure to pesticides and the regulatory response. *Environmental Health Perspectives* **102** (12): 1088–1096.

8 Maroni, M. and Fait, A. (1993). Health effects in man from long-term exposure to pesticides. A review of the 1975-1991 literature. *Toxicology* **78** (1-3): 1–180.

9 Agrawal, S. and Rathore, P. (2014). Nanotechnology pros and cons to agriculture: a review. *International Journal of Current Microbiology and Applied Sciences* **3** (3): 43–55.

10 Chen, H. and Yada, R. (2011). Nanotechnologies in agriculture: new tools for sustainable development. *Trends in Food Science & Technology* **22** (11): 585–594.

11 Chowdappa, P. and Gowda, S. (2013). Nanotechnology in crop protection: status and scope. *Pest Management in Horticultural Ecosystems* **19** (2): 131–151.

12 Kah, M., Beulke, S., Tiede, K., and Hofmann, T. (2013). Nanopesticides: state of knowledge, environmental fate, and exposure modeling. *Critical Reviews in Environmental Science and Technology*.

13 Kah, M. and Hofmann, T. (2014). Nanopesticide research: current trends and future priorities. *Environment International* **63**: 224–235.

14 Sasson, Y., Levy-Ruso, G., Toledano, O., and Ishaaya, I. (2007). Nanosuspensions: emerging novel agrochemical formulations. In: *Insecticides Design Using Advanced Technologies*, 1–39. Springer.

15 Sekhon, B.S. (2014). Nanotechnology in agri-food production: an overview. *Nanotechnology, Science and Applications* **7**: 31.

16 Madhuban, G., Rajesh, K., and Arunava, G. (2012). Nano-pesticides-a recent approach for pest control. *The Journal of Plant Protection Sciences* **4** (2): 1–7.

17 Borm, P.J., Robbins, D., Haubold, S. et al. (2006). The potential risks of nanomaterials: a review carried out for ECETOC. *Particle and Fibre Toxicology* **3** (1): 11.

18 Dolez, P.I. (2015). *Nanoengineering: Global Approaches to Health and Safety Issues*. Elsevier.

19 Frohlich, E. (2013). Cellular targets and mechanisms in the cytotoxic action of non-biodegradable engineered nanoparticles. *Current Drug Metabolism* **14** (9): 976–988.

20 Hussain, S., Hess, K.L., Gearhart, J.M. et al. (2005). in vitro toxicity of nanoparticles in BRL 3A rat liver cells. *Toxicology in vitro* **19** (7): 975–983.

21 Unsworth, J.B., Corsi, C., Van Emon, J.M. et al. (2016). Developing global leaders for research, regulation, and stewardship of crop protection chemistry in the 21st century. *Journal of Agricultural and Food Chemistry* **64** (1): 52–60.

22 García, M., Forbe, T., and Gonzalez, E. (2010). Potential applications of nanotechnology in the agro-food sector. *Food Science and Technology* **30** (3): 573–581.

23 Jampílek, J. and Kráľová, K. (2015). Application of nanotechnology in agriculture and food industry, its prospects and risks. *Ecological Chemistry and Engineering S* **22** (3): 321–361.

24 Pérez-de-Luque, A. and Hermosín, M.C. (2013). Nanotechnology and its use in agriculture. In: *Bio-Nanotechnology: A Revolution in Food, Biomedical and Health Sciences* (ed. D. Bagchi), 383–398. United States: Wiley-Blackwell.

25 Prasad, R., Kumar, V., and Prasad, K.S. (2014). Nanotechnology in sustainable agriculture: present concerns and future aspects. *African Journal of Biotechnology* **13** (6): 705–713.

26 Sonkaria, S., Ahn, S.-H., and Khare, V. (2012). Nanotechnology and its impact on food and nutrition: a review. *Recent Patents on Food, Nutrition & Agriculture* **4** (1): 8–18.

27 Chen, H., Seiber, J.N., and Hotze, M. (2014). ACS select on nanotechnology in food and agriculture: a perspective on implications and applications. *ACS Publications* 62 (6): 1209–1212.

28 Coles, D. and Frewer, L.J. (2013). Nanotechnology applied to European food production–A review of ethical and regulatory issues. *Trends in Food Science & Technology* **34** (1): 32–43.

29 Kuzma, J. and VerHage, P. (2006). Nanotechnology in agriculture and food production: anticipated applications. *Project on Emerging Nanotechnologies*: 1–44.

30 Raliya, R., Tarafdar, J.C., Gulecha, K. et al. (2013). Scope of nanoscience and nanotechnology in agriculture. *Journal of Applied Biology and Biotechnology* **1** (3): 041–044.

31 Thornton, P.K. (2010). Livestock production: recent trends, future prospects. *Transactions of the Royal Society B* 365: 2853–2867.

32 Zhao, M., Liu, L., Her, R. et al. (2012). Nano-sized delivery for agricultural chemicals. In: *NanoFormulation*, 256–265. Cambridge: Royal Society of Chemistry.

33 Neethirajan, S. and Jayas, D.S. (2011). Nanotechnology for the food and bioprocessing industries. *Food and Bioprocess Technology* **4** (1): 39–47.

34 Ragaei, M. and Sabry, A.-k.H. (2014). Nanotechnology for insect pest control. *International Journal of Science, Environment and Technology* **3** (2): 528–545.

35 Bhattacharyya, A., Bhaumik, A., Rani, P.U. et al. (2010). Nano-particles-A recent approach to insect pest control. *African Journal of Biotechnology* **9** (24): 3489–3493.

36 Rai, M. and Ingle, A. (2012). Role of nanotechnology in agriculture with special reference to management of insect pests. *Applied Microbiology and Biotechnology* **94** (2): 287–293.

37 de Oliveira, J.L., Campos, E.V., Bakshi, M. et al. (2014). Application of nanotechnology for the encapsulation of botanical insecticides for sustainable agriculture: prospects and promises. *Biotechnology Advances* **32** (8): 1550–1561.

38 Anjali, C., Khan, S.S., Margulis-Goshen, K. et al. (2010). Formulation of water-dispersible nanopermethrin for larvicidal applications. *Ecotoxicology and Environmental Safety* **73** (8): 1932–1936.

39 Bhagat, D., Samanta, S.K., and Bhattacharya, S. (2013). Efficient management of fruit pests by pheromone nanogels. *Scientific Reports* **3**: 1294.

40 Forim, M.R., Costa, E.S., Da Silva, M.F. et al. (2013). Development of a new method to prepare nano-/microparticles loaded with extracts of Azadirachta indica, their characterization and use in controlling *Plutella xylostella*. *Journal of Agricultural and Food Chemistry* **61** (38): 9131–9139.

41 Lyons, K., Scrinis, G., and Whelan, J. (2011). Nanotechnology, agriculture, and food. *Nanotechnology and Global Sustainability* **117**.

42 Scrinis, G. and Lyons, K. (2007). The emerging nano-corporate paradigm: nanotechnology and the transformation of nature, food and agri-food systems. *International Journal of Sociology of Agriculture and Food* **15** (2): 22–44.

43 Lyons, K. and Scrinis, G. (2009). Under the regulatory radar? Nanotechnologies and their impacts for rural Australia. In: *Tracking Rural Change* (eds. F. Merlan and D. Raftery), 151. Canberra: ANU E Press.

44 Ali, M.A., Rehman, I., Iqbal, A. et al. (2014). Nanotechnology, a new frontier in agriculture. *Advancements in Life Sciences* **1** (3): 129–138.

45 Nair, R., Varghese, S.H., Nair, B.G. et al. (2010). Nanoparticulate material delivery to plants. *Plant Science* **179** (3): 154–163.

46 Pérez-de-Luque, A. and Rubiales, D. (2009). Nanotechnology for parasitic plant control. *Pest Management Science: Formerly Pesticide Science* **65** (5): 540–545.

47 Pereira, A.E., Grillo, R., Mello, N.F. et al. (2014). Application of poly (epsilon-caprolactone) nanoparticles containing atrazine herbicide as an alternative technique to control weeds and reduce damage to the environment. *Journal of Hazardous Materials* **268**: 207–215.

48 Jiang, L.C., Basri, M., Omar, D. et al. (2012). Green nano-emulsion intervention for water-soluble glyphosate isopropylamine (IPA) formulations in controlling *Eleusine indica* (*E. indica*). *Pesticide Biochemistry and Physiology* **102** (1): 19–29.

49 Guo, Y., Yang, Q., Yan, W. et al. (2014). Controlled release of acetochlor from poly (butyl methacrylate-diacetone acrylamide) based formulation prepared by nanoemulsion polymerisation method and evaluation of the efficacy. *International Journal of Environmental Analytical Chemistry* **94** (10): 1001–1012.

50 McGrath, M. (2004). What are fungicides. *The Plant Health Instructor* https://doi.org/10.1094/PHI-I-2004-0825-01.

51 Yadav, A. and Aggarwal, A. (2014). Effect of benomyl and dithane M-45 on mycoflora, mycorrhizae and growth enhancement of *Helianthus annuus L. New York Science Journal* **7** (11): 7–15.

52 Sarlak, N., Taherifar, A., and Salehi, F. (2014). Synthesis of nanopesticides by encapsulating pesticide nanoparticles using functionalized carbon nanotubes and application of new nanocomposite for plant disease treatment. *Journal of Agricultural and Food Chemistry* **62** (21): 4833–4838.

53 Kumar, R., Nair, K.K., Alam, M.I. et al. (2015). Development and quality control of nanohexaconazole as an effective fungicide and its biosafety studies on soil nitifiers. *Journal of Nanoscience and Nanotechnology* 15: 1350–1356.

54 Kumar, R.S., Shiny, P.J., Anjali, C.H. et al. (2013). Distinctive effects of nano-sized permethrin in the environment. *Environmental Science and Pollution Research* **20** (4): 2593–2602.

55 Kanhed, P., Birla, S., Gaikwad, S. et al. (2014). in vitro antifungal efficacy of copper nanoparticles against selected crop pathogenic fungi. *Materials Letters* **115**: 13–17.

56 Anjali, C.H., Sharma, Y., Mukherjee, A., and Chandrasekaran, N. (2012). Neem oil (*Azadirachta indica*) nanoemulsion—a potent larvicidal agent against *Culex quinquefasciatus*. *Pest Management Science* 68 (2): 158–163.

57 Piola, L., Fuchs, J., Oneto, M.L. et al. (2013). Comparative toxicity of two glyphosate-based formulations to *Eisenia andrei* under laboratory conditions. *Chemosphere* **91** (4): 545–551.

58 Nguyen, H., Hwang, I.C., Park, J.W., and Park, H.J. (2012a). Enhanced payload and photo-protection for pesticides using nanostructured lipid carriers with corn oil as liquid lipid. *Journal of Microencapsulation* **29** (6): 596–604.

59 Nguyen, H.M., Hwang, I.C., Park, J.W., and Park, H.J. (2012b). Photoprotection for deltamethrin using chitosan-coated beeswax solid lipid nanoparticles. *Pest Management Science* **68** (7): 1062–1068.

60 Yin, Y.-h., Guo, Q.-m., Han, Y. et al. (2012). Preparation, characterization and nematicidal activity of lansiumamide B nano-capsules. *Journal of Integrative Agriculture* **11** (7): 1151–1158.

61 Brunel, F., El Gueddari, N.E., and Moerschbacher, B.M. (2013). Complexation of copper (II) with chitosan nanogels: toward control of microbial growth. *Carbohydrate Polymers* **92** (2): 1348–1356.

62 Persano, L., Camposeo, A., Tekmen, C., and Pisignano, D. (2013). Industrial upscaling of electrospinning and applications of polymer nanofibers: a review. *Macromolecular Materials and Engineering* **298** (5): 504–520.

63 Xiang, C., Taylor, A.G., Hinestroza, J.P., and Frey, M.W. (2013). Controlled release of nonionic compounds from poly (lactic acid)/cellulose nanocrystal nanocomposite fibers. *Journal of Applied Polymer Science* **127** (1): 79–86.

64 Hellmann, C., Greiner, A., and Wendorff, J.H. (2011). Design of pheromone releasing nanofibers for plant protection. *Polymers for Advanced Technologies* **22** (4): 407–413.

65 Pardeshi, C., Rajput, P., Belgamwar, V. et al. (2012). Solid lipid based nanocarriers: an overview. *Acta Pharmaceutica* **62** (4): 433–472.

66 Bang, S., Yu, Y.M., Hwang, I.C., and Park, H.J. (2009). Formation of size-controlled nano carrier systems by self-assembly. *Journal of Microencapsulation* **26** (8): 722–733.

67 Hwang, I., Kim, T., Bang, S. et al. (2011). Insecticidal effect of controlled release formulations of etofenprox based on nano-bio technique. *Journal of the Faculty of Agriculture, Kyushu University* **56** (1): 33–40.

68 Kang, M.A., Seo, M.J., Hwang, I.C. et al. (2012). Insecticidal activity and feeding behavior of the green peach aphid, *Myzus persicae*, after treatment with nano types of pyrifluquinazon. *Journal of Asia-Pacific Entomology* **15** (4): 533–541.

69 Kah, M. (2015). Nanopesticides and nanofertilizers: emerging contaminants or opportunities for risk mitigation? *Frontiers in Chemistry* **3**: 64.

70 Ávila, D.S., Garcia, S.C., Arbo, M.D. et al. (2018). Use of nanomaterials in agriculture: potential benefits and challenges. In: *Emerging Trends in Agri-Nanotechnology: Fundamental and Applied Aspects* (eds. H.B. Singh, S. Mishra, L.F. Fraceto and R. de Lima), 10. https://doi.org/10.1079/9781786391445.0010.

71 Nicolopoulou-Stamati, P., Maipas, S., Kotampasi, C. et al. (2016). Chemical pesticides and human health: the urgent need for a new concept in agriculture. *Frontiers in Public Health* **4**: 148.

72 Anselmo, A.C. and Mitragotri, S. (2015). A review of clinical translation of inorganic nanoparticles. *The AAPS Journal* **17** (5): 1041–1054.

73 Namasivayam, K.R.S., Aruna, A., and Gokila (2014). Evaluation of silver nanoparticles-chitosan encapsulated synthetic herbicide paraquat (AgNp-CS-PQ) preparation for the controlled release and improved herbicidal activity against *Eichhornia crassipes*. *Research Journal of Biotechnology* **9** (9): 19–27.

74 Simonin, M., Colman, B.P., Tang, W. et al. (2018). Plant and microbial responses to repeated Cu (OH) 2 nanopesticide exposures under different fertilization levels in an agro-ecosystem. *Frontiers in Microbiology* **9**: 1769.

75 Zhao, L., Huang, Y., Adeleye, A.S., and Keller, A.A. (2017). Metabolomics reveals Cu (OH) 2 nanopesticide-activated anti-oxidative pathways and decreased beneficial antioxidants in spinach leaves. *Environmental Science & Technology* **51** (17): 10184–10194.

76 Zhao, L., Ortiz, C., Adeleye, A.S. et al. (2016). Metabolomics to detect response of lettuce (Lactuca sativa) to Cu (OH) 2 nanopesticides: oxidative stress response and detoxification mechanisms. *Environmental Science & Technology* **50** (17): 9697–9707.

77 Aksakal, F.I. and Arslan, H. (2020). Detoxification and reproductive system-related gene expression following exposure to Cu (OH) 2 nanopesticide in water flea (*Daphnia magna* Straus 1820). *Environmental Science and Pollution Research* **27** (6): 6103–6111.

78 Zhang, X., Xu, Z., Wu, M. et al. (2019). Potential environmental risks of nanopesticides: application of Cu (OH) 2 nanopesticides to soil mitigates the degradation of neonicotinoid thiacloprid. *Environment International* **129**: 42–50.

79 Shang, Y., Hasan, M., Ahammed, G.J. et al. (2019). Applications of nanotechnology in plant growth and crop protection: a review. *Molecules* **24** (14): 2558.

80 Naseri, N., Valizadeh, H., and Zakeri-Milani, P. (2015). Solid lipid nanoparticles and nanostructured lipid carriers: structure, preparation and application. *Advanced Pharmaceutical Bulletin* **5** (3): 305.

81 Shen, Y., Zhu, H., Cui, J. et al. (2018). Construction of lambda-cyhalothrin nano-delivery system with a high loading content and controlled-release property. *Nanomaterials* **8** (12): 1016.

82 de Oliveira, J.L., Campos, E.V., Goncalves da Silva, C.M. et al. (2015). Solid lipid nanoparticles co-loaded with simazine and atrazine: preparation, characterization, and evaluation of herbicidal activity. *Journal of Agricultural and Food Chemistry* **63** (2): 422–432.

83 Jacques, M.T., Oliveira, J.L., Campos, E.V. et al. (2017). Safety assessment of nanopesticides using the roundworm *Caenorhabditis elegans*. *Ecotoxicology and Environmental Safety* **139**: 245–253.

84 Gomes, S.I., Scott-Fordsmand, J.J., Campos, E.V. et al. (2019). On the safety of nanoformulations to non-target soil invertebrates–an atrazine case study. *Environmental Science: Nano* **6** (6): 1950–1958.

85 Mohd Firdaus, M.A., Agatz, A., Hodson, M.E. et al. (2018). Fate, uptake, and distribution of nanoencapsulated pesticides in soil–earthworm systems and implications for environmental risk assessment. *Environmental Toxicology and Chemistry* **37** (5): 1420–1429.

86 Clemente, Z., Grillo, R., Jonsson, M. et al. (2014). Ecotoxicological evaluation of poly (ε-caprolactone) nanocapsules containing triazine herbicides. *Journal of Nanoscience and Nanotechnology* **14** (7): 4911–4917.

87 Grillo, R., dos Santos, N.Z., Maruyama, C.R. et al. (2012). Poly (ε-caprolactone) nanocapsules as carrier systems for herbicides: physico-chemical characterization and genotoxicity evaluation. *Journal of Hazardous Materials* **231**: 1–9.

88 de Andrade, L.L., Santo Pereira, A.D., Fraceto, L.F., and dos Reis Martinez, C.B. (2019). Can atrazine loaded nanocapsules reduce the toxic effects of this herbicide on the fish *Prochilodus lineatus*? A multibiomarker approach. *Science of the Total Environment* **663**: 548–559.

89 Oliveira, J.L.d., Campos, E.V., Pereira, A.E. et al. (2018). Zein nanoparticles as eco-friendly carrier systems for botanical repellents aiming sustainable agriculture. *Journal of Agricultural and Food Chemistry* **66** (6): 1330–1340.

90 Pascoli, M., Jacques, M.T., Agarrayua, D.A. et al. (2019). Neem oil based nanopesticide as an environmentally-friendly formulation for applications in sustainable agriculture: an ecotoxicological perspective. *Science of the Total Environment* **677**: 57–67.

91 Özel, R.E., Wallace, K.N., and Andreescu, S. (2011). Chitosan coated carbon fiber microelectrode for selective in vivo detection of neurotransmitters in live zebrafish embryos. *Analytica Chimica Acta* **695** (1-2): 89–95.

92 Lotti, M. (1995). Mechanisms of toxicity and risk assessment. *Toxicology Letters* **77** (1–3): 9–14.

93 Curry, P.B., Iyengar, S., Maloney, P.A., and Maroni, M. (eds.) (2013). *Methods of Pesticide Exposure Assessment*, vol. 19. Springer Science & Business Media.

94 World Health Organization (1996). *Biological Monitoring of Chemical Exposure in the Workplace: Guidelines*. World Health Organization.

95 Tordoir, W.F. and Maroni, M. (1994). Basic concepts in occupational health management of pesticide workers. *Toxicology* 91: 5–14.

96 Edly, K. and Cottle, M.R. (2018). Microcapsules encapsulating lambda-cyhalothrin. PCT Application WO 2018/001895 A1, 4 January 2018.

97 Zedda, A., Ferri, G., and Sala, M. (2004). Preparation of sterically hindered amine ethers. Google Patents.

98 Braig, A., Leppard, D.G., and Wendeborn, F. (2008). Synergistic UV absorber combination. Google Patents.

99 Scher, H.B., Chen J. (2012)., Mazeaud I Microcapsules, U. Patent, Editor.

100 Bosch, H.W., Cooper, E.R., and McGurk, S.L. (2010). Bioadhesive nanoparticulate compositions having cationic surface stabilizers. Google Patents.

101 Phillips, F. and Gillham, E. (1973). Al comparison of sticker performance against rainwashing of microcapsules on leaf surfaces. *Pesticide Science* **4** (1): 51–57.

102 Quong D, Nielson K. (2003). Adherent microcapsules.

103 Preisler, A.C., Pereira, A.E., Campos, E.V. et al. (2020). Atrazine nanoencapsulation improves pre-emergence herbicidal activity against *Bidens pilosa* without enhancing long-term residual effect on Glycine max. *Pest Management Science* **76** (1): 141–149.

104 Wilson, S.L. (2010). Microencapsulated insecticide with enhanced residual activity. Google Patents.

105 Markus, A. and Linder, C. (2015). Encapsulated essential oils. US Patent 9,079,152, 14 Jul 2015.

106 Urch, H., Franz, D., Nolte, M. et al. (2018). Microcapsules comprising cinmethylin in the core and a polyurea derived from diphenylmethane diisocyanate or an oligomer thereof.

107 Europea, C. (2006). The appropriateness of existing methodologies to assess the potential risks associated with engineered and adventitious products of nanotechnologies. European Commission.

108 McNaught, A.D. and Wilkinson, A. (1997). *Compendium of Chemical Terminology*, vol. 1669. Blackwell Science Oxford.

109 Fernandes Fraceto, L., de Castro, V.L.S.S., Grillo, R. et al. (eds.) (2020). *Nanopesticides: From Research and Development to Mechanisms of Action and Sustainable Use in Agriculture*. Switzerland AG: Springer.

16

Eco-friendly Natural Polymers-based Nanotechnology

Twishi Puri[1] *and Yashwant Pathak*[2,3]

[1] *Department of Biology, Worcester State University, Worcester, MA, USA*
[2] *Taneja College of Pharmacy, University of South Florida, Tampa, FL, USA*
[3] *Faculty of Pharmacy, Airlangga University, Surabaya, Indonesia*

16.1 Introduction

Drug delivery is perhaps one of the most important aspects of modern medicine since most of the medical treatments depend on the effective delivery of drugs to the target areas. Previously, drug delivery systems were based on synthetic polymers. Recently, however, the research on drug delivery has shifted its focus from conventional methods to biodegradable and biocompatible polymers. This is because natural polymers have various advantages over synthetic polymers. Such advantages include their intrinsic property of environmental responsiveness, no toxicity, and the ability to be easily modified in or incorporated in oral delivery and bolus matrix delivery systems. Additionally, nanoscale polymer particles have successfully been applied to oral and intramuscular delivery systems for nonviral gene therapy [1].

There are several advantages of polymeric nanoparticles in terms of their use in drug delivery. For instance, they have the ability to increase the stability of volatile drugs, can improve the efficiency and effectiveness of the traditional oral and intravenous methods of drug administration, and can deliver a higher concentration of drug to the target location [2]. Additionally, polymers can also be used to encapsulate biodegradable particles that can act as drug reservoirs. These drug reservoirs can then be used to deliver the desired materials, such as drugs, genes, and vaccines, to the target location. It can also be administered to obtain a local therapeutic effect [3–5].

Henceforth, this chapter focuses on few possible candidates of natural polymers and their properties that are useful for successful drug delivery systems. Additionally, the various important preparation methods of nanomaterial-based polymers are also discussed.

16.2 Natural Polymers

Natural polymers, due to their sustainability, eco-friendly nature, nontoxicity, biodegradability, and biocompatibility, are known as excellent hosting materials for nanoparticles [6]. Any polymer that is used in the process, regardless of their identity, has to meet the following requirements: be able to adapt to the body, should exhibit nontoxic and nonantigenicity performance, and should be biodegradable and biocompatible [7]. Using natural polymers as drug carriers is a challenge because their broad muscular weight distributions and batch-to-batch variability can hinder the process or inhibit the system from successfully delivering the drug [8, 9]. Some of the polymers discussed below work well together with each other. For example, there has been extensive research on chitosan–alginate and its use in targeted drug delivery. The combination has proven to be satisfactory due to chitosan's versatility and alginate's acid-resistant property [10–12]. Polymers, other than chitosan, discussed below have not been explored as extensively as chitosan and alginate. Even though they have similar properties to chitosan and alginate, these polymers

Sustainable Nanotechnology: Strategies, Products, and Applications, First Edition.
Edited by Yashwant V. Pathak, Govindan Parayil, and Jayvadan K. Patel.

have not gone further postclinical trials [13]. This provides opportunity for further research on the incorporation of these polymers in the drug delivery systems.

16.2.1 Chitosan

Chitosan, a bioactive polymer, structurally is a polysaccharide composed of linear β-(1→4)-linked monosaccharides. It is obtained from deacetylation of chitin, a biocompatible polysaccharide [14]. It is closely related to cellulose in structure; however, unlike cellulose, chitosan is composed of 2-amino-2-deoxy-β-D-glucan, which are bonded through glycosidic linkages. The amine group in its composition makes chitosan a useful candidate in pharmaceutical applications, especially drug delivery [15]. By nature, chitosan is a cationic polysaccharide in neutral or basic pH conditions, which provides it the ability to interact with anionic components found on the cell surface such as macromolecules and nucleic acids [16–18]. Additionally, chitosan's ability of biocompatibility with living tissue makes it an excellent molecule for various medical applications. It does not cause any allergic reaction nor is it prone to rejection by the body because it is able to break down into amino sugars which the human body can absorb completely [19].

In terms of developing polymer nanocomposites, chitosan's ability to control the release of drugs, solubility in aqueous acidic solution, presence of free amine groups which are ready for cross-linking, cationic nature, and mucoadhesive character, among many other characteristics, make it a very useful polymer [19]. Most chitosan nanoparticles are prepared in aqueous environment through methods of ionic gelation and complex coacervation, due to their ability to stabilize drugs [20]. In recent research, various modified forms of chitosan have been developed, which include cisplatin-mediated polyampholytic chitosan [21] and folate- and heptamethine-based modification of chitosan for tumor-specific imaging and photodynamic therapy [22]. In terms of drug delivery, modified chitosan has been used for various application. For example, chitosan derivatives have been used for insulin delivery [23]. Additionally, antibody-modified chitosan nanoparticles have been developed to deliver drugs to the brain, an organ to which delivering drugs has never been easy due to the blood bank barrier not permitting cellular uptake to the brain [24].

A particular research investigating chitosan nanoparticles as tumor-targeted carrier for dextran–doxorubicin (DOX) conjugate has shown promising results. Alone, the cancer therapy drug DOX can have many side effects; however, dextran–DOX conjugate has shown to minimize the side effects, and containment of this molecule in a chitosan nanoparticle has shown to increase the success of drug delivery [25]. A similar research done on the use of chitosan to encapsulate DOX and its analogue (*N*-(trifluoroacetyl) DOX) tested three chitosans with different molecular weights (15, 100, and 200 kDa). Out of the three, chitosans with molecular weight of 100 kDa had the highest ability to bind to DOX and its analogue and, hence, is a more suitable carrier for delivery [26]. Chitosan nanoparticles have certainly been a promising development in the field of drug delivery. Furthermore, chitosan's low toxicity, antimicrobial activity, and low immunogenicity have expanded the range of its application in medicine [27].

16.2.2 Gelatin

Gelatin, a natural biopolymer, has wide variety of applications in food, cosmetic, medical, and pharmaceutical industry [28]. It is collected from hydrolysis of collagen, a major component of skin, bones, and connective tissue [29]. Two types of gelatin are produced based on how their source of collagen is treated. Type A gelatin requires collagen that has gone through acid treatment. On the other hand, type B gelatin requires collagen that has been treated with alkaline [30]. Due to its biocompatible, biodegradable, and nontoxic characteristics, gelatin is considered as a potential candidate for preparing colloidal drug delivery systems [29]. Gelatin, on a molecular level, is composed mainly of amino acids glycine, proline, and alanine. The presence of these and many more functional groups gives the molecule the advantage of having multiple functions and being modifiable, both of which can be important when creating drug carriers [31]. Some examples of gelatin modification through the means of chemical derivatization are chitosan-conjugated gelatin [32], poly(D,L-lactide)-grafted gelatin [33], and poly(ethylene glycol)-modified gelatin [31]. All of these have pharmaceutical applications.

For preparation of gelatin particles for nanocomposites, methods such as desolvation can be used. In this method, a desolving agent is added to an aqueous gelatin solution, which dehydrates the gelatin molecules, causing a change in the structure of gelatin, from stretched to coil. This results in an increase in turbidity [34].

16.2.3 Albumin

Albumin already has the reputation of being a protein carrier for drug delivery due to its nontoxic, nonimmunogenic, biocompatible, and biodegradable characteristics [35]. Its easy accessibility and ability to dissolve in water allow it to be delivered via injections, which makes it a promising candidate for nanoparticle preparation [36]. There are three sources of albumin, which results in three different albumins. Ovalbumin (OVA) is a monomeric phosphoglycoprotein consisting of 385 amino acid residues. It is obtained from the whites of eggs [37, 38]. Due to its low cost, high availability, ability to form gel networks, and stabilize emulsions and foams, it is often used in drug delivery systems. In fact, it is an ideal candidate for carrier for controlled drug release [39]. Bovine serum albumin (BSA) is obtained from bovine serum. It also has low cost, high availability, and is easily purifiable. Additionally, its ligand-binding properties, found in no other albumin, have earned it recognition in the pharmaceutical industry [40, 41]. Human serum albumin (HSA) is obtained from human serum and is often a substitute for BSA when avoidance of immunological response is required *in vivo*. It is a soluble globular monomeric protein consisting of 35 cysteinyl residues which form one sulfhydryl group and 17 disulfide brides. Compared to the other two albumin types, HSA is most abundant and has an average half-life of 19 days [42]. Similar to the previous two albumin types, its lack of toxicity and biodegradability make it a promising carrier for drug delivery systems [43, 44]. Specifically, both BSA and HSA are usually used in protein-binding studies and targeted drug delivery applications [45, 46].

Previously conducted studies have proved that albumin's contributions have increased the effectiveness of drug delivery. A major issue with nano drug delivery is the destruction or deactivation of the drug from the body before it is able to reach and affect the target site. When nanoparticles are used in the delivery system, high amounts of plasma proteins emerge and surround themselves around the system, forming a protein corona which hinders the function of the nanoparticles. With this, the results of *in vitro* are affected as much as the ones for *in vivo*. In this system, albumin was used to coat the nanoparticles, which protects the drug during its passage to the target site, hence being a useful candidate in protective delivery [47]. Albumin has also been useful in delivering drugs to the brain. Attachment of albumin to sinapic acid, a biomolecule, allows the acid to cross the blood–brain barrier. Prototype albumin nanoparticles were later developed using β-cyclodextrin-modified glycol chitosan and adamantane-conjugated HSA. The results confirmed that albumin had the ability to target tumors [48]. The results of these studies confirm that albumin has the ability to attach to solid tumors, which could be a useful characteristic in Site-targeted delivery systems for antitumor drugs [49].

16.2.4 Starch

Starch is a polysaccharide mainly found in plants such as corn, potato, rice, tapioca, wheat, sorghum, and barley. Depending on the source, the type of starch obtained can vary. Starch obtained directly from plants is known as "native starch," and starch that has been through enzymatic modification is known as "modified starch." In addition to being biodegradable, starch has the ability to dissolve in cold water due to its polymerized structure and the presence of hydrogen bonds between adjacent chains [50, 51]. It also has the ability to form inter- and intramolecular cross-linked bonds with polyanions such as sodium tripolyphosphate (STPP) [52]. The use of starch in controlled drug delivery has enhanced drug solubility and drug stability, while decreasing toxicity and side effects. Among the starches used, native starch has proved to be less beneficial due to its limitations. This results in starch modification by techniques such as oxidation, cross-linking, hydroxypropylation, and retrogradation [53–58]. An useful alternative to native starch is resistant starch, a type of starch that is not digested by the upper-gastrointestinal tract and stimulates a prebiotic effect in the colon. Fermentation of resistant starch produces short-chain fatty acids, which results in a change in intestinal pH and enhancement of the bioavailability of divalent cations. This provides resistant starch the ability of being incorporated in drug encapsulation material for colon-specific drug delivery [59].

16.3 Preparation

The methods of developing nanotechnology-based polymers play a vital role in achieving the desired properties for a particular application. Generally, polymeric nanoparticles are made from biocompatible and -degradable polymers that are approximately 10–1000 nm in size. This allows for better dissolvement, entrapment, and encapsulation of drugs [60–62]. Additionally, optimization of the physiochemical characteristics of polymeric nanoparticles is required based on the specific application. There are various methods that can be used to produce various nanoparticulate systems with the polymers mentioned above. The two main methods discussed are dispersion of preformed polymers and polymerization of monomers. Each method includes various development methods and the analytical techniques used to evaluate the performance of the polymers used.

16.3.1 Nanoparticles from Dispersion of Preformed Polymer

16.3.1.1 Solvent Evaporation

Among the first methods developed for preparing polymeric nanoparticles, solvent evaporation now has various pharmaceutical applications, especially regarding biodegradable polymers. The method involves polymer solutions to be prepared in volatile solvents. Previously, dichloromethane and chloroform were used as solutions, but studies have replaced them with ethyl acetate which has a far better toxicological profile. The drug that is intended to be delivered is then dispersed in the resulting solution, and emulsification is triggered due to the emulsifying agents present in the nonsolvent water of the polymer solution. The emulsion is converted into a nanoparticle suspension by evaporating the solvent.

Two strategies have been developed for forming emulsions, preparation of single emulsions and double emulsions. Single emulsions can be simplified to the analogy of adding oil in water. Double emulsions, on the other hand, can be described as adding water in oil and adding the resulting mixture in water. Both methods utilize high-speed homogenization or ultrasonication, superseding the evaporation stage which can be achieved by reducing the pressure of continuous magnetic stirring at room temperature. Using ultracentrifuge, the resulting solidified nanoparticles can be collected and washed with distilled water for removal of any molecules present on the surface. The resulting product is then lyophilized [63–65]. Following these methods, 200 nm poly(lactic-co-glycolic acid) (PLGA) nanoparticles were created using 1% dichloromethane as the volatile solvent and poly(vinyl alcohol) (PVA) or Span 40 as the stabilizing agent [66]. Alternatively, 60–200 nm PLGA nanoparticles were also prepared using dichloromethane and acetone in a 8 : 2 ratio as the solvent system and PVA as the stabilizing agent [67]. Both developments concluded that particle size was highly influenced by the type and concentrations of stabilizer, homogenizer speed, and polymer concentration [68]. Additionally, the characteristics of the organic and the aqueous phase, especially their temperature and viscosity, can also influence the size of the nanoparticles [69]. To further reduce the size of the particles, a high-speed homogenization or ultrasonication utilization is advised [60].

16.3.1.2 Nanoprecipitation

Nanoprecipitation, also referred to as the solvent displacement method, involves precipitation of a polymer from an organic solution and the diffusion of the organic solvent in an aqueous medium [70]. A preformed polymer, usually polylactic acid (PLA), is dissolved in a semipolar solvent, miscible in water, which results in nanosphere precipitation. The precipitate, known as the organic phase, is added to a aqueous solution, known as the aqueous phase, containing a stabilizer as a surfactant. Through rapid diffusion of the solution, polymer deposition found between the water and organic solvent forms nanoparticles [71]. A successful preparation of nanoparticles through nanoprecipitation depends on the organic phase injection rate, aqueous phase agitation rate, method of mixing both phases, phase ratio of organic to aqueous phase, and the nature and concentration of their components. The addition of surfactant as a stabilizer is an important part of nanoprecipitation because it prevents the nanoparticles from accumulating during long periods of storage. Additionally, a surfactant's characteristics and concentration can also influence the size of the resulting nanoparticles [72].

16.3.1.3 Salting Out

The salting-out technique is utilized to decrease the use of high-risk organic solvents, surfactants, and chlorinated solvents. The method is based on separation of water-soluble solvents from aqueous solutions. It involves a polymer and drug being dissolved in a water-miscible solvent, usually acetone, and the resulting solution being emulsified in an aqueous gel containing salting-out agents and colloidal stabilizer. The resulting emulsion is then diluted with an aqueous solution, usually water, to enhance the diffusion of acetone into aqueous phase, prompting the formation of nanospheres [64]. Through cross-flow filtration, the solvent and the salting-out agent are removed [73].

There are a variety of salting-out agents available generally in the electrolyte family, such as magnesium chloride, calcium chloride, and magnesium acetate. Some nonelectrolyte salting-out agents include sucrose. Additionally, various colloidal stabilizers, such as polyvinylpyrrolidone and hydroxyethyl cellulose, are used [64]. The reason for the availability of a variety of salting-out agents is that their selection is important because certain characteristics that one agent exhibits can influence the encapsulation efficiency of the drug [73]. There are several advantages of using the salting-out technique for preparing nanoparticles. Mainly, salting-out method minimizes the stress that the protein encapsulants experience [73]. Additionally, it can be used to process heat-sensitive substances because it does not require an increase in temperature in any of its stages [74].

16.3.1.4 Supercritical Fluid Technology (SCF)

Regarded as an eco-friendly approach, supercritical fluids are environmentally friendly solvents that have the reputation of producing polymeric nanoparticles with a gool level of purity. Additionally, it is regarded as being able to overcome the limitations expressed by conventional methods of nanoparticle preparation [75]. Supercritical fluids are generally defined as solvents present at a temperature above their critical temperature. During this it remains in a single phase regardless of the pressure [61]. There are two procedures developed for nanoparticle preparation using supercritical fluids, namely, rapid expansion of supercritical solution into liquid solvent (RESOLV) and rapid expansion of critical solution (RESS).

In RESS, the solute is dissolved in a supercritical fluid, such as supercritical methanol. The solution is then put through the process of rapid expansion across a capillary nozzle or orifice into ambient air. The degree of supersaturation from being dissolved in the solvent and the pressure reduction experiences in expansion produces well-dispersed and solvent-free nanoparticles. A study previously conducted using the RESS method involved the production of poly(perfluoropolyetherdiamide) droplets from CO_2 solution. It was concluded that the concentration and the degree of saturation of the polymer certainly influenced the size and the morphology of the particles that are produces by RESS [61, 76–79].

RESOLV, in theory, is similar to RESS method. In RESS, the supercritical solution was expanded into ambient air; RESOLV, however, requires the supercritical fluid to be expanded into a liquid solution. Previously, RESOLV has been used to produce poly(heptadecafluorodecyl acrylate) nanoparticles with sizes smaller than 50 nm. Results produced from both RESS and RESOLV are crucial to understand the reason behind the development of two methods utilizing supercritical fluids instead of relying on one method. The results obtained from mechanistic studies of the RESS process show that the method produces not only nano-sized particles but also micro-sized particles [76]. RESOLV, on the other hand, is able to produce primarily nanosized particle due to its capability of suppressing the growth of the particles in the expansion jet [80–82]. Regardless of the differences, both supercritical fluid technologies are environmentally friendly and utilizable in mass production. However, one drawback of the technology is its requirement for highly evolved equipment which does not make it cost effective.

16.3.2 Nanoparticles from Polymerization of Monomers

16.3.2.1 Emulsion Polymerization

Even though known as one of the fastest methods for nanoparticle preparation, emulsion polymerization is not widely used today due to the involvement of various toxic chemicals during preparation phases. Based on the characteristics of the phase being utilized, organic or aqueous, emulsion polymerization method can be divided into two categories. If an organic continuous phase is used, the monomer is dispersed into an emulsion, an inverse microemulsion, or an emulsion in which the monomer will not dissolve [64]. Previously, this method was used to produce polyacrylamide nanosphere. As previously stated, surfactants and other protective soluble polymers used to prevent aggregation of the product are considered to be toxic materials and no longer used to form the nanoparticles. Aqueous continuous phase, on the other hand, requires the monomer to be completely dissolved in an aqueous solution. Coincidently, the method eliminates the need to use toxic materials, such as surfactants, making the method more environmentally friendly. There are various ways to initiate polymerization. In aqueous continuous phase, initiation takes place when the monomer dissolved in the phase collides with an ion or radical-free initiator. Additionally, the monomer can also be converted into an initiating radical through various radiations. Formation of nanoparticles can occur before or after completion of the polymerization reaction [64, 83].

16.3.2.2 Interfacial Polymerization

Interfacial polymerization is a well-researched method for developing polymeric nanoparticles [84–86]. In this method, two reactive monomers are dissolved in continuous and dispersed phases, and the reaction occurs at the interface of the two liquids [87]. Interfacial cross-linking reactions such as polyaddition, polycondensation [88–90], and radical polymerization aid in the synthesis of hollow nanoparticles [91, 92]. Using this method, polymerization of monomers at the oil/water interface of oil-in-water microemulsion resulted in the development of capsule-like nanoparticles containing oil [83, 93, 94]. On the other hand, nanocapsules containing water can be developed using the same method while substituting water-in-oil emulsion for oil-in-water emulsion. Polymers found near the water–oil interface were precipitated to form nanocapsule shell [95].

Several aprotic solvents, such as acetone and acetonitrile, can be used to enhance nanocapsules formation. However, the use of protic solvents, such as ethanol, *n*-butanol, and isopropanol, stimulated the formation of nanocapsules as well as nanospheres [96].

16.3.2.3 Controlled/Living Radical Polymerization (C/LRP)

Radical polymerization was previously mentioned in the discussion of aqueous continuous-phase method of emulsion polymerization. This process was developed recently in comparison with the other nanoparticle preparation technologies discussed above. The reason for development of this method was the need for utilizing hydrophilic polymers as well as the requirement for eco-friendly technology. Previously, the use of radical polymerization posed limitations such as the lack of control over the molar mass, the molar mass distribution, the resulting function, and the macromolecule architecture of the polymer. These limitations are the result of fast radical–radical termination reactions which are unavoidable. The use of controlled/living radical polymerization (C/LRP) exhibits the ability to eliminate these limitations. Additionally, C/LRP can be effectively executed at the industrial scale [97–99]. The use of C/LRP has also increased the demand for environmentally friendly and chemically benign solvents, such as water and supercritical carbon dioxide. C/LRP, at the industrial scale, is performed in aqueous dispersed systems and emulsion polymerization. The main goal of the method is to control the characteristics of the polymer, which was previously not possible with radical polymerization. Polymeric nanoparticles produced via C/LRP have a precise particle size as well as a controlled size distribution, crucial characteristics for the future commercial success of this technique [100].

Some of the most successful and extensively researched methods of C/LRP are nitroxide-mediated polymerization (NMP) [101–105], atom-transfer radical polymerization (ATRP) [106–112], and reversible addition and fragmentation transfer chain polymerization (RAFT) [113–115]. The size of the nanoparticles produced through these methods is highly influenced by the characteristics of the control agent, monomer, initiator, and the type of emulsion. Temperature, however, does not influence the particle size.

16.4 Conclusion

The incorporation of nanoscience and technology in drug delivery systems has surely worked toward maximizing their efficiency and safety. With the availability of various polymers and polymeric nanoparticle development methods comes the advancement of medicine, especially drug delivery, toward a cleaner and safer path. Naturally available polymeric nanoparticles have proved to be a good candidate for controlled drug release due to the various beneficial characteristics that they possess. Additionally, the techniques used for preparing these polymeric nanoparticles have also been a success in laboratories. The development method of biodegradable polymers was dependent on the nature of the polymer, the drug being transported, and the application of the prepared nanoparticle. This allowed for a more material-specific development as well as a safe development and drug release process. The next step for this research is to move from a laboratory-led investigation toward an industry-led investigation and evaluate the progress and efficiency of the prepared polymeric nanoparticles [37, 116].

16.5 Future Trends

The availability of various natural biodegradable polymers certainly provides a large ground for development for more safe and efficient drug delivery systems. As discussed previously, many possible polymers, which have been identified as good candidates for drug delivery, have not been through extensive research experiments in comparison to chitosan and alginate. Knowing the advantages each polymer provides toward drug delivery, the future steps should include clinical trials of these polymers and evaluate how they perform when incorporated in targeted and/or controlled delivery systems. Additionally, issues with the search, design, and development of the polymers must be addressed before using them for drug delivery. These issues include biocompatibility, target specificity, biodistribution, containment of sufficient amount of drug, and its safe release, in addition to its stability and shelf life. Lack of improvement in any of these issues could result in a poor delivery system with other health problems [37].

References

1 Dang, J.M. and Leong, K.W. (2006). Natural polymers for gene delivery and tissue engineering. *Advanced Drug Delivery Reviews* 58 (4) https://doi.org/10.1016/j.addr.2006.03.001.

2 Abhilash, M. (2010). Potential applications of nanoparticles. *International Journal of Pharma and Bio Sciences* 1 (1): 1–12.

3 Jahanshahi, M. and Babaei, Z. (2008). Protein nanoparticle: a unique system as drug delivery vehicles. *African Journal of Biotechnology* 7 (25) https://doi.org/10.5897/AJB08.081.

4 Zafar, N., Fessi, H., and Elaissari, A. (2014). Cyclodextrin containing biodegradable particles: from preparation to drug delivery applications. *International Journal of Pharmaceutics* 461 (1–2) https://doi.org/10.1016/j.ijpharm.2013.12.004.

5 Iqbal, M., Zafar, N., Fessi, H., and Elaissari, A. (2015). Double emulsion solvent evaporation techniques used for drug encapsulation. *International Journal of Pharmaceutics* 496 (2) https://doi.org/10.1016/j.ijpharm.2015.10.057.

6 Ali, A. and Ahmed, S. (2018). A review on chitosan and its nanocomposites in drug delivery. *International Journal of*

Biological Macromolecules 109 https://doi.org/10.1016/j.ijbiomac.2017.12.078.

7 Ghosh, P.K. (2000). Hydrophilic polymeric nanoparticles as drug carriers. *Indian Journal of Biochemistry & Biophysics* 37 (5): 273–282.

8 Hu, L., Sun, Y., and Wu, Y. (2013). Advances in chitosan-based drug delivery vehicles. *Nanoscale* 5 (8) https://doi.org/10.1039/c3nr00338h.

9 Park, J.H., Saravanakumar, G., Kim, K., and Kwon, I.C. (2010). Targeted delivery of low molecular drugs using chitosan and its derivatives. *Advanced Drug Delivery Reviews* 62 (1) https://doi.org/10.1016/j.addr.2009.10.003.

10 Azevedo, M.A., Bourbon, A.I., Vicente, A.A., and Cerqueira, M.A. (2014). Alginate/chitosan nanoparticles for encapsulation and controlled release of vitamin B2. *International Journal of Biological Macromolecules* 71 https://doi.org/10.1016/j.ijbiomac.2014.05.036.

11 Biswas, S., Chattopadhyay, M., Sen, K.K., and Saha, M.K. (2015). Development and characterization of alginate coated low molecular weight chitosan nanoparticles as new carriers for oral vaccine delivery in mice. *Carbohydrate Polymers* 121 https://doi.org/10.1016/j.carbpol.2014.12.044.

12 Li, T., Shi, X.W., Du, Y.M., and Tang, Y.F. (2007). Quaternized chitosan/alginate nanoparticles for protein delivery. *Journal of Biomedical Materials Research Part A* 83 (2) https://doi.org/10.1002/jbm.a.31322.

13 Rosenblum, D., Joshi, N., Tao, W. et al. (2018). Progress and challenges towards targeted delivery of cancer therapeutics. *Nature Communications* 9 (1) https://doi.org/10.1038/s41467-018-03705-y.

14 Dodane, V. and Vilivalam, V.D. (1998). Pharmaceutical applications of chitosan. *Pharmaceutical Science & Technology Today* 1 (6) https://doi.org/10.1016/S1461-5347(98)00059-5.

15 Berscht, P.C., Nies, B., Liebendörfer, A., and Kreuter, J. (1994). Incorporation of basic fibroblast growth factor into methylpyrrolidinone chitosan fleeces and determination of the in vitro release characteristics. *Biomaterials* 15 (8) https://doi.org/10.1016/0142-9612(94)90209-7.

16 Sannan, T., Kurita, K., and Iwakura, Y. (1976). Studies on chitin, 2. Effect of deacetylation on solubility. *Die Makromolekulare Chemie* 177 (12) https://doi.org/10.1002/macp.1976.021771210.

17 Jean, M., Alameh, M., De Jesus, D. et al. (2012). Chitosan-based therapeutic nanoparticles for combination gene therapy and gene silencing of in vitro cell lines relevant to type 2 diabetes. *European Journal of Pharmaceutical Sciences* 45 (1–2) https://doi.org/10.1016/j.ejps.2011.10.029.

18 Nagarajan, E., Shanmugasundaram, P., Ravichandiran, V. et al. (2015). Development and evaluation of chitosan based polymeric nanoparticles of an antiulcer drug Lansoprazole. *Journal of Applied Pharmaceutical Science* 5 (4) https://doi.org/10.7324/JAPS.2015.50404.

19 Agnihotri, S.A., Mallikarjuna, N.N., and Aminabhavi, T.M. (2004). Recent advances on chitosan-based micro- and nanoparticles in drug delivery. *Journal of Controlled Release* 100 (1) https://doi.org/10.1016/j.jconrel.2004.08.010.

20 Saboktakin, M.R., Tabatabaie, R.M., Maharramov, A., and Ramazanov, M.A. (2011). Synthesis and in vitro evaluation of carboxymethyl starch-chitosan nanoparticles as drug delivery system to the colon. *International Journal of Biological Macromolecules* 48 (3) https://doi.org/10.1016/j.ijbiomac.2010.10.005.

21 Kang, R.H., Kwon, J.Y., Kim, Y., and Lee, S.M. (2017). Cisplatin-mediated formation of polyampholytic chitosan nanoparticles with attenuated viscosity and pH-sensitive drug release. *Langmuir* 33 (36) https://doi.org/10.1021/acs.langmuir.7b02043.

22 Antoniraj, M.G., Ayyavu, M., Henry, L.J.K. et al. (2018). Cytocompatible chitosan-graft-mPEG-based 5-fluorouracil-loaded polymeric nanoparticles for tumor-targeted drug delivery. *Drug Development and Industrial Pharmacy* 44 (3) https://doi.org/10.1080/03639045.2017.1371741.

23 Wang, J., Tan, J., Luo, J. et al. (2017). Enhancement of scutellarin oral delivery efficacy by vitamin B12-modified amphiphilic chitosan derivatives to treat type II diabetes induced-retinopathy. *Journal of Nanbiotechnology* 15 (1) https://doi.org/10.1186/s12951-017-0251-z.

24 Gu, J., Al-Bayati, K., and Ho, E.A. (2017). Development of antibody-modified chitosan nanoparticles for the targeted delivery of siRNA across the blood-brain barrier as a strategy for inhibiting HIV replication in astrocytes. *Drug Delivery and Translational Research* 7 (4) https://doi.org/10.1007/s13346-017-0368-5.

25 Mitra, S., Gaur, U., Ghosh, P.C., and Maitra, A.N. (2001). Tumour targeted delivery of encapsulated dextran-doxorubicin conjugate using chitosan nanoparticles as carrier. *Journal of Controlled Release* 74 https://doi.org/10.1016/S0168-3659(01)00342-X.

26 Sanyakamdhorn, S., Agudelo, D., and Tajmir-Riahi, H.A. (2013). Encapsulation of antitumor drug doxorubicin and its analogue by chitosan nanoparticles. *Biomacromolecules* 14 (2) https://doi.org/10.1021/bm3018577.

27 Jayakumar, R., Nwe, N., Tokura, S., and Tamura, H. (2007). Sulfated chitin and chitosan as novel biomaterials. *International Journal of Biological*

Macromolecules 40 (3) https://doi.org/10.1016/j.ijbiomac.2006.06.021.

28 Burke, C.J., Hsu, T.A., and Volkin, D.B. (1999). Formulation, stability, and delivery of live attenuated vaccines for human use. *Critical Reviews in Therapeutic Drug Carrier Systems* 16 (1) https://doi.org/10.1615/critrevtherdrugcarriersyst.v16.i1.10.

29 Vandervoort, J. and Ludwig, A. (2004). Preparation and evaluation of drug-loaded gelatin nanoparticles for topical ophthalmic use. *European Journal of Pharmaceutics and Biopharmaceutics* 57 (2) https://doi.org/10.1016/S0939-6411(03)00187-5.

30 Busch, S., Schwarz, U., and Kniep, R. (2003). Chemical and structural investigations of biomimetically grown fluorapatite-gelatin composite aggregates. *Advanced Functional Materials* 13 (3) https://doi.org/10.1002/adfm.200390029.

31 Kaul, G. and Amiji, M. (2002). Long-circulating poly(ethylene glycol)-modified gelatin nanoparticles for intracellular delivery. *Pharmaceutical Research* 19 (7) https://doi.org/10.1023/A:1016486910719.

32 Chen, T., Embree, H.D., Wu, L.Q., and Payne, G.F. (2002). in vitro protein-polysaccharide conjugation: Tyrosinase-catalyzed conjugation of gelatin and chitosan. *Biopolymers* 64 (6) https://doi.org/10.1002/bip.10196.

33 Ma, J., Cao, H., Li, Y., and Li, Y. (2002). Synthesis and characterization of poly(DL-lactide)-grafted gelatins as bioabsorbable amphiphilic polymers. *Journal of Biomaterials Science. Polymer Edition* 13 (1) https://doi.org/10.1163/156856202753525945.

34 Tanaka, N., Takino, S., and Utsumi, I. (1963). A new oral gelatinized sustained-release dosage form. *Journal of Pharmaceutical Sciences* 52 (7) https://doi.org/10.1002/jps.2600520714.

35 Elzoghby, A.O., Samy, W.M., and Elgindy, N.A. (2012). Albumin-based nanoparticles as potential controlled release drug delivery systems. *Journal of Controlled Release* 157 (2) https://doi.org/10.1016/j.jconrel.2011.07.031.

36 Rahimnejad, M., Jahanshahi, M., and Najafpour, G.D. (2006). Production of biological nanoparticles from bovine serum albumin for drug delivery. *African Journal of Biotechnology* 5 (20) https://doi.org/10.5897/AJB2006.000-5072.

37 George, A., Shah, P.A., and Shrivastav, P.S. (2019). Natural biodegradable polymers based nano-formulations for drug delivery: a review. *International Journal of Pharmaceutics* 561 https://doi.org/10.1016/j.ijpharm.2019.03.011.

38 Damodaran, S. and Paraf, A. (2017). *Food Proteins and Their Applications*. CRC Press https://doi.org/10.1201/9780203755617.

39 Wongsasulak, S., Patapeejumruswong, M., Weiss, J. et al. (2010). Electrospinning of food-grade nanofibers from cellulose acetate and egg albumen blends. *Journal of Food Engineering* 98 (3) https://doi.org/10.1016/j.jfoodeng.2010.01.014.

40 Hu, Y.J., Liu, Y., Sun, T.Q. et al. (2006). Binding of anti-inflammatory drug cromolyn sodium to bovine serum albumin. *International Journal of Biological Macromolecules* 39 (4–5) https://doi.org/10.1016/j.ijbiomac.2006.04.004.

41 Tantra, R., Tompkins, J., and Quincey, P. (2010). Characterisation of the de-agglomeration effects of bovine serum albumin on nanoparticles in aqueous suspension. *Colloids and Surfaces B: Biointerfaces* 75 (1) https://doi.org/10.1016/j.colsurfb.2009.08.049.

42 Hirose, M., Tachibana, A., and Tanabe, T. (2010). Recombinant human serum albumin hydrogel as a novel drug delivery vehicle. *Materials Science and Engineering: C* 30 (5) https://doi.org/10.1016/j.msec.2010.02.020.

43 Kratz, F. (2008). Albumin as a drug carrier: design of prodrugs, drug conjugates and nanoparticles. *Journal of Controlled Release* 132 (3) https://doi.org/10.1016/j.jconrel.2008.05.010.

44 Fasano, M., Curry, S., Terreno, E. et al. (2005). The extraordinary ligand binding properties of human serum albumin. *IUBMB Life* 57 (12) https://doi.org/10.1080/15216540500404093.

45 Jahanban-Esfahlan, A., Panahi-Azar, V., and Sajedi, S. (2015). Spectroscopic and molecular docking studies on the interaction between N-acetyl cysteine and bovine serum albumin. *Biopolymers* 103 (11) https://doi.org/10.1002/bip.22697.

46 Jahanban-Esfahlan, A., Panahi-Azar, V., and Sajedi, S. (2016). Interaction of glutathione with bovine serum albumin: spectroscopy and molecular docking. *Food Chemistry* 202 https://doi.org/10.1016/j.foodchem.2016.02.026.

47 Jiang, Y., Lu, H., Chen, F. et al. (2016). PEGylated albumin-based polyion complex micelles for protein delivery. *Biomacromolecules* 17 (3) https://doi.org/10.1021/acs.biomac.5b01537.

48 Wang, N., Sun, P., Lv, M. et al. (2017). Mustard-inspired delivery shuttle for enhanced blood-brain barrier penetration and effective drug delivery in glioma therapy. *Biomaterials Science* 5 (5) https://doi.org/10.1039/c7bm00133a.

49 Merodio, M., Arnedo, A., Renedo, M.J., and Irache, J.M. (2001). Ganciclovir-loaded albumin nanoparticles: characterization and in vitro release properties. *European Journal of Pharmaceutical Sciences* 12 (3) https://doi.org/10.1016/S0928-0987(00)00169-X.

50 Han, S., Li, M., Liu, X. et al. (2013). Construction of amphiphilic copolymer nanoparticles based on gelatin as drug carriers for doxorubicin delivery. *Colloids and Surfaces B: Biointerfaces* 102 https://doi.org/10.1016/j.colsurfb.2012.09.010.

51 Prado, H.J., Matulewicz, M.C., Bonelli, P.R., and Cukierman, A.L. (2009). Preparation and characterization of a novel starch-based interpolyelectrolyte complex as matrix for controlled drug release. *Carbohydrate Research* 344 (11) https://doi.org/10.1016/j.carres.2009.04.026.

52 El-Naggar, M.E., El-Rafie, M.H., El-sheikh, M.A. et al. (2015). Synthesis, characterization, release kinetics and toxicity profile of drug-loaded starch nanoparticles. *International Journal of Biological Macromolecules* 81 https://doi.org/10.1016/j.ijbiomac.2015.09.005.

53 Ding, Y., Zheng, J., Zhang, F., and Kan, J. (2016). Synthesis and characterization of retrograded starch nanoparticles through homogenization and miniemulsion cross-linking. *Carbohydrate Polymers* 151 https://doi.org/10.1016/j.carbpol.2016.06.007.

54 Bel Haaj, S., Magnin, A., Pétrier, C., and Boufi, S. (2013). Starch nanoparticles formation via high power ultrasonication. *Carbohydrate Polymers* 92 (2) https://doi.org/10.1016/j.carbpol.2012.11.022.

55 Kim, H.Y., Park, S.S., and Lim, S.T. (2015). Preparation, characterization and utilization of starch nanoparticles. *Colloids and Surfaces B: Biointerfaces* 126 https://doi.org/10.1016/j.colsurfb.2014.11.011.

56 Moad, G. (2011). Chemical modification of starch by reactive extrusion. *Progress in Polymer Science* 36 (2) https://doi.org/10.1016/j.progpolymsci.2010.11.002.

57 Soares, G.A., de Castro, A.D., BSF, C., and Evangelista, R.C. (2013). Blends of cross-linked high amylose starch/pectin loaded with diclofenac. *Carbohydrate Polymers* 91 (1) https://doi.org/10.1016/j.carbpol.2012.08.014.

58 Yang, J., Huang, Y., Gao, C. et al. (2014). Fabrication and evaluation of the novel reduction-sensitive starch nanoparticles for controlled drug release. *Colloids and Surfaces B: Biointerfaces* 115 https://doi.org/10.1016/j.colsurfb.2013.12.007.

59 Gibson, G.R. and Roberfroid, M.B. (1995). Dietary modulation of the human colonic microbiota: introducing the concept of prebiotics. *The Journal of Nutrition* 125 (6) https://doi.org/10.1093/jn/125.6.1401.

60 Soppimath, K.S., Aminabhavi, T.M., Kulkarni, A.R., and Rudzinski, W.E. (2001). Biodegradable polymeric nanoparticles as drug delivery devices. *Journal of Controlled Release* 70 (1–2) https://doi.org/10.1016/S0168-3659(00)00339-4.

61 Mohanraj, V.J. and Chen, Y. (2007). Nanoparticles – a review. *Tropical Journal of Pharmaceutical Research* 5 (1) https://doi.org/10.4314/tjpr.v5i1.14634.

62 Huang, Y. and Dai, W.-G. (2014). Fundamental aspects of solid dispersion technology for poorly soluble drugs. *Acta Pharmaceutica Sinica B* 4 (1) https://doi.org/10.1016/j.apsb.2013.11.001.

63 Rao, J.P. and Geckeler, K.E. (2011). Polymer nanoparticles: preparation techniques and size-control parameters. *Progress in Polymer Science* 36 (7) https://doi.org/10.1016/j.progpolymsci.2011.01.001.

64 Pinto Reis, C., Neufeld, R.J., Ribeiro, A.J., and Veiga, F. (2006). Nanoencapsulation I. Methods for preparation of drug-loaded polymeric nanoparticles. *Nanomedicine: Nanotechnology, Biology and Medicine* 2 (1) https://doi.org/10.1016/j.nano.2005.12.003.

65 Anton, N., Benoit, J.P., and Saulnier, P. (2008). Design and production of nanoparticles formulated from nano-emulsion templates – a review. *Journal of Controlled Release* 128 (3) https://doi.org/10.1016/j.jconrel.2008.02.007.

66 Lemoine, D. and Préat, V. (1998). Polymeric nanoparticles as delivery system for influenza virus glycoproteins. *Journal of Controlled Release* 54 (1) https://doi.org/10.1016/S0168-3659(97)00241-1.

67 Song, C.X., Labhasetwar, V., Murphy, H. et al. (1997). Formulation and characterization of biodegradable nanoparticles for intravascular local drug delivery. *Journal of Controlled Release* 43 (2–3) https://doi.org/10.1016/S0168-3659(96)01484-8.

68 Zohri, M., Gazori, T., Mirdamadi, S.S. et al. (2009). Polymeric nanoparticles: production, applications and advantage. *The Internet Journal of Nanotechnology* 3.

69 Tice, T.R. and Gilley, R.M. (1985). Preparation of injectable controlled-release microcapsules by a solvent-evaporation process. *Journal of Controlled Release* 2 https://doi.org/10.1016/0168-3659(85)90056-2.

70 Fessi, H., Puisieux, F., Devissaguet, J.P. et al. (1989). Nanocapsule formation by interfacial polymer deposition following solvent displacement. *International Journal of Pharmaceutics* 55 (1) https://doi.org/10.1016/0378-5173(89)90281-0.

71 Quintanar-Guerrero, D., Allémann, E., Fessi, H., and Doelker, E. (1998). Preparation techniques and mechanisms of formation of biodegradable nanoparticles from preformed polymers. *Drug Development and Industrial Pharmacy* 24 (12) https://doi.org/10.3109/03639049809108571.

72 Limayem Blouza, I., Charcosset, C., Sfar, S., and Fessi, H. (2006). Preparation and characterization of spironolactone-loaded nanocapsules for paediatric use. *International Journal of Pharmaceutics* 325 (1–2) https://doi.org/10.1016/j.ijpharm.2006.06.022.

73 Jung, T., Kamm, W., Breitenbach, A. et al. (2000). Biodegradable nanoparticles for oral delivery of peptides:

is there a role for polymers to affect mucosal uptake? *European Journal of Pharmaceutics and Biopharmaceutics* 50 (1) https://doi.org/10.1016/S0939-6411(00)00084-9.

74 Lambert, G., Fattal, E., and Couvreur, P. (2001). Nanoparticulate systems for the delivery of antisense oligonucleotides. *Advanced Drug Delivery Reviews* 47 (1) https://doi.org/10.1016/S0169-409X(00)00116-2.

75 Kawashima, Y. (2001). Panoparticulate systems for improved drug delivery. *Advanced Drug Delivery Reviews* 47 (1): 1–2. https://doi.org/10.1016/S0169-409X(00)00117-4.

76 Weber, M. and Thies, M.C. (2002). Understanding the RESS process. In: *Supercritical Fluid Technology in Materials Science and Engineering*, 1e (ed. Y.-P. Sun), 385–439. New York: Marcel Dekker https://doi.org/10.1201/9780203909362.ch11.

77 Blasig, A., Shi, C., Enick, R.M., and Thies, M.C. (2002). Effect of concentration and degree of saturation on RESS of a CO2-soluble fluoropolymer. *Industrial and Engineering Chemistry Research* 41 (20) https://doi.org/10.1021/ie0201819.

78 Lim, K.T., Ganapathy, H.S., Hwang, H.S. et al. (2005). Novel semiconducting polymer particles by supercritical fluid process. *Macromolecular Rapid Communications* 26 (22) https://doi.org/10.1002/marc.200500438.

79 Sane, A. and Thies, M.C. (2007). Effect of material properties and processing conditions on RESS of poly(l-lactide). *Journal of Supercritical Fluids* 40 (1) https://doi.org/10.1016/j.supflu.2006.04.003.

80 Meziani, M.J., Pathak, P., Hurezeanu, R. et al. (2004). Supercritical-fluid processing technique for nanoscale polymer particles. *Angewandte Chemie, International Edition* 43 (6) https://doi.org/10.1002/anie.200352834.

81 Meziani, M.J., Pathak, P., Wang, W. et al. (2005). Polymeric nanofibers from rapid expansion of supercritical solution. *Industrial and Engineering Chemistry Research* 44 (13) https://doi.org/10.1021/ie048796o.

82 Hutchenson, K. (2002). Organic chemical reactions and catalysis in supercritical fluid media. In: *Supercritical Fluid Technology in Materials Science and Engineering*, 1e (ed. Y.-P. Sun), 88–191. New York: Marcel Dekker https://doi.org/10.1201/9780203909362.ch3.

83 Vauthier, C., Dubernet, C., Fattal, E. et al. (2003). Poly(alkylcyanoacrylates) as biodegradable materials for biomedical applications. *Advanced Drug Delivery Reviews* 55 (4) https://doi.org/10.1016/S0169-409X(03)00041-3.

84 Gaudin, F. and Sintes-Zydowicz, N. (2008). Core-shell biocompatible polyurethane nanocapsules obtained by interfacial step polymerisation in miniemulsion. *Colloids and Surfaces A: Physicochemical and Engineering Aspects* 331 (1–2) https://doi.org/10.1016/j.colsurfa.2008.07.028.

85 Landfester, K., Musyanovych, A., and Mailänder, V. (2010). From polymeric particles to multifunctional nanocapsules for biomedical applications using the miniemulsion process. *Journal of Polymer Science, Part A: Polymer Chemistry* 48 (3) https://doi.org/10.1002/pola.23786.

86 Wu, M., Dellacherie, E., Durand, A., and Marie, E. (2009). Poly(n-butyl cyanoacrylate) nanoparticles via miniemulsion polymerization. 2. PEG-based surfactants. *Colloids and Surfaces B: Biointerfaces* 69 (1) https://doi.org/10.1016/j.colsurfb.2008.10.003.

87 Karode, S.K., Kulkarni, S.S., Suresh, A.K., and Mashelkar, R.A. (1998). New insights into kinetics and thermodynamics of interfacial polymerization. *Chemical Engineering Science* 53 (15) https://doi.org/10.1016/S0009-2509(98)00083-9.

88 Crespy, D., Stark, M., Hoffmann-Richter, C. et al. (2007). Polymeric nanoreactors for hydrophilic reagents synthesized by interfacial polycondensation on miniemulsion droplets. *Macromolecules* 40 (9) https://doi.org/10.1021/ma0621932.

89 Danicher, L., Frère, Y., and Calvé, A.L. (2000). Synthesis by interracial polycondensation of polyamide capsules with various sizes. Characteristics and properties. In: *Macromolecular Symposia*, vol. 151. https://doi.org/10.1002/1521-3900(200002)151:1<387::AID-MASY387>3.0.CO;2-D.

90 Torini, L., Argillier, J.F., and Zydowicz, N. (2005). Interfacial polycondensation encapsulation in miniemulsion. *Macromolecules* 38 (8) https://doi.org/10.1021/ma047808e.

91 Sarkar, D., El-Khoury, J., Lopina, S.T., and Hu, J. (2005). An effective method for preparing polymer nanocapsules with hydrophobic acrylic shell and hydrophilic interior by inverse emulsion radical polymerization. *Macromolecules* 38 (20) https://doi.org/10.1021/ma050661m.

92 Scott, C., Wu, D., Ho, C.C., and Co, C.C. (2005). Liquid-core capsules via interfacial polymerization: a free-radical analogy of the nylon rope trick. *Journal of the American Chemical Society* 127 (12) https://doi.org/10.1021/ja044532h.

93 Gallardo, M., Couarraze, G., Denizot, B. et al. (1993). Study of the mechanisms of formation of nanoparticles and nanocapsules of polyisobutyl-2-cyanoacrylate. *International Journal of Pharmaceutics* 100 (1–3) https://doi.org/10.1016/0378-5173(93)90075-Q.

94 Aboubakar, M., Puisieux, F., Couvreur, P. et al. (1999). Study of the mechanism of insulin encapsulation in poly(isobutylcyanoacrylate) nanocapsules obtained by interfacial polymerization. *Journal of Biomedical Materials Research* 47 (4) https://doi.org/10.1002/(sici)1097-4636(19991215)47:4<568::aid-jbm14>3.3.co;2-o.

95 Watnasirichaikul, S., Davies, N.M., Rades, T., and Tucker, I.G. (2000). Preparation of biodegradable insulin nanocapsules from biocompatible microemulsions. *Pharmaceutical Research* 17 (6) https://doi.org/10.1023/A:1007574030674.

96 Puglisi, G., Fresta, M., Giammona, G., and Ventura, C.A. (1995). Influence of the preparation conditions on poly(ethylcyanoacrylate) nanocapsule formation. *International Journal of Pharmaceutics* 125 (2) https://doi.org/10.1016/0378-5173(95)00142-6.

97 Matyjaszewski, K. and Xia, J. (2001). Atom transfer radical polymerization. *Chemical Reviews* 101 (9) https://doi.org/10.1021/cr940534g.

98 Zetterlund, P.B., Nakamura, T., and Okubo, M. (2007). Mechanistic investigation of particle size effects in TEMPO-mediated radical polymerization of styrene in aqueous miniemulsion. *Macromolecules* 40 (24) https://doi.org/10.1021/ma0712403.

99 Zetterlund, P.B., Kagawa, Y., and Okubo, M. (2008). Controlled/living radical polymerization in dispersed systems. *Chemical Reviews* 108 (9) https://doi.org/10.1021/cr800242x.

100 Nicolas, J., Charleux, B., Guerret, O., and Magnet, S. (2005). Nitroxide-mediated controlled free-radical emulsion polymerization using a difunctional water-soluble alkoxyamine initiator. Toward the control of particle size, particle size distribution, and the synthesis of triblock copolymers. *Macromolecules* 38 (24) https://doi.org/10.1021/ma0511999.

101 Nicolas, J., Ruzette, A.V., Farcet, C. et al. (2007). Nanostructured latex particles synthesized by nitroxide-mediated controlled/living free-radical polymerization in emulsion. *Polymer (Guildf)* 48 (24) https://doi.org/10.1016/j.polymer.2007.09.039.

102 Dire, C., Magnet, S., Couvreur, L., and Charleux, B. (2009). Nitroxide-mediated controlled/living free-radical surfactant-free emulsion polymerization of methyl methacrylate using a poly(methacrylic acid)-based macroalkoxyamine initiator. *Macromolecules* 42 (1) https://doi.org/10.1021/ma802083g.

103 Farcet, C., Nicolas, J., and Charleux, B. (2002). Kinetic study of the nitroxide-mediated controlled free-radical polymerization of n-butyl acrylate in aqueous miniemulsions. *Journal of Polymer Science, Part A: Polymer Chemistry* 40 (24) https://doi.org/10.1002/pola.10531.

104 Farcet, C., Charleux, B., and Pirri, R. (2001). Poly(n-butyl acrylate) homopolymer and poly[n-butyl acrylate-b-(n-butyl acrylate-co-styrene)] block copolymer prepared via nitroxide-mediated living/controlled radical polymerization in miniemulsion [6]. *Macromolecules* 34 (12) https://doi.org/10.1021/ma010011s.

105 Farcet, C., Lansalot, M., Charleux, B. et al. (2000). Mechanistic aspects of nitroxide-mediated controlled radical polymerization of styrene in miniemulsion, using a water-soluble radical initiator. *Macromolecules* 33 (23) https://doi.org/10.1021/ma000952p.

106 Li, W., Matyjaszewski, K., Albrecht, K., and Möller, M. (2009). Reactive surfactants for polymeric nanocapsules via interfacially confined miniemulsion ATRP. *Macromolecules* 42 (21) https://doi.org/10.1021/ma901574y.

107 Siegwart, D.J., Srinivasan, A., Bencherif, S.A. et al. (2009). Cellular uptake of functional nanogels prepared by inverse miniemulsion ATRP with encapsulated proteins, carbohydrates, and gold nanoparticles. *Biomacromolecules* 10 (8) https://doi.org/10.1021/bm9004904.

108 Oh, J.K., Perineau, F., Charleux, B., and Matyjaszewski, K. (2009). AGET ATRP in water and inverse minlemulsion: a facile route for preparation of high-molecular-weight biocompatible brush-like polymers. *Journal of Polymer Science, Part A: Polymer Chemistry* 47 (7) https://doi.org/10.1002/pola.23272.

109 Li, W., Min, K., Matyjaszewski, K. et al. (2008). PEO-based block copolymers and homopolymers as reactive surfactants for AGET ATRP of butyl acrylate in miniemulsion. *Macromolecules* 41 (17) https://doi.org/10.1021/ma800892e.

110 Min, K., Gao, H., Yoon, J.A. et al. (2009). One-pot synthesis of hairy nanoparticles by emulsion ATRP. *Macromolecules* 42 (5) https://doi.org/10.1021/ma8026244.

111 Min, K., Gao, H., and Matyjaszewski, K. (2006). Development of an ab initio emulsion atom transfer radical polymerization: from microemulsion to emulsion. *Journal of the American Chemical Society* 128 (32) https://doi.org/10.1021/ja0629054.

112 Min, K. and Matyjaszewski, K. (2005). Atom transfer radical polymerization in microemulsion. *Macromolecules* 38 (20) https://doi.org/10.1021/ma051675v.

113 Rieger, J., Zhang, W., Stoffelbach, F., and Charleux, B. (2010). Surfactant-free RAFT emulsion polymerization using poly(N,N -dimethylacrylamide) trithiocarbonate macromolecular chain transfer agents. *Macromolecules* 43 (15) https://doi.org/10.1021/ma1009269.

114 Manguian, M., Save, M., and Charleux, B. (2006). Batch emulsion polymerization of styrene stabilized by a hydrophilic macro-RAFT agent. *Macromolecular Rapid Communications* 27 (6) https://doi.org/10.1002/marc.200500807.

115 Zhou, X., Ni, P., and Yu, Z. (2007). Comparison of RAFT polymerization of methyl methacrylate in conventional emulsion and miniemulsion systems. *Polymer (Guildf)* 48 (21) https://doi.org/10.1016/j.polymer.2007.08.053.

116 Bennet, D. and Kim, S. (2014). Polymer nanoparticles for smart drug delivery. In: *Application of Nanotechnology in Drug Delivery* (ed. A.D. Sezer), 257–310. Intech Open https://doi.org/10.5772/58422.

17

Cobalt Oxide-engineered Nanomaterials for Environmental Remediation

Komal Parmar[1] and Jayvadan Patel[2]

[1] *ROFEL, Shri G.M. Bilakhia College of Pharmacy, Gujarat, India*
[2] *Nootan Pharmacy College, Faculty of Pharmacy, Sankalchand Patel University, Gujarat, India*

17.1 Introduction

Industrial development, urbanization, and population growth in short span of time resulted in numerous pollutants being produced and dumped into the atmosphere. Environmental degradation is without question one of the most critical issues which society is facing today. It is the deterioration of the atmosphere by the rapid depletion of the resources such as air, water, and soil. Biodiversity loss, habitat loss, and wild life extinction are few consequences to name resulting from environmental deterioration. Environmental remediation has come out as a necessity to save the mother earth and life on earth. It refers to employment of the techniques and habits in order to reduce the excessive exploitation of natural resources and improper disposal of the used material. Emerging innovations for remediating air, water, and soil pollutants are constantly being explored [1–6]. Remediation and reconstruction of contaminated environments are among the critical challenges tackled by the world community by means of delivering environmental sustainability for future generations. Various pollutants are responsible for the environmental degradation. For instance, extensive mining of minerals together with improper clearance of harmful by-products severely impacts the soil and water [7]. Uncontrolled use of pesticides and chemical fertilizers for exponential growth of agricultural production has perilous influence on soil and water [8]. Waste water effluent that consists of dyes, heavy metals, and organic compounds from chemical industries sternly disturbs the water bodies [9]. The most common heavy metal pollutants are arsenic, cadmium, chromium, copper, nickel, lead, and mercury which are raising concerns over the potential effects on the environment [10]. Nuclear technology which has proved to be the most efficient energy resource has also detrimentally affected the environment. The nuclear radioactive waste, radiation exposure, and accidents in nuclear reactors have potential harmful effects on the life and environment [11]. Figure 17.1 highlights various causes of contaminants polluting natural resources. In light of all the abovementioned issues that constantly challenge our environment, environmental restoration has now become a global priority. The US Market Research Report, 2020, states that the remediation and environmental clean-up services industry has grown at an annualized rate of 0.2% to $17.1 billion over the past five years [12]. This is representative of impressive developments in the production of new technologies and methods for environmental remediation and restoration.

Nanotechnology refers to the area of study and engineering that is concerned with constructing "things" on the scale of atoms and molecules – in general with size ranging in 1–100 nm. In the recent decades, nanotechnology has gained lot of attention due to the exceptional characteristics of nanomaterials. Owing to the small size, the nanoparticles offer high surface-to-volume ratio compared to their counterparts, which is beneficial for efficient adsorption of nanowaste materials [13]. Another advantage of nanomaterial is that it can be fabricated in various shapes such as sphere, rod, fiber, dots, tube, and wires, each having unique properties that help in the removal of harmful pollutants from the environment [14]. In addition to the abovementioned advantages, the surface of nanomaterials can be modified with functional groups/moieties that can

Sustainable Nanotechnology: Strategies, Products, and Applications, First Edition.
Edited by Yashwant V. Pathak, Govindan Parayil, and Jayvadan K. Patel.

Figure 17.1 Various contaminants polluting the environment.

specifically act on targeted pollutants for remediation process with enhanced efficiency. For instance, by combining NiO and PdO/Pd to Co_3O_4s nanoparticles, the catalytic efficiency of functionalized nanocatalysts was increased to 99 and 98%, respectively [15]. Various nanomaterials employed for environmental remediation are reported by different investigators. Table 17.1 briefs about various nanomaterials applied for environmental remediation.

Cobalt is the ferromagnetic metal belonging to the group of the "transition metals that exists in various inorganic forms such as oxides, carbonates, nitrates, sulfates, and chlorides. Due to its unique mechanical, thermal, and magnetic properties, cobalt oxide nanoparticles and their composite-based nanoparticles attract more interest from researchers among all the magnetic nanoparticles. Pure cobalt at room temperature converts into oxides such as CoO, Co_2O_3, and Co_3O_4. Cobalt oxide nanoparticles can be synthesized using various methods. For instance, green synthesis that employs plant extracts or microorganisms has proved to be the environmentally friendly approach of nanoparticle synthesis. Recently, cobalt oxide (Co_3O_4) nanoparticles were synthesized using the leaves extract of *Populus ciliata* (safaida) and cobalt nitrate hexahydrate as a source of cobalt. Further, antibacterial activities of the synthesized Co_3O_4 nanoparticles were evaluated against Gram-negative and -positive bacteria, and they were found active against *Escherichia coli* (*E. coli*), *Klebsiella pneumoniae* (*K. pneumoniae*), *Bacillus subtilis* (*B. subtilis*), and *Bacillus licheniformis* (*B. licheniformis*). Results revealed successful preparation of nanoparticles and its antibacterial application [34]. In another study, novel mycosynthesis method was reported for the fabrication of cobalt oxide nanoparticles using *Aspergillus brasiliensis* ATCC 16404. The prepared nanoparticles had exceptional magnetic properties and exhibited a good antimicrobial activity against some pathogenic microorganisms [35]. Thermal decomposition is another method of preparation of metal oxide nanoparticles. In the process, metal transition from the precursor takes place at high temperature, and metallic oxide nanoparticles are formed. In one such attempt, cobalt oxide nanoparticles were synthesized by thermal decomposition of cobalt carboxamide complex as a precursor. The complex was heated at 500 °C for 2 h in the presence of air, which thereby resulted in cobalt oxide (Co_3O_4) nanoparticles [36]. Various researchers have reported the synthesis of cobalt oxide nanoparticles employing different methods of synthesis (Table 17.2).

17.2 Applications of Cobalt Oxide Nanoparticles for Environmental Remediation

Cobalt oxide nanoparticles are employed in various applications such as drug delivery, electronics, as catalyst, in chemical industries, and sensors. Cobalt oxide

Table 17.1 Application of nanomaterials for environmental remediation.

Nanomaterial	Environmental remediation	Role of nanomaterials	Reference
Silver nanoparticles	Cadmium removal from aqueous solution	Biosorbent	Al-Qahtani [16]
Silver nanoparticles	Removal of pathogenic bacteria from wastewater	Antibacterial	Moustafa [17]
Titanium oxide-gallium nanoparticles	Photocatalytic degradation of organic dye in waste water treatment under UV irradiation	Catalyst	Truong [18]
Titanium dioxide nanoparticles	Remediation of antimony from soil	Adsorbent	Zand et al. [19]
Iron oxide nanoparticles	Mercury removal in waste water	Adsorbent	Velez et al. [20]
Iron nanoparticles	For methylene blue removal from aqueous solutions and textile wastewater treatment	Adsorbent	Hamdy et al. [21]
Melamine-based dendrimer amine-modified magnetic nanoparticles	Removal of lead from waste water	Adsorbent	Sharahi and Shahbazi [22]
Carbon nanotubes	Removal of ozone (gaseous pollutant)	Adsorbent	Yang et al. [23]
Carbon fibers	Removal of sulfur dioxide gas	Adsorbent	Bai et al. [24]
Gold nanoparticle decorated titanium dioxide nanocomposites	Removal of methylene blue from water	Adsorbent	Perera et al. [25]
Nanomaterial (organosilane compound)	Soil treatment	Adsorbent	Ugwu et al. [26]
Graphene oxide nanoparticles	Removal of arsenic, copper, lead, and cadmium from soil	Immobilizing agent	Baragaño et al. [27]
Carbon nanomaterial	Removal of petroleum hydrocarbons from soil	Adsorbent	Apul et al. [28]
Silver-titanium dioxide nanoparticles	For air purification	Photocatalyst	Le et al. [29]
Zinc oxide nanoparticles	Treatment of municipal wastewater	Antibacterial	Elmi et al. [30]
Titanium dioxide nanoparticles	Waste water treatment	Photocatalyst	Araoyinbo et al. [31]
Iron oxide and copper oxide nanoparticles	River water treatment	Adsorbent and catalyst	Goud et al. [32]
Zinc oxide-doped stannic oxide nanoparticles	Photocatalytic degradation of harmful and toxic organic dye	Photocatalyst	Lamba et al. [33]

nanomaterial can prove to be a potential candidate for environmental remediation process. Figure 17.2 demonstrates the possible mechanism of nanoparticles in the removal of contaminants. Various investigators have reported the use of cobalt and its oxides as catalysts for organic pollutant degradation. Recently, Adekunle and coresearchers have demonstrated the potential of cobalt and its oxide nanoparticles catalyst activity toward dye degradation in waste water remediation. Nanoparticles were fabricated using microwave and reductive chemical heating method. Nanoparticles showed efficient photocatalytic activity against murexide dye (43.6%) and Erichrome black-T (39.4%) [67]. In another such study, malachite green dye was removed from waste water using magnetic cobalt oxide nanoparticles. The nanoparticles were prepared using starch-assisted modified self-propagator combustion process. Results revealed maximum Langmuir adsorption capacity of $238.10\,mg\,g^{-1}$. The magnetic characteristics of cobalt oxide nanoparticles further facilitate the rapid removal of nanoparticles after use by employing a handheld magnet [68]. One such study reported the preparation of cobalt oxide nanoparticles by employing green synthesis approach. The leaf extract of *Citrus medica* plant was utilized in the fabrication of nanoparticles. Further, the nanoparticles were characterized and evaluated for photocatalytic activity against methyl orange dye. Results demonstrated 90% dye degradation in the first 60 min at natural pH of 6.5. In addition, it was found that at higher pH values, dye degradation was enhanced due to increase in the formation of OH radicals [69]. Another such example of

Table 17.2 Fabrication methods for cobalt oxide nanoparticles.

Method of preparation	Precursor/additive agent	Reference
Thermal decomposition	Cobalt carbonate	Rahimi-Nasrabadi et al. [37]
	Cobalt chloride and Schiff base (5-bromo-2-hydroxybenzyl-2-furylmethyl)imine	Khalaji [38]
	bis(*N*-cyclohexyl-1-naphthaldehydato)cobalt(II) complex	Xaba et al. [39]
	Cobalt(II) nitrate hexahydrate	Qasem et al. [40]
	Cobalt hydroxy carbonates	Singh et al. [41]
	Cobalt nitrate and sucrose	Makhlouf et al. [42]
	Cobalt nitrate	Alayoglu et al. [43]
	Cobalt hydroxide	Lv et al. [44]
Green synthesis	Leaf extracts of *Sageretia thea*	Khalil et al. [45]
	Aspalathus linearis extract	Diallo et al. [46]
	Piper longum extract	Ranaei Siadat [47]
	Camellia sinensis (L.) Kuntze and *Apium graveolens* (L.) extract	Urabe and Aziz [48]
	Latex of *Calotropis procera*	Dubey et al. [49]
	Punica granatum peel extract	Bibi et al. [50]
	Moringa oleifera extract	Matinise et al. [51]
Sol–gel method	Cobalt (II) nitrate hexahydrate as the precursor and ethylene glycol as the dissolvent	Sardjono and Puspitasari [52]
	Cobalt salt, THF, and ammonia	Jagtap et al. [53]
	Cobalt nitrate and polyvinyl alcohol	Lima-Tenório et al. [54]
	Cobalt salt and starch	Vaya et al. [55]
	Cobalt nitrate and ammonium oxalate	Prabaharan et al. [56]
Coprecipitation method	Cobalt chloride hexahydrate and sodium carbonate	Januja [57]
Pyrolysis	Aqueous solution of cobalt acetate	Yang et al. [58]
	Aqueous solution of cobalt chloride	Abbas et al. [59]
Laser ablation method	Pure cobalt	Ghaem et al. [60]
	Pure cobalt	Rousta et al. [61]
Wet synthesis	Cobalt acetate	Athar et al. [62]
Polymer matrix-assisted synthesis	Cobalt chloride, hyperbranched polyester polyol Boltorn H20, and sodium borohydride	Medvedeva et al. [63]
Polyol method	Cobalt acetate tetrahydrate, polyvinylpyrrolidone, and diethylene glycol	Izu et al. [64]
Microwave-assisted synthesis	Cobalt(II) acetate (tetrahydrated), urea, ethylene glycol, and acetone	Prem Latha et al. [65]
Surfactant-assisted synthesis	Cobalt chloride (hexahydrate), sodium tartarate, and trisodium citrate	Raman et al. [66]

green synthesis of cobalt oxide nanoparticles for environmental remediation process was reported by Samuel and coworkers, wherein the nanoparticles were synthesized using Jumbo muscadine (*Vitis rotundifolia*). The photocatalytic activity of green-synthesized cobalt oxide nanoparticles was estimated by degrading Acid Blue-74 dye. About 98% of dye degradation was achieved at 150 min at pH 10 with 60 mg/100 ml concentration [70]. Adsorption of Rhodamine B dye from waste water was reported using cobalt oxide nanoparticles fabricated using hydrothermal technique. Graphene oxide as a supporting substance and ascorbic acid as a reducing agent were employed in the

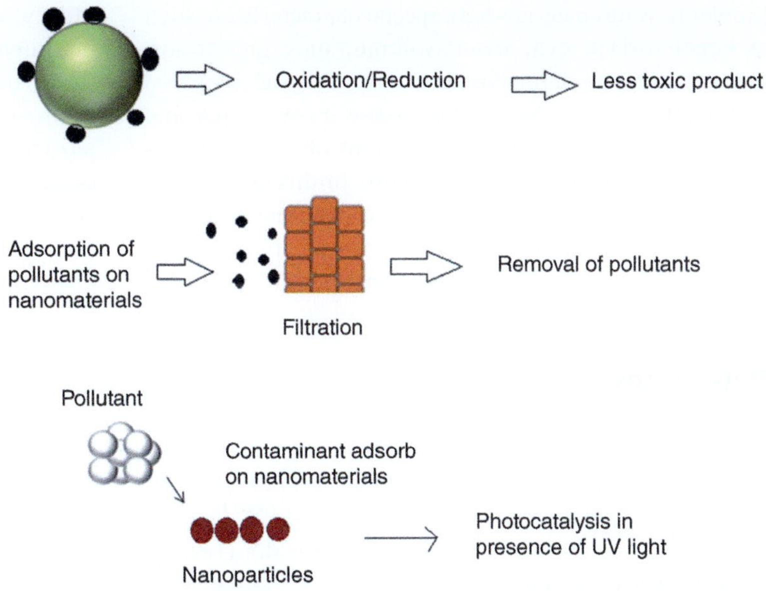

Figure 17.2 Schematic representation of the mechanisms of nanoparticles in the removal of contaminants.

Table 17.3 Applications of cobalt oxide nanomaterials in environmental application.

Type of preparation	Remediation of material	Reference
Cobalt oxide supported on graphene oxide	Degradation of orange II in water	Shi et al. [75]
Spinel-type cobalt oxide nanoparticles	Carbon monoxide oxidation activity	Pandey et al. [76]
Cobalt oxide nanoparticles	Dehydrogenation of cyclohexane	Tyo et al. [77]
Cobalt oxide nanocrystals	Catalytic reduction of nitric oxide by ammonia	Meng et al. [78]
Cobalt oxide nanorod	Reduction of nitric oxide with carbon monoxide	Zhang et al. [79]
Gold-doped cobalt oxide catalyst	Oxidation of propane and toluene	Solsona et al. [80]
Cobalt oxide nanoparticles	Oxidation of chlorinated short-chain alkanes (model: 1,2-dichloroethane)	De Rivas et al. [81]
Cobalt oxide nanowires and nanorods	Catalysis of toluene	Bai et al. [82]
Cobalt oxide nanoflower clusters	Catalytic oxidation of gaseous toluene	Yan et al. [83]
Mesoporous cobalt oxide	Catalytic oxidation of toluene and methanol	Xia et al. [84]
Cobalt oxide nanotubes	Methane combustion	Fei et al. [85]
Cobalt oxide nanobelt and nanocube	Carbon monoxide oxidation	Hu et al. [86]

preparation of the nanoparticles [71]. Ferrous and nickel-doped cobalt oxide nanoparticles were fabricated to study the photocatalytic degradation of Eosin blue dye. About 95% of degradation of the dye in waste water was observed with the modified nanoparticles [72]. Photocatalytic degradation of Methylene blue dye in wastewater management was studied using cobalt oxide nanoparticles. Nanoparticles were prepared using coprecipitation method. The result demonstrated successful photocatalytic degradation of waste water dyes in 120 minutes [73]. In another similar study, cobalt oxide nanoparticles were synthesized using thermal decomposition method for catalytic degradation of methylene blue. Aqua pentaammine-cobalt(III) nitrate was employed as the precursor complex for the synthesis of nanoparticles [74]. Table 17.3 summarizes various other applications of cobalt oxide nanomaterials for environmental remediation.

17.3 Future Perspectives

It is likely that the use of nanomaterials will have a significant impact with greater sensitivity and greater productivity in rectifying the existing declining environmental problems. Nanomaterials have special characteristics, such as larger surface area, area-to-volume ratios, greater surface sensitivity, and tunable semiconductivity, as compared to their bulk equivalents. This makes them unique and innovative resources in the treatment of contaminants to fulfill the functional requirements. In this chapter, we have discussed about metallic oxide nanomaterials and emphasized the method of preparation and application of cobalt oxide nanoparticles in environmental remediation. Mostly, cobalt oxide nanomaterials act as catalysts for decomposition of pollutants owing to their excellent redox properties, although significant efforts are needed to improve the practical applications of cobalt oxide nanoparticles. Despite recent nanotechnological advances in environmental remediation, appropriate assessment and ecosystem-wide studies of these cobalt oxide nanomaterials have to be carried out in order to avoid any detrimental environmental consequences.

References

1 Khan, F.I. and Ghoshal, A.K. (2000). Removal of volatile organic compounds from polluted air. *Journal of Loss Prevention in the Process Industries* 13: 527–545.

2 Marley, M.C., Hazebrouck, D.J., and Walsh, M.T. (1992). The application of in situ air sparging as an innovative soils and ground water remediation technology. *Ground Water Monitoring and Remediation* 12 (2): 137–145.

3 Masciangioli, T. and Zhang, W.X. (2003). Environmental technologies at the nanoscale. *Environmental Science and Technology* 37 (5): 102A–108A. https://doi.org/10.1021/es0323998.

4 Sarkar, B., Tsang, D.C.W., Song, H. et al. (2019). Technological innovation for soil/sediment remediation. *Journal of Soils and Sediments* 19: 3881–3882.

5 Vaseashta, A., Vaclavikova, M., Vaseashta, S. et al. (2007). Nanostructures in environmental pollution detection, monitoring, and remediation. *Science and Technology of Advanced Materials* 8: 47–59.

6 Wang, L., Hou, D., Cao, Y. et al. (2020). Remediation of mercury contaminated soil, water, and air: a review of emerging materials and innovative technologies. *Environment International* 134: 105281. https://doi.org/10.1016/j.envint.2019.105281.

7 Bian, Z., Dong, J., Lei, S. et al. (2009). The impact of disposal and treatment of coal mining wastes on environment and farmland. *Environmental Geology* 58: 625–634.

8 Xudong, L. (2011). Study on environmental pollution of agriculture and countermeasures under the double failure. *Energy Procedia* 5: 204–208.

9 Peschke, K., Capowiez, Y., Kohler, H.R. et al. (2019). Impact of a wastewater treatment plant upgrade on amphipods and other macroinvertebrates: individual and community responses. *Frontiers in Environmental Science* 7: 64. https://doi.org/10.3389/fenvs.2019.00064.

10 Tchounwou, P.B., Yedjou, C.G., Patlolla, A.K., and Sutton, D.J. (2012). Heavy metal toxicity and the environment. *Experientia Supplementum* 101: 133–164.

11 Kautsky, U., Saetre, P., Berglund, S. et al. (2016). The impact of low and intermediate-level radioactive waste on humans and the environment over the next one hundred thousand years. *Journal of Environmental Radioactivity* 151 (2): 395–403.

12 IBISWorld Industry Report (2020). Remediation & environmental cleanup services in the US industry trends (2015–2020). https://www.ibisworld.com/united-states/market-research-reports/remediation-environmental-cleanup-services-industry (accessed September 2020).

13 Sadegh, H., Ali, G.A.M., Gupta, V.K. et al. (2017). The role of nanomaterials as effective adsorbents and their applications in wastewater treatment. *Journal of Nanostructure in Chemistry* 7: 1–14.

14 Khin, M.M., Nair, A.S., Babu, V.J. et al. (2012). A review on nanomaterials for environmental remediation. *Energy and Environmental Science* 5: 8075–8109.

15 Shaheen, I. and Ahmad, K.S. (2020). Green synthesis of doped Co_3O_4 nanocatalysts using organic template for fast azo dye degradation from aqueous environment. *Journal of Chemical Technology and Biotechnology* 95 (11): 2898–2910. https://doi.org/10.1002/jctb.6449.

16 Al-Qahtani, K.M. (2017). Cadmium removal from aqueous solution by green synthesis zero valent silver nanoparticles with Benjamina leaves extract. *The Egyptian Journal of Aquatic Research* 43 (4): 269–274.

17 Moustafa, M.T. (2017). Removal of pathogenic bacteria from wastewater using silver nanoparticles synthesized by two fungal species. *Water Science* 31 (2): 164–176.

18 Truong, V.V. (2012). Photocatalytic degradation of organic dye by sol-gel-derived gallium-doped anatase titanium oxide nanoparticles for environmental remediation. *Journal of Nanomaterials* 2012: 201492. https://doi.org/10.1155/2012/201492.

19 Zand, A.D., Tabrizi, A.M., and Heir, A.V. (2020). Co-application of biochar and titanium dioxide nanoparticles to promote remediation of antimony from soil by Sorghum bicolor: metal uptake and plant response.

Heliyon 6: e04669. https://doi.org/10.1016/j.heliyon.2020.e04669.

20 Velez, E., Campillo, G.E., Morales, G. et al. (2016). Mercury removal in wastewater by iron oxide nanoparticles. *Journal of Physics Conference Series* 687: 012050. https://doi.org/10.1088/1742-6596/687/1/012050.

21 Hamdy, A., Mostafa, M.K., and Nasr, M. (2018). Zero-valent iron nanoparticles for methylene blue removal from aqueous solutions and textile wastewater treatment, with cost estimation. *Water Science and Technology* 78 (2): 367–378.

22 Sharahi, F.J. and Shahbazi, A. (2017). Melamine-based dendrimer amine-modified magnetic nanoparticles as an efficient Pb(II) adsorbent for wastewater treatment: adsorption optimization by response surface methodology. *Chemosphere* 189: 291–300.

23 Yang, S., Nie, J., Wei, F., and Yang, X. (2016). Removal of ozone by carbon nanotubes/quartz fiber film. *Environmental Science and Technology* 50 (17): 9592–9598.

24 Bai, B.C., Lee, Y.S., and Im, J.S. (2019). Activated carbon fibers for toxic gas removal based on electrical investigation: mechanistic study of p-type/n-type junction structures. *Scientific Reports* 9: 14458. https://doi.org/10.1038/s41598-019-50707-x.

25 Perera, M., Wijenayaka, L.A., Siriwardana, K. et al. (2020). Gold nanoparticle decorated titania for sustainable environmental remediation: green synthesis, enhanced surface adsorption and synergistic photocatalysis. *RSC Advances* 10: 29594–29602.

26 Ugwu, O.O., Ogboin, A.S., and Nwoji, C.U. (2018). Characterization of engineering properties of active soils stabilized with nanomaterial for sustainable infrastructure delivery. *Frontiers in Built Environment* 4: 65. https://doi.org/10.3389/fbuil.2018.00065.

27 Baragaño, D., Forján, R., Welte, L., and Gallego, J.L.R. (2020). Nanoremediation of As and metals polluted soils by means of graphene oxide nanoparticles. *Scientific Reports* 10: 1896. https://doi.org/10.1038/s41598-020-58852-4.

28 Apul, O.G., Delgado, A.G., Kidd, J. et al. (2016). Carbonaceous nano-additives augment microwave-enabled thermal remediation of soils containing petroleum hydrocarbons. *Environmental Science: Nano* 3 (5): 997–1002.

29 Le, T.S., Dao, T.H., Nguyen, D.C. et al. (2015). Air purification equipment combining a filter coated by silver nanoparticles with a nano-TiO_2 photocatalyst for use in hospitals. *Advances in Natural Sciences: Nanoscience and Nanotechnology* 6 (1): 015016. https://doi.org/10.1088/2043-6262/6/1/015016.

30 Elmi, F., Alinezhad, H., Moulana, Z. et al. (2014). The use of antibacterial activity of ZnO nanoparticles in the treatment of municipal wastewater. *Water Science and Technology* 70 (5): 763–770.

31 Araoyinbo, A.O., Abdullah, M.M.A.B., Rahmat, A. et al. (2018). Preparation of heat-treated titanium dioxide (TiO_2) nanoparticles for water purification. *IOP Conference Series: Materials Science and Engineering* 374: 012084. https://doi.org/10.1088/1757-899X/374/1/012084.

32 Goud, R.K., Ajay Kumar, V., Reddy, T.R. et al. (2018). Effect of metal oxide nanoparticles on Godavari river water treatment. *AIP Conference Proceedings* 1966 (1): 020003. https://doi.org/10.1063/1.5038682.

33 Lamba, R., Umar, A., Mehta, S.K., and Kansal, S.K. (2015). ZnO doped SnO_2 nanoparticle heterojunction photo-catalyst for environmental remediation. *Journal of Alloys and Compounds* 653: 327–333.

34 Hafeez, M., Shaheen, R., Akram, B. et al. (2020). Green synthesis of cobalt oxide nanoparticles for potential biological applications. *Materials Research Express* 7 (2): 025019. https://doi.org/10.1088/2053-1591/ab70dd.

35 Omran, B.A., Nassar, H.N., Younis, S.A. et al. (2020). Novel mycosynthesis of cobalt oxide nanoparticles using *Aspergillus brasiliensis* ATCC 16404—optimization, characterization and antimicrobial activity. *Journal of Applied Microbiology* 128 (2): 438–457.

36 Meghdadi, S., Amirnasr, M., Zhiani, M. et al. (2017). Facile synthesis of cobalt oxide nanoparticles by thermal decomposition of cobalt(II) carboxamide complexes: application as oxygen evolution reaction electrocatalyst in alkaline water electrolysis. *Electrocatalysis* 8: 122–131.

37 Rahimi-Nasrabadi, M., Naderi, H.R., Karimi, M.S. et al. (2017). Cobalt carbonate and cobalt oxide nanoparticles synthesis, characterization and supercapacitive evaluation. *Journal of Materials Science: Materials in Electronics* 28: 1877–1888.

38 Khalaji, A.D. (2019). Cobalt oxide nanoparticles by solid-state thermal decomposition: synthesis and characterization. *Eurasian Chemical Communications* 1: 113–116. https://doi.org/10.33945/SAMI/ECC.2019.1.10.

39 Xaba, T., Mokgoera, L., Nate, Z., and Shumbula, P.M. (2018). Synthesis and characterization of cubic structured cobalt oxide nanoparticles capped with TOPO through the decomposition of bis(*N*-cyclohexyl-1-naphthaldehydato)cobalt(II) complex as a single source precursor. *Digest Journal of Nanomaterials and Biostructures* 13 (4): 1141–1147.

40 Qasem, M.A.A., Aziz, M.A., Hakeem, A.S., and Onaizi, S.A. (2018). Preparation of nano-Co_3O_4 by direct thermal decomposition of cobalt(II) nitrate hexahydrate for electrochemical water oxidation. *Current Nanoscience* 14

(2): 154–159. https://doi.org/10.2174/1573413713666171201150215.

41 Singh, M., Ralhan, N.K., and Singh, S. (2015). Synthesis, characterization, scale-up and catalytic behaviour of Co_3O_4 nanoparticles. *Bulletin of Materials Science* 38 (2): 297–301.

42 Makhlouf, M.T., Abu-Zied, B.M., and Mansoure, T.H. (2013). Direct fabrication of cobalt oxide nanoparticles employing sucrose as a combustion fuel. *Journal of Nanoparticles* 2013: 384350. https://doi.org/10.1155/2013/384350.

43 Alayoglu, S., Rosenberg, D.J., and Ahmed, M. (2016). Hydrothermal synthesis and characterization under dynamic conditions of cobalt oxide nanoparticles supported over magnesium oxide nano-plates. *Dalton Transactions* 45 (24): 9932–9941.

44 Lv, L., Su, Y.G., Liu, X.Q. et al. (2013). Synthesis of cellular-like Co_3O_4 nanocrystals with controlled structural, electronic and catalytic properties. *Journal of Alloys and Compounds* 553: 163–166.

45 Khalil, A.T., Ovais, M., Ullah, I. et al. (2020). Physical properties, biological applications and biocompatibility studies on biosynthesized single phase cobalt oxide (Co_3O_4) nanoparticles via Sageretia thea (Osbeck.). *Arabian Journal of Chemistry* 13 (1): 606–619.

46 Diallo, A., Beye, A.C., Doyle, T.B. et al. (2015). Green synthesis of Co_3O_4 nanoparticles via *Aspalathus linearis*: physical properties. *Green Chemistry Letters and Reviews* 8 (3–4): 30–36.

47 Ranaei Siadat, S.O. (2015). Green synthesized cobalt nano particles for using as a good candidate for sensing organic compounds. *Journal of Electrochemical Science and Technology* 6 (4): 111–115.

48 Urabe, A.A. and Aziz, W.J. (2019). Biosynthesis of cobalt oxide (Co_3O_4) nanoparticles using plant extract of *Camellia sinensis* (L.) Kuntze and *Apium graveolens* L. as the antibacterial application. *World News of Natural Sciences* 24: 357–365.

49 Dubey, S., Kumar, J., Kumar, A., and Sharma, Y.C. (2018). Facile and green synthesis of highly dispersed cobalt oxide (Co_3O_4) nano powder: characterization and screening of its eco-toxicity. *Advanced Powder Technology* 29 (11): 2583–2590.

50 Bibi, I., Nazar, N., Iqbal, M. et al. (2017). Green and eco-friendly synthesis of cobalt-oxide nanoparticle: characterization and photo-catalytic activity. *Advanced Powder Technology* 28 (9): 2035–2043.

51 Matinise, N., Mayedwa, N., Fuku, X.G. et al. (2018). Green synthesis of cobalt (II, III) oxide nanoparticles using *Moringa oleifera* natural extract as high electrochemical electrode for supercapacitors. *AIP Conference Proceedings* 1962: 040005. https://doi.org/10.1063/1.5035543.

52 Sardjono, S.A. and Puspitasari, P. (2020). Synthesis and characterization of cobalt oxide nanoparticles using sol-gel method. *AIP Conference Proceedings* 2231: 040046. https://doi.org/10.1063/5.0002419.

53 Jagtap, S.V., Tale, A.S., and Thakre, S.D. (2017). Synthesis by sol gel method and characterization of Co_3O_4 nanoparticles. *International Journal of Research in Engineering and Applied Sciences* 7 (8): 1–6.

54 Lima-Tenório, M.K., Ferreira, C.S., Rebelo, Q.H.F. et al. (2018). Pseudocapacitance properties of Co_3O_4 nanoparticles synthesized using a modified sol-gel method. *Materials Research* 21 (2): e20170521. https://doi.org/10.1590/1980-5373-mr-2017-0521.

55 Vaya, D., Meena, and Das, B.K. (2019). Green synthesis of cobalt oxide nanoparticles by a starch-assisted method. *Nanoscience & Nanotechnology-Asia* 9 (3): 362–370.

56 Prabaharan, D.D.M., Sadaiyandi, K., Mahendran, M., and Sagadevan, S. (2017). Precipitation method and characterization of cobalt oxide nanoparticles. *Applied Physics A* 123 (4): 264. https://doi.org/10.1007/s00339-017-0786-8.

57 Januja, M.R.S.A. (2017). Synthesis of Co_3O_4 nano aggregates by co-precipitation method and its catalytic and fuel additive applications. *Open Chemistry* 17 (1): 865–873.

58 Yang, W., Lu, J.E., Zhang, Y. et al. (2019). Cobalt oxides nanoparticles supported on nitrogen-doped carbon nanotubes as high-efficiency cathode catalysts for microbial fuel cells. *Inorganic Chemistry Communications* 105: 69–75.

59 Abbas, T.A.H., Slewa, L.H., Khizir, H.A., and Kakil, S.A. (2017). Synthesis of cobalt oxide (Co_3O_4) thin films by electrostatic spray pyrolysis technique (ESP). *Journal of Materials Science: Materials in Electronics* 28: 1951–1957.

60 Ghaem, E.N., Dorranian, D., and Sari, A.H. (2020). Characterization of cobalt oxide nanoparticles produced by laser ablation method: effects of laser fluence. *Physica E: Low-dimensional Systems and Nanostructures* 115: 113670. https://doi.org/10.1016/j.physe.2019.113670.

61 Rousta, A., Dorranian, D., and Elahi, M. (2020). Electrophoretic deposition of cobalt oxide nanoparticles on aluminium substrate. *Surface Engineering* 36 (9): 919–928.

62 Athar, T., Hakeem, A., Topnani, N., and Hashmi, A. (2012). Wet synthesis of monodisperse cobalt oxide nanoparticles. *International Scholarly Research Notices* 2012: 691032. https://doi.org/10.5402/2012/691032.

63 Medvedeva, O.I., Kambulova, S.S., Bondar, O.V. et al. (2017). Magnetic cobalt and cobalt oxide nanoparticles in hyperbranched polyester polyol matrix. *Journal of*

Nanotechnology 2017: 7607658. https://doi.org/10.1155/2017/7607658.

64 Izu, N., Matsubara, I., Uchida, T. et al. (2017). Synthesis of spherical cobalt oxide nanoparticles by a polyol method. *Journal of the Ceramic Society* 125 (9): 701–704.

65 Prem Latha, K., Prema, C., and Sundar, S.M. (2018). Synthesis and characterization of cobalt oxide nanoparticles. *Journal of Nanoscience and Technology* 4 (5): 475–477.

66 Raman, V., Suresh, S., Savarimuthu, P.A. et al. (2016). Synthesis of Co_3O_4 nanoparticles with block and sphere morphology, and investigation into the influence of morphology on biological toxicity. *Experimental and Therapeutic Medicine* 11: 553–560.

67 Adekunle, A.S., Oyekunle, J.A.O., Durosinmi, L.M. et al. (2020). Potential of cobalt and cobalt oxide nanoparticles as nanocatalyst towards dyes degradation in wastewater. *Nano-Structures & Nano-Objects* 21: 100405. https://doi.org/10.1016/j.nanoso.2019.100405.

68 Mohanta, J., Dey, B., and Dey, S. (2020). Magnetic cobalt oxide nanoparticles: sucrose-assisted self-sustained combustion synthesis, characterization, and efficient removal of malachite green from water. *Journal of Chemical and Engineering Data* 65 (5): 2819–2829.

69 Siddique, M., Khan, N.M., Saeed, M. et al. (2021). Green synthesis of cobalt oxide nanoparticles using *Citrus medica* leaves extract: characterization and photo-catalytic activity. *Zeitschrift für Physikalische Chemie* 235 (6): 663–681. https://doi.org/10.1515/zpch-2019-1583.

70 Samuel, M.S., Selvarajan, E., Mathimani, T. et al. (2020). Green synthesis of cobalt-oxide nanoparticle using jumbo Muscadine (*Vitis rotundifolia*): characterization and photo-catalytic activity of acid Blue-74. *Journal of Photochemistry and Photobiology B: Biology* 211: 112011. https://doi.org/10.1016/j.jphotobiol.2020.112011.

71 Alwan, S.H., Alshamsi, H.A.H., and Jasim, L.S. (2018). Rhodamine B removal on A-rGO/cobalt oxide nanoparticles composite by adsorption from contaminated water. *Journal of Molecular Structure* 1161: 356–365.

72 Koli, P.B., Kapadnis, K.H., Deshpande, U.G., and Patil, M.R. (2018). Fabrication and characterization of pure and modified Co_3O_4 nanocatalyst and their application for photocatalytic degradation of eosine blue dye: a comparative study. *Journal of Nanostructure in Chemistry* 8: 453–463.

73 Mathew, J. and Shetty, N. (2017). Treatment of wastewater using synthesised photocatalyst cobalt oxide CO_3O_4. *International Journal of Civil Engineering and Technology* 8 (4): 1840–1844.

74 Farhadi, S., Javanmard, M., and Nadri, G. (2016). Characterization of cobalt oxide nanoparticles prepared by the thermal decomposition of $[Co(NH_3)_5(H_2O)](NO_3)_3$ complex and study of their photocatalytic activity. *Acta Chimica Slovenica* 63 (2): 335–343.

75 Shi, P., Su, R., Zhu, S. et al. (2012). Supported cobalt oxide on graphene oxide: highly efficient catalysts for the removal of Orange II from water. *Journal of Hazardous Materials* 229–230: 331–339.

76 Pandey, A.D., Jia, C., Schmidt, W. et al. (2012). Size-controlled synthesis and microstructure investigation of Co_3O_4 nanoparticles for low-temperature CO oxidation. *The Journal of Physical Chemistry C* 116 (36): 19405–19412.

77 Tyo, E.C., Yin, C.R., Vece, M.D. et al. (2012). Oxidative dehydrogenation of cyclohexane on cobalt oxide (Co_3O_4) nanoparticles: the effect of particle size on activity and selectivity. *ACS Catalysis* 2: 2409–2423.

78 Meng, B., Zhao, Z.B., Wang, X.Z. et al. (2013). Selective catalytic reduction of nitrogen oxides by ammonia over Co_3O_4 nanocrystals with different shapes. *Applied Catalysis B: Environmental* 129: 491–500.

79 Zhang, S.R., Shan, J.J., Zhu, Y. et al. (2013). Restructuring transition metal oxide nanorods for 100% selectivity in reduction of nitric oxide with carbon monoxide. *Nano Letters* 13: 3310–3314.

80 Solsona, B., Aylón, E., Murillo, R. et al. (2011). Deep oxidation of pollutants using gold deposited on a high surface area cobalt oxide prepared by a nanocasting route. *Journal of Hazardous Materials* 187 (1–3): 544–552.

81 De Rivas, B., López-Fonseca, R., Jiménez-González, C., and GutiérrezOrtiz, J.I. (2012). Highly active behaviour of nanocrystalline Co_3O_4 from oxalate nanorods in the oxidation of chlorinated short chain alkanes. *Chemical Engineering Journal* 184: 184–192.

82 Bai, G.M., Dai, H.X., Deng, J.G. et al. (2013). Porous Co_3O_4 nanowires and nanorods: highly active catalysts for the combustion of toluene. *Applied Catalysis A: General* 450: 42–49.

83 Yan, Q.Y., Li, X.Y., Zhao, Q.D., and Chen, G.H. (2012). Shape-controlled fabrication of the porous Co_3O_4 nanoflower clusters for efficient catalytic oxidation of gaseous toluene. *Journal of Hazardous Materials* 209–210: 385–391.

84 Xia, Y., Dai, H., Jiang, H., and Zhang, L. (2010). Three-dimensional ordered mesoporous cobalt oxides: highly active catalysts for the oxidation of toluene and methanol. *Catalysis Communications* 11 (15): 1171–1175.

85 Fei, Z., He, S., Li, L. et al. (2012). Morphology-directed synthesis of Co_3O_4 nanotubes based on modified Kirkendall effect and its application in CH_4 combustion. *Chemical Communications* 48 (6): 853–855.

86 Hu, L.H., Sun, K.Q., Peng, Q. et al. (2010). Surface active sites on Co_3O_4 nanobelt and nanocube model catalysts for CO oxidation. *Nano Research* 3: 363–368.

18

Eco-friendly Nanotechnology in Agriculture: Opportunities, Toxicological Implications, and Occupational Risks

Layla Muraisi[1], Dewi Melani Hariyadi[2], Umi Athiyah[2], and Yashwant Pathak[1,3]

[1] *Taneja College of Pharmacy, University of South Florida, Tampa, FL, USA*
[2] *Faculty of Pharmacy, Universitas Airlangga, Surabaya, Indonesia*
[3] *Faculty of Pharmacy, Universitas Airlangga, Surabaya, Indonesia*

18.1 Introduction

According to the National Nanotechnology Initiative, nanotechnology is defined as the science, engineering, and technology conducted at the nanoscale, which is about 1–100 nm. In the past few decades, nanotechnology has made its way to the agriculture industry. The ideas of nanotechnology have been applied to create pesticides and fertilizers that are at the nanoscale to reduce the ecological harm they cause and to increase the efficiency of fertilizers and pesticides. With the production of these new products it looked very promising for the agriculture industry, and it has been. Many studies have shown that the use of nanotechnology in agriculture can increase crop yield and nutrient levels in the crop being harvested. These developments have been important because of the recent amount of food scarcity, droughts, and lack of nutritious food around the globe. Attention to these atrocities has brought the concept of global sustainability, which is defined by Cambridge as humans and nature, societies and the biosphere, and the world and the Earth can coexist in ways that enable productive harmony, stability, and resilience to support the present and future generations. This ties in significantly to nanotechnology in agriculture because if nanotechnology can be used to safely fertilize the land, to only target specific organism with pesticides and reduce the damage to the land, it will allow the globe to maintain land where future generation can farm on. However, with the development of these new products it raises a lot of questions as to how safe are the nanopesticides and nanofertilizers? What constitutes a nanopesticide and nanofertilizer? Will they cause more severe harm to the environment than pesticides and fertilizers in use currently? These questions are being explored, but there is no definitive answer to any. When it comes to the safety of nanopesticides and nanofertilizers, a lot of concerns are raised about the digestion of these nanomaterials and their capability to invade our cells, but no decrees by the governments around the world have been made. When it comes to what are considered nanotechnology, governments cannot agree and are relying on old laws for agriculture to apply for the current products in circulation. If laws and regulations can be established for nanotechnology, then this will aid in helping to achieve some of the United Nation Sustainability goals such as GOAL 3: Good Health and Well-being, GOAL 6: Clean Water and Sanitation, GOAL 7: Affordable and Clean Energy, GOAL 8: Decent Work and Economic Growth, GOAL 9: Industry, Innovation, and Infrastructure, GOAL 11: Sustainable Cities and Communities, GOAL 12: Responsible Consumption and Production, GOAL 13: Climate Action, and GOAL 15: Life on Land. Nanotechnology in agriculture is pertinent for the development and progression of the industry.

Sustainable Nanotechnology: Strategies, Products, and Applications, First Edition.
Edited by Yashwant V. Pathak, Govindan Parayil, and Jayvadan K. Patel.

18.2 Nanotechnology-enabled Agrochemicals

Agrochemicals is in reference to fertilizers and pesticides used in the agriculture business. And they are used by a significant amount of people. However, there is a downside to agrochemicals, they can be a health hazard to those who distribute it, can pollute the water nearby, have a negative long-lasting effect on the soil, and lead to eradication of organisms that are not the target (Hasan et al. [1], pp. 75–79). A case study documented that farms in Indonesia use an excessive amount of agrochemicals in intensive farming, a majority of the agrochemicals ended up as waste and having adverse impacts (Mariyono et al. [2], pp. 773–774). To reduce the agrochemical waste and to prevent adverse impacts, nanotechnology can be used. Nanotechnology has the ability to target specific organisms, reduce the harm agrochemicals have, and improve the efficiency of farming. Refer Table 18.1 for more insight.

18.2.1 Nanoscale Carriers

Nanoscale carriers could be utilized for the efficient delivery of fertilizers, pesticides, herbicides, and plant growth regulators. The mechanisms employed by these carriers in the efficient delivery, better storage, and controlled release include encapsulation and entrapment, polymers and dendrimers, and surface ionic and weak bond attachments (Ditta et al. [4], p. 62). These mechanisms help to improve their stability against degradation in the environment and ultimately reduce the amount to be applied, which reduces chemical runoff and alleviates environmental problems [5].

18.2.2 Nanobarcode Technology

In our daily life, identification tags have been applied in wholesale agriculture and livestock products. Due to their small size, nanoparticles have been applied in many fields ranging from advanced biotechnology to agricultural encoding. Nanobarcodes (>1 million) have been applied in multiplexed bioassays and general encoding because of their possibility of formation of a large number of combinations that make them attractive for this purpose (Mathew et al. [6], p. 10). However, it is important to consider the risks of nanobarcoding such as the dependence of nanobarcodes for anticounterfeiting of edible products including food and drugs, which requires even more careful consideration as they are ingested by the end users (Shikha et al. [7], p. 46). For instance, the organic solid-phase change nanoparticles such as stearic acid, palmitic acid, paraffin wax, and polyethylene have been demonstrated to be mixed with drug powder such as acetaminophen (Duong et al. [8], p. 4). Nanobarcoding is also used for the packaging of agriculture goods and husbandry products. This leads to opportunities to help improve traceability in the food trade and be a tool in the promotion of biosafe global agrofood business (Jain et al. [9], pp. 11–12).

18.2.3 Nanofertilizers

Nanofertilizers are nanomaterials responsible for providing one or more types of nutrients to the growing plants and support their growth and improve production (Liu and Lal [10], pp. 121–129). Nanofertilizers provide more surface area and more availability of nutrients to the crop plant which help to increase these quality parameters of the plant (such as protein, oil content, and sugar content) by enhancing the rate of reaction or synthesis process in the plant system (Qureshi et al. [11], pp. 3325–3335). Nanofertilizers are going to be key for global sustainability with their ability to produce greater crop yield. Seeing as there are food shortages in several countries in the world, it is pertinent to use nanofertilizers to aid those in a food crisis such as the one occurring in Yemen where famine is common for many citizens (Tarran [12], p. 28). When looking into nanofertilizers, the outcome desired can determine the type of nanofertilizer used. There are three general categories of nanofertilizers based on plant nutrient requirement: (i) macronanofertilizer, (ii) micronanofertilizer, and (iii) nanoparticulate fertilizer (Chhipa [13], p. 16). Macronanofertilizers are utilized when nutrient is required in large amount in conventional farming practices (Chhipa [13], p. 16). Nitrogen (N), phosphorus (P), potassium (K), magnesium (Mg), sulfur (S), and calcium (Ca) are included as macronutrients for plant growth (Chhipa [13], p. 16). Micronanofertilizers are those used in minute quantity. An example of the use of micronanofertilizers is found in the study "Evaluation of efficacy of ZnO nanoparticles as remedial zinc nanofertilizer for rice," which confirmed the potential of foliar application of zinc micronutrient nanofertilizer for remediation of the Zn-deficiency symptoms in rice [14]. This is important because around the globe rice is a staple food for many households and accounts for 27% of dietary energy for the developing world, according to the United Nations [15]. Nanoparticles increase the efficiency of nutrients use and minimize the costs of environmental protection (Nair et al. [16], pp. 154–163). An example of the use of nanoparticles is seen in the study "Effects of Zn/B nanofertilizer on biophysical characteristics and growth of coffee seedlings in a greenhouse," where it was found that chitosan nanoparticles prepared by ionic gelation are suitable for the synthesis of Zn/B nanofertilizer [17]. The Zn/B nanofertilizer was adsorbed easily by the plant and had positive effects on the growth of coffee seedlings in a greenhouse by enhancing the

Table 18.1 Nanotechnology in agriculture: applications, opportunities, and practical challenges (Iavicoli et al. [3], p. 97).

Nanotechnology in Agriculture			
	Applications	Opportunities	Practical challenges
Nano-enabled agrochemicals	*Nanofertilizers*	✓ Targeted and controlled nutrient release	✓ Nanomaterial phytotoxicity
	✓ Macronutrient nanofertilizers	✓ Increased nutrient availability and plant uptake efficiency	✓ Variability and reactivity of nanomaterials in the environment
	✓ Micronutrient nanofertilizers	✓ Increased enzymatic activity	✓ Possible adverse effects for exposed workers under real application conditions
	✓ Nutrient-loaded nanofertilizers, i.e. nutrient-augmented zeolites	✓ Reduced adverse impact of conventional compounds	
	✓ Plant growth enhancers, i.e. TiO_2-NPs and CNTs		
	Nanopesticides	✓ Greater pesticide solubility, mobility, and durability	✓ Biosafety of nanopesticides
	✓ Nanoemulsions, -dispersions, -spheres, -capsules, and -gels of traditional pesticides	✓ Reduced amount of ingredients via targeted/controlled release	✓ Toxicological profile: interactions with coformulants; environmental fate
	✓ Solid lipid NPs, coated liposomes, or inorganic NPs associated with active ingredients	✓ Reduced resistance and damage to nontarget organisms	✓ Long-term effects on the environment and chronically exposed workers under real application conditions
	✓ Engineered NPs, i.e. Ag- and TiO_2-NPs		
Detection and remediation of environmental pollution	*Sensing devices*	✓ Improved detection limits for real-time monitoring of toxicants	✓ Fabrication and validation of sensitive instruments
	✓ Nanosensors to monitor a variety of pesticides and pathogens as well as to assess crop growth, field conditions, and early biological alterations	✓ Precise farming support: reduction of inputs and sustainable use of resources	✓ Environmental and health consequences of nanomaterial released from devices
	Water and soil remediation	✓ Destruction and transformation of toxic contaminants with high removal effectiveness	✓ Environmental and health effects of nanomaterials and released products
	✓ Zero-valent iron NPs, metal oxide NPs, i.e. TiO_2. Fe_{34}, bimetallic NPs, and carbon-based nanomaterials have been explored for degradation of chlorinated, halogenated compounds and heavy metals	✓ Reduced cleanup time	✓ No knowledge on nanomaterial large-scale application, regeneration, and reutilization
		✓ Cost and energy-efficient processes	
Nanoscale agricultural products	*Water remediation and crop coating*	✓ Good adsorbent properties due to nanocellulose high surface area, mechanical properties, biocompatibility, easily functionalizable surface, sustainable sources, and biodegradability	✓ *in vitro* studies demonstrated some cytotoxic and genotoxic effects of cellulose nanomaterials
	✓ Cellulose nanocrystals or nanofibers derived from plant-based resources are promising adsorbents for water remediation	✓ Improved membrane strength, selectivity, permeability, hydrophilicity, and biofouling resistance	✓ *in vivo* exposure via the respiratory tract induced lung inflammatory and oxidative stress reactions
	✓ Nanocellulose may stabilize reactive NPs or improve membrane properties		✓ Concerns regarding nanocellulose high aspect ratio, stiffness, biodurability
	✓ A cellulose nanofiber-based coating for reducing cherry rain cracking has been recently developed		

Table 18.2 Depicts ways in which nanotechnology can be applied and used in todays' agriculture.[a]

Nanofertilizer	Crop name	Amount of nanofertilizer (ppm)	Impact	References
Au	Pearl millet (*Pennisetum glaucum*)	50	Improved seed germination and growth of seedlings	Parveen et al. [18]
Ca	Peanut (*Arachis hypogaea*)		Nutrient content increased in shoot and root	Liu et al. [19]
CeO_2	Cucumber (*Cucumis sativus*)	400	Improved scratch and globulin content	Zhao et al. [20]
CNTs	Date palm (*Phoenix dactylifera*)	0.05–0.1	Shoot length and number of leaf increased	Taha et al. [21]
	Tobacco	5–500	55–64% plant growth improved	Khodakovskaya et al. [22]
Cu	Lettuce (*Lactuca sativa*)	130–600	Shoot and root length increased	Shah and Belozerova [22]
CuO	Maize (*Zea mays*)	10	51% plant growth	Adhikari et al. [23]
Fe/SiO_2	Barley (*Hordeum vulgare*) and maize (*Zea mays*)	0–25	Improved mean germination time	Najafi Disfani et al. [24]
FeO	Soybean (*Glycine max*)	30–60	Chlorophyll increased	Ghafariyan et al. [25]
	Pea (*Pisum sativum*)	250–500	Seed weight and chlorophyll increased	Delfani et al. [26]
Mg	Cow peas (*Vigna unguiculata*)	2.5	Increment in stem Mg content, plasma membrane stability, and chlorophyll content	Delfani et al. [26]
Mn	Mung bean (*Vigna radiata*)	0.05–1	Shoot length, chlorophyll content, and the photosynthesis rate increased	Pradhan et al. [27]
	Rice (*Oryza sativa*)		Improved Zn uptake 5.66 mg/hill	Yuvaraj and Subramanian [28]
Mo	Chickpea (*Cicer arietinum*)	8	Plant mass and number of modules increased	Taran et al. [29]
P	Soybean (*Glycine max*)	100	Increased the growth rate and seed yield by 32.6 and 20.4%	Liu and Lal [10]
TiO_2	Spinach (*Spinacia oleracea*)	0.25–4	Plant dry weight increased by 73%	Zheng et al. [30]
	Spinach (*Spinacia oleracea*)		N_2 fixation improvement	
	Mung bean (*Vigna radiata*)	10	Improvement in plant growth and nutrient content	
	Cowpea (*Vigna unguiculata*)	125 cc Ha^{-1}	Cowpea yield up to 26–51%	Owolade and Ogunleti [31]
Zn	Ryegrass (*Lolium*)	1–2000	Root elongation	Lin and Xing [32]
ZnO	Mung bean (*Vigna radiata*) and chickpea (*Cicer arietinum*)	1–2000	Plant growth increased at 20 ppm in mung bean and in check pea at 1 ppm	Mahajan et al. [33]
	Cucumber (*Cucumis sativus*)	400–800	Root dry weight and fruit gluten increased	
	Rap seed (*Brassica napus*)	1–2000	Root elongation	Lin and Xing [34]
	Peanut (*Arachis hypogaea*)	1000	34% increment in pod yield per plant	Prasad et al. [35]
	Chickpea (*Cicer arietinum*)	1.5	Improved shoot dry weight and antioxidant activity	Burman et al. [36]
	Maize (*Zea Maize*)	10	Improved plant height and dry weight	Adhikari et al. [23]
	Cluster bean (*Cyamopsis tetragonoloba*)	10	Improvement in plant growth and nutrient content	Raliya and Tarafdar [37]

[a] It also shows how this serves to provide opportunities in the field of agriculture, and the reasonable risks that can occur if the nanotechnology is used. *Source:* Chhipa [13].

nutritional uptake, chlorophyll content, and photosynthesis [17]. If this way of farming could be applied to other crops, it could benefit many developing and developed countries. The benefits of nanofertilizers are listed in Table 18.2.

18.2.3.1 Nanopesticides

According to the European Commission [38], a pesticide is defined as "something that prevents, destroys, or controls a harmful organism (pest) or disease, or protects plants or plant products during production, storage, and transport." The term pesticide is also defined by the US EPA [39] as "any substance or mixture of substances intended for preventing, destroying, repelling, or mitigating any pest." The problem with pesticides is the inadvertent harm they can cause by runoff into water, killing organisms that were not intended to be killed, and causing health problems for those dispersing the pesticides. In 2004, the worldwide consumption of pesticides was about two million tons per year, of which 24% was in the United States alone, 45% in Europe, and 25% in the rest of the world (Gupta [40], pp. 83–90). Due to this large usage of pesticides and the harm it causes, there is an opportunity for nanopesticides to reduce this harm using less chemicals. When it comes to nanopesticides, there are a variety of types, such as nanoemulsion, -dispersion, -sphere, -capsule, -gel, and -fibers. The first type of nanopesticides refers to the pesticides that are loaded with nanomaterials or doped in nanomaterials or directly coated with nanomaterials to put a "nanocoat" on their surface (Sun et al. [41], pp. 1–2). However, most of the emulsifiable concentrates and microemulsions used in nanopesticides are polar solvents such as benzenes, alcohols, and ketones, which have high acute toxicity and easily find their way into the farmland and groundwater [42–44].

18.3 Nanotechnology for Detection and Remediation of Environmental Pollutants

The world currently faces a lot of pollutants, climate change, and environmental harm. By utilizing nanomaterials, trace detection sensors have shown great promise for detection of chemical markers of pollutants because the materials are used to either capture the marker or amplify the signal associated with detection [45, 46].

18.3.1 Environmental Monitoring of Toxicants and Pathogens

Nanoparticles such as colloidal metals and inorganic crystals (e.g. quantum dots (QDs)) are currently being used as labels, markers, or probes for detection of pollutant molecules [47]. One study that used sensitive immunochromatographic strip (ICS) assay, based on QDs, was developed for the qualitative and quantitative detection of acetamiprid in agricultural samples. The whole sample detection could be accomplished within one hour. According to the spiked recovery test, the minimum detectable quantity of acetamiprid in tea was $0.02\,mg\,kg^{-1}$ by QDs-ICS [48]. Studies like this are important to test the efficiency, effectiveness, and efficacious of nanotechnology when it comes to the monitoring of environmental pollutants. Other alternatives include biosensors that can monitor microorganisms, organic compounds, and heavy metals in wastewater. This is relevant because agriculture is the largest consumer of water, where the irrigation of croplands accounts for 70% of water withdrawals [49]. The small size of the sensor enables the integration of several processes into a single device to design real-time monitoring of multiple materials, which is very useful for monitoring the composition of wastewater. Furthermore, the use of nanomaterials has a positive effect on their physical and chemical features and improves the sensitivity, decreasing the analysis time, increasing measurement reliability, and demonstrating high-throughput multiplex detection [50–52]. This is important because for global sustainability what people are consuming and the effect that agriculture has on the land and water today and in the future matter.

18.3.2 Nanotechnology for Water and Soil Remediation

Biological remediation, a process defined as the use of microorganisms or plants to detoxify or remove organic and inorganic xenobiotic compounds from the environment, is a remediation option that offers green technology solution to the problem of environmental degradation [53]. The focus here, however, is how nanotechnology can play a role in water and soil remediation after it has been detected by nanotechnology of toxicants and pollutants in the water and soil. Nanoremediation involves applications of reactive nanomaterials to the contaminated soil for either transformation or detoxification of pollutants [54, 55]. While progress has been shown with nanomaterials serving to remediate soil, the risks are still unknown because not enough time has passed to be able to assess any damages or risks it could potentially cause to the environment, ecosystem, and humans [56].

18.4 Agriculture Product Accessibility

Nanoscale agriculture products are easily accessible and can be purchased on websites such as https://www.americanelements.com/agriculture.html with ease. This website offers services to 32 countries including China, India, Brazil, and the United Kingdom. It is observed that they cater to mostly European and Asian countries, with no countries from Africa on their list, and few from South America and the Middle East. They also have big name cliental such as General Motors, Apple, Nike, and Tesla. This is concerning because there is not a lot of regulation yet for nanomaterials and nanoproducts. This poses a potential danger for the public because the long-term exposure effects of these nanomaterials and products have yet to be discovered because of their newness to the market. While many countries still lack proper regulation on nanotechnology, last year India created a guideline for nanoagriculture products, http://dbtindia.gov.in/sites/default/files/DBT_Draft1-Nano-Agri_Input_nd_Nano-Agri_Products.pdf, titled "Guidelines for Evaluation of Nano-agri Input and Nano-agri Products in India." This shows a progression toward regulation for nanotechnology, which is greatly needed. At this moment, the United States Department of Agriculture has no guidelines or regulations on nanotechnology in agriculture.

18.5 Critical Issues of Occupational Risk in Nanoagriculture Field

When it comes to the usage on nanoagriculture, it is important to assess the risk when using these products. It has been observed that particle surface area dose predicts the better toxic and pathological responses to inhaled particles as compared to the particle mass dose [57]. Specifically, reactive oxygen species are chemically active free-radical species that further lead to an imbalance in downstream pathways by triggering DNA damage, altered cellular signaling, and programmed cell death. Due to the fact that nanomaterials are so small, it is easy for them to invade our macromolecule interactions including DNA, RNA, and proteins which lead to mutations [58]. To further explore the risks of nanotechnology, there will be three categories: (i) risk assessment, (ii) risk governance, and (iii) risk management.

18.5.1 Risk Assessment

It is difficult to assess the risk on nanotechnology in agriculture because there is yet to be an establishment of a standard process of risk assessment for nanotechnology. The United States Environmental Protection Agency currently applies the risk assessment process for pesticides on nanopesticides, so it has been abiding by that process [59, 60]. This can cause problems because nanopesticides have different characteristics than traditional pesticides, so a government standard process to asses these new products would be beneficial for the agriculture industry and for the safety of the consumer and living organism. The European Union also shares these same issues, which has led to companies that either choose to include nano in the name of their pesticides because of the ambiguity when it comes to laws and some companies or choose to forgo the nano because of the negative connotation it has because of the lack of regulatory laws on it [61]. The assessment of nanochemicals can be done by tools such as quantitative structure–activity relationship/quantitative structure property relationship (QSAR/QSPR) devices. The model that would shift from standard QSARs/QSPRs to novel nano-QSARs/QSPRs should take into account that (i) classical QSAR/QSPR descriptors are not appropriate for nanoparticles, and specific nano-QSAR/QSPR descriptors are needed; (ii) comprehensive experimental and empirical data amassed are not yet sufficient to fully assess the endpoints; (iii) nanoparticles are large and structurally limited in variety unlike classic small organic compounds with varied structural forms. In this sense, it becomes essential to have high-quality and well-described datasets for the series of related but differently sized nanoparticles/nanopesticides; and (iv) novel size and structural-dependent descriptors for modeling nanoparticle reactivity are also required [61]. There are also other tools such as dominance-based rough set approach (DRSA), multicriteria decision analysis (MCDA), elimination and choice expressing the reality (ELECTRE), and analytical hierarchy process (AHP) [62]. DRSA is used to deal with missing attribute values and inconsistency problems and is used for the calculation of subvariable weights in each main variable in the proposed model [63]. MCDA application can balance societal benefits against unintended side effects and risks, and how it can also bring together multiple lines of evidence to estimate the likely toxicity and risk of nanomaterials, given limited information on physical and chemical properties. For example, let us consider three hypothetical alternative nanomaterials, each with different social and economic values as well as environmental properties and associated risks and benefits. In this example, Nanomaterial #1 has the most benefits for industry, while Nanomaterial #2 and Nanomaterial #3 are more environmentally friendly, although the knowledge on the potential environmental risks and benefits of these materials is very uncertain [64].

18.5.2 Risk Management

Concerns regarding the management of nanomaterials has been a minimally researched topic these past few decades. In Google scholar, there are only 148000 articles relating to the management of nanomaterials, and 59% of them have been since 2000. When it comes to risk management, the goal is to access all possible risks and try to prevent or minimize their harmful impact. When it comes to nanotechnology in agriculture this is a hard thing to access because as discussed previously there is no set process to access the risk of nanomaterials, and in order to minimize negative impact you have to first assess the situation. The U.S. Environmental Protection Agency even stated in 2011 that, "We found that EPA does not currently have sufficient information or processes to effectively manage the human health and environmental risks of nanomaterials. EPA has the statutory authority to regulate nanomaterials but currently lacks the environmental and human health exposure and toxicological data to do so effectively."

18.5.3 Risk Governance

According to the international risk governance council, risk governance is when principles of good governance are applied to the identification, assessment, management, and communication of risks. There is an established framework for risk governance established by the IRGC as well, and it would be beneficial to apply it to nanotechnology in order to prevent the harm nanomaterials could cause to the environment. Currently, the European Union is funding research to develop and implement a transparent, transdisciplinary, and science-based risk governance framework (RGF) for managing nanotechnology risks regarding social, environmental, and economic benefits [65]. The need for risk management of nanotechnology has been growing, as the use of nanotechnology is increasing.

18.6 IRGC Risk Governance Framework

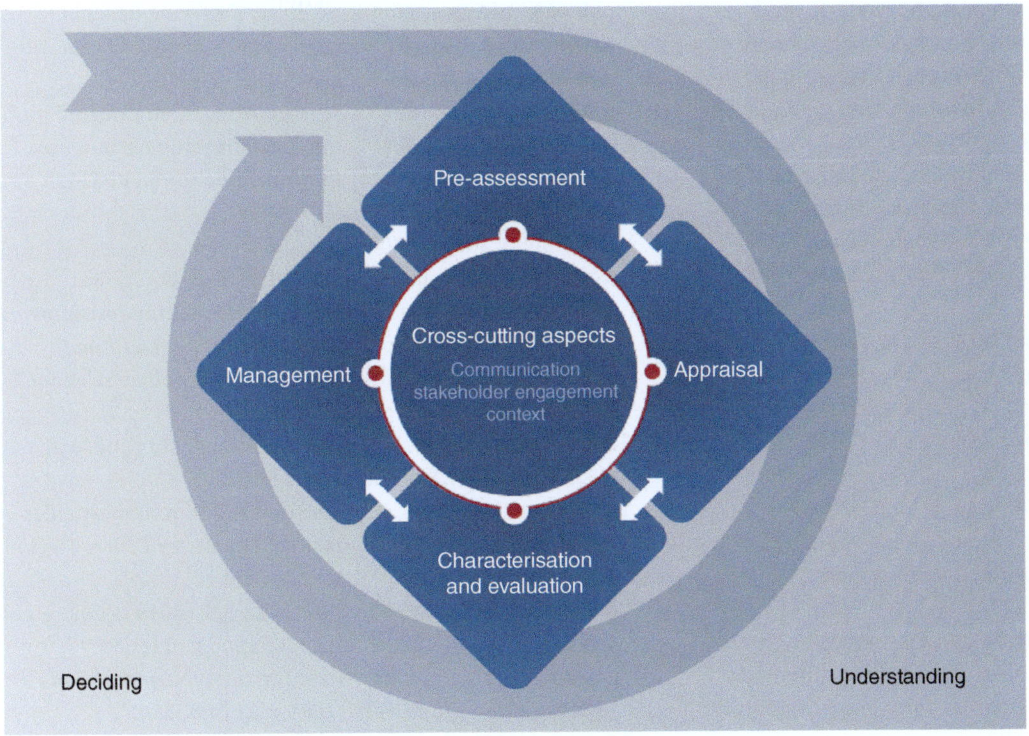

18.6.1 Future Trends

Currently, nanotechnology in agriculture is heavily focused on nanopesticides and nanofertilizers to improve crop yield. However, a recent trend in nanotechnology is focused on nanotechnology for packaging, shipping, remediation, and monitoring. Especially with the aim to be more eco-friendly and globally sustainable, people and countries are looking at nanotechnology to use as alternatives to their traditional pesticides and fertilizers. Specifically, China and India are setting many trends for agriculture and are developing plenty of new articles and case studies. This is because China and

India are both the top agriculture produces in the world for products such as wheat, grains, and rice, according to the data from the United Nations. Nanotechnology, however, is limited in certain aspects because of lack of regulation, laws, and governance. The next few years we will hopefully be seeing more regulations and progression in the field of nanotechnology in agriculture.

18.7 Conclusion

Overall, it has been found that agriculture has a positive impact on the agriculture industry by improving crop yield, increasing nutrients in the crop, reducing the amount of agrochemical waste by targeting specific organisms, and helping it the remediation of water and soil. What was also seen is that while nanotechnology has existed for some time, there is no clear answer to the long-term effects they will have because it has not been long enough to assess that. However, the short-term affects can be assessed and have been suspected of being a health hazard, toxic, and an environmental hazard because these agrochemicals are nano and can invade our gastrointestinal system, our respiratory system, and cells easier because of their small size. Also, nanotechnology in agriculture does not have any specific regulations or guidelines for them, but that some countries and organization such as India and the European Union are working on creating these guidelines and regulations.

References

1 Hasan, M.K., Shahriar, A., and Jim, K.U. (2019). Water pollution in Bangladesh and its impact on public health. *Heliyon* 5 (8): e02145.

2 Mariyono, J., Kuntariningsih, A., Suswati, E., and Kompas, T. (2018). Quantity and monetary value of agrochemical pollution from intensive farming in Indonesia. *Management of Environmental Quality: An International Journal* 29 (4): 759–779.

3 Iavicoli, I., Leso, V., Beezhold, D.H., and Shvedova, A.A. (2017). Nanotechnology in agriculture: opportunities, toxicological implications, and occupational risks. *Toxicology and Applied Pharmacology* 329: 96–111.

4 Ditta, A., Arshad, M., and Ibrahim, M. (2015). Nanoparticles in sustainable agricultural crop production: applications and perspectives. In: *Nanotechnology and Plant Sciences*, 55–75. Cham: Springer.

5 Sawant, R., Hurley, J., Salmaso, S. et al. (2006). “SMART” drug delivery systems: double-targeted pH-responsive pharmaceutical nanocarriers. *Bioconjugate Chemistry* 17 (4): 943–949.

6 Mathew, A.P., Laborie, M.P., and Oksman, K. (2009). Cross-linked chitosan/chitin crystal nanocomposites with improved permeation selectivity and pH stability. *Biomacromolecules* 10: 1627.

7 Shikha, S., Salafi, T., Cheng, J., and Zhang, Y. (2017). Versatile design and synthesis of nano-barcodes. *Chemical Society Reviews* 46 (22): 7054–7093.

8 Duong, B., Liu, H., Ma, L., and Su, M. (2014). Covert thermal barcodes based on phase change nanoparticles. *Scientific Reports* 4: 5170.

9 Jain, S., Kumar, A., Dhewa, T. et al. (2018). Applications of nanotechnology in agriculture—a review: concepts, aspects, prospects, and constraints. In: *Sustainable Biological Systems for Agriculture: Emerging Issues in Nanotechnology, Biofertilizers, Wastewater, and Farm Machines*, 1e (ed. M.R. Goyal). CRC Press.

10 Liu, R. and Lal, R. (2015). Potentials of engineered nanoparticles as fertilizers for increasing agronomic productions. *Science of the Total Environment* 514: 131–139. https://doi.org/10.1016/j.scitotenv.2015.01.104.

11 Qureshi, A., Singh, D.K., and Dwivedi, S. (2018). Nano-fertilizers: a novel way for enhancing nutrient use efficiency and crop productivity. *International Journal of Current Microbiology and Applied Sciences* 7 (2): 3325–3335.

12 Tarran, B. (2019). The crisis in Yemen: short of food, short of data? *Significance* 16 (1): 28.

13 Chhipa, H. (2017). Nanofertilizers and nanopesticides for agriculture. *Environmental Chemistry Letters* 15 (1): 15–22.

14 Bala, R., Kalia, A., and Dhaliwal, S.S. (2019). Evaluation of efficacy of ZnO nanoparticles as remedial zinc nanofertilizer for rice. *Journal of Soil Science and Plant Nutrition* 19 (2): 379–389.

15 FAO (2020).Rice is life. http://www.fao.org/3/Y6618E/Y6618E00.htm (accessed 2 February 2020)

16 Nair, R., Varghese, S.H., Nair, B.G. et al. (2010). Nanoparticulate material delivery to plants. *Plant Science* 179: 154–163.

17 Wang, S.L. and Nguyen, A.D. (2018). Effects of Zn/B nanofertilizer on biophysical characteristics and

growth of coffee seedlings in a greenhouse. *Research on Chemical Intermediates* 44 (8): 4889–4901.

18 Parveen, A., Mazhari, B.B.Z., and Rao, S. (2016). Impact of bio-nanogold on seed germination and seedling growth in *Pennisetum glaucum*. *Enzyme and Microbial Technology* 95: 107–111.

19 Liu, X., Zhang, F., Zhang, S. et al. (2004). Responses of peanut to nano-calcium carbonate. *Plant Nutrition and Fertilizing Science* 11: 385–389.

20 Zhao, L., Peralta-Videa, J.R., Rico, C.M. et al. (2014). CeO_2 and ZnO nanoparticles change the nutritional qualities of cucumber (*Cucumis sativus*). *Journal of Agricultural and Food Chemistry* 62 (13): 2752–2759.

21 Taha, R.A., Hassan, M.M., Ibrahim, E.A. et al. (2016). Carbon nanotubes impact on date palm in vitro cultures. *Plant Cell, Tissue and Organ Culture* 127 (2): 525–534.

22 Shah, V. and Belozerova, I. (2009). Influence of metal nanoparticles on the soil microbial community and germination of lettuce seeds. *Water, Air, and Soil Pollution* 197 (1): 143–148. https://doi.org/10.1007/s11270-008-9797-6.

23 Adhikari, T., Sarkar, D., Mashayekhi, H., and Xing, B. (2016). Growth and enzymatic activity of maize (*Zea mays* L.) plant: solution culture test for copper dioxide nano particles. *Journal of Plant Nutrition* 39 (1): 99–115.

24 Najafi Disfani, M., Mikhak, A., Kassaee, M.Z., and Maghari, A. (2017). Effects of nano Fe/SiO_2 fertilizers on germination and growth of barley and maize. *Archives of Agronomy and Soil Science* 63: 817–826. https://doi.org/10.1080/03650340.2016.1239016.

25 Ghafariyan, M.H., Malakouti, M.J., Dadpour, M.R. et al. (2013). Effects of magnetite nanoparticles on soybean chlorophyll. *Environmental Science & Technology* 47: 10645–10652. https://doi.org/10.1021/es402249b.

26 Delfani, M., Firouzabadi, M.B., Farrokhi, N., and Makarian, H. (2014). Some physiological responses of black-eyed pea to iron and magnesium nanofertilizers. *Communications in Soil Science and Plant Analysis* 45: 11. https://doi.org/10.1080/00103624.2013.863911.

27 Pradhan, S., Patra, P., Das, S. et al. (2013). Photochemical modulation of biosafe manganese nanoparticles on *Vigna radiata*: a detailed molecular, biochemical, and biophysical study. *Environmental Science & Technology* 47: 9. https://doi.org/10.1021/es402659t.

28 Yuvaraj, M. and Subramanian, K.S. (2015). Controlled-release fertilizer of zinc encapsulated by a manganese hollow core shell. *Soil Science & Plant Nutrition* 61 (2): 319–326.

29 Taran, N.Y., Gonchar, O.M., Lopatko, K.G. et al. (2014). The effect of colloidal solution of molybdenum nanoparticles on the microbial composition in rhizosphere of *Cicer arietinum* L. *Nanoscale Research Letters* 9: 289. https://doi.org/10.1186/1556-276X-9-289.

30 Zheng, L., Hong, F., Lu, S., and Liu, C. (2005). Effect of nano-TiO_2 on strength of naturally aged seeds and growth of spinach. *Biological Trace Element Research* 104: 83–91. https://doi.org/10.1385/BTER:104:1:08.

31 Owolade, O. and Ogunleti, D. (2008). Effects of titanium dioxide on the diseases, development and yield of edible cowpea. *Journal of Plant Protection Research* 48 (3): 329–336.

32 Lin, D. and Xing, B. (2008). Root uptake and phytotoxicity of ZnO nanoparticles. *Environmental Science & Technology* 42: 5580–5585.

33 Mahajan, P., Dhoke, S.K., and Khanna, A.S. (2011). Effect of nano-ZnO particle suspension on growth of mung (*Vigna radiata*) and gram (*Cicer arietinum*) seedlings using plant agar method. *Journal of Nanotechnology* 2011: 7. https://doi.org/10.1155/2011/696535.

34 Lin, D. and Xing, B. (2007). Phytotoxicity of nanoparticles: inhibition of seed germination and root growth. *Environmental Pollution* 150: 243–250. https://doi.org/10.1016/j.envpol.2007.01.016.

35 Prasad, T.N.V.K.V., Sudhakar, P., Sreenivasulu, Y. et al. (2012). Effect of nanoscale zinc oxide particles on the germination, growth and yield of peanut. *Journal of Plant Nutrition* 35 (6): 905–927.

36 Burman, U., Saini, M., and Kumar, P. (2013). Effect of zinc oxide nanoparticles on growth and antioxidant system of chickpea seedlings. *Toxicological and Environmental Chemistry* 95 (4): 605–612.

37 Raliya, R. and Tarafdar, J. (2013). ZnO nanoparticle biosynthesis and its effect on phosphorous-mobilizing enzyme secretion and gum contents in cluster bean (*Cyamopsis tetragonoloba* L.). *Agricultural Research* 2: 48–57. https://doi.org/10.1007/s40003-012-0049-z.

38 European Commission (2015). Pesticides. http://ec.europa.eu/food/plant/pesticides/index_en.htm (accessed 2 February 2020).

39 US EPA (2015). Pesticides. http://www.epa.gov/pesticides (accessed 2 February 2020)

40 Gupta, P.K. (2004). Pesticide exposure—Indian scene. *Toxicology* 198 (1–3): 83–90.

41 Sun, Y., Liang, J., Tang, L. et al. (2019). Nano-pesticides: a great challenge for biodiversity? *Nano Today* 28: 100757.

42 Alexander, M. and Lustigman, B. (1966). Effect of chemical structure on microbial degradation of substituted benzenes. *Journal of Agricultural and Food Chemistry* 14: 410–413.

43 Li, M. (2015). *Organic Chemistry of Drug Degradation*. Royal Society of Chemistry.

44 Baertschi, S.W. and Alsante, K.M. (2016). *Stress Testing: The Chemistry of Drug Degradation, Pharmaceutical Stress Testing*, 60–152. CRC Press.

45 Chen, X., Dong, Y., Fan, L., and Yang, D. (2007). Fluorescence for the ultrasensitive detection of peptides with functionalized nano-ZnS. *Analytica Chimica Acta* 582 (2): 281–287.

46 Xu, W., Zhang, J., Zhang, L. et al. (2009). Ultrasensitive detection using surface enhanced Raman scattering from silver nanowire arrays in anodic alumina membranes. *Journal of Nanoscience and Nanotechnology* 9 (8): 4812–4816.

47 Clapp, A.R., Medintz, I.L., Uyeda, H.T. et al. (2005). Quantum dot-based multiplexed fluorescence resonance energy transfer. *Journal of the American Chemical Society* 127 (51): 18212–18221.

48 Liu, Y., Zhao, Y., Zhang, T. et al. (2019). Quantum dots-based immunochromatographic strip for rapid and sensitive detection of acetamiprid in agricultural products. *Frontiers in Chemistry* 7: 76.

49 McDaniel, R.L., Munster, C., and Nielsen-Gammon, J. (2017). Crop and location specific agricultural drought quantification: part III. Forecasting water stress and yield trends. *Transactions of the ASABE* 60 (3): 741–752.

50 Ansari, S. (2017). Combination of molecularly imprinted polymers and carbon nanomaterials as a versatile biosensing tool in sample analysis: recent applications and challenges. *TrAC Trends in Analytical Chemistry* 93: 134–151.

51 Kim, M., Lim, J.W., Kim, H.J. et al. (2015). Chemostat-like microfluidic platform for highly sensitive detection of heavy metal ions using microbial biosensors. *Biosensors and Bioelectronics* 65: 257–264.

52 Rodriguez-Mozaz, S., de Alda, M.J.L., and Barceló, D. (2006). Biosensors as useful tools for environmental analysis and monitoring. *Analytical and Bioanalytical Chemistry* 386 (4): 1025–1041.

53 Peter Abioye, O. (2011). Biological remediation of hydrocarbon and heavy metals contaminated soil. In: *Soil Contamination* (ed. S. Pascucci). IntechOpen https://doi.org/10.5772/24938, https://www.intechopen.com/books/soil-contamination/biological-remediation-of-hydrocarbon-and-heavy-metals-contaminated-soil.

54 Karn, B., Kuiken, T., and Otto, M. (2009). Nanotechnology and in situ remediation: a review of the benefits and potential risks. *Environmental Health Perspectives* 117 (12): 1813–1831.

55 Otto, M., Floyd, M., and Bajpai, S. (2008). Nanotechnology for site remediation. *Remediation Journal: The Journal of Environmental Cleanup Costs, Technologies & Techniques* 19 (1): 99–108.

56 Bakshi, M. and Abhilash, P.C. (2020). Nanotechnology for soil remediation: revitalizing the tarnished resource. In: *Nano-Materials as Photocatalysts for Degradation of Environmental Pollutants*, 345–370. Elsevier.

57 Fadeel, B. and Garcia-Bennett, A.E. (2010). Better safe than sorry: understanding the toxicological properties of inorganic nanoparticles manufactured for biomedical applications. *Advanced Drug Delivery Reviews* 6 (5): 535–542. https://doi.org/10.1016/j.addr.2009.11.008.

58 Lombi, E., Donner, E., Dusinska, M., and Wickson, F. (2019). A one health approach to managing the applications and implications of nanotechnologies in agriculture. *Nature Nanotechnology* 14 (6): 523–531.

59 EPA (2019). Ecological risk assessment for pesticides: technical overview. https://www.epa.gov/pesticide-science-and-assessing-pesticide-risks/ecological-risk-assessment-pesticides-technical (accessed 2 February 2020)

60 EPA (2011). Needs to manage nanomaterial risks more effectively. https://nepis.epa.gov/Exe/ZyPDF.cgi/P100W3ON.PDF?Dockey=P100W3ON.PDF (accessed 2 February 2020)

61 Villaverde, J.J., Sevilla-Moran, B., López-Goti, C. et al. (2018). Considerations of nano-QSAR/QSPR models for nanopesticide risk assessment within the European legislative framework. *Science of the Total Environment* 634: 1530–1539.

62 Windsor, R., Cinelli, M., and Coles, S.R. (2018). Comparison of tools for the sustainability assessment of nanomaterials. *Current Opinion in Green and Sustainable Chemistry* 12: 69–75.

63 Deepa, N. and Ganesan, K. (2019). Hybrid rough fuzzy soft classifier based multi-class classification model for agriculture crop selection. *Soft Computing* 23 (21): 10793–10809.

64 Linkov, I., Satterstrom, F.K., Steevens, J. et al. (2007). Multi-criteria decision analysis and environmental risk assessment for nanomaterials. *Journal of Nanoparticle Research* 9 (4): 543–554.

65 European Commission (2020). Establishing a nanotechnology risk governance framework. https://cordis.europa.eu/project/id/814530 (accessed 1 February 2020)

19

Novel Approaches to Design Eco-Friendly Materials Based on Natural Nanomaterials

Twishi Puri[1] *and Yashwant Pathak*[2,3]

[1] *Department of Biology, Worcester State University, Worcester, MA, USA*
[2] *Taneja College of Pharmacy, University of South Florida, Tampa, FL, USA*
[3] *Faculty of Pharmacy, Airlangga University, Surabaya, Indonesia*

19.1 Introduction

Today, more than any other time, there is a necessary requirement for the production of eco-friendly materials since years of pollution and waste build-up has resulted in degradation of the environment. Environment damage is a threat not only to the survival of humans but also to the other forms of life residing on this planet [1]. To combat this, the marriage between physical, biological, and social sciences, as well as engineering and economic studies, has become not only necessary but also beneficial [2]. Nanomaterials have already been considered as the future of fabrications of new technologies, especially those that lead to global sustainability. However, the use of fabricated nanomaterials has posed a viable risk to the health of the environment. This has led to the utilization of natural nanomaterials to develop more eco-friendly innovations.

Natural nanomaterials have been used in various ways to curate eco-friendly innovation, but these innovations do not necessarily work toward cleaning the environment or making it healthy. Keeping this in mind, this chapter focuses solely on those natural nanoparticle-based materials that have possible applications in the removal of pollutants and hazardous materials from the environment. The three types of material categories discussed below are adsorbents, catalysts, and polymer composites.

19.2 Adsorbents

Adsorption can be defined as the process of transferring a substance from one phase, usually the liquid phase, and attaching it to the surface of another phase, usually a solid, through physical and/or chemical interactions [3]. The separating agent for adsorption is known as an adsorbent [4]. Adsorbents can be key separation tools in the removal of waste products or unwanted substances from useful materials such as water. The general methods of adsorption include removal of adsorbates, substance that is being adsorbed, from a fluid to a fluid-adsorbent interface. The adsorbate diffused through the layer that surrounds the adsorbent molecule. The adsorbate reactants then react with the internal surface of the adsorbent to form adsorbed products [5]. There are several ways to produce eco-friendly adsorbents that can remove hazardous substances from desired materials. The methods discussed below focus on curating adsorbents from natural nanomaterials.

19.2.1 Coal

For the adsorption of toxic heavy metals, such as Cd(II) and Pb(II) ions, from aqueous solutions, adsorbents derived from coal and ogbono (*Irvingia gabonensis*) seed shells have been shown to be a possible eco-friendly alternative to other adsorbents [6]. Coal is the result of plant material degradation. Depending of the age of degradation and the carbon content, coal can be classified as lignite, sub-bituminous, bituminous, and anthracite [7]. The use of coal in the preparation of adsorbents is fairly new compared to the other industrial uses of coal [8]. For this particular study [9], all of the raw materials were collected from Nigeria; lignite coal from Garinmaiganga mine in Gombe state, sub-bituminous coal from Okaba mine in Kogi state, and the *I. gabonensis* seed shells from Obein Enugu state. Before the preparation of the adsorbents, the adsorbate solution was

Sustainable Nanotechnology: Strategies, Products, and Applications, First Edition.
Edited by Yashwant V. Pathak, Govindan Parayil, and Jayvadan K. Patel.

prepared using 2.03 g cadmium chloride, 1.6 g lead nitrate, and 1000 ml of distilled water to make solutions of Cd(II) and Pb(II) ions, each at 1000 $mg\,l^{-1}$ concentration. The first stage of adsorbent preparation involves carbonization of the coals and the seeds. The lignite and sub-bituminous coal were carbonized at 800 °C and the lignite was carbonized again at 400 °C due to its ability for better adsorption efficiency. The lignite carbonized at 400 °C was combined with equal proportion of *I. gabonensis* biomass. This blend was chemically modified with 0.1 M of NaOH and 0.1 M of H_3PO_4 after utilizing the method of Argun and Dursun [10]. After rounds of washing, modification, and drying, the adsorbents were ready for analysis.

Three methods of analysis were used for this particular adsorbent. X-ray fluorescence (XRF) analysis was used to determine the chemical compositions of the two types of coal and seed shell used. The results show that Al_2O_3 and SiO_2 were present in the raw materials and could be a factor contributing to the removal of Cd(II) and Pb(II) ions from the adsorbate solution. Fourier transform infrared analysis was used to test the vibration frequencies of the adsorbents. The results of the analysis indicate that carbonyl and hydroxyl groups are major constituents of lignite coal. This is important because these functional groups contain adsorption centers. Analysis of *I. gabonensis* seed shell indicated the presence of aromatic carbon–hydrogen bonds. Lastly, scanning electron microscopy was used to study the morphology of the prepared adsorbents. The images obtained from the analysis showed that the surface of the adsorbents is irregular and is covered with pores. This particular structure is a classic indication of an adsorbent and contributes to the removal of the metal ions [11]. Additionally, further testing of the lignite coal was done in various settings. A carbonized and unmodified version of lignite coal exhibited optimal performance at pH10, 120 minutes contact time, adsorption temperature of 50 °C, 250 μm adsorbent particle size, and 40 $g\,l^{-1}$ dosage. However, the blend of lignite coal and *I. gabonensis* seed shell modified with NaOH and H_3PO_4performed better at adsorption [9].

19.2.2 CL-LDH (Mg-Al-Cl-Layered Double Hydroxide)

Pharmaceutical and personal care products (PPCPs) are environmental contaminants with the unique ability of being able to induce physiological effects in humans at very low dosage. Pharmaceuticals, especially, have structures that are designed to exhibit maximum biological activity and prolonged action at low concentrations. The bioaccumulation of PPCPs in the environment, especially natural waters, has been confirmed to have disadvantageous effects on both humans and wildlife [12]. PPCPs, however, are in high demand due to their involvement in disease treatment, agriculture, and livestock farming. Tetracycline (TC) in particular is widely used in animal feed, even though it is metabolized by neither animals nor human beings [13]. By 2015, the environmental concentrations of TC were about 1.12 $\mu g\,l^{-1}$. Diclofenac (DF) sodium is another nonsteroidal, anti-inflammatory drug with potential threats to the municipal wastewaters and surface waters. By 2015, it was hypothesized that the environmental concentrations of DF could reach 28.4 $\mu g\,l^{-1}$ [14]. Additionally, chloramphenicol (CAP), a broad-spectrum antibiotic, was used in animal and human medicine, exhibiting severe side effects such as bone marrow suppression, aplastic anemia acute leukemia, and grey syndrome in humans [15]. With severe risks such as these, it has become necessary for elimination of PPCPs from the environment, especially natural waters. As a possible solution, Cl-LDH (Cl-layered double hydroxides) were synthesized and tested for removal of PC, DF, and CAP from aqueous solution. Cl-LDH was synthesized by co-precipitation methods developed in previous studies [16–18]. Two aqueous solutions were prepared for the Cl-LDH synthesis. Solution A was prepared with $MgCl_2{\cdot}6H_2O$ (24.396 g, 0.12 mol) and AlCl3•9H2O (7.2429 g, 0.03 mol) which was dissolved in 150 ml of deionized water. Solution B was prepared with NaOH (12.0 g, 0.03 mol) and NaCl (1.7532 g, 0.03 mol) which was dissolved in 150 ml of boiled deionized water. Each solution was added to the same three-necked flask simultaneously at the rate of two drops per second with a burette. The mixture was stirred to achieve homogeneity and was kept at pH of 10. The mixture was then stored at 70 °C for 12 hours. It was then centrifuged after which the precipitate was collected and washed multiple times with deionized water. The sample was then kept to dry at 110 °C for 24 hours after which it was collected. To prevent carbonate ions from interfering with the process, an inert atmosphere of $N_2$was used.

Three adsorption experiments were conducted with the results studies for further analyses. To prepare for the experiments, 0.015 g Cl-LDH and 25 ml PPCPs (pH of 7) solution were blended in a 50 ml centrifuge tube. For the isotherm study, 0.03 $mmol\,l^{-1}$ to 3 $Mmol\,l^{-1}$ concentrations of TC, DF, and CAP were used. For the pH study, PPCPs with the initial concentration of 2.4 $mmol\,l^{-1}$ with different pH values were prepared. For these two studies, the solutions were shaken at the speed of 150 rpm in a water bath at room temperature for 24 hours. For the kinetic study of PPCPs, the initial concentration used for all three drugs was 2.4 $mmol\,l^{-1}$. The solutions were then shaken at the same speed, same temperature, and in the same environment as the solutions for the previous two studies for 1–1440 minutes. The mixtures from all three studies were then centrifuges at 4500 rpm for 5 minutes.

For analyses, a clear difference between the initial and the final concentration of the solution was the indication of PPCPs removal. To determine the equilibrium concentration of the PPCPs, the supernatant from each solution was analyzed through UV/VIS spectrophotometer. Through Rigaku D/Max-IIIa X-ray diffraction (XRD) analyses diffractometer was used to analyze the structure of the sample. The sharpness and intensity of the peaks present on the XRD pattern suggested that Cl-LDH had a crystalline structure. Analyses of the pattern also showed that TC was adsorbed on the external surface of Cl-LDH and CAP was either adsorbed by the external surface or not adsorbed at all by Cl-LDH. The zeta potential of the material was studied which indicated that the adsorbent had good dispersity in water. The result of this analysis was congruent with the results of XRD regarding the adsorption of CAP by Cl-LDH. Through surface area analyzer, the Brunauer, Emmett and Teller (BET) specific surface area of the sample was determined which showed that the Cl-LDH had strong adsorption characteristics. Additionally, analyses of the FTIR spectra of the samples before and after the adsorption of the drugs also concluded that CAP was not adsorbed by Cl-LDH. It is suggested that the coherent results of no adsorption of CAP could be due to the lack of electrostatic interaction between the Cl-LDH and the antibiotic [19].

19.3 Catalysts

Perhaps one of the most useful materials in a recycle-forward society is catalysts. The transition from a waste generative society to a recycling economy has been remarkable yet it would not have been possible if there had not been ways to rid the environment of the harmful products [20]. Catalysts can be used to modify substances that are composed of environment-harming materials and work toward making them safer and environmental friendly. Below, the discussion focuses on natural nanomaterials that can remove pollutants from fuel and water.

19.3.1 Natural Attapulgiteand Rare Earth Materials

Hydrodesulfurization (HDS) is a common method used to desulfurize fuels such as naphtha, kerosene, and gas oil [21]. Desulfurizing has become necessary from excessive use of fuel has negative consequences on the environment that results in natural phenomena like acid rain and atmospheric haze [22, 23]. Even though HDS's role in desulfurization of fuels is crucial, it does have several energy-based drawbacks such as several energy-based drawbacks such as its requirements of high temperature and pressure and high-end hydrogen [24, 25]. As an alternative, photocatalytic oxidative desulfurization is a hopeful candidate for deep desulfurization due to it being cheaper and more efficient than the other desulfurization strategies studied (extractive desulfurization, biodesulfurization, and adsorption desulfurization) [26–29].

Attapulgite (ATP) is a magnesium aluminum phyllosilicate that can be found in clay soil. It is commonly used in medicine, especially antidiarrheal medicine, due to its ability to physically bind to acids and toxins found in the stomach and the digestive tract [30, 31]. However, it can also be used to aid catalysts due to their large surface area, unique pore structure, and high ability of adsorption [32]. Previous studies have shown that ATP-based nanocomposites, such as attapulgite-CeO_2/MoS_2, CeO_2/ATP/g-C_3N_4, and $BiP_{1-x}V_xO_4$/ATP, have the ability to increase the rate of oxidative desulfurization under irradiation of visible light [33–35]. The use of visible light, however, does not allow all of ATP's abilities to be utilized since ATP can only be stimulated by ultraviolet light [36]. Keeping this in mind, several studies have changed their course and invested their focus on upconversion luminescence materials that are able to absorb and upconvert low-energy light with long wavelength to high-energy light with short wavelength [37]. Among various possibilities, Er^{3+}, metal belonging to the rare earth family, has been considered as an excellent candidate for converting visible light to UV light due to its abundant energy levels and a higher upconversion luminescence efficiency than other materials [38]. In addition, CeO_2, a metal oxide, exhibits high chemical stability and low phonon energy which makes it an excellent matrix material for upconversion. Previously, Er, Yb-CeO_2 hollow spheres have been produced doping Er^{3+} into crystal lattice of CeO_2 for improving the efficiency of dye-sensitized solar cells [39]. However, it was reported that as a result of doping a rare earth upconversion luminescence materials was construction of a heterostructure which causes annihilation of photo-generated charges with lower reduction and oxidation potential [40]. Considering the results of previously done studies, to achieve the strongest emission of UV light, the doping fraction of $Er^{3+}$was adjusted. This adjusted Er^{3+}-doped CeO_2 upconversion luminescence oxide was immobilized on ATP to effectively stimulate it.

Through a one-step method, Er^{3+}:CeO_2/ATP composites were synthesized by dissolving 1 g ATP and ample amount of $Ce(NO_3)_3{\cdot}6H_2O$, $Er(NO_3)_3{\cdot}5H_2O$ in 100 ml deionized water. This was followed by the addition of hexamethylene tetramine (HMT) in a 5 : 1 molar ratio of HMT to $Ce(NO_3)_3$ to the mixture. Meanwhile, the molar fraction of Er was adjusted from 0.5% to 2.5 and the mass ratio of Er^{3+} : CeO_2 to ATP was adjusted from 1 : 10 to 5 : 10. The mixture was then heated at 80 °C in a water bath for 2 hours. The mixture was removed from the water bath and set aside to cool down at room temperature. The precipitate collected was washed thrice using deionized water, dried in a vacuum at 80 °C for 10 hours, and calcined at 300 °C for 2 hours. Analysis of the UV–Vis image of CeO_2, ATP, and Er doping fraction for 1% concluded that ATP can certainly be excited by the upconversion wavelength of 1% $Er^{3+} : CeO_{2+}$. The study concluded that integration of natural clay and rare earth upconversion in catalysts can produce eco-friendly photocatalysts [41].

19.3.2 Zero Valent Iron and Clay Minerals

Amongst the most hazardous material for the environment is wastewater from the textile production. Textile industry is known for its large water consumption and high production of wastewater from printing and dyeing. This specific wastewater has high concentration of chroma and pollutants along with a large water volume [42, 43]. Without proper treatment, if the wastewater is discharged into the environment, its assimilation with bodies of water such as rivers and lakes could result in serious threat to the environment [44]. There are several methods and materials proposed for wastewater treatment. Zero-valent iron (ZVI) has been previously used to treat water such as groundwater, stormwater, and acid mine drainage [45–47]. Additionally, zero-valent iron nanoparticles (nZVI) have been used to treat smelting wastewater due to their ability to remove contaminants such as heavy metals and arsenic [48]. Clay minerals composed of silicon, aluminum, and magnesium are also widely used in wastewater treatment. It has complex pore structures and high specific surface area, two highly beneficial characteristics for the separation and purification of materials [49]. Additionally, due to its abundance and low cost, clay minerals have previously played an important role in protecting the environment as well as in the process of developing new mineral materials [50]. The process of water treatment requires nZVI to be fixated on clay minerals' surface. Most suitable clay minerals for efficient results are kaolinite and sepiolite. Kaolinite has a 1 : 1 layered lattice structure which is efficiently ordered to allow no isomorphous substitution in its structure. This prohibits the minerals from swelling in water [51] as well as allowing them to adsorb organic pollutants such as congo red dye [52], heavy metals [53], and dipalmitoyl lecithin [54]. Sepiolite, on the other hand, has a common 2 : 1 chain layered structure assembled of two layers of silicone tetrahedron and a layer of magnesium octahedron [55]. It also does not exhibit good swelling performance in water; however, it does have the ability to adsorb organic pollutants such as reactive blue 221 [56], 3-aminopropyltriethoxysilane [57], and β-carotene [58]. Both clay minerals are suitable to load nZVI onto their surface because due to their structure they are unable to swell in water.

The following study was conducted using kaolinite and sepiolite-loaded nZVIs to degrade Rhodamine 6G (Rh 6G) under microwave irradiation. Microwaves have been proven to improve the rate of catalytic reactions by being adsorbed by aqueous solutions in the reaction which then begin to self-heat. The nZVI used is Fe^{3+}. nZVI/kaolinite are synthesized by adding 100 ml distilled water, 5 g kaolinite, and 10 g $FeCl_3{\cdot}6H_2O$ in a three-necked flask. The solution is stirred for 12 hours to allow fixation of Fe^{3+} onto the clay minerals. To this, 8 g $NaBH_4$was added and the mixture was stirred for 30 minutes under nitrogen at room temperature. The resulting suspension was then centrifuged and washed thrice three times. The pellet obtained was then freeze-dried. Sepiolite was substituted for kaolinite to make nZVI/sepiolite. A 4000 mg l^{-1} Rh 6G solution was made using 2000 mg Rh 6G dissolved in 500 ml of distilled water and ultrasonic for 15 minutes. 0.1 g of the prepared samples of nZVI loaded clay minerals were added to 10 ml Rh 6G solution. The mixture was placed in a microwave oven for a certain time and under different microwave power to degrade Rh 6G. Following this, the mixture was centrifuged, and the collected supernatant was filtered using 0.22 μm syringe filters. Analysis of degradation results of Rh 6G concluded that nZVI/sepiolite had better removal performance than nZVI/kaolinite. This was due to the difference in the iron loading amount which was the result of the difference in the surface area and zeta potential of the two clay minerals [59].

19.4 Polymer Composites

In the recent years, plastic littering has become a prominent problem for the environment. Its involvement in various industries makes it difficult to eliminate it completely. Alternative biodegradable materials, however, can be developed to prevent the accumulation of harmful plastics. Two such fabricated polymer composites are composed of sodium alginate (SA) and polyurethane. Characteristics, such as biodegradability and biocompatibility, of both materials make them great replacement candidates for the harmful used in industries such as automobile and the food sector.

19.4.1 Sodium Alginate Nanocomposites

Plastics such as polyethylene, polypropylene, and polyethylene terephthalate have taken over the food industry in the form of food packaging material. Their involvement, however, can result in some serious environmental pollution due to their nonbiodegradable structure [60–62]. Continuous production of hazardous food packaging materials requires development of biodegradable films composed of natural, nontoxic, and safe polymers such as proteins, lipids, and polysaccharides [63–65]. The possible use of SA, a natural, biodegradable polymer, in packaging has modified many studies regarding food packaging. SA by nature is biocompatible, has firm-forming abilities, and is composed of various active functional groups [66, 67]. In addition, being the product of marine algae and bacteria, SA can be found in great abundance and has other characteristics such as nontoxicity, water-solubility, and biodegradability [68]. It does, however, have limitations, for instance, fragility and hydrophilicity, that hinder their applications in film materials [69, 70]. However, the involvement of nanoparticles can eliminate its limitations. Previous studies have tested the properties of alginate enhanced by montmorillonite (MMT) nanoparticles. MMT is a 2 : 1-type clay mineral exhibiting a structure composed of two 2D platelets and cations, a detail that is highly relevant to the research of polymer nanocomposites. The conclusion was that the mechanical properties of alginate were significantly enhanced after incorporation of MMT; however, excess MMT leads to agglomeration in the polymer matrix which works toward lessening the mechanical properties of the film [71–73]. On the other hand, rod-like nanoclays make great nanoparticles and are capable of being applied to the development of polymer/nanoclays composites. Additionally, their small surface area, which decreases the amount of contact prone surface, and weaker interaction amongst rods allow easy dispersion in the polymer matrix with less aggregation [74–78]. One specific nanoclay being studied is Palygorskite (PAL) which is a natural 1-D nanorod-like silicate clay mineral [79, 80]. Its characteristics such as high aspect ratio, large specific area, good thermal stability, and high modulus give it an advantage over other materials in terms of applications in development of polymer composites.

Even though two categorical possibilities of nanoclays, 1D and 2D, are available to use in developing SA-based films, research comparing the two types is minimal. To rectify this, the following experiment was developed to study the effect of 1D PAL and 2D MMT on SA/nanoclays nanocomposite film. The method used to prepare SA, SA/MMT, and SA/PAL films was solution casting. A 2 wt% SA-based aqueous solution was prepared using SA powder and distilled water. The two were mixed using a mechanical stirrer for 24 hours until all SA powder was dissolved. To this glycerol was added and the mixture was stirred at 600 rpm. Both PAL and MMT dispersions (2 wt%) were prepared by dispersing the clay minerals in distilled water under mechanical agitation for an hour at 600 rpm, all done at room temperature. the mixture was then dispersed using ultrasound equipment. The equipment was operated at 40 kHz for 10 minutes. These dispersions were added to the prepared SA solution. The resulting mixture was poured into PS dishes and evaporate at atmospheric temperature for 72 hours. The dried films were then kept in a conditioning cabinet with RH of 53%. The same method, barring the addition of Pal and MMT clay minerals, was used to prepare the SA films. The final films had the length of 80 mm and width of 10 mm. The three films were tested for their mechanical properties and moisture uptake. The tensile performance was tested using an AG-IS material testing machine and at room temperature. Based on the results provided by the machine, it was concluded that the tensile strength of SA/PAL nanocomposites had gone from 13.67 to 25.23 MPa, showing an increase by 84.56%. On the other hand, the tensile strength of SA/MMT had initially increased a the MMT content reached 6 wt%, however, as more MMT was added the tensile strength experienced a decrease, producing an increase in strength by 82.96%. The moisture uptake of the films was tested by using small 2×2 cm pieces of the films and drying them at 70 °C for 24 hours. After fully dried, the samples were weighed (W_0) and conditions at 76% RH for 72 hours in a saturated NaCl solution. After conditioning, the films were weighed again. The initial and final difference between the weight of the films is the indication of how much moisture the films can hold. The following equation is used to determine the moisture uptake of the films:

$$\mathrm{MU} = \frac{\left[W_1 - W_0\right]}{W_0} \times 100\% \tag{19.1}$$

The moisture uptake of the SA without any clay minerals was higher than the other two films developed. This suggests that addition of clay minerals affects the film's ability to penetrate water molecules which negatively affects the moisture uptake of the films [81]. The reason for these results is that PAL and MMT form hydrogen bonding networks with the SA films which results in surface roughness and blockage in the diffusion pathway of the water molecules [81–83]. Additionally, PAL and MMT exhibit weaker hydrophilicity than the SA matrix which could also decrease the attraction between water molecules and the film, hence decreasing the moisture uptake. In comparison with each other, PAL had a higher reduction effect on the moisture uptake of the SA films than MMT. The addition of 10 wt% MMT reduced the moisture uptake by only 8.1% whereas 10 wt% PAL was able to reduce it by 12.9%. A good explanation for this is the fact that PAL has better dispersion abilities than MMT when incorporated in large amounts. Based on these results, it was concluded that 1D PAL-based nanorods could be a suitable alternative in the fabrication of other biopolymer-based nanocomposite films. Additionally, the incorporation of these films in food packaging materials certainly could be beneficial to the environment [84].

19.4.2 Elastomer Materials

With progression toward sustainable development comes the modification of the automobile industry, one of the main industries with high contribution to the risks regarding energy and the environment. Tires specifically pose a great risk to the health of the environment by being accountable for 5% of the carbon dioxide released in the environment. Additionally, it is also accountable for consuming 6% of the world's total energy consumption per annum. Even though they are one of the main contributors of environmental pollution, production of tires cannot be halted since tires are a vital part of an automobile by being the only part that is always directly in contact with the road. In 1980, the United Nations, through "The World Conservative Strategy," had proposed the concept of sustainable development which addressed concerns such as the fossil fuel depletion, pollutant emissions, global warming, and sustainable energy development systems [85]. According to the data collected by the European Union (EU), transportation industry is alone responsible for consuming more than 20% of the world's total energy consumption in the decade of the 2000s. Additionally, tires are accountable for 24% of the total carbon dioxide emissions caused by road vehicles. To rectify this, the EU had released the legislation of "Tire Labelling Regulation" in 2009 which led to the following estimated: decreasing the rolling resistance of tires by 20% results in a reduction in fuel consumption by 5% and switching to B-level green tires can save 20 billion liters of fuel and reduce the carbon dioxide emission by 50 million tons every year.

What accounts for the high energy consumption of tires? The answer is simple: rubber. Currently, the processing process of rubber composites that are used to manufacture pneumatic tires demands large amounts of energy. Additionally, the production process results in not only the production of tires but also significant amounts of carbon disulfide and nonmethane hydrocarbons along with other organic waste. Even with high energy consumption and production of hazardous waste products, pneumatic tires are not considered the best since they are prone to punctures, leakage, and may blow out when in use [86]. A good candidate for replacement for rubber is polyurethane elastomer materials, known for the presence of chemically end-linked nanoparticles and exhibiting low dynamic hysteresis loss. Additionally, the urethane groups present on the macromolecular chain of the material created alternating flexible soft and rigid hard segments [87–90]. Polyurethane has been proven to have better wear resistance, high tear strength and elongation, wide hardness range, low rolling resistance, large carrying capacity as well as excellent oil and chemical resistance [91]. The production process of polyurethane is also simple in comparison to rubber which results in less energy consumption. Polyurethane's biodegradable properties also make it easier to recycle, something not possible with rubber [92].

However, even pure polyurethane is not fully suitable to be used in tire production. The material's glass transition temperature can be affected by wet traction caused during rain and it also has poor heat resistance which could cause problems during driving [93]. The following experiment addressed this problem by replacing polyester polyols and polyether polyols of the oft segments present in polyurethane with hydroxyl-terminated solute ion-polymerized styrene-butadiene tubber (HTSSBR). HTSSBR is known to have a wide range of glass transition temperature that is adjustable and possesses better wet-skid resistance. Additionally, 1,5-naphthalene diisocyanate (NDI) is used to synthesize the polyurethane elastomer material due to its high heat resistance which allows the material to be stable at high temperatures caused by friction.

The preparation of hydroxyl-terminated solute ion-polymerized styrene-butadiene tubber polyurethane (HTSSBR-PU) involves two steps: preparation of the prepolymer through reaction of 1,5-naphthalene diisocyanate (NDI) and HTSSBR under the nitrogen atmosphere at 80 °C and creating a mixture of 1.4-butanediol (BDO) (extender), trimethylolpropane (TMP) (cross-linking agent), and the prepared prepolymer at 65 °C. The mixture of the three substances was stirred at high speed for 6–8 minutes. To shape the material, a metal mold was used to contain the mixture, which afterward was cured at 105 °C for 24 hours. Various material characterization tests were performed on the resulting product however the two we will focus on are mechanical performance and thermal properties of solution-polymerized styrene-butadiene rubber (SSBR) and HTSSBR. The mechanical performances of the elastomer were tested using machines such as the universal testing machine (LRX Plus/LLOYD LRX/MTS CMT4104), Akron abrasion tester (MINGZHU MZ-4061), and friction coefficient tester (LEIYUN BM-III). The testing results concluded that the tensile strength of the elastomer (HTSSBR) was 22.3 MPa, 38.5% higher than the tensile strength of just SSBR. Regarding wear resistance, HTSSBR experienced Akron abrasion loss of 10.8 mm^{-3} whole SSBR experiences a loss of 74 mm^{-3}, making HTSSBR's wear resistance better than SSBR's. The thermal properties of the two materials were tested using TGA (Mettler-Toeledo), DSC (Mettler-ToeledoSTARe), TMA (TA Q800), and DMA (TA Q800). It was concluded that the glass transition temperature of HTSSBR was −22.45 °C.

Based on the results of comparison between SSBR and HTSSBR elastomer material, it can be concluded that HTSSBR has various advantages that not only make tires safe but also protect the environment with a low carbon dioxide emission [86].

19.5 Conclusion

Naturally available nanomaterials, or those that have been synthesized from natural resources, have certainly proved themselves to be useful in various industries that have direct and indirect influence on the health of the environment. Actions toward global sustainability and combating global warming now include the use of separation material (adsorbents), catalysts, and polymer composites for environmental remediation. Natural nanomaterials that have been mentioned in this chapter, and many more, have already been proved to exhibit eco-friendly characteristics that their conventional counterparts lack. For example, coal, *I. gabonensis* seed shells, and Cl-type Mg-Al hydrotalcite (Cl-LDH) have been excellent candidates in the process of formulating adsorbents that can be used to rid fuel and water from pollutants. Additionally, the combination of natural clay materials, such as ATP, and rare earth materials, such as Er^{3+} and CeO_2 has been useful for effective desulfurization. ZVI-loaded nanoclay based catalysts have also been effective in degrading Rh 6G under microwave irradiation. Materials such as food packaging and tire have also been made safe with the use of polymer nanocomposites such as 1D-PAL nanorods and HTSSBR-PU. The next step certainly includes further research on the materials and industrial level investigation.

19.6 Future Trends

Environment protection and sustainability are perhaps one of the most important global issues of the twenty-first century. Progression toward achieving a healthy environment has so far been significant however there is a lot more than research can do to reach the optimal goal of eliminating hundred percent of the pollutants from the environment. Adsorbents, catalysts, and polymer composites are just few of the many possibilities of materials that can be modified using eco-friendly, natural nanomaterials. Even though the attention that natural materials have receive from several scientific fields has surged, there is still a great deal to be done regarding the complete replacement of conventional synthetic nanomaterials with natural nanomaterials. Additionally, limitations, both theoretical and technological, must be addressed regarding the design and synthesis of new materials from natural materials. This requires continuous and extensive research of various natural materials available through the environment along with evaluation of their effectiveness and how its success rate can be improved [94].

References

1 Purwendah, E.K., Mangku, D.G.S., and Periani, A. (2019). Dispute settlements of oil spills in the sea towards sea environment pollution. *First International Conference on Progressive Civil Society (ICONPROCS 2019)*. Atlantis Press, 245–248. doi: https://doi.org/10.2991/iconprocs-19.2019.51.

2 Hochella, M.F., Mogk, D.W., Ranville, J. et al. (2019). Natural, incidental, and engineered nanomaterials and their impacts on the Earth system. *Science* 363 (6434) https://doi.org/10.1126/science.aau8299.

3 Kurniawan, T.A., GYS, C., Lo, W.H., and Babel, S. (2006). Physico-chemical treatment techniques for wastewater laden with heavy metals. *Chemical Engineering Journal* 118 (1–2) https://doi.org/10.1016/j.cej.2006.01.015.

4 Yang, R.T. (2003). *Adsorbents: Fundamentals and Applications*. Hoboken, NJ: Wiley-Interscience https://doi.org/10.1002/047144409x.

5 Nabi, S.A. and Naushad, M. (2008). Synthesis, characterization and analytical applications of a new composite cation exchanger cellulose acetate-Zr(IV) molybdophosphate. *Colloids and Surfaces A: Physicochemical and Engineering Aspects* 316 (1-3) https://doi.org/10.1016/j.colsurfa.2007.09.005.

6 Golder, A.K., Chanda, A.K., Samanta, A.N., and Ray, S. (2007). Removal of Cr(VI) from aqueous solution: electrocoagulation vs chemical coagulation. *Separation Science and Technology* 42 (10) https://doi.org/10.1080/01496390701446464.

7 Odesola, I.F., Samuel, E., and Olugasa, T. (2013). Coal development in Nigeria: prospects and challenges. *International Journal of Engineering and Applied Sciences* 4 (1): 64–73. www.eaas-journal.org.

8 Ani, J.U., Nnaji, N.J., Okoye, C.O.B., and Onukwuli, O.D. (2012). Factorial analyses, equilibrium studies,

kinetics and thermodynamics of the removal of suspended particles from an industrial effluent on coal based activated carbon. *International Journal of Chemical Sciences* 10 (3): 1765–1784.

9 Ezeokonkwo, M.A., Ofor, O.F., and Ani, J.U. (2018). Preparation and evaluation of adsorbents from coal and *Irvingia gabonensis* seed shell for the removal of Cd(II) and Pb(II) ions from aqueous solutions. *Frontiers in Chemistry* 5 https://doi.org/10.3389/fchem.2017.00132.

10 Argun, M.E. and Dursun, S. (2006). Removal of heavy metal ions using chemically modified adsorbents. *Journal of International Environmental Application & Science* 1 (1-2): 27–40.

11 Vafakhah, S., Bahrololoom, M.E., Bazarganlari, R., and Saeedikhani, M. (2014). Removal of copper ions from electroplating effluent solutions with native corn cob and corn stalk and chemically modified corn stalk. *Journal of Environmental Chemical Engineering* 2 (1) https://doi.org/10.1016/j.jece.2014.01.005.

12 Ebele, A.J., Abou-Elwafa Abdallah, M., and Harrad, S. (2017). Pharmaceuticals and personal care products (PPCPs) in the freshwater aquatic environment. *Emerging Contaminants* 3 (1) https://doi.org/10.1016/j.emcon.2016.12.004.

13 Ahmadi, M., Ramezani Motlagh, H., Jaafarzadeh, N. et al. (2017). Enhanced photocatalytic degradation of tetracycline and real pharmaceutical wastewater using MWCNT/TiO2 nano-composite. *Journal of Environmental Management* 186 https://doi.org/10.1016/j.jenvman.2016.09.088.

14 Qi, M., Zhihao, Z., Xiuwen, C. et al. (2015). Research progress in the pollution situation and treatment methods for diclofenac in water. *Shangdong Chenm. Indus.* 44: 66–68.

15 Sai, N., Wu, Y., Sun, Z. et al. (2019). A novel photonic sensor for the detection of chloramphenicol. *Arabian Journal of Chemistry* 12 (8) https://doi.org/10.1016/j.arabjc.2016.06.015.

16 Vreysen, S. and Maes, A. (2008). Adsorption mechanism of humic and fulvic acid onto Mg/Al layered double hydroxides. *Applied Clay Science* 38 (3–4) https://doi.org/10.1016/j.clay.2007.02.010.

17 Yue, X., Liu, W., Chen, Z., and Lin, Z. (2017). Simultaneous removal of Cu(II) and Cr(VI) by Mg–Al–Cl layered double hydroxide and mechanism insight. *Journal of Environmental Sciences (China)* 53 https://doi.org/10.1016/j.jes.2016.01.015.

18 Lv, X., Chen, Z., Wang, Y. et al. (2013). Use of high-pressure CO2 for concentrating CrVI from electroplating wastewater by Mg-Al layered double hydroxide. *ACS Applied Materials & Interfaces* 5 (21) https://doi.org/10.1021/am4035009.

19 Li, E., Liao, L., Lv, G. et al. (2018). The interactions between three typical PPCPs and LDH. *Frontiers in Chemistry* 6 https://doi.org/10.3389/fchem.2018.00016.

20 Franken, T., Terreni, J., Borgschulte, A., and Heel, A. (2020). Solid solutions in reductive environment – a case study on improved CO2 hydrogenation to methane on cobalt based catalysts derived from ternary mixed metal oxides by modified reducibility. *Journal of Catalysis* 382 https://doi.org/10.1016/j.jcat.2019.12.045.

21 Safari, A. and Vesali-Naseh, M. (2019). Design and optimization of hydrodesulfurization process for liquefied petroleum gases. *Journal of Cleaner Production* 220 https://doi.org/10.1016/j.jclepro.2019.02.226.

22 Liu, R., Zhang, J., Xu, Z. et al. (2018). Visible light photocatalytic oxidative desulfurization using Ti-MCM-41-loaded iron phthalocyanine combined with ionic liquid extraction. *Journal of Materials Science* 53 (7) https://doi.org/10.1007/s10853-017-1954-0.

23 Zhang, G., Ren, J., Liu, B. et al. (2018). In situ hydrothermal preparation and photocatalytic desulfurization performance of metallophthalocyanine sensitized SnO2. *Inorganica Chimica Acta* 471 https://doi.org/10.1016/j.ica.2017.12.025.

24 Lei, W., Wenya, W., Mominou, N. et al. (2016). Ultra-deep desulfurization of gasoline through aqueous phase in-situ hydrogenation and photocatalytic oxidation. *Applied Catalysis B: Environmental* 193 https://doi.org/10.1016/j.apcatb.2016.04.032.

25 Zeng, X., Xiao, X., Li, Y. et al. (2017). Deep desulfurization of liquid fuels with molecular oxygen through graphene photocatalytic oxidation. *Applied Catalysis B: Environmental* 209 https://doi.org/10.1016/j.apcatb.2017.02.077.

26 Raj, J.J., Magaret, S., Pranesh, M. et al. (2018). Extractive desulfurization of model fuel oil using ester functionalized imidazolium ionic liquids. *Separation and Purification Technology* 196 https://doi.org/10.1016/j.seppur.2017.08.050.

27 Khodadadi Dizaji, A., Mortaheb, H.R., and Mokhtarani, B. (2018). Complete oxidative desulfurization using graphene oxide-based phosphomolybdic acid catalyst: process optimization by two phase mass balance approach. *Chemical Engineering Journal* 335 https://doi.org/10.1016/j.cej.2017.10.129.

28 Agarwal, M., Dikshit, P.K., Bhasarkar, J.B. et al. (2016). Physical insight into ultrasound-assisted biodesulfurization using free and immobilized cells of *Rhodococcus rhodochrous* MTCC 3552. *Chemical Engineering Journal* 295 https://doi.org/10.1016/j.cej.2016.03.042.

29 Yang, Y., Lv, G., Li, J. et al. (2018). Synthesis of ceria nanorods as adsorbent for the adsorption desulfurization of gasoline fuel. *Journal of Alloys and Compounds* 747 https://doi.org/10.1016/j.jallcom.2018.03.026.

30 Zaid, M.R.B., Hasan, M., and Khan, A.K.A. (1995). Attapulgite in the treatment of acute diarrhoea: a double-blind placebo-controlled study. *Journal of Diarrhoeal Diseases Research* 13 (1): 44–46.

31 Nwachukwu, C.E. and Okebe, J.U. (2008). Antimotility agents for chronic diarrhoea in people with HIV/AIDS. *Cochrane Database of Systematic Reviews* 4 https://doi.org/10.1002/14651858.CD005644.pub2.

32 Zhang, G., Wang, H., Guo, S. et al. (2016). Synthesis of Cu/TiO 2 /organo-attapulgite fiber nanocomposite and its photocatalytic activity for degradation of acetone in air. *Applied Surface Science* 362 https://doi.org/10.1016/j.apsusc.2015.11.218.

33 Li, X., Zhang, Z., Yao, C. et al. (2016). Attapulgite-CeO 2 / MoS 2 ternary nanocomposite for photocatalytic oxidative desulfurization. *Applied Surface Science* 364 https://doi.org/10.1016/j.apsusc.2015.12.196.

34 Li, X., Zhu, W., Lu, X. et al. (2017). Integrated nanostructures of CeO2/attapulgite/g-C3N4 as efficient catalyst for photocatalytic desulfurization: mechanism, kinetics and influencing factors. *Chemical Engineering Journal* 326 https://doi.org/10.1016/j.cej.2017.05.131.

35 Li, X., Li, F., Lu, X. et al. (2018). Microwave hydrothermal synthesis of BiP1 − xVxO4/attapulgite nanocomposite with efficient photocatalytic performance for deep desulfurization. *Powder Technology* 327 https://doi.org/10.1016/j.powtec.2018.01.005.

36 Balaji, R., Kumar, S., Reddy, K.L. et al. (2017). Near-infrared driven photocatalytic performance of lanthanide-doped NaYF4@CdS core-shell nanostructures with enhanced upconversion properties. *Journal of Alloys and Compounds* 724 https://doi.org/10.1016/j.jallcom.2017.07.050.

37 Feng, W., Han, C., and Li, F. (2013). Upconversion-nanophosphor-based functional nanocomposites. *Advanced Materials* 25 (37) https://doi.org/10.1002/adma.201301946.

38 Pickering, J.W., Bhethanabotla, V.R., and Kuhn, J.N. (2017). Assessment of mechanisms for enhanced performance of TiO2/YAG:Yb+3,Er+3 composite photocatalysts for organic degradation. *Applied Catalysis B: Environmental* 202 https://doi.org/10.1016/j.apcatb.2016.09.007.

39 Han, G., Wang, M., Li, D. et al. (2017). Novel upconversion Er, Yb-CeO2 hollow spheres as scattering layer materials for efficient dye-sensitized solar cells. *Solar Energy Materials & Solar Cells* 160 https://doi.org/10.1016/j.solmat.2016.10.021.

40 Šutka, A., Vanags, M., Joost, U. et al. (2018). Aqueous synthesis of Z-scheme photocatalyst powders and thin-film photoanodes from earth abundant elements. *Journal of Environmental Chemical Engineering* 6 (2) https://doi.org/10.1016/j.jece.2018.04.003.

41 Wu, F., Li, X., Zhang, H. et al. (2018). Z-scheme photocatalyst constructed by natural attapulgite and upconversion rare earth materials for desulfurization. *Frontiers in Chemistry* 6 https://doi.org/10.3389/fchem.2018.00477.

42 Liu, M., Shadbegian, R., and Zhang, B. (2017). Does environmental regulation affect labor demand in China? Evidence from the textile printing and dyeing industry. *Journal of Environmental Economics and Management* 86 https://doi.org/10.1016/j.jeem.2017.05.008.

43 Zhao, K., Zhao, G., Li, P. et al. (2010). A novel method for photodegradation of high-chroma dye wastewater via electrochemical pre-oxidation. *Chemosphere* 80 (4) https://doi.org/10.1016/j.chemosphere.2010.04.019.

44 Oluwakemi Kehinde, F. and Abdul, A.H. (2016). Classical optimization of process variables in the treatment of real textile wastewater using clinoptilolite. *Journal of Environmental Chemical Engineering* 4 (1) https://doi.org/10.1016/j.jece.2015.08.019.

45 Rangsivek, R. and Jekel, M.R. (2005). Removal of dissolved metals by zero-valent iron (ZVI): kinetics, equilibria, processes and implications for stormwater runoff treatment. *Water Research* 39 (17) https://doi.org/10.1016/j.watres.2005.07.040.

46 Wilkin, R.T. and McNeil, M.S. (2003). Laboratory evaluation of zero-valent iron to treat water impacted by acid mine drainage. *Chemosphere* 53 (7) https://doi.org/10.1016/S0045-6535(03)00512-5.

47 Wilkin, R.T., Su, C., Ford, R.G., and Paul, C.J. (2005). Chromium-removal processes during groundwater remediation by a zerovalent iron permeable reactive barrier. *Environmental Science & Technology* 39 (12) https://doi.org/10.1021/es050157x.

48 Yan, W., Ramos, M.A.V., Koel, B.E., and Zhang, W.X. (2012). As(III) sequestration by iron nanoparticles: study of solid-phase redox transformations with X-ray photoelectron spectroscopy. *Journal of Physical Chemistry C* 116 (9) https://doi.org/10.1021/jp208600n.

49 Steudel, A., Batenburg, L.F., Fischer, H.R. et al. (2009). Alteration of non-swelling clay minerals and magadiite by acid activation. *Applied Clay Science* 44 (1–2) https://doi.org/10.1016/j.clay.2009.02.001.

50 Kelm, U., Sanhueza, V., and Guzman, C. (2003). Filtration and retention of mineral processing slurries with pumice and common clay: low-cost materials for environmental applications in the small-scale mining industry. *Applied Clay Science* 24 (1–2) https://doi.org/10.1016/j.clay.2003.07.004.

51 Ezzatahmadi, N., Ayoko, G.A., Millar, G.J. et al. (2017). Clay-supported nanoscale zero-valent iron composite materials for the remediation of contaminated aqueous solutions: a review. *Chemical Engineering Journal* 312 https://doi.org/10.1016/j.cej.2016.11.154.

52 Vimonses, V., Lei, S., Jin, B. et al. (2009). Adsorption of congo red by three Australian kaolins. *Applied Clay Science* 43 (3–4) https://doi.org/10.1016/j.clay.2008.11.008.

53 Matłok, M., Petrus, R., and Warchoł, J.K. (2015). Equilibrium study of heavy metals adsorption on kaolin. *Industrial and Engineering Chemistry Research* 54 (27) https://doi.org/10.1021/acs.iecr.5b00880.

54 Wallace, W.E., Headley, L.C., and Weber, K.C. (1975). Dipalmitoyl lecithin surfactant adsorption by kaolin dust in vitro. *Journal of Colloid and Interface Science* 51 (3) https://doi.org/10.1016/0021-9797(75)90152-6.

55 Kaviratna, H. and Pinnavaia, T.J. (1994). Acid hydrolysis of octahedral Mg2+ sites in 2:1 layered silicates: an assessment of edge attack and gallery access mechanisms. *Clays and Clay Minerals* 42 (6) https://doi.org/10.1346/CCMN.1994.0420607.

56 Alkan, M., Demirbaş, Ö., and Doğan, M. (2007). Adsorption kinetics and thermodynamics of an anionic dye onto sepiolite. *Microporous and Mesoporous Materials* 101 (3) https://doi.org/10.1016/j.micromeso.2006.12.007.

57 Demirbaş, Ö., Alkan, M., Doğan, M. et al. (2007). Electrokinetic and adsorption properties of sepiolite modified by 3-aminopropyltriethoxysilane. *Journal of Hazardous Materials* 149 (3) https://doi.org/10.1016/j.jhazmat.2007.04.036.

58 Sabah, E., Çinar, M., and Çelik, M.S. (2007). Decolorization of vegetable oils: adsorption mechanism of β-carotene on acid-activated sepiolite. *Food Chemistry* 100 (4) https://doi.org/10.1016/j.foodchem.2005.12.052.

59 Rao, W., Lv, G., Wang, D., and Liao, L. (2018). Enhanced degradation of Rh 6G by zero valent iron loaded on two typical clay minerals with different structures under microwave irradiation. *Frontiers in Chemistry* 6 https://doi.org/10.3389/fchem.2018.00463.

60 Souza, V.G.L., Fernando, A.L., Pires, J.R.A. et al. (2017). Physical properties of chitosan films incorporated with natural antioxidants. *Industrial Crops and Products* 107 https://doi.org/10.1016/j.indcrop.2017.04.056.

61 Costa, M.J., Marques, A.M., Pastrana, L.M. et al. (2018). Physicochemical properties of alginate-based films: effect of ionic crosslinking and mannuronic and guluronic acid ratio. *Food Hydrocolloids* 81 https://doi.org/10.1016/j.foodhyd.2018.03.014.

62 Salama, H.E., Abdel Aziz, M.S., and Sabaa, M.W. (2018). Novel biodegradable and antibacterial edible films based on alginate and chitosan biguanidine hydrochloride. *International Journal of Biological Macromolecules* 116 https://doi.org/10.1016/j.ijbiomac.2018.04.183.

63 Mushi, N.E., Utsel, S., and Berglund, L.A. (2014). Nanostructured biocomposite films of high toughness based on native chitin nanofibers and chitosan. *Frontiers in Chemistry* 2 (NOV) https://doi.org/10.3389/fchem.2014.00099.

64 Wang, S. and Jing, Y. (2017). Effects of formation and penetration properties of biodegradable montmorillonite/chitosan nanocomposite film on the barrier of package paper. *Applied Clay Science* 138 https://doi.org/10.1016/j.clay.2016.12.037.

65 Youssef, A.M. and El-Sayed, S.M. (2018). Bionanocomposites materials for food packaging applications: concepts and future outlook. *Carbohydrate Polymers* 193 https://doi.org/10.1016/j.carbpol.2018.03.088.

66 Shankar, S., Wang, L.F., and Rhim, J.W. (2016). Preparations and characterization of alginate/silver composite films: effect of types of silver particles. *Carbohydrate Polymers* 146 https://doi.org/10.1016/j.carbpol.2016.03.026.

67 Fabra, M.J., Falcó, I., Randazzo, W. et al. (2018). Antiviral and antioxidant properties of active alginate edible films containing phenolic extracts. *Food Hydrocolloids* 81 https://doi.org/10.1016/j.foodhyd.2018.02.026.

68 Wang, W. and Wang, A. (2010). Synthesis and swelling properties of pH-sensitive semi-IPN superabsorbent hydrogels based on sodium alginate-g-poly(sodium acrylate) and polyvinylpyrrolidone. *Carbohydrate Polymers* 80 (4) https://doi.org/10.1016/j.carbpol.2010.01.020.

69 Rhim, J.W. (2004). Physical and mechanical properties of water resistant sodium alginate films. *LWT – Food Science and Technology* 37 (3) https://doi.org/10.1016/j.lwt.2003.09.008.

70 Zhang, N., Xu, J., Gao, X. et al. (2017). Factors affecting water resistance of alginate/gellan blend films on paper cups for hot drinks. *Carbohydrate Polymers* 156 https://doi.org/10.1016/j.carbpol.2016.08.101.

71 Zhang, Y., Zhu, W., Lu, Y. et al. (2014). Nano-scale blocking mechanism of MMT and its effects on the properties of polyisocyanate-modified soybean protein adhesive. *Industrial Crops and Products* 57 https://doi.org/10.1016/j.indcrop.2014.03.027.

72 Block, K.A., Trusiak, A., Katz, A. et al. (2015). Exfoliation and intercalation of montmorillonite by small peptides. *Applied Clay Science* 107 https://doi.org/10.1016/j.clay.2015.01.021.

73 Liu, S., Cai, P., Li, X. et al. (2016). Effect of film multi-scale structure on the water vapor permeability in hydroxypropyl starch (HPS)/Na-MMT nanocomposites.

Carbohydrate Polymers 154 https://doi.org/10.1016/j.carbpol.2016.08.006.

74 Nikolic, M.S., Petrovic, R., Veljovic, D. et al. (2017). Effect of sepiolite organomodification on the performance of PCL/sepiolite nanocomposites. *European Polymer Journal* 97 https://doi.org/10.1016/j.eurpolymj.2017.10.010.

75 Ajmal, A.W., Masood, F., and Yasin, T. (2018). Influence of sepiolite on thermal, mechanical and biodegradation properties of poly-3-hydroxybutyrate-co-3-hydroxyvalerate nanocomposites. *Applied Clay Science* 156 https://doi.org/10.1016/j.clay.2018.01.010.

76 Shankar, S., Kasapis, S., and Rhim, J.W. (2018). Alginate-based nanocomposite films reinforced with halloysite nanotubes functionalized by alkali treatment and zinc oxide nanoparticles. *International Journal of Biological Macromolecules* 118 https://doi.org/10.1016/j.ijbiomac.2018.07.026.

77 Zhang, P., Tian, N., Zhang, J., and Wang, A. (2018). Effects of modification of palygorskite on superamphiphobicity and microstructure of palygorskite@fluorinated polysiloxane superamphiphobic coatings. *Applied Clay Science* 160 https://doi.org/10.1016/j.clay.2018.01.015.

78 Bilotti, E., Zhang, R., Deng, H. et al. (2009). Sepiolite needle-like clay for PA6 nanocomposites: an alternative to layered silicates? *Composites Science and Technology* 69 (15–16) https://doi.org/10.1016/j.compscitech.2009.07.016.

79 Deng, Y., Wang, L., Hu, X. et al. (2012). Highly efficient removal of tannic acid from aqueous solution by chitosan-coated attapulgite. *Chemical Engineering Journal*: 181–182. https://doi.org/10.1016/j.cej.2011.11.082.

80 Zhang, Z., Wang, W., Tian, G. et al. (2018). Solvothermal evolution of red palygorskite in dimethyl sulfoxide/water. *Applied Clay Science* 159 https://doi.org/10.1016/j.clay.2017.06.014.

81 Almasi, H., Ghanbarzadeh, B., and Entezami, A.A. (2010). Physicochemical properties of starch-CMC-nanoclay biodegradable films. *International Journal of Biological Macromolecules* 46 (1) https://doi.org/10.1016/j.ijbiomac.2009.10.001.

82 Wu, H., Liu, C., Chen, J. et al. (2009). Structure and properties of starch/α-zirconium phosphate nanocomposite films. *Carbohydrate Polymers* 77 (2) https://doi.org/10.1016/j.carbpol.2009.01.002.

83 Bidsorkhi, H.C., Soheilmoghaddam, M., Pour, R.H. et al. (2014). Mechanical, thermal and flammability properties of ethylene-vinyl acetate (EVA)/sepiolite nanocomposites. *Polymer Testing* 37 https://doi.org/10.1016/j.polymertesting.2014.05.007.

84 Huang, D., Zhang, Z., Ma, Z., and Quan, Q. (2018). Effect of natural nanostructured rods and platelets on mechanical and water resistance properties of alginate-based nanocomposites. *Frontiers in Chemistry* 6 https://doi.org/10.3389/fchem.2018.00635.

85 Hammond, G.P. (2004). Engineering sustainability: thermodynamics, energy systems, and the environment. *International Journal of Energy Research* 28 (7) https://doi.org/10.1002/er.988.

86 Qin, X., Wang, J., Han, B. et al. (2018). Novel design of eco-friendly super elastomer materials with optimized hard segments micro-structure: toward next-generation high-performance tires. *Frontiers in Chemistry* 6 https://doi.org/10.3389/fchem.2018.00240.

87 Qin, X., Han, B., Lu, J. et al. (2018). Rational design of advanced elastomer nanocomposites towards extremely energy-saving tires based on macromolecular assembly strategy. *Nano Energy* 48 https://doi.org/10.1016/j.nanoen.2018.03.038.

88 Cooper, S.L. and Tobolsky, A.V. (1967). Anomalous depression of rubbery modulus through crosslinking. *Journal of Applied Polymer Science* 11 (8) https://doi.org/10.1002/app.1967.070110801.

89 Liff, S.M., Kumar, N., and McKinley, G.H. (2007). High-performance elastomeric nanocomposites via solvent-exchange processing. *Nature Materials* 6 (1) https://doi.org/10.1038/nmat1798.

90 Jiang, L., Wu, J., Nedolisa, C. et al. (2015). Phase separation and crystallization in high hard block content polyurethane thin films. *Macromolecules* 48 (15) https://doi.org/10.1021/acs.macromol.5b01083.

91 Fu, G., Yuan, L., Liang, G., and Gu, A. (2016). Heat-resistant polyurethane films with great electrostatic dissipation capacity and very high thermally reversible self-healing efficiency based on multi-furan and liquid multi-maleimide polymers. *Journal of Materials Chemistry A* 4 (11) https://doi.org/10.1039/c6ta00953k.

92 Chiou, B.S. and Schoen, P.E. (2002). Effects of crosslinking on thermal and mechanical properties of polyurethanes. *Journal of Applied Polymer Science* 83 (1) https://doi.org/10.1002/app.10056.

93 Yoshida, S., Ejima, H., and Yoshie, N. (2017). Tough elastomers with superior self-recoverability induced by bioinspired multiphase design. *Advanced Functional Materials* 27 (30) https://doi.org/10.1002/adfm.201701670.

94 Wang, A., Wang, W., Cheng, H., and Zhou, C.E. (2019). *Novel Approaches to Design Eco-Friendly Materials Based on Natural Nanomaterials*. Lausanne: Frontiers Media https://doi.org/10.3389/978-2-88945-912-4.

20

Biomedical Applications of Nanofibers

Mehtap Sahiner[1], Saliha B. Kurt[2], and Nurettin Sahiner[2,3,4]

[1] *Canakkale Applied Science, Department of Fashion Design, CanakkaleOnsekiz Mart University, Canakkale, Turkey*
[2] *Faculty of Science and Arts, Department of Chemistry, Canakkale Onsekiz Mart University, Canakkale, Turkey*
[3] *Department of Chemical and Biomolecular Engineering, University of South Florida, Tampa, FL, USA*
[4] *Department of Ophthalmology, Morsani College of Medicine, University of South Florida, Tampa, FL, USA*

20.1 Introduction

In the twenty-first century, nanotechnology is the one of the top fields of science and technology since the industrial revolution. Considering nanotechnology's important role in many different areas such as economy, industry, human-made equipment, and lifestyle, it has improved the quality of human life considerably. Controllable nanosized materials have great impact in a broad range of applications such as biomedical, environmental, and energy Because of the recent developments in nanotechnology, nanosized materials with novel, superior, and unique features are constantly being discovered. In the past decades, nanosized materials with diverse functions, controllable shape, dimensions, composition, and physicochemical properties have been developed for daily use in industry. These nanosized materials, such as self-assembly thin films, quantum dots, nanorods, nanotubes, nanowires, nanofoam, nanocrystals, and nanofibers, were synthesized using different methods and formulations [1]. Among the nanosized materials, nanofibers occupy a place with increasing significance due to their interesting properties, large surface area, tunable porosity, ability for surface functionalization, selectiveness for certain materials, and superior mechanical performance [2]. Nanofibers prepared by the electrospinning technique have great possibilities in versatile platforms including biotechnology, environmental engineering, air and water filtration, textile industry, automotive industry, catalysis, energy conversion and storage, optoelectronics, pharmaceutics, biomedical devices, and health care. However, the most important thing in all these applications is the tunability and versatility of the nanosized fibrillar structure in comparison to other features [3]. To prepare nanofibers, various techniques such as electrospinning, self-assembly, and phase separation methods are applied. Among these, electrospinning is considered to be the front runner as it provides uniform fibers with controllable porosity [4].

In this chapter, the synthesis of nanofibers using some of the most important techniques, e.g. traditional electrospinning, supercritical CO_2 (SC CO_2) extraction with electrospinning, and self-assembling will be examined. Additionally, several important biomedical applications of nanofibers such as drug delivery systems, wound-healing applications, and biosensor applications will be considered.

20.2 Natural Polymeric-based Nanofibrils

Nanofibers are structures that are sometimes expected to mimic the extracellular matrix (ECM). The ECM has three-dimensional networks providing settings for the biochemical environment for the cells. Thus, cells can communicate, differentiate and adhere, proliferate, etc. Therefore, in biomedical applications, ECM-derived proteins such as collagen, gelatin (GEL), laminin, and fibronectin are frequently studied for the preparation of fibers. Besides, these proteins can provide cells with various biological signals through informational amino acid sequences [5]. Apart from these proteins, there are many natural proteins and polysaccharides that are suitable for nanofiber preparation for biomedical applications. Natural polymeric-based nanofibrils (NFs)

Sustainable Nanotechnology: Strategies, Products, and Applications, First Edition.
Edited by Yashwant V. Pathak, Govindan Parayil, and Jayvadan K. Patel.

can be obtained from animal or plant sources. In general, plant-based polymers are expected to be biocompatible, inexpensive, and less immunogenic; however, animal-based polymers are more active in terms of cell behavior and rapid biodegradability [6]. Table 20.1 summarizes some of the natural fibrils based on their sources.

20.2.1 Collagen

Collagen is a fibrillar protein that has a triple-helix structure and is commonly found in skin, tendons, bones, cartilage, and supportive tissues. It has six functional groups: carboxyl, amino, hydroxyl, imidazole, guanidine, and sulfhydryl. These groups provide collagen with the ability to bind various structures and functions. Moreover, ECM is also mainly consists of collagen. To mimic human skin, ECM with different kinds of structures are suggested such as hydrogels, pellets, and foam that contain collagen [7]. Collagen/black phosphorus/poly(ε-caprolactone) nanofibers were synthesized via electrospinning for bone tissue engineering to enhance osteo differentiation [8]. It was also suggested that embedding any collagen NF without cross-linking, collagen fibrils may induce cross-links with different chemical structures such as tannic acid (TA), glutaraldehyde, diisocyanates, and polyepoxide. Furthermore, even photo cross-linking can be used as an alternative in the preparation of collagen-based fibrils [9].

20.2.2 Chitin Nanofibers

Chitin nanofibers (CNFs) are very attractive for biomedical applications because of their inherent properties. CNFs are longer, more flexible, and less degradable than chitin nanocrystals. CNFs can be used as reinforcement materials to improve the mechanical properties of GEL, carboxymethyl cellulose (CMC), and chitosan (Ch). The antibacterial activity of CNFs is very well known, and CNFs can render additional strength to structures and improve the barrier properties of materials that it is incorporated into [10]. The deacetylated form of chitin is Ch, which has reactive hydroxyl groups along with an amino group as linear polyamine. This amine group can readily protonate and bind to the negatively charged cell membrane, enabling mucoadhesion, which is very important for bone tissue engineering. In the inflammatory environment, the generated protons can aid in diminishing pain via protonation of the amino group of D-glucosamine. Furthermore, Ch is biocompatible and biodegradable because of the existence of glyosidic bonds, making this natural polymer an excellent essential element in composite applications [11].

20.2.3 Resilin

In nature, there are three elastomeric proteins known, which are resilin, silk fibrinoid, and elastin. Resilin is found abundantly in insects and arthropods. It has rubber-like elasticity characterized by low stiffness, high extensibility, efficient energy storage, and exceptional resilience and fatigue lifetime [12]. Grove and Regan mentioned that bundles that contain resilin and elastin can be made softer and stimuli responsive in hydrogel formulations [13].

20.2.4 Silk Fibroin

Silk fibroin is a fibril that is a biocompatible protein obtained from silkworms [8]. This is an insoluble structure; however, it was reported that fibroin can be dissolved in concentrated neutral salts such as lithium bromide for the preparation of a fibroin solution and can be used in the preparation of composite films using a mixture of cellulose nanofiber (CNF)/silk fibroin fiber for biomedical applications [14]. The use of organic solvents is uneconomical and environmentally unfavorable from the viewpoint of pollution and environmental concerns [15]. Therefore, a novel preparation method without using organic solvents for silk fibroin solutions and silk fibroin films needs to be explored.

Table 20.1 Natural polymeric fibrils.[a]

	Animal based	Plant based
Protein	Collagen, silk fibroin, keratin, elastin, resilin, casein, laminin	Zein(corn), gluten, soy protein
Polysaccharide	Chitin, chitosan, alginate, hyaluronic acid, heparin	Cellulose and its derivatives, alginate

[a] *Source:* Add the following credit: Based on [6].

20.2.5 Elastin

Because of its' tough, and other beneficial properties, elastin performs special tasks in the body. It acts with tissues for differentiation, migration, and proliferation of cells. The flexibility of skin, lung, and blood vessels relies on elastin peptide. Elastin nanofibers possess great potential for regenerative therapies of salivary glands and other epithelial tissues [16]. The inclusion of elastin improves the biomatching capability of materials, e.g. the addition of elastin to a cellular bioprostheses contributes to matching with native tissues [9].

20.2.6 Keratin

Keratin is a fibrous protein and exist in hair, nail, wool, feather, and horns. Keratin has a cysteine residue that possesses disulfide bonds. It is a biopolymer and studied in the area of hemostasis, blood vessels, wound healing, nerve regeneration, and drug delivery. Keratin and keratose nanofibers were fabricated into electrospun nanofibers in combination with the synthetic polymer polyurethane (PU). Both keratin and keratose structures enhance cytocompatibility, especially of cell attachment [17]. Keratin can be used in fiber preparation along with polymers such as polyethylene terephthalate (PET), making these fibers very useful in environmental applications, i.e. for the removal of Cr(VI) [18].

20.2.7 Casein

Casein is a phosphoprotein that is found in milk. It is cheap, nontoxic, and heat stable. Due to the large polar groups, it can easily coil. Casein is capable of creating intra- and intermolecular hydrogen bonds due to the suitable chemical structure. It is used for drug and enzyme delivery, such as lipase delivery [19]. Poly(ethylene oxide) (PEO) was used as a copolymer with sodium dodecyl sulfate (SDS) along with casein for the electrospinning process, and SDS as a surfactant assumed to break hydrophobic interactions. The use of SDS that decreases surface tension enabled easy fiber preparation [19].

20.2.8 Zein

Zein is a plant protein from corn. It has some inherent properties such as being antimicrobial, antioxidant, and resilient [6]. It was reported that zein is an alcohol-soluble protein and this property makes zein to be electrospun in aqueous solutions of ethanol, acetic acid, and methanol. Weak mechanical properties of zein nanofibers require additional natural polymers such as silk fibroin and GEL [20]. Gallic acid-loaded PEO core/zein shell nanofibers were fabricated via the coaxial electrospinning method for treatment of gallbladder cancer cells. The SEM and TEM images of the core/shell structure of these fibers are shown in Figure 20.1. This core–shell structure provides increased drug release performance and protects unstable drug or drugs which can be protected from harsh environment such as pH and light [21].

20.2.9 Cellulose

Cellulose has different types depending on the functional groups that are attached to the structures such as methyl, ethyl, and propyl and is widely used in the biomedical field. It was reported that cellulose nanofibers can be embedded in hydrogel to prepare composites [22]. TA and CNF are added to polyvinyl alcohol (PVA) and borax to provide a self-healing composite hydrogel and enable these composite hydrogels to adhere to biological surfaces, e.g. finger epidermis. The fabricated hydrogels can render any desired shapes easily several times without changing the rheological properties. Moreover, the stretchable capability was approximately 20-fold. As the hydrogel was cut into two

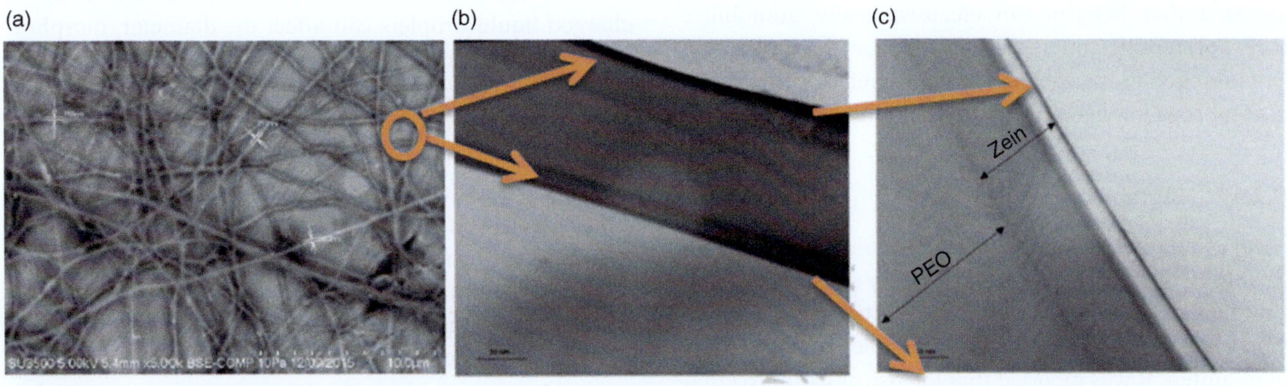

Figure 20.1 (a) SEM, (b) TEM, and (c) magnified TEM images of poly(ethylene oxide) core/zein shell nanofibers. *Source:* Acevedo et al. [21] Fig 01, Pg 44/with permission of Elsevier.

parts, the parts can bind to each other when reconnected. Dynamic borate ester bonds between borate and the other components (PVA, TA, and CNF) provide binding at the cut interfaces. These kinds of self-healing devices and composite hydrogels afford great potential in many biomechanical applications [22]. Bacterial cellulose nanofibers (BCNFs) are also widely used as biomedical fibers, because of the three-dimensional nanofibrillar networks supplying good mechanical and barrier performance, water-holding capacity, and structure-forming potential, as well as excellent film-forming ability. Due to their good crystalline properties, BCNFs can also be used to reinforce biocomposite films with high strength [10].

20.2.10 Alginate

Alginate is an anionic polysaccharide that is obtained from brown algae. The biocompatibility, high moisture-holding capacity, gelation capability, and low cost are the most attractive properties of alginate. Because of these properties, it is widely used in wound healing and dental-impression material. The negative properties of alginates, such as low solubility and high viscosity of high molecular weight alginate, high density of intra- and intermolecular hydrogen bonding, polyelectrolyte nature of aqueous solution, and lack of appropriate organic solvent, prevent it from directly forming electrospun fibers. However, it can be widely used to make fibers by combining with synthetic polymers such as PVA and PEO [4].

20.2.11 Heparin

Heparin is a linear nonbranched polysaccharide obtained from animals, and it is also known as the oldest anticoagulant agent [23]. Heparin nanofibers were studied for burn regeneration. It was reported that heparin has high affinity for growth factors such as brain-derived neurotrophic factor (BDNF) and basic fibroblast growth factor (bFGF). It was suggested that heparin can electrostatically immobilize bFGF. Polycaprolactone/GEL NFs conjugated with heparin were prepared for peripheral nerve regeneration and heparin accounted for vigorous cell proliferation [5, 24, 25].

20.2.12 Gluten

Wheat gluten has two major proteins, namely gliadin and glutenin. Gliadin is the more commonly studied material as biomedical nanofibers due to its elastic properties [6]. Upon changing the solvent, it is possible to make fibers out of gluten. It was reported that the use of trifluoro ethanol enables a convenient method for making gliadin fibers [26].

20.3 Nanofibers with Various Fabrication Techniques

Electrospinning is one of the most convenient, versatile, and also common technique for the synthesis of fibers with tunable size, e.g. micro- and nanodimensions. This method embodies the use of high-voltage electric field to prepare natural or synthetic polymer solutions. The applied electric field charges the liquid droplets and then develops electrostatic repulsion, leading to deactivation of the surface tension of the droplets. Charged droplets stretch in the direction of a collector with the opposite polarity to the droplets. Fibers form due to evaporation of the solvent with the remaining polymer. Electrospinning is one of the facile methods for establishing fluctuating fibers. In particular, needle-based electrospinning is the most common way to fabricate fibers as it is more cost-effective in comparison to the other methods. The needle-based electrospinning instrument usually consists of a syringe containing polymer solution as a reservoir, a piston for polymer solution to flow from the syringe, a metallic needle, and a collector for droplets of the solution formed at the needle to interact with high-voltage electric forces and produce fibers on the collector [3, 4]. To produce fibers, an air-tight closed reservoir containing polymer solution is necessary to minimize and avoid evaporation. Therefore, a broad range of materials including high volatile solvents can be readily processed. In the past decades, the electrospinning technique in combination with a variety of techniques was used in the production of different fibers [27, 28]. The improvement of new methods for versatile fiber production has progressed to commercialize the electrospinning method.

Fibers can be generated with tunable properties by precisely controlling the electrospinning parameters including voltage, distance between the collector and the fiber spraying needle, flow rate, and the viscosity of the solution. The structure of the collector, applied voltage, length between the needle and the collector, and scattering of charged liquid droplets can affect the diameter, morphology, alignment, and molecular placement of the fibers. In addition to the electrospinning techniques and arrangement of parameters, there are several important elements for the preparation of the polymer solution and environmental factors such as humidity and temperature that need to be considered. The concentration of the solution affects the fiber diameter, e.g. the increase in concentration results in an increase in fiber diameter. Viscosity of the solution also greatly influences the fiber thickness. Polymer solution with low viscosity may not render successful fiber formation, while high viscosity of the solution requires higher electric field to form fibers. The optimum viscosity value of the solution may deliver smooth continuous fibers. The

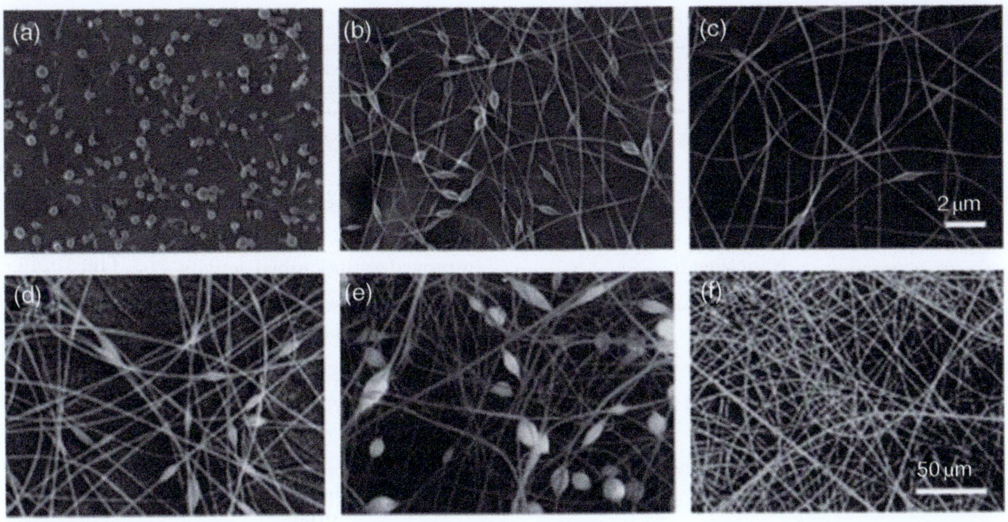

Figure 20.2 The effect of the different viscosities on fiber morphology: (a) 13 mPa s, (b) 160 mPa s, and (c) 527 mPa s. SEM images for the effects of using different grades of dimethylformamide (DMF): (d) DMF2-Aldrich 99%, (e) Sigma 99.8%, and (f) Fluka 98% forming polystyrene (PS) fibers from a 20% (w/v) PS solution. (a, b, c) *Source:* Fong et al. [29] Fig 01, Pg 03/with permission of Elsevier. (d, e, f) *Source:* Uyar et al. [30] Fig 04, Pg 04/with permission of Elsevier.

effects of viscosity of the polymer solution on the quality of fibers are shown in Figure 20.2 [29, 30]. Producing fibers from diverse materials provides remarkable features such as high surface area, complex pore structure, great mechanical stability, and flexibility with variable diameters (micrometer to nanometer scale range) that can be useful in environmental to biomedical applications. It was suggested that decreasing fiber sizes from micro- to nanoscale provides great advantages for certain applications, e.g. biomedical applications [31].

While using the electrospinning method, some other methods can be used simultaneously such as SC CO_2. SC CO_2 is a green technology that contains no toxic solvents. A silk fibroin porous scaffold was produced by SC CO_2 combined with porogen leaching. This structure has potential applications for nerve regeneration [8]. Lupin hull cellulose nanofiber aerogel was prepared by supercritical CO_2 drying [32]. Moreover, SC CO_2 is used to prepare microfibers via traditional electrospray and electrospinning methods to overcome some limitations of surface tension and viscosity control [33].

Self-assembling is another method for the preparation of nanofibers with structures that provide large length-to-diameter ratio [34]. In the body, naturally self-assembling proteins are found, i.e. aptamers assemble DNA and RNA. Furthermore, amino acids are assembled to form proteins. However, accidental misfolding of proteins can also occur. This misfolding during aggregation of proteins causes some neurodegenerative diseases such as Parkinson and Alzheimer. The scientific studies about hindering the misfolding of protein are very important for neurotic disease as the investigation of self-assembling fibrillation can make a huge contribution and provide clues to understand, analyze, hinder, or control the misfolded protein [25]. Additionally, the self-assembling protein mimetic peptide system that can mimic the properties and functions of the ECM provides a new multifunctional, recognized biomedical material.

20.4 Biomedical Applications of Nanofibers

In recent years, researchers have developed many remarkable nanomaterials, e.g. nanofibers with extraordinary features for use in the human body. Accordingly, nanofibers with high surface volume ratio, high porous structure, and improved physicomechanical properties obtained by the electrospinning method support human life. Nanofibers are widely used in drug delivery systems, wound healing, biosensor, and tissue engineering applications. Therefore, the biocompatibility of fibers, e.g. as not causing any toxic effects, immunologic rejection, and not showing any physiological reactivity in human body and its systems, has paramount significance [35]. It is important to use biocompatible nanofibers that do not threaten the body cycle, and it is also expected that the nanofibers will prevent microorganisms and promote the healing process in

Table 20.2 Fibers derived from various polymers for different applications.

Materials	Form	Applications	References
Chitosan–Ag nanoparticles (NPs)	Nanofibers	Antibacterial wound dressing material	[36]
Silk–Ag NPs	Nanofibers	Antibacterial wound dressing material	[37]
Chitin	Nanofibers	Improved antimicrobial activity	[38]
Chitosan–Ag NPs-PVA	Nanofibrous mat	Antimicrobial activity, wound dressing material	[39]
Polyurethane–cellulose acetate–zein	Composite mats	Wound dressing material	[40]
Sodium alginate–poly(ethylene oxide)	Core–shell nanofibrous scaffold	Tissue engineering	[41]
Poly(D,L-lactide-co-trimethylene carbonate) (PLMC), nanohydroxyapatite (HAp)	Fibrous scaffolds	Bone screw hole healing	[42]
Chitosan-PVA-clotrimazole microemulsion	Nanofiber mats	Oral candidiasis	[43]
PEO-chitosan-graphene oxide	Composite nanofibrous scaffolds	Drug delivery	[44]
PVA-curcumin/PVA-cyclodextrin curcumin complex	Nanofibers	Drug delivery	[45]
PVDF-MWCNT-Pt NPs	Nanofibrous membrane	Biosensor	[46]
PLGA-fusidic acid (FA) and rifampicin (RIF)	Nanofibers	Orthopedic implant-related infection	[47]

many circumstances [4, 7–9]. Some applications of the different types of nanofibers are summarized in Table 20.2.

A considerable amount of tissue loss or organ failure is one of the main problems in many wounds in terms of human health. Nanofibers are great candidates for tissue engineering in helping to cure damaged tissue because of their ability to mimic the ECM. Nanofibers for tissue engineering offer a 3-D environment for cell adhesion proliferation. Biocompatible and bioresorbable polymers are preferred for the production of tissue scaffolds that mimic the ECM. There are diverse biomaterials that offer many versatile potentials for the design of nanofibers in tissue engineering that can improve the quality of life, e.g. regeneration of the loss of skin or bone tissues, cartilage, cardiovascular tissue, nerves, bladder, spinal cord, tendons, and ligaments. The three main apex components in tissue engineering are scaffolds as nanofibers, undifferentiated or differentiated cells, and growth factors for maintaining the growth of cells. There are certain polymers such as collagen, chitin, Ch, alginate, and GEL that can be utilized alone or in combination for the production of different scaffolds [2, 48].

Ch nanofibers are perfect materials for nerve regeneration or treatment of demyelinating lesions of the Schwann cells [2]. Both aligned and randomly formed electrospun nanofibers were applied, and the aligned nanofibers showed more cell maturation. As a result of this research, the aligned Ch nanofibers are ideal scaffolds for this case [49]. In another study, hydroxyapatite nanorods-poly(lactide-co-glycolide) (nHA/PLGA) nanofibers were shown to enhance osteoblast growth, proliferation, and differentiation for insulin release in a drug delivery system [50]. Farokhi et al. showed the sustained release of vascular endothelial growth factor for bone tissue engineering using silk fibroin-calcium phosphate and PLGA (SF/Ca/PLGA) nanofibers [51]. Another report illustrated CS/PVA nanofibers as potential nerve growth factors (NGFs) that provide for the attachment and proliferation of SKNMC and U373 neural cell lines [52]. Wu et al. prepared starch-based nanographene oxide nanofibers for potential uses in bone tissue engineering. Under simultaneous mineralization and degradation for one week, the nanofiber showed mimicking features and morphologic similarities with ECM [53]. Biocompatible multilayer scaffolds made up of chlorhexidine-doped-PLGA/polycaprolactone (PCL), PLGA/PCL, and β-tricalcium phosphate-doped-PLGA/PCL synthesized by Qian et al. revealed the osteoconductive behavior and antibacterial activity for regeneration of bone deficiency [54].

20.5 Nanofiber Drug Delivery Systems

New discoveries and innovations in the medical field enabled the development of drug delivery systems with unprecedented increasing rates. The correct drug delivery system is required for achieving the therapeutic effect needed from a drug with specific release profile. Several important arrangements in the release system of these drugs, such as the placement, timing, and rate of release, need to be accomplished for the therapeutic target of the drug. However, it is cumbersome to meet all the requirements of an ideal drug delivery system [55, 56]. Therefore, nanofibers provide an innovative solution because of their tunable morphological structures and sizes ranging from micrometer to nanometer scale. Because of the diverse size range, surface properties, and the morphology of fibers, a broad range of drug release kinetics can be ensured. Especially, the nanosize range of materials that are similar in size to the biological molecules and the building blocks of a living organism allow them to adapt to the biological environments readily. Lately, nanofibers have been used as drug delivery systems with controllable drug-releasing capabilities due to the high surface area-to-volume ratio making it easier to load the intended amount of drug into nanofibers. Additionally, the control of porosity of the nanofibers enables the right amount of drug loading because the fibrous structure has huge resemblance to the extracellular membrane, making nanofibers attractive materials for use in drug delivery systems [57, 58].

Recent advances in the design and preparation of nanofibers make it possible to use them in various applications of drug delivery systems for cancer therapy, antibacterial sheets, surgical treatments, and tissue engineering. For example, the solubility of the drug can be enhanced by increasing the nanofibers. Various drugs, herbal extracts, biological molecules, genes, and bioactive agents, e.g. anticancer drugs, anti-inflammatory drugs, antibiotics, and proteins, can be readily loaded and delivered by means of nanofibers. There are several requirements in choosing materials that carry drugs. Firstly, nanofibers must protect the therapeutic agents or drugs from deterioration in the blood stream. Secondly, the structure of the nanofibers needs to be biodegradable to transport the drug to the target area. Finally, nanofibers must permit the release of the drug in controllable time frame at certain release rate for effective treatment. Therefore, the methods used to prepare nanofibers, such as electrospinning which is a facile, economical, and tunable method, offer broad opportunities in the selection of materials to produce fibers to meet the desired criteria, e.g. high loading capacity and high encapsulation efficiency to ensure a viable drug delivery system. The therapeutic effectiveness of drugs involves drug specificity, efficiency, tolerability, and availability during the period of treatment [59, 60]. When a drug delivery system is designed, many parameters such as controlled release, therapeutic delivery, and targeted delivery need to be carefully accounted for. The benchmarks for designing effective drug delivery systems are shown in Figure 20.3.

Figure 20.3 Benchmarks for designing effective drug delivery systems.

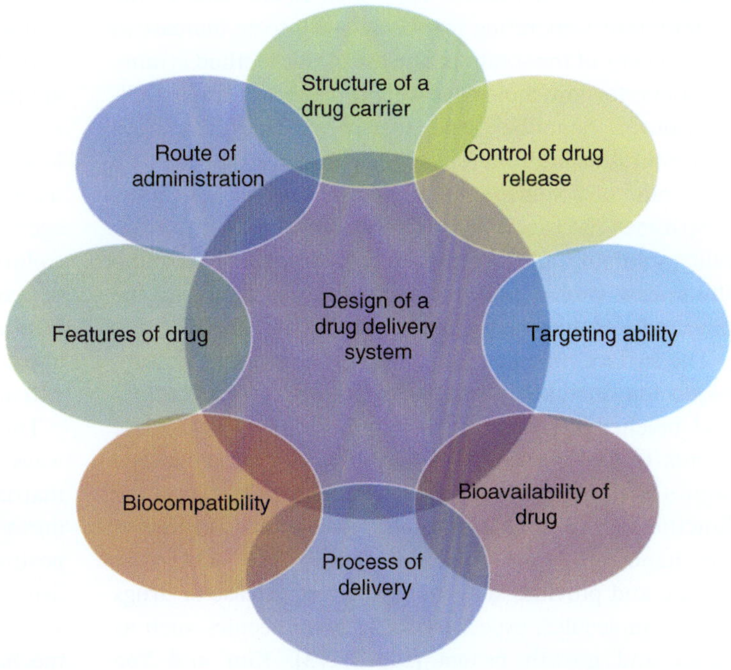

There are a huge number of drugs with different structures and properties to incorporate into a delivery system, so different loading methods are necessary. The drug solubility, targeted region, and sometimes the requirement for multiple drugs obligate employment of several types of drug loading methods in or out of the nanofibers including the use of emulsion, surface modification, blending, adsorption, and conjugation [61–63]. The blending method is one of the most commonly used facile techniques for loading of drugs into fibers. This method is easy as it just involves dissolving or dispersing the drug in the polymer solution to be used for the preparation of the nanofiber. As the blending method is utilized before electrospinning, the nanofibers, physicochemical properties of the drug, and polymer solution need to be evaluated carefully as it affects the drug encapsulation efficiency, drug allocation, and release kinetics [64]. With the general principle of like dissolves like, lipophilic drugs can be readily carried in lipophilic polymer solutions, and the opposite is valid as hydrophilic drugs are soluble in hydrophilic polymer solutions to make a better drug-encapsulated nanofiber. The distribution of the drug is very important; it requires uniform distribution, otherwise the release of the drug can be problematic, e.g. bursts of release, slow release (not therapeutic), and no release etc. are not desired [58, 65, 66].

The emulsion method is defined as emulsifying a drug solution that is immiscible in a polymer solution for the electrospinning process. Generally, this emulsion system is constituted with the polymer dissolved in aqueous solution called a water phase, and the drug dissolved in oil and oil-related solution called a continuous phase; thus, mixing these two solutions makes core–shell fibers. Due to the water phase evaporating very fast, it leads to an increase in the viscosity of the solution. The use of this method is more favorable for low molecular weight drugs to form a well-distributed nanofiber, and it can provide core–shell nanofibers for high molecular weight drugs [67–70]. The right ratio of polymer solution, which is the water phase, determines the success of the process. The structural stability, solubility, and effectiveness of the system determine the release profile from the nanofibers, ascertaining the bioactivity of the encapsulated drugs [57].

The surface modification method is a noteworthy technique that enables integration with tissue for structural and biochemical bonding or conjugation of therapeutic agents to the surface of the nanofibers. This technique allows modulation of the drug release and protection of the functionality of the biomaterials. Surface modification of the nanofibers prevents burst release and short-term release and provides slow and prolonged release of drugs and biomolecules, especially for large molecules such as genes and growth factors [52, 71–73]. Kim and Yoo modified the surface of the fiber and incorporated with small interfering RNA (siRNA) to provide enhanced outcome for gene silencing and wound healing [74]. In another study, the nanofiber was modified to enhance the wetting capacity of the nanofibers of polycaprolactone (PCL) that were prepared via the electrospinning method, and simultaneously curcumin was embedded into the nanofibers [75]. The nanofibers were surface modified with poly(ethylene glycol) methyl ether methacrylate (PEGMA) to increase the wetting capability. The water contact angle of the pristine PCL nanofibers was measured in 120–90 °C. However, the water contact angle was decreased significantly to approximately 55 °C for PCL modified with PEGMA. Similarly, cropped nanofibers were embedded into GEL/Ch for skin regeneration [75]. In another study, the surface modification effort was applied to cellulose nanofibers. Cellulose nanofibers were converted to acetylated cellulose nanofibers. This modification increases the hydrophobicity and thermal stability of CNF, allowing enhanced dispersibility of CNF in hydrophobic polymers [76].

In recent years, there is a new drug delivery method called multidrug delivery, which is a desirable technique and versatile tool that uses multiple drugs with or without similar therapeutic effects when employed with the appropriate polymer solutions [76, 77]. So, different drugs are carried by nanofibers that have the same diffusion pathways and matrix-degradation rate required for successful independent delivery [78, 79]. One of the good examples of multiple drug-loaded fibers was studied by Okuda and his coworkers [80]. The system of incorporating multiple drugs into fibers was done as four layers, two of them were blank polymer fibers used as barrier mesh, and the other two fibers were used as drug-loaded mesh. The process was applied so that the drug-loaded polymer is the first layer, blank polymers are the second and third layers, the drug-loaded polymer is used next, and lastly, blank polymer is used as the final layer. This smart procedure ensured management of release time and the amounts of drugs by controlling the fiber size, the thickness of each fiber layer, and the position of the layers. The technique used in this procedure was very interesting due to biochemical regulation in chemotherapy in which the Okuda research group used two different dyes as antitumor drugs [80].

The drug release process from fibers occurs in different ways. For example, desorption from the nanofiber surface, matrix degradation, and diffusion of drugs through the pores or channels are the most common routes. The composition and arrangement of the nanofibers also affects the drug release profile. Polymers can degrade with enzymes and/or hydrolytically, which can also affect the drug release mechanism [81–83]. The other significant circumstances that affect drug release are the crystallinity of the polymer used for fiber preparation and the molecular weight of the drugs. Usually, semicrystalline polymers exhibit bursts of drug release at first due to the drug position on the surface of the nanofibers and because the narrow water uptake of the semicrystalline polymers leads to interrupted release. The limited water uptake of the highly packed crystalline section of a crystalline polymer induces slower drug release on the surface of the nanofibers compared to amorphous polymers [83, 84]. The state of drugs, molecular weight, and solubility of the drug also affect the release mechanism. A drug on amorphous polymer fibers provides a more sustainable drug release profile than on crystalline polymer fibers. The rate and the total amount of the drug released depend on the molecular weight, with the lowest molecular weight of drug released in higher amounts and faster rates compared to other releases [85–88].

The production of fibers at nanoscale is desired due to their resemblance to biological molecules in human body. Accordingly, the nano range of the fibers make it a prominent option for drug delivery applications in biomedical applications. Diverse drugs and biomolecules including DNA, RNA, and growth factors are used for drug delivery at desired dosages using tunable nanofibers [89].

Disintegration can occur in the blood stream when systemic application is implemented, so the protection of drugs mainly takes places within the nanofibers. The modification of the drug delivery system within the nanofibers offers controllable amounts and long-term release. Nanofibers with constant release kinetics may allow the drug to pass through membranes or barriers such as blood–brain barriers. In recent years, nanofibers were expected to be used as medical devices such as bandages which carry certain drugs. The main reason for the use of nanofibers that are obtained by electrospinning technique as drug delivery systems is due to their controllable and tunable functionality as well as adjustable morphological features at nanometer range [59, 90].

The drugs as therapeutic agents within the nanofibers can be anticancer drugs, antibiotics, proteins, polysaccharides, and growth factors, which can be successfully used in the biomedical field for gene delivery, wound-healing applications, tissue engineering, biosensors, and so on. The quality of the structure of the nanofibers, such as adhesiveness to the mucous layer for a certain part of the body, makes them perfect candidates to improve the healing process [91].

Vitamins can also be used in the delivery system by attaching them to the surface of the nanofibers, which could be very useful to treat skin. Vitamins A and E, which are lipid-soluble molecules, are the most commonly used vitamins to enhance the treatment of acne and other skin syndromes as well as in the cosmetic field. Besides treatment of acne and skin disorders, vitamin A is also used in the treatment of leukemia. Vitamin E has a hydroxyl group in the chromanol ring which makes it possible to display antioxidant features, which means that it can reduce free radicals harming the human body. Recently, vitamins have been loaded on the nanofibers successfully, and the results of the release of these vitamins revealed high potential in wound healing of the skin due to their innate properties, as shown by Fahami and Fathi and Kheradvar et al. [90, 92]. Taepaiboon et al. compared the release of vitamins A and E from a cellulose acetate polymer nanofiber and solvent cast film. The result showed that vitamin A and E-loaded nanofibers have steady increase with cumulative release, whereas the film had burst release [93]. It was shown that as vitamin A is encapsulated with PVA-composited nanofibers, the thermal stability of vitamin A was increased in acidic conditions such as gastric fluids [90]. Vitamin B_{12}, cyanocobalamin is a water-soluble vitamin, and the intake of Vitamin B_{12} can be accomplished in many ways such as pulmonary, transdermal patch, and intradermal injection mode. In a study where Vitamin B_{12}-loaded polycaprolactone nanofibers were used as transdermal patch. Before loading, polycaprolactone was modified with plasma at different time periods to render hydrophilic properties that enabled the modified transdermal patch to be used as nanoreservoir for prolonged release applications [94].

In tumor or cancer treatments, there is a need for targeted drug delivery systems to cover the degenerated tissue region. Thus, the frequency of drug administration will be reduced by use of nanofibers that can be placed on the target tissue, and the arrangement of drug dosage will be provided by the nanofibers. Normally, when a cancer drug is administered orally or intravenously into the body, the efficacy and concentration of the drug is lower around the tumor site. However, incorporation of a cancer drug into a nanofiber increases the drug release concentration and drug efficacy due to delivery of the drugs at the target with the nanofibers [86, 95, 96].

Doxorubicin, paclitaxel, cisplatin, and dichloroacetate are frequently used chemotherapeutic drugs, and these drugs can be readily loaded on nanofibers that are commonly produced from poly(lactic acid) and poly(lactic-co-glycolic acid) polymer solutions. Cancer drugs incorporated with hydrophobic paclitaxel and hydrophilic doxorubicin need a multidrug delivery system. The toxicity of cancer drugs is high when they are taken orally or intravenously, damaging the healthy cells. On the other hand, as the cancer drugs loaded into nanofibers are favorable for administration of these drugs to certain places in the body, they have decreased cytotoxic effects on other cells. The placement of drug-loaded nanofibers is localized, and it directly

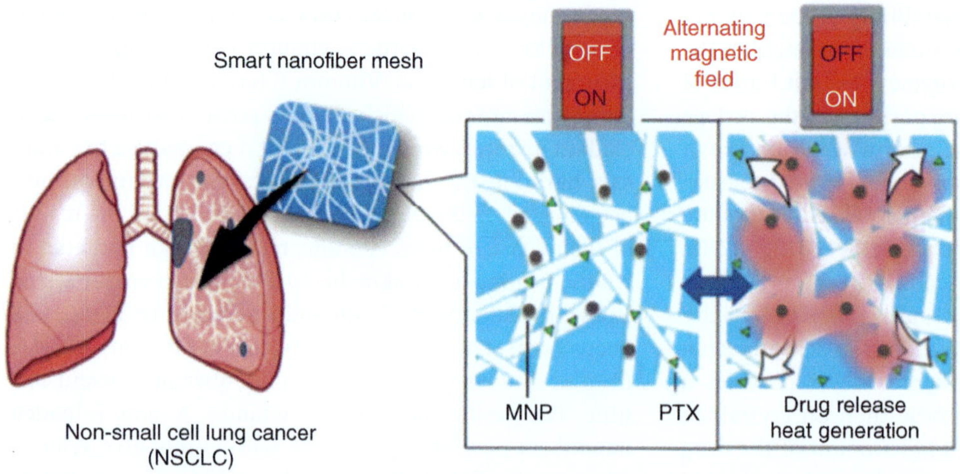

Figure 20.4 Illustration of temperature-responsive nanofiber mesh for thermochemotherapy containing magnetic nanoparticles (MNPs) and paclitaxel (PTX) as an anticancer drug. The system can be switched on and off under an alternating magnetic field. *Source:* Reproduced by permission [100].

targets the cancerous tissue [96–99]. Niiyama et al. fabricated a temperature-responsive nanofiber by chemical cross-linking from poly(NIPAAm-co-HMAAm) with magnetic nanoparticles (MNPs) and paclitaxel (PTX) as anticancer drug, as shown in Figure 20.4. This smart nanofiber mesh has excellent functions with tumor-killing activity against lung adenocarcinoma. Heat can be generated by the external switch of MNPs exposed to alternating magnetic fields, and with this generated heat, PTX can be released from the nanofibers because of the existence of thermoresponsive PNIPAM [100]. This system was proven as a safe and efficient treatment for cancer.

In another study, nanofibers were used as a drug carrier to prevent abdominal adhesion that is an instance occurring after abdominal surgical procedures [101]. Abdominal adhesion is the attachment of organs to the surgical site. Generally, antibiotics are used for the prevention of abdominal adhesion. Biocompatible and degradable nanofibers are used to prevent abdominal adhesion with and without using antibiotics. Nanofibers are resistant against the migration of the tissue cells, otherwise cells are free to move from one place to another. Bölgen et al. delivered the common antibiotic, Biteral, using poly(epsilon-caprolactone) nanofibers and showed that it inhibited abdominal adhesion [102].

20.6 Wound-healing Applications of Nanofibers

Skin is accepted as the first line of defense of the immunologic system against infections. Wound-healing materials are sterilized covers that prevent external microorganisms from entering to prevent infection. After an injury, skin has the ability to heal itself, but under certain circumstances, e.g. after burning of the skin, it is necessary to intervene with medication. Wound-healing applications have attracted much interest because of the health care industry. Wound dressings stimulate or start the healing process of the affected area and also protect from microorganisms. Platelet adhesion is the first step to heal the wound naturally, then to prevent high amounts of blood loss vasoconstriction develops, and lastly, leucocytes migrate around the wound to prevent or fight inflammation [101, 103].

There is a great need for materials to promote tissue growth around the wound, which have the ability to fight infections or external microorganisms [2]. Nanofibers also have great potential to boost hemostasis and gas permeation as they can be embedded in the wound. Usually, bioactive materials containing films, hydrogels, foams, and sponges are used as wound dressings [66]. Nanofibers, on the other hand, are more advantageous in wound-healing applications owing to the high surface area for effective absorption of exudates, flexible nature, and high mechanical strength. Electrospun nanofiber mats can be designed to have similar building architecture to ECM that is found in human cells, e.g. 60–90% porosity. High porosity of these nanofibers can supply enough oxygen for the cells but not enough for bacterial infections and can hold moisture in the wound to accelerate healing without scars on the skin [66, 104–106]. Nanofibers have controllable water-absorbance capacity between 18% and 213% and higher water capacity in comparison to films [107].

For wound healing, bacterial cellulose-based nanofibers have low cell toxicity, and it is converted from stem cell proliferation to adipose tissue [108, 109]. These kinds of nanofibers facilitate new capillary formation and speed up the regeneration of new tissue [110, 111]. The average ulcer treatment takes 81 days; however, using bacterial cellulose-based nanofibers, this time is reduced with the healing process to 11 days as reported by Solway et al. It was reported that bacterial cellulose-based nanofibers used in wound dressings improved the efficacy of the healing process drastically, as a 75% reduction in time was observed compared to the normal process for lower extremity ulcers [110]. In another study, Xu et al. showed that silver nanoparticles embedded in nylon nanofibers provided antimicrobial properties as a wound dressing [111]. Generally, depending on the chronic state, the process of wound healing takes two to three days. Also, Abdelhady et al. proved that nanofibers are better wound dressings compared to other types of wound-dressing materials for treatment and processing [112]. The clinical test outcomes on the human body showed a reduction in pain from dressing removal, airing of the wound, soaking up the exudate, prevention of the formation of infection, and finally, skin tissue regeneration using biodegradable nanofibers, e.g. made up of Ch, as illustrated in Figure 20.5 [113].

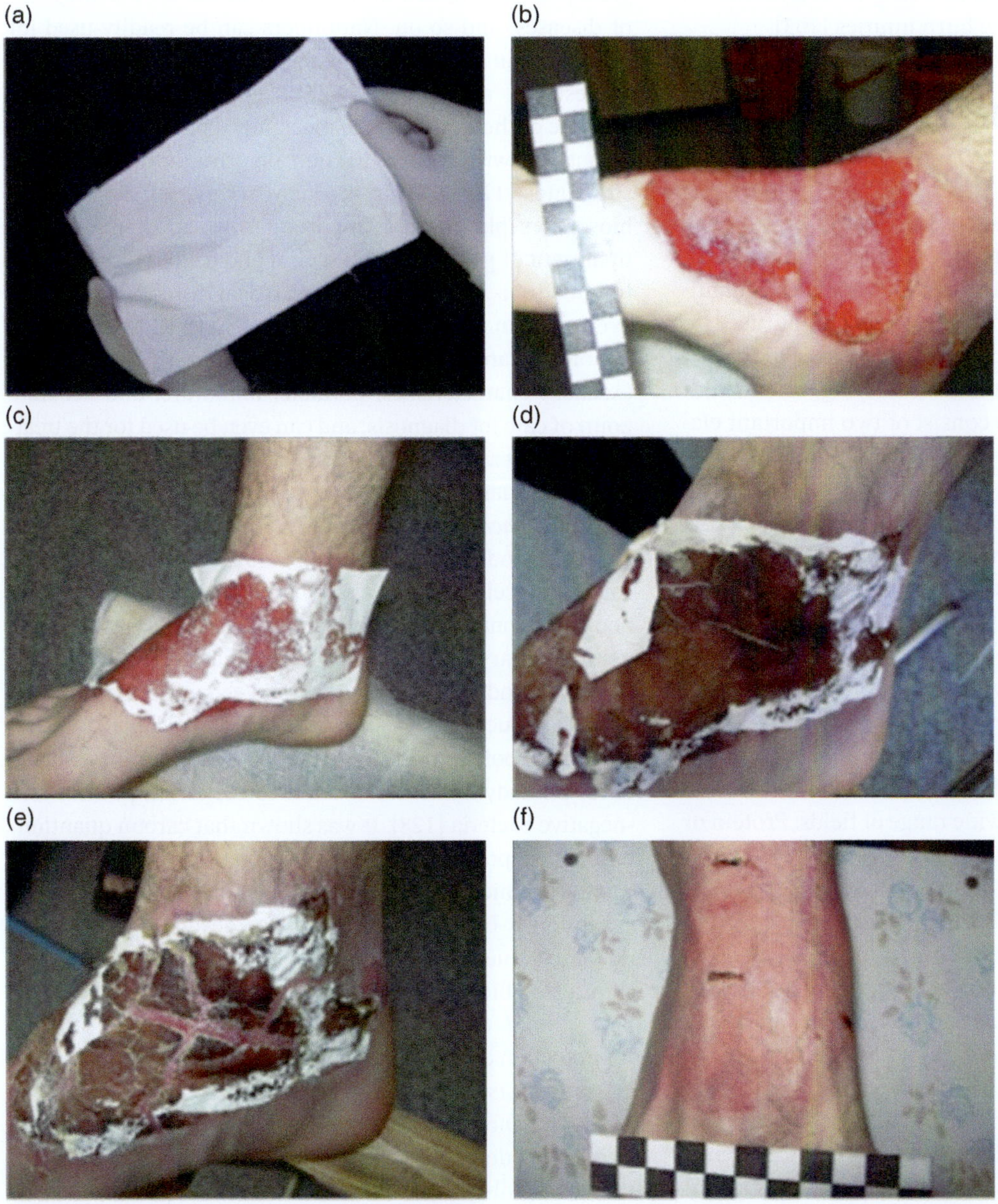

Figure 20.5 Illustration of a burn (a) chitosan nanofiber as wound dressing material, (b) burn before covering with the nanofiber, (c) after covering with nanofiber, (d) 5 days after covering, (e) 10 days after covering, and (f) 14 days after covering [113].
Source: Balusamy et al. [113] Fig 6a, Pg 10/with permission of Elsevier.

A Ch/PVA composite nanofiber membrane cross-linked with glutaraldehyde was reported prepared via the electrospinning method with 2D nanofiber membranes immersed in $NaBH_4$ solution, and H_2 gas output during the hydrolysis reaction ensured that 2D samples expanded to 3-D sponges [114]. The expanded layered nanofiber sponges could accelerate wound healing, prevent scar formation, and ensure faster blood coagulation [114]. Lavender oil and silver nanoparticles loaded into PU nanofibers were synthesized via electrospinning by Sofi et al., resulting in improved hydrophilicity and decreased rigidity, affording antibacterial properties and providing an environment for proliferation of fibroblast cells, and long-term lavender oil release capabilities for chronic burn injuries [103].

20.7 Nanofibers in Biosensors Applications

A biosensor is a self-contained device that is used for the recognition of biological elements such as enzymes, antibodies, and receptors to obtain semiquantitative or quantitative analytical information. Biosensors can be used to detect diverse biological components in many fields such as food security, environmental monitoring, and health care applications. Biosensors consist of two important elements of the bioreceptor and transducer. A bioreceptor is a sensitive element which is immobilized onto a substrate such as an enzyme and an antibody, and the transducer is a device employed to convert the received biochemical signal to an electronic signal due to the interaction between the analyte and the bioreceptor [115]. Sensitivity and limit of detection of a biosensor are very significant, even though many biosensors exist in low amounts and/or extents and concentrations. Biosensors are categorized into four classes of electrochemical, optical, acoustic, and immunosensors. Electrochemical and optical-based biosensors are the most commonly used sensors in a wide range of fields. Protein or enzyme-based biosensors are electrochemical sensors which are also broadly used due to their sensitivity, selectivity, repeatability, and easy and economic production. In addition to electrochemical biosensors, optical biosensors such as fiber optics and fluorescence properties are important examples of other biosensors. Fluorescence is widely used for bioimaging in scanning systems mostly to track drugs in the human body [116]. In the past two decades, new zero-dimensional materials such as carbon dots with size range from few nm to 20 nm with strong florescence properties are considered as promising materials for optical sensors. In addition to the photoluminescent properties of carbon dots, they have excellent characteristics such as being nontoxic, high biocompatibility, inexpensive synthesis methods, high thermal and optical stability, and tunable surface functionality via chemical modification. They are used for broad fluorescence imaging, as nanocarriers for medical diagnosis, and so on [117, 118].

Nanofibers are great candidates for biosensors with superb properties such as high porosity and surface area-to-volume ratio, interconnectivity, and simple production. The incorporation of biosensor materials into nanofibers improves the sensitivity and selectivity to detect troublesome analytes [119]. Furthermore, nanofibers can be functionalized further with surface-modification agents and affording additional properties. Biosensors including nanofibers possess great potential for health care applications, for the diagnosis of diseases, and so on. Nanofibers can be readily used for the determination of some analytes such as glucose, urea, cholesterol, and microRNA in comparison to biofilm because of their remarkable mechanical properties and high bioactivity with immobilization on various materials and substrates [120–123]. In recent years, nanofibers used as biosensors played very important roles in diagnosis and treatment of genetic diseases, viral and bacterial pathogens, in the fight against the bioterrorism threats, and were even used in the discovery of drugs [124]. Biosensor-based nanofibers are the most pertinent materials due to their excellent characteristics; portable, miniature, and accurate point of care for diagnosis; and can even be used for the treatment of diseases and disorders [125]. Some studies showed that these nanofibers allow detection of specific microbes such as *Escherichia coli*, bovine viral diarrhea virus, malarial parasites, and Hepatitis B virus [126, 127]. Luo et al. fabricated nitrocellulose biosensor-based nanofibers with antibody functionalization to detect viral and bacterial pathogens using capillary immunoassay, direct-charge electrical measurement, and integrating magnetic separation [126, 127]. Nie et al. successfully showed that carbon quantum dot-embedded polyacrylonitrile nanofibers mediate the antibacterial photodynamic inactivation of the Gram-positive and -negative bacteria [128]. It was shown that carbon quantum-embedded polyacrylonitrile nanofibers can be used as a photodynamic material against infections [128].

Biosensor-based nanofibers were designed for glucose detection due to their high sensitivity and selectivity. To synthesize biosensor nanofibers for glucose, glucose oxidase is used, and this is the most commonly used enzyme for detection of glucose molecule due to its excellent properties such as good stability at a broad pH range. Ren et al. produced biosensor nanofibers from PVA and glucose oxidase to detect glucose [129]. Tsou et al. reported that the use of silica nanofiber membranes for enzyme-linked immunosorbent assay (ELISA) is more favorable compared to conventional ELISA using polystyrene well plates [130]. It was shown that with the use of silica nanofibers, the limit of detection for ELISA was 1.6 pM, which is 32 times lower than conventional ELISA, and the detection time decreased to only 1 hour from 24 hours [131, 132].

Vitamin B is a very important nutrient for the human body that maintains and regulates metabolism, especially enzymatic cofactors and fatty acid metabolism. For that reason, it is vital to determine vitamin B in different nutritional and dietary conditions. Polypyrrole nanofibers were used to determine water-soluble vitamins such as B_2, B_9, and B_{12} in human urine via solid-phase extraction with HPLC. The determination of vitamin B in urine is valuable, as otherwise obtaining cerebrospinal fluid or plasma samples can be traumatic sometimes. So, the use of polypyrrole fiber-based sensors to determine vitamin B is safe, noninvasive, easy to apply, fast, repeatable, and easy to use with very small amounts of sample [133].

20.8 Conclusions and Outlook

The electrospinning method is considered as the most common route to obtain fibers and has been utilized for decades and now is versatile and tunable with many advantageous for biomedicine applications. Electrospun fibers are used in a wide range of fields due to their attractive properties such as being easy to use, flexibility and adaptability, tunable fiber size and functionality, high porosity, selectivity, and sensitivity for different materials. Diverse polymers can be used with the electrospinning technique to produce micro/nanofibers and for biomedical use in almost any field. In addition, the polymers that are used to fabricate the nanofibers give their characteristic features to the nanofibers, and biological molecules such as DNA, RNA, enzyme, growth factor, and toxins; bioactive molecules such as drugs and phenolic compounds, and particles such as metals, silicate, and carbon dots can be introduced into fibers for versatile use. As many polymers and materials are naturally antimicrobial, sensors, or have good innate features than can be combined with added structure, fiber formulations ensure the combined effect of both constituents for diverse benefits in health-related fields. Thus, these features of nanofibers make them indispensable materials for future applications for drug delivery, wound healing, and biosensors. Moreover, multifuntional nanofibers are still evolving for multitask applications, providing the required skills such as higher precision, higher sensitivity, speed of response, release and recovery, and stability and degradability with superior performances in comparison to their rivals for biomedical applications. It is expected that nanofibers obtained using diverse methods will keep evolving with enduring features, and their use will significantly increase in biomedical fields.

References

1 Chee, B.S. and Nugent, M. (2019). *Electrospun Natural Polysaccharide for Biomedical Application*. Elsevier Inc.

2 Rasouli, R., Barhoum, A., Bechelany, M., and Dufresne, A. (2019). Nanofibers for biomedical and healthcare applications. *Macromolecular Bioscience* **19** (2): 1–27.

3 Li, Y., Abedalwafa, M.A., Tang, L. et al. (2019, 2021). *Electrospun nanofibers for sensors. International Journal of Molecular Sciences* 22 (12): 6357.

4 Taemeh, M.A., Shiravandi, A., Korayem, M.A., and Daemi, H. (2020). Fabrication challenges and trends in biomedical applications of alginate electrospun nanofibers. *Carbohydrate Polymers* **228**: 115419.

5 Ikegami, Y. and Ijima, H. (2019). Development of heparin-conjugated nanofibers and a novel biological signal by immobilized growth factors for peripheral nerve regeneration. *Journal of Bioscience and Bioengineering* 129 (3): 354–362.

6 Babitha, S., Rachita, L., Karthikeyan, K. et al. (2017). Electrospun protein nanofibers in healthcare: a review. *International Journal of Pharmaceutics* **523** (1): 52–90.

7 Sahiner, M., Alpaslan, D., and Bitlisli, B.O. (2014). Collagen-based hydrogel films as drug-delivery devices with antimicrobial properties. *Polymer Bulletin* **71** (11): 3017–3033.

8 Li, Z.-H., Wang, L., Dai, H.-L. et al. (2019). Fabrication, characterization, and in vitro evaluation of biomimetic silk fibroin porous scaffolds via supercritical CO_2 technology. *Journal of Supercritical Fluids* **150**: 86–93.

9 Sallach, R.E. and Chaikof, E. (2008). Electrospun elastin and collagen nanofibers and their application as biomaterials. In: *Natural-Based Polymers for Biomedical Applications* (eds. R.L. Reis, N.M. Neves, J.F. Mano, et al.), Chapter 11, 315–336. Woodhead Publishing.

10 Yang, Y.-N., Lu, K.-Y., Wang, P. et al. (2020). Development of bacterial cellulose/chitin multi-nanofibers based smart films containing natural active microspheres and nanoparticles formed in situ. *Carbohydrate Polymers* **228**: 115370.

11 Balagangadharan, K., Dhivya, S., and Selvamurugan, N. (2017). Chitosan based nanofibers in bone tissue engineering. *International Journal of Biological Macromolecules* **104**: 1372–1382.

12 Li, L. and Kiick, K.L. (2013). Resilin-based materials for biomedical applications. *ACS Macro Letters* **2** (8): 635–640.

13 Grove, T.Z. and Regan, L. (2012). New materials from proteins and peptides. *Current Opinion in Structural Biology* **22** (4): 451–456.

14 Narita, C., Okahisa, Y., and Yamada, K. (2019). A novel technique in the preparation of environmentally friendly cellulose nanofiber/silk fibroin fiber composite films with improved thermal and mechanical properties. *Journal of Cleaner Production* **234**: 200–207.

15 Xie, X., Zhang, T., Gu, J., and Huang, Z. (2018). Water footprint assessment of coal-based fuels in China: exploring the impact of coal-based fuels development on water resources. *Journal of Cleaner Production* **196**: 604–614.

16 Foraida, Z.I., Kamaldinov, T., Nelson, D.A. et al. (2017). Elastin-PLGA hybrid electrospun nanofiber scaffolds for salivary epithelial cell self-organization and polarization. *Acta Biomaterialia* **62**: 116–127.

17 Yang, G., Yao, Y., and Wang, X. (2018). Comparative study of kerateine and keratose based composite nanofibers for biomedical applications. *Materials Science and Engineering: C* **83**: 1–8.

18 Jin, X., Wang, H., Jin, X. et al. (2019). Preparation of keratin/PET nanofiber membrane and its high adsorption performance of Cr(VI). *Science of the Total Environment* 710: 135546.

19 Selvaraj, S., Thangam, R., and Fathima, N.N. (2018). Electrospinning of casein nanofibers with silver nanoparticles for potential biomedical applications. *International Journal of Biological Macromolecules* **120**: 1674–1681.

20 Deng, L., Zhang, X., Li, Y. et al. (2018). Characterization of gelatin/zein nanofibers by hybrid electrospinning. *Food Hydrocolloids* **75**: 72–80.

21 Acevedo, F., Hermosilla, J., Sanhueza, C. et al. (2018). Gallic acid loaded PEO-core/zein-shell nanofibers for chemopreventive action on gallbladder cancer cells. *European Journal of Pharmaceutical Sciences* **119**: 49–61.

22 Ge, W., Cao, S., Shen, F. et al. (2019). Rapid self-healing, stretchable, moldable, antioxidant and antibacterial tannic acid-cellulose nanofibril composite hydrogels. *Carbohydrate Polymers* **224**: 115147.

23 Oduah, E., Linhardt, R., and Sharfstein, S. (2016). Heparin: past, present, and future. *Pharmaceuticals* **9** (3): 38.

24 Yergoz, F., Hastar, N., Cimenci, C.E. et al. (2017). Heparin mimetic peptide nanofiber gel promotes regeneration of full thickness burn injury. *Biomaterials* **134**: 117–127.

25 Zhang, W., Yu, X., Li, Y. et al. (2018). Protein-mimetic peptide nanofibers: motif design, self-assembly synthesis, and sequence-specific biomedical applications. *Progress in Polymer Science* **80**: 94–124.

26 Soares, R.M.D., Patzer, V.L., Dersch, R. et al. (2011). A novel globular protein electrospun fiber mat with the addition of polysilsesquioxane. *International Journal of Biological Macromolecules* **49** (4): 480–486.

27 Nguyen, L.T.H., Chen, S., Elumalai, N.K. et al. (2013). Biological, chemical, and electronic applications of nanofibers. *Macromolecular Materials and Engineering* **298** (8): 822–867.

28 Agrahari, V., Agrahari, V., Meng, J., and Mitra, A.K. (2017). *Electrospun Nanofibers in Drug Delivery: Fabrication, Advances, and Biomedical Applications.* Elsevier.

29 Fong, H., Chun, I., and Reneker, D. (1999). Beaded nanofibers formed during electrospinning. *Polymer (Guildf)* **40** (16): 4585–4592.

30 Uyar, T. and Besenbacher, F. (2008). Electrospinning of uniform polystyrene fibers: the effect of solvent conductivity. *Polymer (Guildf)* **49** (24): 5336–5343.

31 Leung, V. and Ko, F. (2011). Biomedical applications of nanofibers. *Polymers for Advanced Technologies* **22** (3): 350–365.

32 Ciftci, D., Ubeyitogullari, A., Huerta, R.R. et al. (2017). Lupin hull cellulose nanofiber aerogel preparation by supercritical CO_2 and freeze drying. *Journal of Supercritical Fluids* **127**: 137–145.

33 Baldino, L., Cardea, S., and Reverchon, E. (2019). A supercritical CO_2 assisted electrohydrodynamic process used to produce microparticles and microfibers of a model polymer. *Journal of CO_2 Utilization* **33**: 532–540.

34 Huang, Y., Song, J., Yang, C. et al. (2019). Scalable manufacturing and applications of nanofibers. *Materials Today* 28: 98–113.

35 de Moraes Porto, I.C.C. (2012). Polymer biocompatibility. In: *Polymerization* (ed. A.D.S. Gomes), 47–62. IntechOpen. doi: 10.5772/47786. https://www.intechopen.com/chapters/38906.

36 Lee, S.J., Heo, D.N., Moon, J.-H. et al. (2014). Electrospun chitosan nanofibers with controlled levels of silver nanoparticles. Preparation, characterization and antibacterial activity. *Carbohydrate Polymers* **111**: 530–537.

37 Kang, M., Jung, R., Kim, H.-S. et al. (2007). Silver nanoparticles incorporated electrospun silk fibers. *Journal of Nanoscience and Nanotechnology* **7** (11): 3888–3891.

38 Mushi, N.E., Butchosa, N., Salajkova, M. et al. (2014). Nanostructured membranes based on native chitin nanofibers prepared by mild process. *Carbohydrate Polymers* **112**: 255–263.

39 Wang, D., Lu, Q., Wei, M., and Guo, E. (2018). Ultrasmall Ag nanocrystals supported on chitosan/PVA nanofiber mats with bifunctional properties. *Journal of Applied Polymer Science* **135** (28): 46504.

40 Unnithan, A.R., Gnanasekaran, G., Sathishkumar, Y. et al. (2014). Electrospun antibacterial

polyurethane–cellulose acetate–zein composite mats for wound dressing. *Carbohydrate Polymers* **102**: 884–892.

41 Ma, G., Fang, D., Liu, Y. et al. (2012). Electrospun sodium alginate/poly(ethylene oxide) core–shell nanofibers scaffolds potential for tissue engineering applications. *Carbohydrate Polymers* **87** (1): 737–743.

42 Bao, M., Wang, X., Yuan, H. et al. (2016). HAp incorporated ultrafine polymeric fibers with shape memory effect for potential use in bone screw hole healing. *Journal of Materials Chemistry B* **4** (31): 5308–5320.

43 Tonglairoum, P., Ngawhirunpat, T., Rojanarata, T. et al. (2015). Fabrication of a novel scaffold of clotrimazole-microemulsion-containing nanofibers using an electrospinning process for oral candidiasis applications. *Colloids and Surfaces, B: Biointerfaces* **126**: 18–25.

44 Ardeshirzadeh, B., Anaraki, N.A., Irani, M. et al. (2015). Controlled release of doxorubicin from electrospun PEO/chitosan/graphene oxide nanocomposite nanofibrous scaffolds. *Materials Science and Engineering: C* **48**: 384–390.

45 Sun, X.-Z., Williams, G.R., Hou, X.-X., and Zhu, L.-M. (2013). Electrospun curcumin-loaded fibers with potential biomedical applications. *Carbohydrate Polymers* **94** (1): 147–153.

46 Zhang, P., Zhao, X., Zhang, X. et al. (2014). Electrospun doping of carbon nanotubes and platinum nanoparticles into the β-phase polyvinylidene difluoride nanofibrous membrane for biosensor and catalysis applications. *ACS Applied Materials and Interfaces* **6** (10): 7563–7571.

47 Gilchrist, S.E., Lange, D., Letchford, K. et al. (2013). Fusidic acid and rifampicin co-loaded PLGA nanofibers for the prevention of orthopedic implant associated infections. *Journal of Controlled Release* **170** (1): 64–73.

48 Rivero, G., Aldana, A.A., and Abraham, G.A. (2019). *Nanocomposite Electrospun Micro/Nanofibers for Biomedical Applications*. Elsevier Inc.

49 Haider, S., Al-Zeghayer, Y., Ahmed Ali, F.A. et al. (2013). Highly aligned narrow diameter chitosan electrospun nanofibers. *Journal of Polymer Research* 20 (4): 1–11.

50 Haider, A., Gupta, K., and Kang, I.-K. (2014). PLGA/nHA hybrid nanofiber scaffold as a nanocargo carrier of insulin for accelerating bone tissue regeneration. *Nanoscale Research Letters* 9 (1): 314.

51 Farokhi, M., Mottaghitalab, F., Shokrgozar, M.A. et al. (2014). Bio-hybrid silk fibroin/calcium phosphate/PLGA nanocomposite scaffold to control the delivery of vascular endothelial growth factor. *Materials Science and Engineering: C* **35**: 401–410.

52 Mottaghitalab, F., Farokhi, M., Mottaghitalab, V. et al. (2011). Enhancement of neural cell lines proliferation using nano-structured chitosan/poly(vinyl alcohol) scaffolds conjugated with nerve growth factor. *Carbohydrate Polymers* **86** (2): 526–535.

53 Wu, D., Samanta, A., Srivastava, R.K., and Hakkarainen, M. (2017). Starch derived nano-graphene oxide paves the way for electrospinnable and bioactive starch scaffolds for bone tissue engineering. *Biomacromolecules* **18** (5): 1582–1591.

54 Qian, Y., Zhou, X., Sun, H. et al. (2017). Biomimetic domain-active electrospun scaffolds facilitating bone regeneration synergistically with antibacterial efficacy for bone defects. *Applied Materials and Interfaces* **10** (4): 3248–3259.

55 Safari, J. and Zarnegar, Z. (2014). Advanced drug delivery systems: nanotechnology of health design A review. *Journal of Saudi Chemical Society* **18** (2): 85–99.

56 Liechty, W.B., Kryscio, D.R., Slaughter, B.V., and Peppas, N.A. (2010). Polymers for drug delivery systems. *Annual Review of Chemical and Biomolecular Engineering* **1** (1): 149–173.

57 Bhattarai, R.S., Bachu, R.D., Boddu, S.H.S., and Bhaduri, S. (2019). Biomedical applications of electrospun nanofibers: drug and nanoparticle delivery. *Pharmaceutics* **11** (1): 5.

58 Kim, K., Luu, Y.K., Chang, C. et al. (2004). Incorporation and controlled release of a hydrophilic antibiotic using poly(lactide-co-glycolide)-based electrospun nanofibrous scaffolds. *Journal of Controlled Release* **98** (1): 47–56.

59 Akhgari, A., Shakib, Z., and Sanati, S. (2017). A review on electrospun nanofibers for oral drug delivery. *Nanomedicine Journal* **4** (4): 197–207.

60 Sanchez-Alvarado, D.I., Guzmán-Pantoja, J., Páramo-García, U. et al. (2018). Morphological study of chitosan/poly (vinyl alcohol) nanofibers prepared by electrospinning, collected on reticulated vitreous carbon. *International Journal of Molecular Sciences* **19** (6): 1–12.

61 Kim, C., Kim, D., Kang, S. et al. (2006). Structural studies of electrospun cellulose nanofibers. *Polymer* 47: 5097–5107.

62 Online, V.A., Sharma, A., Gupta, A. et al. (2013). Electrospun composite nanofiber-based transmucosal patch for anti-diabetic drug delivery. *Journal of Materials Chemistry B* **1**: 3410–3418.

63 Kataria, K., Sharma, A., Garg, T. et al. (2014). Novel technology to improve drug loading in polymeric nanofibers. *Drug Delivery Letters* **4**: 79–86.

64 Zeng, J., Yang, L., Liang, Q. et al. (2005). Influence of the drug compatibility with polymer solution on the release kinetics of electrospun fiber formulation. *Journal of Controlled Release* **105** (1–2): 43–51.

65 Meng, Z.X., Xu, X.X., Zheng, W. et al. (2011). Preparation and characterization of electrospun PLGA/gelatin nanofibers as a potential drug delivery system. *Colloids and Surfaces, B: Biointerfaces* **84** (1): 97–102.

66 Jannesari, M., Varshosaz, J., Morshed, M., and Zamani, M. (2011). Composite poly(vinyl alcohol)/poly(vinyl acetate) electrospun nanofibrous mats as a novel wound dressing matrix for controlled release of drugs. *International Journal of Nanomedicine* **6**: 993–1003.

67 Shastri, V.P., Sy, J.C., and Klemm, A.S. (2009). Emulsion as a means of controlling electrospinning of polymers. *Advanced Materials* **21** (18): 1814–1819.

68 Yang, Y., Li, X., Qi, M. et al. (2008). Release pattern and structural integrity of lysozyme encapsulated in core-sheath structured poly(dl-lactide) ultrafine fibers prepared by emulsion electrospinning. *European Journal of Pharmaceutics and Biopharmaceutics* **69** (1): 106–116.

69 He, S., Xia, T., Wang, H. et al. (2012). Multiple release of polyplexes of plasmids VEGF and bFGF from electrospun fibrous scaffolds towards regeneration of mature blood vessels. *Acta Biomaterialia* **8** (7): 2659–2669.

70 Yang, Y., Li, X., Cheng, L. et al. (2011). Core-sheath structured fibers with pDNA polyplex loadings for the optimal release profile and transfection efficiency as potential tissue engineering scaffolds. *Acta Biomaterialia* **7** (6): 2533–2543.

71 Zomer Volpato, F., Almodóvar, J., Erickson, K. et al. (2012). Preservation of FGF-2 bioactivity using heparin-based nanoparticles, and their delivery from electrospun chitosan fibers. *Acta Biomaterialia* **8** (4): 1551–1559.

72 Kim, H.S. and Yoo, H.S. (2010). MMPs-responsive release of DNA from electrospun nanofibrous matrix for local gene therapy: in vitro and in vivo evaluation. *Journal of Controlled Release* **145** (3): 264–271.

73 Choi, J.S., Leong, K.W., and Yoo, H.S. (2008). in vivo wound healing of diabetic ulcers using electrospun nanofibers immobilized with human epidermal growth factor (EGF). *Biomaterials* **29** (5): 587–596.

74 Kim, H.S. and Yoo, H.S. (2013). Matrix metalloproteinase-inspired suicidal treatments of diabetic ulcers with siRNA-decorated nanofibrous meshes. *Gene Therapy* **20** (4): 378–385.

75 Ghaee, A., Bagheri-Khoulenjani, S., Amir Afshar, H., and Bogheiri, H. (2019). Biomimetic nanocomposite scaffolds based on surface modified PCL-nanofibers containing curcumin embedded in chitosan/gelatin for skin regeneration. *Composites. Part B, Engineering* 177 (August): 107339.

76 Jamaluddin, N., Kanno, T., Asoh, T.A., and Uyama, H. (2019). Surface modification of cellulose nanofiber using acid anhydride for poly(lactic acid) reinforcement. *Materials Today Communications* 21 (August): 100587.

77 Wang, Y., Wang, B., Qiao, W., and Yin, T. (2010). A novel controlled release drug delivery system for multiple drugs based on electrospun nanofibers containing nanoparticles. *Journal of Pharmaceutical Sciences* **99** (12): 4805–4811.

78 Xu, J., Jiao, Y., Shao, X. et al. (2011). Controlled dual release of hydrophobic and hydrophilic drugs from electrospun poly (l-lactic acid) fiber mats loaded with chitosan microspheres. *Journal of Pharmaceutical Sciences* **65** (4): 1165–1172.

79 Zamani, M., Prabhakaran, M.P., and Ramakrishna, S. (2013). Advances in drug delivery via electrospun and electrosprayed nanomaterials. *International Journal of Nanomedicine* **8**: 2997–3017.

80 Okuda, T., Tominaga, K., and Kidoaki, S. (2010). Time-programmed dual release formulation by multilayered drug-loaded nanofiber meshes. *Journal of Controlled Release* **143** (2): 258–264.

81 Ranganath, S.H. and Wang, C.H. (2008). Biodegradable microfiber implants delivering paclitaxel for post-surgical chemotherapy against malignant glioma. *Biomaterials* **29** (20): 2996–3003.

82 Pillay, V., Dott, C., Choonara, Y.E. et al. (2013). A review of the effect of processing variables on the fabrication of electrospun nanofibers for drug delivery applications. *Journal of Nanomaterials* **2013**: 789289.

83 Meng, Z.X., Zheng, W., Li, L., and Zheng, Y.F. (2011). Fabrication, characterization and in vitro drug release behavior of electrospun PLGA/chitosan nanofibrous scaffold. *Materials Chemistry and Physics* **125** (3): 606–611.

84 Kenawy, E.R., Bowlin, G.L., Mansfield, K. et al. (2002). Release of tetracycline hydrochloride from electrospun poly(ethylene-co-vinylacetate), poly(lactic acid), and a blend. *Journal of Controlled Release* **81** (1–2): 57–64.

85 Goonoo, N., Bhaw-Luximon, A., and Jhurry, D. (2014). Drug loading and release from electrospun biodegradable nanofibers. *Journal of Biomedical Nanotechnology* **10** (9): 2173–2199.

86 Zamani, M., Morshed, M., Varshosaz, J., and Jannesari, M. (2010). Controlled release of metronidazole benzoate from poly ε-caprolactone electrospun nanofibers for periodontal diseases. *European Journal of Pharmaceutics and Biopharmaceutics* **75** (2): 179–185.

87 Natu, M.V., de Sousa, H.C., and Gil, M.H. (2010). Effects of drug solubility, state and loading on controlled release in bicomponent electrospun fibers. *International Journal of Pharmaceutics* **397** (1–2): 50–58.

88 Taepaiboon, P., Rungsardthong, U., and Supaphol, P. (2006). Drug-loaded electrospun mats of poly(vinyl

alcohol) fibres and their release characteristics of four model drugs. *Nanotechnology* **17** (9): 2317–2329.

89 Dhand, C., Dwivedi, N., Sriram, H. et al. (2017). *Nanofiber Composites in Drug Delivery*. Elsevier Ltd.

90 Fahami, A. and Fathi, M. (2018). Development of cress seed mucilage/PVA nanofibers as a novel carrier for vitamin A delivery. *Food Hydrocolloids* **81**: 31–38.

91 Yan, L., Shiya, Z., Yanlin, G., and Yinglei, Z. (2019). Electrospun nanofibers as a wound dressing for treating diabetic foot ulcer. *Asian Journal of Pharmaceutical Sciences* 14 (1): 130–143.

92 Kheradvar, S.A., Nourmohammadi, J., Tabesh, H., and Bagheri, B. (2018). Starch nanoparticle as a vitamin E-TPGS carrier loaded in silk fibroin-poly(vinyl alcohol)-Aloe vera nanofibrous dressing. *Colloids and Surfaces, B: Biointerfaces* **166**: 9–16.

93 Taepaiboon, P., Rungsardthong, U., and Supaphol, P. (2007). Vitamin-loaded electrospun cellulose acetate nanofiber mats as transdermal and dermal therapeutic agents of vitamin A acid and vitamin E. *European Journal of Pharmaceutics and Biopharmaceutics* **67** (2): 387–397.

94 Madhaiyan, K., Sridhar, R., Sundarrajan, S. et al. (2013). Vitamin B12 loaded polycaprolactone nanofibers: a novel transdermal route for the water soluble energy supplement delivery. *International Journal of Pharmaceutics* **444** (1–2): 70–76.

95 Kohane, D.S. and Langer, R. (2010). Biocompatibility and drug delivery systems. *Chemical Science* **1** (4): 441–446.

96 Luo, X., Xie, C., Wang, H. et al. (2012). Antitumor activities of emulsion electrospun fibers with core loading of hydroxycamptothecin via intratumoral implantation. *International Journal of Pharmaceutics* **425** (1–2): 19–28.

97 Aggarwal, U., Goyal, A.K., and Rath, G. (2017). Development and characterization of the cisplatin loaded nanofibers for the treatment of cervical cancer. *Materials Science and Engineering: C* **75**: 125–132.

98 Wang, Y., Ying, X., Xu, H. et al. (2017). The functional curcumin liposomes induce apoptosis in C6 glioblastoma cells and C6 glioblastoma stem cells in vitro and in animals. *International Journal of Nanomedicine* **12**: 1369–1384.

99 Cheng, S., Du, Y., Ma, B., and Tan, D. (2009). Total synthesis of a furostan saponin, timosaponin BII. *Organic and Biomolecular Chemistry* **7** (15): 3112–3118.

100 Niiyama, E., Uto, K., Lee, C.M. et al. (2018). Alternating magnetic field-triggered switchable nanofiber mesh for cancer thermo-chemotherapy. *Polymers (Basel)* **10** (9): 1–13.

101 Andrary, A.L. (2007). Biomedical applications of nanofibers. In: *Science and Technology of Polymer Nanofibers* (ed. A.L. Andrary), 183–223. Hoboken, NJ, USA: Wiley.

102 Bölgen, N., Vargel, İ., Korkusuz, P. et al. (2007). in vivo performance of antibiotic embedded electrospun PCL membranes for prevention of abdominal adhesions. *Journal of Biomedical Materials Research Part B: Applied Biomaterials* **81B** (2): 530–543.

103 Sofi, H.S., Akram, T., Tamboli, A.H. et al. (2019). Novel lavender oil and silver nanoparticles simultaneously loaded onto polyurethane nanofibers for wound-healing applications. *International Journal of Pharmaceutics* 569 (August): 118590.

104 Tian, L., Prabhakaran, M.P., Ding, X. et al. (2012). Emulsion electrospun vascular endothelial growth factor encapsulated poly(l-lactic acid-co-ε-caprolactone) nanofibers for sustained release in cardiac tissue engineering. *Journal of Materials Science* **47** (7): 3272–3281.

105 Boateng, J.S., Matthews, K.H., Stevens, H.N.E., and Eccleston, G.M. (2008). Wound healing dressings and drug delivery systems: a review. *Journal of Pharmaceutical Sciences* **97** (8): 2892–2923.

106 Kanani, G.A. and Bahrami, H.S. (2010). Review on electrospun nanofibers scaffold and biomedical applications. *Biomaterials and Artificial Organs* **24** (2): 93–115.

107 Lee, C.K., Kim, S.I., and Kim, S.J. (2005). The influence of added ionic salt on nanofiber uniformity for electrospinning of electrolyte polymer. *Synthetic Metals* **154** (1–3): 209–212.

108 Fu, L., Zhang, Y., Li, C. et al. (2012). Skin tissue repair materials from bacterial cellulose by a multilayer fermentation method †. *Journal of Materials Chemistry* **22**: 12349–12357.

109 Fu, L., Zhou, P., Zhang, S., and Yang, G. (2013). Evaluation of bacterial nanocellulose-based uniform wound dressing for large area skin transplantation. *Materials Science and Engineering: C* **33** (5): 2995–3000.

110 Solway, D.R., Consalter, M., and Levinson, D.J. (2010). Microbial cellulose wound dressing in the treatment of skin tears in the frail elderly. *Wounds* **22**: 17–19.

111 Xu, Z., Mahalingam, S., Rohn, J.L. et al. (2015). Physio-chemical and antibacterial characteristics of pressure spun nylon nanofibres embedded with functional silver nanoparticles. *Materials Science and Engineering: C* **56**: 195–204.

112 Abdelhady, S., Honsy, K.M., and Kurakula, M. (2015). Electro spun- nanofibrous mats: a modern wound dressing matrix with a potential of drug delivery and therapeutics. *Journal of Engineered Fibers and Fabrics* **10** (4): 155892501501000.

113 Balusamy, B., Senthamizhan, A., and Uyar, T. (2017). *in vivo Safety Evaluations of Electrospun Nanofibers for Biomedical Applications*. Elsevier Ltd.

114 Zhang, K., Bai, X., Yuan, Z. et al. (2019). Layered nanofiber sponge with an improved capacity for promoting blood coagulation and wound healing. *Biomaterials* **204** (March): 70–79.

115 Koyun, A., Ahlatcolu, E., and Koca, Y. (2012). Biosensors and their principles. In: *A Roadmap of Biomedical Engineers and Milestones* (ed. S. Kara). InTech.

116 Mansoori, G.A., George, T.F., Zhang, G., and Assoufid, L. (2007). *Molecular Building Blocks for Nanotechnology: From Diamondoids to Nanoscale Materials and Applications*. New York: Springer-Verlag.

117 Sagbas, S. and Sahiner, N. (2019). Carbon dots: preparation, properties, and application. In: *Nanocarbon and its Composites* (eds. A. Khan, M. Jawaid, Inamuddin and A.M. Asiri), 651–676. Elsevier Ltd.

118 Sahiner, N., Suner, S.S., Sahiner, M., and Silan, C. (2019). Nitrogen and sulfur doped carbon dots from amino acids for potential biomedical applications. *Journal of Fluorescence* **29** (5): 1191–1200.

119 Ahadian, S., Obregón, R., Ramón-Azcón, J. et al. (2017). *Clinical/Preclinical Aspects of Nanofiber Composites*. Elsevier Ltd.

120 Cash, K.J. and Clark, H.A. (2010). Nanosensors and nanomaterials for monitoring glucose in diabetes. *Trends in Molecular Medicine* **16** (12): 584–593.

121 Gomes, S., Rodrigues, G., Martins, G. et al. (2017). Evaluation of nanofibrous scaffolds obtained from blends of chitosan, gelatin and polycaprolactone for skin tissue engineering. *International Journal of Biological Macromolecules* **102**: 1174–1185.

122 Sapountzi, E., Braiek, M., Chateaux, J.F. et al. (2017). Recent advances in electrospun nanofiber interfaces for biosensing devices. *Sensors (Switzerland)* **17** (8): 1887.

123 Zarekhalili, Z., Bahrami, S.H., Ranjbar-Mohammadi, M., and Milan, P.B. (2017). Fabrication and characterization of PVA/Gum tragacanth/PCL hybrid nanofibrous scaffolds for skin substitutes. *International Journal of Biological Macromolecules* **94**: 679–690.

124 Sassolas, A., Blum, L.J., and Leca-Bouvier, B.D. (2011). Homogeneous assays using aptamers. *Analyst* **136** (2): 257–274.

125 Wang, J. (2006). Electrochemical biosensors: towards point-of-care cancer diagnostics. *Biosensors and Bioelectronics* **21** (10): 1887–1892.

126 Luo, Y., Nartker, S., Miller, H. et al. (2010). Surface functionalization of electrospun nanofibers for detecting *E. coli* O157:H7 and BVDV cells in a direct-charge transfer biosensor. *Biosensors and Bioelectronics* 26 (4): 1612–1617.

127 Inamuddin, Khan, R., Mohammad, A., and Asiri, A. (eds.) (2019). *Advanced Biosensors for Health Care Applications*. Elsevier.

128 Nie, X., Wu, S., Mensah, A. et al. (2020). Carbon quantum dots embedded electrospun nanofibers for efficient antibacterial photodynamic inactivation. *Materials Science and Engineering: C* **108**: 110377.

129 Ren, G., Xu, X., Liu, Q. et al. (2006). Electrospun poly(vinyl alcohol)/glucose oxidase biocomposite membranes for biosensor applications. *Reactive and Functional Polymers* **66** (12): 1559–1564.

130 Tsou, P.-H., Chou, C.-K., Saldana, S.M. et al. (2008). The fabrication and testing of electrospun silica nanofiber membranes for the detection of proteins. *Nanotechnology* **19** (44): 445714.

131 Kai, D., Liow, S.S., and Loh, X.J. (2015). Biodegradable polymers for electrospinning: towards biomedical applications. *Materials Science and Engineering: C* **45**: 659–670.

132 Su, X., Wei, J., Ren, X. et al. (2013). A new amperometric glucose biosensor based on one-step electrospun poly(vinyl alcohol)/chitosan nanofibers. *Journal of Biomedical Nanotechnology* **9** (10): 1776–1783.

133 Xie, L., Huang, J., Han, Q. et al. (2019). Solid phase extraction with Polypyrrole nanofibers for simultaneously determination of three water-soluble vitamins in urine. *Journal of Chromatography. A* **1589**: 30–38.

Abbreviations

BCNF	Bacterial cellulose nanofibers
BDNF	Brain-derived neurotrophic factor
bFGF	Basic fibroblast growth factor
CNF	Cellulose nanofibers
CNT	Carbon nanotubes
CNF	Chitin nanofibers
DMF	Dimethylformamide
ECM	Extracellular matrix
ELISA	Enzyme-linked immunosorbent assay
FA	Fusidic acid
Hap	Hydroxyapatite
MWCNT	Multiwalled carbon nanotubes
NP	Nanoparticles
PCL	Polycaprolactone
PEO	Poly(ethylene oxide)
PET	Polyethylene terephthalate
PLMC	Poly(D,L-lactide-co-trimethylene carbonate)
PLGA	Poly(lactic acid-co-glycolic acid)
Pt	Platinum
PU	Polyurethane
PVA	Polyvinyl alcohol
PVDF	Polyvinylidene
RIF	Rifampicin
SC CO_2	Supercritical CO_2
SDS	Sodium dodecyl sulfate
SEM	Scanning electron microscope
TA	Tannic acid
TEM	Transmission electron microscope

21

Environmentally Sustainable and Safe Production of Nanomedicines

Samson A. Adeyemi, Pradeep Kumar, Viness Pillay, and Yahya E. Choonara

Wits Advanced Drug Delivery Platform Research Unit, Department of Pharmacy and Pharmacology, School of Therapeutic Sciences, Faculty of Health Sciences, University of the Witwatersrand, Johannesburg, South Africa
† Professor Viness Pillay (as corresponding author) passed away shortly after the work was produced.

21.1 Introduction: Going Green – Advances in Nanomedicines

Advances in nanotechnology during the past decade have enabled scientists across the globe to invent techniques for the production and characterization of new materials on the nanoscale level such as nanoparticles, nanotubes, and nanolayers. Meanwhile, the design and production of these nanoscale materials having controlled characteristics remains a significant hurdle within nanoscience and nanotechnology. Various scientific experimentations to either develop nascent chemicals or create new physical methods to synthesis nanomaterials often raise the concern about the adverse impact on the environment. Some of the negative effects that emanate from the chemical techniques involved in the synthesis of nanomaterials include toxic solvents, which have the capacity to produce dangerous by-products, and usually require high-energy utilization. Most important is the constant dispute of the potential toxic nature of some of these nanomaterials.

This in essence has led to the renewed consciousness of the need to design and develop clean, safe, and environmentally friendly protocols for the fabrication and assembly of nanoparticles. In this regard, the present paradigm shift in nanoscience has evolved toward the green nanotechnology. Generally, the green nanotechniques involve a wide variety of promising applications. These eco-friendly biofabrication procedures decrease waste products. The green chemistry principles are employed in such a way that they allow for the use of nanomaterials as catalysts for enhanced efficiency in the present manufacturing processes through the reduction or total elimination of toxic materials in the reaction chain. Nanomaterials and nanodevices have been used to decrease pollution (e.g. water and air filters) as well as are viable alternatives for the production of energy (e.g. solar and fuel cells). More recently, researchers explore the possibility of making use of plants as biofactories for green nanomaterials.

The term green chemistry has been defined as "the utilization of a set of principles that reduces or eliminates the use or generation of hazardous substances in the design, manufacture, and application of chemical products" [1]. Anastas and Warner (1998) categorized green chemistry into 12 principles which are now being employed in the design of a number of chemical products and synthesis protocols with the objectives of reducing environmental hazards and toxic chemicals to health, minimizing waste, and stopping pollution. The use of these principles has resulted in drastic reduction in the use of hazardous reagents and solvents and enhanced the material and energy efficiency of the chemical processes. Thus, making use of these principles in nanoscience will aggravate the synthesis and biofabrication of nontoxic nanomaterials and nanostructured devices. The green nanotechnology approach entails the use of green chemistry principles to design nanoscale products, to develop nanomaterial fabrication techniques, and for the application of nanomaterials. In this way, the characteristics of nanomaterials are properly understood including those related to toxicity and environmental hazard. Nanoscale materials are also designed that can be included into high-performance products that show little or no danger to human health and the environment. Furthermore, the methods of synthesis that exonerate the inclusion of dangerous reagents and improve the output of

Sustainable Nanotechnology: Strategies, Products, and Applications, First Edition.
Edited by Yashwant V. Pathak, Govindan Parayil, and Jayvadan K. Patel.

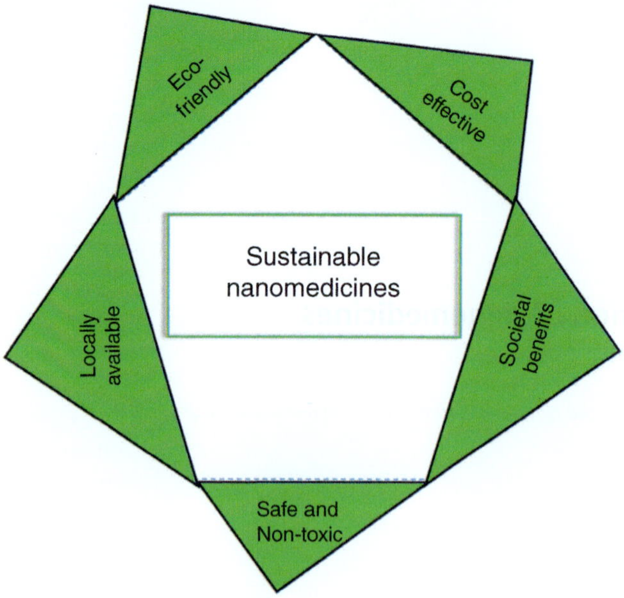

Figure 21.1 Sustainable green nanomedicines.

these methods are discovered and provide the required quantity of pure material in a way that supports and enhances economy management. Using this approach, the synthesized nanomaterials are intrinsically safer by assessing the biological and ecological hazards in line with the design. Lastly, the green chemistry approach applies nanoscience in a way that optimizes benefits to the society and greatly reduces the hazards to the ecosystem (Figure 21.1).

21.2 Convectional Techniques and Production of Polymeric Nanomedicines Using Bio/synthetic Polymers

Polymeric nanoparticles (PNPs) have been defined as the minute particles dispersion or solid particles within a size range of 10–1000 nm [2]. They are often synthesized from biodegradable and biocompatible polymers in which the chemotherapeutic agent is dissolved, entrapped, encapsulated, or attached to the nanoparticle matrix [3]. Polymeric materials including nanospheres or -capsules are generated depending on the method of synthesis [4]. The method for the preparation of PNPs depends to a larger extent on the properties of the PNPs to be optimized and their particular application. As such, finding an appropriate method for PNPs synthesis is a major determinant to obtaining the desired properties to suffix a defined application of interest [4]. The techniques for the preparation of PNPs have been broadly classified into two including preparation from a macromolecule or preformed polymers and polymerization from a monomer. Organic solvents discussed in this review are classified according to the International Conference on Harmonization (ICH) based on their possible risk to human health. These solvents are placed into one of the three categories, namely, category I (solvents to be avoided), category II (solvents to be limited), and category III (solvents with low toxic potential).

21.2.1 Synthesis of Polymeric Nanoparticles from Preformed Polymers

Direct synthesis of PNPs can be achieved from preformed polymers. These polymers are either synthetic or natural, depending on the properties and application of the desired generated PNPs to be harnessed.

21.2.1.1 Preparation of PNPs from Synthetic Preformed Polymers

21.2.1.1.1 Emulsification/Solvent Evaporation

The preparation of PNPs through the emulsification–solvent technique was the first method to be developed. It basically involves two subsequent procedures, namely, (i) the emulsification of the polymer and drug solutions into the aqueous phase and (ii) the evaporation of the polymer solvent, thereby inducing the precipitation of the polymer as nanospheres [5]. The polymer and drug solutions are dissolved in volatile organic solvents such as acetone/dichloromethane (ICH, class 2), chloroform (ICH, class 2), and ethyl acetate (ICH, class 3), and emulsions are formulated [5]. Conventionally, two methods are being employed for the formation of emulsions: (i) the preparation of single emulsions: oil-in-water (o/w), (ii) the preparation of double emulsions: water-in-oil-in-water (w/o)/w [4]. The emulsions are then dispersed into nanodroplets using dispersing agents and high-energy homogenization or ultrasonication [6]. Nanoparticle suspensions are formed from the dispersed emulsions upon the evaporation of the solvent either by increasing the temperature of the solvent under reduced pressure or by continuous magnetic stirring at room temperature [2]. The reaction mixture is then subjected to ultracentrifugation and washed with distilled water to eliminate additives such as surfactants, and solidified nanoparticles are then collected and lyophilized [5, 7]. The size of the PNPs obtained depends largely on the stir rate, the type and amount of the dispersing agent, the viscosity of both the organic and aqueous phases, and the temperature [6].

21.2.1.1.2 Solvent Displacement/Nanoprecipitation and Interfacial Deposition

Both solvent displacement/nanoprecipitation and interfacial deposition are closely related techniques in the preparation of PNPs. These two procedures are based on spontaneous emulsification of the organic internal phase containing the dissolved polymer into the aqueous external medium with or without a surfactant [8]. Meanwhile, the solvent displacement technique forms either nanospheres or nanocapsules, while the interfacial deposition method only produces nanocapsules. The polymer is dissolved in a water-miscible solvent, having an intermediate polarity, which eventually results in the precipitation of nanospheres. This resultant solution is then dropwisely injected into the aqueous solution containing a stabilizer as a surfactant under stirring. Interfacial polymer deposition between the water and the organic solvent, as a result of the fast diffusion of the solvent, causes the instantaneous formation of a colloidal suspension [9]. Phase separation is performed with a solvent which is totally miscible but a nonsolvent of the polymer [10]. This in turn will initiate the formation of colloidal polymer particles during the first step of the procedure. However, the solvent displacement method allows for the preparation of nanocapsules upon the addition of a small amount of nontoxic oil in the organic phase [9]. This technique is mainly applicable and efficient in encapsulating lipophilic drugs rather than hydrophilic drugs due to the miscibility of the solvent with the aqueous phase [9]. The percentage entrapment ratio for lipophilic drugs is usually very high when nanocapsules are prepared compared to nanospheres. This is generally due to the oil-based central cavities of the nanocapsules [9]. Meanwhile, a major setback in the application of this simple technique is that of its usefulness in water-miscible solvents in which the diffusion rate is sufficient to produce spontaneous emulsification [9]. In addition, some water-miscible solvents yield a certain level of instability when mixed in water. Thus, spontaneous emulsification is not observed in cases where the coalescence rate of the formed droplets is very high [11]. Though acetone and dichloromethane (ICH, class 2) are often employed to dissolve and increase drug entrapments, the use of dichloromethane has been shown to increase the mean particulate size of the formed nanoparticles. Meanwhile, dichloromethane is considered toxic [12]. Another difficulty in the application of this technique is the choice of suitable drug/polymer/solvent/nonsolvent system in which particles will be formed and the drug efficiently entrapped. This is because both the solvent and nonsolvent of the polymer must be mutually miscible [13]. Nanospheres are generally formed with an approximate size of 200 nm by the progressive addition of the polymer solution to the nonsolvent [5].

21.2.1.1.3 Emulsification/Solvent Diffusion (ESD)

Emulsification/solvent diffusion (ESD) technique is a modification of the solvent evaporation method [14]. It involves dissolving the polymer to be used for encapsulating the drug in a solvent such as propylene carbonate, which is partially soluble in water. This is then saturated with water in order to allow for the two liquids to be thermodynamically stable. It is important to facilitate the diffusion of the solvent into the dispersed phase by diluting it with sufficient water when the organic solvent is water miscible or with another suitable organic solvent in the opposite case. This will in turn allow for the polymer to precipitate and consequently results in the formation of nanoparticles. Thereafter, the polymer-water saturated solvent phase is emulsified in an aqueous solution containing a stabilizer, which allows the solvent to diffuse to the external phase and the formation of either nanospheres or nanocapsules, depending on the oil-to-polymer ratio. The solvent is then removed either by evaporation or filtration based on its boiling point and the reaction product lyophilized. The advantages of ESD [5] include (i) high encapsulation efficiencies (usually above 70%), (ii) high batch-to-batch reproducibility, (iii) narrow size distribution, easy to scale up, and simple, and (iv) absence of homogenization process. However, the major setbacks in its application are the high volumes of water to be eliminated from the suspension and the leakage of hydrophilic drugs into the saturated-aqueous external phase during emulsification, which reduces its encapsulation efficiency. Similarly, as with other methods, it is efficient in the encapsulation of lipophilic drugs [9].

21.2.1.1.4 Salting-out Technique

The salting-out method for the preparation of PNPs is referred to as a modification of the ESD technique. It relied on the separation of a water-miscible solvent from an aqueous solution through the intermediate effect of a salting-out agent. Both the polymer and drug are initially dissolved in an organic solvent, e.g. acetone. The reaction mixture is then emulsified into an aqueous gel containing the salting-out agent and a colloidal stabilizer. This oil–water emulsion is then diluted with excess volume of water or the aqueous solution to facilitate the diffusion of the organic solvent into the aqueous phase, thereby resulting in the formation of nanospheres. Finally, both the organic solvent and the salting-out agent are then removed by cross-flow filtration [9]. Electrolytes, such as magnesium chloride, calcium chloride, and magnesium acetate, are used as salting-out agents. Meanwhile, nonelectrolytes such as sucrose are also employed as salting-out agents. Colloidal stabilizers such as polyvinylpyrrolidone, hydroxyethylcellulose, span 80, Tween 80, and polyvinylchloride (PVA) are often used in this technique. Their main advantages lie in their ability to minimize stress to the protein encapsulants [15], can be easily scaled-up [9], and the fact that they do not require an increase in temperature, which allows for their application when heat-sensitive substances have to be encapsulated [16]. However, their main disadvantages are their exclusive usage in lipophilic drugs and the rigorous nanoparticle washing steps [17].

21.2.1.1.5 Dialysis

Dialysis technique allows for the preparation of small and evenly distributed PNPs with a simple, effective, and efficient procedure [18, 19]. A suitable organic solvent is used to dissolve the polymer and placed inside a dialysis membrane having an appropriate molecular weight. Dialysis is then carried out against a nonsolvent which is miscible with the initial miscible solvent in which the polymer is dissolved. Homogeneous suspensions of nanoparticles are formed, resulting from the progressive aggregation of the polymer due to the loss of solubility upon the displacement of the solvent within the membrane. The mechanism by which PNPs are formed by dialysis technique is not yet fully elucidated. Meanwhile, a similar mechanism of PNP formation through nanoprecipitation reported by Fessi et al. [18] has been proposed as well for dialysis technique. A simple but versatile osmosis-based method using various sizes of the dialysis membrane has been employed by Chronopoulou et al. [20] to generate different types of PNPs including polystyrene-based nanoparticles.

21.2.1.1.6 Supercritical Fluid Technology

While nearly all the methods for the preparation of PNPs employ the use of organic solvents, some of which are toxic, harmful, and not user friendly, supercritical fluid technology provides an alternative method for an environmentally safer technique using supercritical fluids with the potential to generate PNPs with high purity and without any trace of organic solvent [7, 21]. There are basically two principles designed for the generation of PNPs using supercritical fluids: (i) rapid expansion of supercritical solution (RESS) and (ii) rapid expansion of supercritical solution into liquid solvent (RESOLV).

21.2.1.1.6.1 Rapid Expansion of Supercritical Solution (RESS)

Basically, in the application of RESS, the solute (polymer) is dissolved in a supercritical fluid to form a solution. This is followed by the rapid expansion of the solution across the orifice or a capillary nozzle into the ambient air. Well-dispersed particles are formed as a result of the high degree of supersaturation of the solution which is accompanied by the reduction in the pressure of the expanded solution with homogeneous nucleation in the capillary nozzle. Previous studies using the RESS process for different model solutes have shown that both nanometer- and micrometer-sized range particles are present in the expansion jet [22]. A study on the formation of fine powders of caffeine from the rapid expansion of supercritical fluid, CO_2 solutions, using RESS has been reported, among others [23]. There are three major compartments in the RESS experimental apparatus comprising (i) a high-pressure stainless mixing cell, (ii) a syringe pump, and (iii) a pre-expansion unit. The polymer is dissolved in CO_2 at ambient temperature, pumped into the pre-expansion unit using the syringe pump, and heated isobarically to attain the pre-expansion temperature. The supercritical solution is then allowed to expand through the nozzle at ambient temperature, and the nanoparticles are collected. The solute (polymer) concentration, the pre-expansion and expansion temperatures, and the pressure of extraction have been shown to affect the size and morphology of the formed powder during the RESS process [24]. Though RESS technique does not use organic solvent in its preparation procedures for PNPs production, a major disadvantage in its application is the generation of microscaled products rather than nanoscaled products [7].

21.2.1.1.6.2 Rapid Expansion of Supercritical Solution into Liquid Solvent (RESOLV) In order to circumvent and overcome the drawback encountered in RESS application, a simple but important improvement was implemented in RESOLV. This modification allows for the expansion of the supercritical solution into a liquid solvent instead of the convectional ambient air implemented in RESS [7, 25]. In order to obtain nanosized particles, the liquid solvent tends to suppress the particle growth in the expansion jet despite the increasing availability of different supercritical fluids including carbon monoxide, *n*-pentane, water, and ammonium [7, 26]. A major challenge to overcome in both RESS and RESOLV, though environmentally friendly technologies for producing PNPs, is the poor solubility or at some instances nonsolubility of the polymers in these supercritical fluids.

21.2.1.2 Preparation of PNPs from Natural Macromolecules

Biodegradable hydrophilic polymers such as alginate, agarose, chitosan, gelatin, and albumin have been used to prepare PNPs.

21.2.1.2.1 Alginate Nanoparticles

Sodium alginate is a hydrophilic polymer that forms gels in the presence of multivalent cations such as calcium [27]. Through a gradual dropwise extrusion of sodium alginate solution into calcium chloride solution, alginate particles are formed having a size that is directly dependent on the size of the initial extrusion droplet. Air atomization was employed to produce alginate particles of about 5 to 15 μm [28]. Alginate particles have also been produced through a modified emulsification/internal gelation technique [29]. This method is simple in that it does not require any specialized equipment and can be carried out at ambient temperature. Meanwhile, a major drawback in its application is the need to eliminate the residual oil droplets from the nanoparticles through various washing steps.

21.2.1.2.2 Chitosan Nanoparticles
Chitosan nanoparticles have been developed to serve as cargo and encapsulants for various therapeutic agents. Proteins including bovine serum albumin [30], tetanus and diphtheria toxoid [2], vaccines [31], anticancer agents [32–34], insulin [35], and nucleic acids [36] have been investigated. The absorption of peptides such as insulin and calcitonin across the nasal epithelium is favorably facilitated by chitosan [37]. Different methods have been proposed for the preparation of chitosan nanoparticles [38]. These methods are based on either the spontaneous formation of complexes between chitosan and the polyanions [39] or the gelation of a chitosan solution dispersed in an oil emulsion [40].

21.2.2 Nanoparticles Obtained by Polymerization of a Monomer

The preparation of PNPs from a monomer using the polymerization technique can be subdivided into three major categories, namely, (i) emulsion polymerization, (ii) interfacial polymerization and polycondensation, and (iii) living/controlled radical polymerization. Meanwhile, in this review, we focus principally on emulsion polymerization and its various subclassifications as the major methods currently employed for the preparation of PNPs.

21.2.2.1 Emulsion Polymerization
Emulsion polymerization is the most frequently used technique with a wide range of applications in the production of nanoparticles. This method is highly scalable and is subdivided based on the incorporation of surfactant into the system or not. As such, it is classified into convectional emulsion polymerization and surfactant-free emulsion polymerization (SFMP) [7, 41].

21.2.2.1.1 Convectional Emulsion Polymerization
A larger proportion of the world's polymer production based on emulsion polymerization is achieved though the convectional technique. Water, a monomer of low water solubility (e.g. styrene), a water-soluble initiator (e.g. persulfate), and a surfactant constitute the regimen for the convectional emulsion method [41]. The by-products of the reaction usually have an approximate size of 10^2 nm, each containing many polymer chains. Colloidal stabilizers may be electrostatic, steric, or electrosteric, displaying both the sterilizing mechanisms. When a dissolved monomer molecule in the continuous phase collides with an initiator molecule that may either be an ion or a free radical, the reaction is initiated. Conversely, the monomer molecule can be converted into an initiating radical by high-energy radiation such as Gamma-ray, ultraviolet ray, or strong visible light. Phase separation and formation of solid particles usually occur prior to or after the termination of the polymerization reaction [42]. The size of the PNPs produced depends on the type of the surfactant employed in the polymerization reaction [43]. A number of nanoparticles including polystyrene (PS) [44], poly(methylmethacrylate) (PMMA) [45], poly(vinylcarbazole) [46], poly(ethylcyanoacrylate) (PECA), and poly(butylcyanoacrylate) [47] were produced through the application of the surfactants into the solvents such as cyclohexane, *n*-pentane, and toluene. Anionic polymerization of the monomer has been the main method employed in the preparation of most poly(alkylcyanoacrylate) nanoparticles with well-defined properties [48]. The hydroxyl groups of water or other nucleophiles present in the molecules that are dissolved in the polymerization system serve as the spontaneous initiator of the reaction.

21.2.2.1.2 Surfactant-free Emulsion Polymerization (SFMP)
SFMP is usually referred to as surfactant-free [49], emulsifier-free [50], or soapless [51] emulsion polymerization. As opposed to the convectional emulsion polymerization systems that incorporate varying amounts of surfactants that must be eliminated from the final products even though such surfactants are difficult to remove and time consuming, the advent of SFMP technique has received a considerable mileage and well employed in circumventing the drawbacks observed in the former without the use of stabilizing surfactants [52]. Stabilization of the PNPs is achieved through the use of ionizable initiators or ionic comonomers. SFMP systems consist of deionized water, a water-soluble initiator such as potassium persulfate, and monomers (usually vinyl or acryl monomers). The two main mechanisms proposed for the nucleation and growth of the particles in SFMP include (i) micellar-like nucleation [53] and (ii) homogeneous nucleation based on the aqueous solubility of the monomers [50, 54]. Meanwhile, homogeneous nucleation is much important in SEMP reactions [50]. SFMP has been used in the preparation of PMMA nanoparticles [55], polyacrylate nanoparticles [56], polystyrene nanoparticles [57], poly(hydroxyeth ylmethacrylate) (PHEMA) nanoparticles [58], polyimide nanoparticles [59], and polypyrrole [60] nanospheres. The solution dielectric constant, solubility of monomers, and sonication time were reported to contribute immensely to the size and distribution of the PNPs formed. While SFMP

has emerged as a simple and environment-friendly method for the production of PNPs, several other factors still limit its expanded application such as the preparation of monodisperse and precisely controlled particle size [61].

21.2.2.1.3 Miniemulsion Polymerization

A number of recent research and publication outputs have been documented using the miniemulsion polymerization technique, which is a type of surfactant-based approach of emulsion polymerization. Vital differences between emulsion polymerization and miniemulsion polymerization subtype are the application of low molecular mass compound as the costabilizer and the use of a high-shear device such as ultrasound. However, miniemulsions are highly stable, requiring a high shear to reach a steady state and possess an interfacial tension that is far greater than zero. The choice and combination of different costabilizers and initiators have been shown to drastically influence the nature and formation of various PNPs with a wide range of applications [62]. PMMA [63], poly(*n*-butylacrylate) [64], polyacrylic acid [65], PHEMA [66], and polyacrylonitrile [67] nanoparticles have been produced using the miniemulsion polymerization approach with excellent PNP outputs as discussed in the literatures. A major disadvantage of this technique is that remnant quantity of surfactant is found within the polymer latex even though a very smaller amount of the surfactant is used [68]. High hydrophobic stearyl methacrylate (SMA) and dodecyl methacrylate (DMA) were employed as cosurfactants during the polymerization of styrene to stabilize the miniemulsion in order to overcome this limitation. These cosurfactants were chemically incorporated into the emulsion polymer during polymerization, thereby reducing the presence of unwanted volatile organic compounds in the latex product [68].

21.2.2.1.4 Microemulsion Polymerization

Significant focus has been given of recent to microemulsion polymerization as a new and effective technique for the production of nanosized polymeric particles. Both emulsion and microemulsion polymerization methods seem to be similar since the two techniques can produce colloidal polymeric particles. However, they are distinctively different in terms of their mechanisms of kinetic [69]. While three consecutive reaction rates are exhibited in the emulsion polymerization method, only two are implemented in the microemulsion polymerization system, with the particulate size and average number of chains per particle being considerably smaller when compared [7]. The microemulsion method entails the addition of an initiator, which is usually water soluble, into the aqueous phase of the microemulsion containing swollen micelles in a thermodynamically stable state. The polymerization reaction kick starts from this thermodynamically stable and spontaneous state and depends on the high amount of the surfactant systems, whose interfacial tension at the oil/water interface is close to zero. Also, the particles are wholly submerged into the surfactant due to the high amount utilized in the polymerization process. At the initial state, some droplets of the polymers are formed because the initiation cannot be attained in all the microdroplets. But as the reaction proceeds toward the end, the osmotic and elastic influence of the chains change the fragile microemulsions, which leads to an increase in the particle size, the formation of empty micelles, and secondary nucleation. The final product contains minute latexes of about 5–50 nm in size, which coexists with other several empty micelles [53]. Several experiments have been carried out to investigate the effect of the major critical parameters that influence the kinetics of microemulsion method and the properties of PNPs such as (i) the types of initiator and concentration [70], (ii) the surfactant employed [71], (iii) the monomer type and amount [72], and (iv) the temperature of the reaction process [70]. A major limitation of microemulsion polymerization despite the wide range application of its by-products is due to the fact that its typical polymer formulations are diluted and as such require a large ratio of surfactant to monomer. The amount of surfactants concentration required for the stability of the polymer is usually less as compared to the amount incorporated into the polymerization system [7].

21.2.2.2 Interfacial Polymerization

Interfacial polymerization (IP) has gained a lot of recognition and application as an established technique for the preparation of PNPs [73]. This method involves stepwise polymerization of two reactive monomers or agents in two different phases, namely, (i) the continuous phase and (ii) the dispersed phase. The two reactive monomers are dissolved in the two phases, respectively, and the reaction occurs at the interface of the two mediums [74]. IP technique is easy to use, which has facilitated its application in encapsulating pharmaceutical products [75] as well as in the preparation of conducting polymers [76]. Through the application of polyaddition and polycondensation [77] or radical polymerization [78] as interfacial cross-linking reactions, nanometer-sized hollow polymeric particles have been synthesized. The polymerization of monomers at the oil/water interface of a very fine oil-in-water microemulsion has also been employed to produce oil-containing nanocapsules. The organic solvent, which mixed completely with water, serves as the monomer to facilitate the reaction, and the interfacial polymerization of the monomer was believed to occur at the surface of the oil droplets that is generated during emulsification [10, 79]. Two major classes of solvents, (i) aprotic solvents and (ii) protic solvents, are usually employed in IP reactions, depending on the final by-product desired, either nanocapsules or -spheres. Aprotic solvents such as acetone and acetonitrile were recommended to facilitate the formation of nanocapsules. Meanwhile, the use of protic solvents, such as ethanol, *n*-butanol, and isopropanol, was discovered to form both nanospheres and -capsules [80]. Conversely, nanocapsules containing water as their inner core can be produced by the interfacial polymerization of monomers in water-in-oil micro-emulsions. Using such strategies, the polymer is formed locally at the water–oil interface and subsequently precipitated to form the nanocapsules shell [81].

Spontaneous emulsification has been combined with the traditional IP technique as a new method to produce nanoparticles with wider range of applications. Polyurethane nanocapsules with particle size of 258 ± 29 nm [82], polystyrene nanocapsules [83], poly(alkylcyanoacrylate) nanoparticles [84], polyaniline nanoparticles [85], PANI-PSS nanoparticles [86], and polyindole nanoparticles [87] were produced.

The drawbacks in the precise regulation of the size of PNPs produced using the various IP techniques serve as a major disadvantage of this method. Meanwhile, in order to circumvent this problem, the use of membrane reactors that showed different functions, such as the separation of products from reaction mixtures and the immobilization of catalysts, has been considered recently [88]. These reactors allow for the control of how one monomer (the organic phase) is added to another monomer (the aqueous phase). This improved method has been employed by Charcosset and Fessi [89] to synthesize nanoparticles by investigating the influence of parameters such as (i) the membrane pore radius, (ii) cross-flow rate, and (iii) transmembrane pressure. Though this improvement in IP method for PNPs preparation is quite advantageous, it is quite an expensive process with a huge burden to the environment. Also, to effectively control the size of the products, various membranes with different inner pore sizes are required.

21.2.2.3 Controlled/Living Radical Polymerization (C/LRP)

Radical polymerization on its own is limited with a number of factors including lack of control over the molar mass, the molar mass distribution, the end functionalities, and the macromolecular architecture. These limitations are due to the fast radical–radical termination reactions in the system, which cannot be avoided. Recently, different types of controlled or living radical polymerization (C/LRP) systems have allowed for new strategies based on the old polymerization method [90]. The increase in both the medical and pharmaceutical applications of water-soluble polymers coupled with the increase in global burden for environment-friendly processes accounts for the exploration of these newly improved C/LRP techniques. Industrially, radical polymerization is often carried out in aqueous dispersed systems and particularly in emulsion polymerization. This was aimed at controlling the characteristics of the polymer in terms of its molar mass, molar mass distribution, architecture, and function [91]. Among the most successful and widely applied C/LRP methods are (i) nitroxide-mediated polymerization (NMP) [92], (ii) atom-transfer radical polymerization (ATRP) [93], and (iii) reversible addition and fragmentation transfer chain polymerization (RAFT) [94]. Apart from temperature, the nature and concentration of the control agent, the monomer, the initiator, and the type of emulsion have been shown to be the most important parameters that influence the size of PNPs produced by the C/LRP technique [95].

While the use of C/LRP has received a broad spectrum of opportunities in the production of PNPs that can be used in functional studies, a number of challenges have hampered the quest to fully elucidate its chemistry. Parts of the limitations are central to the various forms of C/LRP, while others are specific to each of the individual technique. Some of the general problems encountered across the different C/LRP types are as compared to the convectional radical polymerizations (i) the presence of a residual control agent in the product and (ii) additional process complexity and cost. The presence of residual control agents in the by-products has aggravated concerns on its effect on the color, odor, stability, and its compliance to the environment. Meanwhile, it is nearly impossible to eliminate the mediating agents from aqueous dispersion than from homogeneous solutions [7].

21.2.3 One-pot Synthesis – An Economic and Time-saving Approach for Nanomedicines Synthesis

A major paradigm shift in the twenty-first century is the greater influence of organic synthesis as the driver of synthetic economy such as atom economy [96, 97], redox economy [98], and step economy [99], among others. Meanwhile, the conceptual approach and instruments for the design and planning of process-oriented synthesis are provided by synthetic economy. To obtain a holistic process efficiency, reducing the synthetic steps, particularly the redox reaction steps, and making use of superfluous atoms, is basic and effective. In order to achieve generalized synthetic methods, combining multiple step reactions in one-pot operations remains a viable approach to increase the efficiency of synthetic production. The one-pot synthetic process has been defined as an economically viable approach in which a number of bond-formation steps that involve the formation of rings and stereogenic centers are condensed into reduced number of unit operations without the removal of the intermediate products.

The one-pot synthesis is viable, promising, and effective due to its ability to enhance unification of several synthetic transformations and bond-forming steps that can be performed in a single pot, while at the same time omitting diverse purification procedures [100]. The increasing demand for an efficient chemical synthesis has resulted in the enormous boost in the use of one-pot procedures over the past decades. The multicomponent reaction (MCR) remains one of the methods employed to achieve one-pot operation [101]. In the MCR process, more than three starting materials are converged in a single reaction vessel to produce a singular product that has portions of all the start-up components. Contrary to the convectional stepwise synthesis, the one-pot MCR technique demands minimal isolation and purification procedures, thereby usually resulting in quantitative yields [102].

Compared to the MCRs are the cascade reactions that enable the efficient production of complex molecules from simple start-up materials [103, 104]. A cascade or domino reaction refers to a process that allows for two or more bond-formation steps to occur governed by the same reaction conditions in a sequential manner. A more interesting phenomenon in cascade reaction is its ability to efficiently produce the expected targets in one reaction vessel without the need to purify the intermediate(s). In this way, cascade reactions enable the drastic reduction of wastage that could occur during the process, improve reaction time, and enhance productivity. The three stages in a cascade reaction are the initiation stage, a relay stage, and a termination stage. The initiation stage can be any reaction type once it is capable of producing the functionality that is perfect for the next relay step(s). The relay stage can add one or more sequential reaction steps, while the termination stage serves as the last step of the sequential event wherein the final product will be inert under the reaction conditions. Meanwhile, when the relay step is singular, it will also serve as the termination step.

The tandem reaction is quite similar to the cascade reaction but with relaxed features that allow for each step in the tandem sequences to be performed under adjustable conditions such as temperature, acid and base, catalyst, and solvent, to mention but a few [105–107]. These two similar reactions have been employed to effectively produce pharmaceutical products in recent years.

One strategy by which the robustness, safety, process efficiency, etc. can be enhanced is through the telescoping technique. A major telescoping procedure in one-pot synthesis is the azeotropic distillation. This technique is mainly important for telescoping reactions that are running in different solvents. Meanwhile, in order to decrease the number of unit operations, it is more viable to conduct one-pot synthesis using a common solvent. Interestingly, both the cascade and the tandem reaction sequence can be accommodated in a telescoped process [102].

21.3 Eco-friendly Nanotechnology for Nanomedicines

In recent years, several promising biosynthetic approaches have evolved using microorganisms including fungi, bacteria, algae, yeast, and viruses (Figure 21.2) or plant extracts as promising alternatives to both physical and chemical synthetic methods.

21.3.1 Nanoparticle Synthesis Using Fungi

The synthesis of nanoparticles using fungi represents a direct and simple biosynthetic method that has been widely engaged for the production of nanoparticles. Using fungi biosynthesis approach, the amount of nanoparticles produced is higher with increased tolerance to metals in terms of higher cell-wall-binding propensity of metal ions with biomass when compared to bacteria [109]. In addition, fungi are relatively easy to manipulate during biomass treatments and downstream processing when compared to viruses and bacteria. Most importantly, fungi exhibit increased bioaccumulation potential with respect to metal ions, thereby yielding an excellent alternative for nanoparticles production with low cost. Meanwhile, nanoparticles biosynthesis is often dependent on the process parameters, which showed significant effects on their bioproduction.

The work of Bhargava et al. investigated the influence of process parameters including pH, salt concentration, and reaction time on both the particle size and yield using the fungi *Cladosporium oxysporum* to generate nanoparticles from gold ion [110]. Their results showed an optimum production of gold nanoparticles (AUNPs) with biomass-to-water ratio of 1 : 5 at a salt concentration of 1 mM and pH of 7. The AUNPs fabricated displayed an effective catalytic property on textile dye, rhodamine B, showing an optimum degradation within 7 minutes. The effect of incubation time and reaction temperature on the biosynthesis of AUNPs from the culture substrate of *Hypocrea lixii* and *Trichoderma viride* was investigated by Mishra et al. [111]. The biosynthesized AUNPs from *T. viride* were fabricated extracellularly within 10 minutes at 30 °C with efficient biocatalytic and antimicrobial properties. In another development, Metuku et al. examined a white rot fungus, *Schizophyllum radiatum*, and discovered its ability to

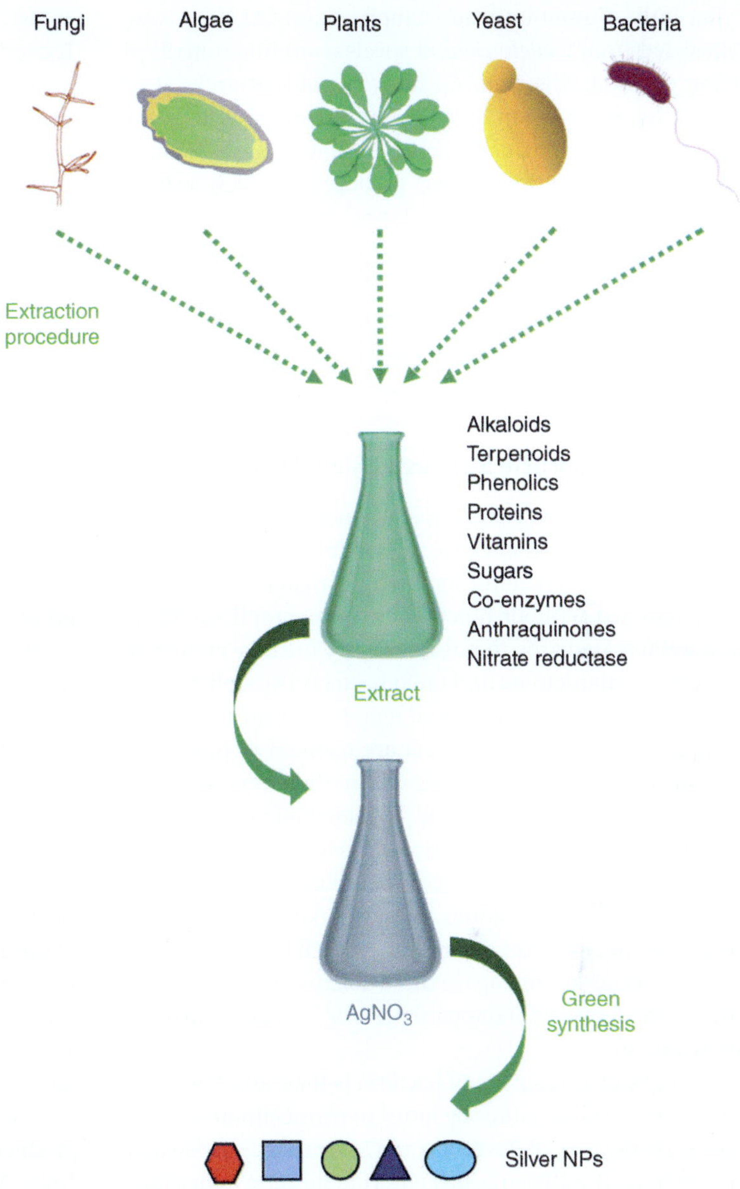

Figure 21.2 Graphical representation of a mechanistic scheme showing the synthesis of metal nanoparticles from microorganisms. Source: This figure has been adapted from Ref. [108] under a Creative Commons Attribution-NonCommercial 3.0 Unported Licence.

generate uniformly dispersed silver nanoparticles (AgNPs) [112]. The outputs established the ability of white rot fungus for the extracellular biosynthesis of AgNPs of size 10–40 nm from silver ions. Interestingly, the biosynthesized AgNPs exhibited versatile antimicrobial activity against diverse infectious Gram-positive and -negative bacteria strains.

The use of extracellular matrix has characterized the majority of the studies reported in recent years for the production of nanomaterials. A major advantage using the extracellular component-enhanced biosynthesis is it freedom from the influence of contaminations through intracellular proteins as well as the nontreatment with detergents and ultrasound. In addition, the need to comprehend the mechanistic protocols of nanoparticles biosynthesis has become critical for the production of dependable applications. In order to bridge this knowledge gap, Rajput et al. investigated diverse fungal strains of *Fusarium oxysporum* for the synthesis of AgNPs and reported the influence of the selection of isolate, temperature, and pH on the morphology of nanoparticles [113]. The need to understand the relationship between the interactions of interfacial and organic layer, which will be of immense assistance in the production of novel applications such as biosensors, was the major finding from their study. Furthermore, Kitching et al. studied the production of bioinspired nanoparticles. They extracted the cell surface proteins of *Rhizopus oryzae* for the *in vitro* synthesis of AuNPs, which were used for biomedical and biocatalytic interventions [114].

Recently, aluminum oxide nanoparticles (ALNPs) were fabricated from *Colletotrichum* species and functionalized using essential oils from *Eucalyptus globulus* and *Citrus medica* by Suryavanshi et al. [115]. It was found that the oil-functionalized nanoparticles exhibited antimicrobial properties against food-borne pathogens, which inhibit food spoilage. In another development, an edible mushroom, *Pleurotus ostreatus*, and two other filamentous fungi, *Penicillium citreonigrum* and *Scopulariopsis brumptii*, have been employed for the fabrication of nanoparticles for both the antimicrobial and -cancer applications, respectively [116].

21.3.2 Nanoparticle Synthesis Using Bacteria

A number of unique characteristics underscored the use of bacteria as the most effective alternative for nanoparticle biosynthesis from metallic ions, which include their ability to be handled without much stress and complications as well as increased growth rates. Contrary to other microbes, it is easy to manipulate and mold bacteria through genetic engineering for the reduction of metal ions into nanoparticles [117]. Quiet often, bacteria are exposed to high concentrations of heavy metal ions within their environments that are nonfriendly and toxic to their biological nature. Nonetheless, they acquired diverse intrinsic defense mechanisms to buffer these stressors including their ability for intracellular sequestration, extracellular precipitation, change in metal ion concentration, and efflux pumps [118]. As such, bacteria can engage these defense mechanisms to effectively fabricate nanoparticles for various types of applications.

A unique characteristic of bacteria is their ability to produce nanoparticles either by intra- or extracellular mechanisms. In the work of Beveridge and Murray, the suspension of unfixed cell wall using gold chloride solution to produce extracellular AuNPs on the cell wall of *Bacillus subtilis* was first reported [119]. Meanwhile, using a silver-resistant strain of *Pseudomonas stutzeri* AG259, nano-range AgNPs with average size of less than 200 nm were synthesized intracellularly using NADH-dependent reductase enzyme as electron donors while itself undergoing oxidation to NAD+ [120]. Silver ions were biologically reduced to AgNPs through the transfer of electrons from NADH. About a decade ago, the work of Srivastava and Constanti showed that a number of metallic nanoparticles including silver, iron, platinum, palladium, rhodium, cobalt, nickel, and lithium were intracellularly synthesized from *Pseudomonas aeruginosa* [121]. Interestingly, their study excludes the use of any external stabilizing and electron-donating agents and omits the pH-modifying step during the biomineralization process of diverse metal ions. As presented in Table 21.1, a number of bacterial strains including *Escherichia coli*, *B. subtilis*, *Bacillus magaterium*, *P. aeruginosa*, *Klebsiella pneumonia*, *Bacillus cereus*, *Alteromonas*, and *Ochrobactrum* have been widely employed for fabrication of nanoparticles. Using *Bacillus cereus* extracted from contaminated soil containing heavy metals, Das et al. reported the production of AgNPs extracellularly at room temperature within 24 hours [126]. The AgNPs produced exhibited surface plasmon resonance characteristics that could be helpful in different applications. In another experiment, Kulkani et al. fabricated AgNPs using radiation-resistant strain of *Deinococcus radiodurans* through the reduction of silver chloride solution [128]. The fabricated AgNPs possessed a wide-spectrum antibiofilm and antibacterial effects against Gram-positive and -negative pathogenic bacteria. Similarly, the AgNPs displayed viable anticancer property against human breast cancer cell lines by inhibiting cell viability and increased cytotoxicity for the proliferation of cancer cells.

The ability of *Deinococcus radiodurans* to survive high radiation and desiccation condition allows for its on-field usage for bioremediation. Recent researches are now targeted toward the fabrication and production of diverse kinds of metallic nanoparticles including platinum, tellurium, and palladium, which directly focus them for different applications. For instance, electrochemically active biofilms of *Shewanella loihica* PV-4 were employed to fabricate ultrasmall nano-sized range 2–7 nm palladium and platinum nanoparticles by Ahmed et al. [130]. These palladium and platinum nanoparticles displayed outstanding catalytic property in the degradation of methyl orange dye. Toxic tellurite oxyanions were converted into important nanoparticles using *Ochrobactrum* sp. as a nanofactory in the study reported by Zonaro et al. [135]. AuNPs were also produced from high-resistant *B. subtilis* isolated from Hatti Gold Mine in India in a study reported by Srinath et al. [136]. The synthesized AuNPs showed biocatalytic activity in the decomposition of methylene blue dye and can be employed for the degradation of other harmful dyes in the environment. Spherical AgNPs with nano-sized range of 41–62 nm were also fabricated using *Bacillus brevis* by Saravanan et al. [133]. In this study, the synthesized AgNPs exhibited viable antibacterial property against multidrug resistant strains of both *Salmonella typhi* and *Staphylococcus aureus*.

A number of parameters have been reported to affect the reduction of metallic ions into nanoparticles. The type of the organic functional molecules present on the microbial cell wall which induce biomineralization and appropriate environmental conditions such as the pH, medium composition, the concentration of metallic salt, and temperature are important factors that influence nanoparticle

Table 21.1 Selective representation of bacteria and actinomycetes employed for the biofabrication of nanoparticles and their usage.[a]

Bacteria	Source of isolation	Metal nanoparticles	Characterization		Applications	Reference
			Size (nm)	Shape		
Stenotrophomonas GSG2	Coral sample collected from Bay of Bengal	Silver and gold	Gold – 10 to 50; silver – 40 to 60	Circular, triangular, and hexagonal	nd	[122]
Alteromonas macleodii	Sediment sample from Kochi back water, India	Silver	70	Spherical	nd	[123]
Bacillus cereus	Leaf of *Garcinia xanthochymus*	Silver	20–40	Spherical	Antibacterial activity	[124]
Kocuria flava	Kanyakumari coast of India	Copper	5–30	Spherical	nd	[125]
Bacillus sp. CS11	Soil samples from Cochin, India	Silver	42–92	Spherical	nd	[126]
Pseudomonas aeruginosa JP-11	Marine water	Cadmium sulfide	20–40	Spherical	Removal of cadmium pollutant from aqueous solution	[127]
Deinococcus radiodurans	American Type Culture Collection, Manassas, USA	Silver	4–50	Spherical	Antibacterial activity, anti-biofouling agent, and anticancer activity	[128]
Ochrobactrum rhizosphaerae	Marine water	Silver	10	Spherical	Antibacterial activity	[129]
Shewanella loihica PV-4	DSMZ, Germany	Palladium and platinum	2–7	Spherical	Degradation of methyl orange dye	[130]
Klebsiella pneumoniae, Escherichia coli, Pseudomonas jessinii	Isolated from tiger nut, carrot juice, and feces	Silver	50–100	Cubic and star/flower-like shapes	nd	[131]
Pseudomonas aeruginosa	Mangrove water sample Mandovi Estuary	Silver	35–60	Spherical and triangular	Antibacterial activity	[132]
Bacillus brevis NCIM 2533	National Collection of Industrial Microorganism (NCL), Pune, India	Silver	41–68	Spherical	Antibacterial activity against multidrug-resistant bacteria	[133]
Alcaligenes faecalis	Coral from Palk Bay located near Mandapam, Gulf of Mannar	Silver	30–50	Spherical	Antimicrobial and antibiofilm activity	[134]

(Continued)

Table 21.1 (Continued)

Bacteria	Source of isolation	Metal nanoparticles	Characterization		Applications	Reference
			Size (nm)	Shape		
Ochrobactrum sp. MPV1	Roasted Arsenopyrites, Tuscany, Italy	Tellurium	nd	Roughly spherical and rods	Reduction of toxic compounds	[135]
Bacillus subtilis	Hutti Gold Mine, India	Gold	20–25	Spherical	Degradation of methylene blue	[136]
Morganella psychrotolerans	—	Silver	2–5	Spherical	—	[137]
Actinomycetes						
Rhodococcus sp. NCIM 2891	National Chemical Laboratory, India	Silver	10–15	Spherical	Antimicrobial activity, catalytic reduction of 4-nitrophenol	[138]
Streptomyces rochei MHM13	Sediment samples along Suez Gulf, Red Sea, Egypt	Silver	22–85	Spherical	Antimicrobial activity and synergistic effect with antibiotics	[139]
Streptomyces sp. LK3	Marine soil sample, Nicobar Island	Silver	5	Spherical	Acaricidal activity	[140]
Rhodococcus NCIM 2891	nd	Silver	10	Spherical	nd	[141]
Streptacidiphilus durhamensis	Acidic forest soil	Silver	8–48	Spherical	Antibacterial and -cancer activity	[142]
Streptomyces capillispiralis Ca-1	Medicinal plant *Convolvulus arvensis*	Copper	3.6–59	Spherical	Antimicrobial activity	[143]
Streptomyces xinghaiensis OF1	Sediment sample of Lonar Crater, Maharashtra, India	Silver	5–20	Spherical	Antimicrobial activity and synergistic effect with antibiotics	[144]
Streptomyces parvulus	DPUA Culture Collection, Brazil	Silver	1–40	nd	Antimicrobial activity	[145]
Streptomyces griseoruber	Soil Sample, Mercara region	Gold	5–50	Spherical, hexagonal, and triangular	Degradation of methylene blue	[146]

[a]Table adapted and modified from Gahlawat and Choudhury [147].

biosynthesis [120]. These environmental factors can tremendously affect the size, morphology, and composition of biometallic nanoparticles [148]. As such, optimizing these parameters is important during the biosynthetic process in order to optimize the overall performance of particles. For instance, AgNPs were fabricated by Ramanathan et al. using *Morganella psychrotolerans* and optimized factors that influence growth kinetics to investigate their effect on the morphology of AgNPs [137]. Spherical AgNPs having an average size between 2 and 5 nm were synthesized at optimum growth temperature of 20 °C. As the temperature increased to 25 °C, mixed morphology was observed for the AgNPs including triangular, hexagonal nanoplates and spherical nanostructures. Reducing the growth temperature from 20 to 15 °C also produced a mixture of nanoplates and spherical nanoparticles. There was a significant increase in the number of nanoplates with scanty spherical nanoparticles when both the physiological activity and bacteria growth decreased due to further reduction in bacteria optimum growth temperature to 4 °C. Interestingly, the spherical nanoparticles produced at lower temperature of 4 °C exhibit larger size between 70 and 100 nm.

In a similar study, Yumei and coworkers synthesized AgNPs using the *Arthrobacter* sp. and reported the influence of metal ion concentration, temperature, and pH as modulating factors during AgNPs biosynthesis [149]. At a lower concentration of 1 mM silver nitrate solution, face-centered-cubic AgNPs were fabricated with an average size between 9 and 72 nm at 70 °C between pH 7 and 8. As the concentration of silver nitrate increased to 3 mM, AgNPs were seen to be aggregated at 70 °C. Meanwhile, AgNPs were not synthesized at pH below 5 and above 8 but within the pH range of 7–8. As the incubation temperature increased from 70 to 90 °C, the synthesis period reduced from 10 to 2 minutes. Concisely, the study reported that both medium metal concentration and pH have significant impact on the fabrication of nanoparticles.

Focus on terrestrial bacteria has taken the center stage by researchers for many years until recently. The use of microbial cultures from the marine environment has gained enormous attention as nanofactories in recent times. AuNPs and AgNPs were biofabricated by Malhotra et al. using the novel marine bacterium, *Strenotrophomonas* [122]. Their findings showed that low molecular weight secretory proteins within the supernatant facilitate the biosynthesis of AuNPs and AgNPs. In a similar experiment, Kaur and associates reported the production of copper nanoparticles using newly discovered marine strain of *Kocuria flava* having a nanoparticulate size within the range of 5–30 nm [125]. The use of the extracellular polymeric substances (EPSs) of bacteria as both bioreductant and capping agents for nanoparticles biosynthesis has gained a lot of attention in the literature [123, 127, 129]. For instance, narrow size distribution of AgNPs was reported by Mehta et al. using the EPS produced by an osmotolerant marine isolate of *Alteromonas macleodii* [123].

The overview of the biosynthesis of nanoparticles from different bacteria and actinomycetes as well as their applications in bioremediation and biomedical field is presented in Table 21.1.

Usually, actinomycetes have been employed for the production of both extracellular enzymes and secondary metabolites [132]. They have also been explored for the biofabrication of nanoparticles due to their unequally high protein content as well as their unique ability to generate different bioactive compounds. While actinomycetes have the capability to produce nanoparticles using both the intra- and extracellular pathways, their extracellular reduction potential remains the most commonly explored pathway with more commercial applications in diverse fields. Otari et al. reported the bioproduction of AgNPs using actinobacteria *Rhodococcus* NCIM 2891 [141] in which the transmission electron microscopy (TEM) graph showed spherical shape AgNPs having an average nanosize of 10 nm. The silver nuclei produced were proposed to be the result of the intracellular biomineralization of silver ions due to the surface enzymes on the cell wall of the microorganism. Also, the marine bacterium, *Streptomycetes* sp. LK-3 was employed by Karthik et al. to reduce silver ions into AgNPs [140]. In an electron-transfer reaction, $NADH^+$-dependent nitrate reductase was reported to account for the reduction of silver ion into stable AgNPs. Their findings showed that the synthesized AgNPs possessed antiparasitic or acaricidal properties when tested on both *Rhipicephalus microplus* and *Haemaphysalis bispinosa*. The work of Buszewski et al. in a recent experiment produced AgNPs using an acidophilic actinobacteria, *Streptacidiphilus durhamensis* [142]. Stable spherical AgNPs were produced from their study having an average range size of 8–48 nm with an antibacterial activity against the trio of *Proteus mirabilis*, *P. aeruginosa*, and *Staphylococcus aureus*.

On a general note, nanoparticles fabricated using green synthesis showed higher antimicrobial activity than those synthesized through the conventional procedure as a result of the influence of other bioactive molecules that partake in capping and stabilizing the biosynthesized nanoparticles. In a later research, 41 actinomycetes isolates, taken from the Gulf, Red Sea, were evaluated by Abd-Elnaby et al. and found only two of the isolates having the potential to produce AgNPs [139]. However, the synthesized AgNPs showed vibrant antibacterial potential against a number of infectious bacteria including *P. aeruginosa*, *Escherichia coli*, *Bacillus cereus*, *B. subtilis*, *Salmonella typhimurium*, *Staphylococcus aureus*, *Vibrio fluvialis*, and *Vibrio damsella*.

Concisely, AgNPs remain the most explored and investigated among the biosynthesized nanoparticles using actinomycetes. Nonetheless, recent works have evolved in which copper and AUNPs have been biosynthesized using both *Streptomyces griseoruber* and *Streptomyces capillispiralis* Ca-1, respectively [143, 146]. It is interesting to note that the strain of *Streptomyces* remains the most widely applied in pharmaceutical and enzymatic processes for antibiotic production among the actinomycetes, since 55% of more than the 10 000 known antibiotics are synthesized by them.

21.3.3 Nanoparticle Synthesis Using Yeasts

Aside bacteria and fungi, several research studies have explored the possibilities of using yeast for the green synthesis of nanomedicines. The innate ability to take in and store increased concentrations of toxic metal ions present in its environment remains a special capability of yeast as a biogenic nanofactory [150]. In order to cope with their toxic and unfavorable environment due to metallic ions' toxicity, yeast cells employ different adaptive detoxification strategies including bioprecipitation, intracellular sequestration, and chelation. A number of researchers have exploited these adaptive mechanisms for the biosynthesis of metallic nanoparticles. For instance, a biomimetic fabrication of AgNPs was investigated using a marine strain of ascomycetous yeast *Yarrowia lipolytica* in a cell-related manner [151]. It was observed that the melanin secreted by the yeast cells facilitates the biodegradation of the silver ions into AgNPs which exhibited an antibiofilm activity against *Salmonella paratyphi* pathogen. In another study, environmentally friendly AgNPs were biosynthesized extracellularly by Waghmare et al. using *Candida utilis* NCIM 3469 [152]. Circular-shaped AgNPs were synthesized with size range between 20 and 80 nm, which displayed antibacterial efficacy against infectious strains of *Escherichia coli*, *Staphylococcus aureus*, and *P. aeruginosa*. Elahian et al. [153] recently employed genetically engineered yeast strain, *Pichia pastoris*, for the biofabrication of AgNPs. The modified *Pichia pastoris* strain produced a high metal-resistant gene, cytochrome b5 reductase enzyme gotten from *Mucor racemosus*, used for the reduction of metal ion into nanoparticles. The metallic nanoparticles generated by the cytochrome b5 reductase enzymatic reaction were stable and uniformly dispersed within the size range of 70–180 nm.

Using *Candida lusitaniae* isolate obtained from the gut of a termite as a nanofactory, Eugenio et al., synthesized nano-ranged AgNPs of 2–10 nm [154]. The green synthesis AgNPs displayed antiproliferative property against *Staphylococcus aureus* and *Klebsiella pneumoniae* as well as showed a viable alternative to the frequently used antibiotics. Nano-sized hexagonal palladium nanoparticles (PdNPs) having a size of 32 nm were biosynthesized by Sriramulu and Sumathi using aqueous extract from *Saccharomyces cerevisiae* [155]. The produced PdNPs exhibited photocatalytic bioremediation of textile azo dye to 98% within 60 minutes postexposure to UV radiation. Interestingly, these literature surveys suggested that the variance in size, shape, and properties exhibited by nanoparticles biofabricated by yeast rests squarely on the unique adaptive mechanisms demonstrated by yeast cells for their synthesis and stability.

21.3.4 Nanoparticle Synthesis Using Algae

The literature is also littered with several reports on the use of algae as a nanofactory for the biofabrication of metal nanoparticles. Spherical-shaped AgNPs with average size range of 9.8 ± 5.8 nm were biosynthesized using the dried unicellular microalga, *Chlorella vulgaris*, by da Silva Ferreira et al. [156]. These nanoparticles were investigated as viable eco-friendly option for biomedical antimicrobial usage. In a subsequent experiment, Arsiya et al. [157] employed an aqueous extract of *Chlorella vulgaris* for the biosynthesis of PdNPs within 10 minutes. Uniform and circularly dispersed PdNPs were shown by the TEM graphical analysis with an average size ranging between 5 and 20 nm. Interestingly, this represents the first biosynthesis of PdNPs in a comparatively shorter time duration using *Chlorella vulgaris*. Meanwhile, the marine alga *Sargassum bovinum*, obtained from the Persian Gulf environment, has also been employed for the biofabrication of PdNPs [158]. Silver chloride nanoparticles (AgclNPs) were synthesized using the aqueous extract of the marine alga *Sargassum plagiophyllum* in a study performed by Dhas et al. [159]. In recent time, an eco-friendly and cheap green method has been employed to produce AgNPs using the marine green alga *Caulerpa racemosa* as a nanofactory [160]. These green nanoparticles displayed a plausible catalytic activity for the reduction of methylene blue. Ramakrishna et al. employed aqueous extracts of brown algae *Sargassum tenerrimum* and *Turbinaria conoides* for the biosynthesis of AuNPs [161]. The nanoparticles exhibited efficient biocatalytic property for the degradation of aromatic nitro compounds as well as organic dyes.

There has been an increased curiosity for the biosynthesis of zinc oxide (ZnO) nanoparticles in recent years among scientists owing to their excellent physiochemical properties and broad applications in biomedicines and food additives, optoelectronics, sunscreens, etc. Algal formulation from two marine brown seaweeds, *Padina tetrastromatica* and *Turbinaria conoides*, were used by Rajeshkumar for the biofabrication of ZnO nanoparticles [162]. Their antimicrobial activity was investigated against fish pathogens. Also, ZnO nanoparticles were produced by Sanaeimehr et al. using the extracts of *Sargassum muticum* [163]. Their antiapoptotic and -angiogenic activities against human liver cancer cell lines were evaluated. It is worthy of note that these literature accounts connote the increasing interest of researchers in the exploration of marine microbes for biogenic synthesis of metal nanoparticles since marine algae possess diverse biologically active compounds and secondary metabolites which avail them to serve as nanofactories [164]. Importantly, these marine algae have numerous applications in biomedicines such as anticancer, antiviral agents, antioxidants, antidiabetic, hepatoprotective, and cardioprotective agents.

21.3.5 Nanoparticle Synthesis Using Viruses

A unique characteristic of viruses that distinguishes them from other microbes is their thick extracellular coating of capsid proteins that serve as appropriate landscape for metallic ion interaction [165]. Such encapsulated protein cages can be genetically engineered to form uniformly dispersed units that are highly robust and malleable. It is possible to fabricate viruses to form templates for material deposition or modified to form three-dimensional vectors for targeted drugs delivery [166]. Another important application of viruses is their usage in the production of nanoconjugates and nanocomposites with metal nanoparticles that are valuable fabricated nanocargos in drug delivery and cancer therapy (Table 21.1).

Mao et al. [167] examined the nucleation and orientation process of semiconductor nanocrystals using M13 bacteriophage. Their work described the biogenic fabrication of genetically controllable nanocrystals semiconductor from zinc and cadmium sulfides. Presently, different nanotechnology applications are investigating the safe usage of plant viruses owing to their biochemical and structural stability, being easy to cultivate, nonpathogenicity, and non-toxicity in human and animal models. In a particular study, reduced concentrations of tobacco mosaic virus (TMV) and bovine papillomavirus (BPV) were employed as additives together with diverse plants extracts such as *Nicotiana benthamiana*, *Avena sativa*, and *Musa paradisiaca* [168]. Interestingly, the addition of TMV and BPV both assisted to decrease the size of the nanoparticles as well as drastically increased their numbers when compared to the nonvirus control. The sustained release of doxorubicin for chemotherapy was investigated by Cao et al. using nanoparticles biosynthesized from red clover necrotic mosaic virus (RCNMV) [169]. The infusion of doxorubicin to the viral capsid through the pore-formation mechanism was because of the unique morphology of RCNMV and the structural changes in response to bivalent cations removal.

In another study, Le et al. examined the delivery of doxorubicin drug for the treatment of cancer using nanoparticles produced from potato virus X [170]. Elongated filamentous nanoparticles were synthesized by potato virus X, which showed efficient tumor homing and penetration power when compared to spherical ones. Nonetheless, nanoparticle production using viruses is still confronted with diverse setbacks such as dependence on the host organism for protein expression, synthetic processes that are not fully developed, and reduced prospects for large-scale application.

21.4 Plant-based Nanomedicines

21.4.1 Plants – The Paradigm Shift for the Green Synthesis of Nanomaterials

Nano- and micro-scaled inorganic materials with varying lengths and sizes have been relatively developed through diverse eco-friendly mechanisms by Mother Earth. These natural processes by the earth present promising and viable alternative for the synthesis of nanomaterials using bio-organisms as biofactories and are aligned to the principles of green chemistry. The green synthesis paradigm presents an alternative mechanism for the fabrication of nanoparticles using nontoxic, environmentally friendly, and safe reagents.

Phytomining refers to the process that employs specialized plant species known as hyperaccumulating plants to absorb metal ions from the soil with viable economic return upon the recovery of the metals from the biomass. Hyperaccumulator species possess unique physiological mechanism that enhances the uptake, storage, and regulation of high metal ion concentration from the soil into their tissues [171]. Stress-tolerant plants have been reported to have increased capacity for the reduction of metal ions into metallic nanoparticles [172]. The concept of phytoremediation in plants has been proposed to be responsible for the mechanism underlining the biosynthesis of metal nanoparticles [173]. While some plants can absorb more heavy metals from their growth environment, others cannot. As such, those plants that have the capacity to accumulate metal up to about 1000 times more than that found in normal plants accumulator are referred to as hyperaccumulators [174]. A number of researchers have investigated the nature of the metal accumulated on plants. Gardea-Torresdey et al. reported that gold ions absorbed and stored in leaf and stem biomass can be biofabricated into uniform and monodispersed nanoparticles of pure metal [175]. Their finding was based on their earlier study in which gold chloride-enriched agar with concentration of 320 mg kg^{-1} was employed to grow alfalfa sprouts and characterized using both transmission electron microscopy (TEM) and X-ray absorption spectroscopy (XAS) [176]. A mixture of clustered particles with sizes between 20 and 40 nm and free-flowing monodispersed nanoparticles of size range of 2–20 nm was seen to spread across some areas within the plant tissue. Subsequently, the alfalfa roots demonstrated the ability to absorb silver ions from the agar medium and later transfer it to the shoot for storage in the same oxidation state. These silver ions aggregate to form nanoparticles that eventually agglomerate to produce larger particles having another organizational architecture [177].

Gold recovery through phytomining or deposit of gold ores in plant has been reported as eco-friendly and economy viable alternative method for the synthesis of AuNPs [175, 178]. This phenomenon of gold accumulation as nanoparticles in plants resulted in the new paradigm in gold production termed gold phytomining, metal extraction from soils or ores for recovery and marketing [179]. The extracts from the trio plants of *Brassica juncea*, *Cinnamomum camphora*, and *Capsicum annum* have been shown to accumulate gold and AgNPs in other research reports that supported their production [180]. The biosynthesis of metal nanoparticles directly from plant leaf extracts and their potential applications have also been reported. The work of Shankar et al. investigated the conversion of silver or chloroaurate ions through oxidative reduction into nanoparticles using an extract from Geranium leaf (*Pelargonium graveolens*) [181] or neem leaf (*Azadirachta indica*) [182].

The work of Paciotti et al. investigated and reported the biosynthesis of triangular-shaped AUNPs in Lemon grass extracts (*Cymbopogon flexuous*) and from the extract of tamarind leaf [183]. In another recent study, nanotriangles of AuNPs and AgNPs were synthesized from Aloe vera plant extracts. The findings showed that gold nanotriangles were formed from the bioreduction of chloroaurate ions by biomolecules with an average molecular weight less than 3 kDa. The plant extracts employed for the synthesis of these AuNPs were broths obtained from boiling fresh plants leaves, while some were obtained from dried mass extracts from sundried *Cinnamonum camphora* leaf. Also, AgNPs were biofabricated from *Capsicum annuum* extract when applied on silver ions. Also, AgNPs were biofabricated from *C. annuum* extract when applied on silver ions [180]. Interestingly, AuNPs were synthesized from dead biomass of hops (*Humulus lupulus*). The extracellular biofabrication of both AuNPs and AgNPs was observed using the fruit extract *Emblica officinalis* as a reducing agent [184].

The enhanced capacity of stress-tolerant plants for the synthesis of metal nanoparticles has also been reported. Record showed that these plants possessed high reducing capacity in converting metallic ions to nanoparticles. The desert willow plant (*Chilopsis linearis*) possessed such high reducing capacity with high storage capacity to yield large amounts of biomass with low water consumption. Ankamwar et al. reported that the average size of AuNPs produced from gold enriched medium is directly proportional to the total gold concentration in the plant tissues [185]. The physicochemical properties of the biofabricated AuNPs were examined using TEM, selected area electron diffraction (SEAD), and X-ray diffraction (XRD) spectra. The evaluation of the crystallinity of the particles revealed their transition from poly to single crystallinity with increasing size as the reaction time was increased [180]. There are a number of peptides and proteins that facilitate the formation of nanoparticles in plants including amino acids such as lysine, cysteine, arginine, and methionine [186–188]. These amino acids have been shown to be capable of binding silver ions [189]. Similarly, in some higher plants, biosilcification also yields metallic nanoparticles. Silicon nanoparticles have been shown to possess unique physiological properties that enhance their entry into plants and change plant metabolic activities [190, 191].

21.4.2 Synthesis Mechanism of Plant-based Eco-friendly Nanomaterials

The principle that governed plant-based nanoparticle synthesis is premised on the mechanism of mineral uptake from the soil. The reservoir of minerals such as gold, silver, and silicon in the soil or the production medium enhances the accumulation of nanoparticles in plants. Meanwhile, there are a lot of doubts and discrepancies concerning the mechanism that governs the synthesis of metallic nanoparticles in plants. A great matter of controversy is that nanoparticles are biofabricated from within living plants as well as from sundried biomass [192]. However, the unilateral fundamental mechanism involves the formation of nanoparticles from their oxidative reduction from metal ions. Interestingly, this reduction mechanism is influenced by the activities of some reducing agents or in-bound enzymes within the plant's cell wall. Several experimental reports proposed that various biomolecules such as proteins are responsible for this process [186]. Report has shown that the amino groups of proteins act as reducing and controlling agents during the synthesis of AgNPs from silver ions. A major change occurred to the secondary structure of the protein upon their interaction with the silver ions [180]. For instance, *C. annuum* proteins play critical role in the synthesis of AgNPs [193]. However, despites all these propositions, there is no clear report that reveals the mechanistic aspects of nanoparticles synthesis in plants till date though some assertions hold that phytoremediation concept in plant might be responsible.

21.5 Conclusion and Future Recommendations

The establishment of nanotechnology is to enhance the objective of preserving the sustainability of the environment. Through this objective, both human environmental issues and human health hazards are managed. The use of green nanotechnology has been investigated in applications such as fabrication of green materials, coatings, and production of biocides in order to circumvent the outlet of toxic substances into the environment. Meanwhile, the potential risk of more refined green nanotechnologies has to be examined. The more the refinement, the greater the risks they pose.

As opposed to the synthesis of nanoparticles through the physical and chemical methods, the biosynthesis of nanoparticles has been proven to be more biocompatible and stable. Meanwhile, major setbacks still exist in the biosynthesis of nanoparticles including low production and yield, poor product quality control, contamination from biological cells, and the inability to effectively separate the nanoparticles from the biological materials.

A number of microorganisms have been investigated over the past few decades for their potential to convert toxic metal and metalloid ions into functional nanoparticles. These are often referred to as efficient eco-friendly nanofactories. While insights into some of the mechanisms for bioreduction remain unknown, a lot have been achieved in discovering reasonable hypothesis that explained the experimental phenomena that give further understanding into the synthesis process of the biofabrication of nanoparticles using microbes. Nanoparticles biosynthesis occurred in the absence of high-temperature treatments or additional chemicals. These biosynthesized nanoparticles have displayed several outstanding characteristics for optical applications, catalytic applications, biosensors, aerospace applications, and as gas sensors.

For future purposes, subsequent studies should focus on a number of prospects for effectiveness. It is important to have a better grasp of the reaction mechanisms underlying the formation of nanoparticles for the reproduction of biomaterials and the possible manipulation of their shape, size, and dispersity. Also, as it stands, more of the studies are mainly on the biosynthesis of Au, Ag, Pd, Te, and Se nanoparticles, while information on other nanoparticles produced from other metallic ions are not well documented. As such, more diversified research is needed by exploring the microbial and plant diversity to discover novel nanofactories for the biosynthesis of nanomaterials for nanomedicines.

References

1 Anastas, P.T. and Warner, J.C. (2000). *Green Chemistry: Theory and Practice*. Oxford University Press.

2 Soppimath, K.S., Aminabhavi, T.M., Kulkarni, A.R., and Rudzinski, W.E. (2001). Biodegradable polymeric nanoparticles as drug delivery devices. *Journal of Controlled Release* 70: 1–20. https://doi.org/10.1016/S0168-3659(00)00339-4.

3 Mohanraj, V.J. and Chen, Y. (2007). Nanoparticles – a review. *Tropical Journal of Pharmaceutical Research* 5: 561–573.

4 Nagavarma, B.V.N., Hemant, K.Y., Ayaz, A. et al. (2012). Different techniques for preparation of polymeric nanoparticles-a review. *Asian Journal of Pharmaceutical and Clinical Research* 5: 16–23.

5 Pinto Reis, C., Neufeld, R.J., Ribeiro, A.J., and Veiga, F. (2006). Nanoencapsulation I. Methods for preparation of drug-loaded polymeric nanoparticles. *Nanomedicine: Nanotechnology, Biology and Medicine* 2: 8–21. https://doi.org/10.1016/j.nano.2005.12.003.

6 Tice, T.R. and Gilley, R.M. (1985). Preparation of injectable controlled-release microcapsules by a solvent-evaporation process. *Journal of Controlled Release* 2: 343–352. https://doi.org/10.1016/0168-3659(85)90056-2.

7 Rao, J.P. and Geckeler, K.E. (2011). Polymer nanoparticles: preparation techniques and size-control parameters. *Progress in Polymer Science* 36: 887–913. https://doi.org/10.1016/j.progpolymsci.2011.01.001.

8 Galindo-Rodriguez, S., Allémann, E., Fessi, H., and Doelker, E. (2004). Physicochemical parameters associated with nanoparticle formation in the salting-out, emulsification-diffusion, and nanoprecipitation methods. *Pharmaceutical Research* 21: 1428–1439. https://doi.org/10.1023/B:PHAM.0000036917.75634.be.

9 Quintanar-Guerrero, D., Allémann, E., Fessi, H., and Doelker, E. (1998). Preparation techniques and mechanisms of formation of biodegradable nanoparticles from preformed polymers. *Drug Development and*

Industrial Pharmacy 24: 1113–1128. https://doi.org/10.3109/03639049809108571.
10 Vauthier, C., Dubernet, C., Fattal, E. et al. (2003). Poly(alkylcyanoacrylates) as biodegradable materials for biomedical applications. *Advanced Drug Delivery Reviews* 55: 519–548. https://doi.org/10.1016/S0169-409X(03)00041-3.
11 Dimitrova, B., Ivanov, I.B., and Nakache, E. (1988). Mass transport effects on the stability of emulsion: emulsion films with acetic acid and acetone diffusing across the interface. *Journal of Dispersion Science and Technology* 9: 321–341.
12 Wehrle, P., Magenheim, B., and Benita, S. (1995). The influence of process parameters on the PLA nanoparticle size distribution, evaluated by means of factorial design. *European Journal of Pharmaceutics and Biopharmaceutics* 41: 19–26.
13 Barichello, J.M., Morishita, M., Takayama, K., and Nagai, T. (1999). Encapsulation of hydrophilic and lipophilic drugs in PLGA nanoparticles by the nanoprecipitation method. *Drug Development and Industrial Pharmacy* 25: 471–476. https://doi.org/10.1081/DDC-100102197.
14 Niwa, T., Takeuchi, H., Hino, T. et al. (1993). Preparations of biodegradable nanospheres of water-soluble and insoluble drugs with d,l-lactide/glycolide copolymer by a novel spontaneous emulsification solvent diffusion method, and the drug release behavior. *Journal of Controlled Release* 25: 89–98. https://doi.org/10.1016/0168-3659(93)90097-O.
15 Jung, T., Kamm, W., Breitenbach, A. et al. (2000). Biodegradable nanoparticles for oral delivery of peptides: is there a role for polymers to affect mucosal uptake? *European Journal of Pharmaceutics and Biopharmaceutics* 50: 147–160. https://doi.org/10.1016/S0939-6411(00)00084-9.
16 Lambert, G., Fattal, E., and Couvreur, P. (2001). Nanoparticulate systems for the delivery of antisense oligonucleotides. *Advanced Drug Delivery Reviews* 47: 99–112. https://doi.org/10.1016/S0169-409X(00)00116-2.
17 Couvreur, P., Dubernet, C., and Puisieux, F. (1995). Controlled drug delivery with nanoparticles: current possibilities and future trends. *European Journal of Pharmaceutics and Biopharmaceutics* 41: 2–13.
18 Fessi, H., Puisieux, F., Devissaguet, J.P. et al. (1989). Nanocapsule formation by interfacial polymer deposition following solvent displacement. *International Journal of Pharmaceutics* 55: R1–R4. https://doi.org/10.1016/0378-5173(89)90281-0.
19 Jeong, Y.-I., Cho, C.-S., Kim, S.-H. et al. (2001). Preparation of poly(DL-lactide-co-glycolide) nanoparticles without surfactant. *Journal of Applied Polymer Science* 80: 2228–2236. https://doi.org/10.1002/app.1326.
20 Chronopoulou, L., Fratoddi, I., Palocci, C. et al. (2009). Osmosis based method drives the self-assembly of polymeric chains into micro- and nanostructures. *Langmuir* 25: 11940–11946. https://doi.org/10.1021/la9016382.
21 Mishima, K. (2008). Biodegradable particle formation for drug and gene delivery using supercritical fluid and dense gas. *Advanced Drug Delivery Reviews* 60: 411–432. https://doi.org/10.1016/j.addr.2007.02.003.
22 Khapli, S. and Jagannathan, R. (2014). Supercritical CO2 based processing of amorphous fluoropolymer Teflon-AF: surfactant-free dispersions and superhydrophobic films. *Journal of Supercritical Fluids* 85: 49–56. https://doi.org/10.1016/j.supflu.2013.10.014.
23 Ksibi, H., Subra, P., and Garrabos, Y. (1995). Formation of fine powders of caffeine by RESS. *Advanced Powder Technology* 6: 25–33. https://doi.org/10.1016/S0921-8831(08)60545-2.
24 Sane, A. and Thies, M.C. (2007). Effect of material properties and processing conditions on RESS of poly(l-lactide). *Journal of Supercritical Fluids* 40: 134–143. https://doi.org/10.1016/j.supflu.2006.04.003.
25 Rollins, H.W., Bunker, C.E., Sun, Y.-P. et al. (2002). Preparation and processing of nanoscale materials by supercritical fluid technology. In: *Supercritical Fluid Technology in Materials Science and Engineering*, 1e (ed. Y.-P. Sun). CRC Press. http://www.crcnetbase.com/doi/abs/10.1201/9780203909362.ch13 (accessed 5 November 2014).
26 Hutchenson, K.W. (2002). Organic chemical reactions and catalysis in supercritical fluid media. In: *Supercritical Fluid Technology in Materials Science and Engineering* (ed. Y.-P. Sun). CRC Press. http://www.crcnetbase.com/doi/abs/10.1201/9780203909362.ch3 (accessed 5 November 2014).
27 Aslani, P. and Kennedy, R.A. (1996). Studies on diffusion in alginate gels. I. Effect of cross-linking with calcium or zinc ions on diffusion of acetaminophen. *Journal of Controlled Release* 42: 75–82. https://doi.org/10.1016/0168-3659(96)01369-7.
28 Kwok, K.K., Groves, M.J., and Burgess, D.J. (1991). Production of 5-15 microns diameter alginate-polylysine microcapsules by an air-atomization technique. *Pharmaceutical Research* 8: 341–344.
29 Paques, J.P., Sagis, L.M.C., van Rijn, C.J.M., and van der Linden, E. (2014). Nanospheres of alginate prepared through w/o emulsification and internal gelation with nanoparticles of CaCO3. *Food Hydrocolloids* 40: 182–188. https://doi.org/10.1016/j.foodhyd.2014.02.024.
30 Adeyemi, S.A., Choonara, Y.E., Kumar, P. et al. (2017). Synthesis and in vitro characterization of a pH-responsive chitosan- polyethylenimine nanosystem for the delivery

of therapeutic proteins. *Journal of Drug Delivery Science and Technology* 39: 266–276. https://doi.org/10.1016/j.jddst.2017.03.022.

31 Vila, A., Sánchez, A., Janes, K. et al. (2004). Low molecular weight chitosan nanoparticles as new carriers for nasal vaccine delivery in mice. *European Journal of Pharmaceutics and Biopharmaceutics* 57: 123–131. https://doi.org/10.1016/j.ejpb.2003.09.006.

32 Janes, K.A., Fresneau, M.P., Marazuela, A. et al. (2001). Chitosan nanoparticles as delivery systems for doxorubicin. *Journal of Controlled Release* 73: 255–267. https://doi.org/10.1016/S0168-3659(01)00294-2.

33 Adeyemi, S.A., Choonara, Y.E., Kumar, P. et al. (2017). Design and characterization of endostatin-loaded nanoparticles for in vitro antiangiogenesis in squamous cell carcinoma. *Journal of Nanomaterials* https://doi.org/10.1155/2017/2539065.

34 Adeyemi, S.A., Choonara, Y.E., Kumar, P. et al. (2019). Folate-decorated, endostatin-loaded, nanoparticles for anti-proliferative chemotherapy in esophaegeal squamous cell carcinoma. *Biomedicine & Pharmacotherapy* 119: 109450. https://doi.org/10.1016/j.biopha.2019.109450.

35 Ma, Z., Lim, T.M., and Lim, L.-Y. (2005). Pharmacological activity of peroral chitosan–insulin nanoparticles in diabetic rats. *International Journal of Pharmaceutics* 293: 271–280. https://doi.org/10.1016/j.ijpharm.2004.12.025.

36 Mansouri, S., Lavigne, P., Corsi, K. et al. (2004). Chitosan-DNA nanoparticles as non-viral vectors in gene therapy: Strategies to improve transfection efficacy. *European Journal of Pharmaceutics and Biopharmaceutics* 57: 1–8. https://doi.org/10.1016/S0939-6411(03)00155-3.

37 Illum, L., Farraj, N.F., and Davis, S.S. (1994). Chitosan as a novel nasal delivery system for peptide drugs. *Pharmaceutical Research* 11: 1186–1189. https://doi.org/10.1023/A:1018901302450.

38 Aktaş, Y., Andrieux, K., Alonso, M.J. et al. (2005). Preparation and in vitro evaluation of chitosan nanoparticles containing a caspase inhibitor. *International Journal of Pharmaceutics* 298: 378–383. https://doi.org/10.1016/j.ijpharm.2005.03.027.

39 Calvo, P., Remuñan-López, C., Vila-Jato, J.L., and Alonso, M.J. (1997). Chitosan and chitosan/ethylene oxide-propylene oxide block copolymer nanoparticles as novel carriers for proteins and vaccines. *Pharmaceutical Research* 14: 1431–1436. https://doi.org/10.1023/A:1012128907225.

40 Tokumitsu, H., Ichikawa, H., and Fukumori, Y. (1999). Chitosan-gadopentetic acid complex nanoparticles for gadolinium neutron-capture therapy of cancer: preparation by novel emulsion-droplet coalescence technique and characterization. *Pharmaceutical Research* 16: 1830–1835.

41 Thickett, S.C. and Gilbert, R.G. (2007). Emulsion polymerization: state of the art in kinetics and mechanisms. *Polymer* 48: 6965–6991. https://doi.org/10.1016/j.polymer.2007.09.031.

42 Kreuter, J. (1982). On the mechanism of termination in heterogeneous polymerization. *Journal of Polymer Science, Polymer Letters Edition* 20: 543–545. https://doi.org/10.1002/pol.1982.130201006.

43 Vranckx, H., Demoustier, M., and Deleers, M. (1996). A new nanocapsule formulation with hydrophilic core: application to the oral administration of salmon calcitonin in rats. *European Journal of Pharmaceutics and Biopharmaceutics* 42: 345–347.

44 Muñoz-Bonilla, A., van Herk, A.M., and Heuts, J.P.A. (2010). Preparation of hairy particles and antifouling films using brush-type amphiphilic block copolymer surfactants in emulsion polymerization. *Macromolecules* 43: 2721–2731. https://doi.org/10.1021/ma9027257.

45 Costa, C., Santos, A.F., Fortuny, M. et al. (2009). Kinetic advantages of using microwaves in the emulsion polymerization of MMA. *Materials Science and Engineering: C* 29: 415–419. https://doi.org/10.1016/j.msec.2008.08.013.

46 Yoon, S.-J., Chun, H., Lee, M.-S., and Kim, N. (2009). Preparation of poly(N-vinylcarbazole) (PVK) nanoparticles by emulsion polymerization and PVK hollow particles. *Synthetic Metals* 159: 518–522. https://doi.org/10.1016/j.synthmet.2008.11.011.

47 El-Samaligy, M.S., Rohdewald, P., and Mahmoud, H.A. (1986). Polyalkyl cyanoacrylate nanocapsules. *The Journal of Pharmacy and Pharmacology* 38: 216–218. https://doi.org/10.1111/j.2042-7158.1986.tb04547.x.

48 Nicolas, J. and Couvreur, P. (2009). Synthesis of poly(alkyl cyanoacrylate)-based colloidal nanomedicines. *Wiley Interdisciplinary Reviews. Nanomedicine and Nanobiotechnology* 1: 111–127. https://doi.org/10.1002/wnan.15.

49 Hearn, J., Wilkinson, M.C., Goodall, A.R., and Chainey, M. (1985). Kinetics of the surfactant-free emulsion polymerization of styrene: -The post nucleation stage. *Journal of Polymer Science, Polymer Chemistry Edition* 23: 1869–1883. https://doi.org/10.1002/pol.1985.170230703.

50 Song, Z. and Poehlein, G.W. (1990). Kinetics of emulsifier-free emulsion polymerization of styrene. *J. Polym. Sci. Part Polym. Chem.* 28: 2359–2392. https://doi.org/10.1002/pola.1990.080280911.

51 Liu, G. and Liu, P. (2010). Synthesis of monodispersed crosslinked nanoparticles decorated with surface carboxyl groups via soapless emulsion polymerization. *Colloids and Surfaces A: Physicochemical and Engineering Aspects* 354: 377–381. https://doi.org/10.1016/j.colsurfa.2009.05.016.

52 Akgöl, S., Öztürk, N., and Denizli, A. (2010). New generation polymeric nanospheres for lysozyme adsorption. *Journal of Applied Polymer Science* 115: 1608–1615. https://doi.org/10.1002/app.31294.

53 Landfester, K. (2001). Polyreactions in Miniemulsions. *Macromolecular Rapid Communications* 22: 896–936. https://doi.org/10.1002/1521-3927(20010801)22:12<896::AID-MARC896>3.0.CO;2-R.

54 Shibuya, K., Nagao, D., Ishii, H., and Konno, M. (2014). Advanced soap-free emulsion polymerization for highly pure, micron-sized, monodisperse polymer particles. *Polymer* 55: 535–539. https://doi.org/10.1016/j.polymer.2013.12.039.

55 Camli, S.T., Buyukserin, F., Balci, O., and Budak, G.G. (2010). Size controlled synthesis of sub-100 nm monodisperse poly(methylmethacrylate) nanoparticles using surfactant-free emulsion polymerization. *Journal of Colloid and Interface Science* 344: 528–532. https://doi.org/10.1016/j.jcis.2010.01.041.

56 Cui, X., Zhong, S., and Wang, H. (2007). Emulsifier-free core–shell polyacrylate latex nanoparticles containing fluorine and silicon in shell. *Polymer* 48: 7241–7248. https://doi.org/10.1016/j.polymer.2007.10.019.

57 Faridi-Majidi, R. and Sharifi-Sanjani, N. (2007). Emulsifier-free miniemulsion polymerization of styrene and the investigation of encapsulation of nanoparticles with polystyrene via this procedure using an anionic initiator. *Journal of Applied Polymer Science* 105: 1244–1250. https://doi.org/10.1002/app.26114.

58 Oztürk, N., Bereli, N., Akgöl, S., and Denizli, A. (2008). High capacity binding of antibodies by poly(hydroxyethyl methacrylate) nanoparticles. *Colloids and Surfaces B: Biointerfaces* 67: 14–19. https://doi.org/10.1016/j.colsurfb.2008.07.005.

59 Frank, H., Ziener, U., and Landfester, K. (2009). Formation of polyimide nanoparticles in heterophase with an ionic liquid as continuous phase. *Macromolecules* 42: 7846–7853. https://doi.org/10.1021/ma901392h.

60 Kim, S.W., Cho, H.G., and Park, C.R. (2009). Fabrication of unagglomerated polypyrrole nanospheres with controlled sizes from a surfactant-free emulsion system. *Langmuir* 25: 9030–9036. https://doi.org/10.1021/la9007872.

61 Zhang, G., Niu, A., Peng, S. et al. (2001). Formation of novel polymeric nanoparticles. *Accounts of Chemical Research* 34: 249–256. https://doi.org/10.1021/ar000011x.

62 Jiang, X.E., Dausend, J., Hafner, M. et al. (2010). Specific effects of surface amines on polystyrene nanoparticles in their interactions with mesenchymal stem cells. *Biomacromolecules* 11: 748–753.

63 Mouran, D., Reimers, J., and Schork, F.J. (1996). Miniemulsion polymerization of methyl methacrylate with dodecyl mercaptan as cosurfactant. *Journal of Polymer Science. Part A, Polymer Chemistry* 34: 1073–1081. https://doi.org/10.1002/(SICI)1099-0518(19960430)34:6<1073::AID-POLA16>3.0.CO;2-4.

64 Leiza, J.R., Sudol, E.D., and El-Aasser, M.S. (1997). Preparation of high solids content poly(n-butyl acrylate) latexes through miniemulsion polymerization. *Journal of Applied Polymer Science* 64: 1797–1809. https://doi.org/10.1002/(SICI)1097-4628(19970531)64:9<1797::AID-APP16>3.0.CO;2-V.

65 Kriwet, B., Walter, E., and Kissel, T. (1998). Synthesis of bioadhesive poly(acrylic acid) nano- and microparticles using an inverse emulsion polymerization method for the entrapment of hydrophilic drug candidates. *Journal of Controlled Release* 56: 149–158. https://doi.org/10.1016/S0168-3659(98)00078-9.

66 Landfester, K., Willert, M., and Antonietti, M. (2000). Preparation of polymer particles in nonaqueous direct and inverse miniemulsions. *Macromolecules* 33: 2370–2376. https://doi.org/10.1021/ma991782n.

67 Landfester, K. and Antonietti, M. (2000). The polymerization of acrylonitrile in miniemulsions: "crumpled latex particles" or polymer nanocrystals. *Macromolecular Rapid Communications* 21: 820–824. https://doi.org/10.1002/1521-3927(20000801)21:12<820::AID-MARC820>3.0.CO;2-9.

68 Chern, C.-S. and Sheu, J.-C. (2001). Effects of carboxylic monomers on the styrene miniemulsion polymerizations stabilized by SDS/alkyl methacrylates. *Polymer* 42: 2349–2357. https://doi.org/10.1016/S0032-3861(00)00608-X.

69 Sosa, N., Peralta, R.D., López, R.G. et al. (2001). A comparison of the characteristics of poly (vinyl acetate) latex with high solid content made by emulsion and semi-continuous microemulsion polymerization. *Polymer* 42: 6923–6928. https://doi.org/10.1016/S0032-3861(01)00157-4.

70 Sosa, N., Zaragoza, E.A., López, R.G. et al. (2000). Unusual free radical polymerization of vinyl acetate in anionic microemulsion media. *Langmuir* 16: 3612–3619. https://doi.org/10.1021/la991065m.

71 Shah, S., Pal, A., Gude, R., and Devi, S. (2010). Synthesis and characterization of thermo-responsive copolymeric nanoparticles of poly(methyl methacrylate-co-N-vinylcaprolactam). *European Polymer Journal* 46: 958–967. https://doi.org/10.1016/j.eurpolymj.2010.01.005.

72 Macías, E.R., Rodríguez-Guadarrama, L.A., Cisneros, B.A. et al. (1995). Microemulsion polymerization of methyl methacrylate with the functional monomer N-methylolacrylamide. *Colloids and Surfaces A: Physicochemical and Engineering Aspects* 103: 119–126. https://doi.org/10.1016/0927-7757(95)03209-V.

73 Landfester, K., Musyanovych, A., and Mailänder, V. (2010). From polymeric particles to multifunctional nanocapsules for biomedical applications using the miniemulsion process. *J. Polym. Sci. Part Polym. Chem.* 48: 493–515. https://doi.org/10.1002/pola.23786.

74 Karode, S.K., Kulkarni, S.S., Suresh, A.K., and Mashelkar, R.A. (1998). New insights into kinetics and thermodynamics of interfacial polymerization. *Chemical Engineering Science* 53: 2649–2663. https://doi.org/10.1016/S0009-2509(98)00083-9.

75 Hirech, K., Payan, S., Carnelle, G. et al. (2003). Microencapsulation of an insecticide by interfacial polymerisation. *Powder Technology* 130: 324–330. https://doi.org/10.1016/S0032-5910(02)00211-5.

76 Sree, U., Yamamoto, Y., Deore, B. et al. (2002). Characterisation of polypyrrole nano-films for membrane-based sensors. *Synthetic Metals* 131: 161–165. https://doi.org/10.1016/S0379-6779(02)00179-0.

77 Crespy, D., Stark, M., Hoffmann-Richter, C. et al. (2007). Polymeric nanoreactors for hydrophilic reagents synthesized by interfacial polycondensation on miniemulsion droplets. *Macromolecules* 40: 3122–3135. https://doi.org/10.1021/ma0621932.

78 Sarkar, D., El-Khoury, J., Lopina, S.T., and Hu, J. (2005). An effective method for preparing polymer nanocapsules with hydrophobic acrylic shell and hydrophilic interior by inverse emulsion radical polymerization. *Macromolecules* 38: 8603–8605. https://doi.org/10.1021/ma050661m.

79 Aboubakar, M., Puisieux, F., Couvreur, P. et al. (1999). Study of the mechanism of insulin encapsulation in poly (isobutylcyanoacrylate) nanocapsules obtained by interfacial polymerization. *Journal of Biomedical Materials Research* 47: 568–576. https://doi.org/10.1002/(SICI)1097-4636(19991215)47:4<568::AID-JBM14>3.0.CO;2-X.

80 Puglisi, G., Fresta, M., Giammona, G., and Ventura, C.A. (1995). Influence of the preparation conditions on poly(ethylcyanoacrylate) nanocapsule formation. *International Journal of Pharmaceutics* 125: 283–287. https://doi.org/10.1016/0378-5173(95)00142-6.

81 Watnasirichaikul, S., Davies, N.M., Rades, T., and Tucker, I.G. (2000). Preparation of biodegradable insulin nanocapsules from biocompatible microemulsions. *Pharmaceutical Research* 17: 684–689. https://doi.org/10.1023/A:1007574030674.

82 Bouchemal, K., Briançon, S., Perrier, E. et al. (2004). Synthesis and characterization of polyurethane and poly(ether urethane) nanocapsules using a new technique of interfacial polycondensation combined to spontaneous emulsification. *International Journal of Pharmaceutics* 269: 89–100. https://doi.org/10.1016/j.ijpharm.2003.09.025.

83 Chen, Y., Liu, H., Zhang, Z., and Wang, S. (2007). Preparation of polymeric nanocapsules by radiation induced miniemulsion polymerization. *European Polymer Journal* 43: 2848–2855. https://doi.org/10.1016/j.eurpolymj.2007.04.013.

84 Krauel, K., Davies, N.M., Hook, S., and Rades, T. (2005). Using different structure types of microemulsions for the preparation of poly(alkylcyanoacrylate) nanoparticles by interfacial polymerization. *Journal of Controlled Release* 106: 76–87. https://doi.org/10.1016/j.jconrel.2005.04.013.

85 Zhang, L. and Liu, P. (2010). Polyaniline micro/nano capsules via facile interfacial polymerization approach. *Soft Materials* 8: 29–38. https://doi.org/10.1080/15394451003598460.

86 Kuo, C.-W. and Wen, T.-C. (2008). Dispersible polyaniline nanoparticles in aqueous poly(styrenesulfonic acid) via the interfacial polymerization route. *European Polymer Journal* 44: 3393–3401. https://doi.org/10.1016/j.eurpolymj.2008.07.018.

87 An, S., Abdiryim, T., Ding, Y., and Nurulla, I. (2008). A comparative study of the microemulsion and interfacial polymerization for polyindole. *Materials Letters* 62: 935–938. https://doi.org/10.1016/j.matlet.2007.07.014.

88 Drioli, E., Criscuoli, A., and Curcio, E. (2003). Membrane contactors and catalytic membrane reactors in process intensification. *Chemical Engineering and Technology* 26: 975–981. https://doi.org/10.1002/ceat.200300012.

89 Charcosset, C. and Fessi, H. (2005). Preparation of nanoparticles with a membrane contactor. *Journal of Membrane Science* 266: 115–120. https://doi.org/10.1016/j.memsci.2005.05.016.

90 Matyjaszewski, K. (2012). Atom transfer radical polymerization: from mechanisms to applications. *Israel Journal of Chemistry* 52: 206–220. https://doi.org/10.1002/ijch.201100101.

91 Nicolas, J., Charleux, B., Guerret, O., and Magnet, S. (2005). Nitroxide-mediated controlled free-radical emulsion polymerization using a difunctional water-soluble alkoxyamine initiator. Toward the control of particle size, particle size distribution, and the synthesis of triblock copolymers. *Macromolecules* 38: 9963–9973. https://doi.org/10.1021/ma0511999.

92 Dire, C., Magnet, S., Couvreur, L., and Charleux, B. (2009). Nitroxide-mediated controlled/living free-radical surfactant-free emulsion polymerization of methyl methacrylate using a poly(methacrylic acid)-based macroalkoxyamine initiator. *Macromolecules* 42: 95–103. https://doi.org/10.1021/ma802083g.

93 Siegwart, D.J., Srinivasan, A., Bencherif, S.A. et al. (2009). Cellular uptake of functional nanogels prepared by inverse miniemulsion ATRP with encapsulated proteins, carbohydrates, and gold nanoparticles. *Biomacromolecules* 10: 2300–2309. https://doi.org/10.1021/bm9004904.

94 Rieger, J., Zhang, W., Stoffelbach, F., and Charleux, B. (2010). Surfactant-free RAFT emulsion polymerization using poly(N,N-dimethylacrylamide) trithiocarbonate macromolecular chain transfer agents. *Macromolecules* 43: 6302–6310. https://doi.org/10.1021/ma1009269.

95 Cao, J., He, J., Li, C., and Yang, Y. (2001). Nitroxide-mediated radical polymerization of styrene in emulsion. *Polymer Journal* 33: 75–80. https://doi.org/10.1295/polymj.33.75.

96 Trost, B.M. (1991). The atom economy-A search for synthetic efficiency. *Science* 254: 1471–1477.

97 Trost, B.M. (1995). Atom economy—a challenge for organic synthesis: homogeneous catalysis leads the way. *Angewandte Chemie International Edition In English* 34: 259–281. https://doi.org/10.1002/anie.199502591.

98 Wender, P.A., Croatt, M.P., and Witulski, B. (2006). New reactions and step economy: the total synthesis of (±)-salsolene oxide based on the type II transition metal-catalyzed intramolecular [4+4] cycloaddition. *Tetrahedron* 62: 7505–7511. https://doi.org/10.1016/j.tet.2006.02.085.

99 Burns, N.Z., Baran, P.S., and Hoffmann, R.W. (2009). Redox economy in organic synthesis. *Angewandte Chemie, International Edition* 48: 2854–2867. https://doi.org/10.1002/anie.200806086.

100 Hayashi, Y. (2016). Pot economy and one-pot synthesis. *Chemical Science* 7: 866–880. https://doi.org/10.1039/C5SC02913A.

101 Armstrong, R.W., Combs, A.P., Tempest, P.A. et al. (1996). Multiple-component condensation strategies for combinatorial library synthesis. *Accounts of Chemical Research* 29: 123–131. https://doi.org/10.1021/ar9502083.

102 Zhao, W. and Chen, F.-E. (2012). One-pot synthesis and its practical application in pharmaceutical industry. *Current Organic Synthesis* 9: 873–897. https://doi.org/10.2174/157017912803901619.

103 Tietze, L.F. (1996). Domino reactions in organic synthesis. *Chemical Reviews* 96: 115–136. https://doi.org/10.1021/cr950027e.

104 Tietze, L.F. and Rackelmann, N. (2004). Domino reactions in the synthesis of heterocyclic natural products and analogs. *Pure and Applied Chemistry* 76: 1967–1983. https://doi.org/10.1351/pac200476111967.

105 Sutton, A.E., Seigal, B.A., Finnegan, D.F., and Snapper, M.L. (2002). New tandem catalysis: preparation of cyclic enol ethers through a ruthenium-catalyzed ring-closing metathesis—olefin isomerization sequence. *Journal of the American Chemical Society* 124: 13390–13391. https://doi.org/10.1021/ja028044q.

106 Denmark, S.E. and Thorarensen, A. (1996). Tandem [4+2]/[3+2] cycloadditions of nitroalkenes. *Chemical Reviews* 96: 137–166. https://doi.org/10.1021/cr940277f.

107 Nicolaou, K.C., Edmonds, D.J., and Bulger, P.G. (2006). Cascade reactions in total synthesis. *Angewandte Chemie International Edition In English* 45: 7134–7186. https://doi.org/10.1002/anie.200601872.

108 Roy, A., Bulut, O., Some, S. et al. (2019). Green synthesis of silver nanoparticles: biomolecule-nanoparticle organizations targeting antimicrobial activity. *RSC Advances* 9: 2673–2702. https://doi.org/10.1039/C8RA08982E.

109 Singh, P., Kim, Y.-J., Zhang, D., and Yang, D.-C. (2016). Biological synthesis of nanoparticles from plants and microorganisms. *Trends in Biotechnology* 34: 588–599. https://doi.org/10.1016/j.tibtech.2016.02.006.

110 Bhargava, A., Jain, N., Khan, M.A. et al. (2016). Utilizing metal tolerance potential of soil fungus for efficient synthesis of gold nanoparticles with superior catalytic activity for degradation of rhodamine B. *Journal of Environmental Management* 183: 22–32. https://doi.org/10.1016/j.jenvman.2016.08.021.

111 Mishra, A., Kumari, M., Pandey, S. et al. (2014). Biocatalytic and antimicrobial activities of gold nanoparticles synthesized by Trichoderma sp. *Bioresource Technology* 166: 235–242. https://doi.org/10.1016/j.biortech.2014.04.085.

112 Gudikandula, K., Vadapally, P., and Singara Charya, M.A. (2017). Biogenic synthesis of silver nanoparticles from white rot fungi: their characterization and antibacterial studies. *OpenNano* 2: 64–78. https://doi.org/10.1016/j.onano.2017.07.002.

113 Rajput, S., Werezuk, R., Lange, R.M., and McDermott, M.T. (2016). Fungal isolate optimized for biogenesis of silver nanoparticles with enhanced colloidal stability. *Langmuir* 32: 8688–8697. https://doi.org/10.1021/acs.langmuir.6b01813.

114 Kitching, M., Choudhary, P., Inguva, S. et al. (2016). Fungal surface protein mediated one-pot synthesis of stable and hemocompatible gold nanoparticles. *Enzyme and Microbial Technology* 95: 76–84. https://doi.org/10.1016/j.enzmictec.2016.08.007.

115 Suryavanshi, P., Pandit, R., Gade, A. et al. (2017). Colletotrichum sp.- mediated synthesis of sulphur and aluminium oxide nanoparticles and its in vitro activity against selected food-borne pathogens. *LWT- Food Science and Technology* 81: 188–194. https://doi.org/10.1016/j.lwt.2017.03.038.

116 Domany, E.B.E., Essam, T.M., Ahmed, A.E., and Farghali, A.A. (2018). Biosynthesis physico-chemical optimization of gold nanoparticles as anti-cancer and synergetic antimicrobial activity using *Pleurotus* *ostreatus* fungus. *Journal of Applied Pharmaceutical Science* 8 (5): 119–128. https://doi.org/10.7324/japs.2018.8516.

117 Faramarzi, M.A. and Sadighi, A. (2013). Insights into biogenic and chemical production of inorganic nanomaterials and nanostructures. *Advances in Colloid and Interface Science* 189–190: 1–20. https://doi.org/10.1016/j.cis.2012.12.001.

118 Iravani, S. (2014). Bacteria in nanoparticle synthesis: current status and future prospects. *International Scholarly Research Notices* 2014: 359316. https://doi.org/10.1155/2014/359316.

119 Beveridge, T.J. and Murray, R.G. (1980). Sites of metal deposition in the cell wall of *Bacillus subtilis*. *Journal of Bacteriology* 141: 876–887.

120 Klaus-Joerger, T., Joerger, R., Olsson, E., and Granqvist, C. (2001). Bacteria as workers in the living factory: metal-accumulating bacteria and their potential for materials science. *Trends in Biotechnology* 19: 15–20. https://doi.org/10.1016/s0167-7799(00)01514-6.

121 Srivastava, S.K. and Constanti, M. (2012). Room temperature biogenic synthesis of multiple nanoparticles (Ag, Pd, Fe, Rh, Ni, Ru, Pt, Co, and Li) by *Pseudomonas aeruginosa* SM1. *Journal of Nanoparticle Research* 14: 831. https://doi.org/10.1007/s11051-012-0831-7.

122 Malhotra, A., Dolma, K., Kaur, N. et al. (2013). Biosynthesis of gold and silver nanoparticles using a novel marine strain of *Stenotrophomonas*. *Bioresource Technology* 142: 727–731. https://doi.org/10.1016/j.biortech.2013.05.109.

123 Mehta, A., Sidhu, C., Pinnaka, A.K., and Roy Choudhury, A. (2014). Extracellular polysaccharide production by a novel osmotolerant marine strain of *Alteromonas macleodii* and its application towards biomineralization of silver. *PLoS One* 9: e98798. https://doi.org/10.1371/journal.pone.0098798.

124 Sunkar, S. and Nachiyar, C.V. (2012). Biogenesis of antibacterial silver nanoparticles using the endophytic bacterium *Bacillus cereus* isolated from *Garcinia xanthochymus*. *Asian Pacific Journal of Tropical Biomedicine* 2: 953–959. https://doi.org/10.1016/S2221-1691(13)60006-4.

125 Kaur, H., Dolma, K., Kaur, N. et al. (2015). Marine microbe as nano-factories for copper biomineralization. *Biotechnology and Bioprocess Engineering* 20: 51–57. https://doi.org/10.1007/s12257-014-0432-7.

126 Das, V.L., Thomas, R., Varghese, R.T. et al. (2014). Extracellular synthesis of silver nanoparticles by the Bacillus strain CS 11 isolated from industrialized area. *3 Biotech* 4: 121–126. https://doi.org/10.1007/s13205-013-0130-8.

127 Raj, R., Dalei, K., Chakraborty, J., and Das, S. (2016). Extracellular polymeric substances of a marine bacterium mediated synthesis of CdS nanoparticles for removal of cadmium from aqueous solution. *Journal of Colloid and Interface Science* 462: 166–175. https://doi.org/10.1016/j.jcis.2015.10.004.

128 Kulkarni, R.R., Shaiwale, N.S., Deobagkar, D.N., and Deobagkar, D.D. (2015). Synthesis and extracellular accumulation of silver nanoparticles by employing radiation-resistant Deinococcus radiodurans, their characterization, and determination of bioactivity. *International Journal of Nanomedicine* 10: 963–974. https://doi.org/10.2147/IJN.S72888.

129 Gahlawat, G., Shikha, S., Chaddha, B.S. et al. (2016). Microbial glycolipoprotein-capped silver nanoparticles as emerging antibacterial agents against cholera. *Microbial Cell Factories* 15: 25. https://doi.org/10.1186/s12934-016-0422-x.

130 Ahmed, E., Kalathil, S., Shi, L. et al. (2018). Synthesis of ultra-small platinum, palladium and gold nanoparticles by Shewanella loihica PV-4 electrochemically active biofilms and their enhanced catalytic activities. *Journal of Saudi Chemical Society* 22: 919–929. https://doi.org/10.1016/j.jscs.2018.02.002.

131 Müller, A., Behsnilian, D., Walz, E. et al. (2016). Effect of culture medium on the extracellular synthesis of silver nanoparticles using *Klebsiella pneumoniae*, *Escherichia coli* and *Pseudomonas jessinii*. *Biocatalysis and Agricultural Biotechnology* 6: 107–115. https://doi.org/10.1016/j.bcab.2016.02.012.

132 Naik, M.M., Prabhu, M.S., Samant, S.N. et al. (2017). Synergistic action of silver nanoparticles synthesized from silver resistant estuarine *Pseudomonas aeruginosa* Strain SN5 with antibiotics against antibiotic resistant bacterial human pathogens. *Thalass* 33: 73–80. https://doi.org/10.1007/s41208-017-0023-4.

133 Saravanan, M., Barik, S.K., MubarakAli, D. et al. (2018). Synthesis of silver nanoparticles from Bacillus brevis (NCIM 2533) and their antibacterial activity against pathogenic bacteria. *Microbial Pathogenesis* 116: 221–226. https://doi.org/10.1016/j.micpath.2018.01.038.

134 Divya, M., Kiran, G.S., Hassan, S., and Selvin, J. (2019). Biogenic synthesis and effect of silver nanoparticles (AgNPs) to combat catheter-related urinary tract infections. *Biocatalysis and Agricultural Biotechnology* 18: 101037. https://doi.org/10.1016/j.bcab.2019.101037.

135 Zonaro, E., Piacenza, E., Presentato, A. et al. (2017). *Ochrobactrum* sp. MPV1 from a dump of roasted pyrites can be exploited as bacterial catalyst for the biogenesis of selenium and tellurium nanoparticles. *Microbial Cell Factories* 16: 215. https://doi.org/10.1186/s12934-017-0826-2.

136 Srinath, B.S., Namratha, K., and Byrappa, K. (2018). Eco-friendly synthesis of gold nanoparticles by *Bacillus subtilis* and their environmental applications. *Advanced Science Letters* 24: 5942–5946. https://doi.org/10.1166/asl.2018.12224.

137 Ramanathan, R., O'Mullane, A.P., Parikh, R.Y. et al. (2011). Bacterial kinetics-controlled shape-directed biosynthesis of silver nanoplates using *Morganella psychrotolerans*. *Langmuir* 27: 714–719. https://doi.org/10.1021/la1036162.

138 Otari, S.V., Patil, R.M., Nadaf, N.H. et al. (2014). Green synthesis of silver nanoparticles by microorganism using organic pollutant: its antimicrobial and catalytic application. *Environmental Science and Pollution Research* 21: 1503–1513. https://doi.org/10.1007/s11356-013-1764-0.

139 Abd-Elnaby, H.M., Abo-Elala, G.M., Abdel-Raouf, U.M., and Hamed, M.M. (2016). Antibacterial and anticancer activity of extracellular synthesized silver nanoparticles from marine *Streptomyces rochei* MHM13. *Egyptian Journal of Aquatic Research* 42: 301–312. https://doi.org/10.1016/j.ejar.2016.05.004.

140 Karthik, L., Kumar, G., Kirthi, A.V. et al. (2014). *Streptomyces* sp. LK3 mediated synthesis of silver nanoparticles and its biomedical application. *Bioprocess and Biosystems Engineering* 37: 261–267. https://doi.org/10.1007/s00449-013-0994-3.

141 Otari, S.V., Patil, R.M., Nadaf, N.H. et al. (2012). Green biosynthesis of silver nanoparticles from an actinobacteria *Rhodococcus* sp. *Materials Letters* 72: 92–94. https://doi.org/10.1016/j.matlet.2011.12.109.

142 Buszewski, B., Railean-Plugaru, V., Pomastowski, P. et al. (2018). Antimicrobial activity of biosilver nanoparticles produced by a novel *Streptacidiphilus durhamensis* strain. *Journal of Microbiology, Immunology, and Infection* 51 (1): 45–54. https://doi.org/10.1016/j.jmii.2016.03.002.

143 Hassan, S.E.-D., Salem, S.S., Fouda, A. et al. (2018). New approach for antimicrobial activity and bio-control of various pathogens by biosynthesized copper nanoparticles using endophytic actinomycetes. *Journal of Radiation Research and Applied Science* 11: 262–270. https://doi.org/10.1016/j.jrras.2018.05.003.

144 Wypij, M., Czarnecka, J., Świecimska, M. et al. (2018). Synthesis, characterization and evaluation of antimicrobial and cytotoxic activities of biogenic silver nanoparticles synthesized from *Streptomyces xinghaiensis* OF1 strain. *World Journal of Microbiology and Biotechnology* 34: 23. https://doi.org/10.1007/s11274-017-2406-3.

145 Silva-Vinhote, N.M., Caballero, N.E.D., de Amorim Silva, T. et al. (2017). Extracellular biogenic synthesis of silver nanoparticles by Actinomycetes from amazonic biome and its antimicrobial efficiency. *African Journal of Biotechnology* 16: 2072–2082. https://doi.org/10.5897/AJB2017.16148.

146 Ranjitha, V.R. and Rai, V.R. (2017). Actinomycetes mediated synthesis of gold nanoparticles from the culture supernatant of *Streptomyces griseoruber* with special reference to catalytic activity. *3 Biotech* 7: 299. https://doi.org/10.1007/s13205-017-0930-3.

147 Gahlawat, G. and Choudhury, A.R. (2019). A review on the biosynthesis of metal and metal salt nanoparticles by microbes. *RSC Advances* 9: 12944–12967. https://doi.org/10.1039/C8RA10483B.

148 Hulkoti, N.I. and Taranath, T.C. (2014). Biosynthesis of nanoparticles using microbes- a review. *Colloids and Surfaces B: Biointerfaces* 121: 474–483. https://doi.org/10.1016/j.colsurfb.2014.05.027.

149 Yumei, L., Yamei, L., Qiang, L., and Jie, B. (2017). Rapid biosynthesis of silver nanoparticles based on flocculation and reduction of an exopolysaccharide from Arthrobacter sp. B4: its antimicrobial activity and phytotoxicity. *Journal of Nanomaterials* 2017: e9703614. https://doi.org/10.1155/2017/9703614.

150 Shah, M., Fawcett, D., Sharma, S. et al. (2015). Green synthesis of metallic nanoparticles via biological entities. *Materials (Basel)* 8: 7278–7308. https://doi.org/10.3390/ma8115377.

151 Apte, M., Sambre, D., Gaikawad, S. et al. (2013). Psychrotrophic yeast *Yarrowia lipolytica* NCYC 789 mediates the synthesis of antimicrobial silver nanoparticles via cell-associated melanin. *AMB Express* 3: 32. https://doi.org/10.1186/2191-0855-3-32.

152 Waghmare, S.R., Mulla, M.N., Marathe, S.R., and Sonawane, K.D. (2015). Ecofriendly production of silver nanoparticles using Candida utilis and its mechanistic action against pathogenic microorganisms. *3 Biotech* 5: 33–38. https://doi.org/10.1007/s13205-014-0196-y.

153 Elahian, F., Reiisi, S., Shahidi, A., and Mirzaei, S.A. (2017). High-throughput bioaccumulation, biotransformation, and production of silver and selenium nanoparticles using genetically engineered *Pichia pastoris*. *Nanomedicine: Nanotechnology, Biology and Medicine* 13: 853–861. https://doi.org/10.1016/j.nano.2016.10.009.

154 Eugenio, M., Müller, N., Frasés, S. et al. (2016). Yeast-derived biosynthesis of silver/silver chloride nanoparticles and their antiproliferative activity against bacteria. *RSC Advances* 6: 9893–9904. https://doi.org/10.1039/C5RA22727E.

155 Sriramulu, M. and Sumathi, S. (2018). Biosynthesis of palladium nanoparticles using *Saccharomyces cerevisiae* extract and its photocatalytic degradation behaviour.

Advances in Natural Sciences: Nanoscience and Nanotechnology 9: 025018. https://doi.org/10.1088/2043-6254/aac506.

156 da Silva Ferreira, V., Conz Ferreira, M.E., Lima, L.M.T.R. et al. (2017). Green production of microalgae-based silver chloride nanoparticles with antimicrobial activity against pathogenic bacteria. *Enzyme and Microbial Technology* 97: 114–121. https://doi.org/10.1016/j.enzmictec.2016.10.018.

157 Arsiya, F., Sayadi, M.H., and Sobhani, S. (2017). Green synthesis of palladium nanoparticles using *Chlorella vulgaris*. *Materials Letters* 186: 113–115. https://doi.org/10.1016/j.matlet.2016.09.101.

158 Momeni, S. and Nabipour, I. (2015). A simple green synthesis of palladium nanoparticles with *Sargassum* alga and their electrocatalytic activities towards hydrogen peroxide. *Applied Biochemistry and Biotechnology* 176: 1937–1949. https://doi.org/10.1007/s12010-015-1690-3.

159 Dhas, T.S., Kumar, V.G., Karthick, V. et al. (2014). Facile synthesis of silver chloride nanoparticles using marine alga and its antibacterial efficacy. *Spectrochimica Acta. Part A, Molecular and Biomolecular Spectroscopy* 120: 416–420. https://doi.org/10.1016/j.saa.2013.10.044.

160 Edison, T.N.J.I., Atchudan, R., Kamal, C., and Lee, Y.R. (2016). Caulerpa racemosa: a marine green alga for eco-friendly synthesis of silver nanoparticles and its catalytic degradation of methylene blue. *Bioprocess and Biosystems Engineering* 39: 1401–1408. https://doi.org/10.1007/s00449-016-1616-7.

161 Ramakrishna, M., Rajesh Babu, D., Gengan, R.M. et al. (2016). Green synthesis of gold nanoparticles using marine algae and evaluation of their catalytic activity. *J. Nanostructure Chem.* 6: 1–13. https://doi.org/10.1007/s40097-015-0173-y.

162 Rajeshkumar, S. (2018). Synthesis of zinc oxide nanoparticles using algal formulation (*Padina tetrastromatica* and *Turbinaria conoides*) and their antibacterial activity against fish pathogens. *Research Journal of Biotechnology* 13: 15–19.

163 Sanaeimehr, Z., Javadi, I., and Namvar, F. (2018). Antiangiogenic and antiapoptotic effects of green-synthesized zinc oxide nanoparticles using *Sargassum muticum* algae extraction. *Cancer Nanotechnology* 9: 3. https://doi.org/10.1186/s12645-018-0037-5.

164 Fawcett, D., Verduin, J.J., Shah, M. et al. (2017). A review of current research into the biogenic synthesis of metal and metal oxide nanoparticles via marine algae and seagrasses. *Journal of Nanoscience* 2017: e8013850. https://doi.org/10.1155/2017/8013850.

165 Kobayashi, M., Tomita, S., Sawada, K. et al. (2012). Chiral meta-molecules consisting of gold nanoparticles and genetically engineered tobacco mosaic virus. *Optics Express* 20: 24856–24863. https://doi.org/10.1364/OE.20.024856.

166 Zeng, Q., Wen, H., Wen, Q. et al. (2013). Cucumber mosaic virus as drug delivery vehicle for doxorubicin. *Biomaterials* 34: 4632–4642. https://doi.org/10.1016/j.biomaterials.2013.03.017.

167 Mao, C., Flynn, C.E., Hayhurst, A. et al. (2003). Viral assembly of oriented quantum dot nanowires. *Proceedings of the National Academy of Sciences of the United States of America* 100: 6946–6951. https://doi.org/10.1073/pnas.0832310100.

168 Love, A.J., Talianski, M.E., Chapman, S.N. and Shaw, J. US Patent No. 9,688,964, US Patent and Trademark Office, Washington, DC, 2017, US Patent, No. 9,688,964. https://www.uspto.gov/web/offices/com/sol/og/2008/week48/TOC.htm (accessed 20 July 2020).

169 Cao, J., Guenther, R.H., Sit, T.L. et al. (2014). Loading and release mechanism of red clover necrotic mosaic virus derived plant viral nanoparticles for drug delivery of doxorubicin. *Small* 10: 5126–5136. https://doi.org/10.1002/smll.201400558.

170 Le, D.H.T., Lee, K.L., Shukla, S. et al. (2017). Potato virus X, a filamentous plant viral nanoparticle for doxorubicin delivery in cancer therapy. *Nanoscale* 9: 2348–2357. https://doi.org/10.1039/C6NR09099K.

171 Rascio, N. and Navari-Izzo, F. (2011). Heavy metal hyperaccumulating plants: how and why do they do it? And what makes them so interesting? *Plant Science* 180: 169–181. https://doi.org/10.1016/j.plantsci.2010.08.016.

172 Corso, M. and de la Torre, V.S.G. (2020). Biomolecular approaches to understanding metal tolerance and hyperaccumulation in plants. *Metallomics* 12: 840–859. https://doi.org/10.1039/D0MT00043D.

173 da Conceição Gomes, M.A., Hauser-Davis, R.A., de Souza, A.N., and Vitória, A.P. (2016). Metal phytoremediation: general strategies, genetically modified plants and applications in metal nanoparticle contamination. *Ecotoxicology and Environmental Safety* 134: 133–147. https://doi.org/10.1016/j.ecoenv.2016.08.024.

174 Dushenkov, V., Kumar, P.B.A.N., Motto, H., and Raskin, I. (1995). Rhizofiltration: the use of plants to remove heavy metals from aqueous streams. *Environmental Science & Technology* 29: 1239–1245. https://doi.org/10.1021/es00005a015.

175 Gardea-Torresdey, J.L., Rodriguez, E., Parsons, J.G. et al. (2005). Use of ICP and XAS to determine the enhancement of gold phytoextraction by Chilopsis linearis using thiocyanate as a complexing agent. *Analytical and Bioanalytical Chemistry* 382: 347–352. https://doi.org/10.1007/s00216-004-2966-6.

176 De la Rosa, G., Torres, J., Parsons, J.G. et al. (2009). X-ray absorption spectroscopy unveils the formation of gold nanoparticles in corn. *Acta Universitaria* 19: 76–81. https://doi.org/10.15174/au.2009.106.

177 Gardea-Torresdey, J.L., Gomez, E., Peralta-Videa, J.R. et al. (2003). Alfalfa sprouts: a natural source for the synthesis of silver nanoparticles. *Langmuir* 19: 1357–1361. https://doi.org/10.1021/la020835i.

178 Anderson, C.W.N., Brooks, R.R., Chiarucci, A. et al. (1999). Phytomining for nickel, thallium and gold. *Journal of Geochemical Exploration* 67: 407–415. https://doi.org/10.1016/S0375-6742(99)00055-2.

179 Lamb, A.E., Anderson, C.W.N., Haverkamp, R.G. (2001). The extraction of gold from plants and its application to phytomining. https://mro.massey.ac.nz/handle/10179/9612 (accessed 20 July 2020).

180 Chandran, S.P., Chaudhary, M., Pasricha, R. et al. (2006). Synthesis of gold nanotriangles and silver nanoparticles using Aloe vera plant extract. *Biotechnology Progress* 22: 577–583. https://doi.org/10.1021/bp0501423.

181 Shankar, S.S., Ahmad, A., and Sastry, M. (2003). Geranium leaf assisted biosynthesis of silver nanoparticles. *Biotechnology Progress* 19: 1627–1631. https://doi.org/10.1021/bp034070w.

182 Shankar, S.S., Ahmad, A., Pasricha, R., and Sastry, M. (2003). Bioreduction of chloroaurate ions by geranium leaves and its endophytic fungus yields gold nanoparticles of different shapes. *Journal of Materials Chemistry* 13 (7): 1822–1826. https://doi.org/10.1039/b303808b.

183 Paciotti, G.F., Myer, L., Weinreich, D. et al. (2004). Colloidal gold: a novel nanoparticle vector for tumor directed drug delivery. *Drug Delivery* 11: 169–183. https://doi.org/10.1080/10717540490433895.

184 Iravani, S. (2011). Green synthesis of metal nanoparticles using plants. *Green Chemistry* 13: 2638–2650. https://doi.org/10.1039/C1GC15386B.

185 Ankamwar, B., Damle, C., Ahmad, A., and Sastry, M. (2005). Biosynthesis of gold and silver nanoparticles using Emblica Officinalis fruit extract, their phase transfer and transmetallation in an organic solution. *Journal of Nanoscience and Nanotechnology* 5: 1665–1671. https://doi.org/10.1166/jnn.2005.184.

186 Makarov, V.V., Love, A.J., Sinitsyna, O.V. et al. (2014). "Green" nanotechnologies: synthesis of metal nanoparticles using plants. *Acta Naturae* 6: 35–44.

187 Das, R.K., Pachapur, V.L., Lonappan, L. et al. (2017). Biological synthesis of metallic nanoparticles: plants, animals and microbial aspects. *Nanotechnology for Environmental Engineering* 2: 18. https://doi.org/10.1007/s41204-017-0029-4.

188 Khandel, P., Yadaw, R.K., Soni, D.K. et al. (2018). Biogenesis of metal nanoparticles and their pharmacological applications: present status and application prospects. *Journal of Nanostructure in Chemistry* 8: 217–254. https://doi.org/10.1007/s40097-018-0267-4.

189 Gruen, L.C. (1975). Interaction of amino acids with silver (I) ions. *Biochimica et Biophysica Acta* 386: 270–274. https://doi.org/10.1016/0005-2795(75)90268-8.

190 Kuppusamy, P., Yusoff, M.M., Maniam, G.P., and Govindan, N. (2016). Biosynthesis of metallic nanoparticles using plant derivatives and their new avenues in pharmacological applications – an updated report. *Saudi Pharmaceutical Journal* 24: 473–484. https://doi.org/10.1016/j.jsps.2014.11.013.

191 Rastogi, A., Tripathi, D.K., Yadav, S. et al. (2019). Application of silicon nanoparticles in agriculture. *3 Biotech* 9: 90. https://doi.org/10.1007/s13205-019-1626-7.

192 Huang, J.W. and Cunningham, S.D. (1996). Lead phytoextraction: species variation in lead uptake and translocation. *The New Phytologist* 134: 75–84. https://doi.org/10.1111/j.1469-8137.1996.tb01147.x.

193 Li, S., Shen, Y., Xie, A. et al. (2007). Green synthesis of silver nanoparticles using *Capsicum annuum* L. extract. *Green Chemistry* 9: 852–858. https://doi.org/10.1039/B615357G.

Index

Sustainable Nanotechnology: Strategies, Products, and Applications, First Edition.
Edited by Yashwant V. Pathak, Govindan Parayil, and Jayvadan K. Patel.

d

h

o

t

Printed and bound by CPI Group (UK) Ltd, Croydon, CR0 4YY